W9-DIF-041

ANALYSIS OF NUMERICAL METHODS

ANALYSIS OF NUMERICAL METHODS

ANALYSIS OF NUMERICAL METHODS

EUGENE ISAACSON
Professor of Mathematics
Chairman
University Computing Center

HERBERT BISHOP KELLER
Professor of Applied Mathematics
Associate Director
A.E.C. Computing and Applied
Mathematics Center

COURANT INSTITUTE OF MATHEMATICAL SCIENCES
NEW YORK UNIVERSITY

JOHN WILEY AND SONS
New York Chichester Brisbane Toronto Singapore

ISBN 0 471 42865 5
Library of Congress Catalog Card Number: 66-17630
Printed in the United States of America

20 19 18

To our understanding wives, Muriel and Loretta

Preface

Digital computers, though mass produced for no more than fifteen years, have become indispensable for much current scientific research. One basic reason for this is that by implementing numerical methods, computers form a universal tool for "solving" broad classes of problems. While numerical methods have always been useful it is clear that their role in scientific research is now of fundamental importance. No modern applied mathematician, physical scientist, or engineer can be properly trained without some understanding of numerical methods.

We attempt, in this book, to supply some of the required knowledge. In presenting the material we stress techniques for the development of new methods. This requires knowing why a particular method is effective on some problems but not on others. Hence we are led to the analysis of numerical methods rather than merely their description and listing.

Certainly the solving of scientific problems should not be and is not the sole motivation for studying numerical methods. Our opinion is that the analysis of numerical methods is a broad and challenging mathematical activity whose central theme is the effective constructibility of various kinds of approximations.

Many numerical methods have been neglected in this book since we do not attempt to be exhaustive. Procedures treated are either quite good and efficient by present standards or else their study is considered instructive (while their use may not be advocated). Unfortunately the limitations of space and our own experience have resulted in the omission of many important topics that we would have liked to include (for example, linear programming, rational approximation, Monte Carlo methods).

The present work, it turns out, could be considered a mathematics text in selected areas of analysis and matrix theory. Essentially no mathematical preparation beyond advanced calculus and elementary linear algebra (or matrix theory) is assumed. Relatively important material on norms in finite-dimensional spaces, not taught in most elementary courses, is included in Chapter 1. Some familiarity with the existence theory for differential equations would be useful, but is not necessary. A cursory knowledge of the classical partial differential equations of mathematical physics would help in Chapter 9. No significant use is made of the theory of functions of a complex variable and our book is elementary in that sense. Deeper studies of numerical methods would also rely heavily on functional analysis, which we avoid here.

The listing of algorithms to concretely describe a method is avoided. Hence some practical experience in using numerical methods is assumed or should be obtained. Examples and problems are given which extend or amplify the analysis in many cases (starred problems are more difficult). It is assumed that the instructor will supplement these with computational problems, according to the availability of computing facilities.

References have been kept minimal and are usually to one of the general texts we have found most useful and compiled into a brief bibliography. Lists of additional, more specialized references are given for the four different areas covered by Chapters 1–4, Chapters 5–7, Chapter 8, and Chapter 9. A few outstanding journal articles have been included here. Complete bibliographies can be found in several of the general texts.

Key equations (and all theorems, problems, and figures) are numbered consecutively by integers within each section. Equations, etc., in other sections are referred to by a decimal notation with explicit mention of the chapter if it is not the current one [that is, equation (3.15) of Chapter 5]. Yielding to customary usage we have not sought historical accuracy in associating names with theorems, methods, etc.

Several different one-semester and two-semester courses have been based on the material in this book. Not all of the subject matter can be covered in the usual one-year course. As examples of some plans that have worked well, we suggest:

Two-semester courses:

(A) Prerequisite—Advanced Calculus and Linear Algebra, Chapters 1–9;
(B) Prerequisite—Advanced Calculus (with Linear Algebra required only for the second semester), Chapters 3, 5–7, 8 (through Section 3), 1, 2, 4, 8, 9.

One-semester courses:

(A) Chapters 3, 5–7, 8 (through Section 3);
(B) Chapters 1–5;

(C) Chapters 8, 9 (plus some material from Chapter 2 on iterative methods).

This book benefits from our experience in trying to teach such courses at New York University for over fifteen years and from our students' reactions. Many of our former and present colleagues at the Courant Institute of Mathematical Sciences are responsible for our education in this field. We acknowledge our indebtedness to them, and to the stimulating environment of the Courant Institute. Help was given to us by our friends who have read and used preliminary versions of the text. In this connection we are happy to thank Prof. T. E. Hull, who carefully read our entire manuscript and offered much constructive criticism; Dr. William Morton, who gave valuable suggestions for Chapters 5–7; Professor Gene Golub, who helped us to improve Chapters 1, 2, and 4. We are grateful for the advice given us by Professors H. O. Kreiss, Beresford Parlett, Alan Solomon, Peter Ungar, Richard Varga, and Bernard Levinger, and Dr. Olof Widlund. Thanks are also due to Mr. Julius Rosenthal and Dr. Eva Swenson who helped in the preparation of mimeographed lecture notes for some of our courses. This book grew from two sets of these notes upon the suggestion of Mr. Earle Brach. We are most grateful to Miss Connie Engle who carefully typed our manuscript and to Mr. Richard Swenson who helped in reading galleys. Finally, we must thank Miss Sallyanne Riggione, who as copy editor made many helpful suggestions to improve the book.

New York and Pasadena
April, 1966

E. ISAACSON AND H. B. KELLER

Contents

1

Norms, Arithmetic, and Well-Posed Computations

0. INTRODUCTION

In this chapter, we treat three topics that are generally useful for the analysis of the various numerical methods studied throughout the book. In Section 1, we give the elements of the theory of *norms* of finite dimensional vectors and matrices. This subject properly belongs to the field of *linear algebra.* In later chapters, we may occasionally employ the notion of the norm of a function. This is a straightforward extension of the notion of a vector norm to the infinite-dimensional case. On the other hand, we shall not introduce the corresponding natural generalization, i.e., the notion of the norm of a linear transformation that acts on a space of functions. Such ideas are dealt with in *functional analysis,* and might profitably be used in a more sophisticated study of numerical methods.

We study briefly, in Section 2, the practical problem of the effect of rounding errors on the basic operations of arithmetic. Except for calculations involving only exact-integer arithmetic, rounding errors are invariably present in any computation. A most important feature of the later analysis of numerical methods is the incorporation of a treatment of the effects of such rounding errors.

Finally, in Section 3, we describe the computational problems that are "reasonable" in some general sense. In effect, a numerical method which produces a solution insensitive to small changes in data or to rounding errors is said to yield a *well-posed computation.* How to determine the sensitivity of a numerical procedure is dealt with in special cases throughout the book. We indicate heuristically that any *convergent* algorithm is a well-posed computation.

1. NORMS OF VECTORS AND MATRICES

We assume that the reader is familiar with the basic theory of linear algebra, not necessarily in its abstract setting, but at least with specific reference to finite-dimensional linear vector spaces over the field of complex scalars. By "basic theory" we of course include: the theory of linear systems of equations, some elementary theory of determinants, and the theory of matrices or linear transformations to about the Jordan normal form. We hardly employ the Jordan form in the present study. In fact a much weaker result can frequently be used in its place (when the divisor theory or invariant subspaces are not actually involved). This result is all too frequently skipped in basic linear algebra courses, so we present it as

THEOREM **1**. *For any square matrix A of order n there exists a non-singular matrix P, of order n, such that*

$$B = P^{-1}AP$$

is upper triangular and has the eigenvalues of A, say $\lambda_j \equiv \lambda_j(A)$, $j = 1, 2, \ldots, n$, on the principal diagonal (i.e., any square matrix is equivalent to a triangular matrix).

Proof. We sketch the proof of this result. The reader should have no difficulty in completing the proof in detail.

Let λ_1 be an eigenvalue of A with corresponding eigenvector $\mathbf{u}_1$.† Then pick a basis for the n-dimensional complex vector space, C_n, with $\mathbf{u}_1$ as the first such vector. Let the independent basis vectors be the columns of a non-singular matrix P_1, which then determines the transformation to the new basis. In this new basis the transformation determined by A is given by $B_1 \equiv P_1{}^{-1}AP_1$ and since $A\mathbf{u}_1 = \lambda_1\mathbf{u}_1$,

$$B_1 = P_1{}^{-1}AP_1 = \begin{pmatrix} \lambda_1 & \alpha_1 & \alpha_2 & \cdots & \alpha_{n-1} \\ 0 & & & & \\ \vdots & & & A_2 & \\ 0 & & & & \end{pmatrix}$$

where A_2 is some matrix of order $n - 1$.

The characteristic polynomial of B_1 is clearly

$$\det(\lambda I_n - B_1) = (\lambda - \lambda_1)\det(\lambda I_{n-1} - A_2),$$

† Unless otherwise indicated, boldface type denotes column vectors. For example, an n-dimensional vector $\mathbf{u}_k$ has the components u_{ik}; i.e.,

$$\mathbf{u}_k \equiv \begin{pmatrix} u_{1k} \\ u_{2k} \\ \vdots \\ u_{nk} \end{pmatrix}.$$

where I_n is the identity matrix of order n. Now pick some eigenvalue λ_2 of A_2 and corresponding $(n-1)$-dimensional eigenvector, $\mathbf{v}_2$; i.e.,

$$A_2\mathbf{v}_2 = \lambda_2\mathbf{v}_2.$$

With this vector we define the independent n-dimensional vectors

$$\hat{\mathbf{u}}_1 = \begin{pmatrix} 1 \\ 0 \\ \vdots \\ 0 \end{pmatrix}, \qquad \hat{\mathbf{u}}_2 = \begin{pmatrix} 0 \\ \mathbf{v}_2 \end{pmatrix}.$$

Note that with the scalar $\alpha = \alpha_1 v_{12} + \alpha_2 v_{22} + \cdots + \alpha_{n-1} v_{n-1,2}$

$$B_1\hat{\mathbf{u}}_1 = \lambda_1\hat{\mathbf{u}}_1, \qquad B_1\hat{\mathbf{u}}_2 = \lambda_2\hat{\mathbf{u}}_2 + \alpha\hat{\mathbf{u}}_1,$$

and thus if we set $\mathbf{u}_1 = P_1\hat{\mathbf{u}}_1$, $\mathbf{u}_2 = P_1\hat{\mathbf{u}}_2$, then

$$A\mathbf{u}_1 = \lambda_1\mathbf{u}_1, \qquad A\mathbf{u}_2 = \lambda_2\mathbf{u}_2 + \alpha\mathbf{u}_1.$$

Now we introduce a new basis of C_n with the first two vectors being $\mathbf{u}_1$ and $\mathbf{u}_2$. The non-singular matrix P_2 which determines this change of basis has $\mathbf{u}_1$ and $\mathbf{u}_2$ as its first two columns; and the original linear transformation in the new basis has the representation

$$B_2 = P_2^{-1}AP_2 = \begin{pmatrix} \lambda_1 & x & x & \cdots & x \\ 0 & \lambda_2 & x & \cdots & x \\ 0 & 0 & & & \\ \vdots & \vdots & & A_3 & \\ 0 & 0 & & & \end{pmatrix},$$

where A_3 is some matrix of order $n - 2$.

The theorem clearly follows by the above procedure; a formal inductive proof could be given. ∎

It is easy to prove the related stronger result of Schur stated in Theorem 2.4 of Chapter 4 (see Problem 2.13(b) of Chapter 4). We turn now to the basic content of this section, which is concerned with the generalization of the concept of distance in n-dimensional linear vector spaces.

The "distance" between a vector and the null vector, i.e., the origin, is a measure of the "size" or "length" of the vector. This generalized notion of distance or size is called a *norm*. In particular, all such generalizations are required to have the following properties:

(0) To each vector $\mathbf{x}$ in the linear space, $\mathscr{V}$, say, a unique real number is assigned; this number, denoted by $\|\mathbf{x}\|$ or $N(\mathbf{x})$, is called the norm of $\mathbf{x}$ iff:

(i) $\|\mathbf{x}\| \geq 0$ for all $\mathbf{x} \in \mathscr{V}$ and $\|\mathbf{x}\| = 0$ iff $\mathbf{x} = \mathbf{o}$;
where $\mathbf{o}$ denotes the *zero vector* (if $\mathscr{V} \equiv C_n$, then $o_i = 0$);

(ii) $\|\alpha\mathbf{x}\| = |\alpha| \cdot \|\mathbf{x}\|$ for all scalars α and all $\mathbf{x} \in \mathscr{V}$;

(iii) $\|\mathbf{x} + \mathbf{y}\| \leq \|\mathbf{x}\| + \|\mathbf{y}\|$, the *triangle inequality*,† for all $\mathbf{x}, \mathbf{y} \in \mathscr{V}$.

Some examples of norms in the complex n-dimensional space C_n are

$$\text{(1a)} \qquad \|\mathbf{x}\|_1 \equiv N_1(\mathbf{x}) \equiv \sum_{j=1}^{n} |x_j|,$$

$$\text{(1b)} \qquad \|\mathbf{x}\|_2 \equiv N_2(\mathbf{x}) \equiv \left(\sum_{j=1}^{n} |x_j|^2\right)^{1/2},$$

$$\text{(1c)} \qquad \|\mathbf{x}\|_p \equiv N_p(\mathbf{x}) \equiv \left(\sum_{j=1}^{n} |x_j|^p\right)^{1/p}, \qquad p \geq 1,$$

$$\text{(1d)} \qquad \|\mathbf{x}\|_\infty \equiv N_\infty(\mathbf{x}) \equiv \max_j |x_j|.$$

It is an easy exercise for the reader to justify the use of the notation in (1d) by verifying that

$$\lim_{p \to \infty} N_p(\mathbf{x}) = N_\infty(\mathbf{x}).$$

The norm, $\|\cdot\|_2$, is frequently called the *Euclidean norm* as it is just the formula for distance in ordinary three-dimensional Euclidean space extended to dimension n. The norm, $\|\cdot\|_\infty$, is called the *maximum norm* or occasionally the *uniform norm.* In general, $\|\cdot\|_p$, for $p \geq 1$ is termed the *p-norm.*

To verify that (1) actually defines norms, we observe that conditions (0), (i), and (ii) are trivially satisfied. Only the triangle inequality, (iii), offers any difficulty. However,

$$\begin{aligned} N_1(\mathbf{x} + \mathbf{y}) &= \sum_{j=1}^{n} |x_j + y_j| \\ &\leq \sum_{j=1}^{n} (|x_j| + |y_j|) = \sum_{j=1}^{n} |x_j| + \sum_{j=1}^{n} |y_j| \\ &= N_1(\mathbf{x}) + N_1(\mathbf{y}); \end{aligned}$$

† For complex numbers x and y the elementary inequality $|x + y| \leq |x| + |y|$ expresses the fact that the length of any side of a triangle is not greater than the sum of the lengths of the other two sides.

and

$$\begin{aligned} N_\infty(\mathbf{x} + \mathbf{y}) &= \max_j |x_j + y_j| \\ &\leq \max_j (|x_j| + |y_j|) \leq \max_j |x_j| + \max_k |y_k| \\ &= N_\infty(\mathbf{x}) + N_\infty(\mathbf{y}), \end{aligned}$$

so (1a) and (1d) define norms.

The proof of (iii) for (1b), the Euclidean norm, is based on the well-known *Cauchy-Schwarz inequality* which states that

$$\text{(2)} \qquad \left| \sum_{j=1}^{n} x_j y_j \right| \leq \left(\sum_{j=1}^{n} |x_j|^2 \right)^{1/2} \left(\sum_{j=1}^{n} |y_j|^2 \right)^{1/2} = N_2(\mathbf{x}) N_2(\mathbf{y}).$$

To prove this basic result, let $|\mathbf{x}|$ and $|\mathbf{y}|$ be the n-dimensional vectors with components $|x_j|$ and $|y_j|$, $j = 1, 2, \ldots, n$, respectively. Then for any real scalar, ξ,

$$0 \leq N_2(\xi|\mathbf{x}| + |\mathbf{y}|) = \xi^2 \sum_{j=1}^{n} |x_j|^2 + 2\xi \sum_{j=1}^{n} |x_j|\,|y_j| + \sum_{j=1}^{n} |y_j|^2.$$

But since the real quadratic polynomial in ξ above does not change sign its discriminant must be non-positive; i.e.,

$$\left(\sum_{j=1}^{n} |x_j| \cdot |y_j| \right)^2 \leq \left(\sum_{j=1}^{n} |x_j|^2 \right) \left(\sum_{j=1}^{n} |y_j|^2 \right).$$

However, we note that

$$\left| \sum_{j=1}^{n} x_j y_j \right|^2 \leq \left(\sum_{j=1}^{n} |x_j|\,|y_j| \right)^2,$$

and (2) follows from the above pair of inequalities.

Now we form

$$\begin{aligned} N_2(\mathbf{x} + \mathbf{y}) &= \left(\sum_{j=1}^{n} |x_j + y_j|^2 \right)^{1/2} = \left(\sum_{j=1}^{n} (x_j + y_j)(\bar{x}_j + \bar{y}_j) \right)^{1/2} \\ &= \left(\sum_{j=1}^{n} |x_j|^2 + \sum_{j=1}^{n} (x_j \bar{y}_j + \bar{x}_j y_j) + \sum_{j=1}^{n} |y_j|^2 \right)^{1/2} \\ &\leq \left(N_2^2(\mathbf{x}) + 2 \sum_{j=1}^{n} |x_j| \cdot |y_j| + N_2^2(\mathbf{y}) \right)^{1/2}. \end{aligned}$$

An application of the Cauchy-Schwarz inequality yields finally

$$N_2(\mathbf{x} + \mathbf{y}) \leq N_2(\mathbf{x}) + N_2(\mathbf{y})$$

and so the Euclidean norm also satisfies the triangle inequality.

The statement that

$$(3) \qquad N_p(\mathbf{x} + \mathbf{y}) \leq N_p(\mathbf{x}) + N_p(\mathbf{y}), \qquad p \geq 1,$$

is know as *Minkowski's inequality*. We do not derive it here as general p-norms will not be employed further. (A proof of (3) can be found in most advanced calculus texts.)

We can show quite generally that all vector norms are continuous functions in C_n. That is,

LEMMA 1. *Every vector norm, $N(\mathbf{x})$, is a continuous function of $x_1, x_2, \ldots, x_n$, the components of* $\mathbf{x}$.

Proof. For any vectors $\mathbf{x}$ and $\boldsymbol{\delta}$ we have by (iii)

$$N(\mathbf{x} + \boldsymbol{\delta}) \leq N(\mathbf{x}) + N(\boldsymbol{\delta}),$$

so that

$$N(\mathbf{x} + \boldsymbol{\delta}) - N(\mathbf{x}) \leq N(\boldsymbol{\delta}).$$

On the other hand, by (ii) and (iii),

$$\begin{aligned} N(\mathbf{x}) &= N(\mathbf{x} + \boldsymbol{\delta} - \boldsymbol{\delta}) \\ &\leq N(\mathbf{x} + \boldsymbol{\delta}) + N(\boldsymbol{\delta}), \end{aligned}$$

so that

$$-N(\boldsymbol{\delta}) \leq N(\mathbf{x} + \boldsymbol{\delta}) - N(\mathbf{x}).$$

Thus, in general

$$|N(\mathbf{x} + \boldsymbol{\delta}) - N(\mathbf{x})| \leq N(\boldsymbol{\delta}).$$

With the unit vectors† $\{\mathbf{e}_k\}$, any $\boldsymbol{\delta}$ has the representation

$$\boldsymbol{\delta} = \sum_{k=1}^{n} \delta_k \mathbf{e}_k.$$

Using (ii) and (iii) repeatedly implies

$$(4a) \qquad \begin{aligned} N(\boldsymbol{\delta}) &\leq \sum_{k=1}^{n} N(\delta_k \mathbf{e}_k) \\ &\leq \sum_{k=1}^{n} |\delta_k| N(\mathbf{e}_k) \\ &\leq \max_k |\delta_k| \sum_{j=1}^{n} N(\mathbf{e}_j) \\ &= M N_\infty(\boldsymbol{\delta}), \end{aligned}$$

† $\mathbf{e}_k$ has the components e_{ik}, where $e_{ik} = 0$, $i \neq k$; $e_{kk} = 1$.

where

$$M \equiv \sum_{j=1}^{n} N(\mathbf{e}_j). \tag{4b}$$

Using this result in the previous inequality yields, for any $\epsilon > 0$ and all $\boldsymbol{\delta}$ with $N_\infty(\boldsymbol{\delta}) \leq \epsilon/M$,

$$|N(\mathbf{x} + \boldsymbol{\delta}) - N(\mathbf{x})| \leq \epsilon.$$

This is essentially the definition of continuity for a function of the n variables $x_1, x_2, \ldots, x_n$. ∎

See Problem 6 for a mild generalization.

Now we can show that all vector norms are equivalent in the sense of

THEOREM 2. *For each pair of vector norms, say $N(\mathbf{x})$ and $N'(\mathbf{x})$, there exist positive constants m and M such that for all $\mathbf{x} \in C_n$:*

$$mN'(\mathbf{x}) \leq N(\mathbf{x}) \leq MN'(\mathbf{x}).$$

Proof. The proof need only be given when one of the norms is N_∞, since N and N' are equivalent if they are each equivalent to N_∞. Let $S \subset C_n$ be defined by

$$S \equiv \{\mathbf{x} \mid N_\infty(\mathbf{x}) = 1, \mathbf{x} \in C_n\}$$

(this is frequently called the surface of the unit ball in C_n). S is a closed bounded set of points. Then since $N(\mathbf{x})$ is a continuous function (see Lemma 1), we conclude by a theorem of Weierstrass that $N(\mathbf{x})$ attains its minimum and its maximum on S at some points of S. That is, for some $\mathbf{x}^0 \in S$ and $\mathbf{x}^1 \in S$

$$N(\mathbf{x}^0) = \min_{\mathbf{x} \in S} N(\mathbf{x}), \qquad N(\mathbf{x}^1) = \max_{\mathbf{x} \in S} N(\mathbf{x})$$

or

$$0 < N(\mathbf{x}^0) \leq N(\mathbf{x}) \leq N(\mathbf{x}^1) < \infty$$

for all $\mathbf{x} \in S$.

For any $\mathbf{y} \neq \mathbf{o}$ we see that $\mathbf{y}/N_\infty(\mathbf{y})$ is in S and so

$$N(\mathbf{x}^0) \leq N\left(\frac{\mathbf{y}}{N_\infty(\mathbf{y})}\right) \leq N(\mathbf{x}^1)$$

or

$$N(\mathbf{x}^0)N_\infty(\mathbf{y}) \leq N(\mathbf{y}) \leq N(\mathbf{x}^1)N_\infty(\mathbf{y}).$$

The last two inequalities yield

$$mN_\infty(\mathbf{y}) \leq N(\mathbf{y}) \leq MN_\infty(\mathbf{y}),$$

where $m \equiv N(\mathbf{x}^0)$ and $M \equiv N(\mathbf{x}^1)$. ∎

A matrix of order n could be treated as a vector in a space of dimension n^2 (with some fixed convention as to the manner of listing its elements). Then matrix norms satisfying the conditions (0)–(iii) could be defined as in (1). However, since the product of two matrices of order n is also such a matrix, we impose an additional condition on matrix norms, namely that

(iv) $\|AB\| \leq \|A\| \cdot \|B\|$.

With this requirement the vector norms (1) do not all become matrix norms (see Problem 2). However, there is a more natural, geometric, way in which the norm of a matrix can be defined. Thus, if $\mathbf{x} \in C_n$ and $\|\cdot\|$ is some vector norm on C_n, then $\|\mathbf{x}\|$ is the "length" of $\mathbf{x}$, $\|A\mathbf{x}\|$ is the "length" of $A\mathbf{x}$, and we define a norm of A, written as $\|A\|$ or $N(A)$, by the maximum relative "stretching,"

$$(5) \qquad \|A\| \equiv \sup_{\mathbf{x} \neq \mathbf{o}} \frac{\|A\mathbf{x}\|}{\|\mathbf{x}\|}.$$

Note that we use the same notation, $\|\cdot\|$, to denote vector and matrix norms; the context will always clarify which is implied. We call (5) a *natural norm* or the *matrix norm induced by the vector norm*, $\|\cdot\|$. This is also known as the *operator norm* in functional analysis. Since for any $\mathbf{x} \neq \mathbf{o}$ we can define $\mathbf{u} = \mathbf{x}/\|\mathbf{x}\|$ so that $\|\mathbf{u}\| = 1$, the definition (5) is equivalent to

$$(6) \qquad \|A\| = \max_{\|\mathbf{u}\|=1} \|A\mathbf{u}\| = \|A\mathbf{y}\|, \quad \|\mathbf{y}\| = 1.$$

That is, by Problems 6 and 7, $\|A\mathbf{u}\|$ is a continuous function of $\mathbf{u}$ and hence the maximum is attained for some $\mathbf{y}$, with $\|\mathbf{y}\| = 1$.

Before verifying the fact that (5) or (6) defines a matrix norm, we note that they imply, for any vector $\mathbf{x}$, that

$$(7) \qquad \|A\mathbf{x}\| \leq \|A\| \cdot \|\mathbf{x}\|.$$

There are many other ways in which matrix norms may be defined. But if (7) holds for some such norm then it is said to be *compatible* with the vector norm employed in (7). The natural norm (5) is essentially the "smallest" matrix norm compatible with a given vector norm.

To see that (5) yields a norm, we first note that conditions (i) and (ii) are trivially verified. For checking the triangle inequality, let $\mathbf{y}$ be such that $\|\mathbf{y}\| = 1$ and from (6),

$$\|(A + B)\| = \|(A + B)\mathbf{y}\|.$$

But then, upon recalling (7),

$$\begin{aligned} \|A + B\| &\leq \|A\mathbf{y}\| + \|B\mathbf{y}\| \\ &\leq \|A\| + \|B\|. \end{aligned}$$

Finally, to verify (iv), let $\mathbf{y}$ with $\|\mathbf{y}\| = 1$ now be such that

$$\|(AB)\| = \|(AB)\mathbf{y}\|.$$

Again by (7), we have

$$\begin{aligned}\|AB\| &\leq \|A\| \cdot \|B\mathbf{y}\| \\ &\leq \|A\| \cdot \|B\|,\end{aligned}$$

so that (5) and (6) do define a matrix norm.

We shall now determine the natural matrix norms induced by some of the vector p-norms ($p = 1, 2, \infty$) defined in (1). Let the nth order matrix A have elements a_{jk}, $j, k = 1, 2, \ldots, n$.

(A) The matrix norm induced by the *maximum norm* (1d) is

$$\|A\|_\infty = \max_j \sum_{k=1}^{n} |a_{jk}|, \tag{8}$$

i.e., the *maximum absolute row sum*. To prove (8), let $\mathbf{y}$ be such that $\|\mathbf{y}\|_\infty = 1$ and

$$\|A\|_\infty = \|A\mathbf{y}\|_\infty.$$

Then,

$$\begin{aligned}\|A\|_\infty &= \max_j \left| \sum_{k=1}^{n} a_{jk} y_k \right| \leq \max_j \left(\sum_{k=1}^{n} |a_{jk}|\,|y_k| \right) \\ &\leq \max_k |y_k| \cdot \max_j \sum_{k=1}^{n} |a_{jk}| = \|\mathbf{y}\|_\infty \cdot \max_j \sum_{k=1}^{n} |a_{jk}| \\ &= \max_j \sum_{k=1}^{n} |a_{jk}|,\end{aligned}$$

so the right-hand side of (8) is an upper bound of $\|A\|_\infty$. Now if the maximum row sum occurs for, say, $j = J$ then let $\mathbf{x}$ have the components

$$x_k = \begin{cases} \bar{a}_{Jk}/|a_{Jk}|, & a_{Jk} \neq 0 \\ 0, & a_{Jk} = 0, \end{cases} \qquad k = 1, 2, \ldots, n.$$

Clearly $\|\mathbf{x}\|_\infty = 1$, if A is non-trivial, and

$$\|A\mathbf{x}\|_\infty = \sum_{k=1}^{n} |a_{Jk}| \leq \|A\|_\infty,$$

so (8) holds. [If $A \equiv O$, property (ii) implies $\|A\| = 0$ for any natural norm.] ∎

(B) Next, we claim that

$$\|A\|_1 = \max_k \sum_{j=1}^{n} |a_{jk}|, \tag{9}$$

i.e., the *maximum absolute column sum*. Now let $\|\mathbf{y}\|_1 = 1$ and be such that

$$\|A\|_1 = \|A\mathbf{y}\|_1.$$

Then,

$$\begin{aligned}\|A\|_1 &= \sum_{j=1}^{n} \left| \sum_{k=1}^{n} a_{jk} y_k \right| \leq \sum_{j=1}^{n} \sum_{k=1}^{n} |a_{jk}| \cdot |y_k| \\ &= \sum_{k=1}^{n} \left(|y_k| \sum_{j=1}^{n} |a_{jk}| \right) \leq \sum_{k=1}^{n} |y_k| \left(\max_m \sum_{j=1}^{n} |a_{jm}| \right) \\ &= \|\mathbf{y}\|_1 \max_m \sum_{j=1}^{n} |a_{jm}| = \max_m \sum_{j=1}^{n} |a_{jm}|,\end{aligned}$$

and the right-hand side of (9) is an upper bound of $\|A\|_1$. If the maximum is attained for $m = K$, then this bound is actually attained for $\mathbf{x} = \mathbf{e}_K$, the Kth unit vector, since $\|\mathbf{e}_K\|_1 = 1$ and

$$\begin{aligned}\|A\mathbf{e}_K\|_1 &= \sum_{j=1}^{n} \left| \sum_{k=1}^{n} a_{jk} \delta_{kK} \right| \\ &= \sum_{j=1}^{n} |a_{jK}|.\end{aligned}$$

Thus (9) is established. ∎

(C) Finally, we consider the Euclidean norm, for which case we recall the notation for the *Hermitian transpose* or *conjugate transpose* of any rectangular matrix $A \equiv (a_{ij})$,

$$A^* \equiv \bar{A}^T,$$

i.e., if $A^* \equiv (b_{ij})$, then $b_{ij} = \bar{a}_{ji}$. Further, the *spectral radius* of any square matrix A is defined by

$$\rho(A) \equiv \max_s |\lambda_s(A)|, \tag{10}$$

where $\lambda_s(A)$ denotes the sth eigenvalue of A. Now we can state that

$$\|A\|_2 = \sqrt{\rho(A^*A)}. \tag{11}$$

To prove (11), we again pick $\mathbf{y}$ such that $\|\mathbf{y}\|_2 = 1$ and

$$\|A\|_2 = \|A\mathbf{y}\|_2.$$

From (1b) it is clear that $\|\mathbf{x}\|_2{}^2 = \mathbf{x}^*\mathbf{x}$, since $\mathbf{x}^* \equiv (\bar{x}_1, \bar{x}_2, \ldots, \bar{x}_n)$. Therefore, from the identity $(A\mathbf{y})^* = \mathbf{y}^*A^*$, we find

$$\begin{aligned}\|A\|_2{}^2 &= \|A\mathbf{y}\|_2{}^2 = (A\mathbf{y})^*(A\mathbf{y}) \\ &= \mathbf{y}^*A^*A\mathbf{y}.\end{aligned} \tag{12}$$

But since A^*A is Hermitian it has a complete set of n orthonormal eigenvectors, say $\mathbf{u}_1, \mathbf{u}_2, \ldots, \mathbf{u}_n$, such that

$$\mathbf{u}_j{}^*\mathbf{u}_k = \delta_{jk}, \tag{13a}$$

$$A^*A\mathbf{u}_s = \lambda_s\mathbf{u}_s. \tag{13b}$$

The multiplication of (13b) by $\mathbf{u}_s{}^*$ on the left yields further

$$\lambda_s = \mathbf{u}_s{}^*A^*A\mathbf{u}_s \geq 0.$$

Every vector has a unique expansion in the basis $\{\mathbf{u}_s\}$. Say in particular that

$$\mathbf{y} = \sum_{s=1}^{n} \alpha_s\mathbf{u}_s,$$

and then (12) becomes, upon recalling (13),

$$\begin{aligned}
\|A\|_2{}^2 &= \sum_{t=1}^{n} \bar{\alpha}_t\mathbf{u}_t{}^*A^*A \sum_{s=1}^{n} \alpha_s\mathbf{u}_s \\
&= \sum_{t=1}^{n} \bar{\alpha}_t\mathbf{u}_t{}^* \sum_{s=1}^{n} \alpha_s\lambda_s\mathbf{u}_s = \sum_{s=1}^{n} \lambda_s|\alpha_s|^2 \\
&\leq \max_s \lambda_s \sum_{t=1}^{n} |\alpha_t|^2 = \max_s \lambda_s = \rho(A^*A).
\end{aligned}$$

Thus $\rho^{1/2}(A^*A)$ is an upper bound of $\|A\|_2$. However, using $\mathbf{y} = \mathbf{u}_s$, where $\lambda_s = \rho(A^*A)$, we get

$$\begin{aligned}
\|A\mathbf{u}_s\|_2 &= (\mathbf{u}_s{}^*A^*A\mathbf{u}_s)^{1/2} \\
&= \rho^{1/2}(A^*A),
\end{aligned}$$

and so (11) follows. ■

We have observed that a matrix of order n can be considered as a vector of dimension n^2. But since every matrix norm satisfies the conditions (0)–(iii) of a vector norm the results of Lemma 1 and Theorem 2 also apply to matrix norms. Thus we have

LEMMA 1′. *Every matrix norm, $\|A\|$, is a continuous function of the n^2 elements a_{ij} of A.* ■

THEOREM 2′. *For each pair of matrix norms, say $\|A\|$ and $\|A\|'$, there exist positive constants m and M such that for all* nth *order matrices A*

$$m\|A\|' \leq \|A\| \leq M\|A\|'.$$ ■

The proofs of these results follow exactly the corresponding proofs for vectors norms so we leave their detailed exposition to the reader.

There is frequently confusion between the spectral radius (10) of a matrix and the Euclidean norm (11) of a matrix. (To add to this confusion, $\|A\|_2$ is sometimes called the *spectral norm* of A.) It should be observed that if A is Hermitian, i.e., $A^* = A$, then $\lambda_s(A^*A) = \lambda_s^2(A)$ and so *the spectral radius is equal to the Euclidean norm for Hermitian matrices.* However, in general this is not true, but we have

LEMMA 2. *For any natural norm, $\|\cdot\|$, and square matrix, A,*

$$\rho(A) \leq \|A\|.$$

Proof. For each eigenvalue $\lambda_s(A)$ there is a corresponding eigenvector, say $\mathbf{u}_s$, which can be chosen to be normalized for any particular vector norm, $\|\mathbf{u}_s\| = 1$. But then for the corresponding natural matrix norm

$$\|A\| = \max_{\|\mathbf{y}\|=1} \|A\mathbf{y}\| \geq \|A\mathbf{u}_s\| = \|\lambda_s\mathbf{u}_s\| = |\lambda_s|.$$

As this holds for all $s = 1, 2, \cdots, n$, the result follows. ∎

On the other hand, for each matrix some natural norm is arbitrarily close to the spectral radius. More precisely we have

THEOREM 3. *For each nth order matrix A and each arbitrary $\epsilon > 0$ a natural norm, $\|A\|$, can be found such that*

$$\rho(A) \leq \|A\| \leq \rho(A) + \epsilon.$$

Proof. The left-hand inequality has been verified above. We shall show how to construct a norm satisfying the right-hand inequality. By Theorem 1 we can find a non-singular matrix P such that

$$PAP^{-1} \equiv B \equiv \Lambda + U$$

where $\Lambda = (\lambda_j(A)\delta_{ij})$ and $U \equiv (u_{ij})$ has zeros on and below the diagonal. With $\delta > 0$, a "sufficiently small" positive number, we form the diagonal matrix of order n

$$D \equiv (\delta^{1-j}\delta_{ij}) = \begin{pmatrix} 1 & & & 0 \\ & \delta^{-1} & & \\ & & \ddots & \\ 0 & & & \delta^{1-n} \end{pmatrix}.$$

Now consider

$$C = DBD^{-1} = \Lambda + E,$$

where $E \equiv (e_{ij}) = DUD^{-1}$ has elements

$$e_{ij} = \begin{cases} 0, & j \leq i \\ u_{ij}\delta^{j-i}, & j > i, \end{cases} \qquad i = 1, 2, \ldots, n.$$

Note that the elements e_{ij} can be made arbitrarily small in magnitude by choosing δ appropriately. Also we have that

$$A = P^{-1}D^{-1}CDP.$$

Since DP is non-singular, a vector norm can be defined by

$$\|\mathbf{x}\| \equiv N_2(DP\mathbf{x}) = (\mathbf{x}^*P^*D^*DP\mathbf{x})^{1/2}.$$

The proof of this fact is left to the reader in Problem 5. The natural matrix norm induced by this vector norm is of course

$$\|A\| \equiv \max_{\|\mathbf{y}\|=1} \|A\mathbf{y}\|.$$

However, from the above form for A, we have, for any $\mathbf{y}$,

$$\|A\mathbf{y}\| = N_2(DPA\mathbf{y}) = N_2(CDP\mathbf{y}).$$

If we let $\mathbf{z} \equiv DP\mathbf{y}$, this becomes

$$\|A\mathbf{y}\| = N_2(C\mathbf{z}) = (\mathbf{z}^*C^*C\mathbf{z})^{1/2}.$$

Now observe that

$$\begin{aligned} C^*C &= (\Lambda^* + E^*)(\Lambda + E) \\ &= \Lambda^*\Lambda + \mathscr{M}(\delta). \end{aligned}$$

Here the term $\mathscr{M}(\delta)$ represents an nth order matrix each of whose terms is $\mathcal{O}(\delta)$.† Thus, we can conclude that

$$\begin{aligned} \mathbf{z}^*C^*C\mathbf{z} &\le \max_s |\lambda_s^2(A)|\, \mathbf{z}^*\mathbf{z} + |\mathbf{z}^*\mathscr{M}(\delta)\mathbf{z}| \\ &\le [\rho^2(A) + \mathcal{O}(\delta)]\mathbf{z}^*\mathbf{z}, \end{aligned}$$

since

$$|\mathbf{z}^*\mathscr{M}(\delta)\mathbf{z}| \le n^2\mathbf{z}^*\mathbf{z}\mathcal{O}(\delta) = \mathbf{z}^*\mathbf{z}\mathcal{O}(\delta).$$

Recalling $\|\mathbf{y}\| = N_2(\mathbf{z})$, we find from $\|\mathbf{y}\| = 1$ that $\mathbf{z}^*\mathbf{z} = 1$. Then it follows that

$$\begin{aligned} \|A\| &\le [\rho^2(A) + \mathcal{O}(\delta)]^{1/2} \\ &= \rho(A) + \mathcal{O}(\delta). \end{aligned}$$

For δ sufficiently small $\mathcal{O}(\delta) < \epsilon$. ■

† A quantity, say f, is said to be $\mathcal{O}(\delta)$, or briefly $f = \mathcal{O}(\delta)$ iff for some constants $K \ge 0$ and $\delta_0 > 0$,

$$\left|\frac{f}{\delta}\right| \le K \quad \text{for} \quad |\delta| \le \delta_0.$$

It should be observed that the natural norm employed in Theorem 3 depends upon the matrix A as well as the arbitrary small parameter ϵ. However, this result leads to an interesting characterization of the spectral radius of any matrix; namely,

COROLLARY. *For any square matrix A*

$$\rho(A) = \inf_{\{N(\cdot)\}} \left(\max_{N(\mathbf{x})=1} N(A\mathbf{x}) \right)$$

where the inf *is taken over all vector norms, $N(\cdot)$; or equivalently*

$$\rho(A) = \inf_{\{\|\cdot\|\}} \|A\|$$

where the inf *is taken over all natural norms, $\|\cdot\|$.*

Proof. By using Lemma 2 and Theorem 3, since $\epsilon > 0$ is arbitrary and the natural norm there depends upon ϵ, the result follows from the definition of inf. ∎

1.1. Convergent Matrices

To study the convergence of various iteration procedures as well as for many other purposes, we investigate matrices A for which

$$\lim_{m\to\infty} A^m = O, \tag{14}$$

where O denotes the *zero matrix* all of whose entries are 0. Any square matrix satisfying condition (14) is said to be *convergent*. Equivalent conditions are contained in

THEOREM 4. *The following three statements are equivalent:*

(a) *A is convergent;*
(b) $\lim_{m\to\infty} \|A^m\| = 0$, *for some matrix norm;*
(c) $\rho(A) < 1$.

Proof. We first show that (a) and (b) are equivalent. Since $\|\cdot\|$ is continuous, by Lemma 1′, and $\|O\| = 0$, then (a) implies (b). But if (b) holds for some norm, then Theorem 2′ implies there exists an M such that

$$\|A^m\|_\infty \leq M\,\|A^m\| \to 0.$$

Hence, (a) holds.

Next we show that (b) and (c) are equivalent. Note that by Theorem 2′ there is no loss in generality if we assume the norm to be a natural norm. But then, by Lemma 2 and the fact that $\lambda(A^m) = \lambda^m(A)$, we have

$$\|A^m\| \geq \rho(A^m) = \rho^m(A),$$

so that (b) implies (c). On the other hand, if (c) holds, then by Theorem 3 we can find an $\epsilon > 0$ and a natural norm, say $N(\cdot)$, such that

$$N(A) \leq \rho(A) + \epsilon \equiv \theta < 1.$$

Now use the property (iv) of matrix norms to get

$$N(A^m) \leq [N(A)]^m \leq \theta^m$$

so that $\lim_{m \to \infty} N(A^m) = 0$ and hence (b) holds. ∎

A test for convergent matrices which is frequently easy to apply is the content of the

COROLLARY. *A is convergent if for some matrix norm*

$$\|A\| < 1.$$

Proof. Again by (iv) we have

$$\|A^m\| \leq \|A\|^m$$

so that condition (b) of Theorem 4 holds. ∎

Another important characterization and property of convergent matrices is contained in

THEOREM 5. (a) *The geometric series*

$$I + A + A^2 + A^3 + \cdots,$$

converges iff A is convergent.

(b) *If A is convergent, then $I - A$ is non-singular and*

$$(I - A)^{-1} = I + A + A^2 + A^3 + \cdots.$$

Proof. A necessary condition for the series in part (a) to converge is that $\lim_{m \to \infty} A^m = O$, i.e., that A be convergent. The sufficiency will follow from part (*b*).

Let A be convergent, whence by Theorem 4 we know that $\rho(A) < 1$. Since the eigenvalues of $I - A$ are $1 - \lambda(A)$, it follows that $\det(I - A) \neq 0$ and hence this matrix is non-singular. Now consider the identity

$$(I - A)(I + A + A^2 + \cdots + A^m) = I - A^{m+1}$$

which is valid for all integers m. Since A is convergent, the limit as $m \to \infty$ of the right-hand side exists. The limit, after multiplying both sides on the left by $(I - A)^{-1}$, yields

$$(I + A + A^2 + \cdots) = (I - A)^{-1}$$

and part (b) follows. ∎

A useful corollary to this theorem is

COROLLARY. *If in some natural norm, $\|A\| < 1$, then $I - A$ is non-singular and*

$$\frac{1}{1 + \|A\|} \leq \|(I - A)^{-1}\| \leq \frac{1}{1 - \|A\|}.$$

Proof. By the corollary to Theorem 4 and part (b) of Theorem 5 it follows that $I - A$ is non-singular. For a natural norm we note that $\|I\| = 1$ and so taking the norm of the identity

$$I = (I - A)(I - A)^{-1}$$

yields

$$\begin{aligned} 1 &\leq \|(I - A)\| \cdot \|(I - A)^{-1}\| \\ &\leq (1 + \|A\|)\|(I - A)^{-1}\|. \end{aligned}$$

Thus the left-hand inequality is established.

Now write the identity as

$$(I - A)^{-1} = I + A(I - A)^{-1}$$

and take the norm to get

$$\|(I - A)^{-1}\| \leq 1 + \|A\| \cdot \|(I - A)^{-1}\|.$$

Since $\|A\| < 1$ this yields

$$\|(I - A)^{-1}\| \leq \frac{1}{1 - \|A\|}. \qquad \blacksquare$$

It should be observed that if A is convergent, so is $(-A)$, and $\|A\| = \|-A\|$. Thus Theorem 5 and its corollary are immediately applicable to matrices of the form $I + A$. That is, if in some natural norm, $\|A\| < 1$, then

$$\frac{1}{1 + \|A\|} \leq \|(I + A)^{-1}\| \leq \frac{1}{1 - \|A\|}.$$

PROBLEMS, SECTION 1

1. (a) Verify that (1b) defines a norm in the linear space of square matrices of order n; i.e., check properties (i)–(iv), for $\|A\|_E^2 = \sum_{ij} |a_{ij}|^2$.

(b) Similarly, verify that (1a) defines a matrix norm, i.e., $\|A\| = \sum_{ij} |a_{ij}|$.

2. Show by example that the maximum vector norm, $\eta(A) = \max_{i,j} |a_{ij}|$, when applied to a matrix, does not satisfy condition (iv) that we impose on a matrix norm.

3. Show that if A is non-singular, then $B \equiv A^*A$ is Hermitian and positive definite. That is, $\mathbf{x}^*B\mathbf{x} > 0$ if $\mathbf{x} \neq \mathbf{o}$. Hence the eigenvalues of B are all positive.

4. Show for any non-singular matrix A and any matrix norm that

$$\|I\| \geq 1 \quad \text{and} \quad \|A^{-1}\| \geq \frac{1}{\|A\|}.$$

[Hint: $\|I\| = \|II\| \leq \|I\|^2$; $\|A^{-1}A\| \leq \|A^{-1}\| \cdot \|A\|$.]

5. Show that if $\eta(\mathbf{x})$ is a norm and A is any non-singular matrix, then $N(\mathbf{x})$ defined by

$$N(\mathbf{x}) \equiv \eta(A\mathbf{x}),$$

is a (vector) norm.

6. We call $\eta(\mathbf{x})$ a *semi-norm* iff $\eta(\mathbf{x})$ satisfies all of the conditions, (0)–(iii), for a norm with condition (i) replaced by the weaker condition

(i′): $\eta(\mathbf{x}) \geq 0$ for all $\mathbf{x} \in \mathscr{V}$.

We say that $\eta(\mathbf{x})$ is *non-trivial* iff $\eta(\mathbf{x}) > 0$ for some $\mathbf{x} \in \mathscr{V}$. Prove the following generalization of Lemma 1:

LEMMA 1″. *Every non-trivial semi-norm, $\eta(\mathbf{x})$, is a continuous function of $x_1, x_2, \ldots, x_n$, the components of $\mathbf{x}$. Hence every semi-norm is continuous.*

7. Show that if $\eta(\mathbf{x})$ is a semi-norm and A any square matrix, then $N(\mathbf{x}) \equiv \eta(A\mathbf{x})$ defines a semi-norm.

2. FLOATING-POINT ARITHMETIC AND ROUNDING ERRORS

In the following chapters we will have to refer, on occasion, to the errors due to "rounding" in the basic arithmetic operations. Such errors are inherent in all computations in which only a fixed number of digits are retained. This is, of course, the case with all modern digital computers and we consider here an example of one way in which many of them do or can do arithmetic; so-called *floating-point arithmetic*. Although most electronic computers operate with numbers in some kind of binary representation, most humans still think in terms of a decimal representation and so we shall employ the latter here.

Suppose the number $a \neq 0$ has the exact decimal representation

$$a = \pm 10^q(.d_1d_2\cdots) \tag{1}$$

where q is an integer and the $d_1, d_2, \ldots,$ are digits with $d_1 \neq 0$. Then the "t-digit floating-decimal representation of a," or for brevity the "floating a" used in the machine, is of the form

$$\text{fl}(a) \equiv \pm 10^q(.\delta_1\delta_2\cdots\delta_t) \tag{2}$$

where $\delta_1 \neq 0$ and $\delta_1, \delta_2, \ldots, \delta_t$ are digits. The number $(.\delta_1\delta_2\cdots\delta_t)$ is called the *mantissa* and q is called the *exponent* of fl(a). There is usually a restriction on the exponent, of the form

$$(3) \qquad -N \leq q \leq M,$$

for some large positive integers N, M. If a number $a \neq 0$ has an exponent outside of this range it cannot be represented in the form (2), (3). If, during the course of a calculation, some computed quantity has an exponent $q > M$ (called *overflow*) or $q < -N$ (called *underflow*), meaningless results usually follow. However, special precautions can be taken on most computers to at least detect the occurrence of such over- or underflows. We do not consider these practical difficulties further; rather, we shall assume that they do not occur or are somehow taken into account.

There are two popular ways in which the floating digits δ_j are obtained from the exact digits, d_j. The obvious *chopping* representation takes

$$(4) \qquad \delta_j = d_j, \qquad j = 1, 2, \ldots, t.$$

Thus the exact mantissa is chopped off after the tth decimal digit to get the floating mantissa. The other and preferable procedure is to *round*, in which case†

$$(5) \qquad \delta_1\delta_2\cdots\delta_t = [d_1d_2\cdots d_t.d_{t+1} + 0.5]$$

and the brackets on the right-hand side indicate the integral part. The error in either of these procedures can be bounded as in

LEMMA 1. *The error in t-digit floating-decimal representation of a number $a \neq 0$ is bounded by*

$$|a - \text{fl}(a)| \leq 5|a|10^{-t}p \quad \begin{cases} p = 1, & \textit{rounded,} \\ p = 2, & \textit{chopped.} \end{cases}$$

Proof. From (1), (2), and (4) we have

$$\begin{aligned} |a - \text{fl}(a)| &= 10^{q-t}(.d_{t+1}d_{t+2}\cdots) \\ &= 10^{q-t}(.d_{t+1}d_{t+2}\cdots)\frac{|a|}{|a|} \\ &= 10^{-t}\frac{(.d_{t+1}d_{t+2}\cdots)}{(.d_1d_2\cdots)}|a|. \end{aligned}$$

† For simplicity we are neglecting the special case that occurs when $d_1 = d_2 = \cdots = d_t = 9$ and $d_{t+1} \geq 5$. Here we would increase the exponent q in (2) by unity and set $\delta_1 = 1$, $\delta_j = 0$, $j > 1$. Note that when $d_{t+1} = 5$, if we were to round up iff d_t is odd, then an *unbiased rounding* procedure would result. Some electronic computers employ an unbiased rounding procedure (in a binary system).

But since $1 \le d_1 \le 9$ and $0.d_{t+1}d_{t+2}\cdots \le 1$ this implies

$$|a - \text{fl}(a)| \le 10^{1-t}|a|,$$

which is the bound for the chopped representation. For the case of rounding we have, similarly,

$$|a - \text{fl}(a)| \le \tfrac{1}{2}\, 10^{q-t} = \tfrac{1}{2}\, 10^{q-t}\frac{|a|}{|a|} \le 5|a|10^{-t}. \qquad \blacksquare$$

We shall assume that our idealized computer performs each basic arithmetic operation correctly to $2t$ digits and then either rounds or chops the result to a t-digit floating number. With such operations it clearly follows from Lemma 1 that

$$\left.\begin{aligned} &\text{(6a)}\quad \text{fl}(a \pm b) = (a \pm b)(1 + \phi 10^{-t}) \\ &\text{(6b)}\quad \text{fl}(ab) \quad\;\; = a\cdot b(1 + \phi 10^{-t}) \\ &\text{(6c)}\quad \text{fl}\left(\frac{a}{b}\right) \quad = \frac{a}{b}(1 + \phi 10^{-t}) \end{aligned}\right\} \quad \begin{aligned} &0 \le |\phi| \le 5 \quad \text{rounding,} \\ &0 \le |\phi| \le 10 \quad \text{chopping.} \end{aligned}$$

In many calculations, particularly those concerned with linear systems, the accumulation of products is required (e.g., the inner product of two vectors). We assume that rounding (or chopping) is done after each multiplication and after each successive addition. That is,

$$\text{(7a)}\qquad \text{fl}(a_1b_1 + a_2b_2) = [a_1b_1(1 + \phi_1 10^{-t}) + a_2b_2(1 + \phi_2 10^{-t})](1 + \theta 10^{-t})$$

and in general

$$\text{(7b)}\qquad \text{fl}\left(\sum_{i=1}^{n} a_ib_i\right) = \text{fl}\left[\text{fl}\left(\sum_{i=1}^{n-1} a_ib_i\right) + \text{fl}(a_nb_n)\right].$$

The result of such computations can be represented as an exact inner product with, say, the a_i slightly altered. We state this as

LEMMA **2**. *Let the floating-point inner product* (7) *be computed with* ***rounding***. *Then if n and t satisfy*

$$\text{(8)}\qquad n10^{1-t} \le 1$$

it follows that

$$\text{(9a)}\qquad \text{fl}\left(\sum_{i=1}^{n} a_ib_i\right) = \sum_{i=1}^{n} (a_i + \delta a_i)b_i$$

where

$$\text{(9b)}\qquad |\delta a_1| \le n|a_1|10^{1-t}, \qquad |\delta a_i| \le (n - i + 2)|a_i|10^{1-t}, \quad i = 2, 3, \ldots, n.$$

Proof. By (6b) we can write

$$\text{fl}(a_k b_k) = a_k b_k(1 + \phi_k 10^{-t}), \qquad |\phi_k| \le 5,$$

since rounding is assumed. Similarly from (6a) and (7b) with $n = k$ we have

$$\text{fl}\left(\sum_{i=1}^{k} a_i b_i\right) = \left[\text{fl}\left(\sum_{i=1}^{k-1} a_i b_i\right) + (a_k b_k)(1 + \phi_k 10^{-t})\right](1 + \theta_k 10^{-t})$$

where

$$\theta_1 = 0; \qquad |\theta_k| \le 5, \qquad k = 2, 3, \ldots .$$

Now a simple recursive application of the above yields

$$\begin{aligned}\text{fl}\left(\sum_{i=1}^{n} a_i b_i\right) &= \sum_{k=1}^{n}\left[a_k b_k(1 + \phi_k 10^{-t}) \prod_{j=k}^{n} (1 + \theta_j 10^{-t})\right] \\ &\equiv \sum_{k=1}^{n} a_k b_k (1 + E_k),\end{aligned}$$

where we have introduced E_k by

$$1 + E_k \equiv (1 + \phi_k 10^{-t}) \prod_{j=k}^{n} (1 + \theta_j 10^{-t}).$$

A formal verification of this result is easily obtained by induction.

Since $\theta_1 = 0$, it follows that

$$(1 - 5\cdot 10^{-t})^{n-k+2} \le 1 + E_k \le (1 + 5\cdot 10^{-t})^{n-k+2}, \qquad k = 2, 3, \ldots, n,$$

and

$$(1 - 5\cdot 10^{-t})^{n} \le 1 + E_1 \le (1 + 5\cdot 10^{-t})^{n}.$$

Hence, with $\epsilon = 5\cdot 10^{-t}$,

$$\begin{aligned}|E_1| &\le (1 + \epsilon)^n - 1, \\ |E_k| &\le (1 + \epsilon)^{n-k+2} - 1, \qquad k = 2, 3, \ldots, n.\end{aligned}$$

But, for $p \le n$, (8) implies that $p\epsilon \le \frac{1}{2}$, so that

$$\begin{aligned}(1 + \epsilon)^p - 1 &\equiv p\epsilon\left(1 + \frac{p-1}{2}\epsilon + \frac{p-1}{2}\frac{p-2}{3}\epsilon^2 + \cdots\right) \\ &\le p\epsilon(1 + \tfrac{1}{2} + (\tfrac{1}{2})^2 + \cdots) \\ &\le 2p\epsilon = p\,10^{1-t}.\end{aligned}$$

Therefore,

$$|E_k| \le (n - k + 2)10^{1-t}, \qquad k = 2, 3, \ldots, n.$$

Clearly for $k = 1$ we find, as above with $k = 2$, that

$$|E_1| \leq n \cdot 10^{1-t}.$$

The result now follows upon setting

$$\delta a_k = a_k E_k.$$

(Note that we could just as well have set $\delta b_k = b_k E_k$.) ∎

Obviously a similar result can be obtained for the error due to chopping if condition (8) is strengthened slightly; see Problem 1.

PROBLEMS, SECTION 2

1. Determine the result analogous to Lemma 2, when "chopping" replaces "rounding" in the statement.

[Hint: The factor 10^{1-t} need only be replaced by $2 \cdot 10^{1-t}$, throughout.]

2. (a) Find a representation for $\mathrm{fl}\left(\sum_{i=1}^{n} c_i\right)$.

(b) If $c_1 > c_2 > \cdots > c_n > 0$, in what order should $\mathrm{fl}\left(\sum_{i=1}^{n} c_i\right)$ be calculated to minimize the effect of rounding?

3. What are the analogues of equations (6a, b, c) in the binary representation:

$$\mathrm{fl}(a) = \pm 2^q(.\delta_1\delta_2 \cdots \delta_t)$$

where $\delta_1 = 1$ and $\delta_j = 0$ or 1?

3. WELL-POSED COMPUTATIONS

Hadamard introduced the notion of well-posed or properly posed problems in the theory of partial differential equations (see Section 0 of Chapter 9). However, it seems that a related concept is quite useful in discussing computational problems of almost all kinds. We refer to this as the notion of a well-posed computing problem.

First, we must clarify what is meant by a "computing problem" in general. Here we shall take it to mean an *algorithm* or equivalently: *a set of rules specifying the order and kind of arithmetic operations* (*i.e., rounding rules*) *to be used on specified data.* Such a computing problem may have as its object, for example, the determination of the roots of a quadratic equation or of an approximation to the solution of a nonlinear partial differential equation. How any such rules are determined for a particular purpose need not concern us at present (this is, in fact, what much of the rest of this book is about).

Suppose the specified data for some particular computing problem are the quantities $a_1, a_2, \ldots, a_m$, which we denote as the m-dimensional vector $\mathbf{a}$. Then if the quantities to be computed are $x_1, x_2, \ldots, x_n$, we can write

$$\mathbf{x} = \mathbf{f}(\mathbf{a}), \tag{1}$$

where of course the n-dimensional function $\mathbf{f}(\cdot)$ is determined by the rules.

Now we will define a computing problem to be *well-posed* iff the algorithm meets **three requirements.** The *first requirement* is that a "solution," $\mathbf{x}$, should *exist* for the given data, $\mathbf{a}$. This is implied by the notation (1). However, if we recall that (1) represents the evaluation of some algorithm it would seem that a solution (i.e., a result of using the algorithm) must always exist. But this is not true, a trivial example being given by data that lead to a division by zero in the algorithm. (The algorithm in this case is not properly specified since it should have provided for such a possibility. If it did not, then the corresponding computing problem is *not well-posed* for data that lead to this difficulty.) There are other, more subtle situations that result in algorithms which cannot be evaluated and it is by no means easy, a priori, to determine that $\mathbf{x}$ is indeed defined by (1).

The *second requirement* is that the computation be *unique*. That is, when performed several times (with the same data) identical results are obtained. This is quite invariably true of algorithms which can be evaluated. If in actual practice it seems to be violated, the trouble usually lies with faulty calculations (i.e., machine errors). The functions $\mathbf{f}(\mathbf{a})$ must be single valued to insure uniqueness.

The *third requirement* is that the result of the computation should depend *Lipschitz continuously on the data with a constant that is not too large*. That is, "small" changes in the data, $\mathbf{a}$, should result in only "small" changes in the computed $\mathbf{x}$. For example, let the computation represented by (1) satisfy the first two requirements for all data $\mathbf{a}$ in some set, say $\mathbf{a} \in D$. If we change the data $\mathbf{a}$ by a small amount $\boldsymbol{\delta}\mathbf{a}$ so that $(\mathbf{a} + \boldsymbol{\delta}\mathbf{a}) \in D$, then we can write the result of the computation with the altered data as

$$\mathbf{x} + \boldsymbol{\delta}\mathbf{x} = \mathbf{f}(\mathbf{a} + \boldsymbol{\delta}\mathbf{a}). \tag{2}$$

Now if there exists a constant M such that for any $\boldsymbol{\delta}\mathbf{a}$,

$$\|\boldsymbol{\delta}\mathbf{x}\| \leq M\|\boldsymbol{\delta}\mathbf{a}\|, \tag{3}$$

we say that the computation depends Lipschitz continuously on the data. Finally, we say (1) is *well-posed* iff the three requirements are satisfied and (3) holds with a not too large constant, $M = M(\mathbf{a}, \eta)$, for some not too small $\eta > 0$ and all $\boldsymbol{\delta}\mathbf{a}$ such that $\|\boldsymbol{\delta}\mathbf{a}\| \leq \eta$. Since the Lipschitz constant

M depends on $(\mathbf{a}, \eta)$ we see that a computing problem or algorithm may be well-posed for some data, $\mathbf{a}$, but not for all data.

Let $\mathscr{P}(\mathbf{a})$ denote the original problem which the algorithm (1) was devised to "solve." This problem is also said to be *well-posed* if it has a unique solution, say

$$\mathbf{y} = \mathbf{g}(\mathbf{a}),$$

which depends Lipschitz continuously on the data. That is, $\mathscr{P}(\mathbf{a})$ is well-posed if for all $\boldsymbol{\delta}\mathbf{a}$ satisfying $\|\boldsymbol{\delta}\mathbf{a}\| \le \zeta$, there is a constant $N = N(\mathbf{a}, \zeta)$ such that

$$\|\mathbf{g}(\mathbf{a} + \boldsymbol{\delta}\mathbf{a}) - \mathbf{g}(\mathbf{a})\| \le N\|\boldsymbol{\delta}\mathbf{a}\|. \tag{4}$$

We call the algorithm (1) *convergent* iff $\mathbf{f}$ depends on a parameter, say ϵ (e.g., ϵ may determine the size of the rounding errors), so that for any small $\epsilon > 0$,

$$\|\mathbf{f}(\mathbf{a} + \boldsymbol{\delta}\mathbf{a}) - \mathbf{g}(\mathbf{a} + \boldsymbol{\delta}\mathbf{a})\| \le \epsilon, \tag{5}$$

for all $\boldsymbol{\delta}\mathbf{a}$ such that $\|\boldsymbol{\delta}\mathbf{a}\| \le \delta$. Now, if $\mathscr{P}(\mathbf{a})$ is well-posed and (1) is convergent, then (4) and (5) yield

$$\begin{aligned} \|\mathbf{f}(\mathbf{a}) - \mathbf{f}(\mathbf{a} + \boldsymbol{\delta}\mathbf{a})\| &\le \|\mathbf{f}(\mathbf{a}) - \mathbf{g}(\mathbf{a})\| + \|\mathbf{g}(\mathbf{a}) - \mathbf{g}(\mathbf{a} + \boldsymbol{\delta}\mathbf{a})\| \\ &\qquad + \|\mathbf{g}(\mathbf{a} + \boldsymbol{\delta}\mathbf{a}) - \mathbf{f}(\mathbf{a} + \boldsymbol{\delta}\mathbf{a})\| \\ &\le \epsilon + N\|\boldsymbol{\delta}\mathbf{a}\| + \epsilon. \end{aligned} \tag{6}$$

Thus, recalling (3), we are led to the heuristic

OBSERVATION **1**. *If* $\mathscr{P}(\mathbf{a})$ *is a well-posed problem, then a necessary condition that* (1) *be a convergent algorithm is that* (1) *be a well-posed computation.*

Therefore we are interested in determining whether a given algorithm (1) is a well-posed computation simply because *only such an algorithm is sure to be convergent* for all problems of the form $\mathscr{P}(\mathbf{a} + \boldsymbol{\delta}\mathbf{a})$, when $\mathscr{P}(\mathbf{a})$ is well-posed and $\|\boldsymbol{\delta}\mathbf{a}\| \le \delta$.

Similarly, by interchanging $\mathbf{f}$ and $\mathbf{g}$ in (6), we may justify

OBSERVATION **2**. *If* $\mathscr{P}$ *is a not well-posed problem, then a necessary condition that* (1) *be an accurate algorithm is that* (1) *be a not well-posed computation.*

In fact, for certain problems of linear algebra (see Subsection 1.2 of Chapter 2), it has been possible to prove that the commonly used algorithms, (1), produce approximations, $\mathbf{x}$, which are exact solutions of slightly perturbed original mathematical problems. In these algebraic cases, the accuracy of the solution $\mathbf{x}$, as measured in (5), is seen to depend on the well-posedness of the original mathematical problem. In algorithms,

(1), that arise from differential equation problems, other techniques are developed to estimate the accuracy of the approximation. For differential equation problems the well-posedness of the resulting algorithms (1) is referred to as the *stability* of the finite difference schemes (see Chapters 8 and 9).

We now consider two elementary examples to illustrate some of the previous notions.

The most overworked example of how a simple change in the algorithm can affect the accuracy of a single precision calculation is the case of determining the smallest root of a quadratic equation. If in

$$x^2 + 2bx + c,$$

$b < 0$ and c are given to t digits, but $|c|/b^2 < 10^{-t}$, then the smallest root, x_2, should be found from $x_2 = c/x_1$, after finding $x_1 = -b + \sqrt{b^2 - c}$ in single precision arithmetic. Using

$$x_2 = -b - \sqrt{b^2 - c}$$

in single precision arithmetic would be disastrous!

A more sophisticated well-posedness discussion, without reference to the type of arithmetic, is afforded by the problem of determining the zeros of a polynomial

$$P_n(z) \equiv z^n + a_{n-1}z^{n-1} + \cdots + a_1 z + a_0.$$

If $Q_n(z) \equiv z^n + b_{n-1}z^{n-1} + \cdots + b_1 z + b_0$, then the zeros of $P_n(z; \epsilon) \equiv P_n(z) + \epsilon Q_n(z)$ are "close" to the zeros of $P_n(z)$. That is, in the theory of functions of a complex variable it is shown that

LEMMA. *If $z = z_1$ is a simple zero of $P_n(z)$, then for $|\epsilon|$ sufficiently small $P_n(z; \epsilon)$ has a zero $z_1(\epsilon)$, such that*

$$\left| z_1(\epsilon) - z_1 + \epsilon \frac{Q_n(z_1)}{P_n'(z_1)} \right| = \mathcal{O}(\epsilon^2).$$

If z_1 is a zero of multiplicity r of $P_n(z)$, there are r neighboring zeros of $P_n(z; \epsilon)$ with

$$\left| z_1(\epsilon) - z_1 - \left[-\frac{r!\,Q_n(z_1)}{P_n^{(r)}(z_1)} \right]^{1/r} \epsilon^{1/r} \right| = \mathcal{O}(\epsilon^{2/r}).$$

Now it is clear that in the case of a simple zero, z_1, the computing problem, to determine the zero, might be well-posed if $P_n'(z_1)$ were not too small and $Q_n(z_1)$ not too large, since then $|z_1(\epsilon) - z_1|/|\epsilon|$ would not be large for small ϵ. On the other hand, the determination of the multiple root would most likely lead to a not well-posed computing problem.

The latter example illustrates Observation (2), that is, a computing problem is not well-posed if the original mathematical problem is not well-posed. On the other hand, the example of the quadratic equation indicates how an ill-chosen formulation of an algorithm may be well-posed but yet *inaccurate* in single precision.

Given an $\epsilon > 0$ and a problem $\mathscr{P}(\mathbf{a})$ we do not, in general, know how to determine an algorithm, (1), that requires the least amount of work to find $\mathbf{x}$ so that $\|\mathbf{x} - \mathbf{y}\| \leq \epsilon$. This is an important aspect of algorithms for which there is no general mathematical theory. For most of the algorithms that are described in later chapters, we estimate the number of arithmetic operations required to find $\mathbf{x}$.

PROBLEM, SECTION 3

1. For the quadratic equation

$$x^2 + 2bx + c = 0,$$

find the small root by using single precision arithmetic in the iterative schemes

(a) $x_{n+1} = -\dfrac{c}{2b} - \dfrac{x_n^2}{2b}$,

and

(b) $x_{n+1} = x_n - \dfrac{x_n^2 + 2bx_n + c}{2x_n + 2b}$.

If your computer has a mantissa with approximately $t = 2p$ digits, use

$$c = 1, \qquad b = -10^p$$

for the two initial values

$$\text{(i) } x_0 = 0; \qquad \text{(ii) } x_0 = -\frac{c}{b}.$$

Which scheme gives the smaller root to approximately t digits with the smaller number of iterations? Which scheme requires less work?

2

Numerical Solution of Linear Systems and Matrix Inversion

0. INTRODUCTION

Finding the solution of a linear algebraic equation system of "large" order and calculating the inverse of a matrix of "large" order can be difficult numerical tasks. While in principle there are standard methods for solving such problems, the difficulties are practical and stem from

(a) the labor required in a lengthy sequence of calculations,

and

(b) the possible loss of accuracy in such lengthy calculations performed with a fixed number of decimal places.

The first difficulty renders manual computation impractical and the second limits the applicability of high speed digital computers with fixed "word" length. Thus to determine the feasibility of solving a particular problem with given equipment, several questions should be answered:

(i) How many arithmetic operations are required to apply a proposed method?

(ii) What will be the accuracy of a solution to be found by the proposed method (*a priori estimate*)?

(iii) How can the accuracy of the computed answer be checked (*a posteriori estimate*)?

The first question can frequently† be answered in a straightforward manner and this is done, by means of an "operational count," for most

† For "direct" methods, the operational count is easily made; while for "indirect" or iterative methods, the operational count is made by multiplying the estimated number of iterations by the operational count per iteration.

of the methods in this chapter. The third question can be easily answered if we have a bound for the norm of the inverse matrix. We therefore indicate, in Subsection 1.3, how such a bound may be obtained if we have an *approximate* inverse. However, the second question has only been recently answered for some methods. After discussing the notions of "well-posed problem" and "condition number" of a matrix, we give an account of Wilkinson's a priori estimate for the Gaussian elimination method in Subsection 1.2. This treatment, in Section 1, of the Gaussian elimination method is followed, in Section 2, by a discussion of some modifications of the procedure. Direct factorization methods, which include Gaussian elimination as a special case, are described in Section 3. Iterative methods and techniques for accelerating them are studied in the remaining three sections.

The matrix inversion problem may be formulated as follows: *Given a square matrix of order n,*

$$A \equiv (a_{ij}) \equiv \begin{pmatrix} a_{11} & a_{12} & \cdots & a_{1n} \\ a_{21} & a_{22} & \cdots & a_{2n} \\ \vdots & \vdots & & \vdots \\ a_{n1} & a_{n2} & \cdots & a_{nn} \end{pmatrix} \tag{1}$$

find its inverse, i.e., a square matrix of order n, say A^{-1}, such that

$$A^{-1}A = AA^{-1} = I \equiv (\delta_{ij}). \tag{2}$$

Here I is the nth order identity matrix whose elements are given by the Kronecker delta:

$$\delta_{ij} \equiv \begin{cases} 0, & \text{if } i \neq j; \\ 1, & \text{if } i = j. \end{cases} \tag{3}$$

It is well known that this problem has one and only one solution iff the determinant of A is non-zero ($\det A \neq 0$), i.e., iff A is non-singular.

The problem of solving a general linear system is formulated as follows: Given a square matrix A and an arbitrary n-component column vector $\mathbf{f}$, find a vector $\mathbf{x}$ which satisfies

$$A\mathbf{x} = \mathbf{f}, \tag{4a}$$

or, in component form,

$$\begin{aligned} a_{11}x_1 + a_{12}x_2 + \cdots + a_{1n}x_n &= f_1, \\ a_{21}x_1 + a_{22}x_2 + \cdots + a_{2n}x_n &= f_2, \\ &\vdots \\ a_{n1}x_1 + a_{n2}x_2 + \cdots + a_{nn}x_n &= f_n. \end{aligned} \tag{4b}$$

Again it is known that this problem has a solution which is unique for *every* inhomogeneous term **f**, iff A is non-singular. [If A is singular the system (4) will have a solution only for *special* vectors **f** and such a solution is not unique. The numerical solution of such "singular" problems is briefly touched on in Section 1 and in Problems 1.3, 1.4 of Chapter 4.]

It is easy to see that the problem of matrix inversion is equivalent to that of solving linear systems. For, let the inverse of A, assumed non-singular, be known and have elements c_{ij}, that is,

$$A^{-1} \equiv (c_{ij}).$$

Then multiplication of (4a) on the left by A^{-1} yields, since $I\mathbf{x} = \mathbf{x}$,

$$\mathbf{x} = A^{-1}\mathbf{f}, \tag{5a}$$

or componentwise,

$$x_i = \sum_{j=1}^{n} c_{ij} f_j, \qquad i = 1, 2, \ldots, n. \tag{5b}$$

Thus when the c_{ij} are known it requires, at most, n multiplications and $(n - 1)$ additions to evaluate each component of the solution, or, for the complete solution, a total of n^2 multiplications and $n(n - 1)$ additions.

On the other hand, assume a procedure is known for solving the non-singular system (4) with an *arbitrary* inhomogeneous term **f**. We then consider the n special systems

$$A\mathbf{x} = \mathbf{e}^{(j)}, \qquad j = 1, 2, \ldots, n, \tag{6}$$

where $\mathbf{e}^{(j)}$ is the jth column of the identity matrix; that is, the elements of $\mathbf{e}^{(j)}$ are $e_i^{(j)} = \delta_{ij}$, $i = 1, 2, \ldots, n$. The solutions of these systems are n vectors which we call $\mathbf{x}^{(j)}$, $j = 1, 2, \ldots, n$; the components of $\mathbf{x}^{(j)}$ are denoted by $x_i^{(j)}$. With these vectors we form the square matrix

$$X \equiv (x_i^{(j)}), \tag{7}$$

in which the jth column is the solution $\mathbf{x}^{(j)}$ of the jth system in (6). Then it follows from the row by column rule for matrix multiplication that

$$AX = (\delta_{ij}) = I. \tag{8}$$

Since A was assumed to be non-singular, we find upon multiplying both sides of (8) on the left by A^{-1} that

$$X = A^{-1}.$$

Thus, by solving the n special systems (6) the inverse may be computed; this is the procedure generally used in practice. The number of operations required is, *at most*, n times that required to solve a single system. However, this number can be reduced by efficiently organizing the computations and by taking account of the special form of the inhomogeneous terms, $\mathbf{e}^{(j)}$, as we shall explain later in Subsection 1.1.

PROBLEMS, SECTION 0

1. If the columns of A form a set of vectors such that at most c of the columns are linearly independent, then we say that the *column rank* of A is c. (Similarly define the *row rank* of A to be r by replacing "columns" by "rows" and "c" by "r.") Prove that the row rank of A equals the column rank of A.

[Hint: Let row rank $(A) \equiv r$; and use a set of r rows of A to define a submatrix B with row rank $(B) = r$. Then show that $c \equiv$ column rank $(A) =$ column rank (B). Hence, since B has r rows, $c \leq r$. Similarly show that $r \leq c$.] Hence define *rank of A* by rank $(A) \equiv r = c$.

2. (*Alternative Principle*) If A is of order n, then either

$$A\mathbf{x} = \mathbf{o} \qquad \text{iff } \mathbf{x} = \mathbf{o},$$

or else

$$r \equiv \text{rank}\,(A) < n$$

and there exist a finite number, p, of linearly independent solutions $\{\mathbf{x}^{(j)}\}$ that span the null space of A, i.e.,

$$A\mathbf{x}^{(j)} = \mathbf{o}, \qquad j = 1, 2, \ldots, p,$$

and if $A\mathbf{x} = \mathbf{o}$ there exist constants $\{\alpha_j\}$ such that $\mathbf{x} = \sum_{j=1}^{p} \alpha_j \mathbf{x}^{(j)}$. Show that $p = n - r$.

3. Observe that

$$A\mathbf{x} = \mathbf{f} \tag{9}$$

has a solution iff $\mathbf{f}$ is a linear combination of the columns of A. Hence show that: (9) has a solution $\mathbf{x}$ iff rank $(A) =$ rank $(A, \mathbf{f})$; (9) has a solution $\mathbf{x}$ iff $\mathbf{y}^T\mathbf{f} = 0$ for all vectors $\mathbf{y} \neq \mathbf{o}$ such that $\mathbf{y}^T A = \mathbf{o}$. (In this problem, A may be rectangular; $(A, \mathbf{f})$ is the *augmented matrix*).

1. GAUSSIAN ELIMINATION

The best known and most widely used method for solving linear systems of algebraic equations and for inverting matrices is attributed to Gauss. It is, basically, the elementary procedure in which the "first" equation is used to eliminate the "first" variable from the last $n - 1$ equations, then the new "second" equation is used to eliminate the "second" variable from the last $n - 2$ equations, etc. If $n - 1$ such eliminations can be performed, then the resulting linear system which is equivalent† to the

† Two linear systems are *equivalent* iff every solution of one is a solution of the other.

original system is triangular and is easily solved. Of course, the ordering of the equations in the system and of the unknowns in the equations is arbitrary and so there is no unique order in which the procedure must be employed. As we shall see, the ordering is important since some orderings may not permit $n - 1$ eliminations, while among the permissible orderings, some are to be preferred since they yield more accurate results.

In order to describe the specific sequence of arithmetic operations used in Gaussian elimination, we will first use the natural order in which the system is given, say

(1a) $$A\mathbf{x} = \mathbf{f},$$

where

(1b) $$A \equiv (a_{ij}), \qquad \mathbf{x} \equiv \begin{pmatrix} x_1 \\ x_2 \\ \vdots \\ x_n \end{pmatrix}, \qquad \mathbf{f} \equiv \begin{pmatrix} f_1 \\ f_2 \\ \vdots \\ f_n \end{pmatrix}.$$

Before the variable x_k is eliminated, we denote the equivalent system (i.e., the reduced system), from which $x_1, x_2, \ldots, x_{k-1}$ have been eliminated, by

(2a) $$A^{(k)}\mathbf{x} = \mathbf{f}^{(k)}, \qquad k = 1, 2, \ldots, n,$$

where

(2b) $$A^{(k)} \equiv (a_{ij}^{(k)}), \qquad \mathbf{f}^{(k)} \equiv \begin{pmatrix} f_1^{(k)} \\ f_2^{(k)} \\ \vdots \\ f_n^{(k)} \end{pmatrix}.$$

For $k = 1$ we have $A^{(1)} \equiv A$, $\mathbf{f}^{(1)} \equiv \mathbf{f}$, and the elements in (2b) for $k = 2, 3, \ldots, n$, are computed recursively by

(3a) $$a_{ij}^{(k)} = \begin{cases} a_{ij}^{(k-1)} & \text{for } i \leq k-1, \\ 0 & \text{for } i \geq k, j \leq k-1, \\ a_{ij}^{(k-1)} - \dfrac{a_{i,k-1}^{(k-1)}}{a_{k-1,k-1}^{(k-1)}} a_{k-1,j}^{(k-1)} & \text{for } i \geq k, j \geq k; \end{cases}$$

(3b) $$f_i^{(k)} = \begin{cases} f_i^{(k-1)} & \text{for } i \leq k-1, \\ f_i^{(k-1)} - \dfrac{a_{i,k-1}^{(k-1)}}{a_{k-1,k-1}^{(k-1)}} f_{k-1}^{(k-1)} & \text{for } i \geq k. \end{cases}$$

These formulae represent the result of multiplying the $(k - 1)$st equation in $A^{(k-1)}\mathbf{x} = \mathbf{f}^{(k-1)}$ by the ratio $(a_{i,k-1}^{(k-1)}/a_{k-1,k-1}^{(k-1)})$ and subtracting the result from the ith equation for all $i \geq k$. In this way, the variable x_{k-1} is eliminated from the last $n - k + 1$ equations. The resulting coefficient matrix and inhomogeneous term have the forms

$$(4) \quad A^{(k)} \equiv \begin{bmatrix} a_{11}^{(1)} & a_{12}^{(1)} & a_{13}^{(1)} & \cdots & a_{1,k-1}^{(1)} & a_{1k}^{(1)} & \cdots & a_{1n}^{(1)} \\ 0 & a_{22}^{(2)} & a_{23}^{(2)} & \cdots & a_{2,k-1}^{(2)} & a_{2k}^{(2)} & \cdots & a_{2n}^{(2)} \\ 0 & 0 & a_{33}^{(3)} & \cdots & \cdot & \cdot & & \cdot \\ & & & & \cdot & \cdot & & \cdot \\ & & 0 & & \cdot & \cdot & & \cdot \\ & & & & a_{k-1,k-1}^{(k-1)} & a_{k-1,k}^{(k-1)} & \cdots & a_{k-1,n}^{(k-1)} \\ & & & & 0 & a_{kk}^{(k)} & \cdots & a_{kn}^{(k)} \\ & & & & 0 & a_{k+1,k}^{(k)} & \cdots & a_{k+1,n}^{(k)} \\ & & & & \vdots & \vdots & & \vdots \\ 0 & 0 & 0 & \cdots & 0 & a_{nk}^{(k)} & \cdots & a_{nn}^{(k)} \end{bmatrix}; \quad \mathbf{f}^{(k)} \equiv \begin{bmatrix} f_1^{(1)} \\ f_2^{(2)} \\ \cdot \\ \cdot \\ \cdot \\ f_{k-1}^{(k-1)} \\ f_k^{(k)} \\ f_{k+1}^{(k)} \\ \vdots \\ f_n^{(k)} \end{bmatrix}.$$

It has been assumed above that the elements $a_{kk}^{(k)} \neq 0$ for $k = 1, 2, \ldots, n$. When this is the case we have

THEOREM **1**. *Let the matrix A be such that the Gaussian elimination procedure defined in (2)–(3) (i.e., in the natural order) yields non-zero diagonal elements $a_{kk}^{(k)}$, $k = 1, 2, \ldots, n$. Then A is non-singular and in fact,*

$$\text{(5a)} \qquad \det A = a_{11}^{(1)} a_{22}^{(2)} \cdots a_{nn}^{(n)}.$$

The final matrix $A^{(n)} \equiv U$ is upper triangular and A has the factorization

$$\text{(5b)} \qquad LU = A,$$

where $L \equiv (m_{ik})$ is lower triangular with the elements

$$\text{(5c)} \qquad m_{ik} = \begin{cases} 0 & \text{for } i < k, \\ 1 & \text{for } i = k, \\ a_{ik}^{(k)}/a_{kk}^{(k)} & \text{for } i > k. \end{cases}$$

The final vector $\mathbf{f}^{(n)} \equiv \mathbf{g}$ is

$$\text{(5d)} \qquad \mathbf{g} = L^{-1}\mathbf{f}.$$

Proof. Once (5b) is established, we have that $\det A = (\det L)(\det U) = \det U$ and so (5a) follows. To verify (5b), let us set $LU = (c_{ij})$. Then, since L and U are triangular and (4) is satisfied for $k = n$,

$$c_{ij} = \sum_{k=1}^{n} m_{ik} a_{kj}^{(n)},$$

$$= \sum_{k=1}^{\min(i,j)} m_{ik} a_{kj}^{(k)}.$$

We recall that $a_{ij}^{(1)} \equiv a_{ij}$ and note from (3a) and (5c) that

$$m_{i,k-1}a_{k-1,j}^{(k-1)} = a_{ij}^{(k-1)} - a_{ij}^{(k)} \qquad \text{for } 2 \leq k \leq i,\ k \leq j.$$

Thus, if $i \leq j$ we get from the above

$$\begin{aligned} c_{ij} &= \sum_{k=1}^{i-1} m_{ik}a_{kj}^{(k)} + a_{ij}^{(i)} \\ &= \sum_{k=1}^{i-1} (a_{ij}^{(k)} - a_{ij}^{(k+1)}) + a_{ij}^{(i)} \\ &= a_{ij}. \end{aligned}$$

This holds also for $i > j$ since $a_{ij}^{(j+1)} = 0$ and so (5b) is verified.

Define $\mathbf{h} \equiv L\mathbf{g}$ so that

$$h_i = \sum_{k=1}^{i} m_{ik}g_k = \sum_{k=1}^{i} m_{ik}f_k^{(k)}.$$

Now from (3b) and (5c)

$$m_{ik}f_k^{(k)} = f_i^{(k)} - f_i^{(k+1)} \qquad \text{for } k < i,$$

and $f_i^{(1)} = f_i$. Thus, we find $h_i = f_i$, and since L is non-singular (5d) follows. ∎

Under the conditions of this theorem, the system (1) can be written as

$$LU\mathbf{x} = L\mathbf{g}.$$

Multiplication on the left by L^{-1} yields the equivalent upper triangular system

$$U\mathbf{x} = \mathbf{g}. \tag{6a}$$

If we write $U \equiv (u_{ij})$, then (5) is easily solved in the order $x_n, x_{n-1}, \ldots, x_1$ to get

$$\text{(6b)} \qquad \begin{aligned} x_n &= \frac{1}{u_{nn}} g_n, \\ x_i &= \frac{1}{u_{ii}}\left(g_i - \sum_{j=i+1}^{n} u_{ij}x_j\right), \qquad i = n-1, n-2, \ldots, 1. \end{aligned}$$

We recall that the elements of $U \equiv A^{(n)}$ and $\mathbf{g} \equiv \mathbf{f}^{(n)}$ are computed by the Gaussian elimination procedure (3), without the explicit evaluation of L^{-1}.

We now consider the generalization in which the order of elimination is arbitrary. Again we set $A^{(1)} \equiv A$ and $\mathbf{f}^{(1)} \equiv \mathbf{f}$. Then we select an arbitrary non-zero element $a_{i_1 j_1}^{(1)}$ called the 1st pivot element. (If this cannot be done then $A \equiv O$ and the system is degenerate, but also trivially in triangular form.) Since $a_{i_1j_1}^{(1)} \neq 0$ is the coefficient of the j_1st variable, x_{j_1}, in the i_1st equation we can eliminate this variable from all of the other equations. To do this, we subtract an appropriate unique multiple of the i_1st equation from each of the other equations; i.e., to eliminate x_{j_1} from the kth equation the multiplier must be $m_{kj_1} = (a_{kj_1}^{(1)}/a_{i_1j_1}^{(1)})$.

The reduced system is written as $A^{(2)}\mathbf{x} = \mathbf{f}^{(2)}$ and it is such that omitting the i_1st equation yields $n - 1$ equations in the $n - 1$ unknowns x_k, $k \neq j_1$. We now proceed with this reduced system and eliminate a second unknown, say x_{j_2}. To do this we must find some element $a_{i_2j_2}^{(2)} \neq 0$ with $i_2 \neq i_1$ and $j_2 \neq j_1$, called the 2nd pivot element. If $a_{rs}^{(2)} = 0$ for all $r \neq i_1$ and $s \neq j_1$ the process is terminated as the remaining equations are degenerate. After this second elimination the resulting system, say, $A^{(3)}\mathbf{x} = \mathbf{f}^{(3)}$, is composed of the i_1st equation of $A^{(1)}\mathbf{x} = \mathbf{f}^{(1)}$, the i_2nd equation of $A^{(2)}\mathbf{x} = \mathbf{f}^{(2)}$ and $n - 2$ remaining equations in only $n - 2$ variables, x_k with $k \neq j_1, j_2$. The general process is now clear and can be used to prove

THEOREM **2**. *Let the matrix A have rank r. Then we can find a sequence of distinct row and column indices $(i_1, j_1), (i_2, j_2), \ldots, (i_r, j_r)$ such that the corresponding pivot elements in $A^{(1)}, A^{(2)}, \ldots, A^{(r)}$ are non-zero and $a_{ij}^{(r)} = 0$ if $i \neq i_1, i_2, \ldots, i_r$. Let us define the permutation matrices, whose columns are unit vectors,*

$$P \equiv (\mathbf{e}^{(i_1)}, \mathbf{e}^{(i_2)}, \ldots, \mathbf{e}^{(i_r)}, \ldots, \mathbf{e}^{(i_n)}),$$

$$Q \equiv (\mathbf{e}^{(j_1)}, \mathbf{e}^{(j_2)}, \ldots, \mathbf{e}^{(j_r)}, \ldots, \mathbf{e}^{(j_n)}),$$

where i_k, j_k, for $1 \leq k \leq r$, are the above pivotal indices and the sets $\{i_k\}$ and $\{j_k\}$ are permutations of $1, 2, \ldots, n$.
Then the system

$$B\mathbf{y} = \mathbf{g},$$

where

$$B \equiv P^T A Q, \qquad \mathbf{y} \equiv Q^T\mathbf{x}, \qquad \mathbf{g} \equiv P^T\mathbf{f},$$

is equivalent to the system (1) *and can be reduced to triangular form by using Gaussian elimination with the natural order of pivots* $(1, 1), (2, 2), \ldots, (r, r)$.

Proof. The generalized elimination alters the matrix $A \equiv A^{(1)}$ by forming successive linear combinations of the rows. Thus, whenever no non-zero pivot elements can be found the null rows must have been linearly dependent upon the rows containing non-zero pivots. The permutations by P and Q simply arrange the order of the equations and unknowns, respectively, so that $b_{\nu\nu} = a_{i_\nu j_\nu}$, $\nu = 1, 2, \ldots, n$. By the first part of the theorem, the reduced matrix $B^{(r)}$ is triangular since all rows after the rth one vanish. ∎

If the matrix A is non-singular, then $r = n$ and Theorem 2 implies that, after the indicated rearrangement of the data, Theorem 1 becomes applicable. This is only useful for purposes of analysis. In actual computations on digital computers it is a simple matter to record the order of the pivot indices (i_ν, j_ν) for $\nu = 1, 2, \ldots, n$, and to do the arithmetic accordingly. Of course, the important problem is to determine some order for the pivots so that the elimination can be completed.

One way to pick the pivots is to require that (i_k, j_k) be the indices of a maximal coefficient in the system of $n - k + 1$ equations that remain at the kth step. This method of selecting *maximal pivots* is recommended as being likely to introduce the least loss of accuracy in the arithmetical operations that are based on working with a finite number of digits. We shall return to this feature in Subsection 1.2. Another commonly used pivotal selection method eliminates the variables $x_1, x_2, \ldots, x_{n-1}$ in succession by requiring that (i_k, k) be the indices of the maximal coefficient of x_k in the remaining system of $n - k + 1$ equations. (This method of *maximal column pivots* is particularly convenient for use on an electronic computer if the large matrix of coefficients is stored by columns since the search for a maximal column element is then quicker than the maximal matrix element search.)

1.1. Operational Counts

If the nth order matrix is non-singular, Gaussian elimination might be employed to solve the n special linear systems (0.6) and thus to obtain A^{-1}. Then to solve any number, say m, of systems with the same coefficient matrix A, we need only perform m multiplications of vectors by A^{-1}. However, we shall show here that *for any value of m this procedure is less efficient than an appropriate application of Gaussian elimination to the m systems in question.* In order to show this, we must count the number of arithmetic operations required in the procedures to be compared. The current convention is to count only multiplications and divisions. This custom arose because the first modern digital computers performed additions and subtractions much faster than they did multiplications and divisions which were done in comparable lengths of time. This variation in the execution time of the arithmetic operations is at present being reduced, but it should be noted that additions and subtractions are about as numerous as multiplications for most methods of this chapter. On the other hand, for some computers, as in the case of a desk calculator, it is possible to accumulate a sequence of multiplications (scalar product of two vectors) in the same time that it takes to perform the multiplications. Hence one is justified in neglecting to count these additions since they do not contribute to the total time of performing the calculation.

Let us consider first the m systems, with arbitrary $\mathbf{f}(j)$,

$$A\mathbf{x} = \mathbf{f}(j), \qquad j = 1, 2, \ldots, m. \tag{7}$$

We assume, for the operational count, that the elimination proceeds in the natural order. The most efficient application then performs the operations in (3a) only once and those in (3b) once for each j; that is, m times. On digital computers, not all of the vectors $\mathbf{f}(j)$ may be available at the same time, and thus the calculations in (3b) may be done later than those in (3a). However, since the final reduced matrix $A^{(n)}$ is upper triangular, we may store the "multipliers"

$$m_{i,k-1} \equiv \frac{a_{i,k-1}^{(k-1)}}{a_{k-1,k-1}^{(k-1)}}, \qquad 2 \le k \le i,$$

in the lower triangular part of the original matrix, A. (That is, $m_{i,k-1}$ is put in the location of $a_{i,k-1}^{(k-1)}$). Thus, no operations in (3a) ever need to be repeated.

From (3) and (4) we see that in eliminating x_{k-1}, a square submatrix of order $n - k + 1$ is determined and the last $n - k + 1$ components of each right-hand side are modified. Each element of the new submatrix and subvectors is obtained by performing a multiplication (and an addition which we ignore), but the quotients which appear as factors in (3) are computed only once. Thus, we find that it requires

$$(n - k + 1)^2 + (n - k + 1) \qquad \text{ops. for (3a)},$$

$$(n - k + 1) \qquad \text{ops. for (3b)}.$$

These operations must be done for $k = 2, 3, \ldots, n$ and hence with the aid of the formulae,

$$\sum_{\nu=1}^{n} \nu = \frac{n(n+1)}{2}, \qquad \sum_{\nu=1}^{n} \nu^2 = \frac{n(n+1)(2n+1)}{6},$$

the total number of operations is found to be:

$$\frac{n(n^2-1)}{3} \qquad \text{ops. to triangularize } A, \tag{8a}$$

$$\frac{n(n-1)}{2} \qquad \text{ops. to modify one inhomogeneous vector, } \mathbf{f}(j). \tag{8b}$$

To solve the resulting triangular system we use (6). Thus, to compute x_i requires $(n - i)$ multiplications and one division. By summing this over $i = 1, 2, \ldots, n$, we get

$$\frac{n(n+1)}{2} \qquad \text{ops. to solve one triangular system.} \tag{8c}$$

Finally, to solve the m systems in (7) we must do the operations in (8b) and (8c) m times while those in (8a) are done only once. Thus, we have

LEMMA **1.**

$$\frac{n^3}{3} + mn^2 - \frac{n}{3} \quad \text{ops.} \tag{9}$$

are required to solve the m systems (7) *by Gaussian elimination.* ■

To compute A^{-1}, we could solve the n systems (0.6) so the above count would yield, upon setting $m = n$,

$$\frac{4n^3}{3} - \frac{n}{3} \quad \text{ops. to compute } A^{-1}.$$

However, the n inhomogeneous vectors $\mathbf{e}^{(j)}$ are quite special, each having only one non-zero component which is unity. If we take account of this fact, the above operational count can be reduced. That is, for any fixed j, $1 \leq j \leq n$, the calculations to be counted in (3b) when $\mathbf{f} \equiv \mathbf{e}^{(j)}$ start for $k = j + 2$. This follows, since $f_\nu^{(\nu)} = 0$ for $\nu = 1, 2, \ldots, j - 1$ and $f_j^{(j)} = 1$. Thus, if $j = n - 1$ or $j = n$, no multiplications are involved and in place of (8b), we have

$$\sum_{k=j+2}^{n} (n - k + 1) = \sum_{\nu=1}^{n-j-1} \nu$$

$$= \tfrac{1}{2}[j^2 - (2n - 1)j + n^2 - n] \quad \text{ops.}$$

to modify the inhomogeneous vector $\mathbf{e}^{(j)}$ for $j = 1, 2, \ldots, n - 2$.

By summing this over the indicated values of j, we find

$$\tfrac{1}{6}(n^3 - 3n^2 + 2n) \quad \text{ops. to modify all } \mathbf{e}^{(j)}, \quad j = 1, 2, \ldots, n.$$

The count in (8c) is unchanged and thus to solve the n resulting triangular systems takes

$$\tfrac{1}{2}(n^3 + n^2) \quad \text{ops.}$$

Upon combining the above with (8a) we find

LEMMA **2.** *It need only require*

$$n^3 \quad \text{ops. } \textit{to compute } A^{-1}. \tag{10}$$ ■

Now let us find the operational count for solving the m systems in (7) by employing the inverse. Since A^{-1} and $\mathbf{f}(j)$ need not have any zero or unit elements, it requires in general

$$n^2 \quad \text{ops. to compute } A^{-1}\mathbf{f}(j).$$

Thus, to solve m systems requires mn^2 ops. and if we include the n^3 operations to compute A^{-1}, we get the result:

$$n^3 + mn^2 \quad \text{ops.} \tag{11}$$

are required to solve the m systems (7), *when using the inverse matrix.*

Upon comparing (9) and (11) it follows that *for any value of m the use of the inverse matrix is less efficient than using direct elimination.*

1.2. A Priori Error Estimates; Condition Number

In the course of actually carrying out the arithmetic required to solve

$$A\mathbf{x} = \mathbf{f} \tag{12}$$

by any procedure, roundoff errors will in general be introduced. But if the numerical procedure is "stable," or if the problem is "well-posed" in the sense of Section 3 of Chapter 1, these errors can be kept within reasonable bounds. We shall investigate these matters for the Gaussian elimination method of solving (12).

We recall that a computation is said to be well-posed if "small" changes in the data cause "small" changes in the solution. For the linear system (12) the data are A and $\mathbf{f}$ while $\mathbf{x}$ is the solution. The matrix A is said to be "well-conditioned" or "ill-conditioned" if the computation is or is not, respectively, well-posed. We shall make these notions somewhat more precise here and introduce a *condition number* for A which serves as a measure of ill-conditioning. Then we will show that the Gaussian elimination procedure yields accurate answers, even for very large order systems, if A is well-conditioned, and single precision arithmetic is used.

Suppose first that the data A and $\mathbf{f}$ in (12) are perturbed by the quantities δA and $\boldsymbol{\delta}\mathbf{f}$. Then if the perturbation in the solution $\mathbf{x}$ of (12) is $\boldsymbol{\delta}\mathbf{x}$ we have

$$(A + \delta A)(\mathbf{x} + \boldsymbol{\delta}\mathbf{x}) = \mathbf{f} + \boldsymbol{\delta}\mathbf{f}. \tag{13}$$

Now an estimate of the relative change in the solution can be given in terms of the relative changes in A and $\mathbf{f}$ by means of

THEOREM 3. *Let A be non-singular and the perturbation δA be so small that*

$$\|\delta A\| < 1/\|A^{-1}\|. \tag{14}$$

Then if $\mathbf{x}$ *and* $\boldsymbol{\delta}\mathbf{x}$ *satisfy* (12) *and* (13) *we have*

$$\frac{\|\boldsymbol{\delta}\mathbf{x}\|}{\|\mathbf{x}\|} \leq \frac{\mu}{1 - \mu\,\|\delta A\|/\|A\|}\left(\frac{\|\boldsymbol{\delta}\mathbf{f}\|}{\|\mathbf{f}\|} + \frac{\|\delta A\|}{\|A\|}\right), \tag{15}$$

where the condition number μ is defined as

$$\mu = \mu(A) \equiv \|A^{-1}\|\cdot\|A\|. \tag{16}$$

Proof. Since $\|A^{-1}\delta A\| \leq \|A^{-1}\| \cdot \|\delta A\| < 1$ by (14) it follows from the Corollary to Theorem 1.5 of Chapter 1 that the matrix $I + A^{-1}\delta A$ is non-singular, and further, that

$$\|(I + A^{-1}\delta A)^{-1}\| \leq \frac{1}{1 - \|A^{-1}\delta A\|} \leq \frac{1}{1 - \|A^{-1}\| \cdot \|\delta A\|}.$$

If we multiply (13) by A^{-1} on the left, recall (12) and solve for $\boldsymbol{\delta}\mathbf{x}$, we get

$$\boldsymbol{\delta}\mathbf{x} = (I + A^{-1}\delta A)^{-1}A^{-1}(\boldsymbol{\delta}\mathbf{f} - \delta A\mathbf{x}).$$

Now take norms of both sides, use the above bound on the inverse, and divide by $\|\mathbf{x}\|$ to obtain

$$\frac{\|\boldsymbol{\delta}\mathbf{x}\|}{\|\mathbf{x}\|} \leq \frac{\|A^{-1}\|}{1 - \|A^{-1}\| \cdot \|\delta A\|} \left(\frac{\|\boldsymbol{\delta}\mathbf{f}\|}{\|\mathbf{x}\|} + \|\delta A\| \right).$$

But from (12) it is clear that we may replace $\|\mathbf{x}\|$ on the right, since

$$\|\mathbf{x}\| \geq \|\mathbf{f}\|/\|A\|,$$

and (15) now easily follows by using the definition (16). ∎

The estimate (15) shows that small relative changes in $\mathbf{f}$ and A cause small relative changes in the solution if the factor

$$\frac{\mu}{1 - \mu\, \|\delta A\|/\|A\|}$$

is not too large. Of course the condition (14) is equivalent to

$$\mu \frac{\|\delta A\|}{\|A\|} < 1.$$

Thus, it is clear that when the condition number $\mu(A)$ is not too large, the system (12) is well-conditioned. Note that we cannot expect $\mu(A)$ to be small compared to unity since

$$\|I\| = \|A^{-1}A\| \leq \mu(A).$$

We can apply Theorem 3 to estimate the effects of roundoff errors committed in solving linear systems by Gaussian elimination and other direct methods. Given any non-singular matrix A, the condition number $\mu(A)$ is determined independently of the numerical procedure. But it is possible to view the computed solution as the exact solution, say $\mathbf{x} + \boldsymbol{\delta}\mathbf{x}$, of a perturbed system of the form (13). The basic problem now is to determine the magnitude of the perturbations, δA and $\boldsymbol{\delta}\mathbf{f}$. This type of approach is called a *backward error analysis*. It is rather clear that there are many perturbations δA and $\boldsymbol{\delta}\mathbf{f}$ which yield the same solution, $\mathbf{x} + \boldsymbol{\delta}\mathbf{x}$, in (13).

Our analysis for the Gaussian elimination method will define δA and $\boldsymbol{\delta}\mathbf{f}$ so that

$$\boldsymbol{\delta}\mathbf{f} \equiv 0.$$

Then the error estimate (15) becomes simply

$$\frac{\|\boldsymbol{\delta}\mathbf{x}\|}{\|\mathbf{x}\|} \leq \frac{\mu\|\delta A\|/\|A\|}{1 - \mu\|\delta A\|/\|A\|}, \tag{17}$$

and it is clear that only $\|\delta A\|$ in this error bound depends upon the round-off errors and method of computation.

In the case of Gaussian elimination we have seen in Theorem 1, that exact calculations yield the factorization (5b),

$$LU = A.$$

Here L and U are, respectively, lower and upper triangular matrices determined by (5c) and (3a). However, with finite precision arithmetic in these evaluations, we do not obtain L and U exactly, but say some triangular matrices $\mathscr{L}$ and $\mathscr{U}$. We define the perturbation E due to these inexact calculations by

$$\mathscr{L}\mathscr{U} \equiv A + E. \tag{18}$$

There are additional rounding errors committed in computing $\mathbf{g}$ defined by (3b) or (5d), and in the final back substitution (6b) in attempting to compute the solution $\mathbf{x}$. With exact calculations, these vectors are defined from (5d) and (6a) as the solutions of

$$L\mathbf{g} = \mathbf{f}, \qquad U\mathbf{x} = \mathbf{g}.$$

The vectors actually obtained can be written as $\mathbf{g} + \boldsymbol{\delta}\mathbf{g}$ and $\mathbf{x} + \boldsymbol{\delta}\mathbf{x}$ which are the exact solutions of, say,

$$(\mathscr{L} + \delta\mathscr{L})(\mathbf{g} + \boldsymbol{\delta}\mathbf{g}) = \mathbf{f}, \tag{19a}$$

$$(\mathscr{U} + \delta\mathscr{U})(\mathbf{x} + \boldsymbol{\delta}\mathbf{x}) = (\mathbf{g} + \boldsymbol{\delta}\mathbf{g}). \tag{19b}$$

Here $\mathscr{L}$ and $\mathscr{U}$ account for the fact that the matrices L and U are not determined exactly, as in (18). The perturbations $\delta\mathscr{L}$ and $\delta\mathscr{U}$ arise from the finite precision arithmetic performed in solving the triangular systems with the coefficients $\mathscr{L}$ and $\mathscr{U}$. Upon multiplying (19b) by $\mathscr{L} + \delta\mathscr{L}$ and using (19a) we have, from (13) with $\boldsymbol{\delta}\mathbf{f} = 0$,

$$(A + \delta A) = (\mathscr{L} + \delta\mathscr{L})(\mathscr{U} + \delta\mathscr{U}).$$

From (18), it follows then that

$$\delta A = E + \mathscr{L}(\delta\mathscr{U}) + (\delta\mathscr{L})\mathscr{U} + (\delta\mathscr{L})(\delta\mathscr{U}). \tag{20}$$

Thus, to apply our error bound (17) we must estimate $\|E\|$, $\|\delta\mathscr{U}\|$, and $\|\delta\mathscr{L}\|$. Since $\mathscr{L}$ and $\mathscr{U}$ are explicitly determined by the computations, their norms can also, in principle, be obtained.

We shall assume that floating-point arithmetic operations are performed with a t-digit decimal mantissa (see Section 2 of Chapter 1) and that the system has been ordered so that the natural order of pivots is used [i.e., as in eq. (3)]. In place of the matrices $A^{(k)} \equiv (a_{ij}^{(k)})$ defined in (2) and (3a), we shall consider the computed matrices $B^{(k)} \equiv (b_{ij}^{(k)})$ with the final such matrix $B^{(n)} \equiv \mathscr{U} = (u_{ij})$, the upper triangular matrix introduced above. Similarly, the lower triangular matrix of computed multipliers $\mathscr{L} \equiv (s_{ij})$ will replace the matrix $L \equiv (m_{ij})$ of (5c). For simplicity, we assume that the given matrix elements $A = (a_{ij})$ can be represented exactly with our floating-decimal numbers.

Now in place of (3a) and (5c), the floating-point calculations yield $b_{ij}^{(k)}$ and s_{ij} which by (2.6) of Chapter 1 can be written as:

for $k = 1$,

(21a) $$b_{ij}^{(1)} = a_{ij}, \qquad i, j = 1, 2, \ldots, n;$$

for $k = 1, 2, \ldots, n - 1$,

(21b) $$b_{ij}^{(k+1)} = \begin{cases} b_{ij}^{(k)}, & i \leq k, \\ 0, & i \geq k + 1,\ j \leq k, \\ [b_{ij}^{(k)} - s_{ik}b_{kj}^{(k)}(1 + \theta_{ij}^{(k)}10^{-t})](1 + \phi_{ij}^{(k)}10^{-t}), & \\ \qquad\qquad i \geq k + 1,\ j \geq k + 1; & \end{cases}$$

and finally

(21c) $$s_{ij} = \begin{cases} 0, & i < j; \\ 1, & i = j; \\ \dfrac{b_{ij}^{(j)}}{b_{jj}^{(j)}}(1 + \psi_{ij}10^{-t}), & i > j. \end{cases}$$

Here the quantities, θ, ϕ, ψ satisfy

$$|\theta_{ij}^{(k)}| \leq 5, \qquad |\phi_{ij}^{(k)}| \leq 5, \qquad |\psi_{ij}| \leq 5,$$

and they account for the rounding procedures in floating-point arithmetic. Of course, the above calculations can be carried out iff the $b_{jj}^{(j)} \neq 0$ for $j \leq n - 1$. However, this can be assured from

LEMMA 1. *If A is non-singular and t sufficiently large, then the Gaussian elimination method, with maximal pivots and floating-point arithmetic (with t-digit mantissas), yields multipliers s_{ij} with $|s_{ij}| \leq 1$ and pivots $b_{jj}^{(i)} \neq 0$.*

Proof. See Problem 8. ∎

It turns out that we require a bound on the growth of the pivot elements for our error estimates. That is, we seek a quantity $G = G(n)$, independent of the a_{ij}, such that

(22a) $$|b_{jj}^{(j)}| \leq G(n)a, \qquad j = 1, 2, \ldots, n;$$

where

(22b) $$a \equiv \max_{i,j} |a_{ij}|.$$

Under the conditions of Lemma 1, it is not difficult to see, by induction in (21b), that

(22c) $$G(n) \leq [1 + (1 + 5 \times 10^{-t})^2]^{n-1} = 2^{n-1} + \mathcal{O}((n - 1)10^{1-t}).$$

This establishes the existence of a bound of the form (22), but it is a tremendous overestimate for large n. In fact for exact elimination (i.e., no roundoff errors) using maximal pivots it can be shown that

(23a) $$|a_{jj}^{(j)}| < g(j)a,$$

where

(23b) $$g(j) \leq 3j^{1/2 + 1/4 \ln j}.$$

The quantity $g(n)$ would be a reasonable estimate for $G(n)$ if the maximal pivots in the sequence $\{B^{(k)}\}$ were located in the same positions as the maximal pivots in $\{A^{(k)}\}$. We know that if A is non-singular and t is sufficiently large, then the indices of the maximal pivotal elements used to find $\{B^{(k)}\}$ are also indices of maximal pivots in an exact Gaussian elimination procedure for A. For two special classes of matrices it is established in Problems 6 and 7 that $g(n) \leq 1$ and $g(n) \leq n$. The best (i.e., lowest) bound for $G(n)$ [or for $g(n)$] is not known at present.

We now turn to estimates of the terms in δA. Our first result is a bound on the elements of E which we state as

THEOREM **4**. *Under the hypothesis of Lemma 1 the Gaussian elimination calculations* (21) *are such that*

$$\mathcal{L}\mathcal{U} = A + E$$

where $E \equiv (e_{ij})$ *satisfies*

(24) $$|e_{ij}| \leq \begin{cases} (i - 1)2aG(n)10^{1-t}, & \text{for } i \leq j; \\ j2aG(n)10^{1-t}, & \text{for } i > j. \end{cases}$$

Here $G(n)$ *is any bound satisfying* (22).

Proof. We write the last line of (21b) as

(25a) $$b_{ij}^{(k+1)} = b_{ij}^{(k)} - s_{ik}b_{kj}^{(k)} + \epsilon_{ij}^{(k+1)}, \qquad i \geq k + 1, \; j \geq k + 1,$$

where from Lemma 1 and (22) it follows that

(25b) $$|\epsilon_{ij}^{(k+1)}| \leq 2aG(n)10^{1-t}, \qquad i \geq k+1, j \geq k+1.$$

Similarly, multiplying the last line of (21c) by $b_{jj}^{(j)}$ and dividing by $1 + \psi_{ij}10^{-t}$, we can write the result as

(26a) $$0 = b_{ij}^{(j)} - s_{ij}b_{jj}^{(j)} + \epsilon_{ij}^{(j+1)}, \qquad i > j;$$

where again we find that

(26b) $$|\epsilon_{ij}^{(j+1)}| \leq 2aG(n)10^{1-t}, \qquad i > j.$$

Upon recalling (21) we have

$$\mathscr{L} \equiv (s_{ij}), \qquad \mathscr{U} \equiv (b_{ij}^{(i)})$$

so that

$$(\mathscr{L}\mathscr{U})_{ij} = \sum_{k=1}^{n} s_{ik}b_{kj}^{(k)}$$

$$= \sum_{k=1}^{\min(i,j)} s_{ik}b_{kj}^{(k)}.$$

Now let $i > j$ and sum (25a) over $k = 1, 2, \ldots, j-1$ and then add (26a) to get, with the aid of (21),

$$0 = b_{ij}^{(1)} - \sum_{k=1}^{j} s_{ik}b_{kj}^{(k)} + \sum_{k=1}^{j} \epsilon_{ij}^{(k+1)}, \qquad i > j.$$

From the last two equations above and the fact that

$$b_{ij}^{(1)} = a_{ij}$$

we see that the elements e_{ij} with $i > j$ of $E = \mathscr{L}\mathscr{U} - A$ are

(27a) $$e_{ij} = \sum_{k=1}^{j} \epsilon_{ij}^{(k+1)}, \qquad i > j.$$

For the elements with $i \leq j$, we just sum (25a) over $k = 1, 2, \ldots, i-1$, recalling that $s_{ii} = 1$, and obtain as before

(27b) $$e_{ij} = \sum_{k=1}^{i-1} \epsilon_{ij}^{(k+1)}, \qquad i \leq j.$$

But now (24) follows from (27) by using the bounds (25b) and (26b). ∎

As a simple corollary of this theorem, we note that since $|e_{ij}| \leq 2(n-1)aG(n)10^{1-t}$, it follows that

(28) $$\|E\|_\infty \leq 2an(n-1)G(n)10^{1-t}.$$

The elements in $\delta\mathscr{L}$ and $\delta\mathscr{U}$ can be estimated from a single analysis of the error in solving any triangular system (with the same arithmetic and rounding). Thus we consider, say,

$$T\mathbf{u} = \mathbf{h} \tag{29a}$$

where $T \equiv (t_{ij})$ is lower triangular and non-singular, i.e.,

$$t_{ii} \neq 0; \qquad t_{ij} = 0, \qquad j > i; \qquad i = 1, 2, \ldots, n. \tag{29b}$$

The exact solution of (29) is easily obtained by recursion and is

$$u_1 = t_{11}^{-1}h_1 \tag{30a}$$

$$u_i = t_{ii}^{-1}\left(h_i - \sum_{k=1}^{i-1} t_{ik}u_k\right), \qquad i = 2, 3, \ldots, n. \tag{30b}$$

For numerical solutions, we have

THEOREM 5. *Let the "solution" of* (29) *be computed by using t-digit floating-point arithmetic to evaluate* (30). *Then the computed solution, say* $\mathbf{v}$, *satisfies*

$$(T + \delta T)\mathbf{v} = \mathbf{h} \tag{31a}$$

where the perturbations are bounded by

$$|\delta t_{ij}| \leq \max\,[2, |i - j + 1|]|t_{ij}|10^{1-t} \leq n|t_{ij}|10^{1-t}. \tag{31b}$$

Here t is required to be so large that $n10^{1-t} \leq 1$.

Proof. In the notation of Section 2 of Chapter 1 the floating-decimal evaluations, v_i, of the formulas (30) are

$$v_1 = \mathrm{fl}\left(\frac{h_1}{t_{11}}\right);$$

$$v_i = \mathrm{fl}\left[\frac{\left(h_i - \sum_{k=1}^{i-1} t_{ik}v_k\right)}{t_{ii}}\right] = \mathrm{fl}\left[\frac{\mathrm{fl}\left(h_i - \sum_{k=1}^{i-1} t_{ik}v_k\right)}{t_{ii}}\right], \qquad i = 2, 3, \ldots, n.$$

Then by using (2.6c) of Chapter 1, we must have from the above

$$v_1 = \frac{h_1}{t_{11}}(1 + \phi_1 10^{-t}), \qquad |\phi_1| \leq 5; \tag{32a}$$

$$v_i = \frac{\mathrm{fl}\left(h_i - \sum_{k=1}^{i-1} t_{ik}v_k\right)}{t_{ii}}(1 + \phi_i 10^{-t}), \qquad |\phi_i| \leq 5,\ i = 2, 3, \ldots, n. \tag{32b}$$

If, in the floating-point evaluation of the numerator in (32b) the sum is first accumulated and then subtracted from h_i, we can write, with the use of (2.6a) and Lemma 2.2 of Chapter 1,

$$\text{fl}\left(h_i - \sum_{k=1}^{i-1} t_{ik}v_k\right) = \left[h_i - \sum_{k=1}^{i-1} (t_{ik} + \delta t_{ik})v_k\right](1 + \theta_i 10^{-t}),$$
$$i = 2, 3, \ldots, n.$$

Here $|\theta_i| \leq 5$ and

$$|\delta t_{ik}| \leq \begin{cases}(i - k + 1)|t_{ik}|10^{1-t}, & 2 \leq k \leq i - 1, \\ 2|t_{i,\,i-1}|10^{1-t}, & k = 1, \ i = 2, 3, \ldots, n.\end{cases}$$

From the above and (32) we now obtain, solving for the h_j,

$$t_{11}v_1(1 + \phi_1 10^{-t})^{-1} = h_1,$$

$$t_{ii}v_i(1 + \phi_i 10^{-t})^{-1}(1 + \theta_i 10^{-t})^{-1} + \sum_{k=1}^{i-1} (t_{ik} + \delta t_{ik})v_k = h_i,$$
$$i = 2, 3, \ldots, n.$$

However, if we write

$$t_{11} + \delta t_{11} = t_{11}(1 + \phi_1 10^{-t})^{-1}$$
$$t_{ii} + \delta t_{ii} = t_{ii}(1 + \phi_i 10^{-t})^{-1}(1 + \theta_i 10^{-t})^{-1}$$

then it follows from $|\phi_i| \leq 5$, $|\theta_i| \leq 5$, and $n10^{1-t} \leq 1$ that

$$|\delta t_{11}| \leq |t_{11}|10^{1-t},$$
$$|\delta t_{ii}| \leq 2|t_{ii}|10^{1-t}, \qquad i > 1. \qquad ■$$

We are now able to obtain estimates of the elements in $\delta\mathscr{L}$ and $\delta\mathscr{U}$ or more importantly those in δA. These results are contained in the basic

THEOREM 6 (WILKINSON). *Let the* nth *order matrix* A *be non-singular and employ Gaussian elimination with maximal pivots and* t*-digit floating-point arithmetic to solve* (12). *Let* t *be so large that Lemma* 1 *applies and that*

$$n10^{1-t} \leq 1.$$

Then the computed solution, say $\mathbf{x} + \boldsymbol{\delta}\mathbf{x}$, *satisfies*

$$(A + \delta A)(\mathbf{x} + \boldsymbol{\delta}\mathbf{x}) = \mathbf{f}$$

where

$$|\delta a_{ij}| \leq (2n + 3n^2)G(n)a10^{1-t} \tag{33}$$

and $G(n)$ *is any bound satisfying* (22).

Proof. We have already shown that δA is given by (20) where the elements of E are estimated in (24). Since $\delta\mathscr{L}$ is the perturbation (19a) in solving the lower triangular system $\mathscr{L}\mathbf{g} = \mathbf{f}$, we can apply Theorem 5. We note that the elements of $\mathscr{L} \equiv (s_{ij})$ are given by (21c). Since maximal pivots are used $|b_{jj}^{(j)}| \geq |b_{ij}^{(j)}|$ so that $|s_{ij}| \leq 1$ and we easily get from (31b) in this case

$$|(\delta\mathscr{L})_{ij}| \equiv |\delta s_{ij}| \leq n10^{1-t}.$$

The elements of $\delta\mathscr{U}$ are the perturbations in solving a system of the form $\mathscr{U}\mathbf{y} = \mathbf{z}$ with $\mathscr{U} \equiv (u_{ij}) \equiv (b_{ij}^{(i)})$ where the $b_{ij}^{(i)}$ are defined by (21b). This system is, of course, upper triangular but the estimates of Theorem 5 still apply. Since maximal pivots are used we have, recalling (22),

$$|u_{ij}| \equiv |b_{ij}^{(i)}| \leq |b_{ii}^{(i)}| \leq G(n)a, \qquad i \leq j, \qquad i = 1, 2, \ldots, n;$$

and now (31b) yields

$$|(\delta\mathscr{U})_{ij}| \equiv |\delta u_{ij}| \leq nG(n)a10^{1-t}.$$

From (20) we have

$$\delta a_{ij} = e_{ij} + \sum_{k=1}^{\min(i,j)} (s_{ik}\delta u_{kj} + \delta s_{ik}u_{kj} + \delta s_{ik}\delta u_{kj}).$$

By taking absolute values and using the above bounds on $|\delta u_{ij}|$, $|\delta s_{ij}|$, $|u_{ij}|$, $|s_{ij}|$, as well as (24), we easily obtain

$$|\delta a_{ij}| \leq (2n + 2n^2 + n^3 10^{1-t})G(n)a10^{1-t}.$$

However, since it was required that $n10^{1-t} < 1$, the result in (33) follows. ∎

From this theorem, it follows that the computed solution is the exact solution of a system only slightly perturbed from the original if enough figures are used, i.e., t sufficiently large. Appropriate values for t depend upon n and the bound $G(n)$. If, as indicated in (22c), $G(n)$ were of the order 2^{n-1}, only relatively small order systems could be treated effectively. On the other hand, if $G(n) \cong g(n) \leq 3n^{\frac{1}{2} + \frac{1}{4}\ln n}$, as in (23), then quite large order systems can be treated with the number of digits used on modern digital computers (say $t \geq 8$). It is generally believed, however, that even this latter estimate for $G(n)$ is a generous overestimate, when using maximal pivots. It should be observed that essentially all of the previous analysis is valid if only partial pivoting, say maximal column pivoting, is employed since then $|s_{ij}| \leq 1$ is maintained. However, the growth factor $G(n)$ for this procedure cannot be estimated well in general. In fact, it is possible that the upper bound (22c) which still applies may be attained. In spite of this, partial pivoting is found to be effective in practice but the absence of *any* type of maximal pivoting strategy frequently leads to catastrophic growth of rounding errors.†

From (33) we easily find that

$$\|\delta A\|_\infty \le (2n^2 + 3n^3)G(n)a10^{1-t}, \tag{34}$$

and this can be employed in (17) to obtain maximum norm bounds on the relative error. It is clear that this relative error in $\mathbf{x}$ may not be small even though the relative perturbation, $\|\delta A\|/\|A\|$, is small. In such a case, A would be ill-conditioned. By (17), the relative error $\|\boldsymbol{\delta}\mathbf{x}\|/\|\mathbf{x}\|$ is small if $\|A^{-1}\|\ \|\delta A\|$ is small. For a given $G(n)$ and $\mu(A)$ equation (34) may be used to find the value of t that assures a solution with a prescribed accuracy.

Finally, we recall that a computed inverse of A can be obtained by solving the n systems (0.6). If we denote the matrix obtained as $A^{-1} + F$, then as above we can show that each column vector of $A^{-1} + F$, i.e., $(A^{-1} + F)_{.j}$, satisfies an equation of the form, for some perturbation matrix $\delta A^{(j)}$,

$$(A + \delta A^{(j)})(A^{-1} + F)_{.j} = \mathbf{e}^{(j)}, \qquad j = 1, 2, \ldots, n. \tag{35a}$$

Under the assumptions of Theorem 6, the estimates (33) and (34) also apply to the current perturbations, $\delta A^{(j)}$. Then, if $\|\delta A^{(j)}\| \le 1/\|A^{-1}\|$ we have, almost as in the proof of Theorem 3,

$$\frac{\|F_{.j}\|}{\|A^{-1}_{.j}\|} \le \frac{\mu(A)\|\delta A^{(j)}\|/\|A\|}{1 - \mu(A)\|\delta A^{(j)}\|/\|A\|}. \tag{35b}$$

Thus, as was to be expected, the columns of the inverse matrix are obtained to within the same relative error [i.e., compare (17)] as is the solution of any particular system.

1.3. A Posteriori Error Estimates

Although we do not advocate inverting a matrix to solve linear systems, it is of interest to consider error estimates related to computed inverses.

† Experience indicates that we usually achieve greater accuracy in the single precision solution, if we first *scale* the matrix A. That is, if with $B = D_1AD_2$, we solve

$$B\mathbf{y} = D_1 f$$

for $\mathbf{y}$; and then determine $\mathbf{x}$ from $D_2\mathbf{y} = \mathbf{x}$. Here D_1 and D_2 are some diagonal matrices chosen so that the n columns and the n rows of the matrix B have approximately equal norms. A complete mathematical explanation of this phenomenon is not available.

Let A be the matrix to be inverted and let C be the computed or alleged inverse. The error in the inverse is defined by

$$F = C - A^{-1}; \tag{36a}$$

we also use another measure of error called the *residual matrix*:

$$R = AC - I. \tag{36b}$$

We have first

THEOREM 7. *If* $\|R\| < 1$ *then*:

$$A \text{ and } C \text{ are non-singular}; \tag{37a}$$

$$\|A^{-1}\| \leq \|C\|/(1 - \|R\|); \tag{37b}$$

$$\|F\| \leq \|C\| \cdot \|R\|/(1 - \|R\|). \tag{37c}$$

Proof. We write (36b) as

$$AC = I + R,$$

and use the corollary to Theorem 1.5 of Chapter 1 and $\|R\| < 1$ to deduce that AC is non-singular. Part (37a) then follows. Take the inverse of both sides in the above equation and multiply on the left by C to find

$$A^{-1} = C(I + R)^{-1}.$$

Now (37b) follows by taking norms and by again using the corollary to Theorem 1.5 of Chapter 1. From (36) we see that $F = A^{-1}R$ and so, $\|F\| \leq \|A^{-1}\| \cdot \|R\|$, and (c) follows by an application of (b). ∎

Note that we may just as well consider A to be an approximation to the inverse of C. Thus we obtain the

COROLLARY. *Under the hypothesis of Theorem* 7,

$$\|C^{-1}\| \leq \|A\|/(1 - \|R\|), \tag{37d}$$

$$\|A - C^{-1}\| \leq \|A\| \cdot \|R\|/(1 - \|R\|). \tag{37e}$$ ∎

Since A and C are presumed known, we could actually compute $\|C\|$, $\|A\|$, and $\|R\|$ in the estimates (37). This, of course, is what is meant by a posteriori estimates. In general, n^3 multiplications are required to form AC and this computation, as well as that of the norms, is subject to simply estimated roundoff errors. In contrast, the quantity δA entering the a priori estimates (17) and (34) cannot be computed. It is hardly necessary to point out that $G(n)$ is determined easily after the elimination process has been completed.

It is of interest to note that, under the hypothesis of Theorem 7, with C an approximate inverse of A we can find the perturbation δA, so that C is the exact inverse of $A + \delta A$. That is, set

$$(A + \delta A)C = I.$$

Hence,

$$-\delta A = (AC - I)C^{-1}$$
$$= RC^{-1}.$$

Upon taking norms and using (37d), we then have

$$\|\delta A\| \leq \frac{\|R\| \cdot \|A\|}{1 - \|R\|}. \tag{38}$$

Finally, we observe that the computed inverse can also be used to estimate the error in solving a linear system. We state this result as

THEOREM **8.** *Let an approximate solution* $\mathbf{y}$ *of*

$$A\mathbf{x} = \mathbf{f}$$

have the residual vector

$$\mathbf{r} = A\mathbf{y} - \mathbf{f}. \tag{39}$$

Then, if an approximate inverse C *of* A *satisfies* $\|R\| \equiv \|AC - I\| < 1$, *we have*

$$\|\mathbf{y} - \mathbf{x}\| \leq \frac{\|\mathbf{r}\| \cdot \|C\|}{1 - \|R\|}. \tag{40}$$

Proof. From Theorem 7 it follows that A is non-singular and so from (39)

$$\mathbf{y} = A^{-1}(\mathbf{r} + \mathbf{f}).$$

Subtract $\mathbf{x} = A^{-1}\mathbf{f}$ from this to find, after taking norms,

$$\|\mathbf{y} - \mathbf{x}\| \leq \|A^{-1}\| \cdot \|\mathbf{r}\|.$$

The result (40) then follows from (37b). ∎

The determination of the *residual vector* $\mathbf{r}$ is the first step in an iterative procedure to improve upon the accuracy of the solution (see Subsection 4.3).

It should be noted that the result in this theorem is independent of the manner in which the approximate solution, $\mathbf{y}$, is obtained. Thus, it was not assumed that $\mathbf{y} = C\mathbf{f}$. This suggests, in fact, that the sole purpose for computing C and R might be to use them in error estimates of the form (40). That is, once the constant $M \equiv \|C\|/(1 - \|R\|)$ is known, it requires only n^2 multiplications to compute $\mathbf{r}$ for each approximate solution $\mathbf{y}$ of a system with coefficient matrix A. If one wished to use (40), after finding $\mathbf{y}$ by Gaussian elimination, then an approximate inverse C could be obtained by using the approximate factorization of A. This, by (8a) and (10), would require twice as much labor as was already expended to find $\mathbf{y}$.

PROBLEMS, SECTION 1

1. Show that $A^{(k)}$ is non-singular iff A is non-singular, for the Gaussian elimination method.

2. Describe how the maximal pivot scheme permits the completion of the elimination method, when A is singular.

3. Prove the following corollary to Theorem 2: If interchanges of rows and of columns are made and $r = n$, then

$$\det A = (-1)^J a^{(1)}_{i_1 j_1} a^{(2)}_{i_2 j_2} \dots a^{(n)}_{i_n j_n},$$

where $a^{(k)}_{i_k j_k}$ are the successive pivotal elements in the Gaussian elimination scheme and $(-1)^J = \det P \det Q$.

4. If A is symmetric and positive definite (that is, $\mathbf{x}^* A\mathbf{x} \geq 0$ and $\sum_{i,j=1}^{n} a_{ij}x_i x_j = 0$ only if $x_i = 0$ for all i; a_{ij} and x_i real), show that

(a) $a_{ii} > 0$

(b) $\max_i a_{ii} = \max_{i,j} |a_{ij}|$. [Hint: For (b), if $|a_{rs}| = \max_{i,j} |a_{ij}|$, then with $x_i = 0$ for $i \neq r, s$

$$\sum_{i,j=1}^{n} a_{ij}x_i x_j \equiv a_{rr}x_r^2 + 2a_{rs}x_r x_s + a_{ss}x_s^2 = 0$$

for non-trivial x_r, x_s if $a_{rr}a_{ss} < a_{rs}^2$.]

5. If A is symmetric, positive definite, then the submatrices $(a^{(k)}_{ij})$ for $k \leq i,j \leq n$ are symmetric, positive definite. [Hint: Use mathematical induction on k. Symmetry from

$$a^{(2)}_{ij} = a^{(1)}_{ij} - \frac{a^{(1)}_{i1}}{a^{(1)}_{11}} a^{(1)}_{1j}.$$

Positive definiteness from Problem (4a) and

$$\sum_{i,j=2}^{n} a^{(2)}_{ij}x_i x_j \equiv \sum_{i,j=1}^{n} a^{(1)}_{ij}x_i x_j - a^{(1)}_{11}\left[x_1 + \sum_{i=2}^{n} \frac{a^{(1)}_{j1}}{a^{(1)}_{11}} x_i\right]^2.$$

That is, if $(a^{(2)}_{ij})$ is not positive definite, then $(a^{(1)}_{ij})$ is not.]

6. (*Von Neumann-Goldstine.*) If A is symmetric, positive definite, then $a^{(k)}_{ii} \leq a^{(k-1)}_{ii}$ for $k \leq i \leq n$, $k = 2, 3, \dots, n$. (Hence by Problem (4b), $\max_{i,j} |a^{(n)}_{ij}| \leq \max_{i,j} |a_{ij}|$.)

7. (*Wilkinson.*) If A is a *Hessenberg matrix* (i.e., $a_{ij} = 0$ for $i \geq j + 2$),

$$\max_{i,j} |a^{(n)}_{ij}| \leq n \max_{i,j} |a_{ij}|,$$

if maximal column pivots are used. [Hint: Only one row is changed in passing from $A^{(k-1)}$ to $A^{(k)}$.]

8. Prove Lemma 1.

9. If A is symmetric, use the first part of Problem 5 to show: the number of operations to solve (1) by Gaussian elimination, with diagonal pivots, is $n^3/6 + \mathcal{O}(n^2)$.

10. If in (1), A and $\mathbf{f}$ are complex, show that (1) may be converted to the solution of a real system of order $2n$.

2. VARIANTS OF GAUSSIAN ELIMINATION

There are many methods for solving linear systems that are slight variations of the Gaussian elimination method. None of these methods has succeeded in reducing the number of operations required, but some have eliminated much of the intermediate storage or recording requirements. Caution should be taken in applying any variation that does not allow for the selection of some sort of maximal pivots, which is generally necessary to prevent the growth of rounding errors.

The modification due to Jordan circumvents the final back substitution. This is accomplished by additional computations which serve to eliminate the variable x_k from the first $k - 1$ equations as well as from the last $n - k$ equations at the kth stage of the reduction. In other words, the coefficients above the diagonal are also reduced to zero and the final coefficient matrix which results is a diagonal matrix. The obvious modifications which are required for this Gauss-Jordan elimination are contained in

$$\text{(1a)} \qquad a_{ij}^{(1)} = a_{ij}$$

$$\text{(1b)} \qquad a_{ij}^{(k)} = a_{ij}^{(k-1)} - \frac{a_{i,k-1}^{(k-1)}}{a_{k-1,k-1}^{(k-1)}} a_{k-1,j}^{(k-1)} \qquad \text{for } i \neq k-1, j \geq k-1$$

$$\text{(1c)} \qquad a_{k-1,j}^{(k)} = a_{k-1,j}^{(k-1)} \qquad \text{for } j \geq k-1$$

$$\text{(1d)} \qquad a_{ij}^{(k)} = a_{ij}^{(k-1)} \qquad \text{for } j < k-1.$$

$$\text{(2a)} \qquad f_i^{(1)} = f_i$$

$$\text{(2b)} \qquad f_i^{(k)} = f_i^{(k-1)} - \frac{a_{i,k-1}^{(k-1)}}{a_{k-1,k-1}^{(k-1)}} f_{k-1}^{(k-1)} \qquad \text{for } i \neq k-1$$

$$\text{(2c)} \qquad f_{k-1}^{(k)} = f_{k-1}^{(k-1)} \qquad \text{for } k = 2, \ldots, n.$$

The solution is then

$$x_i = \frac{f_i^{(n)}}{a_{ii}^{(n)}} = \frac{f_i^{(n)}}{a_{ii}^{(i)}}.$$

It is clear that pivoting on the maximal element in the remaining square submatrix may be retained in this procedure. Hence, multipliers for

$i < k - 1$ may exceed unity. Furthermore, the number of operations is somewhat greater than in the ordinary Gaussian elimination with back substitution; it is now

$$\frac{n^3}{2} + n^2 - \frac{n}{2} \text{ ops.} \tag{3}$$

Thus, there does not seem to be any great advantage in using the Gauss-Jordan elimination in actual calculations with automatic computing equipment.

Another variation is the so-called *Crout reduction.* This method is applicable if the rows and columns are so arranged that *no column interchanges are required* in the Gaussian elimination (as in the case of symmetric, positive definite matrices; see Theorem 3.3). Thus, in general, the pivots will not be the maximal elements. Hence, errors may grow very rapidly in the Crout method and it is not recommended unless the system is of relatively small order or if it can be determined that the error growth will not be catastrophic. (In practice one may apply the method and test the accuracy of the solution a posteriori.) On the other hand, the Crout method is specifically designed to reduce the number of intermediate quantities which must be retained. Thus, for hand computations and digital computers with small storage capacities it may be of great value. The Crout method may be modified to use maximal column pivots, by incorporating row interchanges as described in Theorem 3.1 (or see Theorem 1.2).

This "compact" elimination procedure is based on the fact that only those elements $a_{ij}^{(k)}$, in the Gaussian elimination, for which $j \geq i$ and $i \leq k$, are required for the final back substitution.

Thus, we seek a recursive method of defining the columns of L (lower triangular matrix of multipliers) and rows of U (upper triangular matrix). From Theorem 1.1, we know that

$$LU = A.$$

Hence, let us write the formula for a_{kj} from the rule for matrix multiplication, after a simple algebraic transformation, in the form

$$u_{kj} = a_{kj} - \sum_{p=1}^{k-1} m_{kp} u_{pj}, \qquad \text{if } k \leq j; \tag{4}$$

and the formula for a_{ik} in the form

$$m_{ik} = \frac{1}{u_{kk}} \left[a_{ik} - \sum_{p=1}^{k-1} m_{ip} u_{pk} \right], \qquad \text{if } i > k. \tag{5}$$

We now may use (4) and (5) for $k \geq 2$ to find first the elements of the kth row of U and then the elements of the kth column of L, provided that we know the previous rows and columns, respectively, of U and L. Hence, we need only define the first row

$$(4)_1 \qquad u_{1j} = a_{1j} \qquad \text{for } 1 \leq j \leq n$$

and first column

$$(5)_1 \qquad m_{i1} = \frac{a_{i1}}{a_{11}} \qquad \text{for } 2 \leq i \leq n.$$

If we define $u_{1,\,n+1} = f_1$; $a_{i,\,n+1} = f_i$ and use (4) for $j = n + 1$, we find a column $(u_{i,\,n+1})$ which is the vector $\mathbf{g}$ of Theorem 1.1.

We then use the back substitution to solve $U\mathbf{x} = \mathbf{g}$ as before [where U represents the first n columns of (u_{ij})]. The operational count, for producing L, U, and $\mathbf{g}$, is easily found to be $(2n^3 + 3n^2 - 5n)/6$. It is not surprising that this is the same as the number of operations required by the conventional Gaussian elimination scheme to produce L, U, and $\mathbf{g}$ (since we merely avoid writing down the intermediate elements but have ultimately to do the same multiplications and divisions).

We could show now, if the inner products in (4) and (5) are accumulated in *double precision* before the sum is rounded, that the effect of rounding errors is appreciably diminished. In fact, the estimate in place of (1.34) becomes

$$\|\delta A\|_\infty = \mathcal{O}(n^2 G(n) a 10^{-t}).$$

PROBLEM, SECTION 2

1. Verify the operational count for the Crout method: $(2n^3 + 3n^2 - 5n)/6$.

3. DIRECT FACTORIZATION METHODS

The final forms, (2.4) and (2.5), that are used in the Crout method, suggest a more general study of the direct triangular decomposition

$$(1) \qquad LU = A,$$

in which the diagonal elements of L are not necessarily unity. In fact, if we consider $L \equiv (l_{ij})$ then (1) implies

$$(2a) \qquad l_{kk}u_{kk} = a_{kk} - \sum_{p=1}^{k-1} l_{kp}u_{pk}, \qquad \text{for } k \geq 2$$

(2b) $$u_{kj} = \frac{1}{l_{kk}}\left(a_{kj} - \sum_{p=1}^{k-1} l_{kp}u_{pj}\right), \qquad \text{for } j > k \geq 2$$

(2c) $$l_{ik} = \frac{1}{u_{kk}}\left(a_{ik} - \sum_{p=1}^{k-1} l_{ip}u_{pk}\right), \qquad \text{for } i > k \geq 2.$$

[Equations (2a, b, c) hold for $k = 1$, if we remove the $\sum$ term.] Equation (2a) determines the product $l_{kk}u_{kk}$ in terms of data in previous rows of U and columns of L. Once l_{kk} and u_{kk} are chosen to satisfy (2a), we then use (2b) and (2c) to determine the remaining elements in the kth row and column.

If $l_{kk}u_{kk} = 0$, the factorization is not possible, unless all of the brackets vanish in (2b), for $j > k$, or all of the brackets vanish in (2c), for $i > k$. If A is non-singular, then the use of maximal column pivots results in the sequence $(i_1, 1), (i_2, 2), \ldots, (i_n, n)$ as pivotal elements. Hence, the Gaussian elimination process shows that the triangular decomposition

$$LU = P^T A,$$

is possible, where P is defined in Theorem 1.2. In fact, if A is non-singular, one of the bracketed expressions in (2c) does not vanish, for some $i \geq k$. Therefore, one of the bracketed expressions in (2c) is of maximum absolute value for $i \geq k$, say for $i = i_k \geq k$. We may then move the elements of the row i_k in both A and in the part of L that has already been found up to row k. (The rows $k, k + 1, \ldots, i_k - 1$ are moved down in both L and A to fill the gap.) Hence, if A is non-singular we may, with row interchanges, employ (2a, b, and c) to achieve a triangular factorization. We summarize these facts in

THEOREM 1. *If A is non-singular, a triangular decomposition, $LU = A$, may not be possible. But a permutation of the rows of A can be found such that $B \equiv P^T A = LU$, where $P = (p_{rs})$ and*

$$p_{rs} = \begin{cases} 0, & r \neq i_s \\ 1, & r = i_s. \end{cases}$$

In fact, the P may be found so that

$$|l_{kk}| \geq |l_{ik}| \qquad \textit{for } i > k; \qquad k = 1, 2, \ldots, n - 1. \qquad \blacksquare$$

Note that in this result, in contrast to that in Theorem 1.2, we have only employed row interchanges.

A symmetric choice $l_{kk} = u_{kk}$ may lead to imaginary numbers, if the right-hand side of (2a) is negative; a less symmetric choice $l_{kk} = |u_{kk}|$ keeps the arithmetic real if A is real (see Problem 1).

As in the Crout method, we may consider $\mathbf{f}$ as an additional column of A (i.e., $a_{i,n+1} \equiv f_i$) and use (2b) for $j = n + 1$ to define the elements $g_i \equiv u_{i,n+1}$ such that

$$U\mathbf{x} = L^{-1}\mathbf{f} = \mathbf{g}.$$

In Subsections 3.1, 3.2, and 3.3, we consider special applications of this procedure.

3.1. Symmetric Matrices (Cholesky Method)

We begin with

THEOREM 2. *Let A be symmetric. If the factorization $LU = A$ is possible, then the choice $l_{kk} = u_{kk}$ implies $l_{ik} = u_{ki}$, that is, $LL^T = A$.*

Proof. Use (2) and induction on k. ∎

A simple, non-singular, symmetric matrix for which the factorization is not possible is

$$\begin{pmatrix} 0 & 1 \\ 1 & 0 \end{pmatrix}.$$

On the other hand, if the symmetric matrix A is positive definite (i.e., $\mathbf{x}^*A\mathbf{x} > 0$ if $\mathbf{x}^*\mathbf{x} > 0$), then the factorization is possible. We have

THEOREM 3. *Let A be symmetric, positive definite. Then, A can be factored in the form*

$$LL^T = A.$$

Proof. Problems 4 and 5 of Section 1 show that the Gaussian elimination method can be carried out, without any interchanges, to give the factorization $(m_{ij})(b_{ij}) = A$, where $b_{ii} > 0$. But if we define

$$l_{kk} = u_{kk} = \sqrt{b_{kk}}\,,$$

then by Problem 1, we will obtain from (2b, c) the elements in the factorization

$$LU = A,$$

where

$$l_{ik} = u_{ki}.$$ ∎

A count of the arithmetic operations can be made if we remark that only the elements l_{ik} defined by (2a, c) are involved. If we count the square root operations separately, we have

$$\frac{n^3}{6} + n^2 - \frac{n}{6} \text{ ops.} + n \text{ square roots} = \text{no. of ops. to find } L \text{ and } \mathbf{g}.$$

In addition, to find $\mathbf{x}$ we must solve a triangular system which requires $(n^2 + n)/2$ operations. Thus, *to solve one system using the Cholesky method requires $n^3/6 + 3n^2/2 + n/3$ operation plus n square roots.*

To apply our previous error analysis, we deduce from

$$a_{ii} = \sum_{k=1}^{i} l_{ik}u_{ki} = \sum_{k=1}^{i} l_{ik}^2,$$

the bounds

$$|l_{ik}|^2 \leq a_{ii} \leq a = \max_{i,j} |a_{ij}|.$$

For single precision square roots and $G(n) = 1$, we could prove (as we do in Theorem 1.6)

THEOREM 4. *If A is symmetric and positive definite, then the approximate solution of $A\mathbf{x} = \mathbf{f}$ obtained by factorization and floating-point arithmetic with t digits satisfies*

$$(A + \delta A)\mathbf{y} = \mathbf{f},$$

where for t sufficiently large

$$\|\delta A\|_\infty \leq a10^{1-t}(2n^2 + 3n^3). \qquad ■$$

COROLLARY. *Under the hypothesis of Theorem 4, if inner products are accumulated exactly, prior to a final rounding, then for t sufficiently large*

$$\|\delta A\|_\infty = \mathcal{O}(n^2 a10^{-t}). \qquad ■$$

3.2. Tridiagonal or Jacobi Matrices

A coefficient matrix which frequently occurs is the tridiagonal or Jacobi form, in which $a_{ij} = 0$ if $|i - j| > 1$. That is,

$$(3) \qquad A = \begin{pmatrix} a_1 & c_1 & & & & \\ b_2 & a_2 & c_2 & & & \\ & \cdot & \cdot & \cdot & & \\ & & \cdot & \cdot & \cdot & \\ & & & \cdot & \cdot & \cdot \\ & & & b_{n-1} & a_{n-1} & c_{n-1} \\ & & & & b_n & a_n \end{pmatrix}.$$

Assume this matrix can be factored in the bidiagonal form

$$A = LU \equiv \begin{bmatrix} \alpha_1 & & & & \\ b_2 & \alpha_2 & & & \\ & b_3 & \alpha_3 & & \\ & & \cdot & \cdot & \\ & & & \cdot & \cdot \\ & & & & \cdot \quad \cdot \\ & & & & b_n \quad \alpha_n \end{bmatrix} \begin{bmatrix} 1 & \gamma_1 & & & \\ & 1 & \gamma_2 & & \\ & & 1 & \cdot & \\ & & & \cdot & \cdot \\ & & & & \cdot \quad \cdot \\ & & & & \cdot \quad \gamma_{n-1} \\ & & & & 1 \end{bmatrix}.$$

Then we find,

(4a) $\alpha_1 = a_1, \qquad \gamma_1 = c_1/\alpha_1;$

(4b) $\alpha_i = a_i - b_i\gamma_{i-1}, \qquad i = 2, 3, \ldots, n;$

(4c) $\gamma_i = c_i/\alpha_i, \qquad i = 2, 3, \ldots, n - 1.$

Thus, if none of the α_i vanish, the factorization is accomplished by evaluating the recursions in (4). The "intermediate" solution $\mathbf{g}$ of $L\mathbf{g} = \mathbf{f}$ becomes

(5a) $g_1 = f_1/\alpha_1;$

(5b) $g_i = (f_i - b_i y_{i-1})/\alpha_i, \qquad i = 2, \ldots, n;$

and the final solution $\mathbf{x}$ of $U\mathbf{x} = \mathbf{g}$ is given by

(6a) $x_n = g_n,$

(6b) $x_j = g_j - \gamma_j x_{j+1}, \qquad j = n - 1, n - 2, \ldots, 1.$

In many of the applications of this procedure, the elements (3) of A satisfy

(7a) $|a_1| > |c_1| > 0;$

(7b) $|a_i| \geq |b_i| + |c_i|, \qquad b_i c_i \neq 0, \qquad i = 2, 3, \ldots, n - 1;$

(7c) $|a_n| > |b_n| > 0.$

In such cases, the quantities α_i and γ_i can be shown to be nicely bounded and in fact A is non-singular. We state this as

THEOREM 5. *If the elements of A in* (3) *satisfy* (7) *then* det $A \neq 0$ *and the quantities in* (4) *are bounded by*

$$\text{(a) } |\gamma_i| < 1; \qquad \text{(b) } |a_i| - |b_i| < |\alpha_i| < |a_i| + |b_i|.$$

Proof. From (4a) and (7a), it follows that $|\gamma_1| < 1$. Assume $|\gamma_i| < 1$ for $i = 1, 2, \ldots, j - 1$. Then by (4b, c)

$$\gamma_j = \frac{c_j}{a_j - b_j\gamma_{j-1}},$$

and thus

$$|\gamma_j| \le \frac{|c_j|}{||a_j| - |b_j|\,|\gamma_{j-1}||} < \frac{|c_j|}{|a_j| - |b_j|},$$

by the inductive assumption. Finally, by using (7b, c) in the above, it follows that $|\gamma_j| < 1$ and hence (a) is proved. Using this result and (4b) it follows that

$$|a_i| + |b_i| > |\alpha_i| > |a_i| - |b_i| \; (\ge |c_i|),$$

which concludes the proof of the inequalities (a) and (b). But then

$$\det A = (\det L)\cdot(\det U) = \prod_{j=1}^{n} \alpha_j \ne 0. \qquad \blacksquare$$

It should be noted that, when the conditions (7) hold, the procedure defined in (4) must be valid. Further if $b_i c_i = 0$ for some $i \ne 1, n$, then the system can be reduced to two systems which are essentially uncoupled. Similarly, if $c_1 = 0$ or $b_n = 0$ then x_1 or x_n, respectively, can be eliminated to get a reduced system.

The operational count for this procedure is somewhat striking:

(4) requires $2(n - 1)$ ops.

(5) requires $1 + 2(n - 1)$ ops.

(6) requires $n - 1$ ops.

or a total of

$$5n - 4 \text{ ops.} \tag{8}$$

to solve a single system. If there are m such systems to be solved, the quantities α_i, γ_i in (4) need be computed only once and (5) and (6) are then each done m times for a total of

$$(3n - 2)m + 2n - 2 \text{ ops.} \tag{9}$$

to solve m systems. Consequently, the inverse can be obtained, although it should never be used in such circumstances to solve the system, in not more than

$$3n^2 - 2 \text{ ops.}$$

The low operational counts in (8) and (9) are due to the fact that the zero elements of A have been accounted for in performing the calculations.

It should be observed that the factorization computed in (4) is not unique. Thus, for instance, we could try the form

$$A = LU \equiv \begin{bmatrix} 1 & & & & & & \\ \beta_2 & 1 & & & & & \\ & \beta_3 & 1 & & & & \\ & & \cdot & \cdot & & & \\ & & & \cdot & \cdot & & \\ & & & & \cdot & \cdot & \\ & & & & & \beta_n & 1 \end{bmatrix} \begin{bmatrix} \alpha_1 & c_1 & & & & \\ & \alpha_2 & c_2 & & & \\ & & \cdot & \cdot & & \\ & & & \cdot & \cdot & \\ & & & & \cdot & \cdot \\ & & & & \cdot & c_{n-1} \\ & & & & & \alpha_n \end{bmatrix}.$$

The reader should derive the recursions analogous to (4)–(6) for this case and prove the corresponding version of Theorem 5. We give the reader leave to develop a treatment and operational count for the general *band matrix*. A matrix (c_{ij}) of order n is called a band matrix of *width* (b, a) iff

$$c_{ij} = 0 \quad \text{for} \quad j - i \geq a \quad \text{or} \quad i - j \geq b.$$

3.3. Block-Tridiagonal Matrices

Another form which is encountered frequently, especially in the numerical solution of partial differential equations and integral equations, is the so-called *block-tridiagonal matrix*

$$(10) \qquad A = \begin{bmatrix} A_1 & C_1 & & & & & & & \\ B_2 & A_2 & C_2 & & & & & & \\ & \cdot & \cdot & \cdot & & & & & \\ & & \cdot & \cdot & \cdot & & & & \\ & & & \cdot & \cdot & \cdot & & & \\ & & & & B_i & A_i & C_i & & \\ & & & & & \cdot & \cdot & \cdot & \\ & & & & & & \cdot & \cdot & \cdot \\ & & & & & & & \cdot & \cdot & \cdot \\ & & & & & & & & \cdot & \cdot & \cdot \\ & & & & & & & & & B_n & A_n \end{bmatrix}.$$

Here, each of the A_i represents a square matrix, of order m_i, and each of the B_i and C_i are rectangular matrices which just "fit" the indicated pattern. That is, B_i must have m_i rows and m_{i-1} columns, and C_i must have m_i rows and m_{i+1} columns. Note that if all $m_i = m$, then all the submatrices are square and of order m. The order of the matrix A is $\sum_{i=1}^{n} m_i$, or again if all $m_i = m$ then the order is (mn).

A system with coefficient matrix of the form (10) may be solved by a procedure formally analogous to the previous factorization of a Jacobi matrix. Thus, let the system be

$$A\mathbf{x} = \mathbf{f} \tag{11}$$

where now

$$\mathbf{x} = \begin{pmatrix} \mathbf{x}^{(1)} \\ \vdots \\ \mathbf{x}^{(n)} \end{pmatrix}, \qquad \mathbf{f} = \begin{pmatrix} \mathbf{f}^{(1)} \\ \vdots \\ \mathbf{f}^{(n)} \end{pmatrix}, \tag{12}$$

and each $\mathbf{x}^{(i)}$ and $\mathbf{f}^{(i)}$ are m_i-component column vectors. That is, the components of the vector $\mathbf{x}$ are grouped into subsets, $\mathbf{x}^{(i)}$, and these subsets are to be "eliminated," as in the Gaussian procedure, a group at a time. Thus, the method to be described is a special case of more general methods known as *group-* or *block-elimination.*

Exactly as in Subsection 3.2 we seek a factorization of the form

$$A = LU = \tag{13}$$

$$\begin{bmatrix} \mathrm{A}_1 & & & & & \\ B_2 & \mathrm{A}_2 & & & & \\ & B_3 & \mathrm{A}_3 & & & \\ & & \cdot & \cdot & & \\ & & & \cdot & \cdot & \\ & & & & \cdot & \cdot \\ & & & & B_n & \mathrm{A}_n \end{bmatrix} \begin{bmatrix} I_1 & \Gamma_1 & & & & \\ & I_2 & \Gamma_2 & & & \\ & & I_3 & \cdot & & \\ & & & \cdot & \cdot & \\ & & & & \cdot & \cdot \\ & & & & \cdot & \Gamma_{n-1} \\ & & & & & I_n \end{bmatrix}$$

where the I_j are identity matrices of order m_j, the A_j are square matrices of order m_j, and the Γ_j are rectangular matrices with m_j rows and m_{j+1} columns. Proceeding formally, we find that

$$\mathrm{A}_1 = A_1, \qquad \Gamma_1 = A_1^{-1}C_1; \tag{14a}$$

$$\mathrm{A}_i = A_i - B_i\Gamma_{i-1}, \qquad i = 2, 3, \ldots, n; \tag{14b}$$

$$\Gamma_i = \mathrm{A}_i^{-1}C_i, \qquad i = 2, 3, \ldots, n - 1. \tag{14c}$$

From the definitions of the matrices involved, it is clear that each Γ_i is rectangular of the indicated order and that the product $B_i\Gamma_{i-1}$ and hence A_i is square of order m_i. The system (11) is now equivalent to

$$L\mathbf{y} = \mathbf{f}, \qquad U\mathbf{x} = \mathbf{y} \tag{15}$$

where $\mathbf{y}$ also has the compound form indicated in (12). We thus obtain formally, from (13) in (15),

(16)
$$\mathbf{y}^{(1)} = A_1^{-1}\mathbf{f}^{(1)}$$
$$\mathbf{y}^{(i)} = A_i^{-1}(\mathbf{f}^{(i)} - B_i\mathbf{y}^{(i-1)}), \qquad i = 2, 3, \ldots, n,$$

and

(17)
$$\mathbf{x}^{(n)} = \mathbf{y}^{(n)}$$
$$\mathbf{x}^{(i)} = \mathbf{y}^{(i)} - \Gamma_i\mathbf{x}^{(i+1)}, \qquad i = n-1, n-2, \ldots, 1.$$

This method requires (or rather seems to require) the inversion of the n matrices A_i and the formation of the $2(n-1)$ matrix products $A_i^{-1}C_i$, $B_i\Gamma_{i-1}$. To estimate the total number of operations used, we consider the cases where all $m_i = m$. Then with Gaussian elimination to obtain the inverses, we require from (1.10) (see discussion below on improving efficiency by not computing inverses explicitly),

$$nm^3 \text{ ops. for all } A_i^{-1}.$$

The product of two square matrices of order m requires m^3 operations, hence, we have

$$2(n-1)m^3 \text{ ops. for all } A_i^{-1}C_i \text{ and } B_i\Gamma_{i-1}.$$

Thus, the evaluation of (14) involves not more than

(18)
$$(3n-2)m^3 \text{ ops.}$$

The evaluation of (16) and (17) involves only products of m-component vectors by square matrices and we find

$$(16) \text{ requires } (2n-1)m^2 \text{ ops.};$$
$$(17) \text{ requires } (n-1)m^2 \text{ ops.}$$

The total for (14), (16), and (17) is thus

(19)
$$(3n-2)(m^3 + m^2) \text{ ops.},$$

to solve the system (11) *with coefficient matrix* (10).

Notice that this number is much superior to estimates of the form $\frac{1}{3}(nm)^3$ which are appropriate for direct elimination methods applied to arbitrary systems of order (nm). In fact, if $n = m$ the block-elimination scheme requires about $3m^4$ operations, while from (1.9) straightforward Gaussian elimination uses on the order of $\frac{1}{3}m^6$ operations. The great gain in economy of operations is again due to the careful account taken of the large number of zero elements in A. In fact, even greater efficiency is attained if each Γ_i is computed by solving the m linear systems, $A_i\Gamma_i = C_i$, and not by computing A_i^{-1}; and if similarly (15) is solved for $\mathbf{y}^{(i)}$.

The count in (19) is then reduced from the order of $3nm^3$ operations to the order of $\frac{5}{3}nm^3$ operations when we do not compute inverses.

The justification of the block-factorization method is given in

THEOREM **6.** *If the leading diagonal submatrices*

$$A^{(k)} \equiv \begin{bmatrix} A_1 & C_1 & & & & \\ B_2 & A_2 & \cdot & & & \\ & \cdot & \cdot & \cdot & & \\ & & \cdot & \cdot & \cdot & \\ & & & \cdot & \cdot & C_{k-1} \\ & & & & B_k & A_k \end{bmatrix}, \qquad k = 1, 2, \ldots, n,$$

of the original matrix (10) *are non-singular, then the block-factorization in* (14) *may be carried out* (*i.e., the* A_i *are non-singular*).

Proof. This is left to Problem 2. ∎

PROBLEMS, SECTION 3

1. If $LU \equiv A$ is a factorization of A satisfying (2), show that $l_{ik}u_{kj}$ is independent of the choice of l_{kk} and u_{kk} that satisfy (2a).

2. Prove Theorem 6.

4. ITERATIVE METHODS

The previous direct methods for solving general systems of order n require about $n^3/3$ operations. In addition, it has been indicated that, in practical computations with these methods, the errors which are necessarily introduced through rounding may become quite large for large n. Now we consider iterative methods in which an approximate solution is sought by using fewer operations per iteration. In general, these may be described as methods which proceed from some initial "guess," $\mathbf{x}^{(0)}$, and define a sequence of successive approximations $\mathbf{x}^{(1)}, \mathbf{x}^{(2)}, \ldots$ which, in principle, converge to the exact solution. If the convergence is sufficiently rapid, the procedure may be terminated at an early stage in the sequence and will yield a good approximation. One of the intrinsic advantages of such methods is the fact that errors, due to roundoff or even blunders, may be damped out as the procedure continues. In fact, special iterative methods are frequently used to improve "solutions" obtained by direct methods (see Subsection 4.3).

A large class of iterative methods may be defined as follows: Let the system to be solved be

$$(1) \qquad Ax = \mathbf{f}$$

where $\det |A| \neq 0$. Then the coefficient matrix can be *split*, in an infinite number of ways, into the form

$$(2) \qquad A = N - P$$

where N and P are matrices of the same order as A. The system (1) is then written as

$$(3) \qquad N\mathbf{x} = P\mathbf{x} + \mathbf{f}.$$

Starting with some *arbitrary* vector $\mathbf{x}^{(0)}$, we define a sequence of vectors $\{\mathbf{x}^{(\nu)}\}$, by the recursion

$$(4) \qquad N\mathbf{x}^{(\nu)} = P\mathbf{x}^{(\nu-1)} + \mathbf{f}, \qquad \nu = 1, 2, \ldots.$$

It is now clear that one of the restrictions to be placed on the *splitting* (2) is that

$$(5) \qquad \det |N| \neq 0,$$

in which case the recursions (4) define a unique sequence of vectors for all $\mathbf{x}^{(0)}$ and $\mathbf{f}$. As a practical matter, it is also clear that N should be chosen such that a system of the form

$$(6) \qquad N\mathbf{y} = \mathbf{z}$$

can be "easily" solved. Furthermore, if greater accuracy is desired, it would be better to calculate with (4) in the equivalent form

$$N(\mathbf{x}^{(\nu)} - \mathbf{x}^{(\nu-1)}) = \mathbf{f} - A\mathbf{x}^{(\nu-1)}.$$

This point will be discussed further in Subsection 4.3.

The convergence of the sequence $\{\mathbf{x}^{(\nu)}\}$ to the solution $\mathbf{x}$ of (1) is studied by introducing the matrix

$$(7) \qquad M = N^{-1}P$$

and the error vectors

$$(8) \qquad \mathbf{e}^{(\nu)} = \mathbf{x}^{(\nu)} - \mathbf{x}, \qquad \nu = 0, 1, 2, \ldots.$$

Subtract (3) from (4) to obtain, upon multiplication by N^{-1},

$$(9) \qquad \begin{aligned} \mathbf{e}^{(\nu)} &= M\mathbf{e}^{(\nu-1)} \\ &= M^2\mathbf{e}^{(\nu-2)} \\ &\ \vdots \\ &= M^\nu\mathbf{e}^{(0)}, \qquad \nu = 1, 2, \ldots, \end{aligned}$$

where $\mathbf{e}^{(0)}$ is the *arbitrary* initial error. Thus, it is clear that *a sufficient condition for convergence*, i.e., that $\lim_{\nu\to\infty} \mathbf{e}^{(\nu)} = \mathbf{o}$, *is that* $\lim_{\nu\to\infty} M^\nu = O$, and this is also *necessary if the method is to converge for all* $\mathbf{e}^{(0)}$.

A matrix, M, that satisfies this condition is called a *convergent matrix*. The basic results characteristizing convergent matrices have been established in Chapter 1, Theorem 1.4 and its Corollary, which we restate here as

THEOREM **1**. *The matrix M is convergent, i.e.,*

$$\lim_{\nu\to\infty} M^\nu = O,$$

iff all eigenvalues of M are less than one in absolute value. ■

(This condition is frequently stated as $\rho(M) < 1$ where $\rho(M)$ is the *spectral radius* of M defined by

$$\rho(M) \equiv \max_i |\lambda_i|$$

where the λ_i are the eigenvalues of M.)

COROLLARY **1**. *The matrix M is convergent if, for any matrix norm,* $\|M\| < 1$. ■

It is, in general, difficult to verify the conditions of Theorem 1. However, Corollary 1 may frequently be used to show that $\rho(M) < 1$. We have (see Chapter 1, Section 1, for the notation)

COROLLARY **2**. *The matrix* $M \equiv (m_{ij})$ *is convergent if either*

$$\|M\|_\infty \equiv \max_i \sum_{j=1}^{n} |m_{ij}| < 1; \tag{10a}$$

or

$$\|M\|_1 \equiv \max_j \sum_{i=1}^{n} |m_{ij}| < 1. \tag{10b}$$

Proof. We have shown in Chapter 1, Section 1, that $\|M\|_\infty$ and $\|M\|_1$ are matrix norms. ■

Let us return to the iteration scheme (4)–(9) and assume it to be a convergent one. We introduce the notion of the rate of convergence, R, of the iterative scheme by setting

$$R \equiv -\log \rho(M). \tag{11}$$

The significance of this quantity is most easily seen if we recall the corollary to Theorem 1.3 of Chapter 1, which states that: $\rho(M) = \underset{\{\| \|\}}{g.l.b.} \|M\|$ (here $\{\| \|\}$ is the set of natural norms). Given the initial error, $\mathbf{e}^{(0)}$, (9) permits the estimate in terms of any natural norm

$$\|\mathbf{e}^{(\nu)}\| \leq \|M\|^{\nu}\|\mathbf{e}^{(0)}\|.$$

Then for a given $\epsilon > 0$, there is some norm such that

$$\|\mathbf{e}^{(\nu)}\| \leq [\rho(M) + \epsilon]^{\nu}\|\mathbf{e}^{(0)}\|. \tag{12}$$

On the other hand, again from (9), if $\mathbf{e}^{(0)}$ is an eigenvector of M corresponding to the largest eigenvalue, $\|\mathbf{e}^{(\nu)}\| = [\rho(M)]^{\nu}\|\mathbf{e}^{(0)}\|$. Let it be required to reduce the amplitude of the error by a factor of at least 10^{-m}, $m > 0$. From (12), we see that, in some norm, the error amplitude is reduced by a factor close to $[\rho(M)]^{\nu}$. The number of iterations required is the least value of ν for which

$$[\rho(M)]^{\nu} \leq 10^{-m}.$$

By taking logs and recalling that $0 \leq \rho(M) < 1$, we obtain

$$\nu \geq \frac{m}{-\log \rho(M)} = \frac{m}{R}. \tag{13}$$

Thus, the number of iterations required to reduce the initial error by the factor 10^{-m} is inversely proportional to R, the rate of convergence.

4.1. Jacobi or Simultaneous Iterations

A special case (attributed to *Jacobi*) of the previous general theory is

$$N \equiv (a_{ii}\delta_{ij}), \qquad P \equiv N - A = (a_{ii}\delta_{ij} - a_{ij}). \tag{14}$$

From (14) in (4), it is seen that the components $x_i^{(\nu)}$ of the νth iterate are simply computed with $\mathbf{x}^{(0)}$ arbitrary by

$$x_i^{(\nu)} = \frac{1}{a_{ii}}\left(f_i - \sum_{\substack{j=1 \\ (j \neq i)}}^{n} a_{ij}x_j^{(\nu-1)}\right) \tag{15}$$

$$i = 1, 2, \ldots, n; \qquad \nu = 1, 2, \ldots.$$

Thus, this procedure may be employed provided only that $a_{ii} \neq 0$ for all $i = 1, 2, \ldots, n$. However, for the convergence of these iterations, Theorem 1 requires that all roots of $\det|\lambda I - N^{-1}P| = 0$ satisfy $|\lambda| < 1$. This equation can be written as, assuming $\det|N| \neq 0$,

$$(16)\qquad \det|\lambda N - P| = \det\begin{vmatrix} \lambda a_{11} & a_{12} & \cdots & a_{1n} \\ a_{21} & \lambda a_{22} & & \vdots \\ \vdots & & \ddots & a_{n-1,n} \\ a_{n1} & a_{n2} & & \lambda a_{nn} \end{vmatrix} = 0.$$

In general, the roots of such an equation, for large n, are not easily obtained and so we seek simpler sufficient conditions for convergence, as given in Corollary 2. The relevant matrix M is easily obtained since $N^{-1} = (a_{ii}{}^{-1}\delta_{ij})$ and thus

$$(17)\qquad M \equiv N^{-1}P = \left(\delta_{ij} - \frac{a_{ij}}{a_{ii}}\right)$$

Now conditions (10a) and (10b) of Corollary 2 become

$$(18a)\qquad \|M\|_\infty = \max_i \sum_{\substack{j=1 \\ (j\neq i)}}^{n} \left|\frac{a_{ij}}{a_{ii}}\right| < 1,$$

$$(18b)\qquad \|M\|_1 = \max_j \sum_{\substack{i=1 \\ (i\neq j)}}^{n} \left|\frac{a_{ij}}{a_{ii}}\right| < 1.$$

These tests are easily applied in practice. Since $\rho(M) \leq \|M\|$ we obtain a lower bound on the rate of convergence

$$(19)\qquad \begin{aligned} R &= \log\frac{1}{\rho(M)} \\ &\geq \log\frac{1}{\min(\|M\|_1, \|M\|_\infty)}. \end{aligned}$$

The operational count for the Jacobi iteration is simply obtained from (15); it is

$$(20)\qquad n^2 \text{ ops. per iteration.}$$

Thus by (13), if these iterations converge they require a total of about

$$\frac{m \times n^2}{R}\ \text{ops.},$$

to reduce the initial error by at least 10^{-m}. We see that if such an iterative method is to be at least as efficient as the direct elimination method it should have a rate of convergence and required accuracy factor, say 10^{-m}, satisfying

$$\frac{m}{R} \leq \frac{n}{3}.$$

(We assume here that m has been so chosen that the iterative solution will have an accuracy comparable to the accuracy obtained by the direct elimination method using the same number of digits in the arithmetic.)

4.2. Gauss-Seidel or Successive Iterations

It is clear from (15) that in the ordinary Jacobi iterations some components of $\mathbf{x}^{(\nu)}$ are known, but not used, while computing the remaining components. The Gauss-Seidel method is a modification of the Jacobi method in which all of the latest known components are used. The term "successive" which is frequently applied to this method refers to the fact that "new" components are successively used as they are obtained. (In contrast, the previous scheme was called "simultaneous" since new components were not employed as found, i.e., the "new" components were introduced simultaneously at the end of the iterative cycle.)

The obvious modification of (15) suggested by the above remarks is, with $\mathbf{x}^{(0)}$ arbitrary,

$$(21) \qquad x_i^{(\nu)} = \frac{1}{a_{ii}}\left(f_i - \sum_{j=1}^{i-1} a_{ij}x_j^{(\nu)} - \sum_{j=i+1}^{n} a_{ij}x_j^{(\nu-1)}\right),$$

$$i = 1, 2, \ldots, n; \qquad \nu = 1, 2, \ldots.$$

The splitting of A that yields this iterative scheme is

$$(22) \qquad N \equiv \begin{bmatrix} a_{11} & & & \\ a_{21} & a_{22} & & \\ \cdot & \cdot & & \\ \cdot & & \cdot & \\ \cdot & & & \cdot \\ a_{n1} & a_{n2} & \cdots & a_{nn} \end{bmatrix}, \qquad P \equiv N - A.$$

Since N is triangular, $\det |N| \neq 0$ is assured again if $a_{ii} \neq 0$; $i = 1, 2, \ldots, n$. The characteristic equation, whose roots must be in absolute value less than unity for convergence, is now of the form

$$(23) \qquad \det \begin{vmatrix} \lambda a_{11} & a_{12} & \cdots & a_{1n} \\ \lambda a_{21} & \lambda a_{22} & & \\ \vdots & & & \vdots \\ & & & a_{n-1,n} \\ \lambda a_{n1} & \lambda a_{n2} & \cdots & \lambda a_{nn} \end{vmatrix} = 0.$$

The roots of this equation are just as difficult to find as are those of (16), but the sufficient conditions of Corollary 2 are now much more complicated than those for the Jacobi iteration. However, a simple sufficient condition for convergence of the Gauss-Seidel method can be obtained. To derive this condition, we introduce the error vectors defined in (8) and find from (1) and (21) that the components of these vectors must satisfy

$$e_i^{(\nu)} = -\sum_{j=1}^{i-1} \frac{a_{ij}}{a_{ii}} e_j^{(\nu)} - \sum_{j=i+1}^{n} \frac{a_{ij}}{a_{ii}} e_j^{(\nu-1)}, \tag{24}$$

$$i = 1, 2, \ldots, n; \qquad \nu = 1, 2, \ldots.$$

The result to be proved may now be stated as

LEMMA 1. *Let the vectors* $\mathbf{e}^{(\nu)}$, $\nu = 1, 2, \ldots$, *be defined by* (24) *with* $\mathbf{e}^{(0)}$ *arbitrary. Define the maximum norm,* $\|\cdot\|_\infty$, *and factors,* r_i, *by*

$$\|\mathbf{e}^{(\nu)}\|_\infty \equiv \max_j |e_j^{(\nu)}|, \tag{25a}$$

$$r_i \equiv \sum_{\substack{j=1 \\ (j \neq i)}}^{n} \left|\frac{a_{ij}}{a_{ii}}\right|; \tag{25b}$$

and let the matrix A *satisfy*

$$r \equiv \max_i r_i < 1. \tag{26}$$

Then

$$\|\mathbf{e}^{(\nu)}\|_\infty \leq r^\nu \|\mathbf{e}^{(0)}\|_\infty, \tag{27}$$

and $\mathbf{e}^{(\nu)} \rightarrow \mathbf{o}$ *as* $\nu \rightarrow \infty$.

Proof. The lemma clearly follows from the inequalities

$$\|\mathbf{e}^{(\nu)}\|_\infty \leq r \|\mathbf{e}^{(\nu-1)}\|_\infty, \qquad \nu = 1, 2, \ldots; \tag{28}$$

which we shall prove by induction (on the components of $\mathbf{e}^{(\nu)}$). From (24) with $i = 1$ we obtain, using (25) and (26),

$$\begin{aligned} |e_1^{(\nu)}| &\leq \sum_{j=2}^{n} \left|\frac{a_{1j}}{a_{11}}\right| \cdot |e_j^{(\nu-1)}| \\ &\leq \|\mathbf{e}^{(\nu-1)}\|_\infty \sum_{j=2}^{n} \left|\frac{a_{1j}}{a_{11}}\right| = \|\mathbf{e}^{(\nu-1)}\|_\infty \cdot r_1 \\ &\leq r \|\mathbf{e}^{(\nu-1)}\|_\infty. \end{aligned}$$

Now assume $|e_k^{(\nu)}| \leq r\|\mathbf{e}^{(\nu-1)}\|_\infty$ for $k = 1, 2, \ldots, i-1$. Then again from (24), recalling that $r < 1$

$$
\begin{aligned}
|e_i^{(\nu)}| &\leq \sum_{j=1}^{i-1} \left|\frac{a_{ij}}{a_{ii}}\right| \cdot |e_j^{(\nu)}| + \sum_{j=i+1}^{n} \left|\frac{a_{ij}}{a_{ii}}\right| \cdot |e_j^{(\nu-1)}| \\
&\leq \|\mathbf{e}^{(\nu-1)}\|_\infty \left\{\sum_{j=1}^{i-1} r\left|\frac{a_{ij}}{a_{ii}}\right| + \sum_{j=i+1}^{n} \left|\frac{a_{ij}}{a_{ii}}\right|\right\} \\
&\leq \|\mathbf{e}^{(\nu-1)}\|_\infty \sum_{\substack{j=1 \\ (j \neq i)}}^{n} \left|\frac{a_{ij}}{a_{ii}}\right| = r_i\|\mathbf{e}^{(\nu-1)}\|_\infty \\
&\leq r\|\mathbf{e}^{(\nu-1)}\|_\infty.
\end{aligned}
$$

Thus, the induction argument is complete and since the above inequality is valid for all $i = 1, 2, \ldots, n$, it follows by (25) that (28) is valid. ∎

The convergence test of this lemma is easily applied and is formally identical to that of (18a) for the Jacobi method. However, it is not generally true that if the Gauss-Seidel method converges then the Jacobi method will converge, nor is the converse generally true.

See Subsection 4.4 for other convergence tests.

4.3. Method of Residual Correction

This iterative scheme improves upon the accuracy of the approximate solution of (1) (obtained for example by the Gaussian elimination method), by using the approximate numerical triangular factorization of A. That is, the triangularization of (1), performed with t digits, produces $\mathscr{L}$ (lower), $\mathscr{U}$ (upper), and $\mathbf{x}^{(0)}$. Now define

$$
\begin{aligned}
N &\equiv \mathscr{L}\mathscr{U}, \qquad P \equiv N - A, \\
\mathbf{r}^{(0)} &\equiv \mathbf{f} - A\mathbf{x}^{(0)}.
\end{aligned} \tag{29}
$$

Observe that N is easily invertible, or rather that the equation

$$\mathscr{L}\mathscr{U}\mathbf{y} = \mathbf{z}$$

may be readily "solved," since $\mathscr{L}$ and $\mathscr{U}$ are triangular, by using $n(n + 1)$ operations. [If $\mathscr{L} \equiv (s_{ij})$ has $s_{ii} = 1$ for all i, then the number of operations used to solve $\mathscr{L}\mathbf{w} = \mathbf{z}$ is $n(n - 1)/2$, while the number for solving $\mathscr{U}\mathbf{y} = \mathbf{w}$ is $n(n + 1)/2$. Hence, in this case n^2 is the operational count for solving $N\mathbf{y} = \mathbf{z}$.]

Now, the iteration scheme given by (4) is convergent if M, defined by (7), satisfies

$$\|M\| \equiv \|I - N^{-1}A\| < 1.$$

This inequality is satisfied if $\|P\| \cdot \|A^{-1}\| < \frac{1}{2}$ (see Problem 5). In practice, (4) is not solved in the form

$$\mathscr{L}\mathscr{U}\mathbf{x}^{(\nu)} = (\mathscr{L}\mathscr{U} - A)\mathbf{x}^{(\nu-1)} + \mathbf{f}. \tag{30}$$

Rather, we introduce the *change* in the iterate by

$$\boldsymbol{\delta}\mathbf{x}^{(\nu-1)} \equiv \mathbf{x}^{(\nu)} - \mathbf{x}^{(\nu-1)}$$

and the residual of the iterate by

$$\mathbf{r}^{(\nu-1)} \equiv \mathbf{f} - A\mathbf{x}^{(\nu-1)}. \tag{31}$$

Then (30) can be written simply as

$$\mathscr{L}\mathscr{U}[\boldsymbol{\delta}\mathbf{x}^{(\nu-1)}] = \mathbf{r}^{(\nu-1)}, \tag{32}$$

and the computations are done with these equations.

The evaluation of $\mathbf{r}^{(\nu)}$ involves n^2 operations; hence each iteration step, (31) and (32), requires $n(2n + 1)$ operations (only $2n^2$ operations if $s_{ii} = 1$ for all i). By using (1) and (31) the error satisfies

$$\|\mathbf{x}^{(\nu)} - \mathbf{x}\| = \|A^{-1}\mathbf{r}^{(\nu)}\| \leq \|A^{-1}\| \, \|\mathbf{r}^{(\nu)}\|. \tag{33}$$

But from (32), the definition of M, and the corollary to Theorem 1.5 of Chapter 1,

$$\|A^{-1}\mathbf{r}^{(\nu)}\| = \|A^{-1}N\boldsymbol{\delta}\mathbf{x}^{(\nu)}\| = \|(I - M)^{-1}\boldsymbol{\delta}\mathbf{x}^{(\nu)}\|$$
$$\leq \frac{\|\boldsymbol{\delta}\mathbf{x}^{(\nu)}\|}{1 - q},$$

provided $q \equiv \|M\| = \|N^{-1}(N - A)\| < 1$.

As described in Subsection 1.2, the numerical solution of (32) produces a vector $\boldsymbol{\delta}\mathbf{x}^{(\nu-1)}$ that satisfies

$$\mathscr{L}_{\nu-1}\mathscr{U}_{\nu-1}\boldsymbol{\delta}\mathbf{x}^{(\nu-1)} = \mathbf{r}^{(\nu-1)}, \tag{34}$$

where

$$\mathscr{L}_{\nu-1} \equiv \mathscr{L} + \delta\mathscr{L}_{\nu-1}, \qquad \mathscr{U}_{\nu-1} \equiv \mathscr{U} + \delta\mathscr{U}_{\nu-1}.$$

The perturbations $\delta\mathscr{L}_\nu$ and $\delta\mathscr{U}_\nu$ are small relative to $\mathscr{L}$ and $\mathscr{U}$ respectively if the number of digits carried in the arithmetic calculations is large enough. Set

$$N_\nu \equiv \mathscr{L}_\nu\mathscr{U}_\nu, \qquad P_\nu \equiv N_\nu - A$$

and

$$M_\nu = N_\nu^{-1}P_\nu.$$

Then the error

$$\mathbf{e}^{(\nu)} \equiv \mathbf{x}^{(\nu)} - \mathbf{x},$$

satisfies

$$\begin{aligned}\mathbf{e}^{(\nu)} &= M_{\nu-1}\mathbf{e}^{(\nu-1)} \\ &= M_{\nu-1}M_{\nu-2}\mathbf{e}^{(\nu-2)} \\ &\vdots \\ &= M_{\nu-1}M_{\nu-2}\cdots M_0\mathbf{e}^{(0)}.\end{aligned}$$

If $\|M_i\| \leq q < 1$ for all i, then $\|\mathbf{e}^{(\nu)}\| \leq q^\nu\|\mathbf{e}^{(0)}\|$, and the scheme is convergent for any $\mathbf{e}^{(0)}$.

As a practical matter, from equations (31) and (32), we see that $\mathbf{r}^\nu \rightarrow \mathbf{o}$ may occur only if the right-hand side of (31) is calculated to ever higher precision as ν increases. On the other hand, equation (32) or equation (34) requires only single precision accuracy for $\mathbf{r}^{(\nu-1)}$, in order to determine $\delta\mathbf{x}^{(\nu-1)}$ by using single precision arithmetic.

4.4. Positive Definite Systems

Many of the large order linear systems that arise in practice have real symmetric matrices which are positive definite. In such cases we can show that a quite general class of iteration methods converges. We state this result as

THEOREM 2. *Let A be Hermitian (of order n) and N be any non-singular matrix (of order n) for which*

$$Q \equiv N + N^* - A \tag{35}$$

is positive definite. Then the matrix

$$M \equiv I - N^{-1}A$$

is convergent iff A is positive definite.

Proof. For any eigenvalue, λ, and corresponding eigenvector, $\mathbf{u}$, of M we have

$$M\mathbf{u} = \lambda\mathbf{u}.$$

But since N is non-singular this implies

$$A\mathbf{u} = (1 - \lambda)N\mathbf{u}, \tag{36}$$

and so $\lambda = 1$ iff $A\mathbf{u} = \mathbf{o}$.

Now let A be positive definite (i.e., $\mathbf{v}^*A\mathbf{v} > 0$ if $\mathbf{v} \neq \mathbf{o}$) and $\mathbf{u}$ be any eigenvector of M. Then $A\mathbf{u} \neq \mathbf{o}$ so that the corresponding eigenvalue λ

of M satisfies $\lambda \neq 1$. By taking the complex inner product of each side of (36) with $\mathbf{u}$ we then obtain

$$\frac{1}{1-\lambda} = \frac{\mathbf{u}^* N \mathbf{u}}{\mathbf{u}^* A \mathbf{u}}.$$

The complex conjugate of this expression is, since $\mathbf{u}^* A \mathbf{u}$ is real,

$$\frac{1}{1-\bar{\lambda}} = \frac{\mathbf{u}^* N^* \mathbf{u}}{\mathbf{u}^* A \mathbf{u}}.$$

If we add these two equations we get

$$2 \operatorname{Re} \frac{1}{1-\lambda} = \frac{\mathbf{u}^*(N + N^*)\mathbf{u}}{\mathbf{u}^* A \mathbf{u}}.$$

Now set $\lambda = \alpha + i\beta$ and recall (35) to write this as

$$\frac{2(1-\alpha)}{(1-\alpha)^2 + \beta^2} = 1 + \frac{\mathbf{u}^* Q \mathbf{u}}{\mathbf{u}^* A \mathbf{u}} > 1,$$

since by hypothesis Q is positive definite. Hence, we have the inequality

$$|\lambda|^2 = \alpha^2 + \beta^2 < 1.$$

The sufficiency is thus demonstrated. The necessity part of the proof is indicated in Problems 1 and 2. ■

As an immediate corollary of this theorem, we have a result on the convergence of the Gauss-Seidel method for Hermitian matrices.

COROLLARY 1. *Let A be Hermitian with positive diagonal elements. Then the Gauss-Seidel method for this matrix converges iff A is positive definite.*

Proof. By the hypothesis on A it can be written as

$$A = D + E + E^*$$

where D is a diagonal matrix of positive diagonal elements and E is strictly lower triangular (i.e., zeros on and above the diagonal). The Gauss-Seidel method applied to A, see (22), is equivalent to the splitting

$$N \equiv D + E, \qquad P = -E^*.$$

However, with this choice for N we have

$$Q \equiv N + N^* - A = D$$

which is clearly positive definite. Thus the hypothesis of Theorem 2 applies and the proof is concluded. ■

Similarly, we obtain a result on the convergence of the Jacobi iterations as a special case of

COROLLARY **2.** *Let $D = D^*$ be non-singular and*

$$D - (E + E^*)$$

be positive definite. Then

$$D^{-1}(E + E^*)$$

is convergent iff $A \equiv D + (E + E^*)$ is positive definite.

Proof. Use $N \equiv D$ in the theorem. ∎

In the special case that D is a diagonal matrix, Corollary 2 yields the convergence of the Jacobi method for the matrix A.

4.5. Block Iterations

There are other splittings of A which in many important cases yield rapidly convergent iterations. In particular, since tridiagonal and block-tridiagonal systems are easily solved, it is natural to consider iterations in which N has either of these forms. Many of the large order systems which arise in the finite difference methods for partial differential equations suggest such block iterations. More generally, if the elements "close" to the diagonal of a matrix are large compared to the other elements, it is usually advantageous to include all of these large elements in N (assuming, of course, that the resulting systems which determine the iterates are still easily solved). Of course, in all applications of these block methods, attempts should be made to prove the convergence of the method and, if possible, to estimate the rate of convergence.

PROBLEMS, SECTION 4

1. Let the sequence $\{\mathbf{v}_\nu\}$ be defined, with $\mathbf{v}_0$ arbitrary, by

$$\mathbf{v}_{\nu+1} = M\mathbf{v}_\nu, \qquad \nu = 0, 1, \ldots,$$

where $M \equiv I - N^{-1}A$ and A is Hermitian. Then
(a) Verify the identity

$$\mathbf{v}_\nu{}^*A\mathbf{v}_\nu - \mathbf{v}_{\nu+1}^*A\mathbf{v}_{\nu+1} = (\mathbf{v}_\nu - \mathbf{v}_{\nu+1})^*Q(\mathbf{v}_\nu - \mathbf{v}_{\nu+1})$$

where $Q \equiv N + N^* - A$;
(b) If Q is positive definite show that $\{\mathbf{v}_\nu{}^*A\mathbf{v}_\nu\}$ is a non-increasing sequence. (In fact, the sequence is strictly decreasing if 1 is not an eigenvalue of M.)

2. Use part (b) of Problem 1 to show that if M is convergent then A is positive definite.

[Hint: Use proof by contradiction; assume $\mathbf{v}_0{}^*A\mathbf{v}_0 \le 0$ for some $\mathbf{v}_0 \ne \mathbf{o}$. Then, since M is convergent, $\mathbf{v}_1 \ne \mathbf{v}_0$. Therefore,

$$\mathbf{v}_\nu{}^*A\mathbf{v}_\nu \le \mathbf{v}_1{}^*A\mathbf{v}_1 < \mathbf{v}_0{}^*A\mathbf{v}_0 \le 0.$$

This is a contradiction, since the convergence of M implies $\mathbf{v}_\nu \to \mathbf{o}$.]

3. Analyze the convergence of the Jacobi and the Gauss-Seidel iterative methods for the second order matrix

$$A \equiv \begin{pmatrix} 1 & \rho \\ \rho & 1 \end{pmatrix}, \qquad |\rho| < 1, \mathbf{x}_0 \ne \mathbf{o}.$$

4. Determine when the Jacobi iterative method converges for the compound matrix

$$A \equiv \begin{pmatrix} I & S \\ S^T & I \end{pmatrix}, \text{ with } I \text{ and } S$$

of order n.

[Hint: Work with the compound vectors $\begin{pmatrix} \mathbf{x}_\nu \\ \mathbf{y}_\nu \end{pmatrix}$. Define the compound error vectors

$$\begin{pmatrix} \mathbf{e}_\nu \\ \mathbf{g}_\nu \end{pmatrix} \equiv \begin{pmatrix} \mathbf{x}_\nu \\ \mathbf{y}_\nu \end{pmatrix} - \begin{pmatrix} \mathbf{x} \\ \mathbf{y} \end{pmatrix}.$$

Find a recursion formula for $\{\mathbf{e}_\nu\}$ that doesn't involve $\{\mathbf{g}_\nu\}$.]

5. The convergence of the residual correction scheme defined by (30) is assured if $\|I - N^{-1}A\| < 1$. Verify that this inequality holds if

$$\|P\| \cdot \|A^{-1}\| < \tfrac{1}{2}.$$

[Hint: Let $B = A^{-1}N$. Then

$$\begin{aligned} I - N^{-1}A &= I - B^{-1} = B^{-1}(B - I) \\ &= B^{-1}(A^{-1}P). \end{aligned}$$

Note that $B = A^{-1}P + I$ and therefore, by the remark following the corollary to Theorem 1.5 of Chapter 1, we have $\|B^{-1}\| < 2$.]

5. THE ACCELERATION OF ITERATIVE METHODS

Given any iteration procedure, for a specific system of equations, it may be possible to improve its rate of convergence by a simple device. Such modifications, which we call *acceleration*, are frequently termed "*extrapolation*," "*over-relaxation*," or various other names depending upon the problem to which they are applied or perhaps upon the particular form of device which is used. In any event, the general principle common to almost all acceleration procedures is the introduction of a splitting, similar to (4.2), which depends upon some real parameter, say α, in an "appropriate" manner. The splitting may be denoted by

$$A = N(\alpha) - P(\alpha)$$

and is still subject to the requirement that

$$\det |N(\alpha)| \neq 0.$$

(This will place some restriction on the permissible values of α.) Now, as has been shown in Section 4, an iteration scheme based on the above splitting will converge, for arbitrary initial vectors, iff all eigenvalues of

$$M(\alpha) \equiv N^{-1}(\alpha)P(\alpha),$$

are in absolute value less than unity.

Let these eigenvalues be denoted by

$$\lambda_i(\alpha), \qquad i = 1, 2, \ldots, n;$$

where, as indicated, their values may depend upon the choice of the parameter α. Now if a value of α can be determined such that

$$\rho[M(\alpha)] \equiv \max_i |\lambda_i(\alpha)| < 1,$$

the scheme will converge. Furthermore, since the rate of convergence is

$$R(\alpha) = \log \frac{1}{\rho[M(\alpha)]},$$

the convergence is "best" for the value $\alpha = \alpha_*$ such that

$$\rho[M(\alpha_*)] = \min_\alpha \rho[M(\alpha)].$$

The selection of an *optimal* α_* is the most important feature of acceleration procedures.

Some acceleration procedures that are commonly used are described as follows: Let some definite splitting, (4.2), be given by

$$A = N_0 - P_0, \tag{1}$$

where N_0 and P_0 are fixed matrices with $\det |N_0| \neq 0$. Let the relevant eigenvalues of this scheme, i.e., those of

$$M_0 = N_0^{-1}P_0, \tag{2a}$$

be

$$\lambda_i, \qquad i = 1, 2, \ldots, n. \tag{2b}$$

Then we introduce the one-parameter family of splittings

$$
\begin{aligned}
N(\alpha) &= (1 + \alpha)N_0, \\
P(\alpha) &= (1 + \alpha)N_0 - A = P_0 + \alpha N_0.
\end{aligned} \tag{3}
$$

In order that det $|N(\alpha)| \neq 0$, we need only require $\alpha \neq -1$. Then if the eigenvalues of $M(\alpha) \equiv N^{-1}(\alpha)P(\alpha)$ are denoted by $\mu_i(\alpha)$, $i = 1, 2, \ldots, n$, we claim that

$$
\mu_i(\alpha) = \frac{\lambda_i + \alpha}{1 + \alpha}, \qquad i = 1, 2, \ldots, n. \tag{4}
$$

The verification of (4) requires only a simple application of the definitions of eigenvalue and eigenvector. Specifically from (2) and (3) we have

$$
\begin{aligned}
M(\alpha) = N^{-1}(\alpha)P(\alpha) &= \frac{1}{1 + \alpha} N_0^{-1}(P_0 + \alpha N_0) \\
&= \frac{\alpha}{1 + \alpha} I + \frac{1}{1 + \alpha} M_0.
\end{aligned}
$$

Thus, if $\mathbf{u}$ is any eigenvector of M_0 belonging to the eigenvalue λ, that is $M_0\mathbf{u} = \lambda\mathbf{u}$, we obtain from the above

$$
M(\alpha)\mathbf{u} = \frac{\alpha}{1 + \alpha}\mathbf{u} + \frac{\lambda}{1 + \alpha}\mathbf{u} = \frac{\lambda + \alpha}{1 + \alpha}\mathbf{u}.
$$

That is, $\mathbf{u}$ must also be an eigenvector of $M(\alpha)$ belonging to the eigenvalue $(\lambda + \alpha)/(1 + \alpha)$. Conversely, if $M(\alpha)\mathbf{v} = \mu\mathbf{v}$ we obtain

$$
\mu\mathbf{v} = M(\alpha)\mathbf{v} = \frac{\alpha}{1 + \alpha}\mathbf{v} + \frac{1}{1 + \alpha} M_0\mathbf{v},
$$

or, since $1 + \alpha \neq 0$ by assumption,

$$
M_0\mathbf{v} = [\mu(1 + \alpha) - \alpha]\mathbf{v}.
$$

Thus every eigenvector of $M(\alpha)$ is an eigenvector of M_0 and (4) is established for all $\alpha \neq -1$.

In order to determine convergent schemes of the form (3), we must study the relation (4). This is done first for a very important class of special cases in which the "best" such scheme can be obtained. These results may be stated as

THEOREM 1. *Let N_0 and P_0 be such that the eigenvalues λ_i of $N_0^{-1}P_0$ are all real and satisfy*

$$
\lambda_1 \leq \lambda_2 \leq \cdots \leq \lambda_n < 1. \tag{5}
$$

Then the scheme (3) *will converge for any α such that*

$$
\alpha > -\frac{1 + \lambda_1}{2} > -1. \tag{6}
$$

Furthermore, the largest rate of convergence for these schemes is obtained when

(7) $$\alpha = \alpha_* \equiv -\frac{\lambda_1 + \lambda_n}{2}$$

for which value

(8) $$\rho[M(\alpha_*)] \equiv \min_\alpha \rho[M(\alpha)] = \min_\alpha \left(\max_{i=1}^{n} |\mu_i(\alpha)|\right) = \frac{\lambda_n - \lambda_1}{2 - \lambda_1 - \lambda_n} < 1.$$

Proof. A scheme of the form (3) will converge if $|\mu_i(\alpha)| < 1$, $i = 1, 2, \ldots, n$. Let us introduce

(9) $$x \equiv \frac{1}{1 + \alpha}; \qquad m_j \equiv \lambda_j - 1, \qquad j = 1, 2, \ldots, n.$$

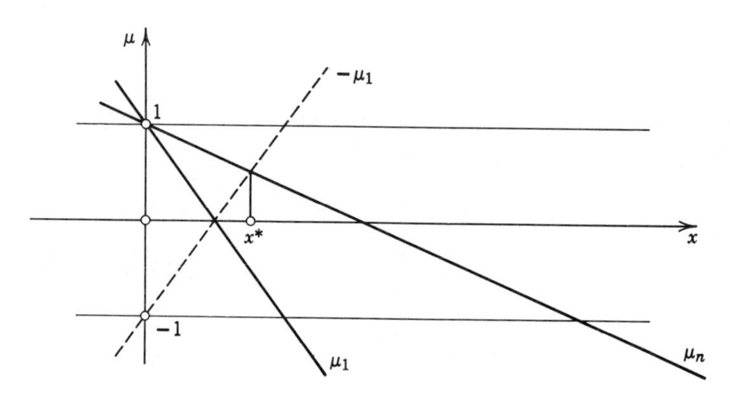

Figure 1

Then (4) can be written as

(10) $$\mu_i = m_i x + 1, \qquad i = 1, 2, \ldots, n,$$

where by (5), all $m_i < 0$. The equations (10), for the μ_i as functions of x, represent n straight lines with negative slopes. Let us assume that the λ_i have been ordered as in (5). Then by (9) we have

$$m_1 \leq m_2 \leq \cdots \leq m_n < 0,$$

and all the lines (10) are bounded by those for $i = 1$ and $i = n$ (see Figure 1). Thus, we have for

(11) $$\begin{aligned} &x > 0\colon \mu_1 = m_1 x + 1 \leq \mu_i \leq m_n x + 1 = \mu_n; \\ &x < 0\colon \mu_n = m_n x + 1 \leq \mu_i \leq m_1 x + 1 = \mu_1, \qquad i = 1, 2, \ldots, n. \end{aligned}$$

Clearly then, all $\mu_i < 1$ iff $x > 0$. Similarly, all $\mu_i > -1$ iff $\mu_1 > -1$ or equivalently $x < -2/m_1$. Thus, $|\mu_i| < 1$ iff $0 < x < -2/m_1$, and using (9) we obtain (6).

For $x > 0$ we have by (11)

$$\mu = \max_{i=1}^{n} |\mu_i| = \max(|m_1 x + 1|, |m_n x + 1|).$$

From Figure 1 it is then clear that

$$\min_{x>0} \mu = |m_1 x_* + 1| = |m_n x_* + 1| = \frac{m_1 - m_n}{m_1 + m_n},$$

where $x_* = -2/(m_1 + m_n)$. Upon applying (9) again, we obtain (7) and (8) and the proof is complete. ■

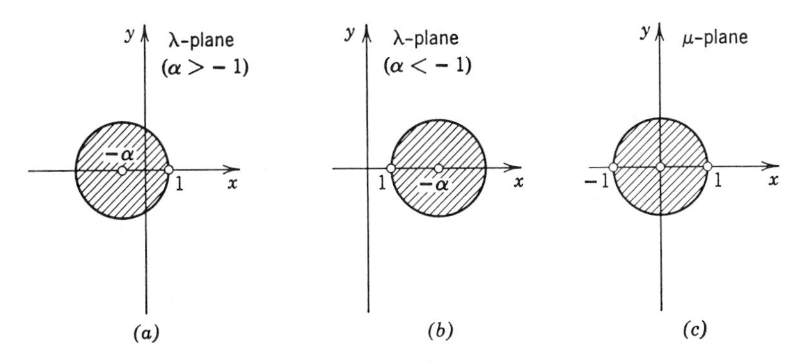

Figure 2

By an exactly analogous proof similar results can be obtained for the case where all $\lambda_i > 1$ (see Problem 1).

In the general case, the λ_j and hence also the $\mu_j(\alpha)$ will be complex. Then the schemes (3) will be convergent if the complex numbers $\mu_j(\alpha) \equiv \xi_j(\alpha) + i\eta_j(\alpha)$ all lie in the interior of the unit circle

$$|\mu|^2 \equiv \xi^2 + \eta^2 = 1, \tag{12}$$

of the (ξ, η)-plane. The relations (4) can now be considered as special points of the mapping of the $\lambda = x + iy$ plane into the μ-plane

$$\mu = \frac{\lambda + \alpha}{1 + \alpha}, \qquad \alpha \neq -1. \tag{13}$$

This is a special case of the well-known Möbius transformations studied in function theory. If α is real, we can easily verify directly that the unit circle (12) in the μ-plane corresponds to a circle in the λ-plane given by

$$(x + \alpha)^2 + y^2 = (1 + \alpha)^2. \tag{14}$$

It can also be shown that any interior point of the circle (14) is mapped by (13) into an interior point of (12). The transformation (13) is illustrated in Figure 2 for $\alpha > -1$ and $\alpha < -1$.

From this figure it can be seen that convergent schemes can be found if the eigenvalues λ_i satisfy either

$$\text{Re}\,(\lambda_i) < 1, \qquad i = 1, 2, \ldots, n, \tag{15a}$$

or

$$\text{Re}\,(\lambda_i) > 1, \qquad i = 1, 2, \ldots, n. \tag{15b}$$

That is, a "convergent" value of α can be obtained corresponding to any circle in the λ-plane which has the properties:

(i) the center is on the real axis;
(ii) it passes through the point (1, 0);
(iii) all eigenvalues λ_i are interior to it.

If such a circle exists, then we call the coordinates of its center $(-\alpha, 0)$ and this value of α yields a convergent scheme. However, now it is not a simple matter to determine the best value of α.

5.1. Practical Application of Acceleration Methods

It is assumed that the basic scheme determined by (1) can be efficiently computed. That is, to solve

$$A\mathbf{x} = \mathbf{f}$$

we consider the iterates, with $\mathbf{x}^{(0)}$ arbitrary, given by

$$N_0\mathbf{x}^{(\nu)} = P_0\mathbf{x}^{(\nu-1)} + \mathbf{f}, \qquad \nu = 1, 2, \ldots. \tag{16}$$

We assume that such systems can be solved in an efficient manner. Now the iterates, $\mathbf{x}^{(\nu)}$, satisfy the system of equations

$$\mathbf{x}^{(\nu)} = M_0\mathbf{x}^{(\nu-1)} + \mathbf{g}, \qquad \nu = 1, 2, \ldots, \tag{17}$$

where $\mathbf{g} \equiv N_0^{-1}\mathbf{f}$ and M_0 is defined in (2a).

The acceleration (3) corresponding to this procedure yields with $\mathbf{y}^{(0)}$ arbitrary

$$\mathbf{y}^{(\nu)} = M(\alpha)\mathbf{y}^{(\nu-1)} + \frac{1}{1 + \alpha}\mathbf{g}, \qquad \nu = 1, 2, \ldots, \tag{18}$$

since $N^{-1}(\alpha)\mathbf{f} = 1/(1 + \alpha)N_0^{-1}\mathbf{f}$. In terms of M_0 these iterates can be written as

$$\mathbf{y}^{(\nu)} = \frac{\alpha}{1 + \alpha}\mathbf{y}^{(\nu-1)} + \frac{1}{1 + \alpha}(M_0\mathbf{y}^{(\nu-1)} + \mathbf{g}), \qquad \nu = 1, 2, \ldots. \tag{19}$$

A comparison of (17) and (19) yields some insight into the relationship between the basic scheme and its accelerated version:

If $\alpha = 0$ the basic scheme results. If $\alpha > 0$ then for any real numbers a and b,

$$\min(a, b) \leq \frac{\alpha}{1+\alpha} a + \frac{1}{1+\alpha} b \leq \max(a, b).$$

So in this case, the acceleration scheme yields a vector on the line segment joining the previous iterate and what would have been the next iterate of the basic scheme. The term "interpolated iteration" is frequently employed to describe this type of acceleration. If $-1 < \alpha < 0$ then

$$\frac{\alpha}{1+\alpha} a + \frac{1}{1+\alpha} b \begin{cases} \leq b & \text{if } b \leq a, \\ \geq b & \text{if } b \geq a; \end{cases}$$

and the acceleration scheme yields vectors with components whose values are definitely not between those of $\mathbf{y}^{(\nu-1)}$ and $M_0\mathbf{y}^{(\nu-1)} + \mathbf{g}$. The scheme is now termed an "extrapolated iteration." Similarly, the remaining case $\alpha < -1$ is such an extrapolation method.

To compute using the scheme (19), we define the vectors $\mathbf{z}^{(\nu)}$ by

$$\text{(20a)} \qquad N_0\mathbf{z}^{(\nu)} = P_0\mathbf{y}^{(\nu-1)} + \mathbf{f}, \qquad \nu = 1, 2, \ldots,$$

and then write (18) as

$$\text{(20b)} \qquad \mathbf{y}^{(\nu)} = \frac{\alpha}{1+\alpha}\mathbf{y}^{(\nu-1)} + \frac{1}{1+\alpha}\mathbf{z}^{(\nu)}.$$

Thus, as in the basic scheme, the calculations only require the solution of systems of the form (20a). [Note that in (20), $\mathbf{z}^{(\nu)}$ is defined by a recursion which is similar, but not identical, to (16).]

In general, the eigenvalues λ_j, or in particular λ_1 and λ_n, of the basic scheme will not be known. But it may be possible to approximate the value $\alpha = \alpha_*$ which yields the fastest convergence. This is accomplished by some test calculations that are easily performed:

Since the rate of convergence is independent of the inhomogeneous term, $\mathbf{f}$, we seek the best scheme for solving

$$\text{(21)} \qquad A\mathbf{y} = \mathbf{o}.$$

Since it is assumed that $\det|A| \neq 0$, this system has the unique solution $\mathbf{y} = \mathbf{o}$. Apply the scheme (18) with some fixed value of $\alpha = \alpha_1$ to (21) and compute [actually use (20)]

$$\text{(22)} \qquad \mathbf{y}^{(\nu)}(\alpha_1) = M^{\nu}(\alpha_1)\mathbf{y}^{(0)}, \qquad \nu = 1, 2, \ldots,$$

where $\mathbf{y}^{(0)}$ is an arbitrary but fixed initial vector. If the value α_1 yields a convergent iteration, compute until

$$\|\mathbf{y}^{(\nu)}(\alpha_1)\| \equiv \max_j |y_j^{(\nu)}(\alpha_1)| \leq 10^{-m},$$

where m is a fixed positive number. This requires some minimum number of iterations which we may call $\nu(\alpha_1)$. Repeat this procedure with the same $\mathbf{y}^{(0)}$ and m, for a sequence of values of $\alpha = \alpha_2, \alpha_3, \ldots$, to obtain the corresponding sequence $\{\nu(\alpha_i)\}$. The approximate value for α_* can be obtained by plotting the points $(\nu(\alpha_i), \alpha_i)$ and choosing that α which seems to minimize the "function" $\nu(\alpha)$.

An obvious alternative is to compute the sequence $\{\|\mathbf{y}^{(N)}(\alpha_i)\|\}$ using a fixed number, say N, of iterations. Then an approximation to α_* is that value which minimizes $\|\mathbf{y}^{(N)}(\alpha)\|$.

5.2. Generalizations of the Acceleration Method

There are numerous generalizations of the acceleration method which in fact are more powerful than the scheme described by (3). The simplest type of generalization proceeds from a single basic splitting of the form (1)–(2) but employs cyclically a fixed sequence of acceleration parameters, say $\alpha_1, \alpha_2, \ldots, \alpha_r$. Specifically for $i = 1, 2, \ldots, r$, define $N(\alpha_i)$ and $P(\alpha_i)$ as in (3) and the corresponding matrices $M(\alpha_i)$ by

$$(23) \qquad M(\alpha_i) \equiv N^{-1}(\alpha_i)P(\alpha_i) = \frac{\alpha_i}{1+\alpha_i} I + \frac{1}{1+\alpha_i} M_0, \qquad i = 1, 2, \ldots, r.$$

The iterations are defined as follows, with $\mathbf{x}^{(0)}$ arbitrary, for $\nu = 1, 2, \ldots$:

$$(24a) \qquad \mathbf{y}^{(\nu, 0)} = \mathbf{x}^{(\nu-1)};$$

$$(24b) \qquad \mathbf{y}^{(\nu, s)} = M(\alpha_s)\mathbf{y}^{(\nu, s-1)} + N^{-1}(\alpha_s)\mathbf{f}, \qquad s = 1, 2, \ldots, r;$$

$$(24c) \qquad \mathbf{x}^{(\nu)} = \mathbf{y}^{(\nu, r)}.$$

Again, each of the r vectors of (24b) can be obtained by solving a linear system of the form

$$N(\alpha_s)\mathbf{y}^{(\nu, s)} = P(\alpha_s)\mathbf{y}^{(\nu, s-1)} + \mathbf{f}.$$

With this notation, one iteration of this generalized acceleration scheme requires the same number of computations as r iterations in the ordinary acceleration scheme. The convergence of this method can be analyzed by means of the equivalent formulation

$$(25) \qquad \mathbf{x}^{(\nu)} = M(\alpha_1, \alpha_2, \ldots, \alpha_r)\mathbf{x}^{(\nu-1)} + \mathbf{g}, \qquad \nu = 1, 2, \ldots,$$

where by (24) we find that

$$(26)\qquad \begin{aligned} M(\alpha_1, \alpha_2, \ldots, \alpha_r) &\equiv M(\alpha_r)\cdots M(\alpha_2)M(\alpha_1); \\ \mathbf{g} &\equiv [N^{-1}(\alpha_r) + M(\alpha_r)N^{-1}(\alpha_{r-1}) + \cdots \\ &\qquad + M(\alpha_r)\cdots M(\alpha_2)N^{-1}(\alpha_1)]\mathbf{f}. \end{aligned}$$

As in the proof of (4), the eigenvalues of $M(\alpha_1, \alpha_2, \ldots, \alpha_r)$ can be determined, by using (23), in terms of the eigenvalues λ_i of M_0. We find in fact, that

$$(27)\qquad \mu_i(\alpha_1, \alpha_2, \ldots, \alpha_r) = \prod_{j=1}^{r} \frac{\lambda_i + \alpha_j}{1 + \alpha_j}, \qquad i = 1, 2, \ldots, n,$$

are the relevant eigenvalues. Now if we define the rth degree polynomial

$$(28)\qquad P_r(\lambda) \equiv \prod_{j=1}^{r} \frac{\lambda + \alpha_j}{1 + \alpha_j},$$

then convergence is implied by $|P_r(\lambda_i)| < 1$ for $i = 1, 2, \ldots, n$. In particular, if all the eigenvalues of M_0 are real and lie in the interval

$$a \leq \lambda \leq b,$$

then convergence is implied by

$$|P_r(\lambda)| < 1, \qquad a \leq \lambda \leq b.$$

In this case, the fastest convergence can be expected for that polynomial which has the smallest absolute magnitude in the indicated interval. Such problems are considered in Chapter 5, Section 4, and it is found that the Chebyshev polynomials can be used to find the polynomials of "least deviation from zero." Hence, in principle, if a and b are known, the best acceleration parameters $\alpha_1, \alpha_2, \ldots, \alpha_r$ can be determined (see Problem 2).

Another type of generalization of the acceleration method is obtained by employing a sequence of different basic splittings, say

$$A = N_i - P_i, \qquad i = 1, 2, \ldots, r;$$

and their corresponding accelerated forms

$$N_i(\alpha_i), \qquad P_i(\alpha_i).$$

An application of this technique is contained in subsection 2.2 of Chapter 9.

PROBLEMS, SECTION 5

1. State and prove a theorem analogous to Theorem 1, for the case that (5) is replaced by

$$1 < \lambda_1 \leq \lambda_2 \leq \cdots \leq \lambda_n.$$

2. If (5) is replaced by

$$-1 \leq \lambda_i \leq b < 1,$$

compare the efficiency of
(a) the method using three acceleration parameters $\{\alpha_i\}$ such that $\{-\alpha_i\}$ are the zeros of the Chebyshev polynomial of third degree for the interval $[-1, b]$, with
(b) the method using the single parameter $\alpha = (1 - b)/2$.

6. MATRIX INVERSION BY HIGHER ORDER ITERATIONS

The previous methods of this chapter have been primarily concerned with solving linear systems. Of course, as demonstrated in Section 0, they can all be employed to determine the inverse of any given non-singular matrix, A. We consider now an iterative method for directly computing A^{-1}. This method is a means for improving the accuracy of an approximate inverse, say R_0, obtained by other procedures. However, in many cases the present method is feasible when the initial approximate inverse is assumed to have the simple form $R_0 = \omega I$. Because of the large number of operations involved in matrix multiplication, these schemes are not generally used.

Assume that R_0 is any approximation to A^{-1} and define the error in this approximation by

$$E_0 = I - AR_0. \tag{1}$$

Clearly, if $R_0 = A^{-1}$ then $E_0 = 0$. Now with R_0 as the initial approximation, we define a sequence of approximate inverses by

$$R_\nu = R_{\nu-1}(I + E_{\nu-1} + E_{\nu-1}^2 + \cdots + E_{\nu-1}^{p-1}), \qquad \nu = 1, 2, \ldots, \tag{2a}$$

$$E_\nu \equiv I - AR_\nu, \qquad \nu = 1, 2, \ldots. \tag{2b}$$

Here, p is an arbitrary fixed integer not less than two. (This method is usually described for the case $p = 2$ but, as will be shown, the "best" value for this integer is $p = 3$.) From (2) we obtain

$$\begin{aligned} E_\nu &= I - AR_\nu \\ &= I - AR_{\nu-1}(I + E_{\nu-1} + \cdots + E_{\nu-1}^{p-1}) \\ &= I - (I - E_{\nu-1})(I + E_{\nu-1} + \cdots + E_{\nu-1}^{p-1}) \\ &= E_{\nu-1}^p, \qquad \nu = 1, 2, \ldots. \end{aligned} \tag{3}$$

Thus, in one iteration, the error matrix is raised to the pth power and the method is consequently called a pth order method. Apply (3) recursively, to find that

$$E_\nu = E_0^{p^\nu}, \qquad \nu = 1, 2, \ldots. \tag{4}$$

Also from (2b) and the above we have

$$(5)\qquad \begin{aligned} A^{-1} - R_\nu &= A^{-1}E_\nu \\ &= A^{-1}E_0^{p^\nu} \\ &= R_0(I - E_0)^{-1}E_0^{p^\nu}; \end{aligned}$$

where we have used (1) and the assumption that $\det |I - E_0| \neq 0$. It is now clear that the iterations converge when E_0 is a convergent matrix (see Section 4).

Let us assume that E_0 is a convergent matrix. Then its eigenvalues $\lambda_i(E_0)$ satisfy $|\lambda_i| < 1$, $i = 1, 2, \ldots, n$, from Theorem 4.1. Let

$$\rho \equiv \rho(E_0) = \max_i |\lambda_i|.$$

Then, since the eigenvalues of $E_0{}^p$ are $\lambda_i{}^p$, the error E_ν of (4) vanishes like ρ^{p^ν}.†

We now pose the problem of determining the "best" value of p to be used for any convergent E_0. By "best" we shall mean that procedure which for the least amount of computation yields an approximate inverse of desired accuracy. Alternatively, the best scheme could be defined as the one for which a given amount of computation yields the most accurate inverse. Adopting our usual convention we find from (2), since the product of two matrices requires n^3 operations, that ν iterations of a pth order scheme require

$$\nu p n^3 \text{ ops.}$$

If only K operations are to be permitted the number of iterations allowed is

$$\nu = \frac{K}{pn^3},$$

where we assume $K/(pn^3)$ is an integer. Thus, the principal eigenvalue is reduced to

$$\rho^{p^\nu} = \rho^{(p)K/(pn^3)} = \rho^{(p^{1/p})K/n^3}.$$

Since K, n, and $\rho < 1$ are independent of p, we find that the error is minimized when $p^{1/p}$ is a maximum. Now it is easily shown that the maximum of $x^{1/x}$ is at $x = e = 2.718\ldots$. But a simple calculation (pointed out by M. Altman) shows that for integers p the maximum is at $p = 3$.

In order to apply the procedure (2), we must have an initial estimate R_0 such that $E_0 = I - AR_0$ is convergent. For a very important class of

† This is only rigorously true if the elementary divisors of E_0 are simple. But, by the corollary to Theorem 1.3 of Chapter 1, the statement isn't very wrong.

matrices such an estimate is easily found. This result is contained in

THEOREM **1.** *Let A have real eigenvalues in the interval*

$$0 < m \leq \lambda_j \leq M, \qquad j = 1, 2, \ldots, n.$$

Then if $R_0(\omega) \equiv \omega I$ and $E_0(\omega) \equiv I - \omega A$, $E_0(\omega)$ will be convergent for all ω in

$$(6) \qquad 0 < \omega < \frac{2}{M}.$$

Further if $\rho(\omega)$ is the spectral radius of $E_0(\omega)$, i.e., $\rho(\omega) \equiv \rho[E_0(\omega)]$, *then*

$$(7) \qquad \rho(\omega_*) = \min_{0<\omega<2/M} \rho(\omega) = \frac{M - m}{M + m}, \qquad \omega_* = \frac{2}{M + m}.$$

Proof. This theorem is essentially a restatement of Theorem 5.1. If we make the association $(x, m_j) \leftrightarrow (\omega, -\lambda_j)$ in the proof of that theorem, the above follows. ∎

PROBLEMS, SECTION 6

1. Newton's method for improving R_0, the approximate inverse of A, is formally obtained by setting

$$\begin{aligned} A &= (R_0 + \delta R_0)^{-1} \\ &= [R_0(I + R_0^{-1}\delta R_0)]^{-1} \\ &= (I + R_0^{-1}\delta R_0)^{-1} R_0^{-1}. \end{aligned}$$

Therefore,

$$AR_0 = (I + R_0^{-1}\delta R_0)^{-1} \cong I - R_0^{-1}\delta R_0.$$

Solve for δR_0. Does this formula fit into the iteration scheme (2) for $p = 2$?

2. Show that if A is non-singular, the choice $R_0 \equiv aA^*$ with $a \equiv 1/\text{tr}(AA^*)$ produces a convergent matrix E_0 in (1).

3

Iterative Solution of Non-Linear Equations

0. INTRODUCTION

In this chapter, we consider iterative methods for determining the roots of equations

$$\mathbf{f}(\mathbf{x}) = \mathbf{0}$$

where $\mathbf{f}$ and $\mathbf{x}$ are vectors of the same dimension k: i.e., if $k = 1$ we have a single equation; if $k = n$ we have a system of n equations. Most of the iterative methods can be written in the form $\mathbf{x}_{n+1} = \mathbf{g}(\mathbf{x}_n)$ for some suitable function $\mathbf{g}$ and initial approximation $\mathbf{x}_0$. The convergence of this iteration process is assured if the mapping $\mathbf{g}(\mathbf{x})$ carries a closed and bounded set $S \subset C_k$ into itself and if the *mapping is contracting*, i.e., if $\|\mathbf{g}(\mathbf{x}) - \mathbf{g}(\mathbf{y})\| \leq M\|\mathbf{x} - \mathbf{y}\|$ for some norm, "Lipschitz" constant $M < 1$, and all $\mathbf{x}, \mathbf{y} \in S$. Such an iteration scheme is sometimes referred to as the *Picard iteration method*, or as a *functional iteration method*. It can be easily shown under these conditions that $\mathbf{g}(\mathbf{x})$ has a unique *fixed point* $\boldsymbol{\alpha}$ in S satisfying

$$\boldsymbol{\alpha} = \mathbf{g}(\boldsymbol{\alpha}).$$

We shall study this contracting mapping theorem in one or more dimensions and the related results which are basic to many of the iterative methods of this chapter.

Usually the iterative methods are valid for real and complex roots. However, in the latter case complex arithmetic must be incorporated into the appropriate digital computer codes and the initial estimate of the root must usually be complex (see Subsection 4.4 for an exception). The iterative methods require at least one initial estimate or guess at the location of the root being sought. If this initial estimate, say $\mathbf{x}_0$, is "sufficiently close" to a root, then, in general, the procedures will converge.

The problem of how to obtain such a "good" $\mathbf{x}_0$ is unresolved in general. Frequently, a good estimate of the root is known to the problem formulator (i.e., the engineer, physicist, mathematician, or other scientist who is interested in the solution) or can be found by an analytical study. For many purposes merely graphical accuracy (about two decimal figures) is needed for the initial value. In these cases, one may tabulate the function and plot the data in one or two variables or "fit" linear forms $a_{i0} + \sum_{j=1}^{k} a_{ij}x_j$ to $f_i(\mathbf{x})$ to find the approximate starting values. If a digital computer is to be employed, this plotting method is quite convenient since all of the required function evaluations will be contained in the eventual machine code for the problem.

As a general empirical rule, the schemes which converge more rapidly (i.e., higher order methods) require closer initial estimates. In practice, these higher order schemes may require the use of more significant digits in order that they converge as theoretically predicted. Thus, it is frequently a good idea to use a simple method to start with and then, when fairly close to the root, to use some higher order method for just a few iterations.

For polynomial equations in one variable we know much about the roots. While the general iteration schemes apply to them there are also special methods which can be used to obtain the zeros of polynomials. Such considerations are to be found in Section 4.

1. FUNCTIONAL ITERATION FOR A SINGLE EQUATION

Let us consider a scalar equation of the special form

$$x - g(x) = 0, \qquad \text{or} \qquad x = g(x). \tag{1}$$

[It is clear that any equation $f(x) = 0$ can be written equivalently in this form by defining $g(x) \equiv x - f(x)$.] If x_0 is some initial estimate of a root of (1), a scheme naturally suggested is to form the sequence

$$x_{\nu+1} = g(x_\nu), \qquad \nu = 0, 1, \ldots. \tag{2}$$

An important result concerning the convergence of this procedure and a proof of the existence of a unique root is contained in

THEOREM 1. *Let $g(x)$ satisfy the Lipschitz condition*

$$|g(x) - g(x')| \leq \lambda|x - x'|, \tag{3a}$$

for all values x, x' in the closed† *interval $I \equiv [x_0 - \rho, x_0 + \rho]$ where the*

† Unless otherwise specified: $[a, b]$ denotes the *closed* interval, $a \leq x \leq b$; (a, b) denotes the *open* interval, $a < x < b$; $(a, b]$ and $[a, b)$ denote respectively the *half-open* intervals $a < x \leq b$ and $a \leq x < b$.

Lipschitz constant, λ, satisfies

$$0 \le \lambda < 1. \tag{3b}$$

Let the initial estimate, x_0, be such that

$$|x_0 - g(x_0)| \le (1 - \lambda)\rho. \tag{4}$$

Then

(i) *all the iterates x_ν, defined by* (2), *lie within the interval I; i.e.,*

$$x_0 - \rho \le x_\nu \le x_0 + \rho, \tag{5}$$

(ii) (existence) *the iterates converge to some point, say,*

$$\lim_{\nu \to \infty} x_\nu = \alpha, \qquad (\text{in fact, } |x_\nu - \alpha| \le \lambda^\nu \rho)$$

which is a root of (1), *and*

(iii) (uniqueness) α *is the only root in* $[x_0 - \rho, x_0 + \rho]$.

Proof. We prove (i) by induction. Since $x_1 = g(x_0)$, we have by (3b) and (4)

$$|x_0 - x_1| \le (1 - \lambda)\rho \le \rho \tag{6}$$

and hence x_1 is in the interval (5). Assume this true for the iterates $x_1, x_2, \ldots, x_\nu$. Then from (2)

$$|x_{\nu+1} - x_\nu| = |g(x_\nu) - g(x_{\nu-1})|$$

and by the inductive assumption x_ν and $x_{\nu-1}$ are in the interval (5). Thus, by (3a), the Lipschitz condition yields

$$\begin{aligned} |x_{\nu+1} - x_\nu| &\le \lambda|x_\nu - x_{\nu-1}| \\ &\le \lambda^2|x_{\nu-1} - x_{\nu-2}| \\ &\;\;\vdots \\ &\le \lambda^\nu|x_1 - x_0| \\ &\le \lambda^\nu(1 - \lambda)\rho. \end{aligned} \tag{7}$$

Here we have used (2) and (3a) recursively and then applied (6). However,

$$\begin{aligned} |x_{\nu+1} - x_0| &= |(x_{\nu+1} - x_\nu) + (x_\nu - x_{\nu-1}) + \cdots + (x_1 - x_0)| \\ &\le |x_{\nu+1} - x_\nu| + |x_\nu - x_{\nu-1}| + \cdots + |x_1 - x_0| \\ &\le (\lambda^\nu + \lambda^{\nu-1} + \cdots + 1)(1 - \lambda)\rho = (1 - \lambda^{\nu+1})\rho \\ &\le \rho, \end{aligned}$$

which completes the proof of (i).

To prove part (ii), we first show that the sequence $\{x_\nu\}$ is a Cauchy sequence. Thus, for arbitrary positive integers m and p, we consider

$$\begin{aligned}
|x_m - x_{m+p}| &= |(x_m - x_{m+1}) + (x_{m+1} - x_{m+2}) + \cdots \\
&\qquad\qquad + (x_{m+p-1} - x_{m+p})| \\
&\leq |x_m - x_{m+1}| + |x_{m+1} - x_{m+2}| + \cdots \\
&\qquad\qquad + |x_{m+p-1} - x_{m+p}| \\
&\leq (\lambda^m + \lambda^{m+1} + \cdots + \lambda^{m+p-1})(1 - \lambda)\rho \\
&\leq (1 - \lambda^p)\rho\lambda^m.
\end{aligned} \tag{8}$$

Here we have used the inequalities (7) which are valid since (i) has been proved. Now given any $\epsilon > 0$, since λ in $0 \leq \lambda < 1$ is fixed, we can find an integer $N(\epsilon)$ such that $|x_m - x_{m+p}| < \epsilon$ for all $m > N(\epsilon)$ and $p > 0$ (we need only take N such that $\lambda^N < \epsilon/\rho$). Hence the sequence $\{x_\nu\}$ is a Cauchy sequence and has a limit, say α, in I. Since the function $g(x)$ is continuous in the interval I, the sequence $\{g(x_\nu)\}$ has the limit $g(\alpha)$ and by (2) this limit must also be α; that is, $\alpha = g(\alpha)$. Now $|x_\nu - \alpha| = |g(x_{\nu-1}) - g(\alpha)| \leq \lambda|x_{\nu-1} - \alpha|$; hence $|x_\nu - \alpha| \leq \lambda^\nu|x_0 - \alpha| \leq \rho\lambda^\nu$.

For part (iii), the uniqueness, let β be another root in $[x_0 - \rho, x_0 + \rho]$. Then, since α and β are both in this interval, (3) holds and we have, if $|\alpha - \beta| \neq 0$,

$$|\alpha - \beta| = |g(\alpha) - g(\beta)| \leq \lambda|\alpha - \beta| < |\alpha - \beta|.$$

This contradiction implies that $\alpha = \beta$ and the proof of the theorem is concluded. ■

COROLLARY. *If $|g'(x)| \leq \lambda < 1$ for $|x - x_0| \leq \rho$ and* (4) *is satisfied, then the conclusion of Theorem* 1 *is valid.*

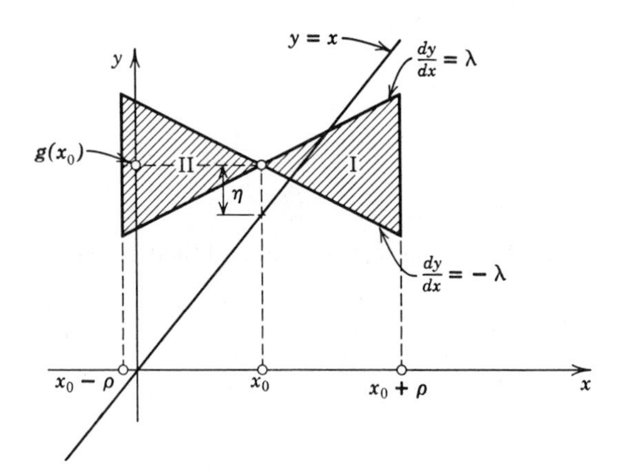

Figure 1

Proof. The mean value theorem implies $g(x_1) - g(x_2) = g'(\xi)(x_1 - x_2)$. Whence λ may serve as the Lipschitz constant in (3a and b). ■

A geometric interpretation of Theorem 1 is suggested by Figure 1. This illustrates the case with $g(x_0) - x_0 = \eta > 0$ and the triangles I and II, determined by lines with slope $\pm\lambda$ through $(x_0, g(x_0))$, are the regions in which the values of $g(x)$ lie for $x_0 - \rho \leq x \leq x_0 + \rho$. It is easy to verify that

(a) if $\lambda \geq 1$, the line $y = x$ will not intersect the upper boundary of triangle I or if
(b) $\lambda < 1$ and $\eta > (1 - \lambda)\rho$ that the line $y = x$ will not intersect the upper boundary of triangle I and hence may not intersect an admissible function $g(x)$ in the interval $[x_0 - \rho, x_0 + \rho]$.

In other words, the conditions $\lambda < 1$, $|\eta| \leq (1 - \lambda)\rho$ are necessary to insure the existence of a root for every function $g(x)$ satisfying conditions (3a and b).

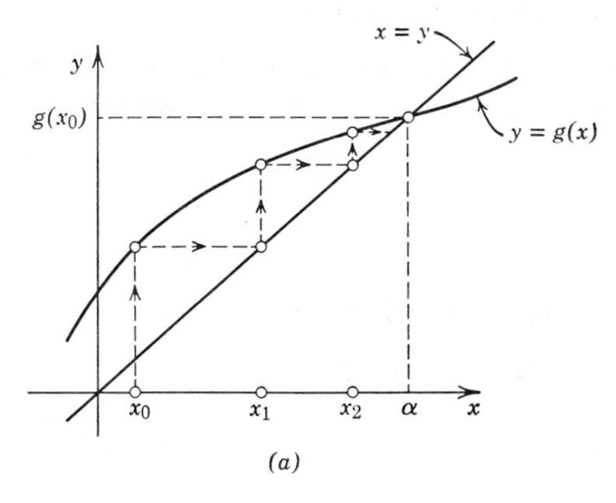

(a)

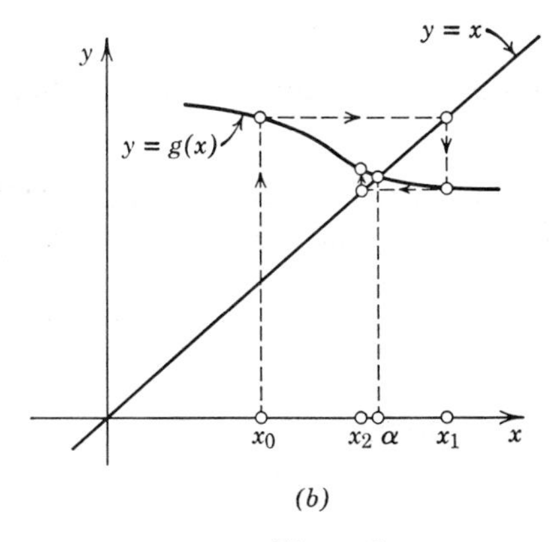

(b)

Figure 2

Figures 2a and 2b illustrate convergent iterative sequences for functions $g(x)$ with positive and negative slope, respectively. Note that the sequence $\{x_n\}$ converges to α monotonically for $g(x)$ of positive slope and converges with values alternately above and below α for $g(x)$ of negative slope.

Another convergence theorem is

THEOREM **2.** *If* $x = g(x)$ *has a root at* $x = \alpha$ *and in the interval*

$$|x - \alpha| < \rho \tag{9a}$$

$g(x)$ *satisfies*

$$|g(x) - g(\alpha)| \leq \lambda |x - \alpha|, \tag{9b}$$

with $\lambda < 1$, *then for any* x_0 *in* (9a):

(i) *all the iterates* x_ν *of* (2) *lie in the interval* (9a),
(ii) *the iterates* x_ν *converge to* α,
(iii) *the root* α *is unique in this interval.*

Proof. Part (i) is again proved by induction. By hypothesis x_0 is in (9a) and we assume $x_1, x_2, \ldots, x_{\nu-1}$ are also. Then since $\alpha = g(\alpha)$ we have from (2)

$$\begin{aligned} |\alpha - x_\nu| &= |g(\alpha) - g(x_{\nu-1})| \\ &\leq \lambda |\alpha - x_{\nu-1}|, \end{aligned} \tag{10a}$$

whence $\lambda < 1$ implies (i). Furthermore,

$$\begin{aligned} |\alpha - x_\nu| &\leq \lambda |\alpha - x_{\nu-1}|, \\ &\leq \lambda^2 |\alpha - x_{\nu-2}|, \\ &\vdots \\ &\leq \lambda^\nu |\alpha - x_0|. \end{aligned} \tag{10b}$$

By letting $\nu \to \infty$, we see that $x_\nu \to \alpha$, since $\lambda < 1$. The uniqueness follows as in Theorem 1. ∎

Notice that condition (9b) is weaker than the general Lipschitz condition for the interval (9a), since the one point α is fixed. This feature is applicable in Problem (1).

We can now prove a corollary (with a hypothesis which is oftentimes more readily verifiable).

COROLLARY. *If we replace* (9b) *by*

$$|g'(x)| \leq \lambda < 1, \tag{9b$'$}$$

then the conclusions (i), (ii), *and* (iii) *follow.*

Proof. From the mean value theorem and (2)

$$\alpha - x_\nu = g(\alpha) - g(x_{\nu-1}) = g'(\xi_{\nu-1})(\alpha - x_{\nu-1}). \tag{10a)'$$

Hence (10a) follows from (9b)′. Therefore (10b) and the rest of the proof of Theorem 2 apply. ■

It is clear from (10a)′ that, if the iterations converge, $\xi_\nu \rightarrow \alpha$ and thus "asymptotically" (as $\nu \rightarrow \infty$)

$$|\alpha - x_{k+\nu}| \approx |g'(\alpha)|^\nu |\alpha - x_k|, \tag{11}$$

for large enough k. The quantity

$$\hat{\lambda} = |g'(\alpha)| \tag{12}$$

is frequently called the (asymptotic) *convergence factor* and in analogy with the iterative solution of linear systems

$$R \equiv \log \frac{1}{\hat{\lambda}} \tag{13}$$

may be called the rate of convergence (if $\hat{\lambda} < 1$). The number of additional iterations required to reduce the error at the kth step by the factor 10^{-m} is then, asymptotically,

$$\nu = \frac{m}{R}. \tag{14}$$

We assume, in these definitions, that $\hat{\lambda} = |g'(\alpha)| \neq 0$ and define such an iteration scheme (2) to be a *first order method*; higher order methods are considered in Subsection 1.2.

1.1. Error Propagation

In actual computations it may not be possible, or practical, to evaluate the function $g(x)$ exactly (i.e., only a finite number of decimals may be retained after rounding or $g(x)$ may be given as the numerical solution of a differential equation, etc.). For any value x we may then represent our approximation to $g(x)$ by $G(x) = g(x) + \delta(x)$ where $\delta(x)$ is the error committed in evaluating $g(x)$. Frequently we may know a bound for $\delta(x)$, i.e., $|\delta(x)| < \delta$. Thus the actual iteration scheme which is used may be represented as

$$X_{\nu+1} = g(X_\nu) + \delta_\nu, \qquad \nu = 0, 1, 2, \ldots, \tag{15}$$

where the X_ν are the numbers obtained from the calculations and the $\delta_\nu \equiv \delta(X_\nu)$ satisfy

$$|\delta_\nu| \leq \delta, \qquad \nu = 0, 1, \ldots. \tag{16}$$

We cannot expect the computed iterates X_ν of (15) to converge. However, under proper conditions, it should be possible to approximate a root to an accuracy determined essentially by the accuracy of the computations, δ.

For example, from Figure 3 it is easy to see that for the special case of $g(x) \equiv \alpha + \lambda(x - \alpha)$, the uncertainty in the root α is bounded by $\pm\,\delta/(1 - \lambda)$. We note that, if the slope λ is close to unity the problem is not "properly posed." We now establish Theorem 3 which states quite generally that when the functional iteration scheme is convergent, the presence of errors in computing $g(x)$, of magnitudes bounded by δ, causes the scheme to estimate the root α with an uncertainty bounded by $\pm\,\delta/(1 - \lambda)$.

THEOREM 3. *Let $x = g(x)$ satisfy the conditions of Theorem 2. Let X_0 be any point in the interval*

$$|\alpha - x| \le \rho_0, \tag{17a}$$

where

$$0 < \rho_0 \le \rho - \frac{\delta}{1 - \lambda}. \tag{17b}$$

Then the iterates X_ν of (15), *with the errors bounded by* (16), *lie in the interval*

$$|\alpha - X_\nu| \le \rho,$$

and

$$|\alpha - X_k| \le \frac{\delta}{1 - \lambda} + \lambda^k\left(\rho_0 - \frac{\delta}{1 - \lambda}\right), \tag{18}$$

where $\lambda^k \to 0$ as $k \to \infty$.

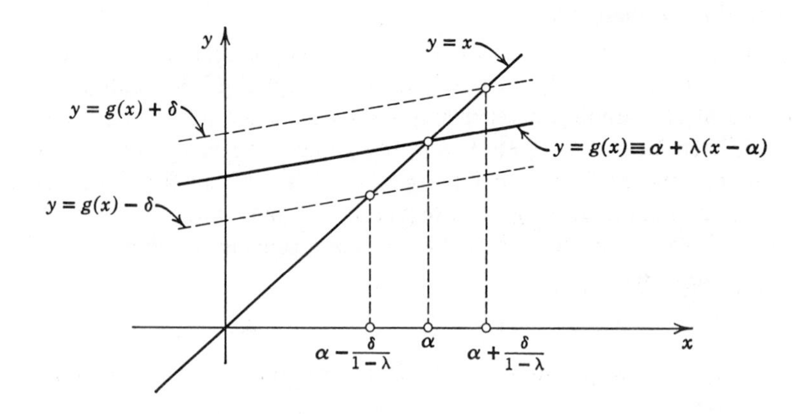

Figure 3

Proof. It is clear that $|\alpha - X_0| \leq \rho_0 \leq \rho_0 + \delta/(1 - \lambda) \leq \rho$. Then for an inductive proof assume $X_0, X_1, \cdots, X_{\nu-1}$ are in $|\alpha - x| \leq \rho$. By (15) and (16)

$$|\alpha - X_\nu| \leq |[g(\alpha) - g(X_{\nu-1})] - \delta_{\nu-1}| \leq |g(\alpha) - g(X_{\nu-1})| + \delta.$$

From (9b), we then have

$$\begin{aligned}
|\alpha - X_\nu| &\leq \lambda|\alpha - X_{\nu-1}| + \delta \\
&\leq \lambda^2|\alpha - X_{\nu-2}| + \lambda\delta + \delta \\
&\leq \lambda^3|\alpha - X_{\nu-3}| + \lambda^2\delta + \lambda\delta + \delta \\
&\vdots \\
&\leq \lambda^\nu|\alpha - X_0| + \lambda^{\nu-1}\delta + \cdots + \lambda\delta + \delta \\
&\leq \lambda^\nu\rho_0 + \frac{1 - \lambda^\nu}{1 - \lambda}\delta \\
&\leq \lambda^\nu\rho_0 + \frac{\delta}{1 - \lambda} - \lambda^\nu\frac{\delta}{1 - \lambda} \\
&\leq \rho_0 + \frac{\delta}{1 - \lambda} \\
&\leq \rho.
\end{aligned}$$

Thus all iterates lie in $|\alpha - x| \leq \rho$ and the iteration process is defined. Moreover, from the last inequality involving ν we find the estimate (18) which completes the proof. ■

Theorem 3 shows that the method is "as convergent as possible," that is, the computational errors which arise from the evaluations of $g(x)$ may cumulatively produce an error of magnitude at most $\delta/(1 - \lambda)$. It is also clear that such errors limit the size of the error bound *independently of the number of iterations.* Thus in actual calculations, it is not worthwhile to iterate until $\lambda^\nu\rho_0 \ll \delta/(1 - \lambda)$. In fact, if reasonable estimates of λ, δ, and ρ_0 are known, it is an efficient procedure to have the two types of error term in (18) of the same magnitude; i.e.,

$$\lambda^\nu\rho_0 \cong \frac{\delta}{1 - \lambda}.\dagger$$

The number of iterations required is then about

$$\begin{aligned}
\nu &\cong \log\left[\frac{\delta}{(1 - \lambda)\rho_0}\right] \Big/ \log\lambda \\
&\cong \log\left[\frac{(1 - \lambda)\rho_0}{\delta}\right] \Big/ \log\frac{1}{\lambda}.
\end{aligned} \tag{19}$$

† $\cong$ reads "approximately equals" and we have tacitly assumed that $\delta/(1 - \lambda) \ll \rho_0$.

Of course, if the acceptable error is much greater than $\delta/(1 - \lambda)$ the number of iterations given by (13) and (14) is the relevant estimate. It is essential to estimate δ from the arithmetic calculations involved in evaluating $g(x)$. Also, note that since the X_ν need not converge any tests for the termination of the iterations should allow for this roundoff effect.

1.2. Second and Higher Order Iteration Methods

It is clear from the corollary to Theorem 2, that if at the root $x = \alpha$,

$$g'(\alpha) = 0, \tag{20}$$

then the convergence should be quite rapid. Let this be the case and assume further that $g''(x)$ exists and is bounded in some interval, $|\alpha - x| \leq \rho$, in which (9) is satisfied. Then for any x in this interval we have by Taylor's theorem:

$$\begin{aligned} g(x) &= g(\alpha) + 0 + \frac{(x - \alpha)^2}{2} g''(\xi), \\ &= \alpha + \frac{(x - \alpha)^2}{2} g''(\xi). \end{aligned}$$

Here ξ is some value between x and α. By using this result we obtain for any iterate (2) (assuming $|\alpha - x_0| < \rho$ and $\frac{1}{2}\rho|g''(x)| \leq \lambda < 1$):

$$|x_\nu - \alpha| = |g(x_{\nu-1}) - g(\alpha)| = |\tfrac{1}{2}g''(\xi_{\nu-1})| \cdot |x_{\nu-1} - \alpha|^2, \qquad \nu = 1, 2, 3, \ldots. \tag{21}$$

Thus, the error in any iterate is proportional to the *square* of the previous error and if $g''(\alpha) \neq 0$ the procedure (2) is now called a *second order method.*

Let the bound on $g''(x)$ be denoted by

$$|g''(x)| \leq 2M, \qquad |\alpha - x| < \rho. \tag{22}$$

Then from (21)

$$\begin{aligned} |x_\nu - \alpha| &\leq M|x_{\nu-1} - \alpha|^2 \\ &\leq M \cdot M^2|x_{\nu-2} - \alpha|^4 \\ &\leq M \cdot M^2 \cdot M^4|x_{\nu-3} - \alpha|^8 \\ &\;\vdots \\ &\leq M^{(2^\nu-1)}|x_0 - \alpha|^{2^\nu} \\ &\leq (M|x_0 - \alpha|)^{2^\nu-1}|x_0 - \alpha|. \end{aligned} \tag{23}$$

Thus, if $M|x_0 - \alpha| < 1$, the second order method converges and reduces the initial error by at least 10^{-m} when

$$(M|x_0 - \alpha|)^{2^\nu-1} \cong 10^{-m}.$$

The number of iterations required is now obtained from

$$(24)\qquad \begin{aligned} 2^{\nu} &\cong \frac{-m}{\log (M|x_0 - \alpha|)}, \\ \nu &\cong \frac{1}{\log 2} \log \frac{m}{\log 1/(M|x_0 - \alpha|)}. \end{aligned}$$

A comparison with first order schemes is possible, i.e., the estimates in (12)–(14), if we assume $\hat{\lambda} = M|x_0 - \alpha|$. That is, by letting $\nu_{(1)}$ and $\nu_{(2)}$ represent the exponents ν in (10b) and (23), respectively, we have equal reduction of the error if

$$(25)\qquad 2^{\nu_{(2)}} = 1 + \nu_{(1)}.$$

For instance, 130 iterations of the first order scheme are equivalent, under the above assumptions, to about 7 iterations of a second order scheme! A further striking property of second order schemes can be obtained by assuming for all $\nu \geq \nu_0$ that

$$|x_\nu - \alpha| = 10^{-p_\nu}, \qquad p_\nu > 0;$$

i.e., p_ν is essentially the number of correct decimals in the νth iterate. Then from the first line of (23):

$$10^{-p_\nu} \leq M 10^{-2p_{\nu-1}},$$

and upon taking the logarithm of both sides

$$(26)\qquad p_\nu \geq 2p_{\nu-1} - \log M.$$

Thus, if $M < 1$, then $-\log M > 0$ and the number of correct decimals *more than doubles* on each iteration. (If $M > 1$ the number does not quite double but, since $p_\nu \gg \log M$ for large ν, this doubling is at least asymptotically true.)

Schemes which are more quickly convergent than second order ones are now easily described. Let us assume that at a root $x = \alpha$ of (1):

$$(27\text{a})\qquad g'(\alpha) = g''(\alpha) = \cdots = g^{(n-1)}(\alpha) = 0;$$

$$(27\text{b})\qquad g^{(n)}(\alpha) \neq 0; \qquad |g^{(n)}(x)| \leq n!M \qquad \text{in} \qquad |x - \alpha| \leq \rho.$$

Then by Taylor's theorem,

$$g(x) = \alpha + \frac{(x - \alpha)^n}{n!} g^{(n)}(\xi), \qquad |x - \alpha| < \rho$$

where ξ is between x and α. Again from (2) and the above

$$\begin{aligned} |x_{\nu+1} - \alpha| &= \frac{|g^{(n)}(\xi_\nu)|}{n!} |x_\nu - \alpha|^n \\ &\leq M \cdot |x_\nu - \alpha|^n. \end{aligned} \tag{28}$$

The method (2) under conditions (27) is now called an nth order procedure and one can easily deduce the results corresponding to (23)–(26) for such methods. In the event that $g(x)$ is calculated with an error of magnitude δ as in (15), the root α may be determined only to within an uncertainty of at best $\pm\delta$. This conclusion follows by letting $\lambda \to 0$ in (18).

PROBLEMS, SECTION 1

1. Given $g(x) \equiv x^2 - 2x + 2$. For what values x_0 does (2) converge? [Hint: Use Theorem 2.] What is the order of the convergence? Sketch a graph analogous to Figures 2a and b.

2. For $g(x) \equiv \cos x$, show that $x_{n+1} = g(x_n)$ defines a convergent sequence for arbitrary x_0. Calculate the root $\alpha = \cos\alpha$ to three decimal places.

2. SOME EXPLICIT ITERATION PROCEDURES

The general problem to which the previous iteration methods are to be applied is that of finding the root (or roots) of

$$f(x) = 0 \tag{1}$$

in some interval, say $a \leq x \leq b$. Let $\phi(x)$ be any function such that

$$0 < |\phi(x)| < \infty, \qquad a \leq x \leq b. \tag{2}$$

Then the equation

$$x = g(x) \equiv x - \phi(x)f(x), \tag{3}$$

has roots which coincide with those of (1) in the interval $[a, b]$ and no others. Many of the standard iterative methods are obtained for special choices of $\phi(x)$.

Another procedure for defining the function $g(x)$ is to use

$$g(x) \equiv x - F(f(x)), \tag{4}$$

where $F(y)$ is a function such that

$$F(0) = 0; \qquad F(y) \neq 0, \qquad y \neq 0.$$

Such methods more naturally describe many higher order schemes.

2.1. The Simple Iteration or Chord Method (First Order)

The simplest choice for $\phi(x)$ in (3) is to take

$$\phi(x) \equiv m \neq 0. \tag{5}$$

If $f(x)$ is differentiable, we note that

$$g'(x) = 1 - mf'(x), \tag{6}$$

and the scheme will be convergent, by the corollary to Theorem 1.2, in some interval about α provided that m is chosen such that

$$0 < mf'(\alpha) < 2. \tag{7}$$

Thus m must have the same sign as $f'(\alpha)$, while if $f'(\alpha) = 0$, (7) cannot be satisfied.

The choice (5) yields the iteration equations

$$x_{\nu+1} = x_\nu - mf(x_\nu).$$

These iterates have a geometric realization in which the value $x_{\nu+1}$ is the x intercept of the line with slope $1/m$ through $(x_\nu, f(x_\nu))$. (See Figure 1.) The inequality (7) implies that this slope should be between ∞ (i.e., vertical) and $\frac{1}{2}f'(\alpha)$ [i.e., half the slope of the tangent to the curve $y = f(x)$ at the root]. It is from this geometric description that the name chord method is derived—the next iterate is determined by a chord of constant slope joining a point on the curve to the x-axis.

2.2. Newton's Method (Second Order)

If the slope of the chord is changed at each iteration so that

$$g'(x_\nu) = 1 - m_\nu f'(x_\nu) = 0, \tag{8}$$

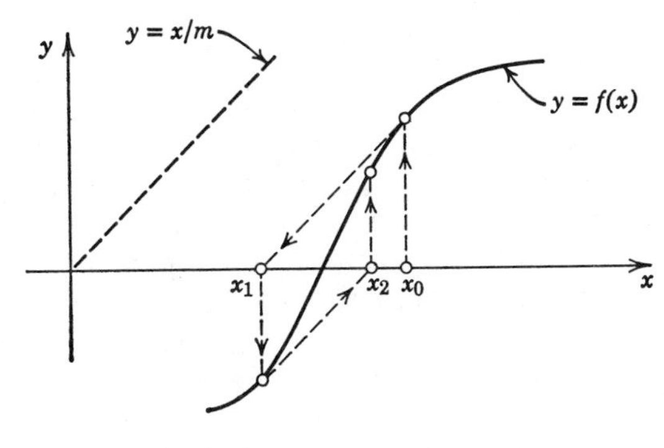

Figure 1

then a second order procedure may be obtained. From (8) we find

$$m_\nu = \frac{1}{f'(x_\nu)}, \tag{9}$$

which suggests the choice in (3) of

$$\phi(x) = \frac{1}{f'(x)}, \quad \text{or} \quad g(x) \equiv x - \frac{f(x)}{f'(x)}. \tag{10}$$

The resulting iteration procedure is now

$$x_{\nu+1} = x_\nu - \frac{f(x_\nu)}{f'(x_\nu)}, \tag{11}$$

and it is at least of second order if $f'(\alpha) \neq 0$ and $f''(x)$ exists, since

$$g'(\alpha) = \frac{f(\alpha)f''(\alpha)}{[f'(\alpha)]^2} = 0. \tag{12}$$

The geometrical interpretation of the scheme (11) simply replaces the chord in Figure 1 by the tangent line to $y = f(x)$ at $(x_\nu, f(x_\nu))$.

In applying Newton's method, we are required to evaluate $f'(x_\nu)$ as well as $f(x_\nu)$ at each step of the procedure. For sufficiently simple functions, which are given explicitly, this may offer no serious difficulty. (This is especially true for polynomials whose derivatives are easily evaluated by synthetic division; see Subsection 4.1.) However, if $f(x)$ is known only implicitly (say as the solution of some differential equation in which x is a parameter in the initial data), it may be impractical to evaluate $f'(x_\nu)$ at each iteration. In such cases the derivative may be approximated by various methods, the most obvious approximation being

$$f'(x_\nu) \cong \frac{f(x_\nu) - f(x_{\nu-1})}{x_\nu - x_{\nu-1}}. \tag{13}$$

If this approximation is used, the procedure is no longer Newton's method but is the method of *false position* discussed in the next subsection.

A useful observation on the application of Newton's method, or the false position variation of it, is based on the fact that as the iterations converge, $f'(x_\nu)$ or its approximations converge to $f'(\alpha)$. Thus, for all iterates $\nu \geq \nu_0$, say, it may suffice to use $f'(x_{\nu_0})$ in place of $f'(x_\nu)$ in (11). The iteration method from this point on is then just the chord method with $1/m = f'(x_{\nu_0})$.

It should be noted that Newton's method may be undefined and condition (2) violated if $f'(x) = 0$ for some x in $[a, b]$. In particular, if at the root $x = \alpha$, $f'(\alpha) = 0$, the procedure may no longer be of second order since (12) is not satisfied. To examine this case we assume that

$$f(x) \equiv (x - \alpha)^p h(x), \qquad p > 1 \tag{14a}$$

where the function $h(x)$ has a second derivative and

$$h(\alpha) \neq 0. \tag{14b}$$

From (14) in (10) we find that

$$g'(x) = \frac{\left(1 - \frac{1}{p}\right) + (x - \alpha)\dfrac{2h'(x)}{ph(x)} + (x - \alpha)^2 \dfrac{h''(x)}{p^2 h(x)}}{\left[1 + (x - \alpha)\dfrac{h'(x)}{ph(x)}\right]^2}.$$

Thus for x_0 sufficiently close to α we have $|g'(x)| < 1$ for $x \in [x_0, \alpha]$ and the iterations (11) will converge. The asymptotic convergence factor is now

$$|g'(\alpha)| = 1 - \frac{1}{p}.$$

So *only in the case of a linear root, i.e., $p = 1$, is Newton's method second order*, but *it will converge as a first order method in the general case* (14). If the order of the root, p, is known (or can be closely estimated) quadratic convergence can be retained or approximated by the modification

$$g(x) \equiv x - p\frac{f(x)}{f'(x)}.$$

The details of this procedure are left to the reader. A convergence proof for Newton's method which *does not require $f'(\alpha) \neq 0$* is contained in Theorem 3.3 [see also Problems (3)–(6) of Section 3].

2.3. Method of False Position (Fractional Order)

If the difference quotient approximation to the derivative, given by (13), is employed in (11) we obtain the iterative procedure:

$$x_{\nu+1} = x_\nu - f(x_\nu)\frac{x_\nu - x_{\nu-1}}{f(x_\nu) - f(x_{\nu-1})}; \qquad \nu = 1, 2, \ldots. \tag{15}$$

It should be noted that two successive iterates, x_0 and x_1, must be estimated before the recursion formula can be used. However, only one function evaluation, $f(x_\nu)$, is required at each step since the previous value, $f(x_{\nu-1})$, may be retained. [This is an advantage over Newton's method where two evaluations, $f(x_\nu)$ and $f'(x_\nu)$, are required.] The order of this procedure cannot be deduced by the analysis of Section 1 since (15) cannot be written in the scalar form $x_{\nu+1} = g(x_\nu)$. To examine this question let $x = \alpha$ be a root of $f(x) = 0$. Then we may write, by subtracting each side of (15) from α,

$$\begin{aligned}\alpha - x_{\nu+1} &= (\alpha - x_\nu) + f(x_\nu)\frac{x_\nu - x_{\nu-1}}{f(x_\nu) - f(x_{\nu-1})} \\ &= (\alpha - x_\nu)\frac{f[x_{\nu-1}, x_\nu] - f[x_\nu, \alpha]}{f[x_{\nu-1}, x_\nu]}\end{aligned}$$

where we define $f[a, b] \equiv [f(b) - f(a)]/(b - a)$. This can be further simplified to the form

$$\alpha - x_{\nu+1} = (\alpha - x_\nu)(\alpha - x_{\nu-1})\left\{-\frac{f[x_{\nu-1}, x_\nu, \alpha]}{f[x_{\nu-1}, x_\nu]}\right\}, \tag{16}$$

by introducing

$$f[x_{\nu-1}, x_\nu, \alpha] \equiv \frac{f[x_{\nu-1}, x_\nu] - f[x_\nu, \alpha]}{x_{\nu-1} - \alpha}.$$

Here we have anticipated the divided difference notation to be studied in Chapter 6 and in Problems 2 and 3 of this section. If the function $f(x)$ has a continuous second derivative in an interval including the points x_ν, $x_{\nu-1}$, and α, then it is shown in Theorem 1.1 of Chapter 6 that

$$\begin{aligned}f[x_{\nu-1}, x_\nu] &= f'(\xi_\nu) \\ f[x_{\nu-1}, x_\nu, \alpha] &= \tfrac{1}{2}f''(\eta_\nu)\end{aligned} \tag{17}$$

for some points ξ_ν and η_ν in the obvious intervals. (See also Problem 3.) Thus we deduce that

$$\alpha - x_{\nu+1} = \frac{-f''(\eta_\nu)}{2f'(\xi_\nu)}(\alpha - x_\nu)(\alpha - x_{\nu-1}); \qquad \nu = 1, 2, \ldots. \tag{18}$$

Let us assume that all the iterates are confined to some interval about the root α and that for all ξ, η in this interval

$$\left|\frac{f''(\eta)}{2f'(\xi)}\right| \leq M. \tag{19}$$

Then by setting $M|\alpha - x_\nu| \equiv e_\nu$ we obtain the inequalities

$$e_{\nu+1} \leq e_\nu e_{\nu-1}; \qquad \nu = 1, 2, \ldots, \tag{20}$$

upon multiplication of (18) with M and the use of (19). If we define $\max(e_0, e_1) = \delta$ the inequalities (20) imply

$$\begin{aligned}e_2 &\leq \delta^2 \\ e_3 &\leq \delta^3 \\ e_4 &\leq \delta^5 \\ &\vdots \\ e_\nu &\leq \delta^{m_\nu}\end{aligned}$$

where $m_0 = m_1 = 1$ and $m_{\nu+1} = m_\nu + m_{\nu-1}$, $\nu = 1, 2, \ldots$. The numbers m_ν form what is known as a *Fibonacci sequence*. It may be shown (see Problem 1) that

$$(21) \qquad m_\nu = \frac{1}{\sqrt{5}}(r_+^{\nu+1} - r_-^{\nu+1}), \qquad r_\pm = \frac{1 \pm \sqrt{5}}{2}.$$

Thus for large ν:

$$m_\nu \approx \frac{1}{\sqrt{5}}(r_+)^{\nu+1} \cong 0.447(1.618)^{\nu+1}.$$

If $\delta < 1$, then the initial error is reduced by 10^{-m} when $\delta^{(m_\nu - 1)} \cong 10^{-m}$ and we may compare this number, ν, of iterations with the corresponding numbers $\nu_{(2)}$ for the second order method and $\nu_{(1)}$ for the first order method (see equation 1.25) in the case $\delta = M|x_0 - \alpha|$ by noting that

$$(22) \qquad m_\nu - 1 = 2^{\nu_{(2)}} - 1 = \nu_{(1)},$$

or

$$\frac{1}{\sqrt{5}}(r_+)^{\nu+1} \cong 2^{\nu_{(2)}}.$$

Hence

$$(23) \qquad \nu = c + d\nu_{(2)}$$

where

$$c = \frac{\log\sqrt{5}}{\log r_+} - 1 \cong 0.672 \qquad \text{and} \qquad d = \frac{\log 2}{\log r_+} \cong 1.440.$$

We see that somewhat more of the current iterations are needed for a given accuracy than is the case for the second order methods (but it should be recalled that only one function evaluation per iteration is used).

If we were to postulate that as $\nu \to \infty$: $|\alpha - x_{\nu+1}| \approx K|\alpha - x_\nu|^r$, then (18), with the coefficient $|f''/(2f')| = M$, would yield $K = M^{1/r}$, $r = r_+$. In other words, we might say that the false position method is of order $\cong 1.618$. Hence, two steps of *Regula Falsi* have an order $\cong (1.6)^2 > 2.5$ and require only two evaluations of $f(x)$.

A geometric interpretation of the scheme (15) is easily given as follows: in the $x, f(x)$ plane let the line through $(x_\nu, f(x_\nu))$ and $(x_{\nu-1}, f(x_{\nu-1}))$ intersect the x-axis at a point called $x_{\nu+1}$. In other words $f(x)$ is approximated by a linear function through the indicated pair of points and the zero of this linear function is taken as the next approximation to the desired root. Depending upon the location of the points in question this procedure may be an interpolation or an extrapolation at each iteration.

In the *classical Regula Falsi* method, the point $(x_0, f(x_0))$ is used in (15) in place of $(x_{\nu-1}, f(x_{\nu-1}))$ for all $\nu = 1, 2, \ldots$. The geometric interpretation of this scheme is quite clear, i.e., all lines pass through the original estimate, which is a poor strategy in general. Again either interpolation or extrapolation may occur.

Finally, we may use in (15), in place of $(x_{\nu-1}, f(x_{\nu-1}))$, the latest point for which the function value has sign opposite that of $f(x_\nu)$. In this latter method, only interpolation occurs, and furthermore, upper and lower bounds for the root are obtained, which is ideal for estimating the error. However, to start this scheme we must initially obtain such upper and lower estimates of the root and, of course, it is only applicable if $f(x)$ changes sign at the root in question. This latter variation requires some additional testing and storage of data and hence is slightly more complicated to employ on a digital computer.

From the geometric description of the method of false position a natural generalization is suggested. That is, we set

$$x_{\nu+1} = P_{k,\nu}(0), \qquad \nu = k, k+1, \ldots.$$

where $P_{k,\nu}(f)$ is the polynomial in f of degree k which passes through the $k + 1$ points $(f(x_\nu), x_\nu), (f(x_{\nu-1}), x_{\nu-1}), \ldots, (f(x_{\nu-k}), x_{\nu-k})$. Clearly, for $k = 1$, this is just the scheme (15). The construction of such interpolation polynomials is, in general, treated in Chapter 6, and Section 2 of that chapter is particularly suited for the present purpose. [We must interchange x and f or else use inverse interpolation. Also, it is assumed that the function values $f(x_j)$ are distinct.] It can be shown that these "multipoint" methods have orders η_k which increase monotonically with k and that $\lim_{k\to\infty} \eta_k = 2$. We have seen that $\eta_1 \cong 1.618$ so that no great improvement over the method of false position can be obtained. For $k = 2$ or 3, the orders are close to 2.

Another possibility along the above lines is to use

$$x_{\nu+1} = P_{\nu,\nu}(0), \qquad \nu = 1, 2, \ldots;$$

that is a νth degree polynomial in f through all the previous iterates $(f(x_j), x_j)$, $j = 0, 1, \ldots, \nu$ is used to determine the $(\nu + 1)$st iterate. Again, the iterative linear interpolation scheme of Chapter 6, Section 2, can be used for this purpose and it can be shown that the order of convergence is now 2 (for simple roots).

2.4. Aitken's δ^2-Method (Arbitrary Order)

This procedure is frequently presented as a means for *accelerating* the convergence of the functional iteration method based on (3). The method can be described and motivated as follows: If x_ν is any number approximating a root of (1) or (3), let $\hat{x}_{\nu+1}$ be defined by

$$\hat{x}_{\nu+1} = g(x_\nu). \tag{24}$$

Then a measure of the "errors" in these two approximations, x_ν and $\hat{x}_{\nu+1}$, can be defined by

$$\begin{aligned} e_\nu &\equiv g(x_\nu) - x_\nu = \hat{x}_{\nu+1} - x_\nu, \\ \hat{e}_{\nu+1} &\equiv g(\hat{x}_{\nu+1}) - \hat{x}_{\nu+1}. \end{aligned} \tag{25}$$

Since for a root this error should vanish, i.e., $e(\alpha) \equiv g(\alpha) - \alpha = -\phi(\alpha)f(\alpha) = 0$, we may seek $x_{\nu+1}$ by "extrapolating the errors to zero." That is, the line segment joining the points (x_ν, e_ν) and $(\hat{x}_{\nu+1}, \hat{e}_{\nu+1})$ is extended to intersect the x-axis and the point of intersection is taken as $x_{\nu+1}$. This yields the expression

$$x_{\nu+1} = \frac{x_\nu \hat{e}_{\nu+1} - \hat{x}_{\nu+1} e_\nu}{\hat{e}_{\nu+1} - e_\nu} \equiv G(x_\nu). \tag{26a}$$

For actual calculations (26a) is usually written as

$$x_{\nu+1} = x_\nu - \frac{e_\nu^2}{\hat{e}_{\nu+1} - e_\nu}, \tag{26b}$$

and the evaluations proceed by using (24), (25), and (26b).

From (24)–(26) we see that the δ^2-*method* can be viewed as functional iteration applied to

$$x = G(x), \tag{27a}$$

where

$$G(x) \equiv \frac{xg(g(x)) - g^2(x)}{g(g(x)) - 2g(x) + x}. \tag{27b}$$

[That is, from x_0 we obtain the same sequence of iterates x_ν by the procedure described in (24)–(26) as is obtained from $x_{\nu+1} = G(x_\nu)$.]

The functional iteration scheme applied to (27) is sometimes known as *Steffensen's method.* In fact, Aitken's δ^2-method† was originally proposed to convert any convergent sequence (no matter how generated), $\{x_n\}$, into a more rapidly convergent sequence, $\{x_n'\}$, by using

$$x_n' \equiv x_n - \frac{(x_{n+1} - x_n)^2}{x_{n+2} - 2x_{n+1} + x_n}. \tag{28}$$

† The denominator in (28) suggests the second difference notation δ^2. See equation (3.16a) of Chapter 6.

Several general applications of the δ^2-*process* are illustrated in Problems 6 and 7.

The function (27b) is indeterminate at the root $x = \alpha$ since $g(\alpha) = \alpha$. However, its value there is easily found by an application of L'Hospital's rule, assuming $g(x)$ to be differentiable at the root and $g'(\alpha) \neq 1$:

$$\begin{aligned} G(\alpha) &= \frac{g(g(\alpha)) + \alpha g'(g(\alpha))g'(\alpha) - 2g(\alpha)g'(\alpha)}{g'(g(\alpha))g'(\alpha) - 2g'(\alpha) + 1} \\ &= \frac{\alpha + \alpha[g'(\alpha)]^2 - 2\alpha g'(\alpha)}{[g'(\alpha)]^2 - 2g'(\alpha) + 1} \\ &= \alpha. \end{aligned}$$

The case $g'(\alpha) = 1$ corresponds to a multiple root of (1) at $x = \alpha$. However, in this case too, it can be shown from (33d) that $\alpha = G(\alpha)$. Thus, it follows that (27a) has roots wherever (3) has them. To show further that all roots of (27) are also roots of (3), assume that x is any finite root of (27). Then there are two cases, either $g(g(x)) - 2g(x) + x$ vanishes or not. If not, then clearing fractions in (27) is legitimate and yields

$$[g(x) - x]^2 = 0.$$

Thus, x is also a root of (3). If the denominator in (27b) vanishes, the numerator must also vanish (since x was assumed finite). Now observe that since the denominator vanishes, we may use

$$xg(g(x)) = 2xg(x) - x^2$$

and substitute in the numerator to find that again $[g(x) - x]^2 = 0$. In other words (27) *has the same roots as* (3).

The order of the δ^2-method is simply related to the order of the functional iteration applied to $x = g(x)$. To derive this result, we assume that $x = \alpha$ is a root and that:

(29a) $$g'(\alpha) = g''(\alpha) = \cdots = g^{(p-1)}(\alpha) = 0;$$

(29b) $$g^{(p)}(\alpha) = p!A \neq 0;$$

(29c) $$g^{(p+1)}(x) \text{ exists in } |x - \alpha| \leq \rho.$$

These conditions imply that $g(x)$ determines a pth order method. By Taylor's theorem and (29) for every ϵ such that $|\epsilon| \leq \rho$:

$$g(\alpha + \epsilon) = g(\alpha) + A\epsilon^p + \frac{g^{(p+1)}(\alpha + \theta\epsilon)}{(p + 1)!}\epsilon^{p+1}, \qquad 0 < \theta < 1;$$

(30a) $$\begin{aligned} &= \alpha + A\epsilon^p + B\epsilon^{p+1} \\ &= \alpha + \delta. \end{aligned}$$

Here we have introduced

$$B \equiv \frac{g^{(p+1)}(\alpha + \theta\epsilon)}{(p+1)!}, \qquad \delta \equiv (A + B\epsilon)\epsilon^p. \tag{31a}$$

Since A and B are bounded, we can pick ϵ sufficiently small such that $|\delta| \leq \rho$ and then as in (30a)

$$g(\alpha + \delta) = \alpha + A\delta^p + B'\delta^{p+1} \tag{30b}$$

where

$$B' = \frac{g^{(p+1)}(\alpha + \phi\delta)}{(p+1)!}, \qquad 0 < \phi < 1. \tag{31b}$$

From (30) in (27b) we obtain, with $x = \alpha + \epsilon$ and $\epsilon \neq 0$,

$$\begin{aligned} G(\alpha + \epsilon) &= \frac{(\alpha + \epsilon)g(\alpha + \delta) - (\alpha + \delta)^2}{g(\alpha + \delta) - 2(\alpha + \delta) + (\alpha + \epsilon)} \\ &= \alpha - \frac{\delta^2 - A\epsilon\delta^p - B'\epsilon\delta^{p+1}}{\epsilon - 2\delta + A\delta^p + B'\delta^{p+1}}. \end{aligned} \tag{32}$$

There are two cases, $p \geq 2$ and $p = 1$, to be considered. First, with $p \geq 2$ equation (32) can be written as

$$G(\alpha + \epsilon) = \alpha - \epsilon^{2p-1}(A + B\epsilon)^2$$
$$\cdot \left\{\frac{1 - A(A + B\epsilon)^{p-2}\epsilon^{(p-1)^2} - B'(A + B\epsilon)^{p-1}\epsilon^{(p-1)^2+p}}{1 - 2(A + B\epsilon)\epsilon^{p-1} + A(A + B\epsilon)^p\epsilon^{p^2-1} + B'(A + B\epsilon)^{p+1}\epsilon^{p^2+p-1}}\right\}.$$

It is clear that the bracketed expression approaches 1 as ϵ approaches 0, and so the above may be written as

$$G(\alpha + \epsilon) = \alpha - A^2\epsilon^{2p-1} + \mathcal{O}(\epsilon^{2p}), \qquad p \geq 2. \tag{33a}$$

For the case $p = 1$, (32) becomes

$$G(\alpha + \epsilon) = \alpha - \epsilon^2(A + B\epsilon) \tag{33b}$$
$$\cdot \left\{\frac{(B - B'A) - B'B\epsilon}{(1 - A)^2 - (2B - BA - B'A^2)\epsilon + 2B'BA\epsilon^2 + B'B^2\epsilon^3}\right\}.$$

Now in general, if $A \neq 1$, the bracketed expression approaches $B^*/(1 - A)$ as ϵ approaches 0 since B' and B approach $B^* \equiv g''(\alpha)/2$ and so (33b) can be written as

$$\mathfrak{z}(\alpha + \epsilon) = \alpha - \frac{AB^*}{1 - A}\epsilon^2 + \mathcal{O}(\epsilon^3); \qquad p = 1, \ g'(\alpha) = A \neq 1. \tag{33c}$$

But, if $A = g'(\alpha) = 1$ and α has multiplicity† m, then by Problem 4:

(33d)
$$G(\alpha + \epsilon) = \alpha + \left(1 - \frac{1}{m}\right)\epsilon + \mathcal{O}(\epsilon^2),$$

for

$$p = 1, \quad g'(\alpha) = 1,$$
$$g''(\alpha) = \cdots = g^{(m-1)}(\alpha) = 0, \quad g^{(m)}(\alpha) \neq 0;$$
$$\text{for} \quad m = 2, 3, \ldots.$$

We now invoke a lemma which shall enable us to determine the orders and convergence properties in the cases represented in (33a–d).

LEMMA 1. *Let $G(x)$ be a function, with $q + 1$ derivatives in a neighborhood of $x = \alpha$, such that*

$$G(\alpha) = \alpha,$$

and for any ϵ sufficiently small

(34)
$$G(\alpha + \epsilon) = \alpha + C\epsilon^q + \mathcal{O}(\epsilon^{q+1}).$$

Then

$$G'(\alpha) = G''(\alpha) = \cdots = G^{(q-1)}(\alpha) = 0, \qquad G^{(q)}(\alpha) = q!C.$$

Proof. By Taylor's theorem we have, for sufficiently small ϵ,

(35)
$$G(\alpha + \epsilon) = \alpha + \frac{\epsilon}{1!}G'(\alpha) + \cdots + \frac{\epsilon^q}{q!}G^{(q)}(\alpha) + \frac{\epsilon^{q+1}}{(q+1)!}G^{(q+1)}(\alpha + \theta\epsilon), \qquad 0 < \theta < 1.$$

The lemma follows by comparing, in the order $k = 1, 2, \ldots, q$, the values obtained from (34) and (35) of:

$$\lim_{\epsilon \to 0}\left[\frac{G(\alpha + \epsilon) - \alpha}{\epsilon^k}k!\right]. \qquad ■$$

By applying this lemma in (33a–d) we deduce the following

THEOREM 2. (i) *If functional iteration applied to* (3) *is of order $p \geq 2$*

† If the functions in (1), (2), and (3) have m derivatives, we may verify the equivalence of the statements:

(i) α is a root of $f(x)$ of multiplicity $m \geq 2$;
(ii) $f(\alpha) = f'(\alpha) = \cdots = f^{(m-1)}(\alpha) = 0$, $f^{(m)}(\alpha) \neq 0$;
(iii) $g(\alpha) = \alpha$, $g'(\alpha) = 1$, $g''(\alpha) = \cdots = g^{(m-1)}(\alpha) = 0$, $g^{(m)}(\alpha) \neq 0$.

Statements (i) and (ii) are equivalent by definition. The equivalence of (ii) and (iii) follows from Leibnitz' rule,

$$(\phi f)^{(k)} = \phi^{(k)}f + k\phi^{(k-1)}f' + \cdots + k\phi' f^{(k-1)} + \phi f^{(k)},$$

and the fact that $\phi \neq 0$, by induction on $k = 1, 2, \ldots, m$.

for some root α of (1), *then the δ^2-method* (24)–(26) *is of order $2p - 1$ for this root.*

(ii) *If functional iteration in* (3) *is of first order (but not necessarily convergent) for a simple root α of* (1), *then the δ^2-method is of second order for this root.*

(iii) *If as in* (ii), *the root α of* (1) *has multiplicity $m \geq 2$, then the δ^2-method is first order with asymptotic convergence factor* $1 - 1/m$.

Proof. Part (i) follows from (29), Lemma 1, and (33a). Part (ii) follows from Lemma 1 and (33c) since $g'(\alpha) \neq 1$ is equivalent to $f'(\alpha) \neq 0$ (i.e., that the root is simple). Finally, (iii) follows from Lemma 1 and (33d) since an m-fold† root of $f(x) = 0$ at $x = \alpha$ implies $g'(\alpha) = 1, g''(\alpha) = \cdots = g^{(m-1)}(\alpha) = 0$ and $g^{(m)}(\alpha) \neq 0$. (This proof has assumed that $f(x)$, $g(x)$ and $G(x)$ have as many derivatives as required.) ∎

From this theorem, it follows that in all cases Aitken's δ^2-method converges if $|\alpha - x_0|$ is sufficiently small. Furthermore, *it is always at least of second order for simple roots.* It is clear that this method can be quite effective and it, or generalizations of it described below, may be very profitably used in practice.

Iterations which converge even faster than the δ^2-method are naturally suggested by the above "derivation" of (26). One such generalization is to consider the set of more than two errors associated with $x_\nu, \hat{x}_{\nu+1}, \ldots, \hat{x}_{\nu+\mu}$ as defined in (24) and (25), say

$$(36) \qquad e_\nu, \hat{e}_{\nu+1}, \ldots, \hat{e}_{\nu+\mu}, \qquad \mu > 1;$$

and then determining $x_{\nu+1}$ such that this set of errors is "extrapolated" to zero. The details of such a procedure require a knowledge of polynomial interpolation which is discussed in Chapter 6 (see Section 2 in particular). The main point in the correct application of this procedure is to consider the x_ν as functions of the e_ν (i.e., inverse interpolation) in which case the approximation $x_{\nu+1}$ can be computed directly by evaluating at $e = 0$ the polynomial of μth degree in e that takes on the values x_ν at e_ν, and $\hat{x}_{\nu+k}$ at $\hat{e}_{\nu+k}$ for $1 \leq k \leq \mu$. Other generalizations of these procedures can be obtained by successively increasing the value of μ. These considerations are, in fact, the same as those in Subsection 2.3 where generalizations of false position were discussed. Another obvious type of modification is described by introducing $G^{(0)}(x) \equiv g(x)$, $G^{(1)}(x) = G(x)$ and then forming $G^{(n)}(x)$ by recursive application of (27).

It should be noted that the correction in (26b), $-e_\nu^2/(\hat{e}_{\nu+1} - e_\nu)$, is the quotient of very small quantities. The denominator, being a difference

† See previous footnote.

of small quantities, may require multiple precision evaluation of $g(g(x_\nu))$ and $g(x_\nu)$, in order not to lose too many significant figures, especially if $g'(\alpha) \cong 1$ (i.e., if the root is multiple or nearly so). For these reasons it is important to determine an appropriate δ which can be used in Theorem 1.3 in estimating the effect of errors in the δ^2-method.

PROBLEMS, SECTION 2

1. Solve the recursion $m_{\nu+1} = m_\nu + m_{\nu-1}$, $\nu = 1, 2, \ldots$, where $m_0 = m_1 = 1$. (Try a solution of the form $m_\nu = r^\nu$ which leads to a quadratic with roots $r_\pm$. Then set $m_\nu = ar_+{}^\nu + br_-{}^\nu$ and determine a and b from $\nu = 0, 1$.)

2. The second divided difference of a function $f(x)$ is defined by:

$$f[x_1, x_2, x_3] \equiv \frac{\dfrac{f(x_1) - f(x_2)}{x_1 - x_2} - \dfrac{f(x_2) - f(x_3)}{x_2 - x_3}}{x_1 - x_3}.$$

The first divided difference is just the difference quotient. Use these definitions to verify the derivation of (16).

3. Verify (17).

[Hint: If $f''(x)$ is continuous in an interval containing x_1, x_2, and x_3, then the second result in (17) can be derived by the expansion, via Taylor's formula with remainder, of $f(x_1)$ and $f(x_3)$ about x_2 plus the fact that a continuous function takes on all values between any two of its values. (Assume, with no loss in generality, that $x_1 < x_2 < x_3$ in the definition above. That is, it is easy to verify $f[x_1, x_2, x_3] = f[x_i, x_j, x_k]$ where (i, j, k) is any permutation of (1, 2, 3). This is a special case of a property established in Chapter 6, Section 1, namely that the divided difference is a symmetric function of its arguments.)]

4. Let $g(\alpha) = \alpha$, $g'(\alpha) = 1$, $g''(\alpha) = g'''(\alpha) = \cdots = g^{m-1}(\alpha) = 0$, and $g^m(\alpha) \neq 0$. Then if we assume g has derivatives of order $2m$, $g(\alpha + \epsilon) = \alpha + \epsilon + B\epsilon^2$ where $m \geq 2$ and

$$B(\epsilon) = \frac{g^{(m)}(\alpha)}{m!}\epsilon^{m-2} + \frac{g^{(m+1)}(\alpha)}{(m+1)!}\epsilon^{m-1} + \cdots$$

and similarly, with $\delta \equiv \epsilon + B\epsilon^2$ then $g(\alpha + \delta) = \alpha + \delta + B'\delta^2$, where

$$B'(\delta) = \frac{g^m(\alpha)}{m!}\delta^{m-2} + \frac{g^{(m+1)}(\alpha)}{(m+1)!}\delta^{m-1} + \cdots.$$

Now observe that

$$\delta = \epsilon + B\epsilon^2 = \epsilon + \frac{g^{(m)}(\alpha)}{m!}\epsilon^m + \cdots$$

and therefore

$$B'B = \left[\frac{g^{(m)}(\alpha)}{m!}\right]^2 \epsilon^{2m-4} + \cdots$$

$$B' - B = (m-2)\left[\frac{g^{(m)}(\alpha)}{m!}\right]^2 \epsilon^{2m-3} + \cdots.$$

Hence, show that formula (33b) yields the results of (33d) for $m = 1, 2, 3, \ldots$.

5. Verify that the functional iteration scheme is divergent for (a) $g(x) \equiv x + x^3$ and (b) $g(x) \equiv 2x + x^3$. Nevertheless, as stated in part (ii) of Theorem 2, the Aitken δ^2-method is convergent and

$$\text{(a)} \qquad G(x) = x - \frac{x}{3 + 3x^2 + x^4} = \tfrac{2}{3}x + \mathcal{O}(x^3)$$

$$\text{(b)} \qquad G(x) = 6x^3 + \mathcal{O}(x^5).$$

6. (*Aitken's δ^2-process*). Let $\{x_n\}$, $n = 0, 1, 2, \ldots$, converge to α; so that, for some constant b,

$$r_n \equiv x_n - \alpha \neq 0, \qquad n \geq N;$$

$$r_{n+1} = (b + \epsilon_n)r_n, \qquad |b| < 1, \epsilon_n = \text{o}(1).\dagger$$

Show that (28) is meaningful for $n \geq N$, i.e.,

$$x_{n+2} - 2x_{n+1} + x_n \neq 0 \qquad \text{for } n \geq N;$$

and that

$$\lim_{n \to \infty} \frac{x_n' - \alpha}{x_n - \alpha} = 0.$$

[Hint: Verify that

$$\begin{aligned} x_{n+2} - 2x_{n+1} + x_n &= r_{n+2} - 2r_{n+1} + r_n \\ &= r_n[(b - 1)^2 + \text{o}(1)]. \end{aligned}$$

Also show from (28) that

$$\begin{aligned} x_n' - \alpha &= r_n - r_n \frac{[b - 1 + \text{o}(1)]^2}{(b - 1)^2 + \text{o}(1)} \\ &= r_n \text{o}(1).] \end{aligned}$$

7. Apply Aitken's δ^2-process (28) to the sequence

$$x_n = \alpha + b\rho_1{}^n + c\rho_2{}^n, \qquad n = 0, 1, 2, \ldots,$$

where $|\rho_2| < |\rho_1| < 1$. Show that

$$x_n' = \alpha + \mathcal{O}(\rho_2{}^n) + \mathcal{O}(\rho_1{}^{2n}).$$

What improvement results by applying the δ^2-process to the sequence $\{x_n'\}$?

3. FUNCTIONAL ITERATION FOR A SYSTEM OF EQUATIONS

Let $\mathbf{x}$ be an n-dimensional column vector with components $x_1, x_2, \ldots, x_n$ and $\mathbf{g}(\mathbf{x})$ an n-dimensional vector valued function, i.e., a column vector with components $g_1(\mathbf{x}), g_2(\mathbf{x}), \ldots, g_n(\mathbf{x})$. Then the system to be solved is

$$\text{(1)} \qquad \mathbf{x} = \mathbf{g}(\mathbf{x}).$$

† We write $\delta_n = \text{o}(1)$ iff there is some number N such that δ_n is defined for all $n \geq N$ and $\lim_{n \to \infty} \delta_n = 0$. In the text, we also use o(1) as a generic symbol to represent the members of any sequence which tends to zero.

The solution (or root) is some vector, say $\boldsymbol{\alpha}$, with components $\alpha_1, \alpha_2, \ldots, \alpha_n$ which is, of course, some point in the n-dimensional space. Starting with a point $\mathbf{x}^{(0)} = [x_1^{(0)}, x_2^{(0)}, \ldots, x_n^{(0)}]^T$, the exact analog of the functional iteration of Section 1 is

$$\mathbf{x}^{(\nu+1)} = \mathbf{g}(\mathbf{x}^{(\nu)}), \qquad \nu = 0, 1, 2, \ldots. \tag{2}$$

The first result is analogous to Theorem 1.1. But where absolute values were used previously, we must now use some vector norm (see Chapter 1, Section 1). For example, we may choose any one of the norms

$$\begin{aligned} \|\mathbf{x}\|_\infty &\equiv \max_{1\le i\le n} |x_i|, \\ \|\mathbf{x}\|_1 &\equiv \sum_{i=1}^{n} |x_i|, \\ \|\mathbf{x}\|_2 &\equiv \sqrt{\sum_{i=1}^{n} |x_i|^2}. \end{aligned} \tag{3}$$

THEOREM **1**. *Let* $\mathbf{g}(\mathbf{x})$ *satisfy*

$$\|\mathbf{g}(\mathbf{x}) - \mathbf{g}(\mathbf{y})\| \le \lambda\|\mathbf{x} - \mathbf{y}\| \tag{4a}$$

for all vectors $\mathbf{x}$, $\mathbf{y}$ *such that* $\|\mathbf{x} - \mathbf{x}^{(0)}\| \le \rho$, $\|\mathbf{y} - \mathbf{x}^{(0)}\| \le \rho$ *with the Lipschitz constant*, λ, *satisfying*

$$0 \le \lambda < 1. \tag{4b}$$

Let the initial iterate, $\mathbf{x}^{(0)}$, *satisfy*

$$\|\mathbf{g}(\mathbf{x}^{(0)}) - \mathbf{x}^{(0)}\| \le (1 - \lambda)\rho. \tag{5}$$

Then: (i) *all iterates*, (2), *satisfy*

$$\|\mathbf{x}^{(\nu)} - \mathbf{x}^{(0)}\| \le \rho;$$

(ii) *the iterates converge to some vector, say*

$$\lim_{\nu\to\infty} \mathbf{x}^{(\nu)} = \boldsymbol{\alpha},$$

which is a root of (1);

(iii) $\boldsymbol{\alpha}$ *is the only root of* (1) *in* $\|\mathbf{x} - \mathbf{x}^{(0)}\| \le \rho$.

Proof. Duplicate the proof of Theorem 1.1 with the replacement of absolute value signs by norm symbols. ■

As a consequence of this proof, it is also seen that the iterates converge geometrically, and at least as fast as $\lambda^\nu \to 0$. Of course, it is more difficult to verify (4), the Lipschitz continuity of a vector valued function, than it is in the case of a scalar function.

However, again as in Section 1, a more useful result can be obtained if we are willing to place more restrictions on $\mathbf{g}(\mathbf{x})$ and assume the existence of a root. We immediately see that Theorem 1.2 and its proof hold if absolute value signs are replaced by norms. Furthermore, the corollary to Theorem 1.2 becomes

THEOREM **2.** *Let* (1) *have a root* $\mathbf{x} = \boldsymbol{\alpha}$. *Let the components* $g_i(\mathbf{x})$ *have continuous first partial derivatives and satisfy*

$$\left|\frac{\partial g_i(\mathbf{x})}{\partial x_j}\right| \leq \frac{\lambda}{n}, \qquad \lambda < 1; \tag{6}$$

for all $\mathbf{x}$ *in*

$$\|\mathbf{x} - \boldsymbol{\alpha}\|_\infty \leq \rho. \tag{7}$$

Then: (i) *For any* $\mathbf{x}^{(0)}$ *satisfying* (7) *all the iterates* $\mathbf{x}^{(\nu)}$ *of* (2) *also satisfy* (7).
(ii) *For any* $\mathbf{x}^{(0)}$ *satisfying* (7) *the iterates* (2) *converge to the root* $\boldsymbol{\alpha}$ *of* (1) *which is unique in* (7).

Proof. For any two points $\mathbf{x}$, $\mathbf{y}$ in (7) we have by Taylor's theorem:

$$g_i(\mathbf{x}) - g_i(\mathbf{y}) = \sum_{j=1}^{n} \frac{\partial g_i(\boldsymbol{\xi}^{(i)})}{\partial x_j}(x_j - y_j), \qquad i = 1, 2, \ldots, n; \tag{8}$$

where $\boldsymbol{\xi}^{(i)}$ is a point on the open line segment joining $\mathbf{x}$ and $\mathbf{y}$. Thus, $\boldsymbol{\xi}^{(i)}$ is in (7), and using (3) and (6) yields

$$\begin{aligned} |g_i(\mathbf{x}) - g_i(\mathbf{y})| &\leq \sum_{j=1}^{n} \left|\frac{\partial g_i(\boldsymbol{\xi}^{(i)})}{\partial x_j}\right| \cdot |x_j - y_j|, \\ &\leq \|\mathbf{x} - \mathbf{y}\|_\infty \sum_{j=1}^{n} \left|\frac{\partial g_i(\boldsymbol{\xi}^{(i)})}{\partial x_j}\right|, \\ &\leq \lambda \|\mathbf{x} - \mathbf{y}\|_\infty. \end{aligned}$$

Since the inequality holds for each i, we have

$$\|\mathbf{g}(\mathbf{x}) - \mathbf{g}(\mathbf{y})\|_\infty \leq \lambda \|\mathbf{x} - \mathbf{y}\|_\infty, \tag{9}$$

and thus we have proven that $\mathbf{g}(\mathbf{x})$ is Lipschitz continuous in the domain (7), with respect to the indicated norm. Now note that for any $\mathbf{x}^{(0)}$ in (7),

$$\begin{aligned} \|\mathbf{x}^{(1)} - \boldsymbol{\alpha}\|_\infty &= \|\mathbf{g}(\mathbf{x}^{(0)}) - \mathbf{g}(\boldsymbol{\alpha})\|_\infty \\ &\leq \lambda \|\mathbf{x}^{(0)} - \boldsymbol{\alpha}\|_\infty \\ &\leq \lambda \rho, \end{aligned}$$

and so $\mathbf{x}^{(1)}$ is also in (7). By an obvious induction we have then

$$\begin{aligned} \|\mathbf{x}^{(\nu)} - \boldsymbol{\alpha}\|_\infty &= \|\mathbf{g}(\mathbf{x}^{(\nu-1)} - \mathbf{g}(\boldsymbol{\alpha})\|_\infty \\ &\le \lambda\|\mathbf{x}^{(\nu-1)} - \boldsymbol{\alpha}\|_\infty \\ &\quad\vdots \\ &\le \lambda^\nu\|\mathbf{x}^{(0)} - \boldsymbol{\alpha}\|_\infty \\ &\le \lambda^\nu \rho \end{aligned} \tag{10}$$

and hence all $\mathbf{x}^{(\nu)}$ lie in (7). The convergence immediately follows from (10) since $\lambda < 1$. The uniqueness follows as before. ∎

The crucial point in the preceding proof is the derivation of (9). It is clear from this derivation that (6) could be replaced by a number of conditions which are perhaps less restrictive and the theorem would still remain valid. One such condition is

$$\max_i \sum_{j=1}^{n} |g_{ij}(\mathbf{x})| \le \lambda < 1, \qquad \text{for all } \|\mathbf{x} - \boldsymbol{\alpha}\|_\infty < \rho, \tag{11}$$

where we have introduced the elements $g_{ij}(\mathbf{x}) = \partial g_i(\mathbf{x})/\partial x_j$. If we define the matrix $G(\mathbf{x}) \equiv (g_{ij}(\mathbf{x}))$ then (11) may be written as $\|G(\mathbf{x})\|_\infty \le \lambda < 1$ in which case we mean the natural matrix norm induced by the maximum vector norm (see Chapter 1, Section 1).

If the function $\mathbf{g}(\mathbf{x})$ is such that at a root

$$G(\boldsymbol{\alpha}) \equiv \left(\frac{\partial g_i(\boldsymbol{\alpha})}{\partial x_j}\right) = 0, \qquad i, j = 1, 2, \ldots, n \tag{12}$$

and these derivatives are continuous near the root, then (6) as well as (11) can be satisfied for some $\rho > 0$. If, in addition, the second derivatives

$$\frac{\partial^2 g_i(\mathbf{x})}{\partial x_j\, \partial x_k}$$

all exist in a neighborhood of the root, then again by Taylor's theorem

$$g_i(\mathbf{x}) - g_i(\boldsymbol{\alpha}) = \frac{1}{2}\sum_{j=1}^{n}\sum_{k=1}^{n} \frac{\partial^2 g_i(\boldsymbol{\xi}^i)}{\partial x_j\, \partial x_k}(x_j - \alpha_j)(x_k - \alpha_k).$$

Now in the iteration, (2), we find

$$\|\mathbf{x}^{(\nu)} - \boldsymbol{\alpha}\|_\infty \le M\|\mathbf{x}^{(\nu-1)} - \boldsymbol{\alpha}\|_\infty^2,$$

where M is chosen such that

$$\max_{i,j,k}\left|\frac{\partial^2 g_i(\mathbf{x})}{\partial x_j\, \partial x_k}\right| \le \frac{2M}{n^2}.$$

Thus, quadratic convergence can occur in solving systems of equations by iteration.

3.1. Some Explicit Iteration Schemes for Systems

In the general case, the system to be solved is of the form

$$\mathbf{f}(\mathbf{x}) = \mathbf{o} \tag{13}$$

where $\mathbf{f}(\mathbf{x}) = [f_1(\mathbf{x}), f_2(\mathbf{x}), \ldots, f_n(\mathbf{x})]^T$ is an n-component column vector. Such a system can be written in the form (1) in a variety of ways; we examine here the choice

$$\mathbf{g}(\mathbf{x}) \equiv \mathbf{x} - A(\mathbf{x})\mathbf{f}(\mathbf{x}), \tag{14}$$

where $A(\mathbf{x})$ is an nth order square matrix with components $a_{ij}(\mathbf{x})$. The equations (1) and (13) will have the same set of solutions if $A(\mathbf{x})$ is non-singular [since in that case $A(\mathbf{x})\mathbf{f}(\mathbf{x}) = \mathbf{o}$ implies $\mathbf{f}(\mathbf{x}) = \mathbf{o}$].

The simplest choice for $A(\mathbf{x})$ is

$$A(\mathbf{x}) \equiv A, \tag{15}$$

a constant non-singular matrix. If we introduce the matrix

$$J(\mathbf{x}) \equiv \left(\frac{\partial f_i(\mathbf{x})}{\partial x_j}\right), \tag{16}$$

whose determinant is the Jacobian of the functions $f_i(\mathbf{x})$, then from (14)–(16) we have

$$G(\mathbf{x}) \equiv \left(\frac{\partial g_i(\mathbf{x})}{\partial x_j}\right) = I - AJ(\mathbf{x}). \tag{17}$$

Thus by Theorem 2, or its modification in which (11) replaces (6), the iterations determined by using

$$\mathbf{x}^{(\nu+1)} = \mathbf{x}^{(\nu)} - A\mathbf{f}(\mathbf{x}^{(\nu)})$$

will converge, for $\mathbf{x}^{(0)}$ sufficiently close to $\boldsymbol{\alpha}$, if the elements in the matrix (17) are sufficiently small, for example, as in the case that $J(\boldsymbol{\alpha})$ is non-singular and A is approximately the inverse of $J(\boldsymbol{\alpha})$. This procedure is the analog of the chord method and it naturally suggests a modification which is again called Newton's method.

In Newton's method (15) is replaced by the choice

$$A(\mathbf{x}) \equiv J^{-1}(\mathbf{x}), \tag{18}$$

with the assumption of course that $\det |J(\mathbf{x})| \neq 0$ for $\mathbf{x}$ in $\|\mathbf{x} - \boldsymbol{\alpha}\| \leq \rho$. In actually using the above procedure, an inverse need not be computed at each iteration; instead, a linear system of order n has to be solved. To see this, and at the same time gain some insight into the method, we note that by using (18) in (14) the iterations for Newton's method are:

$$\text{(19a)} \qquad \begin{aligned} \mathbf{x}^{(\nu+1)} &= \mathbf{g}(\mathbf{x}^{(\nu)}), \\ &= \mathbf{x}^{(\nu)} - J^{-1}(\mathbf{x}^{(\nu)})\mathbf{f}(\mathbf{x}^{(\nu)}). \end{aligned}$$

From this we obtain

$$\text{(19b)} \qquad J(\mathbf{x}^{(\nu)})(\mathbf{x}^{(\nu)} - \mathbf{x}^{(\nu+1)}) = \mathbf{f}(\mathbf{x}^{(\nu)}),$$

which is the system to be solved for the vector $(\mathbf{x}^{(\nu)} - \mathbf{x}^{(\nu+1)})$.

To show that this method is of second order we must verify that (12) is satisfied when (18) is used in (14). The jth column of $G(\mathbf{x})$ is then given by

$$\begin{aligned} \frac{\partial \mathbf{g}(\mathbf{x})}{\partial x_j} &= \frac{\partial \mathbf{x}}{\partial x_j} - \frac{\partial}{\partial x_j}[J^{-1}(\mathbf{x})\mathbf{f}(\mathbf{x})], \\ &= \frac{\partial \mathbf{x}}{\partial x_j} - J^{-1}(\mathbf{x})\frac{\partial \mathbf{f}(\mathbf{x})}{\partial x_j} - \frac{\partial J^{-1}(\mathbf{x})}{\partial x_j}\mathbf{f}(\mathbf{x}). \end{aligned}$$

By setting $\mathbf{x} = \boldsymbol{\alpha}$ in the above and recalling that $\mathbf{f}(\boldsymbol{\alpha}) = \mathbf{o}$ and $J = (\partial f_i/\partial x_j)$ we get

$$G(\boldsymbol{\alpha}) = I - J^{-1}(\boldsymbol{\alpha})J(\boldsymbol{\alpha}) - O = O.$$

To determine $\partial J^{-1}(\mathbf{x})/\partial x_j$, note that

$$\frac{\partial (J^{-1}J)}{\partial x_j} = J^{-1}\frac{\partial J}{\partial x_j} + \frac{\partial J^{-1}}{\partial x_j}J = \frac{\partial I}{\partial x_j} = O$$

and hence

$$\frac{\partial J^{-1}(\mathbf{x})}{\partial x_j} = -J^{-1}(\mathbf{x})\frac{\partial J(\mathbf{x})}{\partial x_j}J^{-1}(\mathbf{x}).$$

Thus, we need only require that $\mathbf{f}(\mathbf{x})$ *have two derivatives and* $J(\mathbf{x})$ *be non-singular at the root, and then the convergence of Newton's method is quadratic.*

For a geometric interpretation of Newton's method we consider a system of two equations and drop subscripts by using

$$\begin{pmatrix} x_1 \\ x_2 \end{pmatrix} \equiv \begin{pmatrix} x \\ y \end{pmatrix}, \qquad \begin{pmatrix} f_1(x_1, x_2) \\ f_2(x_1, x_2) \end{pmatrix} \equiv \begin{pmatrix} f(x, y) \\ g(x, y) \end{pmatrix}.$$

Then

$$J(x) = \begin{pmatrix} f_x & f_y \\ g_x & g_y \end{pmatrix}$$

and the system (19) can be written as:

$$\text{(20a)} \qquad (x_{\nu+1} - x_\nu)f_x(x_\nu, y_\nu) + (y_{\nu+1} - y_\nu)f_y(x_\nu, y_\nu) + f(x_\nu, y_\nu) = 0$$

$$\text{(20b)} \qquad (x_{\nu+1} - x_\nu)g_x(x_\nu, y_\nu) + (y_{\nu+1} - y_\nu)g_y(x_\nu, y_\nu) + g(x_\nu, y_\nu) = 0$$

In the (x, y, z)-space the equations

$$z = (x - x_\nu)f_x(x_\nu, y_\nu) + (y - y_\nu)f_y(x_\nu, y_\nu) + f(x_\nu, y_\nu), \tag{21a}$$

$$z = (x - x_\nu)g_x(x_\nu, y_\nu) + (y - y_\nu)g_y(x_\nu, y_\nu) + g(x_\nu, y_\nu), \tag{21b}$$

each represent planes. The plane (21a) is tangent to the surface $z = f(x, y)$ at the point $(x_\nu, y_\nu, f(x_\nu, y_\nu))$, and the plane (21b) is tangent to $z = g(x, y)$ at the point $(x_\nu, y_\nu, g(x_\nu, y_\nu))$. Clearly, the point $(x_{\nu+1}, y_{\nu+1})$ determined from (20) is the point of intersection of these two planes with the plane $z = 0$, i.e., the (x, y)-plane. Thus, in passing from one dimension (Section 2.2) to two dimensions, Newton's method is generalized by replacing tangent lines with tangent planes. In the more general case of n dimensions the obvious interpretation, using tangent hyperplanes, is valid. Each of the equations

$$z = \sum_{k=1}^{n} (x_k - x_k^{(\nu)}) \frac{\partial f_i(\mathbf{x}^{(\nu)})}{\partial x_k} + f_i(\mathbf{x}^{(\nu)}), \qquad i = 1, 2, \ldots, n,$$

represents a hyperplane in the $(x_1, x_2, \ldots, x_n, z)$ space of $n + 1$ dimensions which is tangent at the point $(x_1^{(\nu)}, x_2^{(\nu)}, \ldots, x_n^{(\nu)})$ to the corresponding hypersurface

$$z = f_i(x_1, x_2, \ldots, x_n).$$

The difficulties which may arise in the solution of systems using Newton's method can be interpreted by means of these geometric considerations.

3.2. Convergence of Newton's Method

If the initial iterate $\mathbf{x}^{(0)}$ is sufficiently close to the root $\boldsymbol{\alpha}$ of $\mathbf{f}(\mathbf{x}) = \mathbf{o}$, then Theorem 2 can be used to prove that the Newton iterates, $\mathbf{x}^{(\nu)}$, defined in (19) converge to the root. In addition, if the Jacobian $J(\mathbf{x})$ is non-singular at the root, $\mathbf{x} = \boldsymbol{\alpha}$, and differentiable there, then the convergence is second order. However, we do not know from these results if a given initial iterate $\mathbf{x}^{(0)}$ is close enough to the unknown root, $\boldsymbol{\alpha}$. We shall develop a sufficient condition, under which Newton's scheme converges, with the property that this condition may be explicitly checked without a knowledge of $\boldsymbol{\alpha}$. In fact, the theorem to be established also proves the existence of a unique root of $\mathbf{f}(\mathbf{x})$ in an appropriate interval about the initial iterate, $\mathbf{x}^{(0)}$. Thus, we have an alternative to Theorem 1 which we state as

THEOREM 3. *Let the initial iterate* $\mathbf{x}^{(0)}$ *be such that the Jacobian matrix* $J(\mathbf{x}^{(0)})$ *defined in* (16) *has an inverse with norm bounded by*

$$\|J^{-1}(\mathbf{x}^{(0)})\| \leq a. \tag{22a}$$

Let the difference of the first two Newton iterates be bounded by

(22b) $$\|\mathbf{x}^{(1)} - \mathbf{x}^{(0)}\| = \|J^{-1}(\mathbf{x}^{(0)})\mathbf{f}(\mathbf{x}^{(0)})\| \leq b.$$

Let the components of $\mathbf{f}(\mathbf{x})$ *have continuous second derivatives which satisfy*

(22c) $$\sum_{k=1}^{n} \left|\frac{\partial^2 f_i(\mathbf{x})}{\partial x_j\, \partial x_k}\right| \leq \frac{c}{n},$$

for all $\mathbf{x}$ *in* $\|\mathbf{x} - \mathbf{x}^{(0)}\| \leq 2b$; $i, j = 1, 2, \ldots, n$. *If the constants a, b, and c are such that*

(22d) $$abc \leq \tfrac{1}{2}$$

then: (i) *the Newton iterates* (19) *are uniquely defined and lie in the "2b-sphere" about* $\mathbf{x}^{(0)}$:

$$\|\mathbf{x}^{(\nu)} - \mathbf{x}^{(0)}\| \leq 2b;$$

(ii) *the iterates converge to some vector, say* $\lim_{\nu \to \infty} \mathbf{x}^{(\nu)} = \boldsymbol{\alpha}$, *for which* $\mathbf{f}(\boldsymbol{\alpha}) = \mathbf{o}$ *and*

(23) $$\|\mathbf{x}^{(\nu)} - \boldsymbol{\alpha}\| \leq \frac{2b}{2^\nu}.$$

[All vector norms in the statement and proof of this theorem are maximum norms, i.e., $\|\mathbf{x}\| = \max_i |x_i|$, and matrix norms are the corresponding induced natural norm, i.e., $\|A\| = \max_i \left(\sum_{j=1}^{n} |a_{ij}|\right)$.]

Proof. The proof proceeds by a somewhat lengthy induction. For convenience, we use the notation $J_\nu \equiv J(\mathbf{x}^{(\nu)})$ for the Jacobian matrices (16) and show for all $\nu = 0, 1, 2, \ldots$ that with $A_{\nu+1} \equiv I - J_\nu^{-1} J_{\nu+1}$,

(24a) $$\|\mathbf{x}^{(\nu+1)} - \mathbf{x}^{(\nu)}\| \leq \frac{b}{2^\nu},$$

(24b) $$\|\mathbf{x}^{(\nu+1)} - \mathbf{x}^{(0)}\| \leq 2b,$$

(24c) $$\|A_{\nu+1}\| \equiv \|J_\nu^{-1}(J_\nu - J_{\nu+1})\| \leq \tfrac{1}{2},$$

(24d) $$\|J_{\nu+1}^{-1}\| = \|(I - A_{\nu+1})^{-1} J_\nu^{-1}\| \leq 2^{\nu+1} a.$$

From the hypothesis (22b) it trivially follows that (24a, b) are satisfied for $\nu = 0$. Now when (24b) is established up to and including any value ν then $\mathbf{x}^{(\nu+1)}$ and $\mathbf{x}^{(\nu)}$ are in the $2b$-sphere about $\mathbf{x}^{(0)}$ in which we are assured that the second derivatives of the $f_i(\mathbf{x})$ are continuous. Then we can apply Taylor's theorem to the components of $J_{\nu+1}$ to obtain

$$\frac{\partial f_i(\mathbf{x}^{(\nu+1)})}{\partial x_j} = \frac{\partial f_i(\mathbf{x}^{(\nu)})}{\partial x_j} + \sum_{k=1}^{n} (x_k^{(\nu+1)} - x_k^{(\nu)}) \frac{\partial^2 f_i\,[\mathbf{x}^{(\nu)} + \theta_i(\mathbf{x}^{(\nu+1)} - \mathbf{x}^{(\nu)})]}{\partial x_j\, \partial x_k},$$
$$0 < \theta_i < 1.$$

Since $\mathbf{x}^{(\nu+1)}$ and $\mathbf{x}^{(\nu)}$ are in $\|\mathbf{x} - \mathbf{x}^{(0)}\| \leq 2b$, so is the point $\mathbf{x}^{(\nu)} + \theta(\mathbf{x}^{(\nu+1)} - \mathbf{x}^{(\nu)})$, and (22c) applies. This gives from the above

$$\|J_{\nu+1} - J_\nu\| \leq c\|\mathbf{x}^{(\nu+1)} - \mathbf{x}^{(\nu)}\|. \tag{25}$$

At the present stage in the proof this is valid only for $\nu = 0$. But then using this and (22a, b, d) in (24c) with $\nu = 0$ yields

$$\begin{aligned}\|A_1\| &\leq \|J_0^{-1}\| \cdot \|J_1 - J_0\| \\ &\leq ac\|\mathbf{x}^{(1)} - \mathbf{x}^{(0)}\| \\ &\leq abc \\ &\leq \tfrac{1}{2}.\end{aligned}$$

Now (24a, b, c) have been established for $\nu = 0$.

If for any ν the matrix J_ν is non-singular, then we have the identity

$$J_{\nu+1} = J_\nu(I - A_{\nu+1}),$$

where, as in (24c), $A_{\nu+1} \equiv J_\nu^{-1}(J_\nu - J_{\nu+1})$. But from the Corollary to Theorem 1.5 of Chapter 1 it follows that *if* $\|A_{\nu+1}\| < 1$ *then* $J_{\nu+1}$ *is non-singular* and

$$\|J_{\nu+1}^{-1}\| \leq \frac{\|J_\nu^{-1}\|}{1 - \|A_{\nu+1}\|}. \tag{26}$$

Since (24c) is valid for $\nu = 0$ we can use this in (26) to get

$$\|J_1^{-1}\| \leq 2a.$$

Thus (24) has been verified for $\nu = 0$.

Let us now make the inductive assumption that (24) is valid for all $\nu \leq k - 1$ and proceed to show that it is also valid for $\nu = k$. Since J_k is non-singular, the $(k + 1)$st Newton iterate, $\mathbf{x}^{(k+1)}$, is uniquely defined and we have from (19a):

$$\begin{aligned}\|\mathbf{x}^{(k+1)} - \mathbf{x}^{(k)}\| &= \|J_k^{-1}\mathbf{f}(\mathbf{x}^{(k)})\| \\ &\leq \|J_k^{-1}\| \cdot \|\mathbf{f}(\mathbf{x}^{(k)})\|.\end{aligned} \tag{27}$$

However, since (24b) is valid for $\nu = k - 1$, the point $\mathbf{x}^{(k)}$ is in the $2b$-sphere about $\mathbf{x}^{(0)}$. Then by Taylor's theorem, with remainder term $\mathbf{R}$, and (19b) with $\nu = k - 1$:

$$\begin{aligned}\mathbf{f}(\mathbf{x}^{(k)}) &= \mathbf{f}(\mathbf{x}^{(k-1)}) + J_{k-1}[\mathbf{x}^{(k)} - \mathbf{x}^{(k-1)}] + \mathbf{R}(\mathbf{x}^{(k)}, \mathbf{x}^{(k-1)}) \\ &= \mathbf{R}(\mathbf{x}^{(k)}, \mathbf{x}^{(k-1)}).\end{aligned}$$

Using (22c), we can bound the above remainder term to yield

$$\|\mathbf{f}(\mathbf{x}^{(k)})\| = \max_i |R_i(\mathbf{x}^{(k)}, \mathbf{x}^{(k-1)})|$$

$$= \max_i \left| \sum_{j=1}^{n} \sum_{l=1}^{n} \frac{(x_j^{(k)} - x_j^{(k-1)})(x_l^{(k)} - x_l^{(k-1)})}{2!} \right.$$

$$\left. \times \frac{\partial^2 f_i}{\partial x_j\, \partial x_l} [\mathbf{x}^{(k-1)} + \phi_i(\mathbf{x}^{(k)} - \mathbf{x}^{(k-1)})] \right|, \qquad 0 < \phi_i < 1,$$

$$\text{(28)} \qquad \leq \frac{c}{2} \|\mathbf{x}^{(k)} - \mathbf{x}^{(k-1)}\|^2.$$

Again, we have used the fact that $\mathbf{x}^{(k-1)} + \phi(\mathbf{x}^{(k)} - \mathbf{x}^{(k-1)})$ is in the $2b$-sphere about $\mathbf{x}^{(0)}$ since $\mathbf{x}^{(k)}$ and $\mathbf{x}^{(k-1)}$ are in it. Now using (28) in (27) and recalling that (24) is assumed valid for all $\nu \leq k - 1$ we get

$$\text{(29)} \qquad \|\mathbf{x}^{(k+1)} - \mathbf{x}^{(k)}\| \leq \frac{c}{2} \|J_k^{-1}\| \cdot \|\mathbf{x}^{(k)} - \mathbf{x}^{(k-1)}\|^2$$

$$\leq \frac{c}{2} (2^k a)\left(\frac{b}{2^{k-1}}\right)^2 = \frac{ab^2c}{2^{k-1}}$$

$$\leq \frac{b}{2^k}.$$

Thus (24a) is established for $\nu = k$. Then since

$$\|\mathbf{x}^{(k+1)} - \mathbf{x}^{(0)}\| = \left\| \sum_{l=0}^{k} (\mathbf{x}^{(l+1)} - \mathbf{x}^{(l)}) \right\|$$

$$\leq \sum_{l=0}^{k} \|\mathbf{x}^{(l+1)} - \mathbf{x}^{(l)}\|$$

$$\leq b \sum_{l=0}^{k} \frac{1}{2^l}$$

$$\leq 2b,$$

we have also established (24b) for $\nu = k$. But then $\mathbf{x}^{(k+1)}$ is in the $2b$-sphere about $\mathbf{x}^{(0)}$ and so (25) is valid with $\nu = k$. This gives

$$\|A_{k+1}\| = \|J_k^{-1}(J_k - J_{k+1})\|$$

$$\leq \|J_k^{-1}\| \cdot \|J_{k+1} - J_k\|$$

$$\leq c\|J_k^{-1}\| \cdot \|\mathbf{x}^{(k+1)} - \mathbf{x}^{(k)}\|$$

$$\leq abc$$

$$\leq \tfrac{1}{2}.$$

Thus (24c) is valid with $\nu = k$ and implies that J_{k+1} is non-singular. Then using (26) with $\nu = k$ yields (24d), and the inductive proof of (24) is complete.

Part (i) of the theorem follows from (24b, d). The convergence of the $\mathbf{x}^{(\nu)}$ follows from (24a) since they form a Cauchy sequence: i.e.,

$$(30) \qquad \|\mathbf{x}^{(\nu+m)} - \mathbf{x}^{(\nu)}\| = \left\| \sum_{l=\nu}^{\nu+m-1} (\mathbf{x}^{(l+1)} - \mathbf{x}^{(l)}) \right\|$$

$$\leq \sum_{l=\nu}^{\nu+m-1} \|\mathbf{x}^{(l+1)} - \mathbf{x}^{(l)}\| \leq b \sum_{l=\nu}^{\nu+m-1} \frac{1}{2^l}$$

$$\leq \frac{b}{2^{\nu-1}}.$$

Calling the limit vector $\boldsymbol{\alpha}$, we use (24a), (28) and the continuity of $\mathbf{f}(\mathbf{x})$ to deduce that

$$\|\mathbf{f}(\mathbf{x}^{(k)})\| \leq \frac{2b^2c}{4^k},$$

and $\lim_{k\to\infty} \mathbf{f}(\mathbf{x}^{(k)}) = \mathbf{f}(\boldsymbol{\alpha}) = \mathbf{o}$. Letting $m \to \infty$, (30) implies

$$\|\boldsymbol{\alpha} - \mathbf{x}^{(\nu)}\| \leq \frac{2b}{2^\nu},$$

and so, part (ii) is established, concluding the proof of the theorem. ∎

This theorem is valid if $n = 1$. The hypothesis permits the case that $J(\boldsymbol{\alpha})$ is singular. Hence, it is reasonable that the conclusion (ii) shows at most linear convergence. But (ii), moreover, implies that the convergence factor is at most $\frac{1}{2}$. This seems to contradict the fact shown earlier for the scalar case ($n = 1$), that the convergence factor is only $1 - 1/p$ if $f(\alpha)$ is a zero of order $p > 1$. The contradiction doesn't exist because, as we show in Problem 6, the requirement $abc \leq \frac{1}{2}$ can only be satisfied if $p \leq 2$. [For example, $f(x) \equiv x^p$ and $x_0 \neq 0$ satisfy

$$a \cdot b \cdot c \geq \left| \frac{1}{f'(x_0)} \cdot \frac{f(x_0)}{f'(x_0)} \cdot f''(x_0) \right| \equiv 1 - 1/p.]$$

On the other hand, if $h \equiv abc < \frac{1}{2}$, then Kantorovich has shown in general that

$$\|\mathbf{x}^{(\nu)} - \boldsymbol{\alpha}\| \leq \frac{(2h)^{2^\nu - 1}}{2^{\nu-1}} b,$$

i.e., quadratic convergence. See Problems 3, 4, 5, and 6 for further results in the scalar case.

3.3. A Special Acceleration Procedure for Non-Linear Systems

If the non-linear system to be solved is written in the form

$$\mathbf{x} = \mathbf{g}(\mathbf{x}), \tag{31}$$

then the obviously implied iterations

$$\mathbf{x}^{(0)} = \text{arbitrary}; \qquad \mathbf{x}^{(\nu+1)} = \mathbf{g}(\mathbf{x}^{(\nu)}), \qquad \nu = 0, 1, \ldots, \tag{32}$$

may or may not converge. However, as with linear systems, we can alter the procedure (32) in a manner which will generally improve the rate of convergence or may even yield a convergent scheme when the basic one diverges. The acceleration procedure is defined by:

$$\begin{aligned} \mathbf{x}^{(0)} &= \text{arbitrary} \\ \mathbf{x}^{(\nu+1)} &= \Theta\mathbf{g}(\mathbf{x}^{(\nu)}) + (I - \Theta)\mathbf{x}^{(\nu)}, \qquad \nu = 0, 1, \ldots, \end{aligned} \tag{33a}$$

where Θ is a diagonal matrix given by

$$\Theta \equiv (\theta_i \delta_{ij}), \qquad \det[\Theta] = \theta_1\theta_2\ldots\theta_n \neq 0. \tag{33b}$$

Of course, if $\Theta = I$, then the basic scheme (32) results. The scalar form of the ith equation in (33a) is clearly

$$x_i^{(\nu+1)} = \theta_i g_i(\mathbf{x}^{(\nu)}) + (1 - \theta_i)x_i^{(\nu)}, \qquad 1 \leq i \leq n,$$

and so the iterations are easily evaluated in an explicit manner.

Let us assume that (31) has a solution, say $\mathbf{x} = \boldsymbol{\alpha}$, and that in some ρ-sphere about this solution, $\|\mathbf{x} - \boldsymbol{\alpha}\| \leq \rho$, the vector function $\mathbf{g}$ has continuous first partial derivatives, $g_{ij}(\mathbf{x}) \equiv \partial g_i(\mathbf{x})/\partial x_j$, which satisfy the conditions

$$|1 - g_{ii}(\mathbf{x})| > \sum_{j \neq i} |g_{ij}(\mathbf{x})|, \qquad 1 \leq i \leq n. \tag{34}$$

Under these conditions it can be shown that the iterations (33) will converge, for some choice of the θ_i, to a solution of (31) for any initial guess $\mathbf{x}^{(0)}$ in $\|\mathbf{x}^{(0)} - \boldsymbol{\alpha}\| \leq \rho$. In fact, under slightly different assumptions we could even demonstrate the existence of a solution if $\|\mathbf{g}(\mathbf{x}^{(0)}) - \mathbf{x}^{(0)}\|$ is sufficiently small. However, we shall not present such specific theorems but instead shall indicate the relevant arguments and concern ourselves with the determination of the appropriate θ_i to be used in (33). These considerations, in turn, suggest a modification in which the θ_i are changed at each iteration.

If the error vector after the νth iteration is denoted by

$$\mathbf{e}^{(\nu)} = \mathbf{x}^{(\nu)} - \boldsymbol{\alpha},$$

then from (33a):

$$\mathbf{e}^{(\nu+1)} = \Theta[\mathbf{g}(\mathbf{x}^{(\nu)}) - \mathbf{g}(\boldsymbol{\alpha})] + (I - \Theta)\mathbf{e}^{(\nu)}.$$

However, by Taylor's theorem we then have

$$(35)\qquad \mathbf{e}^{(\nu+1)} = (I - \Theta + \Theta G_\nu)\mathbf{e}^{(\nu)} \equiv M_\nu \mathbf{e}^{(\nu)}$$

where $G_\nu = (g_{ij})$ and the ith row of G_ν is evaluated at some point $\boldsymbol{\xi}^{(i,\nu)} = \mathbf{x}^{(\nu)} - \phi_i(\mathbf{x}^\nu - \boldsymbol{\alpha})$, $0 < \phi_i < 1$, for $i = 1, 2, \ldots, n$. Clearly, the iterations will converge if the coefficient matrices in (35) satisfy $\|M_\nu\| \le q < 1$ for all ν. Now if we use the maximum vector norm and corresponding induced matrix norm, these inequalities are satisfied if

$$(36)\quad R_i(\theta_i) \equiv |1 - \theta_i[1 - g_{ii}(\boldsymbol{\xi}^{(i,\nu)})]| + |\theta_i| \sum_{j \ne i} |g_{ij}(\boldsymbol{\xi}^{(i,\nu)})| \le q < 1,$$
$$1 \le i \le n; \quad \nu = 0, 1, 2, \ldots.$$

Thus, we are led to consider inequalities of the form

$$R(\theta) \equiv |1 - \theta a| + |\theta| b < 1, \qquad b \ge 0.$$

It is easily shown that $R(\theta) < 1$ if $|a| > b$ and θ is in the interval:

$$(37a)\qquad 0 < \theta < \frac{2}{a + b} \qquad \text{if } a > b;$$

$$(37b)\qquad \frac{2}{a - b} < \theta < 0 \qquad \text{if } -a > b.$$

Furthermore, in each of these intervals the minimum value of $R(\theta)$ is attained at

$$(37c)\qquad \theta^* = \frac{1}{a}, \qquad \text{and} \qquad R^* = R(\theta^*) = \left|\frac{b}{a}\right|.$$

These results are easily deduced by considering the graphs of $|\theta|b$ and $|1 - \theta a|$ as functions of θ. It is also clear from such graphs that underestimates of $|\theta^*|$ produce a smaller R than do overestimates (see Figure 1 for the case $1 > a > b > 0$).

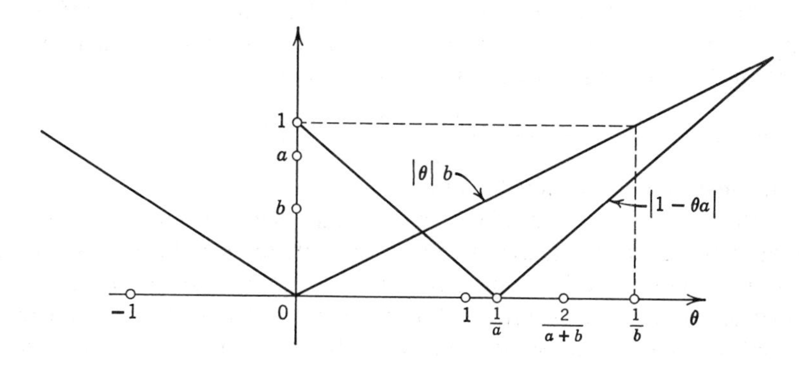

Figure 1

Employing (37) in (36) we find that: *The scheme* (33) *converges if* (34) *is satisfied and the acceleration parameters* θ_i *lie in the intervals*

(38a) $$0 < \theta_i < \frac{2}{1 - g_{ii}(\boldsymbol{\xi}) + r_i(\boldsymbol{\xi})}, \qquad \text{if } 1 - g_{ii}(\boldsymbol{\xi}) > r_i(\boldsymbol{\xi});$$

(38b) $$\frac{2}{1 - g_{ii}(\boldsymbol{\xi}) - r_i(\boldsymbol{\xi})} < \theta_i < 0, \qquad \text{if } 1 - g_{ii}(\boldsymbol{\xi}) < -r_i(\boldsymbol{\xi});$$

where $r_i(\boldsymbol{\xi}) \equiv \sum_{j \neq i} |g_{ij}(\boldsymbol{\xi})|$. [It should be noted that by the assumed continuity of the $g_{ij}(\mathbf{x})$ and (34), if $r_i(\mathbf{x}) \neq 0$ then only one of the ranges (38) can apply for each $i = 1, 2, \ldots, n$.]

From a graph of any $R_i(\theta_i)$, it is clear that the value of θ_i should not lie near the end points of the intervals in (38). In fact, from (37c) it is suggested that $\theta_i = 1/[1 - g_{ii}(\boldsymbol{\xi})]$ is the "best" value for θ_i. However, since this value depends upon $\boldsymbol{\xi}$, a safe choice, θ_i^*, which can be rigorously justified is that for which (38) holds

(39) $$|\theta_i^*| = \min_{\boldsymbol{\xi}} \frac{1}{|1 - g_{ii}(\boldsymbol{\xi})|}, \qquad |\boldsymbol{\xi} - \boldsymbol{\alpha}| \leq \rho.$$

These considerations suggest a modification of the scheme (33) in which an approximation to the best θ_i is used at each step of the iteration. In fact, if $\mathbf{x}^{(\nu)}$ is close to a solution then the values

(40) $$\theta_i^{(\nu)} = \frac{1}{1 - g_{ii}(\mathbf{x}^{(\nu)})}, \qquad i = 1, 2, \ldots, n,$$

can be used in (33), to replace the constants θ_i, and this practical scheme is of the form:

(41) $$\mathbf{x}^{(\nu+1)} = \Theta^{(\nu)}\mathbf{g}(\mathbf{x}^{(\nu)}) + (I - \Theta^{(\nu)})\mathbf{x}^{(\nu)}, \qquad \nu = 0, 1, \ldots.$$

In carrying out this procedure, we need only evaluate the n partial derivatives, $g_{ii} = \partial g_i/\partial x_i$, at each step to predict the appropriate $\theta_i^{(\nu)}$. If these derivatives are not easily obtained or evaluated, one may frequently use difference approximations which just require one extra evaluation of each of the functions $g_i(\mathbf{x})$, i.e., since $g_i(\mathbf{x}^{(\nu)})$ is known we use

$$g_{ii}(\mathbf{x}^{(\nu)}) \approx \frac{g_i(x_1^{(\nu)}, \ldots, x_{i-1}^{(\nu)}, x_i^{(\nu)} + h, x_{i+1}^{(\nu)}, \ldots, x_n^{(\nu)}) - g_i(\mathbf{x}^{(\nu)})}{h}$$

with some suitably small value of h.

Although the conditions (34) seem severe and perhaps unusual they are frequently satisfied in practice. In fact, many difference methods for solving non-linear boundary value problems in ordinary and partial differential equations result in such systems. For many of these systems, convergence can even be obtained for some choice $\theta_1 = \theta_2 = \cdots = \theta_n = \theta$ (for example, see a related scheme in Chapter 8, Subsection 7.2).

PROBLEMS, SECTION 3

1. State and prove a generalization of Theorem 1.3 for systems of equations.

2. State and prove a version of Theorem 2 which employs a different norm (say $\| \|_2$ or $\| \|_1$).

3. For $n = 1$, use the hypothesis of Theorem 3, with $h \equiv abc \leq \frac{1}{2}$ and show directly that there exists a root α with $|\alpha - x_0| \leq 2b$.

[Hint: Use Taylor's theorem,

$$f(x) = f(x_0) + (x - x_0)f'(x_0) + \frac{(x - x_0)^2}{2} f''(\xi)$$

to show that

$$\frac{f(x_0 - 2b)}{f'(x_0)} \leq 0 \leq \frac{f(x_0 + 2b)}{f'(x_0)}.]$$

4. Under the same assumptions as in Problem 3, show that the root α with $|\alpha - x_0| \leq 2b$ is unique.

[Hint: If not unique, there exists η with $|\eta - x_0| < 2b$ and $f'(\eta) = 0$. But from Taylor's formula

$$f'(\eta) = f'(x_0)\left[1 + \frac{(\eta - x_0)f''(\zeta)}{f'(x_0)}\right],$$

show that $f'(\eta) \neq 0$.]

5. Under assumptions of Problem 3, with $h < \frac{1}{2}$, show that $f'(\alpha) \neq 0$ (use hint of Problem 4).

6. Under the assumptions of Problem 3, if $f'(\alpha) = 0$, show that $|\alpha - x_0| = 2b$ (by hint of Problem 4). Furthermore, if $f'(\alpha) = 0$, show that $f''(\alpha) \neq 0$.

[Hint: $f''(\alpha) = \lim_{\eta \to \alpha} f'(\eta)/(\eta - \alpha)$, but by hint of Problem 4 and $\eta < \alpha$,

$$|f'(\eta)| > |f'(x_0)|\left(1 - \frac{|\eta - x_0|}{2b}\right) = |f'(x_0)| \frac{|\eta - \alpha|}{2b}.]$$

4. SPECIAL METHODS FOR POLYNOMIALS

All of the previous schemes for single equations can be employed to compute the roots of polynomials. Complex roots can be obtained by simply using complex arithmetic and complex initial estimates. Or, by reducing the evaluation of a polynomial at a complex point to its real and imaginary parts, the iterative methods for real systems (of order two in this case) could also be used to obtain the complex roots of polynomials. However, it is possible to devise special iterative methods which are frequently more advantageous than the general methods. We shall consider some of these polynomial methods in this section.

It is of interest to note, first, that a very simple a posteriori test of the accuracy of an approximate root of a polynomial is frequently quite effective. Let the nth degree polynomial $P_n(x)$ be

$$P_n(x) \equiv a_0x^n + a_1x^{n-1} + \cdots + a_{n-1}x + a_n \tag{1}$$

with roots $r_1, r_2, \ldots, r_n$. Then since

$$P_n(x) = a_0(x - r_1)\cdots(x - r_n) \tag{2}$$

we have the well-known result that

$$\frac{a_n}{a_0} = (-1)^n r_1 r_2 \cdots r_n. \tag{3}$$

Now let σ be an approximate root of $P_n(x)$ which satisfies the test

$$|P_n(\sigma)| \leq \epsilon. \tag{4}$$

Then from (2) and (3) it follows that (we assume $a_n \neq 0$)

$$\left|\frac{P_n(\sigma)}{a_n}\right| = \left|1 - \frac{\sigma}{r_1}\right| \cdot \left|1 - \frac{\sigma}{r_2}\right| \cdots \left|1 - \frac{\sigma}{r_n}\right| \leq \frac{\epsilon}{|a_n|}.$$

Taking the nth root we now conclude, since

$$\left|\frac{P_n(\sigma)}{a_n}\right| \geq \min_j \left|1 - \frac{\sigma}{r_j}\right|^n,$$

that

$$\min_j \left|1 - \frac{\sigma}{r_j}\right| \leq \left(\frac{\epsilon}{|a_n|}\right)^{1/n}. \tag{5}$$

Thus we obtain an *exact bound on the relative error* of σ as an approximation to some root of $P_n(x)$. In many of the methods to be studied, $P_n(\sigma)$ is already computed and so no extra calculations are required to employ this test. Note that the roots and approximations in this test may be real or complex.

4.1. Evaluation of Polynomials and Their Derivatives

An interesting special feature of polynomials is the ease with which they may be evaluated. Let us write (1) as

$$\begin{aligned} P_n(x) &= a_0x^n + a_1x^{n-1} + \cdots + a_{n-1}x + a_n \\ &= \{\cdots[(a_0x + a_1)x + a_2]x + \cdots\}x + a_n. \end{aligned} \tag{6}$$

The usual way to evaluate $P_n(x)$ is by means of this "*nesting*" procedure. More explicitly, to calculate $P_n(\xi)$ we form:

$$\begin{aligned} b_0 &= a_0, \\ b_1 &= b_0\xi + a_1, \\ b_2 &= b_1\xi + a_2, \\ &\vdots \\ b_\nu &= b_{\nu-1}\xi + a_\nu, \qquad \nu = 1, 2, \ldots, n; \end{aligned} \tag{7}$$

and note that $b_n = P_n(\xi)$. These operations are just those employed in the elementary process of *synthetic division*. In fact, if we write

$$P_n(x) = (x - \xi)Q_{n-1}(x) + R_0,$$

then clearly, $R_0 = P_n(\xi) = b_n$ and it is easily verified (by multiplying out and equating coefficients of like powers of x) that, using the quantities in (7):

$$Q_{n-1}(x) = b_0 x^{n-1} + b_1 x^{n-2} + \cdots + b_{n-1}. \tag{8}$$

Dividing again by $(x - \xi)$ we get, say,

$$Q_{n-1}(x) = (x - \xi)Q_{n-2}(x) + R_1$$

and hence

$$P_n(x) = (x - \xi)^2 Q_{n-2}(x) + (x - \xi)R_1 + R_0.$$

Differentiating the last expression, we find that $R_1 = P_n'(\xi)$. Thus by performing synthetic division of $Q_{n-1}(x)$ by $x - \xi$, we could determine $Q_{n-2}(x)$ and $P_n'(\xi)$. Clearly this procedure can be continued to yield finally

$$P_n(x) = R_n(x - \xi)^n + \cdots + R_1(x - \xi) + R_0. \tag{9}$$

The successive calculations to determine the R_ν and coefficients of the intermediate polynomials $Q_\nu(x)$ can be indicated by the array in Table 1.

Table 1

	0	0			0
a_0	b_0	c_0	$\cdots$		R_n
a_1	b_1	c_1		R_{n-1}	
$\vdots$	$\vdots$	$\vdots$	$\cdot$		
a_{n-2}	b_{n-2}	c_{n-2}	$\cdot$		
a_{n-1}	b_{n-1}	R_1			
a_n	R_0				

Any entry of Table 1 not in the first row or column (which are given initially), is computed by multiplying the entry above by ξ and adding the entry to the left. It follows from (9) by differentiation that

$$R_\nu = \frac{1}{\nu!} \frac{d^\nu P_n(x)}{dx^\nu}\bigg|_{x=\xi}; \qquad \nu = 0, 1, \ldots, n.$$

From (9) it also follows easily that *the polynomial with coefficients* R_ν, *say*

$$R(y) \equiv R_n y^n + R_{n-1} y^{n-1} + \cdots + R_1 y + R_0,$$

has as roots all those of $P(x)$ *reduced by the amount* ξ.

Finally we note that the evaluation of the entire Table 1 requires $\frac{1}{2}(n^2 + n)$ multiplications and as many additions since the evaluation of the first column requires only n of each operation. But in computing the entries of the table, significant figures may be lost in any of the additions; see eqs. (7). Often this necessitates the use of multiple precision arithmetic.

To employ Newton's method on polynomial equations, only the first two columns in Table 1 need be computed. This is easily accomplished by means of the recursions (7) and a similar set with a_ν, b_ν replaced by b_ν, c_ν for $\nu = 0, 1, \ldots, n - 1$. An interesting application of Newton's method to polynomials is found in Problem 1.

4.2. Sturm Sequences

It would be very desirable to obtain successive upper and lower bounds on the *real roots* of a polynomial equation or indeed of any equation, as then the error in approximating the root is easily estimated. This could be done if the number of real roots of the equation in any interval could be determined, and this can in fact be done by means of the so-called Sturm sequences which we proceed to define. Let the equation to be solved be $f_0(x) = 0$ where $f_0(x)$ is differentiable in $[a, b]$. Then the continuous functions

$$f_0(x), f_1(x), \ldots, f_m(x)$$

form a Sturm sequence on $[a, b]$ if they satisfy there:

(i) $f_0(x)$ has at most simple roots in $[a, b]$;
(ii) $f_m(x)$ does not vanish in $[a, b]$;
(iii) if $f_\nu(\alpha) = 0$, then $f_{\nu-1}(\alpha)f_{\nu+1}(\alpha) < 0$ for any root $\alpha \in [a, b]$;
(iv) if $f_0(\alpha) = 0$, then $f_0'(\alpha)f_1(\alpha) > 0$ for any root $\alpha \in [a, b]$.

For every such sequence there follows Sturm's

THEOREM 1. *The number of zeros of* $f_0(x)$ *in* (a, b) *is equal to the difference between the number of sign variations in* $\{f_0(a), f_1(a), \ldots, f_m(a)\}$ *and in* $\{f_0(b), f_1(b), \ldots, f_m(b)\}$ *provided that* $f_0(a)f_0(b) \neq 0$ *and* $\{f_0(x), f_1(x), \ldots, f_m(x)\}$ *form a Sturm sequence on* $[a, b]$.

Proof. The number of sign variations can change as x goes from a to b only by means of some function changing sign in the interval. By (ii) it cannot be $f_m(x)$. Assume that at some $\hat{x} \in (a, b)$, $f_\nu(\hat{x}) = 0$ for some ν in $0 < \nu < m$. In a neighborhood of $\hat{x}$, the signs must be either

x	$f_{\nu-1}(x)$	$f_\nu(x)$	$f_{\nu+1}(x)$
$\hat{x}-\epsilon$	$+$	$\pm$	$-$
$\hat{x}$	$+$	0	$-$
$\hat{x}+\epsilon$	$+$	$\pm$	$-$

or

x	$f_{\nu-1}(x)$	$f_\nu(x)$	$f_{\nu+1}(x)$
$\hat{x}-\epsilon$	$-$	$\pm$	$+$
$\hat{x}$	$-$	0	$+$
$\hat{x}+\epsilon$	$-$	$\pm$	$+$

The signs in the row $x = \hat{x}$ follow from (iii). The signs in the first and last columns then follow by continuity for sufficiently small ϵ. The sign possibilities in the middle columns are general. But we see from these tables that as x passes through $\hat{x}$, there are no changes in the number of sign variations in the Sturm sequence. We now examine the signs near a zero $x = \hat{x}$ of $f_0(x)$:

x	$f_0(x)$	$f_1(x)$
$\hat{x}-\epsilon$	$+$	$-$
$\hat{x}$	0	$-$
$\hat{x}+\epsilon$	$-$	$-$

or

x	$f_0(x)$	$f_1(x)$
$\hat{x}-\epsilon$	$-$	$+$
$\hat{x}$	0	$+$
$\hat{x}+\epsilon$	$+$	$+$

The $f_0(x)$ columns represent the two possible cases for a simple zero. The sign of $f_1(\hat{x})$ then follows from (iv) and the continuity implies the other signs in the last columns. Clearly, there is now a decrease of one change in the number of sign variations as x increases through a zero of $f_0(x)$. When these results are combined the theorem follows. ∎

It is easy to construct a Sturm sequence when $f_0(x) \equiv P_n(x)$ is a polynomial, say of degree n. We define

$$f_1(x) \equiv f_0'(x)$$

so that (iv) is satisfied at simple roots. Divide $f_0(x)$ by $f_1(x)$ and call the remainder $-f_2(x)$. Then divide $f_1(x)$ by $f_2(x)$ and call the remainder $-f_3(x)$. Continue this procedure until it terminates [which it must since each polynomial $f_\nu(x)$ is of lower degree than its predecessor, $f_{\nu-1}(x)$]. We thus have the array:

$$
\begin{aligned}
f_0(x) &= q_1(x)f_1(x) - f_2(x),\\
f_1(x) &= q_2(x)f_2(x) - f_3(x),\\
&\;\vdots\\
f_{m-2}(x) &= q_{m-1}(x)f_{m-1}(x) - f_m(x),\\
f_{m-1}(x) &= q_m(x)f_m(x).
\end{aligned}
\tag{10}
$$

This procedure is well known as the Euclidean algorithm for determining the highest common factor, $f_m(x)$, of $f_0(x)$ and $f_1(x)$. [It is easily seen that

$f_m(x)$ is a factor of all the $f_\nu(x)$, $\nu = 0, 1, \ldots, m-1$; conversely any common factor of $f_0(x)$ and $f_1(x)$ must be a factor of $f_\nu(x)$, $\nu = 2, 3, \ldots, m$.] Thus all multiple roots of $f_0(x)$ are also roots of $f_m(x)$ with multiplicity reduced by one. If $f_m(x)$ is not a constant (i.e., $f_0(x)$ has multiple roots) then we may divide all the $f_\nu(x)$ by $f_m(x)$ and (denoting these quotients by their numerators) obtain the sequence (10) in which $f_0(x)$ has only simple roots and $f_m(x)$ is a constant. It now follows that this reduced sequence $\{f_0(x), f_1(x), \ldots, f_m(x)\}$ is a Sturm sequence. Only (iii) requires proof: If $f_\nu(\hat{x}) = 0$ then, by (10), at this point $f_{\nu-1}(\hat{x}) = -f_{\nu+1}(\hat{x})$; but if $f_{\nu-1}(\hat{x}) = 0$, then also $f_0(\hat{x}) = f_1(\hat{x}) = 0$, a contradiction. Simple formulas for computing the coefficients of the $f_\nu(x)$ can be obtained from (10); we present this as Problem 2.

The usual way to employ a Sturm sequence is with successive bisections of initially chosen intervals. In this way, with each new evaluation of the sequence, the error in determining a real root is at least halved. Thus this procedure converges with an asymptotic convergence factor of at least $\frac{1}{2}$. The value of Sturm sequences is clearly not in rapid convergence properties but rather in the ability to obtain good estimates of all real roots. When the desired root or roots have been located it is more efficient to employ a more rapidly converging iteration, say false position. Or, in fact, when a root is known to lie in the given interval, (a, b), because $f(a)\cdot f(b) < 0$, we may simply calculate $f((a + b)/2)$. Now, $(a + b)/2$ may be a root. If $(a + b)/2$ is not a root, then we would know from the sign of $f((a + b)/2)$ in which of the two sub-intervals, $(a, (a + b)/2)$ or $((a + b)/2, b)$, $f(x)$ does have a root. In this way, we may continue to bisect successive sub-intervals in which we know $f(x)$ to have a root. This procedure is known as the *bisection method*. It has the convergence factor $\frac{1}{2}$. Each step of the bisection method requires fewer calculations than does the evaluation of the Sturm sequence. At each step of the bisection method we have upper and lower bounds for a real root.

4.3. Bernoulli's Method

Consider the polynomial equation

$$P(x) \equiv x^n + a_1 x^{n-1} + \cdots + a_{n-1}x + a_n = 0 \tag{11}$$

and assume that its roots, r_j, are distinct and ordered by

$$|r_1| > |r_2| > \cdots > |r_n|. \tag{12}$$

This equation and its roots bear an important relationship to the difference equation or recursion (see Chapter 8, Section 4):

$$g_\nu = -(a_1 g_{\nu-1} + \cdots + a_n g_{\nu-n}), \qquad \nu = n, n+1, \ldots \tag{13a}$$

$$g_0 = c_0, \ldots, g_{n-1} = c_{n-1}. \tag{13b}$$

Given the n starting values, c_j, the g_ν are easily evaluated (with n multiplications and $n - 1$ additions per step). By seeking a formal solution of (13a) of the form $g_\nu = r^\nu$, we find that any root of (11) yields a solution. Furthermore, since the difference equations are linear it easily follows that any linear combination of the powers of the roots of (11) is also a formal solution of (13a): thus we may write:

$$g_\nu = b_1 r_1^\nu + b_2 r_2^\nu + \cdots + b_n r_n^\nu; \qquad \nu = 0, 1, 2, \ldots. \tag{14}$$

The polynomial (11) is called the *characteristic equation* of the *difference equation* (13a). The conditions (13b) yield a linear system of n equations for the determination of the b_j. The determinant of the coefficient matrix of this system is a Vandermonde determinant which, by (12), does not vanish [see (2.4) in Chapter 5]. It can also be shown that (14) is the most general solution of (13a), and hence with the b_j as determined above, (14) is the unique solution of (13a, b) (see Problem 3).

For $\nu \gg 1$ we have, recalling (12), $g_\nu \approx b_1 r_1^\nu$, and hence we are led to consider the sequence

$$\begin{aligned} \sigma_\nu \equiv \frac{g_{\nu+1}}{g_\nu} &= r_1 \frac{b_1 + b_2\left(\dfrac{r_2}{r_1}\right)^{\nu+1} + \cdots + b_n\left(\dfrac{r_n}{r_1}\right)^{\nu+1}}{b_1 + b_2\left(\dfrac{r_2}{r_1}\right)^{\nu} + \cdots + b_n\left(\dfrac{r_n}{r_1}\right)^{\nu}} \\ &= r_1 + \mathcal{O}\left(\left|\frac{r_2}{r_1}\right|^\nu\right). \end{aligned} \tag{15}$$

Here we have assumed that $b_1 \neq 0$ and, in this case, clearly $\lim_{\nu\to\infty} \sigma_\nu = r_1$. The rapidity of the approach is determined by the ratio $|r_2/r_1|$. If this ratio is not near unity, r_1 is easily obtained and can be eliminated from the original polynomial by synthetic division as in Subsection 4.1 and then a new recursion is evaluated to determine r_2, etc. (In practice, elimination of the roots in decreasing order of magnitude may produce considerably larger errors than elimination in increasing order of magnitude. See Problem 6.)

If $b_1 = 0$ for an unfortunate choice of starting values (13b), it would seem that the ratios then converge to r_2. This is theoretically correct but for actual computations the roundoff introduced in the successive evaluations of (13a) has the effect of altering the exact b_j which should occur in (14). Thus after a few recursions there will be some perhaps small but nonzero component b_1 present and in subsequent steps the dominant root r_1 may still be determined. In like manner, if some error or blunder is committed in the course of these Bernoulli iterations the subsequent steps, performed correctly, will obliterate the error and yield the correct result.

The expansion in (15) shows that the convergence of σ_ν to r_1 is geometric with ratio $|r_2/r_1|$. To test the convergence of the sequence $\{\sigma_\nu\}$, a number of devices can be employed. Perhaps the most frequently used procedure in practice is to compute the differences $|\sigma_{\nu+1} - \sigma_\nu|$ and to stop when these are less than some predetermined small quantity. Another possibility which yields more precise information at the cost of some extra computation is to use the test (4). Of course, if sufficiently many steps have been performed, $\sigma = \sigma_\nu$ is closest to r_1, the dominant root, and so $j = 1$ in (5). Test (4) should not be made after every iteration but say every several steps to reduce the computational effort.

The conditions (12) are most likely not satisfied for a polynomial since in general some conjugate complex roots are to be expected. Suppose then that the dominant real roots have been eliminated and (11) is the reduced equation with r_1 and r_2 conjugate complex roots, $r_2 = \bar{r}_1$, satisfying

$$(16) \qquad |r_1| = |r_2| > |r_3| > \cdots > |r_n|.$$

The solution of (13) is again of the form (14) where the b_j may be complex and while (15) is valid the σ_ν do not converge to r_1 since $|r_2/r_1| = 1$. For large ν we now expect that

$$g_\nu \approx b_1 r_1^{\ \nu} + b_2 r_2^{\ \nu}.$$

Here we note that $r_2 = \bar{r}_1$ and $b_2 = \bar{b}_1$ since the g_ν must be real (assuming that the a_j and c_j are real). A simple calculation now reveals that

$$A_\nu \equiv g_{\nu+1}g_{\nu-1} - g_\nu^{\ 2} \approx |b_1|^2(r_1 - r_2)^2(r_1r_2)^{\nu-1},$$

$$B_\nu \equiv g_{\nu+2}g_{\nu-1} - g_{\nu+1}g_\nu \approx |b_1|^2(r_1 - r_2)^2(r_1r_2)^{\nu-1}(r_1 + r_2).$$

Thus we expect that with $s_\nu \equiv B_\nu/A_\nu$ and $p_\nu \equiv A_{\nu+1}/A_\nu$:

$$\lim_{\nu\to\infty} s_\nu = r_1 + r_2 \equiv s, \qquad \lim_{\nu\to\infty} p_\nu = r_1r_2 \equiv p,$$

and the roots r_1 and r_2 are those of the quadratic equation

$$\xi^2 - s\xi + p = 0.$$

Recalling (16) we can now estimate the error in this procedure for complex roots. Let $|r_1| = |r_2| \equiv r$, $|r_3|/r \equiv \delta$ and we find as above:

$$g_\nu = b_1r_1^{\ \nu} + b_2r_2^{\ \nu} + \mathcal{O}(r^\nu\delta^\nu),$$

$$A_\nu = |b_1|^2(r_1 - r_2)^2(r_1r_2)^{\nu-1} + \mathcal{O}(r^{2\nu}\delta^\nu),$$

$$B_\nu = |b_1|^2(r_1 - r_2)^2(r_1r_2)^{\nu-1}(r_1 + r_2) + \mathcal{O}(r^{2\nu}\delta^\nu).$$

Thus by the usual expansions

$$(17) \qquad s_\nu = s + \mathcal{O}(\delta^\nu), \qquad p_\nu = p + \mathcal{O}(\delta^\nu).$$

The equation actually solved is the quadratic

$$\xi^2 - s_\nu \xi + p_\nu = 0. \tag{18}$$

It is easily shown, since $s^2 - 4p \neq 0$, that the roots of this equation differ from those for $\nu = \infty$ by terms $\mathcal{O}(\delta^\nu)$.

If the dominant roots are equal but of opposite sign, then the above procedure for complex roots is still applicable where now $r_2 = -r_1$. If the dominant root is real but multiple, then the original procedure is applicable but converges more slowly (see Problem 5).

4.4. Bairstow's Method

A much better scheme (than Bernoulli's) for determining quadratic factors of a polynomial, $P_n(x)$, is based on a generalization of synthetic division and Newton's method for systems of equations. This procedure also avoids the use of complex arithmetic. In brief, we seek real numbers, say s and p, such that the quadratic

$$x^2 + sx + p \tag{19}$$

is an exact factor of $P_n(x)$. The division of $P_n(x)$ by this factor may be indicated as:

$$P_n(x) = (x^2 + sx + p)Q_{n-2}(x) + [xR_1(s, p) + R_0(s, p)]. \tag{20}$$

Here, R_1 and R_2 are the coefficients in the remainder which is at most linear in x. As indicated, these coefficients are functions of s and p, the parameters in the quadratic (19). In order that the remainder be zero, s and p must satisfy

$$R_1(s, p) = 0, \qquad R_0(s, p) = 0. \tag{21}$$

This is a system of two equations in two unknowns which we solve by Newton's method. For this purpose we must evaluate the four derivatives

$$\begin{matrix} \dfrac{\partial R_1}{\partial s}, & \dfrac{\partial R_0}{\partial s} \\[2ex] \dfrac{\partial R_1}{\partial p}, & \dfrac{\partial R_0}{\partial p}. \end{matrix} \tag{22}$$

We determine the quantities in (22) indirectly. Let $Q_{n-2}(x)$ in (20) be divided by the quadratic (19) to yield

$$Q_{n-2}(x) = (x^2 + sx + p)Q_{n-4}(x) + [xR_3(s, p) + R_2(s, p)]. \tag{23}$$

We note that the specific values of the $R_j(s, p)$ in (20) and (23) for any fixed (s, p) are easily obtained by carrying out the two indicated divisions.

Using (23) in (20) we have the identity:

$$\begin{aligned}(24)\qquad P_n(x) &= (x^2 + sx + p)^2 Q_{n-4}(x) \\ &\quad + (x^2 + sx + p)[xR_3(s, p) + R_2(s, p)] \\ &\quad + [xR_1(s, p) + R_0(s, p)].\end{aligned}$$

Since $P_n(x)$ is independent of s and p, we can differentiate (24) with respect to s and p and evaluate the result at a root $x = z$ of $z^2 + sz + p = 0$ to get:

$$z(zR_3 + R_2) + \left(z\frac{\partial R_1}{\partial s} + \frac{\partial R_0}{\partial s}\right) = 0,$$

$$(zR_3 + R_2) + \left(z\frac{\partial R_1}{\partial p} + \frac{\partial R_0}{\partial p}\right) = 0.$$

Since $z^2 = -(sz + p)$ these equations can be written as

$$z\left(\frac{\partial R_1}{\partial s} + R_2 - sR_3\right) + \left(\frac{\partial R_0}{\partial s} - pR_3\right) = 0,$$

$$z\left(\frac{\partial R_1}{\partial p} + R_3\right) + \left(\frac{\partial R_0}{\partial p} + R_2\right) = 0.$$

If the quadratic is not a perfect square (in which case the roots would be real and equal), the above must hold for two distinct roots, z, and hence each coefficient in parentheses must vanish. Thus we deduce that:

$$(25)\qquad \begin{aligned}\frac{\partial R_1}{\partial s} &= sR_3 - R_2, \qquad & \frac{\partial R_0}{\partial s} &= pR_3; \\ \frac{\partial R_1}{\partial p} &= -R_3, \qquad & \frac{\partial R_0}{\partial p} &= -R_2.\end{aligned}$$

The iteration scheme proceeds from an initial estimate (s_0, p_0) which is such that $p_0 \neq s_0^2/4$ and defines recursively the sequence (s_ν, p_ν) by Newton's method of solving (21); i.e.,

$$(26)\qquad \begin{aligned}(s_{\nu+1} - s_\nu)\frac{\partial R_1(s_\nu, p_\nu)}{\partial s} + (p_{\nu+1} - p_\nu)\frac{\partial R_1(s_\nu, p_\nu)}{\partial p} &= -R_1(s_\nu, p_\nu), \\ (s_{\nu+1} - s_\nu)\frac{\partial R_0(s_\nu, p_\nu)}{\partial s} + (p_{\nu+1} - p_\nu)\frac{\partial R_0(s_\nu, p_\nu)}{\partial p} &= -R_0(s_\nu, p_\nu).\end{aligned}$$

The coefficients and inhomogeneous terms in (26) are obtained by carrying out the two divisions indicated in (20) and (23) with the quadratic factor $x^2 + s_\nu x + p_\nu$ and then using (25). The divisions can be reduced to the evaluation of simple recursions by equating the coefficients on both sides of the equality signs in (20) and (23); we leave the derivation of this generalized synthetic division as an exercise. During the course of the iterations it should be checked that $p_\nu \neq s_\nu^2/4$, so that (25) is valid. To test the convergence we may employ (4) and (5), after noting that if z is a root of $x^2 + sx + p = 0$ then, by (20),

$$P_n(z) = zR_1(s, p) + R_0(s, p).$$

Usually the coefficients (s_ν, p_ν) converge quickly, and so a direct test on the roots need only be applied when the iterations are about to be terminated.

PROBLEMS, SECTION 4

1. The square roots of a positive number β are the zeros of the polynomial $f(x) \equiv x^2 - \beta$.

(a) Apply Newton's method to obtain the sequence of approximations

$$x_{\nu+1} = \frac{1}{2}\left(x_\nu + \frac{\beta}{x_\nu}\right).$$

This procedure is known as the Newton-Raphson method for computing square roots; it is frequently employed on high-speed digital computers.

(b) If $x_0 > \sqrt{\beta}$ show that: $x_{\nu+1} < x_\nu$, $\nu = 0, 1, \ldots$ (assuming exact calculations).

(c) Derive and study the analogous procedure for the nth root of any positive number where n is an integer.

2. Derive a recursion formula for finding the coefficients of the Sturm sequence (10). Assume $f_0(x) = \sum_{j=0}^{n} a_{j,0}x^{n-j}$.

3. Show that every solution of (13a, b) is unique and hence can be represented in the form (14).

4. If the coefficients a_j in the polynomial $P_n(x) \equiv x^n + a_1x^{n-1} + \cdots + a_n$ are altered by an amount at most ϵ, then the polynomial $P_{n,\epsilon}(x) \equiv P_n(x) + \epsilon g_{n-1}(x)$ is obtained where $g_{n-1}(x) \equiv b_1x^{n-1} + b_2x^{n-2} + \cdots + b_{n-1}$ and $|b_i| \leq 1$.

Show that to each simple real root r_i of $P_n(x)$, corresponds a simple real root $r_{i,\epsilon}$ of $P_{n,\epsilon}(x)$ such that $r_{i,\epsilon} - r_i = \mathcal{O}(\epsilon)$ as $\epsilon \to 0$.

[Hint: Plot $P_{n,\epsilon}(x)$ in a neighborhood of r_i for sufficiently small ϵ.]

5. Show how to modify Bernoulli's method for the case of a dominant real multiple root. Estimate the convergence rate.

6. If $P_n(x) \equiv x^n + a_1x^{n-1} + \cdots + a_{n-1}x + a_n$ with $a_n \neq 0$, then let $Q_n(z) \equiv a_nz^n + a_{n-1}z^{n-1} + \cdots + a_1z + 1$. Show that the roots $\{z_k\}$ of $Q_n(z) = 0$ and roots $\{x_k\}$ of $P_n(x) = 0$ are related by $z_k = 1/x_k$; $k = 1, 2, \ldots, n$. Hence, show how the Bernoulli method may be used to find the zeros of $P_n(x)$ in ascending order of magnitude.

4

Computation of Eigenvalues and Eigenvectors

0. INTRODUCTION

The *eigenvalue-eigenvector problem* for a square matrix $A \equiv (a_{ij})$ of order n is that of determining a scalar λ and vector $\mathbf{x}$ such that

(1) $$A\mathbf{x} = \lambda\mathbf{x}, \qquad \mathbf{x} \neq \mathbf{o}.$$

The problem is clearly non-linear since both λ and $\mathbf{x}$ are unknowns. In fact, as is well known, the *eigenvalues*, λ, are the n roots of the *characteristic equation*

(2) $$\det(\lambda I - A) \equiv p_A(\lambda) = 0.$$

Thus the eigenvalues can, in principle, be found as the zeros of $p_A(\lambda)$ without recourse to any of the eigenvectors. Given some eigenvalue, λ, then the corresponding *eigenvector* is a non-trivial solution of the homogeneous linear system (1). On the other hand, if some particular eigenvector $\mathbf{x}$ is known, then the eigenvalue to which $\mathbf{x}$ belongs can be obtained by taking the inner product of (1) with $\mathbf{x}$ to get

(3) $$\lambda = \frac{\mathbf{x}^* A\mathbf{x}}{\mathbf{x}^*\mathbf{x}}.$$

Alternatively, we could use any non-zero component x_i to get, from the ith row of (1),

$$\lambda = \frac{1}{x_i} \sum_{j=1}^{n} a_{ij}x_j.$$

If some of the n eigenvalues of A are not distinct [i.e., if $p_A(\lambda)$ has multiple roots] then there *may* be fewer than n eigenvectors. For example,

$$A \equiv \begin{pmatrix} 1 & 1 \\ 0 & 1 \end{pmatrix}$$

has the repeated eigenvalue $\lambda = 1$, and only one eigenvector,

$$\mathbf{x} = \begin{pmatrix} 1 \\ 0 \end{pmatrix}.$$

We consider, in Section 1, the well-posedness of the eigenvalue problem and a posteriori error bounds.

In Section 2, we study the power method and its ramifications. These yield a sequence of scalars and vectors that converge (when the procedure works) to some particular eigenvalue and its corresponding eigenvector. In order to compute other eigenpairs, the iteration scheme must be modified. These simple methods yield successively only a few of the eigenvalues and vectors with acceptable accuracy, but for many applications this will be all that is needed.

The methods studied in Section 3 use a finite or infinite sequence of matrix transformations to find a similar matrix $B = P^{-1}AP$, such that the evaluation of det $(\lambda I - B)$ is simpler than the calculation of det $(\lambda I - A)$. We then find the eigenvalues as the zeros of $p_B(\lambda)$, by means of an iterative scheme, which does not explicitly use the coefficients of p_B. In the methods based on the use of an infinite sequence of matrix transformations, the eigenvalues themselves are usually explicitly exhibited in B. A special calculation is then required to obtain the eigenvectors. In summary, these methods may yield all of the eigenvalues without determining any eigenvectors.

[We emphasize the practical importance of not finding explicitly the coefficients of $p_B(\lambda)$ or $p_A(\lambda)$ in order to evaluate the polynomials. All experienced practitioners are aware of the large errors that may result from the use of the approximate coefficients of $p_B(\lambda)$ or $p_A(\lambda)$ for the calculation of the zeros of the characteristic polynomials. We do not study the methods based on finding the coefficients of $p_A(\lambda)$.]

1. WELL-POSEDNESS, ERROR ESTIMATES

A simple criterion for *localizing eigenvalues* is given in

THEOREM 1 (GERSCHGORIN). *Let $A \equiv (a_{ij})$ have eigenvalues λ and define the absolute row and column sums by*

$$r_i \equiv \sum_{\substack{j=1 \\ j \neq i}}^{n} |a_{ij}|, \qquad c_j \equiv \sum_{\substack{i=1 \\ i \neq j}}^{n} |a_{ij}|.$$

Then,

(a) *each eigenvalue lies in the union of the row circles* R_i, $i = 1, 2, \ldots, n$, *where*

$$R_i \equiv \{z \mid |z - a_{ii}| \leq r_i\};$$

(b) *each eigenvalue lies in the union of the column circles* C_j, $j = 1, 2, \ldots, n$, *where*

$$C_j \equiv \{z \mid |z - a_{jj}| \leq c_j\};$$

(c) *each component (maximal connected union of circles) of* $\bigcup_i R_i$ *or* $\bigcup_j C_j$ *contains exactly as many eigenvalues as circles. (The eigenvalues and the circles are counted with their multiplicities.)*

Proof. (a) If λ is an eigenvalue of A, then there exists an eigenvector $\mathbf{x} \neq \mathbf{o}$ such that

$$A\mathbf{x} = \lambda\mathbf{x},$$

or

$$(\lambda - a_{ii})x_i = \sum_{\substack{j=1 \\ j \neq i}}^{n} a_{ij}x_j, \qquad i = 1, 2, \ldots, n.$$

Pick an i such that $|x_i| = \|\mathbf{x}\|_\infty \neq 0$. Then

$$|\lambda - a_{ii}| \leq \sum_{j \neq i} |a_{ij}| \left|\frac{x_j}{x_i}\right| \leq r_i.$$

(b) Since the eigenvalues of A and A^T are identical, (b) follows from (a).

(c) Here we apply a simpler form of the basic lemma of the theory of functions of a complex variable quoted in Chapter 1, Section 3, namely:

LEMMA 1. *The n zeros of a polynomial*

$$p(x) \equiv x^n + a_1x^{n-1} + \cdots + a_{n-1}x + a_n$$

are continuous functions of the coefficients $\{a_j\}$.

Consider the one-parameter family of matrices $A(t) = D + tB$, where $D \equiv (a_{ii}\delta_{ij})$ is a diagonal matrix and $B = A - D$. Consider, corresponding to $A(t)$, the circles $R_i(t)$ and $C_j(t)$ with respective radii tr_i and tc_j about respective centers a_{ii} and a_{jj}. Now, $A(0) = D$ and clearly (c) holds for this diagonal matrix. The eigenvalues $\{\lambda(t)\}$ of $A(t)$ are the zeros of $p_{A(t)}(\lambda) \equiv \det [\lambda I - A(t)]$. As t increases continuously to $t = 1$, the integer number of circles and, by Lemma 1, of eigenvalues in each component of $\bigcup_i R_i(t)$ [or of $\bigcup_j C_j(t)$] varies continuously, and hence is constant, except for a finite number of values, $t = t^l$, at which the number of circles in a component increases. At such values $t = t^l$, (c) holds because of Lemma 1. In other words, when the number of circles in a component increases, the number of eigenvalues in the component increases by the same amount. ■

Gerschgorin's theorem has many applications; the first we make is in the treatment of the well-posedness of the eigenvalue problem. In this connection, it is necessary to introduce the notions of left and right eigenvectors.

The eigenvector $\mathbf{x}$ that satisfies (0.1),

$$A\mathbf{x} = \lambda\mathbf{x}, \qquad \mathbf{x} \neq \mathbf{o},$$

is more properly called a *right eigenvector* of A. Correspondingly a *left eigenvector*, $\mathbf{y}$, of A is a vector that satisfies

$$\mathbf{y}^*A = \mu\mathbf{y}^*, \qquad \mathbf{y} \neq \mathbf{o}. \tag{1}$$

In other words, $\mathbf{y}$ is a left eigenvector of A iff $\mathbf{y}$ is a right eigenvector of A^*, i.e., by *starring* both sides (by taking the complex conjugate transpose of both sides),

$$A^*\mathbf{y} = \bar{\mu}\mathbf{y}.$$

($\mathbf{y}^*$ is also called a *row eigenvector* of A.) The sets of "left" $\{\mu\}$ and "right" $\{\lambda\}$ eigenvalues are identical, since

$$\det(\lambda I - A) = \overline{\det(\bar{\lambda} I - \bar{A})} = \overline{\det(\bar{\lambda} I - A^*)},$$

where the last equality follows from $\det B = \det B^T$.

Now, when the matrix A is Hermitian, the left and right eigenvectors are also identical; otherwise, when A is not Hermitian, the left and right eigenvectors are distinct, in general.

LEMMA 2. *Any left eigenvector,* $\mathbf{y}$, *and any right eigenvector,* $\mathbf{x}$, *of the matrix* A, *corresponding to any pair of distinct eigenvalues, are orthogonal, i.e.,* $\mathbf{y}^*\mathbf{x} = 0$.

Proof. Given as Problem 1. ■

We say that the left and right eigenvectors are *biorthogonal.*

We may now ask, "How well-posed is the eigenvalue problem (0.1) or (1)?" The answer is given in a special case by

THEOREM 2. *Let* A *be of order* n *and have* n *linearly independent eigenvectors. For any fixed matrix* C, *with* $\|C\| = \|A\|$, *define the perturbed matrix*

$$A(\epsilon) \equiv A + \epsilon C. \tag{2}$$

Then, if λ is any eigenvalue of A, there is an eigenvalue $\lambda(\epsilon)$ of $A(\epsilon)$ such that

$$|\lambda(\epsilon) - \lambda| = \mathcal{O}(\epsilon). \tag{3}$$

Moreover, if λ is simple

$$\lim_{|\epsilon|\to 0} \frac{\lambda(\epsilon) - \lambda}{\epsilon} = \frac{\mathbf{y}^*C\mathbf{x}}{\mathbf{y}^*\mathbf{x}}, \tag{4}$$

where $\mathbf{x}$ and $\mathbf{y}$ are, respectively, right and left eigenvectors of A corresponding to λ.

Proof. Let P be the matrix whose columns are the right eigenvectors of A, i.e.,

$$AP = P\Lambda \tag{5}$$

where $\Lambda \equiv (\lambda_{ij})$ is a diagonal matrix, such that $\{\lambda_{ii}\}$ are the eigenvalues of A in some order. Therefore, from (5) and (2),

$$\begin{aligned} P^{-1}A(\epsilon)P &= \Lambda + \epsilon P^{-1}CP \equiv \Lambda + \epsilon E, \\ E &\equiv (e_{ij}) \equiv P^{-1}CP. \end{aligned} \tag{6}$$

The estimate (3) now follows from Gerschgorin's theorem part (a), since the eigenvalues of $A(\epsilon)$ are unchanged by the similarity transformation.

We can now improve upon (3), if the eigenvalue $\lambda = \lambda_{ii}$ is simple. That is, we observe that the circle $R_i(\epsilon)$ of the matrix $P^{-1}A(\epsilon)P$ will not intersect any other circles $R_j(\epsilon)$, if $|\epsilon|$ is small enough, since λ_{ii} is distinct from all other λ_{jj}. Therefore, there is a unique simple eigenvalue $\lambda(\epsilon)$ in $R_i(\epsilon)$ and therefore also a unique corresponding eigenvector $\mathbf{x}(\epsilon)$ of $A(\epsilon)$.

Now, the eigenvalues λ, $\lambda(\epsilon)$ and eigenvectors $\mathbf{x}$, $\mathbf{x}(\epsilon)$ satisfy

$$\begin{aligned} (A + \epsilon C)\mathbf{x}(\epsilon) &= \lambda(\epsilon)\mathbf{x}(\epsilon) \\ A\mathbf{x} &= \lambda\mathbf{x}. \end{aligned} \tag{7}$$

Therefore by subtraction,

$$A[\mathbf{x}(\epsilon) - \mathbf{x}] + \epsilon C\mathbf{x}(\epsilon) = [\lambda(\epsilon) - \lambda]\mathbf{x}(\epsilon) + \lambda[\mathbf{x}(\epsilon) - \mathbf{x}]. \tag{8}$$

If we left-multiply both sides of (8) by the row vector $\mathbf{y}^*$, that satisfies $\mathbf{y}^*A = \lambda\mathbf{y}^*$, then

$$\epsilon\mathbf{y}^*C\mathbf{x}(\epsilon) = [\lambda(\epsilon) - \lambda]\mathbf{y}^*\mathbf{x}(\epsilon). \tag{9}$$

The conclusion (4) follows if we show that

$$\mathbf{y}^*\mathbf{x} \neq 0$$

and that we may select $\mathbf{x}(\epsilon)$ and $\mathbf{x}$ so that, as $\epsilon \to 0$,

$$\mathbf{x}(\epsilon) \to \mathbf{x}.$$

The fact that $\mathbf{y} \neq \mathbf{o}$ is orthogonal to each of the $n - 1$ other right eigenvectors of A, establishes the non-orthogonality of $\mathbf{y}$ and $\mathbf{x}$. From the relation (3), equation (7), and the assumed simplicity of λ, we know that the matrices $A - \lambda I$ and $A + \epsilon C - \lambda(\epsilon)I$ have rank $n - 1$ and that we can delete the same row and the same column from each to form non-singular submatrices, if $|\epsilon|$ is sufficiently small. Hence, if the deleted column is the jth, we may set $x_j(\epsilon) = 1$ and then, from Problem 3, $\mathbf{x}(\epsilon)$ will converge as $|\epsilon| \to 0$, to an eigenvector $\mathbf{x}$ satisfying (7). ∎

COROLLARY (BAUER-FIKE). *Under the same hypothesis and definitions as are used to establish* (3), *each eigenvalue* $\lambda(\epsilon)$ *of* $A + \epsilon C$ *satisfies*

$$(3') \qquad \min_{\lambda} |\lambda(\epsilon) - \lambda| \leq |\epsilon| \, \|C\|_p \|P^{-1}\|_p \|P\|_p,$$

for any p-norm, $\|\mathbf{x}\|_p = \left(\sum_i |x_i|^p\right)^{1/p}$, *with* $1 \leq p \leq \infty$.

Proof. Given as Problem 5. ∎

Observe that the algebraic Lemma of Section 3 in Chapter 1 suggests only that (3) holds if λ is a simple zero of $p_A(\lambda)$; but if λ has multiplicity m, then $|\lambda(\epsilon) - \lambda| = \mathcal{O}(\epsilon^{1/m})$. On the other hand, although (9) holds for the general case, (4) may not hold because $\mathbf{y}^*\mathbf{x}$ may vanish as shown in Problem 2.

We see that when (4) holds, the *well-posedness* of the eigenvalue problem for determining λ depends on the magnitude of

$$\left|\frac{\mathbf{y}^* C \mathbf{x}}{\mathbf{y}^*\mathbf{x}}\right|.$$

If we normalize the vectors, so that $\|\mathbf{y}\|_2 = \|\mathbf{x}\|_2 = 1$, and use the Schwarz inequality (see Problem 6), there results

$$(10) \qquad \left|\frac{\mathbf{y}^* C \mathbf{x}}{\mathbf{y}^*\mathbf{x}}\right| \leq \frac{\|C\|_2}{|\mathbf{y}^*\mathbf{x}|}.$$

In the special case that $C = \|A\|_2 U$, where U is a unitary matrix (i.e., $U^* = U^{-1}$) such that $U\mathbf{x} = \mathbf{y}$ (see Problems 9 and 12), we find that $\|C\|_2 = \|A\|_2$ and furthermore

$$(10') \qquad \left|\frac{\mathbf{y}^* C \mathbf{x}}{\mathbf{y}^*\mathbf{x}}\right| = \|A\|_2 \left|\frac{\mathbf{y}^*\mathbf{y}}{\mathbf{y}^*\mathbf{x}}\right| = \frac{\|A\|_2}{|\mathbf{y}^*\mathbf{x}|} = \frac{\|C\|_2}{|\mathbf{y}^*\mathbf{x}|}.$$

Now, it is possible that the problem of finding λ, an eigenvalue of A, is not well-posed, although the problem of finding the same λ as an eigenvalue of $B = P^{-1}AP$ is well-posed (this fact is illustrated in Problem 13).

The reader may on first impulse think that this statement is contradicted by Theorem 3 given in Problem 11. But a closer examination of Theorem 3 indicates that although the perturbed matrices $A + \epsilon C$ and $B + \epsilon D$ are in the one to one correspondence $D = P^{-1}CP$, the magnitudes of the corresponding matrices, $\|D\|$ and $\|C\|$, may differ considerably unless $\|P\| \cdot \|P^{-1}\|$ is not very large, since (see Problem 7)

$$\frac{\|C\|}{\|P^{-1}\| \, \|P\|} \leq \|P^{-1}CP\| \leq \|C\| \, \|P^{-1}\| \, \|P\|. \tag{11}$$

On the other hand, in the test for well-posedness of λ, it is implied that we consider only perturbations such that $\|C\| = \|A\|$ or such that $\|D\| = \|B\|$. Hence no contradiction is involved.

We may then justifiably say that when (4) holds for all C and $|\mathbf{y}^*\mathbf{x}|$ is not small (for say $\|\mathbf{y}\|_2 = \|\mathbf{x}\|_2 = 1$), the eigenvalue problem for λ of A is *well-posed*, since from (4), (10), and (10′), with $\|C\|_2 = \|A\|_2$,

$$\max_{\{C\}} \lim_{\epsilon \to 0} \left| \frac{\lambda(\epsilon) - \lambda}{\epsilon} \right| = \frac{\|A\|_2}{|\mathbf{y}^*\mathbf{x}|}.$$

We have the immediate consequence of (10).

THEOREM 4. *If A is Hermitian and λ any eigenvalue of A, then the eigenvalue problem for λ is well-posed.*

Proof. If λ is not simple, use (7) and Problem 4 to find $\mathbf{x}(\epsilon)$ and $\mathbf{x}$ with $\|\mathbf{x}(\epsilon) - \mathbf{x}\| \to 0$. Now set $\mathbf{y} = \mathbf{x}$ in (9) and note that $\mathbf{y}^*\mathbf{x}(\epsilon) \to 1$. ∎

Fortunately, the most common matrix transformation methods use orthogonal or unitary similarity transformations of A to produce a simpler matrix B which, by accounting for roundoff errors, can be written as

$$B(\epsilon) = P^{-1}(A + \epsilon C)P.$$

The matrix P is unitary and, depending on the kind of arithmetic used,

$$\|C\|_2 \leq 1; \qquad |\epsilon| = \mathcal{O}(n^p 10^{-t} a), \tag{12}$$

where $a = \max_{i,j} |a_{ij}|$, $p \leq 2$, and t represents the number of figures used in single precision arithmetic. A priori error estimates of the form (12) have been obtained by Givens and Wilkinson. We shall not pursue this topic but rather give an account of some a posteriori error estimates.

1.1. A Posteriori Error Estimates

In the case of a Hermitian matrix, A, we can give a simple error estimate for a computed eigenvalue, λ, in terms of the *residual vector*, $\boldsymbol{\eta}$.

THEOREM 5. *Let A be a Hermitian matrix of order n and have eigenvalues $\{\lambda_i\}$. If*

$$\mathbf{(13)} \qquad A\mathbf{x} - \lambda\mathbf{x} \equiv \boldsymbol{\eta}, \qquad \mathbf{x} \neq \mathbf{o},$$

then

$$\min_i |\lambda - \lambda_i| \leq \frac{\|\boldsymbol{\eta}\|_2}{\|\mathbf{x}\|_2}.$$

Proof. Since A is Hermitian there exists, in C_n, an orthonormal basis of eigenvectors $\{\mathbf{u}_i\}$ such that

$$A\mathbf{u}_i = \lambda_i \mathbf{u}_i, \qquad i = 1, 2, \ldots, n.$$

Therefore, we can express $\mathbf{x}$ as a linear combination

$$(14) \qquad \mathbf{x} = \sum_{i=1}^{n} a_i \mathbf{u}_i,$$

with $a_i = \mathbf{u}_i^* \mathbf{x}$. From (13) and (14),

$$\boldsymbol{\eta} = \sum_{i=1}^{n} a_i(\lambda_i - \lambda)\mathbf{u}_i.$$

Hence

$$\boldsymbol{\eta}^*\boldsymbol{\eta} = \sum_{i=1}^{n} |a_i|^2(\lambda_i - \lambda)^2.$$

Therefore

$$(15) \qquad \frac{\boldsymbol{\eta}^*\boldsymbol{\eta}}{\mathbf{x}^*\mathbf{x}} = \sum_{i=1}^{n} b_i(\lambda_i - \lambda)^2,$$

with

$$b_i = \frac{|a_i|^2}{\sum_{j=1}^{n} |a_j|^2}.$$

Note that

$$b_i \geq 0, \qquad \sum_{i=1}^{n} b_i = 1,$$

whence

$$\min_i (\lambda_i - \lambda)^2 \leq \sum_{i=1}^{n} b_i(\lambda_i - \lambda)^2.$$

The conclusion

$$\min_i |\lambda - \lambda_i| \leq \frac{\|\boldsymbol{\eta}\|_2}{\|\mathbf{x}\|_2}$$

is thus established. ∎

An obvious way to improve upon λ when given an approximate eigenvector, $\mathbf{x}$, is suggested by Theorem 5, namely

THEOREM 6. *Given A, a Hermitian matrix, and $\mathbf{x}$, the residual vector $\boldsymbol{\eta}$ defined by* (13) *is minimal for*

$$\lambda = \lambda_R \equiv \frac{\mathbf{x}^* A \mathbf{x}}{\mathbf{x}^* \mathbf{x}}. \tag{16}$$

Proof. The quadratic expression in λ

$$\|\boldsymbol{\eta}\|_2^2 = (\mathbf{x}^* A - \lambda \mathbf{x}^*)(A\mathbf{x} - \lambda \mathbf{x}) = \mathbf{x}^* A^2 \mathbf{x} - 2\lambda \mathbf{x}^* A \mathbf{x} + \lambda^2 \mathbf{x}^* \mathbf{x}$$

assumes its minimum at $\lambda = \lambda_R$. That is,

$$\min_{\lambda} \boldsymbol{\eta}^* \boldsymbol{\eta} = \mathbf{x}^* A^2 \mathbf{x} - \lambda_R^2 \mathbf{x}^* \mathbf{x}. \qquad \blacksquare$$

The quantity λ_R defined by (16) is called the *Rayleigh quotient* and will be referred to later on in the discussion of iterative methods.

For the Hermitian matrix A, with eigenvalues $\{\lambda_i\}$ and corresponding eigenvectors $\{\mathbf{u}_i\}$, let U_r denote the linear space spanned by the eigenvectors $\mathbf{u}_i$, $i = 1, 2, \ldots, r$. If we know something about the spacing of the eigenvalues, we may be able to estimate the error, $\|\mathbf{x} - U_r\|_2$, of an approximate eigenvector $\mathbf{x}$. (We use here the notation for *distance between a vector* $\mathbf{x}$ *and a set* S:

$$\|\mathbf{x} - S\| = \underset{\mathbf{y} \in S}{\text{g.l.b.}} \; \|\mathbf{x} - \mathbf{y}\|.)$$

That is,

THEOREM 7. *If $|\lambda_i - \lambda| \leq \|\boldsymbol{\eta}\|_2$ for $i = 1, 2, \ldots, r$, with λ, $\boldsymbol{\eta}$, and $\mathbf{x}$ that satisfy* (13), (14)*; $|\lambda_i - \lambda| \geq d > 0$ for $i = r + 1, r + 2, \ldots, n$, then*

$$\|\mathbf{x} - U_r\|_2 \leq \|\mathbf{x} - \sum_{i=1}^{r} a_i \mathbf{u}_i\|_2 \leq \frac{\|\boldsymbol{\eta}\|_2}{d}. \tag{17}$$

Proof. From the above definition and (14),

$$\|\mathbf{x} - U_r\|_2^2 \leq \|\mathbf{x} - \sum_{i=1}^{r} a_i \mathbf{u}_i\|_2^2 = \sum_{i=r+1}^{n} |a_i|^2. \tag{18}$$

Now by (13) and (14)

$$\|\boldsymbol{\eta}\|_2^2 = \sum_{i=1}^{n} |a_i|^2 (\lambda_i - \lambda)^2 \geq \sum_{i=r+1}^{n} |a_i|^2 (\lambda_i - \lambda)^2.$$

Hence by the hypothesis on the spacing of eigenvalues,

$$d^2 \sum_{i=r+1}^{n} |a_i|^2 \leq \|\boldsymbol{\eta}\|_2^2,$$

or

$$\sum_{i=r+1}^{n} |a_i|^2 \leq \frac{\|\boldsymbol{\eta}\|_2^2}{d^2}.$$

Now, from (18), the estimate (17) follows. ∎

Theorems 5 and 7 give us a posteriori estimates for the error in an approximate eigenvalue and eigenvector of a Hermitian matrix. These estimates do not require detailed information about all of the eigenvalues and eigenvectors of the matrix. Unfortunately, for a non-Hermitian matrix, the situation is more complicated and we state

THEOREM 8 (FRANKLIN). *Let A be of order n, and have a set of n linearly independent eigenvectors $\{\mathbf{u}_i\}$, eigenvalues $\{\lambda_i\}$, which satisfy $AU = U\Lambda$, with $U \equiv (\mathbf{u}_1, \mathbf{u}_2, \ldots, \mathbf{u}_n)$, $\Lambda \equiv (\lambda_i \delta_{ij})$.*

If for some $\epsilon > 0$,

$$\|A\mathbf{x} - \lambda\mathbf{x}\|_2 \leq \epsilon\|A\mathbf{x}\|_2, \tag{19}$$

then

$$\min_{\lambda_j \neq 0}\left|1 - \frac{\lambda}{\lambda_j}\right| \leq \epsilon\|U\|_2\|U^{-1}\|_2. \tag{20}$$

Proof. Define $\mathbf{b} \equiv U^{-1}\mathbf{x}$, so that

$$\mathbf{x} = U\mathbf{b},$$

(21)

$$A\mathbf{x} - \lambda\mathbf{x} = U(\Lambda - \lambda I)\mathbf{b}.$$

Now $\mathbf{y} = U^{-1}(U\mathbf{y})$, implies $\|\mathbf{y}\| \leq \|U^{-1}\|\,\|U\mathbf{y}\|$. Therefore

$$\|U\mathbf{y}\| \geq \|U^{-1}\|^{-1}\|\mathbf{y}\|. \tag{22}$$

With $\mathbf{y} \equiv (\Lambda - \lambda I)\mathbf{b}$, (22) and (21) imply

$$\|A\mathbf{x} - \lambda\mathbf{x}\| \geq \|U^{-1}\|^{-1}\|(\Lambda - \lambda I)b\|. \tag{23}$$

But from (19) and (23)

$$\epsilon\|U\Lambda\mathbf{b}\|_2 = \epsilon\|A\mathbf{x}\|_2 \geq \|A\mathbf{x} - \lambda\mathbf{x}\|_2 \geq \|U^{-1}\|_2^{-1}\|(\Lambda - \lambda I)\mathbf{b}\|_2.$$

Hence

$$\epsilon\|U\|_2\|\Lambda\mathbf{b}\|_2 \geq \|U^{-1}\|_2^{-1}\|(\Lambda - \lambda I)\mathbf{b}\|_2.$$

Therefore, since $A\mathbf{x} \neq \mathbf{o}$ implies $\Lambda\mathbf{b} \neq \mathbf{o}$,

$$\frac{\|(\Lambda - \lambda I)\mathbf{b}\|_2}{\|\Lambda\mathbf{b}\|_2} \leq \epsilon\|U^{-1}\|_2\|U\|_2. \tag{24}$$

Now, let $\mathbf{b}$ have components (b_i), $i = 1, 2, \ldots, n$. Clearly with the norm $\|\cdot\|_2$

$$\|(\Lambda - \lambda I)\mathbf{b}\|_2^2 = \sum_{i=1}^{n} |\lambda_i - \lambda|^2|b_i|^2 \geq \sum_{\lambda_i \neq 0} |\lambda_i - \lambda|^2|b_i|^2;$$

$$\|\Lambda\mathbf{b}\|_2^2 = \sum_{j=1}^{n} |\lambda_j|^2|b_j|^2 = \sum_{\lambda_j \neq 0} |\lambda_j|^2|b_j|^2.$$

Therefore,

$$\frac{\|(\Lambda - \lambda I)\mathbf{b}\|_2^2}{\|\Lambda\mathbf{b}\|_2^2} \geq \sum_{\lambda_i \neq 0} \left|1 - \frac{\lambda}{\lambda_i}\right|^2 \frac{|\lambda_i b_i|^2}{\alpha^2}$$

where $\alpha^2 \equiv \sum_{\lambda_j \neq 0} |\lambda_j|^2|b_j|^2$. Hence

$$\frac{\|(\Lambda - \lambda I)\mathbf{b}\|_2^2}{\|\Lambda\mathbf{b}\|_2^2} \geq \min_{\lambda_i \neq 0} \left|1 - \frac{\lambda}{\lambda_i}\right|^2 \tag{25}$$

since $|\lambda_i b_i|^2/\alpha^2 \geq 0$ and $\sum_{\lambda_i \neq 0} |\lambda_i b_i|^2/\alpha^2 = 1$.

By combining the inequalities (24) and (25), the estimate (20) results. ∎

We have also the

THEOREM 9. *The hypothesis of Theorem* 8, *with* (19) *replaced by*

$$\|A\mathbf{x} - \lambda\mathbf{x}\|_2 \leq \epsilon\|\mathbf{x}\|_2, \tag{19'}$$

implies

$$\min_i |\lambda - \lambda_i| \leq \epsilon\|U^{-1}\|_2\|U\|_2. \tag{20'}$$

Proof. Left as Problem 8. ∎

Unfortunately, Theorems 7 and 8 require information about the matrix of eigenvectors, which is not generally available in problems where we are only interested in obtaining a few of the eigenvalues and eigenvectors. In the special case of a Hermitian matrix, A, the matrix of eigenvectors, U, is unitary (see Problem 9), whence $\|U\|_2 = \|U^{-1}\|_2 = 1$. In this case, the estimate (20′) of Theorem 9 reduces to the estimate given in Theorem 5. If, on the other hand, for the case that the eigenvectors $\{\mathbf{u}_i\}$ of A form a basis of C_n, we have a set of vectors $\{\mathbf{v}_i\}$ that approximate the eigenvectors, then let $P \equiv (\mathbf{v}_1, \mathbf{v}_2, \ldots, \mathbf{v}_n)$.

Clearly, if we define R and ϵ by

$$P^{-1}AP \equiv \Lambda + \epsilon R$$

where $\Lambda \equiv (\lambda_i\delta_{ij})$, $\|R\| = 1$, then ϵ is small when P is a good approximation of U. The Gerschgorin circles may now be small enough to give close estimates for the eigenvalues.

PROBLEMS, SECTION 1

1. If **y** is a left eigenvector and **x** is a right eigenvector of A, corresponding to istinct eigenvalues, show that $\mathbf{y}^*\mathbf{x} = 0$.

2. (a) Show that the left and right eigenvectors corresponding to $\lambda = 1$, are orthogonal for

$$A \equiv \begin{pmatrix} 1 & 1 \\ 0 & 1 \end{pmatrix}.$$

(b) Find the eigenvalues $\lambda(\epsilon)$ for $A(\epsilon) \equiv A + \epsilon C$,

$$C \equiv \begin{pmatrix} 0 & 0 \\ 1 & 0 \end{pmatrix},$$

and verify that

$$|\lambda(\epsilon) - \lambda| = \mathcal{O}(\sqrt{\epsilon}).$$

(c) Find the eigenvectors $\mathbf{x}(\epsilon)$ of $A(\epsilon)$ and verify by substitution that (9) holds and that both the eigenvectors $\mathbf{x}(\epsilon)$ converge to the eigenvector of A.

3. Let B and $B(\epsilon)$ be of order n, have rank $n - 1$, and

$$B(\epsilon) \to B \quad \text{as} \quad \epsilon \to 0.$$

Show that for $|\epsilon|$ sufficiently small there exists a vector $\mathbf{x}(\epsilon)$ in the null space of $B(\epsilon)$, i.e.,

$$B(\epsilon)\mathbf{x}(\epsilon) = \mathbf{o},$$

such that

$$\|\mathbf{x}(\epsilon)\|_\infty \geq 1$$

and

$$\mathbf{x}(\epsilon) \to \mathbf{x}(0) \quad \text{as} \quad \epsilon \to 0,$$

where

$$B\mathbf{x}(0) = \mathbf{o}.$$

[Hint: Since $B(\epsilon) \to B$, we see that for all sufficiently small $|\epsilon|$, the $(n - 1)$st order square submatrices B_{ij} of B and $B_{ij}(\epsilon)$ of $B(\epsilon)$ found by deleting the ith row and jth column from each of B and $B(\epsilon)$ for some pair of indices (i, j), are non-singular and $B_{ij}(\epsilon) \to B_{ij}$. Hence, the Gaussian elimination method may be used to triangularize B_{ij}. The same pivotal elements may be used to triangularize $B_{ij}(\epsilon)$ for $|\epsilon|$ sufficiently small. Set $x_j(\epsilon) = 1$ and solve $B(\epsilon)\mathbf{x}(\epsilon) = \mathbf{o}$ by using the above triangularization.]

4. Let B and $B(\epsilon)$ be of order n, B have rank $n - r$, $B(\epsilon)$ have rank $< n$, and

$$B(\epsilon) \to B \quad \text{as} \quad \epsilon \to 0.$$

Let the null space of B be $S \equiv \{\mathbf{x} \mid B\mathbf{x} = \mathbf{o}\}$.

Show that for $|\epsilon|$ sufficiently small, there exists a vector $\mathbf{x}(\epsilon)$ in the null space of $B(\epsilon)$, such that

$$\|\mathbf{x}(\epsilon)\|_\infty \geq 1,$$

$$\min_{\mathbf{x} \in S} \|\mathbf{x}(\epsilon) - \mathbf{x}\| \to 0 \quad \text{as} \quad \epsilon \to 0.$$

5. Carry out the proof of (3′) (see corollary to Theorem 2).

[Hint: Let $\mathbf{x}(\epsilon) \neq \mathbf{o}$ be any eigenvector satisfying

$$(A + \epsilon C)\mathbf{x}(\epsilon) = \lambda(\epsilon)\mathbf{x}(\epsilon).$$

With the matrices P and Λ used in the proof of Theorem 1,

$$[\lambda(\epsilon)I - \Lambda]P^{-1}\mathbf{x}(\epsilon) = \epsilon P^{-1}CP[P^{-1}\mathbf{x}(\epsilon)].$$

If $\lambda(\epsilon)I - \Lambda$ is singular, (3′) holds. If $\lambda(\epsilon)I - \Lambda$ is not singular,

$$P^{-1}\mathbf{x}(\epsilon) = \epsilon[\lambda(\epsilon)I - \Lambda]^{-1}P^{-1}CP[P^{-1}\mathbf{x}(\epsilon)].$$

Take any norm $\|\cdot\|_p$ of both sides to get (3′).]

6. Apply Schwarz' inequality (i.e., $|\mathbf{y}^*\mathbf{x}| \leq \|\mathbf{y}\|_2\|\mathbf{x}\|_2$) to show that

$$|\mathbf{y}^*C\mathbf{x}| \leq \|C\|_2\|\mathbf{y}\|_2\|\mathbf{x}\|_2.$$

7. Show that

$$\frac{\|C\|}{\|P^{-1}\|\,\|P\|} \leq \|P^{-1}CP\| \leq \|C\|\,\|P^{-1}\|\,\|P\|.$$

[Hint: $C = P(P^{-1}CP)P^{-1}$.]

8. Prove Theorem 9.

[Hint: Estimate

$$\frac{\|(\Lambda - \lambda I)\mathbf{b}\|_2^{\,2}}{\sum \|\mathbf{b}\|_2^{\,2}} = \frac{\sum_{i=1}^{n} |\lambda_i - \lambda|^2|b_i|^2}{\sum_{j=1}^{n} |b_j|^2} \geq \min_i |\lambda_i - \lambda|^2.]$$

9. If U is unitary, show that $\|U\|_2 = \|U^{-1}\|_2 = 1$.

10. If $P_m(\lambda) \equiv c_0\lambda^m + c_1\lambda^{m-1} + \cdots + c_m$, and A has the eigenvalues $\{\lambda_i\}$ and the eigenvectors $\{\mathbf{u}_i\}$, then $P_m(A)$ has the eigenvalues $\{P_m(\lambda_i)\}$ with the same eigenvectors.

If A is Hermitian, show that the Rayleigh quotient, defined by (16) for any $\mathbf{x} \neq \mathbf{o}$, satisfies

$$\lambda_1 \leq \lambda_R \leq \lambda_n,$$

where λ_1 and λ_n are the smallest and largest eigenvalues of A.

11. Prove

THEOREM 3. *Let* $\mathbf{x}$ *be a right eigenvector and* $\mathbf{y}$ *a left eigenvector of* A *for the eigenvalue* λ. *For the similarity transformation given by any non-singular* P, *set*

$$B \equiv P^{-1}AP, \qquad B(\epsilon) \equiv P^{-1}A(\epsilon)P.$$

Then $\mathbf{y}^*C\mathbf{x}/\mathbf{y}^*\mathbf{x}$ *is invariant under* P.

[Hint: $\mathbf{u} \equiv P^{-1}\mathbf{x}$ is the right eigenvector of B corresponding to $\mathbf{x}$; $\mathbf{v} \equiv P^*\mathbf{y}$ is the left eigenvector of B corresponding to $\mathbf{y}$; $\epsilon P^{-1}CP$ is the perturbation corresponding to ϵC. Hence, with $D \equiv P^{-1}CP$,

$$\frac{\mathbf{v}^*D\mathbf{u}}{\mathbf{v}^*\mathbf{u}} = \frac{\mathbf{v}^*P^{-1}CP\mathbf{u}}{\mathbf{v}^*\mathbf{u}} = \frac{\mathbf{y}^*C\mathbf{x}}{\mathbf{y}^*\mathbf{x}}.]$$ ■

12. Given $\mathbf{x}$ and $\mathbf{y}$ with $\|\mathbf{x}\|_2 = \|\mathbf{y}\|_2 = 1$, construct a unitary matrix C such that $C\mathbf{x} = \mathbf{y}$.

13. Construct the matrix A which has the eigenvalues λ_j and the corresponding eigenvectors $\mathbf{x}_j$ where $x_{1k} = 1$ for $1 \leq k \leq n$, $x_{kk} = \delta$ for

$2 \le k \le n$, and $x_{ij} = 0$ otherwise. Show that the left unit eigenvectors $\mathbf{y}_j$ are determined by the biorthogonality property. Hence verify that

$$|y_{11}| = \mathcal{O}\left(\frac{|\delta|}{\sqrt{n-1}}\right), \quad \text{and} \quad |\mathbf{y}^*\mathbf{x}| = \mathcal{O}\left(\frac{\sqrt{n-1}}{|\delta|}\right).$$

(Therefore, for small δ, the eigenvalue problem for finding λ_1 is not well-posed. But the eigenvalue problem for $B \equiv P^{-1}AP$ is exceedingly well-posed if $P \equiv (\mathbf{x}_1, \mathbf{x}_2, \ldots, \mathbf{x}_n)$.)

[Hint: Start with the diagonal matrix $B \equiv (\lambda_i \delta_{ij})$ and P. Construct P^{-1} and then $A = PBP^{-1}$. Sketch a picture of $\mathbf{x}_j$ and $\mathbf{y}_j$ in the three dimensional case.]

2. THE POWER METHOD

The power method, in its basic form, is conceptually the simplest iterative procedure for approximating the largest or *principal eigenvalue* and eigenvector of a matrix. Let us assume, throughout this subsection, that the nth order matrix A has real elements (a_{ij}), n linearly independent eigenvectors $\{\mathbf{u}_j\}$, $j = 1, 2, \ldots, n$, and a unique eigenvalue of maximum magnitude, i.e., the eigenvalues satisfy

$$|\lambda_1| > |\lambda_2| \ge |\lambda_3| \ge \cdots \ge |\lambda_n|.$$

Since the λ_j are the roots of a characteristic polynomial with real coefficients, the complex eigenvalues occur in complex conjugate pairs. Hence λ_1 is real. Since $\mathbf{u}_1$ satisfies $A\mathbf{u}_1 = \lambda_1 \mathbf{u}_1$, the components (u_{i1}) of $\mathbf{u}_1$ may be taken to be real.

Let $\mathbf{x}_0$ be an arbitrarily chosen real n-dimensional vector and form the sequence of vectors

$$\text{(1)} \qquad \mathbf{x}_\nu = A\mathbf{x}_{\nu-1} = A^\nu \mathbf{x}_0; \qquad \nu = 1, 2, \ldots.$$

Since A has a complete set of eigenvectors $\{\mathbf{u}_j\}$, say with components (u_{ij}), there exist n scalars a_j such that

$$\text{(2)} \qquad \mathbf{x}_0 = \sum_{j=1}^{n} a_j \mathbf{u}_j.$$

Then the sequence (1) has the representation

$$\text{(3)} \qquad \begin{aligned} \mathbf{x}_\nu &= \sum_{j=1}^{n} \lambda_j^\nu a_j \mathbf{u}_j \\ &= \lambda_1^\nu \left[a_1 \mathbf{u}_1 + \sum_{j=2}^{n} \left(\frac{\lambda_j}{\lambda_1}\right)^\nu a_j \mathbf{u}_j \right]; \qquad \nu = 1, 2, \ldots. \end{aligned}$$

Now clearly, since $|\lambda_j/\lambda_1| < 1$ for all $j \ge 2$, the directions of the vectors $\mathbf{x}_\nu$ converge to that of $\mathbf{u}_1$ as $\nu \to \infty$, provided only that $a_1 \ne 0$. Of course,

λ_1^ν, in general, either converges to zero or becomes unbounded, and so the sequence (1) may not be practical for computations. However, a simple scaling of the iterates $\mathbf{x}_\nu$, to be introduced later, will remedy this defect.

Since $\mathbf{x}_\nu$, for large ν, may be a close approximation to the eigenvector belonging to λ_1, we can employ the methods indicated in the introduction to approximate the eigenvalue. Thus let us form the ratio of, say, the kth components of $\mathbf{x}_\nu$ and $A\mathbf{x}_\nu = \mathbf{x}_{\nu+1}$. Call the ratio $\sigma_{\nu+1}$ and get from (3),

$$\sigma_{\nu+1} = \frac{x_{k,\nu+1}}{x_{k\nu}} = \lambda_1 \frac{a_1 u_{k1} + \sum_{j=2}^{n} \left(\frac{\lambda_j}{\lambda_1}\right)^{\nu+1} a_j u_{kj}}{a_1 u_{k1} + \sum_{j=2}^{n} \left(\frac{\lambda_j}{\lambda_1}\right)^{\nu} a_j u_{kj}}. \tag{4}$$

If $a_1 \neq 0$ and k is chosen such that $u_{k1} \neq 0$, then for ν so large that $|\lambda_2/\lambda_1|^\nu \ll 1$, (4) yields

$$\sigma_{\nu+1} = \lambda_1 + \mathcal{O}\left(\left|\frac{\lambda_2}{\lambda_1}\right|^\nu\right). \tag{5}$$

Thus, in approximating the eigenvalues, the growth or decay of λ_1^ν causes no theoretical difficulties. The convergence of σ_ν to λ_1 is seen to be at least of first order with ratio at most $|\lambda_2/\lambda_1|$. As in our previous studies of iteration schemes (e.g., Chapter 2, Section 4; Chapter 3, Section 1), we may define the rate of convergence as

$$R = \ln\left|\frac{\lambda_1}{\lambda_2}\right|. \tag{6}$$

Difficulties in convergence may be expected if the first two eigenvalues (in magnitude) are "close."

Another way to approximate λ_1 is by means of the Rayleigh quotient indicated in (0.3). Thus we define

$$\sigma'_{\nu+1} = \frac{\mathbf{x}_\nu^* A\mathbf{x}_\nu}{\mathbf{x}_\nu^*\mathbf{x}_\nu} = \frac{\mathbf{x}_\nu^*\mathbf{x}_{\nu+1}}{\mathbf{x}_\nu^*\mathbf{x}_\nu}. \tag{7}$$

Then from (3) and (7) we have

$$\sigma'_{\nu+1} = \lambda_1 \frac{\sum_{i=1}^{n}\sum_{j=1}^{n} \left(\frac{\bar{\lambda}_i}{\lambda_1}\right)^{\nu}\left(\frac{\lambda_j}{\lambda_1}\right)^{\nu+1} \bar{a}_i a_j \mathbf{u}_i^*\mathbf{u}_j}{\sum_{i=1}^{n}\sum_{j=1}^{n} \left(\frac{\bar{\lambda}_i}{\lambda_1}\right)^{\nu}\left(\frac{\lambda_j}{\lambda_1}\right)^{\nu} \bar{a}_i a_j \mathbf{u}_i^*\mathbf{u}_j}.$$

A calculation reveals that σ'_ν converges to λ_1 just as does σ_ν [i.e., as in equation (5)]. However, *if the vectors* $\mathbf{u}_i$ *are mutually orthogonal* or say for convenience *orthonormal*, i.e., $\mathbf{u}_i^*\mathbf{u}_j = \delta_{ij}$, then by Problem 1, A is symmetric and has real eigenvalues and eigenvectors, so that

$$\sigma'_{\nu+1} = \lambda_1 \frac{|a_1|^2 + \sum_{j=2}^{n} \left(\frac{\lambda_j}{\lambda_1}\right)^{2\nu+1} |a_j|^2}{|a_1|^2 + \sum_{j=2}^{n} \left|\frac{\lambda_j}{\lambda_1}\right|^{2\nu} |a_j|^2}. \tag{8}$$

Hence,

$$\sigma'_{\nu+1} = \lambda_1 + \mathcal{O}\left(\left|\frac{\lambda_2}{\lambda_1}\right|^{2\nu}\right), \tag{9}$$

when A is symmetric, and the rate of convergence in this case is twice that of the scheme used in (4). Of course, this gain in using (7) is only achieved when the matrix A is symmetric. An interesting motivation for using the approximation (7) in general is furnished in Problem 2; this result should be compared with that of Theorem 1.6.

In order to terminate the iterative computations (1) and (4) or (7), a variety of different tests can be suggested. For instance, if the quotients in (4) agree for several values of k (i.e., ratios of several components) then a fairly good approximation to λ_1 has been obtained. Usually the obvious test of little or no change in the eigenvalue iterates σ_ν or σ'_ν for several successive values of ν may be successfully employed. However, a quantitative test, based on Theorems 1.8 or 1.9, requires little additional computing and yields very precise information when A is symmetric. That is, pick an arbitrary $\epsilon > 0$ and iterate until

$$\|A\mathbf{x}_\nu - \sigma\mathbf{x}_\nu\|_2 \leq \epsilon\|A\mathbf{x}_\nu\|_2 \equiv \epsilon\|\mathbf{x}_{\nu+1}\|_2. \tag{10}$$

Here $\sigma = \sigma_{\nu+1}$ or $\sigma'_{\nu+1}$. From Theorem 1.8, we then find

$$\min_j \left|1 - \frac{\sigma}{\lambda_j}\right| \leq \epsilon\|U\|_2 \cdot \|U^{-1}\|_2. \tag{11}$$

For sufficiently large ν we can be assured that the minimum in (11) is attained for $j = 1$ [assuming as usual that $a_1 \neq 0$ in (2)].

Thus, a bound on the relative error in approximating λ_1 is attained. However, the quantities $\|U\|_2$ and $\|U^{-1}\|_2$ cannot be estimated in the general case. But if A is symmetric, the matrix U is unitary and

$$\|U\|_2 = \|U^{-1}\|_2 = 1.$$

Thus for the symmetric case the condition (10) implies the precise bound

$$\left|1 - \frac{\sigma}{\lambda_1}\right| \leq \epsilon. \tag{12}$$

The present iterative scheme usually requires that the iterates $\mathbf{x}_\nu$ be scaled at each step of the process. Thus in place of the sequence (1), we actually calculate, with arbitrary $\mathbf{y}_0$,

$$\text{(13)} \qquad \boldsymbol{\xi}_\nu = A\mathbf{y}_{\nu-1}, \qquad \mathbf{y}_\nu = \frac{1}{s_\nu}\boldsymbol{\xi}_\nu; \qquad \nu = 1, 2, \ldots.$$

The sequence of scale factors, s_ν, can be chosen in a variety of ways. The two most commonly used choices are

$$\text{(14a)} \qquad s_\nu = \max_j |\xi_{j,\nu}| = \|\boldsymbol{\xi}_\nu\|_\infty,$$

and

$$\text{(14b)} \qquad s'_\nu = \|\boldsymbol{\xi}_\nu\|_2.$$

The choice (14b) requires the taking of a square root at each step, but the Rayleigh quotient estimate (7) of the eigenvalue becomes, since $\|\mathbf{y}_\nu\|_2 = 1$,

$$(\sigma'_{\nu+1})^2 = \frac{\mathbf{y}_\nu{}^* A\mathbf{y}_\nu}{\mathbf{y}_\nu{}^*\mathbf{y}_\nu} = \mathbf{y}_\nu{}^*\boldsymbol{\xi}_{\nu+1}.$$

The ratio estimate (4) is now computed as

$$\sigma_{\nu+1} = \frac{\xi_{k,\nu+1}}{y_{k\nu}}.$$

The convergence test (10) now takes the form

$$\|\boldsymbol{\xi}_{\nu+1} - \sigma\mathbf{y}_\nu\|_2 \leq \epsilon\|\boldsymbol{\xi}_{\nu+1}\|_2.$$

If the normalization (14b) is employed the computations for this test can be simplified to

$$(s'_{\nu+1})^2 + \sigma^2 - 2\sigma\boldsymbol{\xi}^*_{\nu+1}\mathbf{y}_\nu \leq \epsilon^2(s'_{\nu+1})^2.$$

In any event, the convergence rates of (5) or (9) still apply in the appropriate cases as do the error estimates (11) or (12).

The power method as presented here is frequently adequate for approximating a simple principal eigenvalue and eigenvector. In the event that the principal eigenvalues are $\lambda_1 = \bar{\lambda}_2$, i.e., complex conjugate, but simple, then the numbers λ_1, λ_2 will be approximated by the roots of a quadratic equation found by examining three successive iterates $\mathbf{x}_n$, $\mathbf{x}_{n+1}$, and $\mathbf{x}_{n+2}$ (see Problem 6). Similarly, if the principal eigenvalue has a known multiplicity, the scheme for approximating λ_1 can be suitably modified (see Problem 7).

Of course, if the matrix A has no zero components, the operational count for each iteration (1) is n^2, in general. We now turn to modifications of the power method which improve its rate of convergence.

2.1. Acceleration of the Power Method

Convergence to the principal eigenvalue by the power method has been shown to be geometric with convergence factor $|\lambda_2/\lambda_1|$ (or by using the Rayleigh quotient for a symmetric matrix, the eigenvalue convergence factor is $|\lambda_2/\lambda_1|^2$). This ratio is frequently too near unity for practical computations. Thus we seek modifications, analogous to those employed in Chapter 2, Section 5, and Chapter 3, Subsection 3.3, to accelerate the convergence.

Assume A to be *symmetric* with *unique principal eigenvalue* λ_1 and consider the power method for the modified real symmetric matrix

$$B \equiv A + \alpha I. \tag{15}$$

The eigenvalues μ_i of B are

$$\mu_i = \lambda_i + \alpha, \qquad i = 1, 2, \ldots, n.$$

If μ_1 is the unique principal eigenvalue of B, the rate of convergence is now determined by

$$\max_{j \neq 1} \left| \frac{\lambda_j + \alpha}{\lambda_1 + \alpha} \right|.$$

We minimize this ratio with respect to α and find, if the λ_i are now ordered by

$$\lambda_1 > \lambda_2 \geq \cdots \geq \lambda_n \qquad (\text{or } \lambda_1 < \lambda_2 \leq \cdots \leq \lambda_n),$$

that the *optimal value* of α is

$$\alpha = -\frac{\lambda_2 + \lambda_n}{2}. \tag{16}$$

The proof is a simple modification of that of Theorem 5.1 in Chapter 2.

In order to apply this improvement, estimates of λ_2 and λ_n are required. Such estimates may require auxiliary computations which is one of the disadvantages of the proposed acceleration procedure. For example, if a crude estimate σ of λ_1 is known (obtained perhaps by the ordinary power method) then the principal eigenvalue of $C \equiv A - \sigma I$ will be $\lambda_n - \sigma$. Thus the power method applied to C will yield an estimate of λ_n. Good estimates of λ_2 are more difficult to compute. However, if $\mathbf{u}$ is an approximation to $\mathbf{u}_1$, then using any $\mathbf{x}_0$ such that $\mathbf{x}_0^*\mathbf{u} = 0$ as the initial vector in (1) for a few iterations may yield a reasonable estimate of λ_2 (assuming, of course, that $|\lambda_2| > |\lambda_n|$). The reason is that, if $\mathbf{x}_0$ is almost orthogonal to $\mathbf{u}_1$, then a_1 in (3) will be quite small. Whence, for appropriately "small" values of ν, we may have $|\lambda_1^\nu a_1| \ll |\lambda_2^\nu a_2|$ and the ratio $x_{k,\nu+1}/x_{k\nu}$ will be a better approximation to λ_2 than to λ_1.

On the other hand, if we are given α and β such that, for example,

$$\alpha \leq \lambda_n \leq \cdots \leq \lambda_2 \leq \beta < \lambda_1, \tag{17}$$

then more efficient acceleration schemes for finding λ_1 can be devised. Let $P_m(\lambda)$ be a polynomial of degree m, say

$$P_m(\lambda) \equiv c_0\lambda^m + c_1\lambda^{m-1} + \cdots + c_m. \tag{18}$$

By $P_m(A)$ we indicate the matrix which is the corresponding polynomial in the matrix A. In Problem 10, it is shown that the eigenvalues of $P_m(A)$ are $\{P_m(\lambda_i)\}$. We now consider the power method applied to the matrix $P_m(A)$ with the initial vector $\mathbf{x}_0$ of (2). In place of (3), we now get after τ iterations

$$\begin{aligned} \mathbf{x}_\tau &= P_m(A)\mathbf{x}_{\tau-1} \\ &= P_m{}^\tau(A)\mathbf{x}_0, \end{aligned} \tag{19}$$

$$\mathbf{x}_\tau = P_m{}^\tau(\lambda_1)\left[a_1\mathbf{u}_1 + \sum_{j=2}^{n}\left(\frac{P_m(\lambda_j)}{P_m(\lambda_1)}\right)^\tau a_j\mathbf{u}_j\right].$$

To evaluate $\mathbf{x}_\tau$, we do not actually form the matrix $P_m(A)$ but recursively compute the vectors $A\mathbf{x}_{\tau-1}, \ldots, A^m\mathbf{x}_{\tau-1}$. Thus the number of operations performed in one iteration of (19) is equivalent to that for m iterations of (1). We then can compare the convergence rate by examining

$$\max_{j\neq 1}\left|\frac{\lambda_j}{\lambda_1}\right|^m \quad \text{and} \quad \max_{j\neq 1}\left|\frac{P_m(\lambda_j)}{P_m(\lambda_1)}\right|.$$

In fact, the "best" polynomial (18) of degree $\leq m$ to employ is that for which $\max_z |P_m(z)/P_m(\lambda_1)|$ is a minimum on $\alpha \leq z \leq \beta$. This problem has previously been met in Chapter 2, Section 5, in a similar context. The determination of this best polynomial is described in Chapter 5, Section 4, where we study the Chebyshev polynomials.

When the iterate $\mathbf{x}_\tau$ has been computed the approximations $\sigma_{\tau+1}$ or $\sigma'_{\tau+1}$ are formed as before by using $A\mathbf{x}_\tau$ and $\mathbf{x}_\tau$. The convergence test (10) may still be employed.

The convergence of the eigenvalue and eigenvector iterates can also be improved by the δ^2-process described in Chapter 3, Subsection 2.4 (see Chapter 3, Problems 2.6, 2.7), when A has a unique principal eigenvalue and a complete set of eigenvectors (see Problem 5).

2.2. Intermediate Eigenvalues and Eigenvectors (Orthogonalization, Deflation, Inverse Iteration)

In the previous section, we have indicated procedures whereby the power method could be modified to obtain estimates of λ_2 and λ_n. The careful application of these methods can be made to yield accurate approximations to several eigenvalues and their eigenvectors. In principle, all the values (of a real symmetric matrix) could be determined by these procedures, but in practice much accuracy may be lost in the later stages of the process.

Assume that A is symmetric, with principal eigenvalue λ_1, and that the ordering is

$$\lambda_1 > \lambda_2 \geq \cdots \geq \lambda_{n-1} > \lambda_n.$$

The matrix $A - \sigma I$ has eigenvalues $\lambda_j - \sigma$ and, for real σ, has principal eigenvalue either $\lambda_1 - \sigma$ or $\lambda_n - \sigma$ (since the above ordering is preserved). If $\sigma \geq \lambda_1$ then $\lambda_n - \sigma$ is the principal eigenvalue of $A - \sigma I$ and by the ordinary power method λ_n and $\mathbf{u}_n$ can be accurately approximated. (Note here that Theorem 1.1 provides bounds for $\{\lambda_i\}$, whence such a σ may be obtained.)

However, this device cannot readily be used to yield the intermediate eigenvalues.

The *orthogonalization method* is suitable for determining intermediate eigenvalues and eigenvectors and will be described next. Once $\mathbf{u}_1$ has been accurately determined, we may form a vector $\mathbf{x}_0$ which is orthogonal to $\mathbf{u}_1$. Such a vector has the eigenvector expansion (2) in which $a_1 = 0$. For example, if $\mathbf{z}$ is any vector then

$$\mathbf{x}_0 = \mathbf{z} - \frac{\mathbf{u}_1{}^*\mathbf{z}}{\mathbf{u}_1{}^*\mathbf{u}_1}\mathbf{u}_1 \tag{20}$$

has the property, $\mathbf{x}_0{}^*\mathbf{u}_1 = 0$. Now the sequence $\mathbf{x}_\nu$ is formed and used to compute λ_2 (assuming $|\lambda_2| > |\lambda_j|$, $j = 3, 4, \ldots, n$) and $\mathbf{u}_2$. However, after several iterations, roundoff errors will usually introduce a small but non-zero component of $\mathbf{u}_1$ in the $\mathbf{x}_\nu$ and subsequent iterations will magnify this component. This *contamination by roundoff* may be reduced by removing the $\mathbf{u}_1$ component periodically; i.e., say after every r steps $\mathbf{x}_0$ is recomputed by using $\mathbf{x}_r$ in place of $\mathbf{z}$ in (20). When λ_2 and $\mathbf{u}_2$ have been determined in this manner the procedure can in principle be continued to determine λ_3 and $\mathbf{u}_3$.

In general, if λ_i and $\mathbf{u}_i$ are known for $i = 1, 2, \ldots, s$ then we form, for arbitrary $\mathbf{z}$,

$$\mathbf{x}_0 = \mathbf{z} - \sum_{i=1}^{s} \frac{\mathbf{u}_i{}^*\mathbf{z}}{\mathbf{u}_i{}^*\mathbf{u}_i}\mathbf{u}_i. \tag{21}$$

Since the $\mathbf{u}_i$ are orthogonal, with $\mathbf{z} = \sum_{j=1}^{n} a_j\mathbf{u}_j$, we find that

$$\mathbf{x}_0 = a_{s+1}\mathbf{u}_{s+1} + \cdots + a_n\mathbf{u}_n,$$

and the power method applied to this $\mathbf{x}_0$ will yield λ_{s+1} and $\mathbf{u}_{s+1}$. The roundoff contamination is more pronounced for larger values of s and hence (21) must be frequently reapplied with $\mathbf{z}$ replaced by the current iterate, $\mathbf{x}_\nu$.

If the matrix A were not symmetric, but did have a complete set of eigenvectors, then the corresponding biorthogonalization process could be carried out. For example, the unique principal left and right eigenvectors $\mathbf{v}_1$ and $\mathbf{u}_1$ could be found: $\mathbf{u}_1$ by the iteration scheme (1); and $\mathbf{v}_1$ by the scheme, with $\mathbf{y}_0$ arbitrary,

(1*) $$\mathbf{y}_\nu^* = \mathbf{y}_{\nu-1}^* A, \qquad \nu = 1, 2, \ldots.$$

The unique next largest eigenvalue λ_2 could be found by selecting, for $\mathbf{z}$ arbitrary,

(22a) $$\mathbf{x}_0 = \mathbf{z} - \frac{\mathbf{v}_1{}^*\mathbf{z}}{\mathbf{v}_1{}^*\mathbf{v}_1}\mathbf{v}_1,$$

or

(22b) $$\mathbf{y}_0 = \mathbf{z} - \frac{\mathbf{u}_1{}^*\mathbf{z}}{\mathbf{u}_1{}^*\mathbf{u}_1}\mathbf{u}_1$$

and then applying the power methods (1) or (1*) respectively. The vectors $\{\mathbf{x}_n\}$ and $\{\mathbf{y}_n\}$ are orthogonal to $\mathbf{v}_1$ and $\mathbf{u}_1$ respectively (see Problem 8). Of course, in practice, the effect of rounding errors would have to be removed by periodically re-biorthogonalizing the vectors $\{\mathbf{x}_n\}$ and $\{\mathbf{y}_n\}$. To Problem 9, we leave the development of the analog of (21) for the determination of λ_{s+1}, $\mathbf{u}_{s+1}$, and $\mathbf{v}_{s+1}$.

A method for *roughly* approximating λ_2 and $\mathbf{u}_2$ for the symmetric matrix A is motivated as follows. With (1), (2), and (3), we define

(23) $$\mathbf{z}_\nu \equiv \mathbf{x}_\nu - \lambda_1{}^\nu a_1\mathbf{u}_1 = \sum_{j=2}^{n} \lambda_j{}^\nu a_j\mathbf{u}_j, \qquad \nu = 1, 2, \ldots.$$

By taking ratios of, say, the kth components of z_ν and $z_{\nu+1}$ we have, assuming $|\lambda_2| > |\lambda_3| \geq \cdots \geq |\lambda_n|$,

$$\sigma_{\nu+1} \equiv \frac{z_{k,\nu+1}}{z_{k\nu}} = \lambda_2 + \mathcal{O}\left(\left|\frac{\lambda_3}{\lambda_2}\right|^\nu\right).$$

Similarly, by forming the Rayleigh quotient we have

$$\sigma'_{\nu+1} = \frac{\mathbf{z}_\nu{}^*\mathbf{z}_{\nu+1}}{\mathbf{z}_\nu{}^*\mathbf{z}_\nu} = \lambda_2 + \mathcal{O}\left(\left|\frac{\lambda_3}{\lambda_2}\right|^{2\nu}\right).$$

Also as $\nu \to \infty$ the direction of $\mathbf{z}_\nu$ converges to that of $\mathbf{u}_2$.

Unfortunately, the process defined by (23) with

$$\lambda_1{}^{\nu} a_1 = \frac{\mathbf{u}_1{}^*\mathbf{x}_\nu}{\mathbf{u}_1{}^*\mathbf{u}_1},$$

and

$$\mathbf{u}_1 = \lim_{n\to\infty} \mathbf{x}_n,$$

has the feature that many significant figures are lost in (23) as ν gets large. But these values $\sigma_{\nu+1}$ or $\sigma'_{\nu+1}$ are good approximations of λ_2 only for large ν. Hence we must choose ν judiciously in order to get a reasonable estimate of λ_2 and $\mathbf{u}_2$. This procedure can be readily adapted for the sequence of normalized iterates $\{y_\nu\}$ defined in (13). It should be noted that the vectors $\{\mathbf{x}_\nu\}$ (or $\{\mathbf{y}_\nu\}$) which are required have already been computed in determining λ_1 and $\mathbf{u}_1$. Thus, with little extra computation, we have found a rough approximation for λ_2 and $\mathbf{u}_2$.

These two procedures removed components of the known eigenvectors from the iterated vectors. However, it is also possible to alter the real symmetric matrix A so that the known eigenvectors then correspond to zero eigenvalues. Iteration on an arbitrary vector with this altered matrix then automatically eliminates the known components. Thus, suppose $\mathbf{u}_1$ and λ_1 are known and that $\mathbf{u}_1$ is normalized by $\|\mathbf{u}_1\|_2{}^2 = \mathbf{u}_1{}^*\mathbf{u}_1 = 1$. Then we may form the matrix A_1 as follows

$$A_1 = A - \lambda_1 \mathbf{u}_1 \mathbf{u}_1{}^*. \tag{24}$$

Since A is symmetric, so is A_1. We also note that $A_1\mathbf{u}_1 = \mathbf{o}$. For any other eigenvector, $\mathbf{u}_j$, belonging to an eigenvalue, λ_j, $j > 1$, it follows from the orthogonality of the eigenvectors that $A_1\mathbf{u}_j = \lambda_j\mathbf{u}_j$. Thus A_1 has all the eigenvectors of A and all its eigenvalues except λ_1 which is replaced by zero. A simple calculation and proof by induction reveals that (see Problem 10)

$$\begin{aligned} \mathbf{z}_\nu \equiv A_1{}^\nu \mathbf{z} &= A^\nu \mathbf{z} - \lambda_1{}^\nu \mathbf{u}_1 \mathbf{u}_1{}^* \mathbf{z} \\ &= A^\nu \mathbf{z} - \lambda_1{}^\nu (\mathbf{u}_1{}^* \mathbf{z}) \mathbf{u}_1. \end{aligned} \tag{25}$$

A comparison of this result and the sequence $\{\mathbf{x}_\nu\}$ with $\mathbf{x}_0$ given by (20) shows that, *for exact computation* with normalized eigenvectors, *the present method is exactly equivalent to the orthogonalization method.* The cancellation errors of (23) do not occur now but instead an error grows due to the fact that the λ_1 and $\mathbf{u}_1$ employed are not an exact eigenvalue and eigenvector respectively. Thus the computed A_1 does not satisfy (25) exactly. However, iterations with A_1 are usually more accurate than the more economical computations in (23).

If the matrix A were *sparse* (many zero elements), then we would not recommend using (24) since it would, in general, produce a full matrix A_1. When λ_2 and $\mathbf{u}_2$ have been determined from A_1, the procedure can be repeated by forming $A_2 = A_1 - \lambda_2\mathbf{u}_2\mathbf{u}_2^*$, etc., to determine the remaining eigenvalues and eigenvectors.

In the above modifications of the matrix A, the resulting matrices A_k are still of order n. It is, in principle, possible to successively alter the matrix A, so that the resulting matrices B_k are of order $n - k$, $k = 1, 2, \ldots, n - 1$, and have the same remaining eigenvalues. Such procedures are called *deflation methods*, by analogy with the process of dividing out the roots of a polynomial as they are successively found. For example, the simplest such scheme is based on the method used in Theorem 1.1 of Chapter 1 to show that every matrix is similar to a triangular matrix. (The deflated matrices are the A_k defined there.)

We now describe another scheme, which has the additional feature that when the matrix A is Hermitian, the deflated matrices are also Hermitian.

Let $A\mathbf{u} = \lambda\mathbf{u}$, $\mathbf{u}^*\mathbf{u} = 1$, $u_1 \geq 0$; set

$$P = I - 2\boldsymbol{\omega}\boldsymbol{\omega}^*, \tag{26}$$

where $\boldsymbol{\omega}$ is defined, with $\mathbf{e}_1 \equiv (\delta_{i1})$, by the properties

$$\begin{aligned} \boldsymbol{\omega} &\equiv (\omega_i), \\ \boldsymbol{\omega}^*\boldsymbol{\omega} &= 1, \qquad \omega_1 \geq 0, \\ P\mathbf{e}_1 &= \mathbf{u}. \end{aligned} \tag{27}$$

In Problems 11 and 12, we show that P is unitary and that the components (ω_i) of $\boldsymbol{\omega}$ are defined by

$$\omega_1 = \left(\frac{1 - u_1}{2}\right)^{\frac{1}{2}}, \qquad \omega_k = -\frac{u_k}{2\omega_1}, \qquad k = 2, 3, \ldots, n. \tag{28}$$

Now it is easy to see that

$$AP\mathbf{e}_1 = \lambda P\mathbf{e}_1,$$

whence

$$P^{-1}AP\mathbf{e}_1 = \lambda\mathbf{e}_1. \tag{29}$$

Equation (29) shows that $\mathbf{e}_1$ is an eigenvector of

$$B_1 \equiv P^{-1}AP = PAP. \tag{30}$$

Therefore, the first column of B_1 is $\lambda\mathbf{e}_1$. In other words, A has been deflated.† This process can be continued with the matrix A_1 of order $n - 1$

† In practice the evaluation of B_k could be performed economically by adapting the procedure described in equations (3.16_k).

consisting of the last $n - 1$ rows and columns of B_1. Let the matrix, Q, of order $n - 1$ be of the form (26), and satisfy (27) and (28) relative to the matrix A_1 with an eigenvalue μ and eigenvector $\mathbf{v}$ (of order $n - 1$). Then set

$$P_1 \equiv \begin{pmatrix} 1 & \mathbf{o}^* \\ \mathbf{o} & Q \end{pmatrix},$$

where $\mathbf{o}$ is of order $n - 1$. It is easy to verify that P is unitary, in fact, $P_1^{-1} = P_1^* = P_1$. Hence the matrix

$$B_2 = P_1 PAPP_1$$

has the form

$$B_2 = \begin{bmatrix} \lambda & a_2 & a_3 & \cdots & a_n \\ 0 & \mu & b_3 & \cdots & b_n \\ 0 & 0 & & & \\ \vdots & \vdots & & A_2 & \\ 0 & 0 & & & \end{bmatrix}$$

where A_2 is of order $n - 2$. This process may be continued to provide a proof of

THEOREM 1 (SCHUR). *The matrix A, of order n, is unitarily similar to a triangular matrix.*

Proof. Left as Problem 13. ∎

COROLLARY. *The Hermitian matrix A is unitarily similar to a diagonal matrix.* ∎

Finally, we describe another iteration scheme for determining the intermediate eigenvalues and eigenvectors. This procedure is called *inverse iteration* and is based upon solving

$$(31) \qquad (A - \sigma I)\mathbf{x}_n = \mathbf{x}_{n-1}, \qquad n = 1, 2, \ldots,$$

with $\mathbf{x}_0$ arbitrary and σ a constant.

Clearly, (31) is equivalent to the power method for the matrix $(A - \sigma I)^{-1}$. We may use Gaussian elimination and (31) to calculate $\mathbf{x}_n$.

Of course, the procedure will produce the principal eigenvalue of $(A - \sigma I)^{-1}$, i.e., $1/(\lambda_k - \sigma)$, where

$$|\lambda_k - \sigma| = \min_i |\lambda_i - \sigma|,$$

provided that σ is closer to the simple eigenvalue λ_k of A than to any other eigenvalue of A. Each iteration step, after the first triangularization of $A - \sigma I$, requires about n^2 operations. The first iteration step requires about $n^3/3$ operations [see Chapter 2, equation (1.9)]. The inverse iteration method is very useful for improving upon the accuracy of an approximation to any eigenvalue.

Other iteration schemes based on matrix transformations have been devised to approximate the normal forms given in Theorem 1 and the corollary. We treat such schemes in the next section.

PROBLEMS, SECTION 2

1. Prove that a real square matrix A of order n is symmetric iff it has n orthogonal eigenvectors.

2. Let A, $\mathbf{y}$, and σ be real. Show that the scalar σ such that $\sigma\mathbf{y}$ is the best mean square approximation to the vector $A\mathbf{y}$ is given by

$$\sigma = \frac{1}{2}\frac{\mathbf{y}^*(A + A^*)\mathbf{y}}{\mathbf{y}^*\mathbf{y}} = \frac{\mathbf{y}^*A\mathbf{y}}{\mathbf{y}^*\mathbf{y}}.$$

[Hint: $(\mathbf{y}^*A^*\mathbf{y}) = (\mathbf{y}^*A\mathbf{y})^*$; $\|\boldsymbol{\eta}\|_2{}^2 = \boldsymbol{\eta}^*\boldsymbol{\eta}$.]

3. Show that if A is Hermitian, then with $A = S + iK$, where S is real symmetric, K real skew-symmetric, the eigenvector $\mathbf{u} = \mathbf{x} + i\mathbf{y}$ and eigenvalue λ satisfy

$$\begin{pmatrix} S & -K \\ K & S \end{pmatrix}\begin{pmatrix} \mathbf{x} \\ \mathbf{y} \end{pmatrix} = \lambda\begin{pmatrix} \mathbf{x} \\ \mathbf{y} \end{pmatrix}.$$

Verify that if λ is a simple eigenvalue of A, then λ is a double eigenvalue of the compound matrix.

4. The power method for computing left eigenvectors is based on the sequence

$$\mathbf{z}_\nu{}^* = \mathbf{z}_{\nu-1}^*A = \mathbf{z}_0{}^*A^\nu, \qquad \nu = 1, 2, \ldots.$$

Then, with the use of the sequence (1), we may approximate λ_1 by

$$\sigma''_{\nu+1} = \frac{\mathbf{z}_\nu{}^*\mathbf{x}_{\nu+1}}{\mathbf{z}_\nu{}^*\mathbf{x}_\nu}.$$

Show that, if the matrix A has a complete set of eigenvectors and a unique principal eigenvalue, then σ''_ν converges with the ratio $|\lambda_2/\lambda_1|^2$. (Note, however, that twice as many computations are required to evaluate each iteration step here.)

5. Use Problems 2.6 and 2.7 of Chapter 3 to find the improvement in eigenvalues and eigenvectors obtained by applying the δ^2-process to $\sigma_{\nu+1}$ of (4) or $\sigma'_{\nu+1}$ of (7), when A is real symmetric and has a unique principal eigenvalue.

6. Show how to find the coefficients s, p of the quadratic equation $\lambda^2 - s\lambda + p = 0$, that is satisfied by the unique complex conjugate pair of simple principal eigenvalues λ_1, λ_2 of the real matrix A.

[Hint: Given $\lambda_1 = \bar{\lambda}_2$; $|\lambda_1| = |\lambda_2| > |\lambda_i|$, $i = 3, 4, \ldots, n$; assume that for the corresponding eigenvectors $\mathbf{u}_1$ and $\mathbf{u}_2 = \bar{\mathbf{u}}_1$, the respective first components u_{11} and u_{12} are maximal. Then apply the technique of Bernoulli's method, Chapter 3, equations (4.16)–(4.18).]

7. When the maximal eigenvalue has multiplicity m, how can the power method be used? (See Chapter 3, Problem 4.5.)

8. Verify that the sequences $\{\mathbf{x}_n\}$ and $\{\mathbf{y}_n\}$ defined by (1), (1*), and (22) are orthogonal to $\mathbf{v}_1$ and $\mathbf{u}_1$ respectively. That is, show

$$\mathbf{v}_1{}^*\mathbf{x}_n = \mathbf{u}_1{}^*\mathbf{y}_n = 0.$$

[Hint: Use induction with respect to n.]

9. Develop the biorthogonal analog of (21), for the case that A has a complete set of eigenvectors. Describe the power method for the determination of the unique intermediate eigenvalues when $|\lambda_1| > |\lambda_2| > \cdots > |\lambda_n|$.

[Hint: Generalize (22) and use Problem 8.]

10. Prove (25), i.e., if A is real symmetric, $A\mathbf{u} = \lambda\mathbf{u}$, and $\|\mathbf{u}\|_2 = 1$, then for all $\mathbf{z}$ and $\nu = 1, 2, \ldots,$

$$(A - \lambda\mathbf{u}\mathbf{u}^*)^\nu\mathbf{z} = A^\nu\mathbf{z} - \lambda^\nu(\mathbf{u}^*\mathbf{z})\mathbf{u}.$$

11. Prove that the Hermitian matrix

$$P = I - 2\boldsymbol{\omega}\boldsymbol{\omega}^*$$

is unitary, in fact $P^{-1} = P^* = P$, iff

$$\boldsymbol{\omega}^*\boldsymbol{\omega} = 1.$$

12. With

$$\mathbf{u} = \begin{pmatrix} u_1 \\ u_2 \\ \vdots \\ u_n \end{pmatrix}, \qquad u_1 \geq 0,\ \mathbf{u}^*\mathbf{u} = 1,$$

show that if the matrix P of (26) satisfies (27), then

$$\boldsymbol{\omega} = \begin{pmatrix} \omega_1 \\ \omega_2 \\ \vdots \\ \omega_n \end{pmatrix}, \quad \text{and} \quad \begin{aligned} \omega_1 &= \left(\frac{1 - u_1}{2}\right)^{1/2} \\ \omega_k &= -\frac{u_k}{2\omega_1}, \qquad k = 2, 3, \ldots, n. \end{aligned}$$

13. (a) Give a complete proof by induction on n of Theorem 1 (Schur), along the line indicated in text.

(b) Give another proof of Theorem 1 by making use of Theorem 1.1 of Chapter 1.

[Hint: Construct $B = P^{-1}AP$ where B is triangular. Since P is non-singular, construct a matrix Q whose columns are an orthonormal set of vectors formed successively from the columns of P so that $P = QR$, where R is upper triangular and non-singular. Therefore $RBR^{-1} = Q^{-1}AQ$. Show that the product of two upper triangular matrices and the inverse of an upper triangular matrix are upper triangular matrices.]

3. METHODS BASED ON MATRIX TRANSFORMATIONS

The methods of Section 2 are suitable for the determination of a few eigenvalues and eigenvectors of the matrix A. However, when we are interested in finding all of the eigenvalues and eigenvectors then the methods based on transforming the matrix A seem more efficient. Attempts to implement Schur's theorem, by approximating a unitary matrix H which triangularizes by similarity the given general matrix A, have been recently made. We shall describe one of these later. In the case of a Hermitian matrix A, however, the classical *Jacobi's method* does work well to produce a unitarily similar diagonal matrix. The methods of *Givens* and *Householder* produce either a similar tridiagonal matrix for a Hermitian A or a similar matrix of *Hessenberg*† *form* for a general matrix A, with the use of a unitary similarity transformation.

We will first describe Jacobi's, Givens', and Householder's methods for treating the Hermitian matrix A. In this case, since the computational problem in practice is usually reduced to that of finding the eigenvalues of a real symmetric matrix (see Problem 2.3), we will assume that A is real symmetric. All of these methods, however, require only simple modifications to be directly applicable to the Hermitian case. Jacobi's method reduces the real symmetric matrix A by an *infinite* sequence of simple orthogonal similarity transformations (two dimensional rotations) to diagonal form. The following lemmas provide the basis of the procedure.

LEMMA 1. *Let $B \equiv P^{-1}AP$. If P is orthogonal (unitary), then*

$$\text{trace}(B) = \text{trace}(A),$$
$$\text{trace}(B^*B) = \text{trace}(A^*A). \tag{1}$$

Proof. Since B and A are *similar* matrices, the eigenvalues of B are the same as the eigenvalues of A. By definition,

$$\text{trace}(A) = \sum_{i=1}^{n} a_{ii}.$$

The eigenvalues of A are the roots of

$$p_A(\lambda) \equiv \det(\lambda I - A) = 0.$$

By partially expanding the determinant, the coefficient of λ^n in $p_A(\lambda)$ is seen to be unity while the coefficient of λ^{n-1} is $-\text{trace}(A)$. Hence, we find that

$$\text{trace}(A) = \sum_{i=1}^{n} \lambda_i = \text{trace}(B).$$

The orthogonality of P, i.e., $P^* = P^{-1}$, implies

$$B^*B = (P^*A^*P)(P^*AP) = P^*A^*AP.$$

Again, because the eigenvalues are unchanged under a similarity transformation, the trace of A^*A is unchanged. ∎

† See the definition of (upper) Hessenberg form in Problem 1.7 of Chapter 2.

LEMMA **2.** *Let*

$$A_{ij} \equiv \begin{pmatrix} a_{ii} & a_{ij} \\ a_{ji} & a_{jj} \end{pmatrix}, \qquad a_{ij} \neq 0,$$

be formed from elements of a real symmetric matrix A and let

$$R \equiv \begin{pmatrix} r_{ii} & r_{ij} \\ r_{ji} & r_{jj} \end{pmatrix}, \tag{2a}$$

with $r_{ii} = r_{jj} = \cos\phi$, $r_{ij} = -r_{ji} = \sin\phi$, *where*

$$\tan 2\phi = \frac{2a_{ij}}{a_{jj} - a_{ii}}, \qquad -\frac{\pi}{4} \leq \phi \leq \frac{\pi}{4}. \tag{2b}$$

Then

$$R^* = R^{-1}$$

and

$$B_{ij} \equiv R^* A_{ij} R$$

is a diagonal matrix.

Proof. Let

$$B_{ij} \equiv \begin{pmatrix} b_{ii} & b_{ij} \\ b_{ji} & b_{jj} \end{pmatrix}.$$

Equation (2), by Problem 1, guarantees that $b_{ij} = b_{ji} = 0$. ■

By Lemma 1, since A_{ij} and B_{ij} are orthogonally similar,

$$b_{ii}^2 + b_{jj}^2 = a_{ii}^2 + 2a_{ij}^2 + a_{jj}^2. \tag{3}$$

We say that the matrix R *reduced to zero* the element a_{ij}. We now construct an orthogonal matrix of order n, to reduce to zero the element a_{ij} of A. Let

$$P \equiv (p_{st}) \tag{4}$$

where

$$p_{ii} = r_{ii}, \qquad p_{ij} = r_{ij}, \qquad p_{ji} = r_{ji}, \qquad p_{jj} = r_{jj},$$

and

$$p_{st} = \delta_{st} \qquad \text{otherwise.}$$

With the elements r_{ii}, r_{ij}, r_{ji}, r_{jj} defined in (2), $P^*P = PP^* = I$.

We call such a matrix P a *two dimensional rotation*. Now set,

$$B \equiv P^*AP \equiv (b_{ij}).$$

Clearly, $b_{ij} = b_{ji} = 0$. Hence the matrix P reduces to zero the element a_{ij} of A. We now can show that P reduces the sum of the squares of the off-diagonal elements of A. From the definition of trace (A^*A), it is easy to verify that

$$\text{trace}\,(A^*A) = \sum_{i,j=1}^{n} a_{ij}^2.$$

Now, by Lemmas 1 and 2 and equation (3), we find, since $b_{kk} = a_{kk}$ for $k \neq i, j$,

$$\begin{aligned} \sum_{r \neq s} b_{rs}^2 &= \text{trace}\,(B^*B) - \sum_{s=1}^{n} b_{ss}^2 \\ &= \text{trace}\,(A^*A) - \sum_{s=1}^{n} a_{ss}^2 - 2a_{ij}^2 \\ &= \sum_{r \neq s} a_{rs}^2 - 2a_{ij}^2. \end{aligned} \tag{5}$$

If we pick (i, j) so that $a_{ij}^2 = \max_{k \neq s} \{a_{ks}^2\}$, then

$$a_{ij}^2 \geq \frac{\sum_{k \neq s} a_{ks}^2}{(n^2 - n)} = \frac{\text{trace}\,(A^*A) - \sum_{i=1}^{n} a_{ii}^2}{n^2 - n}. \tag{6}$$

In fact, (6) is satisfied by any a_{ij}^2 that is not less than the average of the squares of the off-diagonal elements of A.

By substituting the inequality (6) in (5), we have

$$\sum_{r \neq s} b_{rs}^2 \leq \left(1 - \frac{2}{n^2 - n}\right) \sum_{r \neq s} a_{rs}^2. \tag{7}$$

Therefore, each two dimensional rotation defined in (2) and (4), and such that (6) holds, reduces the sum of the squares of the off-diagonal elements of a symmetric matrix A by a factor not greater than

$$1 - \frac{2}{n^2 - n} < 1 \qquad \text{for } n \geq 2.$$

In addition, we observe from (3) and the phrase preceding (5) that

$$\sum_{s=1}^{n} b_{ss}^2 = \sum_{s=1}^{n} a_{ss}^2 + 2a_{ij}^2. \tag{8}$$

We therefore have the basis for proving

THEOREM **1** (JACOBI). *Let A be real symmetric, with eigenvalues $\{\lambda_i\}$; let the matrices $\{P_m\}$ be two dimensional rotation matrices of the form* (4) *defined successively so that an above-average element of*

$$B_0 \equiv A$$

is reduced to zero by P_1, and thereafter P_m reduces to zero an above-average element of B_{m-1} in the similarity transformation defining B_m,

$$B_m \equiv P_m^* B_{m-1} P_m, \qquad m = 1, 2, \ldots.$$

Let

(9) $$Q_m \equiv P_1P_2\cdots P_m.$$

Then, as $m \to \infty$,

$$Q_m{}^*AQ_m \to \Lambda,$$

where $\Lambda \equiv (\lambda_i\delta_{ij})$, *for some ordering of the* $\{\lambda_i\}$.

Proof. For m sufficiently large, by (7) and Gerschgorin's theorem

$$B_m \equiv Q_m{}^*AQ_m$$

is approximately the diagonal matrix Λ, for some ordering of $\{\lambda_i\}$. Now, the angle of rotation ϕ, determined for P_{m+1} from B_m, is close to zero, unless the two diagonal elements of B_m used to determine ϕ correspond to a pair of identical eigenvalues. Hence it is easy to verify that

$$Q_m{}^*AQ_m \to \Lambda. \qquad \blacksquare$$

Jacobi's method is the scheme in which P_{k+1} is determined so as to reduce a *maximal* off-diagonal element of B_k to zero, $k = 0, 1, 2, \ldots$. In practice, by listing the magnitude and column index of the maximal off-diagonal element in each row of B_k, the Jacobi scheme is easily carried out. That is, the only elements that change in going from B_k to B_{k+1} are the elements in the rows and columns of index i or j. Hence, the list of maximal elements in each row of B_{k+1} is easily constructed from the list for B_k by making at most $2n - 1$ comparisons. A common variation of the Jacobi method consists in examining the off-diagonal elements of B_m in a systematic *cyclic order* given by the indices $(1, 2)$, $(1,3), \ldots, (1, n)$, $(2, 3)$, $(2, 4), \ldots, (2, n)$, $(3, 4), \ldots, (n - 1, n)$. The indices (i, j) used to determine P_{m+1} correspond to the first element $b_{ij}^{(m)}$ of B_m that satisfies

$$|b_{ij}^{(m)}| \geq t_m,$$

where $\{t_m\}$ is a prescribed decreasing sequence of positive numbers called *thresholds*. Such an iteration procedure is called a *threshold scheme*. A bolder approach, namely to rotate the off-diagonal elements in sequence, irrespective of size, is called the *cyclic Jacobi scheme*. Surprisingly enough, with only a minor change in the definition of the angle ϕ, when $\phi \cong \pm\pi/4$, the *cyclic Jacobi scheme* has been shown (Forsythe-Henrici) to be convergent also. In fact, if a comparison of the residual off-diagonal sum of squares, $\sum_{r \neq s} (b_{rs}^{(m)})^2$, is made after each complete cycle of $q \equiv (n^2 - n)/2$ rotations, then it has been shown that the cyclic Jacobi scheme converges and, in fact, that it *converges quadratically for m large enough*, i.e.,

$$\sum_{r \neq s} (b_{rs}^{(m+q)})^2 \leq K\left[\sum_{r \neq s} (b_{rs}^{(m)})^2\right]^2$$

for m large enough.

In these Jacobi schemes, the eigenvalues of A are approximated by the diagonal elements of B_m, for sufficiently large m. Furthermore, the corresponding eigenvectors of B_m are then approximately the unit vectors $\{\mathbf{e}_i\}$. Hence, the eigenvectors of A are approximated by the columns of

$$Q_m = P_1 P_2 \cdots P_m.$$

Since only two columns of Q_m are changed in going from Q_m to Q_{m+1}, only $4n$ multiplications are involved in this step. On the other hand, the elements of the matrix B_m that are unaffected by the rotation P_{m+1} are those that have indices (r, s) with $r \neq i, j$ and $s \neq i, j$. Therefore, because of symmetry, approximately $4n$ multiplications are needed to carry out the rotation of B_m into B_{m+1} (if we neglect to count the square root operations necessary to determine $\cos \phi$ and $\sin \phi$). Hence, about $8n$ multiplications are required to determine B_{m+1} and Q_{m+1} from B_m and Q_m.

We now consider the *Givens transformation*. Here a finite sequence of

$$M \equiv \frac{(n-2)(n-1)}{2}$$

rotations are employed to reduce the real symmetric matrix A to *tridiagonal form*.

That is, we successively construct a sequence of matrices $\{P_k\}$, $k = 1, 2, \ldots, M$, and define

$$\text{(10)} \qquad \begin{aligned} B_0 &\equiv A, \\ B_k &\equiv P_k^* B_{k-1} P_k, \qquad 1 \leq k \leq M. \end{aligned}$$

The matrices $\{P_m\}$ are two dimensional rotations of the form (4) constructed so that the first k not-codiagonal elements of B_k are zero. That is, we say $\{a_{ii}\}$ and $\{a_{i,i+1}\}$ are the *codiagonal* elements of A. For a symmetric matrix, we refer to the not-codiagonal indices listed cyclically in the order

$$\text{(11)} \quad (1, 3), (1, 4), \ldots, (1, n), (2, 4), (2, 5), \ldots, (2, n), \ldots, (n-2, n).$$

The first k not-codiagonal elements of B_k are the elements of B_k whose indices are among the first k in the list (11). The facts are summarized in

THEOREM 2 (GIVENS). *Let A be real symmetric; $B_0 \equiv A$. Let $(i-1, j)$ be the* kth *pair of indices in the cyclic sequence* (11), *and B_{k-1} have elements* $(b_{rs}^{(k-1)})$. *Let $P_k \equiv I$, if $b_{i-1,j}^{(k-1)} = 0$; otherwise, set $P_k \equiv (p_{rs}^{(k)})$ with*

$$\text{(12)} \qquad \begin{aligned} p_{ii}^{(k)} &= p_{jj}^{(k)} = \frac{b_{i-1,i}^{(k-1)}}{\sqrt{(b_{i-1,i}^{(k-1)})^2 + (b_{i-1,j}^{(k-1)})^2}}, \\ p_{ij}^{(k)} &= -p_{ji}^{(k)} = -\frac{b_{i-1,j}^{(k-1)}}{\sqrt{(b_{i-1,i}^{(k-1)})^2 + (b_{i-1,j}^{(k-1)})^2}}, \\ p_{rs}^{(k)} &= \delta_{rs} \qquad \textit{for other } (r, s). \end{aligned}$$

Let the matrices $\{B_k\}$ *and* $\{P_k\}$ *be defined by* (10) *and* (12) *for* $k = 1, 2, \ldots, M$,

$$M \equiv \frac{(n-2)(n-1)}{2}.$$

Then, B_k *is real symmetric; the first* k *not-codiagonal elements of* B_k *are zero,* $k = 1, 2, \ldots, M$; B_M *is tridiagonal.*

Proof. It is a simple matter to verify that if the first $k - 1$ not-codiagonal elements of B_{k-1} are zero, then the corresponding elements of B_k also vanish. [This property of preserving zeros is *not valid* in general for the rotations of the type used in Jacobi's method! In Jacobi's method the matrix P_k, a two dimensional rotation in the (i, j) coordinates, annihilated the (i, j) element; but in Givens' method the matrix P_k annihilates the $(i - 1, j)$ element]. Furthermore, by Problem 2, the definition (12) of P_k ensures that $b_{i-1,j}^{(k)} = b_{j,i-1}^{(k)} = 0$. Hence, by using mathematical induction the proof may be completed. ∎

Aside from the calculation of the non-trivial elements of P_k, the calculation of the non-zero elements of B_k in (10) involves approximately $4(n - i)$ multiplications. Now, in order to reduce to zero the elements in row $i - 1$, this process must be carried out for $j = i + 1, i + 2, \ldots, n$. That is, $4(n - i)^2$ multiplications are required to put zeros in all of the not-codiagonal elements in row $i - 1$. Therefore, for the complete reduction to tridiagonal form, we have the

COROLLARY. *The Givens method requires*

$$\tfrac{4}{3}n^3 + \mathcal{O}(n^2) \textit{ operations}$$

to transform the real symmetric matrix A *to tridiagonal form.* ∎

We shall complete the description of Givens' method for determining the eigenvalues and eigenvectors after we study *Householder's method* for reducing the matrix A to tridiagonal form. Householder's scheme uses a sequence of $n - 2$ orthogonal similarity transformations of the form

$$(13) \qquad P = I - 2\boldsymbol{\omega}\boldsymbol{\omega}^*; \qquad \boldsymbol{\omega}^*\boldsymbol{\omega} = 1,$$

with suitably chosen vectors $\boldsymbol{\omega}$. In Problem 3, $P^*P = PP^* = I$ is verified.

We now describe how the matrices P_k are defined. Let the rows $i = 1, 2, \ldots, k - 1 \leq n - 3$ of the symmetric matrix B_{k-1} have the reduced form

$$b_{rs} = 0 \qquad \text{for } 1 \leq r \leq k - 1 \text{ and } r + 2 \leq s.$$

If

$$\sum_{t=2}^{n-k} b_{k,k+t}^2 = 0,$$

set

$$P_k \equiv I;$$

if

$$\sum_{t=2}^{n-k} b_{k,k+t}^2 \neq 0,$$

set

$$(14)_k \qquad P_k = I - 2\boldsymbol{\omega}\boldsymbol{\omega}^*,$$

with

$$\boldsymbol{\omega} = \beta \mathbf{v}, \qquad \mathbf{v} = (v_i),$$

where

$$v_i = 0, \qquad i = 1, 2, \ldots, k,$$

$$v_{k+1} = 2y^2 S,$$

$$v_i = b_{ki}, \qquad i = k+2, k+3, \ldots, n;$$

$$S^2 = \sum_{t=1}^{n-k} b_{k,k+t}^2,$$

and if $b_{k,k+1} \neq 0$,

$$S = \text{sign}\,(b_{k,k+1})\sqrt{S^2},$$

$$y = \frac{(b_{k,k+1} + S)}{2K}, \qquad \beta = \frac{1}{2yS},$$

$$2K^2 = S^2 + b_{k,k+1}S.$$

We then have

THEOREM 3 (HOUSEHOLDER). *Let A be real symmetric. Set $B_0 \equiv A$, define*

$$(15)_k \qquad B_k = P_k^* B_{k-1} P_k, \qquad k = 1, 2, \ldots, n-2,$$

by means of $(14)_k$. *Then $\boldsymbol{\omega}^*\boldsymbol{\omega} = 1$; B_k is real symmetric; all of the not-codiagonal elements of B_k, in the rows $i = 1, 2, \ldots, k$, are zero; B_{n-2} is tridiagonal.*

Proof. We leave the verification to Problem 4. ∎

Now, we note that the practical evaluation of B_k in $(15)_k$ can be carried out in the following fashion:

$$\begin{aligned} B_k &= P_k^* B_{k-1} P_k \\ &= (I - 2\boldsymbol{\omega}\boldsymbol{\omega}^*) B_{k-1} (I - 2\boldsymbol{\omega}\boldsymbol{\omega}^*) \\ &= B_{k-1} - 2\boldsymbol{\omega}\boldsymbol{\omega}^* B_{k-1} - 2B_{k-1}\boldsymbol{\omega}\boldsymbol{\omega}^* + 4\boldsymbol{\omega}\boldsymbol{\omega}^* B_{k-1}\boldsymbol{\omega}\boldsymbol{\omega}^*. \end{aligned}$$

Therefore,

$$(16)_k \qquad B_k = B_{k-1} - 2\beta^2(\mathbf{v}\mathbf{u}^* + \mathbf{u}\mathbf{v}^*),$$

where

$$\mathbf{u} \equiv \boldsymbol{\xi} - a\mathbf{v},$$

with

$$\boldsymbol{\xi} \equiv B_{k-1}\mathbf{v},$$

$$a \equiv \beta^2 \mathbf{v}^*\boldsymbol{\xi}.$$

Observe that in $(16)_k$, K need not be evaluated, and therefore only one square root operation is required in each stage of Householder's method. Furthermore, given $\mathbf{v}$, the evaluation of $\boldsymbol{\xi}$ requires

$$(n - k)(n - k - 1) \text{ multiplications;}$$

of a requires

$$n - k \text{ multiplications;}$$

of $\mathbf{u}\mathbf{v}^*$ requires

$$(n - k)^2 \text{ multiplications.}$$

Hence we have shown that the number of multiplications required to produce B_k is approximately $2(n - k)^2$.

COROLLARY. *Householder's method reduces a real symmetric matrix to tridiagonal form with the use of* $\frac{2}{3}n^3 + \mathcal{O}(n^2)$ *multiplications.*

Proof. The result follows from the formula

$$\sum_{k=1}^{n-2} k^2 = \frac{n^3}{3} + \mathcal{O}(n^2). \qquad \blacksquare$$

We now remark that both Givens' method and Householder's method can be employed to reduce any real matrix A to *lower Hessenberg form.* The matrix $B \equiv (b_{ij})$ is in lower Hessenberg form iff $b_{i,s} = 0$ for $i + 2 \leq s \leq n$.

Finally, we give the treatment of Givens for finding the eigenvalues of the symmetric tridiagonal matrix B.

Let $B \equiv (b_{ij})$ be real symmetric and tridiagonal, i.e.,

$$b_{ij} = b_{ji}, \qquad 1 \leq i, j \leq n;$$

$$b_{ij} = 0, \qquad j \neq i - 1, i, i + 1; \qquad 1 \leq i \leq n.$$

Set

$$a_i \equiv b_{ii}, \qquad 1 \leq i \leq n;$$

$$(17) \qquad c_i = b_{i,i+1} = b_{i+1,i}, \qquad 1 \leq i \leq n - 1.$$

Recall that

$$p_B(\lambda) \equiv \det(\lambda I - B).$$

Givens observed that the principal minor determinants of $\lambda I - B$ formed a sequence of polynomials having properties similar to those of a Sturm sequence (see Chapter 3, Subsection 4.2). That is, let

$$
\begin{aligned}
&p_0(\lambda) \equiv 1, \\
&p_1(\lambda) \equiv (\lambda - a_1), \\
(18)\quad &p_2(\lambda) \equiv (\lambda - a_2)(\lambda - a_1) - c_1^2, \\
&\vdots \\
&p_i(\lambda) \equiv (\lambda - a_i)p_{i-1}(\lambda) - (c_{i-1})^2 p_{i-2}(\lambda); \qquad i = 3, 4, \ldots, n.
\end{aligned}
$$

In Problem 5 it is shown that $p_n(\lambda) \equiv p_B(\lambda)$. If any $c_i = 0$, then the determination of the eigenvalues of B is reduced to the eigenvalue problem for a smaller matrix. Hence we assume $c_i \neq 0$, $1 \leq i \leq n - 1$. We have then,

THEOREM 4 (GIVENS). *Let the tridiagonal, real symmetric matrix B be defined by* (17), *with all $c_i \neq 0$. Then the zeros of each $p_i(\lambda)$, $i = 2, 3, \ldots, n$, are distinct and are separated by the zeros of $p_{i-1}(\lambda)$; and, if $p_n(\gamma) \neq 0$, the number of eigenvalues of B that are greater than γ is equal to the number of sign variations in the sequence $p_n(\gamma), p_{n-1}(\gamma), \ldots, p_1(\gamma)$, 1.*

Proof. Since $c_i \neq 0$, no two successive polynomials $p_i(\lambda)$ and $p_{i-1}(\lambda)$ can have a common zero. Otherwise, from (18), $p_{i-2}(\lambda), \ldots, p_0(\lambda)$ would also have that zero. By mathematical induction, we can now prove the separation property. That is, a simple plot of $p_2(\lambda)$ shows that the simple zero of $p_1(\lambda)$ separates the two simple zeros of $p_2(\lambda)$. Assume that the $i - 2$ simple zeros of $p_{i-2}(\lambda)$ separate the $i - 1$ simple zeros of $p_{i-1}(\lambda)$. Now, from (18), at each zero of $p_{i-1}(\lambda)$, the sign of $p_i(\lambda)$ is opposite to the sign of $p_{i-2}(\lambda)$. But, by the induction hypothesis, $p_{i-2}(\lambda)$ changes sign between each pair of neighboring zeros of p_{i-1}. Therefore, $p_i(\lambda)$ also changes sign and hence has a zero *between* each neighboring pair of zeros of $p_{i-1}(\lambda)$. Now

$$p_i(+\infty) = +\infty, \qquad p_i(-\infty) = (-1)^i\infty, \qquad i = 1, 2, \ldots.$$

Therefore $p_i(\lambda)$ has a zero to the right of the largest zero of $p_{i-1}(\lambda)$ and a zero to the left of the smallest zero of $p_{i-1}(\lambda)$. On the other hand, $p_i(\lambda)$ can have no more than i zeros. Therefore, we have shown that the $i - 1$ simple zeros of $p_{i-1}(\lambda)$ separate the i simple zeros of $p_i(\lambda)$. This *separation property* is all that is needed to verify the rest of the theorem's conclusion, by the kind of argument we used for treating Sturm sequences in Chapter 3, Subsection 4.2 (see Problem 6). ∎

The evaluation of the characteristic polynomial of B, once we have calculated $\{c_i^2\}$, is then carried out by using (18). Thus a sequence of $2n - 3$ multiplications is required for each determination of $p_n(\lambda)$. If all of the eigenvalues of B are in the interval $[a, b]$, then we may locate the eigenvalues more precisely by halving the interval and using the theorem to find out how many zeros of $p_B(\lambda)$ are in each half, i.e., pick $\gamma = (a + b)/2$. In this way, after t halvings, we will know the location of an eigenvalue to within $\pm(a + b)2^{-(t+1)}$, at the expense of about $2tn$ multiplications.

Once an eigenvalue λ of B is found, a corresponding eigenvector can be determined by using the fact that

$$B - \lambda I$$

has rank $\leq n - 1$.

If none of the off-diagonal terms c_i vanish,† then the equations

$$(B - \lambda I)\mathbf{x} = \mathbf{o}$$

may be solved simply by applying Gaussian elimination. That is, with the use of maximal column pivots, we proceed to eliminate $x_1, x_2, \ldots, x_n$. We neglect the last r equations of the essentially triangular system that results, if the rank is $n - r$. We give arbitrary non-zero values to the variables $x_{n-r+1}, \ldots, x_n$, that appear in the last r equations; and solve the first $n - r$ equations of the triangular system. If the maximal column pivots do not occur in order along the diagonal, we list the pivotal equations in the order that we find them. In this case, the then resulting upper triangular matrix, U, may not be codiagonal. That is, the only non-trivial elements in row i may be the elements u_{ii}, $u_{i,i+1}$, and $u_{i,i+2}$.

In Problem 8 we see that $\max_{i,j} |u_{ij}| \leq 5b$ where $b = \max_{i,j} |b_{ij}|$. Hence by the theory of a priori estimates in Subsection 1.2, Chapter 2, though the matrix $B - \lambda I$ is singular, the first $n - r$ equations of U, when computed with finite precision, give an accurate representation of the coefficients of the unknowns $x_1, x_2, \ldots, x_{n-r}$ that would arise by exact elimination. Furthermore, the precise determination of r is not necessary!

If $\mathbf{x}$ is an eigenvector of B corresponding to the eigenvalue λ then

$$\mathbf{y} = P\mathbf{x},$$

is the corresponding eigenvector of A, since

$$B = P^{-1}AP.$$

For determining all of the eigenvalues and all of the eigenvectors of B, experience has shown that the QR method (see end of section) for the symmetric, tridiagonal matrix B is a more efficient procedure than Givens' method.

† If any $c_i = 0$, the system splits into two disjoint systems.

We now turn our attention to the case of the general real matrix A. If we are interested in determining all of the eigenvalues of A, then *a preliminary simplification to a similar Hessenberg form B is appropriate, since the iterative operations on B will then require fewer calculations.* For example, as remarked after the corollary to Theorem 3, Givens' or Householder's orthogonal similarity transformations might be used to effect the reduction.

Review the discussion, between Theorem 3 and its corollary, of the number of operations involved in using Householder's method. Since the matrix B_{k-1} is not symmetric when A is not symmetric, we see that the operational count for the vector ξ becomes $(n-k)n + \mathcal{O}(n)$ while the other counts remain the same. Hence, *the number of multiplications required to reduce the general real matrix A to Hessenberg form is* $\frac{5}{6}n^3 + \mathcal{O}(n^2)$.

A convenient and practical technique to evaluate $p_B(\lambda)$, when $B \equiv (b_{ij})$ is in lower Hessenberg form, uses

LEMMA 3 (HYMAN). *Let B be in lower Hessenberg form. If $b_{i,i+1} \neq 0$, $i = 1, 2, \ldots, n-1$, define the sequence of polynomials $m_i(\lambda)$*

$$\begin{aligned} &m_0 \equiv 1 \\ (19) \qquad &-b_{i,i+1}m_i \equiv b_{i1}m_0 + b_{i2}m_1 + \cdots + b_{i,i-1}m_{i-2} + (b_{ii} - \lambda)m_{i-1}, \\ &\qquad\qquad i = 1, 2, \ldots, n-1. \end{aligned}$$

Then

$$\begin{aligned} &g_B(\lambda) \equiv \det(B - \lambda I) = (-1)^{n+1}b_{12}b_{23}\cdots b_{n-1,n}g(\lambda), \\ (20) \qquad &g(\lambda) \equiv b_{n1}m_0 + b_{n2}m_1 + \cdots + b_{n,n-1}m_{n-2} + (b_{nn} - \lambda)m_{n-1}. \end{aligned}$$

Proof. Since $b_{i,i+1} \neq 0$, we can successively add multiples of the columns of $B - \lambda I$ to the first column in order to annihilate the first $n - 1$ elements of the first column. This process defines the polynomials $\{m_i\}$, $i = 1, 2, \ldots, n-1$, and does not change the value of the determinant. But the $(n, 1)$ element is $g(\lambda)$ and the expansion of the determinant with respect to the elements of the first column results in (20). ∎

Clearly, if any $b_{i,i+1} = 0$, the $\det(B - \lambda I)$ can be written as the product of two determinants of submatrices of $B - \lambda I$.

In the case $b_{i,i+1} \neq 0$, formulas (19) and (20) may be used to calculate $g(\lambda)$, with the use of $n^2/2 + \mathcal{O}(n)$ multiplications. Similarly, by differentiating the formulas (19) with respect to λ, a recursive evaluation of $g'(\lambda)$ (or higher derivatives) is also simply carried out. Hence we may apply any of the standard iterative methods for finding the roots of the polynomial $g(\lambda)$ (without evaluating its coefficients). A matter of considerable practical importance for the evaluation of the polynomial $g(\lambda)$ is the double precision accumulation of the inner products in (19) and (20).

Another family of methods for finding the eigenvalues of the lower Hessenberg matrix A are called *factorization methods.* First, we note that $C \equiv A^T$ and A have the same eigenvalues, but that C is of upper Hessenberg form. The first factorization method we describe is *Rutishauser's LR method.* That is, the LR method consists in constructing (when possible) the factorization of the matrix $C_1 \equiv C$ in the form

$$C_1 = L_1R_1, \tag{21}$$

where L is lower triangular (with unit diagonal elements) and R is upper triangular. Then Rutishauser considers

$$L_1^{-1}C_1L_1 = R_1L_1 \equiv C_2. \tag{22}$$

Next find

$$C_2 = L_2R_2, \tag{23}$$

again a lower unit and upper triangular factorization.

Now

$$L_2^{-1}C_2L_2 = L_2^{-1}L_1^{-1}C_1L_1L_2 = R_2L_2 \equiv C_3.$$

In general, a sequence of *similar* matrices $\{C_k\}$ is constructed and their L_kR_k factorization via Gaussian elimination is also found, so that

$$\begin{aligned} C_k &= L_kR_k, \\ C_{k+1} &= L_k^{-1}C_kL_k \\ &= L_k^{-1}\cdots L_1^{-1}C_1L_1\cdots L_k. \end{aligned} \tag{24}$$

But if we define

$$P_k = L_1\cdots L_k,$$

$$Q_k = R_k\cdots R_1,$$

then

$$P_kC_{k+1} = C_1P_k,$$

and therefore

$$\begin{aligned} P_kQ_k &= P_{k-1}C_kQ_{k-1} = C_1P_{k-1}Q_{k-1} \\ &= C_1^2P_{k-2}Q_{k-2} \\ &\;\vdots \\ &= C_1^k. \end{aligned} \tag{25}$$

Hence, P_kQ_k is the LR factorization of $C_1{}^k$. The fact that the matrix C_k converges to an upper triangular form is shown under special assumptions (which may be weakened considerably).

THEOREM **5** (RUTISHAUSER). *Let $C_1X = X\Lambda$, where $\Lambda \equiv (\lambda_i\delta_{ij})$. Assume $|\lambda_1| > |\lambda_2| > \cdots > |\lambda_n| > 0$. Set*

$$Y \equiv X^{-1}.$$

Assume X and Y can be factored in the form

$$X = L_XR_X, \qquad Y = L_YR_Y,$$

where L_X and L_Y are lower unit triangular and R_X and R_Y are upper triangular. Then with $\{C_k\}$ defined by (21), *we have the result: C_k converges to upper triangular form.*

Proof (*Wilkinson*). Clearly,

$$C_1{}^k = X\Lambda^kY = X\Lambda^kL_YR_Y$$
$$= X(\Lambda^kL_Y\Lambda^{-k})(\Lambda^kR_Y).$$

But by the strict inequalities satisfied by $\{\lambda_i\}$, the lower triangular matrix E_k, defined by

$$\Lambda^kL_Y\Lambda^{-k} - I \equiv E_k \equiv (e_{ij}^{(k)})$$

satisfies

$$E_k \rightarrow O; \qquad e_{ii}^{(k)} = 0, \qquad i = 1, 2, \ldots, n.$$

Therefore

$$C_1{}^k = L_XR_X(I + E_k)\Lambda^kR_Y$$
$$= L_X(I + R_XE_kR_X{}^{-1})R_X\Lambda^kR_Y.$$

But $R_XE_kR_X{}^{-1} \rightarrow O$. Therefore, the LR factors of $I + R_XE_kR_X{}^{-1}$ both converge to I.

Hence, since $R_X\Lambda^kR_Y$ is upper triangular, we see that the lower triangular factor of $C_1{}^k$, which by (25) is P_k, converges to L_X. That is,

$$P_k \rightarrow L_X.$$

Hence $L_k = P_{k-1}^{-1}P_k$ converges to I.

But then, since

$$L_k{}^{-1}C_k = R_k,$$

is upper triangular, it follows that C_k must converge to upper triangular form. ∎

It is easy to verify that the LR method preserves the Hessenberg form of the matrices $\{C_i\}$. The LR method can be made to converge much more rapidly by introducing a shift

$$D_k = C_k - s_kI,$$

and then continuing with the factorization of D_k.

But we shall not pursue this avenue further. We merely remark that the *LR* method does not always work as well as we have described, because the *LR* factorization, even if possible, may give rise to ever increasing magnitudes of numbers.

Another factorization method which seems not to suffer from the above defect is the *QR* factorization of *Francis* and *Kublanovskaja*. That is, the upper Hessenberg matrix C_1 is written

$$C_1 = Q_1 R_1,$$

whence

$$Q_1^{-1} C_1 Q_1 = R_1 Q_1 \equiv C_2,$$

where R_1 is upper triangular and Q_1 is unitary, i.e.,

$$Q_1^* = Q_1^{-1}.$$

We then factor

$$C_2 = Q_2 R_2.$$

In general, with

$$C_k = Q_k R_k,$$

we set

$$Q_k^{-1} C_k Q_k \equiv C_{k+1} = R_k Q_k.$$

All the matrices Q_k are unitary and all the matrices R_k are upper triangular. Again, the Hessenberg form of $\{C_i\}$ is preserved.

Francis and Kublanovskaja have given proofs of convergence of C_k to upper triangular form, in special cases. Wilkinson has given a simpler proof using techniques similar to those used in proving Theorem 5.

An important feature of Francis' work is that he shows how to use real arithmetic and maintain the real Hessenberg form, even when some eigenvalues are complex and the accelerating shifts indicated above (for the *LR* method) are complex numbers. Since he works with real numbers, when the eigenvalues are distinct but some are complex conjugate, in pairs

$$\lambda_i = \bar{\lambda}_{i+1},$$

then the matrices C_k converge to a form that is not triangular (i.e., the limiting form has second order matrix blocks on the diagonal).†

The *QR* factorization is unique when C_1 is non-singular. The Gram-Schmidt orthogonalization process is not recommended for carrying out the *QR* factorization. Rather, left-multiplications may be performed upon the matrix C_1, by unitary matrices of the form $I - 2\boldsymbol{\omega}\boldsymbol{\omega}^*$, in order to successively reduce the columns of C_1. That is,

$$(I - 2\boldsymbol{\omega}_{n-1}\boldsymbol{\omega}_{n-1}^*)\cdots(I - 2\boldsymbol{\omega}_1\boldsymbol{\omega}_1^*)C_1 = R_1.$$

† These real matrices C_k are real orthogonally similar and therefore can only be expected to converge to the *Murnaghan-Wintner* canonical form rather than the Schur form of Theorem 2.1.

The matrix Q_1 is given by the product of the unitary matrices,

$$Q_1 = (I - 2\boldsymbol{\omega}_1\boldsymbol{\omega}_1^*)(I - 2\boldsymbol{\omega}_2\boldsymbol{\omega}_2^*)\cdots(I - 2\boldsymbol{\omega}_{n-1}\boldsymbol{\omega}_{n-1}^*).$$

When C_1 is in Hessenberg form, these unitary matrices are simple two dimensional rotations.

PROBLEMS, SECTION 3

1. With A_{ij}, R, and B_{ij} as defined in Lemma 2, show that $b_{ij} = b_{ji} = 0$.

2. Give the details of the proof of Theorem 2.

[Hint: The only elements of B_{k-1} that are transformed by P_k are in the rows and columns with indices i or j. It is necessary only to examine the components (where $j > i \geq 2$)

$$b_{is}^{(k)}, \qquad \text{if } s \leq i - 2;$$

and

$$b_{sj}^{(k)}, \qquad \text{if } 1 \leq s \leq i - 1.]$$

3. Verify that for any matrix P of the form (13), $P^* = P$ and

$$P^*P = PP^* = I.$$

4. Carry out the verification of Theorem 3.

[Hint: Rows $i = 1, 2, \ldots, k - 1$ are unaffected by $(15)_k$. Check that row k is properly reduced.]

5. Verify that $p_n(\lambda)$ given by (18) satisfies $p_n(\lambda) \equiv p_B(\lambda)$.

6. Complete the proof of Theorem 4. That is, use the hypothesis and assume the root separation property to prove the recipe for counting eigenvalues.

7. What recurrence relation is satisfied by the functions $\{p_i'(\lambda)\}$, where $\{p_i(\lambda)\}$ is defined in (18)?

8. Let B be tridiagonal, of form (17); $c_i \neq 0$, $i = 1, 2, \ldots, n - 1$; and λ be an eigenvalue of B. Use Gaussian elimination with maximal column pivots to solve $(B - \lambda I)\mathbf{x} = \mathbf{o}$. Let $U \equiv (u_{ij})$ be the matrix of the resulting equivalent upper triangular system. If $\max_{i,j} (|a_i|, |c_j|) = b$, then $\max_{i,j} |u_{ij}| \leq 5b$ (Wilkinson).

[Hint: By Gerschgorin's theorem $|\lambda| \leq 3b$. Use induction and examine the two equations involved in eliminating x_i.]

9. Let A be Hermitian of order n and μ be an approximation to the simple eigenvalue λ. Let the eigenvector $\mathbf{x}$ correspond to λ. For simplicity, suppose $\max_i |x_i| = x_n = 1$. The usual way to approximate $\mathbf{x}$ consists in solving the $n - 1$ equations

$$(26) \qquad (\hat{A} - \mu\hat{I})\hat{\mathbf{x}} = -\hat{\mathbf{c}},$$

where $\hat{A}$, $\hat{I}$, $\hat{\mathbf{x}}$, $\hat{\mathbf{c}}$ consist in deleting the last row and column of A and I and deleting the last component both of $\mathbf{x}$ and of the last column $\mathbf{c}$ of A, respectively. Then define the residual

$$(27) \qquad f(\mu) = \hat{\mathbf{c}}^*\hat{\mathbf{x}} + a_{nn} - \mu.$$

(a) From (26), since $\hat{\mathbf{x}}$ is a function of μ, verify by differentiation with respect to μ that

$$(\hat{A} - \mu\hat{I})\frac{d\hat{\mathbf{x}}}{d\mu} - \hat{\mathbf{x}} = \hat{\mathbf{o}}.$$

If μ is close enough to λ, and $\hat{A} - \lambda\hat{I}$ is non-singular, then

$$\frac{d\hat{\mathbf{x}}}{d\mu} = (\hat{A} - \mu\hat{I})^{-1}\hat{\mathbf{x}}.$$

From (27),

$$\frac{df}{d\mu} = \hat{\mathbf{c}}^*(\hat{A} - \mu\hat{I})^{-1}\hat{\mathbf{x}} - 1.$$

(b) Verify, then, that in Newton's method for improving μ

$$\Delta\mu = -\frac{f(\mu)}{f'(\mu)} = \frac{\mathbf{x}^*\boldsymbol{\eta}}{\mathbf{x}^*\mathbf{x}},$$

where $\eta_i = 0$ for $1 \leq i \leq n - 1$; $\eta_n = f(\mu)$.

10. Define the *circulant matrix* $A \equiv (a_{ij})$, of order n, generated from $(a_1, a_2, \ldots, a_n)$ by

$$a_{ij} \equiv \begin{cases} a_{j-i+1}, & i \leq j, \\ a_{n+j-i+1}, & i > j. \end{cases}$$

Show that the eigenvectors $\{\mathbf{u}_j\}$ and eigenvalues $\{\lambda_j\}$ are given in terms of the n roots of unity $\{\omega_j\}$ [i.e., $(\omega_j)^n = 1$] by

$$\mathbf{u}_j \equiv (1, \omega_j, \omega_j^2, \ldots, \omega_j^{n-1}),$$

$$\lambda_j \equiv a_1 + a_2\omega_j + a_3\omega_j^2 + \cdots + a_n\omega_j^{n-1}.$$

11. For the circulant matrix generated by $a_i = 1$, $1 \leq i \leq 6$, use Householder's method and Jacobi's method to find the eigenvalues and eigenvectors.

12. Show that Householder's method reduces a real skew-symmetric matrix to a real tridiagonal skew-symmetric matrix. Carry out this procedure for the circulant matrix generated by $(a_1, a_2, a_3, a_4, a_5, a_6) \equiv (0, 1, 1, 0, -1, -1)$.

13. If C_1 is of Hessenberg form, then show that each stage of the LR transformation requires $n^2 + \mathcal{O}(n)$ operations, while each stage of the QR transformation requires $4n^2 + \mathcal{O}(n)$ operations.

5

Basic Theory of Polynomial Approximation

0. INTRODUCTION

There are numerous reasons for seeking approximations to functions. The type of approximation sought depends upon the application intended as well as the ease or difficulty with which it can be obtained. In any event, the "simplest"† approximating functions would seem to be polynomials, and we devote much of our attention to them in this chapter. Some consideration is also given to approximation by trigonometric functions. We shall study the approximation of continuous (possibly differentiable) functions, in a closed bounded interval.

In general, a polynomial, say $P_n(x)$ of degree at most n, may be said to be an approximation to a function, $f(x)$, in an interval $a \leq x \leq b$ if some measure of the deviation of the polynomial from the function in this interval is "small." This notion of approximation becomes precise only when the measure of deviation and magnitude of smallness have been specified.

To this end, we recapitulate the definition of *norm*, this time for a linear space (not necessarily finite dimensional) whose elements are functions $\{f(x)\}$ (see Chapter 1, Section 1). The norm, written

$$\text{Norm}(f) \equiv N(f) \equiv \|f\|,$$

is an assignment of a real number to each element of the linear space such that:

† We do not study the theory of approximation by *rational functions* (i.e., quotient of polynomials), even though rational functions are easy to evaluate and, in certain cases, are more efficient than polynomials.

(i) $\|f\| \geq 0$,
(ii) $\|f\| = 0$ iff $f(x) \equiv 0$,
(iii) $\|cf\| = |c| \cdot \|f\|$ for any constant c,
(iv) $\|f + g\| \leq \|f\| + \|g\|$, the *triangle inequality*.

The notion of linear spaces of functions is of basic importance in the analysis of many approximation procedures. In particular, for present applications we wish to examine not the polynomial approximation of a particular function but, in fact, the properties of such approximations for any function in an appropriate linear space.

A *measure of the deviation* or *error* in the approximation of $f(x)$ by $P_n(x)$ will be denoted generically by

$$\mathbf{|} f(x) - P_n(x) \mathbf{|}$$

(sometimes with an appropriate subscript or superscript). This measure of deviation will be required to satisfy the properties (i), (iii), and (iv) of a *norm*, but not necessarily property (ii), because $\mathbf{|} f \mathbf{|} = 0$ may not imply $f(x) \equiv 0$. Such a measure is actually called a *semi-norm*. If we were to introduce here equivalence classes of functions, i.e., identify $f(x)$ and $g(x)$ if $\mathbf{|} f - g \mathbf{|} = 0$, then the measure $\mathbf{|} \cdot \mathbf{|}$ becomes a norm in a natural way in the linear space composed of these equivalence classes. For simplicity, we refer to $\mathbf{|} \cdot \mathbf{|}$ as a norm in this chapter, even though we do not formally introduce the equivalence classes of functions. Once such a norm has been defined three questions are naturally suggested:

(a) Does a polynomial exist, of a specified maximum degree, which minimizes the error?
(b) If such a polynomial exists, is it unique?
(c) If such a polynomial exists, how can it be determined?

With the convention that, unless otherwise specified, $P_n(x)$ represents a polynomial of degree at most n, it is clear that

$$\text{(1)} \qquad \underset{\{P_n(x)\}}{\text{g.l.b.}} \; \mathbf{|} f(x) - P_n(x) \mathbf{|} \equiv d_n \geq 0,$$

is a monotonic non-increasing function of n. If there exists a unique polynomial $P_n(x)$ for which the minimum error is attained, we may then investigate methods for determining $P_n(x)$ and the magnitude of d_n (or any other norm of the deviation). In particular, we are most interested in those approximation methods for which $d_n \to 0$ as $n \to \infty$.

An example for which questions (a), (b), and (c) are easily answered is furnished by a well-known polynomial approximation: the first $m + 1$ terms in the Taylor expansion of $f(x)$ about x_0. That is, we consider the linear space composed of functions $f(x)$ which have derivatives of order n at $x = x_0$. For any given $f(x)$ in this space we define

$$(2) \qquad P_m(x) \equiv f(x_0) + \frac{(x - x_0)}{1!} f^{(1)}(x_0) + \frac{(x - x_0)^2}{2!} f^{(2)}(x_0) + \cdots + \frac{(x - x_0)^m}{m!} f^{(m)}(x_0), \qquad m \leq n.$$

This polynomial clearly minimizes the measure of error

$$(3) \qquad \| g(x) \|^{(n)} \equiv \sum_{k=0}^{n} |g^{(k)}(x_0)|,$$

with $g(x) \equiv f(x) - Q_m(x)$ among all polynomials $\{Q_m\}$ of degree at most m, since

$$\| f(x) - P_m(x) \|^{(n)} = \begin{cases} 0, & \text{if } m = n, \\ \sum_{k=m+1}^{n} |f^{k}(x_0)|, & \text{if } m < n. \end{cases}$$

Thus existence and explicit construction of a best approximating polynomial for this somewhat contrived norm† are demonstrated. Uniqueness of $P_m(x)$ for a given $m \leq n$ can be proven by assuming that there is some other polynomial of degree at most m, say $Q_m(x)$, which also minimizes the error. By expressing $Q_m(x)$ as a polynomial in $(x - x_0)$ we find that the coefficients $Q_m^{(k)}(x_0)/k!$ must be identical with those of $P_m(x)$ given in (2), since otherwise, $\| f - Q_m \|^{(n)} > \| f - P_m \|^{(n)}$.

Now, if $f(x)$ has an $(n + 1)$st derivative in some interval about x_0, say $|x - x_0| \leq a$, then by Taylor's theorem the remainder in the expansion (or we may call it the pointwise error in the approximation) is given by

$$(4) \qquad R_n(x) \equiv f(x) - P_n(x) = \frac{(x - x_0)^{n+1}}{(n + 1)!} f^{(n+1)}(\xi),$$

where $|x - x_0| \leq a$ and $\xi = \xi(x)$ is some point in the open interval (x, x_0). For the special function $f(x) \equiv 1/(1 + x)$ in the interval $-\frac{1}{2} \leq x \leq 2$ and $x_0 = 0$, we find $P_n(x) \equiv 1 - x + \cdots + (-1)^n x^n$ and

$$R_n(x) \equiv \frac{(-1)^{n+1} x^{n+1}}{1 + x}.$$

In this case, although $\| R_n(x) \|^{(n)} = 0$, we note that for the *maximum norm*

$$\| R_n(x) \|_\infty \equiv \underset{-\frac{1}{2} \leq x \leq 2}{\text{l.u.b.}} |R_n(x)| \geq \frac{2^{n+1}}{3}$$

† Note that we may have $\| g(x) \|^{(n)} = 0$ but $g(x) \not\equiv 0$, e.g., $g(x) \equiv (x - x_0)^{n+1}$. Thus, on the indicated space, (3) is an example of a semi-norm which is not a norm.

and hence some caution must be used. What may be a good approximation when measured in one norm [i.e., in $\|\cdot\|^{(n)}$ of (3)] may be a very poor approximation in another norm (i.e., $\|\cdot\|_\infty$). But, in this example, if the interval were $-\frac{1}{2} \le x \le \frac{1}{2}$, then

$$\|R_n(x)\|_\infty \equiv \underset{-\frac{1}{2} \le x \le \frac{1}{2}}{\text{l.u.b.}} |R_n(x)| \le 2^{-n}$$

and we find that the Taylor series, $P_n(x)$, converges uniformly with respect to x for this function. In fact, the series converges uniformly in any closed interval contained in the open interval $(-1, 1)$. The latter property of Taylor series is typical and is more fully treated in the study of analytic functions of a complex variable.

The questions (a), (b), and (c) can be answered in many other specific cases and we do this for several different norms in this chapter. However, question (a), of existence, can be given an affirmative answer quite generally. We do this in Theorem 1. The answer to question (b), on uniqueness, is a qualified yes given in Theorem 2. (For all the specific approximation problems treated in this chapter the answer is yes.) A general answer to question (c) is not known but we show how to construct the minimizing polynomial for several norms.

For the general results to be presented we assume that the polynomials and the functions, $f(x)$, to be approximated are in a linear space $C[a, b]$ of functions defined on the closed bounded interval, $[a, b]$. Then we have

THEOREM 1. *Let the measure of deviation* $\|\cdot\|$ *be defined in* $C[a, b]$, *and let there exist positive numbers* m_n *and* M_n *which satisfy*

$$0 < m_n \le \left\| \sum_{j=0}^{n} b_j x^j \right\| \le M_n, \qquad n = 0, 1, \ldots, \tag{5}$$

for all $\{b_j\}$ *such that*

$$\sum_{j=0}^{n} |b_j|^2 = 1.$$

Then for any integer n *and* $f(x)$ *in* $C[a, b]$ *there exists a polynomial of degree at most* n *for which*

$$\|f(x) - P_n(x)\|$$

attains its minimum over all such polynomials.

Proof. Write the general nth degree polynomial as

$$P_n(x) = a_0 + a_1 x + \cdots + a_n x^n,$$

and consider the function of the $n + 1$ coefficients

$$\phi(a_0, a_1, \cdots, a_n) \equiv \|f(x) - P_n(x)\|.$$

By the properties (iii) and (iv) of norms and the hypothesis (5), we obtain the continuity of ϕ, as follows:

$$\begin{aligned}\phi(a_0 + \epsilon_0, a_1 + \epsilon_1, \ldots, a_n + \epsilon_n) &= \left\| f(x) - P_n(x) - \sum_{j=0}^{n} \epsilon_j x^j \right\|, \\ &\leq \phi(a_0, a_1, \ldots, a_n) + \left\| \sum_{j=0}^{n} \epsilon_j x^j \right\|, \\ &\leq \phi(a_0, a_1, \ldots, a_n) + \sum_{j=0}^{n} |\epsilon_j| \cdot \|x^j\|, \\ &\leq \phi(a_0, a_1, \ldots, a_n) + M_n \sum_{j=0}^{n} |\epsilon_j|.\end{aligned}$$

Similarly, we find

$$\begin{aligned}\phi(a_0, a_1, \ldots, a_n) &= \left\| f(x) - P_n(x) - \sum_{j=0}^{n} \epsilon_j x^j + \sum_{j=0}^{n} \epsilon_j x^j \right\| \\ &\leq \phi(a_0 + \epsilon_0, a_1 + \epsilon_1, \ldots, a_n + \epsilon_n) + M_n \sum_{j=0}^{n} |\epsilon_j|.\end{aligned}$$

Hence for any $\{a_j\}$ and $\{\epsilon_j\}$

$$(6) \qquad |\phi(a_0 + \epsilon_0, a_1 + \epsilon_1, \ldots, a_n + \epsilon_n) - \phi(a_0, a_1, \ldots, a_n)| \leq M_n \sum_{j=0}^{n} |\epsilon_j|.$$

This demonstrates that $\phi(a_0, a_1, \ldots, a_n)$ is a continuous function of the coefficients $(a_0, a_1, \ldots, a_n)$. (Compare Lemma 1.1 of Chapter 1.)

Since $\phi(a_0, a_1, \ldots, a_n) \geq 0$, the "minimum deviation" in (1) can be characterized as:

$$(7) \qquad \underset{(a_0, a_1, \ldots, a_n)}{\text{g.l.b.}} \ \phi(a_0, a_1, \ldots, a_n) \equiv d_n \geq 0.$$

Thus, the existence problem is reduced to showing that there is a set of coefficients, say $(\hat{a}_0, \hat{a}_1, \ldots, \hat{a}_n)$, such that

$$\phi(\hat{a}_0, \hat{a}_1, \ldots, \hat{a}_n) = d_n.$$

However, since ϕ is continuous, the result will follow from a theorem of Weierstrass if we can show that d_n is the g.l.b. of ϕ in an appropriate closed bounded domain of the coefficients. That is, we will show that for some $R > 0$,

$$(8) \qquad d_n = \underset{\sum_{j=0}^{n} |a_j|^2 \leq R^2}{\text{g.l.b.}} \ \{\phi(a_0, a_1, \ldots, a_n)\}.$$

Then since the continuous function $\phi(a_0, a_1, \ldots, a_n)$ attains its minimum d_n on the closed bounded set $\sum_{j=0}^{n} |a_j|^2 \leq R^2$, the theorem follows.

To verify (8) we observe, using (iii) and (iv), that

$$\|f - g\| \leq \|f\| + \|g\|,$$

and setting $f \equiv u + g$ we get

$$\|u + g\| \geq \|u\| - \|g\|. \tag{9}$$

Therefore, for any constant $\mu > 0$, by (9) and (iii),

$$\begin{aligned} \|f - P_n(x)\| &\geq \|P_n(x)\| - \|f\|, \\ &\geq \mu \left\| \frac{1}{\mu} P_n(x) \right\| - \|f\|. \end{aligned}$$

Let us pick μ such that $\sum_{j=0}^{n} |a_j/\mu|^2 = 1$. Then by (5) in the hypothesis of the theorem

$$\left\| \frac{1}{\mu} P_n(x) \right\| = \left\| \sum_{j=1}^{n} \frac{a_j}{\mu} x^j \right\| \geq m_n,$$

and the previous inequality implies

$$\|f - P_n(x)\| \geq \mu m_n - \|f\|.$$

So, if μ satisfies

$$\mu \geq \frac{\|f\| + d_n + 1}{m_n},$$

then

$$\|f - P_n(x)\| \geq d_n + 1.$$

Thus, we conclude that (8) is valid with the choice $R = (\|f\| + d_n + 1)/m_n$ and the proof is ended. ∎

Observe that the function $f(x)$ need not be continuous. Furthermore, note that this theorem gives no estimate of the magnitude of d_n. The semi-norm (3) satisfies condition (5) for $x_0 = 0$, with $m_n = (n + 1)^{-1/2}$, $M_n = \left(\sum_{j=0}^{n} (j!)^2 \right)^{1/2}$; see Problem 2.

For the uniqueness result, we require the measure of deviation to be *strict*. By definition *a norm* $\|\cdot\|$ *is strict* if

$$\|f + g\| = \|f\| + \|g\|$$

implies there exist constants α, β such that $|\alpha| + |\beta| \neq 0$ and

$$\alpha f(x) + \beta g(x) \equiv 0.$$

Now we state

THEOREM 2. *To the hypothesis of Theorem* 1 *add the requirement that* $\|\cdot\|$ *is strict. Then the minimizing polynomial, say* $P_n(x)$, *is unique.*

Proof. Assume, for a given $f(x)$, that $P_n(x)$ and $Q_n(x)$ both minimize (1). Then from (7), (iii), and (iv), we find

$$d_n \leq \|f - \tfrac{1}{2}(P_n + Q_n)\| \leq \tfrac{1}{2}\|f - P_n\| + \tfrac{1}{2}\|f - Q_n\| = d_n.$$

Hence, equality holds throughout, and since $\|\cdot\|$ is strict there exist non-trivial constants α and β such that

$$\frac{\alpha}{2}[f(x) - P_n(x)] + \frac{\beta}{2}[f(x) - Q_n(x)] \equiv 0.$$

Now if $\alpha = -\beta \neq 0$ then $P_n(x) \equiv Q_n(x)$. Otherwise, $\alpha \neq -\beta$ and then $f(x)$ must be a polynomial of degree at most n. In this case, $d_n = 0$, and using (5) it follows that $P_n(x) \equiv f(x) \equiv Q_n(x)$. ∎

Again, observe that the function $f(x)$ need not be continuous, as remarked after the proof of Theorem 1. This theorem is valid for a semi-norm which satisfies the appropriate additional conditions [i.e., (5) and strictness]. Of course, the minimizing polynomial may be unique even though the norm is not strict. Such an instance is furnished by the semi-norm (3) which is not strict. Further examples follow.

PROBLEMS, SECTION 0

1. Show that if $\|\cdot\|$ is a norm defined for polynomials and satisfying (i)–(iv), then

$$\underset{\sum_{j=0}^{n}|b_j|^2=1}{\text{g.l.b.}} \left\|\sum_{j=0}^{n} b_j x^j\right\| \equiv m_n > 0.$$

[Hint: Prove that

$$\psi(a_0, a_1, \ldots, a_n) \equiv \left\|\sum_{j=0}^{n} a_j x^j\right\|$$

is a continuous function of $(a_0, a_1, \ldots, a_n)$.]

2. Verify that for the semi-norm

$$\|g(x)\|^{(n)} \equiv \sum_{k=0}^{n} |g^{(k)}(0)|,$$

(5) is satisfied with

$$m_n = (n + 1)^{-1/2}, \qquad M_n = \left[\sum_{j=0}^{n} (j!)^2\right]^{1/2}.$$

[Hint: Note that some b_j satisfies $|b_j| \geq (n + 1)^{-1/2}$. On the other hand, apply Schwarz' inequality to estimate $\sum_{j=0}^{n} (j!)|b_j|$.]

1. WEIERSTRASS' APPROXIMATION THEOREM AND BERNSTEIN POLYNOMIALS

We are justified in seeking a close polynomial approximation to a continuous function throughout a finite interval because of the fundamental

WEIERSTRASS' APPROXIMATION THEOREM. *Let $f(x)$ be any function continuous in the (closed) interval $[a, b]$. Then for any $\epsilon > 0$ there exists an integer $n = n(\epsilon)$ and a polynomial $P_n(x)$ of degree at most n such that*

$$|f(x) - P_n(x)| < \epsilon,$$

for all x in $[a, b]$.

This theorem guarantees that arbitrarily close polynomial approximations are possible throughout a closed bounded interval provided only that the function being approximated is continuous. The statement of the theorem is one of existence and gives no hint about constructing such approximations. However, a simple and elegant constructive proof of this result is due to Bernstein and we shall present it in Theorem 1.

First, a basic notion in analysis must be recalled and some preliminary identities will be introduced. If $f(x)$ is a continuous function in a closed interval, say $[0, 1]$,† then the *modulus of continuity of $f(x)$ in $[0, 1]$ is defined as*

$$\omega(f; \delta) \equiv \underset{\left\{\substack{x,\, x' \text{ in } [0,1] \\ |x - x'| \leq \delta}\right\}}{\text{l.u.b.}} |f(x) - f(x')|. \tag{1}$$

Since $f(x)$ is continuous in a closed interval, and hence uniformly continuous, it follows that

$$\lim_{\delta \to 0} \omega(f; \delta) = 0.$$

If, in addition, $f(x)$ satisfies a *Lipschitz condition* in $[0, 1]$, i.e., if

$$|f(x) - f(x')| \leq \lambda |x - x'|,$$

for x, x' in $[0, 1]$ and some constant λ, then from (1):

$$\omega(f; \delta) \leq \lambda\delta.$$

The concept of a modulus of continuity is generally useful in analysis and its use will recur in our study.

† We need only consider this case since:
An arbitrary finite interval $a \leq y \leq b$ is mapped 1–1 onto the unit interval $0 \leq x \leq 1$ by the continuous change of variable: $x = (a - y)/(a - b)$ or $y = (b - a)x + a$. Hence, if $g(y)$ is continuous in $[a, b]$, $f(x) \equiv g((b - a)x + a)$ is continuous in $[0, 1]$. Now, if $P_n(x)$ approximates $f(x)$ in $[0, 1]$ to within ϵ, then $Q_n(y) \equiv P_n((a - y)/(a - b))$ is a polynomial of degree at most n that is within ϵ of $g(y)$ in $[a, b]$.

The identities required all follow from the well-known binomial expansion

$$(a + b)^n = \sum_{j=0}^{n} \binom{n}{j} a^j b^{n-j}. \tag{2a}$$

Upon forming $\partial(a + b)^n/\partial a$ and $\partial^2(a + b)^n/\partial a^2$ we obtain the further identities

$$a(a + b)^{n-1} = \sum_{j=0}^{n} \frac{j}{n} \binom{n}{j} a^j b^{n-j}, \tag{2b}$$

$$\left(1 - \frac{1}{n}\right) a^2 (a + b)^{n-2} = \sum_{j=0}^{n} \left(\frac{j^2}{n^2} - \frac{j}{n^2}\right) \binom{n}{j} a^j b^{n-j}. \tag{2c}$$

Now set $a = x$, and $b = 1 - x$; define the nth degree polynomials

$$\beta_{n,j}(x) \equiv \binom{n}{j} x^j (1 - x)^{n-j}, \qquad j = 0, 1, \ldots, n; \tag{3}$$

and the identities (2) become

$$\sum_{j=0}^{n} \beta_{n,j}(x) = 1, \tag{4a}$$

$$\sum_{j=0}^{n} \frac{j}{n} \beta_{n,j}(x) = x, \tag{4b}$$

$$\sum_{j=0}^{n} \frac{j^2}{n^2} \beta_{n,j}(x) = \left(1 - \frac{1}{n}\right) x^2 + \frac{1}{n} x. \tag{4c}$$

It should be noted that for x in $[0, 1]$, $\beta_{n,j}(x) \geq 0$.

Let the unit interval $[0, 1]$ be subdivided into n equal subintervals with the endpoints

$$x_j = \frac{j}{n}, \qquad j = 0, 1, \ldots, n. \tag{5}$$

We finally introduce the *Bernstein polynomial of degree n for the function* $f(x)$ *on* $[0, 1]$ by the definition:

$$B_n(f; x) \equiv \sum_{j=0}^{n} f(x_j) \beta_{n,j}(x). \tag{6}$$

This is, from (3), clearly a polynomial of degree at most n and has coefficients depending upon the values of $f(x)$ at $n + 1$ equally spaced points in $[0, 1]$.

THEOREM 1. *Let $f(x)$ be any continuous function defined on* [0, 1]. *Then for all x in* [0, 1], *and any positive integer n*,

$$|f(x) - B_n(f; x)| \leq \tfrac{9}{4}\omega(f; n^{-1/2}). \tag{7}$$

Proof. From (4a) and (6) we may write

$$f(x) - B_n(f; x) = \sum_{j=0}^{n} [f(x) - f(x_j)]\beta_{n,j}(x) \equiv S_1(x) + S_2(x),$$

where we define, for any $\delta > 0$,

$$S_1(x) \equiv \sum_{j:|x-x_j|\leq\delta} \cdots, \qquad S_2(x) \equiv \sum_{j:|x-x_j|>\delta} \cdots.$$

Thus by the definition of $\omega(f; \delta)$ and (4a),

$$|S_1(x)| \leq \omega(f; \delta) \sum_{j:|x-x_j|\leq\delta} \beta_{n,j}(x)$$

$$\leq \omega(f; \delta) \sum_{j=0}^{n} \beta_{n,j}(x) = \omega(f; \delta).$$

For the remaining sum, since $|x - x_j| > \delta$, we note

$$f(x) - f(x_j) = [f(x) - f(\xi_1)] + [f(\xi_1) - f(\xi_2)] + \cdots$$
$$+ [f(\xi_{p-1}) - f(\xi_p)] + [f(\xi_p) - f(x_j)],$$

where $p \equiv [|x - x_j|/\delta]$,† and $\xi_1, \xi_2, \ldots, \xi_p$ are p points inserted uniformly between (x, x_j) where each of the $p + 1$ successive intervals is of length $|x - x_j|/(p + 1) < \delta$. Hence,

$$|f(x) - f(x_j)| \leq (p + 1)\omega(f; \delta) \leq \left(1 + \frac{|x - x_j|}{\delta}\right)\omega(f; \delta).$$

Therefore,

$$|S_2(x)| \leq \omega(f; \delta)\left[\sum_{j:|x-x_j|>\delta} \beta_{n,j}(x) + \frac{1}{\delta}\sum_{j:|x-x_j|>\delta} |x - x_j|\beta_{n,j}(x)\right]$$

$$\leq \omega(f; \delta)\left[1 + \frac{1}{\delta^2}\sum_{j:|x-x_j|>\delta} (x - x_j)^2\beta_{n,j}(x)\right]$$

$$\leq \omega(f; \delta)\left[1 + \frac{1}{\delta^2}\sum_{j=0}^{n} (x - x_j)^2\beta_{n,j}(x)\right].$$

From (5) and (4)

$$\sum_{j=0}^{n} (x - x_j)^2\beta_{n,j}(x) = \frac{x(1 - x)}{n} \leq \frac{1}{4n}.$$

† $p \equiv [x]$ for $x > 0$, means p is the largest integer satisfying $p \leq x$.

Therefore,

$$|S_2(x)| \le \omega(f, \delta)\left(1 + \frac{1}{4n\delta^2}\right)$$

and finally,

$$|f(x) - B_n(f; x)| \le |S_1(x)| + |S_2(x)|$$
$$\le \omega(f, \delta)\left(2 + \frac{1}{4n\delta^2}\right).$$

The error estimate (7) is obtained, if we choose $\delta = n^{-1/2}$. ■

Weierstrass' theorem follows by picking n so large that $\omega(f; n^{-1/2}) < 4\epsilon/9$.

If $f(x)$ satisfies a Lipschitz condition, we find easily:

COROLLARY. *Let $f(x)$ satisfy a Lipschitz condition*

$$|f(x) - f(y)| \le \lambda|x - y|,$$

for all x, y in [0, 1]. *Then for all x in* [0, 1],

$$(8) \qquad |f(x) - B_n(f; x)| \le \tfrac{9}{4}\lambda n^{-1/2}.$$ ■

It can be shown that the approximation given by $B_n(f; x)$ may be better than is implied in this result (see Problem 1). However, in general, even if $f(x)$ has p derivatives, the convergence is, at best, of order $1/n^2$. In fact, it can be shown that

$$\lim_{n\to\infty} n^2[B_n(f; x) - f(x)] = \tfrac{1}{2}f''(x)x(1 - x), \qquad \text{if } p \ge 2.$$

As such convergence is quite slow compared to that of many other polynomial approximation methods (see Theorem 9 of Section 3), the Bernstein polynomials are seldom used in practice. It should be emphasized, however, that they converge (uniformly) for any continuous function when many of the other polynomial approximations do not.

The Weierstrass approximation theorem is valid for functions of several variables which are continuous on appropriate sets. In fact, the Bernstein polynomials can again be employed to yield a constructive proof in various cases (see Problem 2).

PROBLEMS, SECTION 1

1. If $f(x)$ satisfies a Lipschitz condition with constant λ in [0, 1], show $E_n \equiv |f(x) - B_n(f; x)| \le (\lambda/2)n^{-1/2}$.

[Hint: Use Schwarz' inequality to get

$$E_n = |\Sigma\, [f(x) - f(x_j)](\beta_{n,j}(x))^{1/2}(\beta_{n,j}(x))^{1/2}|$$
$$\le \{\Sigma\, [f(x) - f(x_j)]^2\beta_{n,j}(x)\}^{1/2}.$$

Now estimate

$$\sum [f(x) - f(x_j)]^2\beta_{n,j}(x) \leq \lambda^2 \sum (x - x_j)^2\beta_{n,j}(x)$$
$$\leq \frac{\lambda^2}{4n}.]$$

2. Let $f(x, y)$ be continuous on the closed unit square: $0 \leq x \leq 1$, $0 \leq y \leq 1$. Then prove Weierstrass' theorem for this case by employing the polynomials:

$$B_{m,n}(f; x, y) \equiv \sum_{j=0}^{m} \sum_{k=0}^{n} f\left(\frac{j}{m}, \frac{k}{n}\right)\beta_{m,j}(x)\beta_{n,k}(y).$$

Show how to extend this theorem to functions of more variables continuous in arbitrary "cubes" with faces parallel to the coordinate planes.

[Hint: Let $R(x, y) \equiv f(x, y) - B_{m,n}(f; x, y)$. Show that

$$R(x, y) = \sum_{j,k} \left[f(x, y) - f\left(\frac{j}{m}, \frac{k}{n}\right)\right]\beta_{m,j}(x)\beta_{n,k}(y),$$

$$|R| \leq \Bigg| \sum_{j,k:\left\{\substack{|x - x_j| \leq \delta \\ \text{and} \\ |y - y_k| \leq \delta}\right\}} \Bigg| + \Bigg| \sum_{j:|x - x_j| > \delta} \Bigg| + \Bigg| \sum_{k:|y - y_k| > \delta} \Bigg|$$

and use the reasoning of the one variable case.

To prove the theorem when $f(x, y)$ is continuous in a square, $0 \leq x - a \leq c$, $0 \leq y - b \leq c$: define $g(u, v) \equiv f(cu + a, cv + b)$ for $0 \leq u, v \leq 1$. Then construct $B_{m,n}(g; u, v)$. Finally, set

$$P_{m,n}(x, y) \equiv B_{m,n}\left(g; \frac{x - a}{c}, \frac{y - b}{c}\right)$$

and observe that $|f(x, y) - P_{m,n}(x, y)|$ can be made small.]

3.* If $f(x)$ has a continuous first derivative in $[0, 1]$, show that the first derivatives of the Bernstein polynomials which approximate $f(x)$ converge to $f'(x)$ uniformly on $[0, 1]$.

[Hint: Verify that

$$\beta'_{n,j} = n(\beta_{n-1,j-1} - \beta_{n-1,j}) \qquad \text{for } j = 1, \ldots, n - 1,$$

$$\beta'_{n,n} = n\beta_{n-1,n-1}, \qquad \beta'_{n,0} = -n\beta_{n-1,0}.$$

Then regroup the sum in terms of $\beta_{n-1,k}$ for $k = 0, 1, \ldots, n - 1$.]

2. THE INTERPOLATION POLYNOMIALS

An approximation polynomial which is equal to the function it approximates at a number of *specified* points is called an *interpolation polynomial*. Given the $n + 1$ *distinct* points x_i, $i = 0, 1, \ldots, n$ and corresponding function values $f(x_i)$, the interpolation polynomial of degree at most n minimizes the norm:†

$$\|f(x) - P_n(x)\|_{\mathscr{I}} \equiv \sum_{i=0}^{n} |f(x_i) - P_n(x_i)|. \tag{1}$$

We shall show that such a polynomial exists (by explicit construction) and is unique; in fact, that the minimum value of the norm in (1) is 0.

The least possible value of $\|f - P_n\|_{\mathscr{I}}$ is, of course, zero. Thus, we seek a polynomial

$$Q_n(x) = \sum_{k=0}^{n} a_k x^k, \tag{2}$$

for which $Q_n(x_i) = f(x_i)$. By considering the coefficients a_k in (2) as unknowns, we have the system of $n + 1$ linear equations

$$Q_n(x_i) = a_0 + a_1 x_i + \cdots + a_n x_i^n = f(x_i), \qquad i = 0, 1, \ldots, n. \tag{3}$$

This system has a unique solution if the coefficient matrix is non-singular. The determinant of this matrix is called a *Vandermonde* determinant and can be easily evaluated (see Problem 1) to yield

$$\begin{vmatrix} 1 & x_0 & \cdots & x_0^n \\ 1 & x_1 & \cdots & x_1^n \\ \vdots & \vdots & & \vdots \\ 1 & x_n & \cdots & x_n^n \end{vmatrix} = \prod_{i>j} (x_i - x_j) \equiv \prod_{j=0}^{n-1} \left[\prod_{i=j+1}^{n} (x_i - x_j) \right]. \tag{4}$$

Since the $\{x_i\}$ are distinct points, the determinant does not vanish and (3) may be uniquely solved for the a_i to determine the interpolation polynomial. Another proof of uniqueness is given in Lemma 1.

Rather than solve the system (3), we may use an alternative procedure to obtain the interpolation polynomial directly. Set

$$P_n(x) = \sum_{j=0}^{n} f(x_j)\phi_{n,j}(x), \tag{5a}$$

where the $n + 1$ functions $\phi_{n,j}(x)$ are nth degree polynomials. We note that $P_n(x_i) = f(x_i)$ if the polynomials $\phi_{n,j}(x)$ satisfy:

$$\phi_{n,j}(x_i) = \delta_{ij}, \qquad i, j = 0, 1, \ldots, n.$$

† This is only a semi-norm. Of course, Theorem 0.1 shows that there is a $P_n(x)$ which minimizes the norm in (1), but we prove more here, namely that $d_n = 0$. We also prove uniqueness which is not covered by Theorem 0.2, since (1) is not strict.

Such polynomials are easily constructed, since the $\{x_i\}$ are distinct, i.e.,

$$\text{(6a)}\quad \phi_{n,j}(x) = \frac{(x - x_0)(x - x_1)\cdots(x - x_{j-1})(x - x_{j+1})\cdots(x - x_n)}{(x_j - x_0)(x_j - x_1)\cdots(x_j - x_{j-1})(x_j - x_{j+1})\cdots(x_j - x_n)},$$

$$j = 0, 1, \ldots, n.$$

By introducing $\omega_n(x) \equiv (x - x_0)(x - x_1)\cdots(x - x_n)$ we find that (6a) can be written in the brief form [where $\omega_n'(x_j) = (d\omega_n(x)/dx)_{x=x_j}$]:

$$\text{(6b)}\qquad \phi_{n,j}(x) = \frac{\omega_n(x)}{(x - x_j)\omega_n'(x_j)}.$$

The interpolation polynomial, especially when in the form (5), is called the *Lagrange interpolation polynomial* and the polynomials (6) are called the *Lagrange interpolation coefficients.* We can use the product notation for ϕ which yields

$$\text{(5b)}\qquad P_n(x) = \sum_{j=0}^{n} f(x_j) \prod_{\substack{k=0\\k\neq j}}^{n} \frac{x - x_k}{x_j - x_k}.$$

That the Lagrange interpolation polynomial is identical to the polynomial defined by (2) and (3) is a consequence of the following

LEMMA 1. *Let $P_n(x)$ and $Q_n(x)$ be any two polynomials, of degree at most n, for which*

$$P_n(x_i) = Q_n(x_i), \qquad i = 0, 1, 2, \ldots, n,$$

where the $n + 1$ points $\{x_i\}$ are distinct. Then $P_n(x) \equiv Q_n(x)$.

Proof. Define the polynomial

$$D_n(x) \equiv P_n(x) - Q_n(x),$$

which is of degree at most n. This polynomial has at least $n + 1$ distinct roots:

$$D_n(x_i) = 0, \qquad i = 0, 1, \ldots, n.$$

However, the only polynomial of degree at most n with more than n roots is the identically vanishing "polynomial" $D_n(x) \equiv 0$. ∎

In summary, there is only one polynomial of degree at most n for which (1) vanishes and it is given by (5) and (6). Of course, there are many other ways of representing this polynomial; since such considerations are of great practical interest they form a large part of the next chapter.

2.1. The Pointwise Error in Interpolation Polynomials

The *pointwise error* between a function, $f(x)$, and some polynomial approximation to it, $P_n(x)$, is defined as

$$\text{(7)}\qquad R_n(x) \equiv f(x) - P_n(x).$$

It is, of course, quite useful to have an explicit expression for this error and, if possible, simple bounds on it. Such information may yield, as in the example of Section 0, an estimate of the rapidity of convergence of $P_n(x)$ to $f(x)$ as $n \to \infty$ (or of divergence). Further, it facilitates a comparison of the different types of polynomial approximation.

For interpolation polynomials, a useful representation of $R_n(x)$ is readily obtained. This result may be stated as

THEOREM 1. *Let $f(x)$ have an $(n + 1)$st derivative, $f^{(n+1)}(x)$, in an interval $[a, b]$. Let $P_n(x)$ be the interpolation polynomial for $f(x)$ with respect to $n + 1$ distinct points x_i, $i = 0, 1, \ldots, n$ in the interval $[a, b]$ (i.e., $P_n(x_i) = f(x_i)$ and $x_i \in [a, b]$). Then for each $x \in [a, b]$ there exists a point $\xi = \xi(x)$ in the open interval:*

$$\text{(8)} \qquad \min(x_0, x_1, \ldots, x_n, x) < \xi < \max(x_0, x_1, \ldots, x_n, x),$$

such that

$$\text{(9)} \quad f(x) - P_n(x) \equiv R_n(x) = \frac{(x - x_0)(x - x_1)\cdots(x - x_n)}{(n + 1)!} f^{(n+1)}(\xi)$$

$$\equiv \frac{\omega_n(x)}{(n + 1)!} f^{(n+1)}(\xi).$$

Proof. Since

$$R_n(x_0) = R_n(x_1) = \cdots = R_n(x_n) = 0,$$

we define $S_n(x)$, for any $x \neq x_i$, by setting

$$\text{(10)} \qquad R_n(x) \equiv (x - x_0)(x - x_1)\cdots(x - x_n)S_n(x) = \omega_n(x)S_n(x).$$

Considering x to be fixed as above we also define a function $F(z)$ by

$$F(z) \equiv f(z) - P_n(z) - \omega_n(z)S_n(x).$$

Clearly, this function and its derivatives with respect to z are defined and continuous wherever $f(z)$ and its derivatives are defined and continuous; thus, $F^{(n+1)}(z)$ is defined in $[a, b]$. (See Problem 5 for a mild generalization.)

We see that $F(z)$ vanishes at $n + 2$ distinct points in $[a, b]$, namely

$$F(x_0) = F(x_1) = \cdots = F(x_n) = F(x) = 0.$$

Thus, there are $n + 1$ adjacent intervals in $[a, b]$ at whose endpoints $F(z)$ vanishes. Rolle's theorem is now applicable, since $F'(z)$ is defined in $[a, b]$. Therefore, in the interior of each of these intervals, there is at least one point at which $F'(z)$ vanishes. Thus, there are at least $n + 1$ distinct points in the interval (8) at which $F'(z) = 0$. They form at least n intervals such that in the interior of each, by another application of Rolle's theorem, the derivative of $F'(z)$ vanishes. That is, $F''(z) = 0$ for at least n distinct points in (8). By continuing this process we find that there is some point, say ξ, in (8) at which the $(n + 1)$st derivative of $F(z)$ vanishes.

However, since $P_n(z)$ is an nth degree polynomial,

$$\frac{d^{n+1}P_n(z)}{dz^{n+1}} = 0,$$

and a simple calculation yields

$$\frac{d^{n+1}}{dz^{n+1}}[\omega_n(z)S_n(x)] = (n + 1)!S_n(x).$$

Thus,

$$F^{(n+1)}(z) = f^{(n+1)}(z) - (n + 1)!S_n(x),$$

and since $F^{(n+1)}(\xi) = 0$ we obtain

$$S_n(x) = \frac{1}{(n + 1)!}f^{(n+1)}(\xi), \qquad \text{for } x \neq x_0, x_1, \ldots, x_n. \tag{11}$$

With this in (10) the theorem follows. It should be pointed out that, although $S_n(x)$ is not defined for $x = x_i$, the final result (9) is valid for all x in $[a, b]$ [in fact, since $R_n(x_i) = 0$, ξ for these values of x may be picked arbitrarily]. ∎

If the maximum and minimum of $f^{(n+1)}(x)$ in $[a, b]$ can be determined, (9) will yield bounds on the error. It should be noted that the error (9) for interpolation polynomials is similar to the remainder in Taylor's expansion (0.4). In fact, we might naïvely assume that if $|x - x_i| < |x - x_0|$ for $i = 1, 2, \ldots, n$ then the interpolation polynomial error is smaller than the error in Taylor's expansion about the point x_0. This assumption is not always justified since the terms $f^{(n+1)}(\xi)$ in (0.4) and (9) are not evaluated at the same point ξ for a given x.

Does the sequence of interpolation polynomials $\{P_n(x)\}$ converge to $f(x)$ in $[a, b]$ if $\{(x_0^{(n)}, \ldots, x_n^{(n)})\}$ covers $[a, b]$? This is a question that is not completely answered. In the case of uniform spacing [i.e., $x_0^{(n)} = a$, $x_j^{(n)} = x_0 + jh_n$, $h_n = (b - a)/n$], we illustrate the fact that divergence is to be expected by studying Runge's example, $f(x) = 1/(1 + x^2)$ over $[-5, 5]$ (see Chapter 6, Subsection 3.4). On the other hand, in Corollary 2, Theorem 2 of Section 5, we exhibit a sequence of non-uniformly spaced points, $\{(x_0^{(n)}, \ldots, x_n^{(n)})\}$, for which uniform convergence of $\{P_n(x)\}$ to $f(x)$ may be established for any function $f(x)$ with continuous second derivatives.†

† Amazingly, for any sequence $\{(x_0^{(n)}, x_1^{(n)}, \ldots, x_n^{(n)})\}$, there exists a *continuous* function $f(x)$ for which $|P_n(x) - f(x)| \nrightarrow 0$!

From the remainder theorem we can deduce some interesting and useful *identities* satisfied by the Lagrange interpolation coefficients. Since $d^{n+1}x^m/dx^{n+1} = 0$ for $m = 0, 1, \ldots, n$, the interpolation polynomials of degree n represent exactly all polynomials of degree at most n. Thus, with $f(x) \equiv x^m$ in (5), equation (9) yields

$$\sum_{j=0}^{n} x_j{}^m \phi_{n,j}(x) = x^m, \qquad m = 0, 1, \ldots, n. \tag{12}$$

The case $m = 0$ is particularly useful. [Compare (12) with equation (1.4) for the Bernstein polynomials.]

2.2. Hermite or Osculating Interpolation

The *osculating polynomial*, a generalization of the interpolation polynomial, is obtained by requiring agreement at the distinct points of interpolation, x_j, with the first $r_j - 1$ derivatives of $f(x)$. (This polynomial arises also as the limit of the interpolation polynomials when r_j points of ordinary interpolation approach each other at the point x_j.) This procedure contains, as special cases, Taylor's expansion and ordinary interpolation. The number of combinations is boundless, but, in fact, a representation of the osculating polynomial can be found together with an expression for the pointwise error (see Chapter 6, Section 1, Problem 10). We shall consider in detail the case in which the function and its first derivative are to be assigned at each point of interpolation. This special procedure is usually called *Hermite* or *osculatory interpolation*.

The problem is to find a polynomial of least degree, say $H_{2n+1}(x)$, such that:

$$\begin{aligned} f(x_j) &= H_{2n+1}(x_j), \\ f'(x_j) &= H'_{2n+1}(x_j), \end{aligned} \qquad j = 0, 1, \ldots, n. \tag{13}$$

By counting the data (i.e., $2n + 2$ conditions), we find that a polynomial of degree $2n + 1$ has the required number of undetermined coefficients. Thus, in analogy with the Lagrange interpolation formula (5), we seek a representation in the form

$$H_{2n+1}(x) = \sum_{j=0}^{n} f(x_j)\psi_{n,j}(x) + \sum_{j=0}^{n} f'(x_j)\Psi_{n,j}(x). \tag{14}$$

Here the polynomials $\psi_{n,j}(x)$ and $\Psi_{n,j}(x)$ are required to be of degree at most $2n + 1$ and to satisfy

$$\begin{aligned} \psi_{n,j}(x_i) &= \delta_{ij}, & \Psi_{n,j}(x_i) &= 0, \\ \psi'_{n,j}(x_i) &= 0, & \Psi'_{n,j}(x_i) &= \delta_{ij}, \end{aligned} \qquad i, j = 0, 1, \ldots, n. \tag{15}$$

Such polynomials are given in terms of the Lagrange interpolation coefficients, $\phi_{n,j}(x)$, as:

$$
(16) \qquad \begin{aligned} \psi_{n,j}(x) &\equiv [1 - 2\phi'_{n,j}(x_j)(x - x_j)]\phi^2_{n,j}(x), \\ \Psi_{n,j}(x) &\equiv (x - x_j)\phi^2_{n,j}(x). \end{aligned}
$$

The error in using (14) to approximate $f(x)$ is

$$
(17) \qquad f(x) - H_{2n+1}(x) = \frac{\omega_n^2(x)}{(2n+2)!} f^{(2n+2)}(\xi),
$$

provided $f(x)$ has a continuous derivative of order $2n + 2$. The point $\xi = \xi(x)$ is again in the interval determined by the points $x, x_0, \ldots, x_n$. The proof of formula (17) is left to Problem 3.

From equation (17) we easily deduce, in analogy with (12), the *identities:*

$$
(18) \qquad \sum_{j=0}^{n} x_j^m \psi_{n,j}(x) + m \sum_{j=0}^{n} x_j^{m-1} \Psi_{n,j}(x) = x^m,
$$
$$
m = 0, 1, \ldots, 2n + 1.
$$

PROBLEMS, SECTION 2

1. Evaluate the Vandermonde determinant to verify (4).

[Hint: Let each x_j, $j = 0, 1, \ldots, n$, in order, be considered variable and determine all the roots of the resulting polynomial. The remaining scalar factor is obtained by evaluating the coefficient of any specific term, say $(1 \cdot x_1 \cdot x_2^2 \cdot \cdots \cdot x_n^n)$. An alternative proof could be given by using mathematical induction and expanding the determinant with respect to the elements of the last column.]

2. Prove that the system (3) is non-singular by assuming that the homogeneous system has a non-trivial solution and using Lemma 1 to obtain a contradiction.

3. Derive the error formula, equation (17), for Hermite interpolation if $f(x)$ is sufficiently differentiable.

[Hint: Proceed exactly as in the derivation of the interpolation error and define: $F(z) \equiv f(z) - H_{2n+1}(z) - \omega_n^2(z)S_n(x)$. After the first application of Rolle's theorem, however, $F'(z)$ will have $2n + 2$ distinct zeros.]

4. Formulate the definition of the Hermite interpolation polynomial as a minimizing polynomial for the appropriate semi-norm. Does this semi-norm satisfy hypothesis (0.5) of Theorem 0.1? Is it strict?

5. Show that the conclusion of Theorem 1 follows under the weaker assumption: $f(x)$ is continuous in the closed interval $[a, b]$, but has the requisite derivatives only in the open interval (a, b).

3. LEAST SQUARES APPROXIMATION

A property which is frequently used to determine an approximating polynomial,

$$Q_n(x) = a_0 + a_1x + \cdots + a_nx^n, \tag{1}$$

is that the L_2 norm, or mean square error,

$$\|f(x) - Q_n(x)\|_2 \equiv \left\{ \int_a^b [f(x) - Q_n(x)]^2\, dx \right\}^{1/2} \tag{2}$$

be a minimum. For the general polynomial (1) and any appropriate† function $f(x)$, we define the function of $n + 1$ variables,

$$\phi(a_0, a_1, \ldots, a_n) \equiv \int_a^b [f(x) - Q_n(x)]^2\, dx. \tag{3}$$

The least squares polynomial approximation to $f(x)$ of degree at most n is then determined by finding a point $(\hat{a}_0, \hat{a}_1, \ldots, \hat{a}_n)$ in the $n + 1$ dimensional space for which ϕ is a minimum.

THEOREM 1. *For each appropriate function $f(x)$, there is a unique least squares polynomial approximation of degree at most n which minimizes* (2).

Proof. The hypotheses of Theorems 0.1 and 0.2 are satisfied by $\|\cdot\|_2$ (see Problem 10), whence existence and uniqueness are established. ∎

We now give an analytical description of a method for calculating the coefficients $\hat{\mathbf{a}} \equiv (\hat{a}_0, \hat{a}_1, \ldots, \hat{a}_n)$ of the polynomial that minimizes $\phi(\mathbf{a})$ in (3).

Since

$$\phi(a_0, a_1, \ldots, a_n) = \int_a^b f^2(x)\, dx - 2\sum_{i=0}^{n} a_i \int_a^b x^i f(x)\, dx + \sum_{i=0}^{n}\sum_{j=0}^{n} a_i a_j \int_a^b x^{i+j}\, dx, \tag{4}$$

ϕ is a quadratic function in the variables a_i. Now, at the minimum of ϕ the coefficients $\hat{\mathbf{a}}$ must satisfy

$$\left.\frac{\partial\phi(a_0, a_1, \ldots, a_n)}{\partial a_k}\right|_{\mathbf{a}=\hat{\mathbf{a}}} = 0, \qquad k = 0, 1, \ldots, n.$$

† The given function, $f(x)$, for this purpose need only be restricted so that it and its square are integrable over $[a, b]$. The *complete* linear space for which (2) is a norm consists of all such functions, if we identify two functions which differ only on a set of measure zero in $[a, b]$, and use the Lebesgue integral. But we do not pursue this avenue of generalization and shall, unless otherwise noted, consider only functions that are continuous, except at a finite number of points, where certain conditions will be specified.

From (4) this necessary condition becomes

$$0 = \left.\frac{\partial \phi}{\partial a_k}\right|_{\hat{\mathbf{a}}} = 0 - 2\int_a^b x^k f(x)\,dx + \sum_{i=0}^{n} \hat{a}_i \int_a^b x^{i+k}\,dx$$

(5a)
$$+ \sum_{j=0}^{n} \hat{a}_j \int_a^b x^{k+j}\,dx$$

$$= 2\left[\sum_{i=0}^{n} \hat{a}_i \int_a^b x^{i+k}\,dx - \int_a^b x^k f(x)\,dx\right],$$

$$k = 0, 1, \ldots, n.$$

A system of $n + 1$ linear equations for the determination of the $\{\hat{a}_i\}$ is defined in (5a); it is frequently called the *normal system.*

We write the normal system (5a) in the form:

(5b)
$$\sum_{j=0}^{n} h_{ij}\hat{a}_j = c_i, \qquad i = 0, 1, \ldots, n;$$

where the coefficient matrix and right-hand side are given by

(5c)
$$H_{n+1}(a, b) \equiv (h_{ij}), \qquad h_{ij} \equiv \int_a^b x^{i+j}\,dx; \; c_i \equiv \int_a^b x^i f(x)\,dx.$$

Now we have

THEOREM 2. *The coefficient matrix $H_{n+1}(a, b)$ is non-singular.*

Proof. For a given arbitrary vector $(c_0, c_1, \ldots, c_n)$ it is possible to find a polynomial $f(x)$, such that

$$\int_a^b x^k f(x)\,dx = c_k, \qquad k = 0, 1, \ldots, n.$$

In fact, the polynomial can be of degree at most n and we leave this construction to Problem 11. If

$$f(x) \equiv \sum_{i=0}^{n} a_i x^i,$$

then (5a) has the solution $\hat{a}_i = a_i$, $i = 0, 1, \ldots, n$. Therefore, the system (5b) has at least one solution for any right-hand side and this implies that the system is non-singular. ∎

In the special case $[a, b] \equiv [0, 1]$, we get from (5c):

(6)
$$H_{n+1}(0, 1) \equiv \begin{pmatrix} \frac{1}{1} & \frac{1}{2} & \cdots & \frac{1}{n+1} \\ \frac{1}{2} & \frac{1}{3} & \cdots & \frac{1}{n+2} \\ \vdots & \vdots & & \vdots \\ \frac{1}{n+1} & \frac{1}{n+2} & \cdots & \frac{1}{2n+1} \end{pmatrix} \equiv (h_{ij}),$$

where $h_{ij} = 1/(i + j - 1)$ and $i, j = 1, 2, \ldots, n + 1$, a so-called *Hilbert segment* matrix. It is difficult to solve numerically a system of equations with the matrix $H_n(0, 1)$ by any of the standard procedures of Chapter 2 (see, for example, Problem 2.2 of Chapter 2). However, it is possible to find the explicit inverse of $H_n(0, 1)$ with the aid of some properties of the Lagrange interpolation coefficients (see Problems 1, 2). As a consequence of the non-singularity of $H_n(a, b)$, and the fact that (5) is a necessary condition for a minimum of $\phi(a_0, a_1, \ldots, a_n)$, we have another proof that the least squares polynomial approximation of degree n is unique.

We may, by the linear change of variable $x = a + (b - a)y$, transform the problem of fitting $f(x)$ by $Q_n(x)$ in $[a, b]$, to that of fitting $f(a + (b - a)y) \equiv g(y)$ by $P_n(y)$ in $[0, 1]$. Afterwards, by setting

$$Q_n(x) \equiv P_n\left(\frac{x - a}{b - a}\right),$$

we have the least squares polynomial that minimizes the norm in (2).

The least squares polynomial can be determined in a way which avoids the difficulties inherent in directly solving the system (5). (This alternative procedure is but a special case of the general theory of approximation by *orthogonal functions*.)

Consider a set of $n + 1$ polynomials $\{P_k(x)\}$, $k = 0, 1, \ldots, n$, where $P_k(x)$ is of degree† k in x. Then without loss of generality, we may let $Q_n(x)$ be a linear combination of these polynomials, say

$$Q_n(x) = \sum_{j=0}^{n} c_j P_j(x). \tag{7}$$

Now the mean square error (2) defines a function

$$J(c_0, c_1, \ldots, c_n) = \int_a^b [f(x) - Q_n(x)]^2\, dx, \tag{8}$$

of the $n + 1$ variables $\{c_k\}$. As before, this function is quadratic in the c_k and at a minimum we must have

$$0 = \frac{\partial J}{\partial c_k} = 0 - 2\int_a^b P_k(x)f(x)\, dx + 2\sum_{j=0}^{n} c_j \int_a^b P_j(x)P_k(x)\, dx;$$

or the normal system

$$\sum_{j=0}^{n} c_j \int_a^b P_j(x)P_k(x)\, dx = \int_a^b P_k(x)f(x)\, dx, \qquad k = 0, 1, \ldots, n. \tag{9}$$

† Here we require that $P_k(x)$ have *exactly* degree k, in order that the set $\{P_k(x)\}$ for $k = 0, 1, 2, \ldots,$ be linearly independent.

There seems to be no apparent gain in replacing the system (5) by (9). However, the main point of the expansion in polynomials, rather than in powers of x, is to choose appropriate $P_k(x)$ so that the system (9) is easily solved. In fact, the simplest choice would be one which makes the coefficient matrix diagonal, or, better still, the identity matrix. This requires

$$(10) \qquad \int_a^b P_j(x)P_k(x)\,dx = \delta_{jk}.$$

In the next subsection, we define such a sequence $\{P_k(x)\}$. A set of polynomials (or any functions) which satisfy (10) are called *orthonormal over* $[a, b]$. The coefficients c_j are now simply given by

$$(11) \qquad c_j = \int_a^b P_j(x)f(x)\,dx, \qquad j = 0, 1, \ldots, n.$$

An additional advantage of the expansion in orthonormal polynomials is that the accuracy of the approximation (7) can be improved by adding an additional term, $c_{n+1}P_{n+1}(x)$, without having to recompute the previously determined coefficients, $c_0, c_1, \ldots, c_n$. [It is also clear that (7), with the coefficients (11), represents an approximation of least mean square error for any set of *orthonormal functions*, not necessarily polynomials, which satisfy (10).] For the approximation determined by (7) and (11), it easily follows from (10) in (8) that

$$(12a) \qquad J(c_0, c_1, \ldots, c_n) = \int_a^b f^2(x)\,dx - \sum_{j=0}^{n} c_j^2 \geq 0.$$

If we let $n \to \infty$ it follows from (12a) that $\sum_{j=0}^{\infty} c_j^2$ converges. Hence, we deduce that $\lim_{j\to\infty} c_j = 0$. This is a conclusion about the integrals of form (11) for general orthonormal functions, $P_j(x)$. The inequality (12a) is known as *Bessel's inequality*.

Convergence in the mean of the least squares polynomial approximation to a continuous function is easily demonstrated. Specifically we have

THEOREM 3. *Let $f(x)$ be continuous on $[a, b]$ and $Q_n(x)$, $n = 0, 1, 2, \ldots$, be the least squares polynomial approximations to $f(x)$ on $[a, b]$ determined by* (7) *and* (11). *Then*

$$\lim_{n\to\infty} J_n \equiv \lim_{n\to\infty} \int_a^b [f(x) - Q_n(x)]^2\,dx = 0,$$

and we have Parseval's equality

$$(12b) \qquad \int_a^b f^2(x)\,dx = \sum_{j=0}^{\infty} c_j^2.$$

Proof. For a proof by contradiction assume that $\lim_{n\to\infty} J_n = \delta > 0$. Then we pick $\epsilon > 0$ such that $\epsilon^2 = \delta/[2(b-a)]$ and by the Weierstrass theorem there is some polynomial $P_m(x)$ such that $|f(x) - P_m(x)| \leq \epsilon$ in $a \leq x \leq b$. For this polynomial,

$$\int_a^b [f(x) - P_m(x)]^2\,dx \leq \epsilon^2(b-a) = \frac{\delta}{2}.$$

However, by (12a) J_n is a non-increasing function of n; hence, the least squares approximation of degree m, say $Q_m(x)$, satisfies

$$\delta/2 \geq \int_a^b [f(x) - Q_m(x)]^2\,dx \geq \delta.$$

This is a contradiction unless $\delta = 0$. Of course, this mean convergence implies (12b), the Parseval equality. ∎

Unfortunately, these simple results yield no information about the pointwise approximation of $f(x)$ by the least squares approximation $Q_n(x)$. In order to estimate $R_n(x) \equiv f(x) - \sum_{j=0}^{n} c_j P_j(x)$, with c_j defined by (11) and $\{P_j(x)\}$ orthonormal, we write

$$\begin{aligned} R_n(x) &= f(x) - \sum_{j=0}^{n} P_j(x) \int_a^b P_j(\xi) f(\xi)\,d\xi \\ &= f(x) - \int_a^b G_n(x, \xi) f(\xi)\,d\xi, \end{aligned}$$

where

$$G_n(x, \xi) \equiv \sum_{j=0}^{n} P_j(x) P_j(\xi). \tag{13a}$$

From the orthogonality property, we observe that

$$\int_a^b G_n(x, \xi)\,d\xi = 1.$$

Therefore, we may rewrite $R_n(x)$ as

$$R_n(x) = \int_a^b G_n(x, \xi)[f(x) - f(\xi)]\,d\xi. \tag{13b}$$

Now, the rate at which $R_n(x) \to 0$, as $n \to \infty$, depends on the nature of the *kernel*, $G_n(x, \xi)$, and on the function $f(x)$.

A direct verification of convergence is possible if the sequence $Q_n(x) \equiv \sum_{j=0}^{n} c_j P_j(x)$ converges in the mean to $f(x)$ and simultaneously converges uniformly in $[a, b]$. That is, if we define

$$g(x) \equiv \lim_{n\to\infty} Q_n(x),$$

the function $g(x)$ will be continuous in $[a, b]$, since it is the uniform limit of a sequence of continuous functions. On the other hand, because of the uniformity of convergence, we may pass to the limit under the integral sign, in the statement of mean convergence, to find

$$\int_a^b [f(x) - g(x)]^2 \, dx = 0.$$

Therefore, $f(x) = g(x)$, since they are both continuous.

Now, it is possible to show that the sequence $Q_n(x)$ converges uniformly if $f(x)$ has two continuous derivatives† in $[a, b]$. We carry out the details for the interval $[a, b] \equiv [-1, 1]$, in Subsection 3.4. On the other hand, if the function $f(x)$ is merely continuous, the sequence $Q_n(x)$ need not converge.

3.1. Construction of Orthonormal Functions

The method by which a set of orthonormal polynomials $\{P_k(x)\}$ can be determined is a special case of a general procedure in which an orthonormal set of functions is constructed from an arbitrary linearly independent set.†† This process is known as the *Gram-Schmidt orthonormalization method* and is described as follows.

We begin by defining the *inner product* of any pair of real valued functions $f(x)$, $g(x)$ by

$$(f, g) = (g, f) = \int_a^b f(x)g(x) \, dx. \tag{14}$$

Now, let $\{g_i(x)\}$, $i = 0, 1, \ldots, n$, be $n + 1$ linearly independent and square integrable functions over $[a, b]$. Consider the functions

$$\begin{aligned} f_0(x) &= d_0[g_0(x)], \\ f_1(x) &= d_1[g_1(x) - c_{01} f_0(x)], \\ &\vdots \\ f_n(x) &= d_n[g_n(x) - c_{0n} f_0(x) - \cdots - c_{n-1,n} f_{n-1}(x)]. \end{aligned} \tag{15}$$

† This requirement may be weakened.

†† In analogy with the definition of linear independence for vectors, the set $\{g_i(x)\}$, $i = 0, 1, \ldots, n$, of functions are linearly independent in some interval $[a, b]$ if and only if the only linear combination $\sum_{i=0}^{n} a_i g_i(x)$ that vanishes identically in $[a, b]$ has $a_i = 0$, $i = 0, 1, \ldots, n$.

We seek coefficients d_k, c_{jk} such that the set $\{f_i(x)\}$ is orthonormal over $[a, b]$; i.e., using the inner product notation:

$$(f_i, f_j) = \delta_{ij}, \qquad i, j = 0, 1, \ldots, n.$$

To normalize $f_0(x)$ we need only require

$$(f_0, f_0) = d_0^2(g_0, g_0) = 1.$$

Since $g_0(x) \not\equiv 0$, or the set $\{g_i(x)\}$ could not be linearly independent, we may define

$$d_0 = \frac{1}{\sqrt{(g_0, g_0)}}. \tag{16$_0$}$$

In order that

$$0 = (f_0, f_1) = d_1(f_0, g_1 - c_{01}f_0)$$
$$= d_1[(f_0, g_1) - c_{01}],$$

we require

$$c_{01} = (f_0, g_1). \tag{16$_{01}$}$$

To normalize f_1 we set

$$(f_1, f_1) = d_1^2(g_1 - c_{01}f_0, g_1 - c_{01}f_0) = 1.$$

The inner product on the right cannot vanish, by the assumed linear independence of the $\{g_i(x)\}$, and thus d_1 is determined to within its sign. As in $(16)_0$ we adopt the convention of using the positive square root.

In general, then, if $(f_i, f_j) = \delta_{ij}$ for all $i, j = 0, 1, \ldots, k - 1$, we find $(f_j, f_k) = 0$ for all $j = 0, 1, \ldots, k - 1$, when we define f_k as in (15) with

$$c_{jk} = (f_j, g_k), \qquad j \le k - 1. \tag{16$_{jk}$}$$

The normalization constant, d_k, is easily obtained as before by setting $(f_k, f_k) = 1$.

To apply the Gram-Schmidt procedure to the problem of obtaining orthonormal polynomials $\{P_j(x)\}$ over an interval $[a, b]$, we observe that the $n + 1$ polynomials

$$g_j(x) \equiv x^j, \qquad j = 0, 1, \ldots, n,$$

are linearly independent over any interval. The proof of their independence follows from the fundamental theorem of algebra used in the proof of Lemma 2.1. As in (15) we form

$$P_0(x) = d_0(1),$$
$$P_1(x) = d_1[x - c_{01}P_0(x)],$$
$$\vdots$$
$$P_n(x) = d_n[x^n - c_{0n}P_0(x) - c_{1n}P_1(x) - \cdots - c_{n-1,n}P_{n-1}(x)].$$

Then

$$1 = \int_a^b P_0^2(x)\,dx = d_0^2 \int_a^b dx = d_0^2(b - a),$$

or

$$(17)_0 \qquad d_0 = \frac{1}{\sqrt{b-a}}.$$

By either repeating the previous derivation or simply applying the formulae $(16)_{jk}$ we have

$$(17)_{01} \quad c_{01} = \int_a^b d_0 x\,dx = \frac{1}{\sqrt{b-a}}\frac{b^2 - a^2}{2} = \sqrt{b-a}\left(\frac{b+a}{2}\right).$$

Normalizing $P_1(x) = d_1[x - c_{01}d_0]$ yields

$$1 = \int_a^b P_1^2(x)\,dx = d_1^2 \int_a^b \left[x^2 - (b+a)x + \frac{(b+a)^2}{4}\right] dx$$

$$= d_1^2\left[\frac{b^3 - a^3}{3} - (b+a)\frac{b^2 - a^2}{2} + \frac{(b+a)^2}{4}(b-a)\right]$$

$$= \frac{d_1^2}{12}(b-a)^3$$

or explicitly

$$d_1 = 2\sqrt{3}\,(b-a)^{-3/2}.$$

The first two polynomials are thus

$$P_0(x) = \frac{1}{\sqrt{b-a}},$$

$$P_1(x) = 2\sqrt{3}\,(b-a)^{-3/2}\left(x - \frac{b+a}{2}\right),$$

and any number of them can be obtained by continuing this procedure. Let us denote by $P_n(x; a, b)$, the sequence of orthonormal polynomials over $[a, b]$. Then it is easily verified that

$$P_n(x; a, b) \equiv \sqrt{\frac{2}{b-a}}\,P_n\left(\frac{2x - (a+b)}{b-a}; -1, 1\right)$$

is their representation in terms of the polynomials orthonormal over $[-1, 1]$. In Problem 6, we verify that

$$P_n(x; -1, 1) = (n + \tfrac{1}{2})^{1/2}\frac{1}{n!\,2^n}\frac{d^n}{dx^n}(x^2 - 1)^n.$$

The polynomials

$$Q_n(x) \equiv (n + \tfrac{1}{2})^{-1/2} P_n(x; -1, 1)$$

are called the *Legendre polynomials.*

3.2. Weighted Least Squares Approximation

The mean square measure of approximation defined in (2) gives equal "weight" at each point in $[a, b]$ to the deviation of the approximating polynomial from the function, $f(x)$. For some purposes, it may be required that the approximation be better over some parts of the interval $[a, b]$ than it is over other parts. This suggests a natural generalization of (2) which is:

$$(18) \qquad \|f(x) - Q_n(x)\|_{2,w} \equiv \left\{\int_a^b [f(x) - Q_n(x)]^2 w(x)\, dx\right\}^{1/2},$$

where $w(x) \geq 0$ in $[a, b]$ and

$$(19) \qquad \int_a^b w(x)\, dx > 0.$$

The non-negative function $w(x)$ is called the *weight function*; clearly, if $w(x) \equiv 1$, the usual least squares approximation results. For convenience, we require that $w(x)$ *be continuous in* (a, b) *and have at most isolated zeros* in this interval. By choosing an appropriate function $w(x)$ and finding the corresponding $Q_n(x)$ which minimizes (18), we may obtain an approximation with good relative accuracy in a specified region of $[a, b]$. An extreme example of this is illustrated by the fact that the interpolation problem can be formulated as a special limiting case of (18) (see Problem 4).

If $Q_n(x)$ is assumed of the form (1) then, as before, we find a system of equations for the determination of the coefficients $\{a_i\}$:

$$\sum_{i=0}^{n} \left[\int_a^b x^{i+k} w(x)\, dx\right] a_i = \int_a^b x^k f(x) w(x)\, dx, \qquad k = 0, 1, \ldots, n.$$

Again, the necessity for solving this system can be eliminated by introducing a set of polynomials having appropriate properties. Specifically we call a set of functions $\{P_j(x)\}$ *orthonormal over* $[a, b]$ *with respect to the weight* $w(x)$ if

$$(20) \qquad \int_a^b P_j(x) P_k(x) w(x)\, dx = \delta_{jk}.$$

Then to minimize (18) with a polynomial of the form (7), we find that

$$(21) \qquad c_j = \int_a^b P_j(x) f(x) w(x)\, dx.$$

The construction of orthonormal functions with respect to a weight $w(x)$ can be accomplished by the procedure of the previous subsection. That is, we introduce, in place of (14), a new definition of inner product

$$(f, g) = (g, f) = \int_a^b f(x)g(x)w(x)\,dx. \tag{22}$$

The generalizations of Bessel's inequality (12a), the mean convergence proof for polynomial approximations of continuous functions and Parseval's relation (12b) are valid in the present case with essentially no change in argument. An important special example $w(x) \equiv (1 - x^2)^{-1/2}$, $[a, b] \equiv [-1, 1]$, gives rise to the Chebyshev polynomials (see Problem 9). The pointwise convergence of weighted mean square approximations is briefly considered in Subsection 3.4. A proof of convergence for the Chebyshev expansion of sufficiently smooth functions is given there.

3.3. Some Properties of Orthogonal Polynomials

Let the polynomials $P_n(x)$, $n = 0, 1, 2, \ldots$, be orthogonal over $a \leq x \leq b$ with respect to the non-negative weight function $w(x)$. Then we have

THEOREM 4. *The roots x_j, $j = 1, 2, \ldots, n$ of $P_n(x) = 0$, $n = 1, 2, \ldots$, are all real and simple and lie in the open interval $a < x_j < b$.*

Proof. Let those roots of $P_n(x) = 0$ in (a, b) be $x_1, x_2, \ldots, x_r$, where any multiple root is repeated the appropriate number of times. Then the polynomial

$$Q_r(x) = (x - x_1)(x - x_2)\cdots(x - x_r)$$

has sign changes wherever $P_n(x)$ does in (a, b) and it is of degree $r \leq n$. Thus, $P_n(x)Q_r(x)$ is of one sign in (a, b) and so

$$\int_a^b P_n(x)Q_r(x)w(x)\,dx \neq 0.$$

This can only be true if $r = n$, since $P_n(x)$ is orthogonal to all polynomials of lower degree. Now assume some root, say x_1, is multiple. Then we can write

$$P_n(x) = (x - x_1)^2 p_{n-2}(x),$$

where $p_{n-2}(x)$ is of degree $n - 2$. But

$$P_n(x)p_{n-2}(x) = \left(\frac{P_n(x)}{x - x_1}\right)^2 \geq 0$$

and hence

$$\int_a^b P_n(x)p_{n-2}(x)w(x)\,dx > 0.$$

But this is a contradiction since $P_n(x)$ is orthogonal to any lower order polynomial. Hence multiple roots cannot occur. ∎

The orthonormal polynomials satisfy a simple recursion formula which is stated in

THEOREM 5. *Any three consecutive orthonormal polynomials are related by*

$$P_{n+1}(x) = (A_n x + B_n)P_n(x) - C_n P_{n-1}(x). \tag{23}$$

If a_k and b_k represent the coefficients of the terms of degree k and $k - 1$ in $P_k(x)$ then

$$A_n = \frac{a_{n+1}}{a_n}, \qquad B_n = \frac{a_{n+1}}{a_n}\left(\frac{b_{n+1}}{a_{n+1}} - \frac{b_n}{a_n}\right), \qquad C_n = \frac{a_{n+1}a_{n-1}}{a_n^2}. \tag{24}$$

Proof. With A_n given by (24) it follows that

$$P_{n+1}(x) - A_n x P_n(x) \equiv Q_n(x)$$

is a polynomial of degree at most n. Hence, $Q_n(x)$ can be expanded as

$$Q_n(x) = \alpha_n P_n(x) + \cdots + \alpha_0 P_0(x).$$

By the orthogonality, however, we find that

$$\begin{aligned}\alpha_k &= \int_a^b Q_n(x)P_k(x)w(x)\,dx \\ &= \int_a^b P_{n+1}(x)P_k(x)w(x)\,dx - A_n \int_a^b P_n(x)P_k(x)xw(x)\,dx \\ &= 0, \qquad \text{for } k = 0, 1, \ldots, n-2.\end{aligned}$$

Thus, the form in (23) follows upon setting $\alpha_n = B_n$ and $\alpha_{n-1} = -C_n$.
Now we may write

$$xP_{n-1}(x) \equiv \frac{a_{n-1}}{a_n} P_n(x) + q_{n-1}(x),$$

where $q_{n-1}(x)$ is of degree at most $n - 1$. Then it follows that

$$\begin{aligned}C_n &= A_n \int_a^b P_n(x)P_{n-1}(x)xw(x)\,dx, \\ &= A_n \frac{a_{n-1}}{a_n} \int_a^b P_n^2(x)w(x)\,dx + A_n \int_a^b P_n(x)q_{n-1}(x)w(x)\,dx \\ &= A_n \frac{a_{n-1}}{a_n}.\end{aligned}$$

The coefficient B_n is easily obtained by equating coefficients of the terms of degree n in (23) and the proof is completed. We observe that (23) and (24) are valid for $n = 0$ if we define $a_{-1} \equiv P_{-1}(x) \equiv 0$. ∎

The result in Theorem 5 can be used to derive what is known as the *Christoffel-Darboux relation.* We state this as

THEOREM 6. *The orthonormal polynomials satisfy*

$$(25)\quad \frac{a_n}{a_{n+1}}[P_{n+1}(x)P_n(\xi) - P_{n+1}(\xi)P_n(x)] = (x - \xi)\sum_{j=0}^{n} P_j(x)P_j(\xi).$$

Proof. Multiply the recursion formula (23) by $P_n(\xi)$ to get:

$$P_n(\xi)P_{n+1}(x) = (A_n x + B_n)P_n(\xi)P_n(x) - C_n P_n(\xi)P_{n-1}(x).$$

Since this is an identity, it holds if we interchange the arguments x and ξ. Subtracting this interchanged form from the original form and multiplying by A_n^{-1} yields, with the aid of (24),

$$\begin{aligned}(x - \xi)P_n(x)P_n(\xi) = {} & A_n^{-1}[P_{n+1}(x)P_n(\xi) - P_{n+1}(\xi)P_n(x)] \\ & - A_{n-1}^{-1}[P_n(x)P_{n-1}(\xi) - P_n(\xi)P_{n-1}(x)].\end{aligned}$$

We now sum these identities over $0, 1, \ldots, n$ and the theorem follows (for $n = 0$, we use the convention $a_{-1} = 0$) since $A_n^{-1} = a_n/a_{n+1}$. ∎

Theorem 6 gives a convenient representation of the kernel $G_n(x, \xi)$ defined in (13a).

3.4. Pointwise Convergence of Least Squares Approximations

We first consider the ordinary least squares approximation over $[-1, 1]$ in which case the orthonormal polynomials [essentially the Legendre polynomials, see (29)] can be defined as

$$(26)\quad P_n(x; -1, 1) \equiv P_n(x) \equiv \frac{\sqrt{n + \frac{1}{2}}}{n!\,2^n}\frac{d^n}{dx^n}(x^2 - 1)^n; \qquad n = 0, 1, 2, \ldots.$$

The derivation of this representation is contained in Problem 6. Given $f(x)$, we find the least squares polynomial approximation of degree at most n to be as in (7) and (11)

$$(27a)\qquad Q_n(x) \equiv \sum_{j=0}^{n} c_j P_j(x);$$

$$(27b)\qquad c_j \equiv \int_{-1}^{1} f(x)P_j(x)\,dx.$$

If $f(x)$ is continuous on $[-1, 1]$, it follows from Theorem 3 that

(28a) $$\lim_{n\to\infty} \int_{-1}^{1} [f(x) - Q_n(x)]^2\, dx = 0,$$

(28b) $$\lim_{n\to\infty} c_n = 0.$$

If $f(x)$ satisfies additional smoothness conditions, we can deduce uniform convergence of $Q_n(x)$ to $f(x)$. In fact, we have

THEOREM 7. *Let $Q_n(x)$ be defined by (27) and let $f(x)$ have a continuous second derivative on $[-1, 1]$. Then for all $x \in [-1, 1]$ and any $\epsilon > 0$*

$$|f(x) - Q_n(x)| \le \frac{\epsilon}{\sqrt{n}}$$

provided n is sufficiently large.

Proof. We introduce the Legendre polynomials

(29) $$p_n(x) = (n + \tfrac{1}{2})^{-1/2} P_n(x) = \frac{1}{n!\, 2^n} \frac{d^n}{dx^n} (x^2 - 1)^n; \qquad n = 0, 1, 2, \ldots$$

some of whose properties are described in Problems 6–8. If we set $u = x^2 - 1$ in (29), it easily follows that

$$\begin{aligned} 2^{n+1}(n + 1)!\, p'_{n+1}(x) &= \frac{d^{n+2}}{dx^{n+2}} u^{n+1} \\ &= \frac{d^n}{dx^n} [2(n + 1)u^{n-1}(u + 2nx^2)], \\ &= 2(n + 1) \frac{d^n}{dx^n} [(2n + 1)u^n + 2nu^{n-1}], \\ &= 2^{n+1}(n + 1)!\, [(2n + 1)p_n(x) + p'_{n-1}(x)]. \end{aligned}$$

Thus we have deduced the relation

(30) $$p'_{n+1}(x) = (2n + 1)p_n(x) + p'_{n-1}(x), \qquad n = 1, 2, \ldots;$$

which by (29) can be rewritten for the $P_n(x)$ as:

(31) $$(n + \tfrac{3}{2})^{-1/2} P'_{n+1}(x) - (n - \tfrac{1}{2})^{-1/2} P'_{n-1}(x) = (2n + 1)(n + \tfrac{1}{2})^{-1/2} P_n(x); \qquad n = 1, 2, \ldots.$$

Now we introduce the notation

(32) $$c_k' \equiv \int_{-1}^{1} f'(x) P_k(x)\, dx, \qquad c_k'' \equiv \int_{-1}^{1} f''(x) P_k(x)\, dx; \qquad k = 0, 1, \ldots,$$

and use integration by parts to deduce

$$\text{(33)} \qquad \begin{aligned} c'_{k\pm 1} &= \int_{-1}^{1} f'(x) P_{k\pm 1}(x)\, dx, \\ &= [f(x)P_{k\pm 1}(x)]\Big|_{-1}^{1} - \int_{-1}^{1} f(x)P'_{k\pm 1}(x)\, dx. \end{aligned}$$

Let us assume for the present that $f(\pm 1) = f'(\pm 1) = 0$. Then (33) simplifies and with (31) and (27b) yields

$$\begin{aligned} (k + \tfrac{3}{2})^{-1/2} c'_{k+1} - (k - \tfrac{1}{2})^{-1/2} c'_{k-1} & \\ &= (2k + 1)(k + \tfrac{1}{2})^{-1/2} \int_{-1}^{1} f(x)P_k(x)\, dx, \\ &= (2k + 1)(k + \tfrac{1}{2})^{-1/2} c_k. \end{aligned}$$

From this, it follows that

$$\text{(34)} \qquad \begin{aligned} c_k &= A_k \frac{c'_{k+1}}{k} - B_k \frac{c'_{k-1}}{k}, \\ A_k &\equiv \left(\frac{2k+1}{2k+3}\right)^{1/2} \frac{k}{2k+1}, \quad B_k \equiv \left(\frac{2k+1}{2k-1}\right)^{1/2} \frac{k}{2k+1}. \end{aligned}$$

But since $f'(x)$ is continuous we may use (28) for $f'(x)$ and c_n' to conclude from (34) that, since $c_k' \to 0$, $A_k \to \frac{1}{2}$ and $B_k \to \frac{1}{2}$,

$$\text{(35)} \qquad \lim_{k\to\infty} kc_k = 0.$$

This argument can be repeated with the function $w(x) \equiv f'(x)$. By the hypothesis, it follows that $w'(x) = f''(x)$ is continuous and in place of (35) we get (having assumed that $w(\pm 1) = f'(\pm 1) = 0$)

$$\lim_{k\to\infty} kc_k' = 0.$$

However, by using this result in (34), we find

$$\text{(36)} \qquad \lim_{k\to\infty} k^2 c_k = 0.$$

From the property

$$|p_n(x)| \le 1$$

exhibited in Problem 8 we deduce from (29) that for $x \in [-1, 1]$

$$\text{(37)} \qquad |P_k(x)| \le \sqrt{k + \tfrac{1}{2}}.$$

By (36) we can pick any $\epsilon > 0$ and find n sufficiently large so that $k^2 c_k < \epsilon$

for all $k \geq n$. Then from (27) and (37) we have for any $m > n$ and $x \in [-1, 1]$:

$$
\begin{aligned}
|Q_m(x) - Q_n(x)| &= \left| \sum_{k=n+1}^{m} \frac{1}{k^2} (k^2 c_k) P_k(x) \right|, \\
&\leq \epsilon \sum_{k=n+1}^{m} \frac{1}{k^2} |P_k(x)|, \\
&\leq \epsilon \sum_{k=n+1}^{m} \frac{\sqrt{k + \frac{1}{2}}}{k^2} = \epsilon \sum_{k=n+1}^{m} \frac{1}{k^{3/2}} \sqrt{1 + \frac{1}{2k}}, \\
&\leq \sqrt{2}\, \epsilon \sum_{k=n+1}^{\infty} \frac{1}{k^{3/2}}. \\
&\leq \sqrt{2}\, \epsilon \int_{n}^{\infty} \frac{1}{\xi^{3/2}}\, d\xi,
\end{aligned}
$$

$$
(38) \quad |Q_m(x) - Q_n(x)| \leq \frac{2\sqrt{2}\, \epsilon}{\sqrt{n}}.
$$

Here we have used $k^{-3/2} < \int_{k-1}^{k} \xi^{-3/2}\, d\xi$. It follows from (38) that the least squares polynomials $\{Q_n(x)\}$ form a Cauchy sequence and converge uniformly on $[-1, 1]$.

If we call

$$
\lim_{n \to \infty} Q_n(x) = g(x),
$$

then $g(x)$ is continuous on $[-1, 1]$ since it is the uniform limit of continuous functions. Again, by the uniformity we may take the limit under the integral sign in (28a) to find, since $f(x)$ and $g(x)$ are continuous, that $f(x) \equiv g(x)$. Finally, letting $m \to \infty$ in (38) and replacing ϵ by $\epsilon/(2\sqrt{2})$ we get the result stated in the theorem.

To complete the proof we must eliminate the requirement that $f(\pm 1) = f'(\pm 1) = 0$. To do this, we construct, for any $f(x)$, the Hermite interpolation polynomial, $h_3(x)$, for which

$$
h_3(\pm 1) = f(\pm 1), \qquad h_3'(\pm 1) = f'(\pm 1).
$$

Then $g(x) \equiv f(x) - h_3(x)$ satisfies all the requirements of the theorem. However, since $h_3(x)$ has degree at most 3, it follows from (27b) and the orthogonality of the polynomials $P_n(x)$, that the c_j are unchanged for $j \geq 4$ if $f(x)$ is replaced by $g(x)$. ∎

By using the technique of the above proof, we find

THEOREM **8**. *Let $f(x)$ have a continuous rth derivative on* $[-1, 1]$ *where* $r \geq 2$. *Then with $Q_n(x)$ and c_n defined in* (27): $\lim_{k\to\infty} k^r c_k = 0$ *and*

$$|f(x) - Q_n(x)| = \mathcal{O}(n^{-r+3/2}), \qquad \textit{for all } x \in [-1, 1].$$

Proof. See Problem 12. ∎

A simple linear change of variable yields the

COROLLARY. *If $f(x)$ has $r \geq 2$ continuous derivatives in $[a, b]$ then there exists a polynomial approximation, $q_n(x)$, of degree at most n such that* $|f(x) - q_n(x)| = \mathcal{O}(n^{-r+3/2})$, *for all* $x \in [a, b]$.

Proof. See Problem 13. ∎

Analogous results can be obtained for various weighted least squares approximations. If the weight function is $w(x)$ and the interval is $[a, b]$, then the approximation to $f(x)$ is

$$Q_n(x) = \sum_{j=0}^{n} c_j P_j(x)$$

where the c_j are defined in (21) and the orthonormal polynomials $P_n(x)$ satisfy (20). A proof of the pointwise convergence of $Q_n(x)$ to a sufficiently smooth $f(x)$ can be given if

(i) the $P_n(x)$ are the *eigenfunctions* of a *regular second order* differential operator, say

$$\mathscr{L}[P_n(x)] \equiv a(x)P_n''(x) + b(x)P_n'(x) = \lambda_n P_n(x),$$

whose *eigenvalues*, λ_n, satisfy

$$\lim_{n\to\infty} \lambda_n n^{-2} = \text{const.};$$

(ii) the $P_n(x)$ are bounded by

$$|P_n(x)| = \mathcal{O}(n^{1/2}) \qquad \text{for all } x \in [a, b].$$

In particular, we shall sketch the proof for the case in which the $P_n(x)$ are related to the *Chebyshev polynomials*. These polynomials are orthogonal over $[-1, 1]$ with respect to the weight

$$w(x) \equiv \frac{1}{\sqrt{1 - x^2}}. \tag{39}$$

In Problem 9 they are determined as (see Section 5):

$$P_n(x) = \sqrt{\frac{2}{\pi}} \cos(n \cos^{-1} x), \qquad n = 1, 2, \ldots \tag{40}$$

and they are solutions of

(41) $$(1 - x^2)P_n''(x) - xP_n'(x) = -n^2P_n(x).$$

Thus in this case $\lambda_n = n^2$. We require $f(x)$ to have a continuous second derivative on $[-1, 1]$. By the remarks at the end of the proof of Theorem 7, we may assume without loss of generality that $f(\pm 1) = f'(\pm 1) = 0$ in the present proof. [Since $(P_i(x), P_j(x)) = \delta_{ij}$, the c_n of (21) are unchanged if $f(x)$ is replaced by $f(x) - h_3(x)$, see last paragraph of Theorem 7.]

From (39) and (41) in (21) with $[a, b] = [-1, 1]$ we have

$$c_n = -\frac{1}{n^2}\int_{-1}^{1} \frac{f(x)}{\sqrt{1 - x^2}} [(1 - x^2)P_n''(x) - xP_n'(x)]\, dx.$$

Integrating by parts, all derivatives can be removed from $P_n(x)$ to get

(42a) $$c_n = -\frac{1}{n^2}\int_{-1}^{1} [\alpha(x)f(x) + \beta(x)f'(x) + \gamma(x)f''(x)]P_n(x)w(x)\, dx,$$

where

(42b) $$\alpha(x) \equiv -\frac{1}{1 - x^2}, \qquad \beta(x) \equiv -3x, \qquad \gamma(x) \equiv 1 - x^2.$$

Since $f(\pm 1) = f'(\pm 1) = 0$ and $f''(x)$ is continuous, we note that $\alpha(x)f(x)$, $\beta(x)f'(x)$ and $\gamma(x)f''(x)$ are continuous on $[-1, 1]$. Thus the coefficients in the expansion of the sum $[\alpha f + \beta f' + \gamma f'']$ tend to zero as $n \to \infty$ (by the analog of Theorem 3 for weighted polynomials). Using this fact in (42a) implies

(43) $$\lim_{n\to\infty} n^2 c_n = 0.$$

A sharper bound than that in (ii) is easily obtained for the Chebyshev polynomials. Clearly, from the representation (40),

(44) $$|P_n(x)| \leq \sqrt{\frac{2}{\pi}}.$$

From (43) and (44) we find, as in the proof of Theorem 7, that the $Q_n(x)$ converge uniformly. However, by using this fact and the mean convergence we easily find that

(45) $$|f(x) - Q_n(x)| = \mathcal{O}\left(\frac{1}{n}\right) \qquad \text{for all } x \in [-1, 1].$$

Note that the error estimate here is smaller by $\mathcal{O}(1/\sqrt{n})$ than that for the Legendre polynomial expansion deduced in Theorem 7.

This argument is easily extended to give

THEOREM 9. *Let $f(x)$ have a continuous rth derivative on $[-1, 1]$ where $r \geq 2$. Then the mean square Chebyshev approximations, $Q_n(x)$, with coefficients c_j defined by* (39) *and* (40) *in* (21) *satisfy*

$$|f(x) - Q_n(x)| = \mathcal{O}(n^{1-r}), \qquad \textit{for all } x \in [-1, 1]. \qquad \blacksquare$$

3.5. Discrete Least Squares Approximation

For a fixed set $(x_0, x_1, \ldots, x_M)$ of distinct points we might seek to minimize

$$\|f(x) - Q_n(x)\|_D = \sqrt{\sum_{i=0}^{M} [f(x_i) - Q_n(x_i)]^2} \tag{46}$$

over all polynomials of degree at most n. Here M is usually much larger than the degree n of the class of approximating polynomials. The fact that $\|\cdot\|_D$ is a norm and essentially strict, is easily shown in Problems 15 and 16. It then follows as in Theorem 0.1, that a minimizing polynomial exists. Further, the minimizing $Q_n(x)$ is uniquely determined if $n \leq M$ (see Problem 16).

To actually determine the discrete least squares approximation, we again use the notation

$$\phi(a_0, \ldots, a_n) \equiv \|f - Q_n(x)\|_D^2, \tag{47}$$

where

$$Q_n(x) = a_0 + a_1 x + \cdots + a_n x^n.$$

This function ϕ is quadratic in the a_i since

$$\phi(a_0, a_1, \ldots, a_n) = \sum_{i=0}^{M} f^2(x_i) - 2 \sum_{k=0}^{n} a_k \sum_{i=0}^{M} x_i^k f(x_i) + \sum_{k=0}^{n} \sum_{j=0}^{n} a_k a_j \sum_{i=0}^{M} x_i^{k+j}. \tag{48}$$

At a minimum, $\{\hat{a}_j\}$ of $\phi(\mathbf{a})$ we have the necessary conditions

$$\left.\frac{\partial \phi}{\partial a_k}\right|_{\mathbf{a}=\hat{\mathbf{a}}} = 0, \qquad k = 0, 1, \ldots, n,$$

which yield the *normal system*

$$\sum_{j=0}^{n} a_j \sum_{i=0}^{M} x_i^{k+j} = \sum_{i=0}^{M} x_i^k f(x_i); \qquad k = 0, 1, \ldots, n. \tag{49}$$

If $n \leq M$, the right-hand side of (49) may take on any preassigned values $(c_0, c_1, \ldots, c_n)$ by suitably picking $[f(x_0), f(x_1), \ldots, f(x_M)]$ (e.g., if $M = n$, the Vandermonde determinant $|x_i^k|$ is non-singular and if $M > n$, there are fewer restrictions than variables $f(x_j)$ to determine). Hence, the system (49) is solvable for any right-hand side and therefore non-singular, if $M \geq n$.

But, we may avoid the necessity of having to solve the general normal system (49) if, in analogy with (10), we can construct a sequence of polynomials $P_n(x)$, $n = 0, 1, \ldots, M$, which is orthonormal relative to summation over x_i, $i = 0, 1, \ldots, M$. To this end, we define an *inner product* by

$$(50) \qquad (f, g) = (g, f) \equiv \sum_{i=0}^{M} f(x_i)g(x_i).$$

With this inner product in the Gram-Schmidt process of Subsection 3.1, we can orthonormalize the independent set $\{x^k\}$ for $k = 0, 1, 2, \ldots, M$. The result is a set of polynomials $\{P_k(x)\}$ for which

$$(51) \qquad (P_r(x), P_s(x)) = \sum_{i=0}^{M} P_r(x_i)P_s(x_i) = \delta_{rs}; \qquad r, s = 0, 1, \ldots, M.$$

Now the unique polynomial $Q_n(x)$ of degree at most $n \leq M$ that minimizes (46) can be written as (see Problem 14)

$$(52a) \qquad Q_n(x) = \sum_{k=0}^{n} d_k P_k(x),$$

where

$$(52b) \qquad d_k = \sum_{j=0}^{M} f(x_j)P_k(x_j).$$

We now consider the determination of polynomials which satisfy (51) over various sets of points. As indicated, the Gram-Schmidt procedure could be used, but for the special cases to be treated it is not required. First, we consider uniformly spaced points $\{x_j\}$ in $[-1, 1]$; say,

$$x_0 = -1, \quad x_j = x_0 + jh, \quad h = \frac{2}{M}; \qquad j = 0, 1, \ldots.$$

The corresponding orthonormal polynomials that satisfy (51) can be written as

$$(53) \qquad P_n(x) = C_n\Delta^n(u^{[n]}(x)v^{[n]}(x)), \qquad n = 0, 1, \ldots, M.$$

Here the forward difference operators Δ^n are defined by:

$$\begin{aligned}
&\Delta^0 f(x) \equiv f(x)\\
&\Delta f(x) \equiv f(x+h) - f(x),\\
(54)\qquad &\Delta^2 f(x) \equiv \Delta(\Delta f(x)) = f(x+2h) - 2f(x+h) + f(x),\\
&\quad\vdots\\
&\Delta^n f(x) \equiv \Delta[\Delta^{n-1} f(x)] = \sum_{j=0}^{n} (-1)^j \tbinom{n}{j} f(x + (n-j)h);
\end{aligned}$$

the coefficients C_n are constants given in (58) and

$$\begin{aligned}
&u^{[0]}(x) = v^{[0]}(x) = 1\\
(55)\qquad &u^{[n]}(x) = (x - x_0)(x - x_1)\cdots(x - x_{n-1})\\
&v^{[n]}(x) = (x - x_{M+1})(x - x_{M+2})\cdots(x - x_{M+n}).
\end{aligned}$$

Formula (53) is the discrete analog of the formula (26) for the Legendre polynomials.

The fact that these polynomials satisfy (51) may be verified by the use of a formula for *summation by parts*, which is analogous to integration by parts. To derive this formula, we note the identity

$$\Delta(FG) = F\Delta G + G\Delta F + (\Delta F)(\Delta G) \tag{56a}$$

which can be written as

$$G\Delta F = \Delta(FG) - F\Delta G - (\Delta F)(\Delta G). \tag{56b}$$

Now assume $r > s$ and let

$$F(x) \equiv \Delta^{r-1}(u^{[r]}(x)v^{[r]}(x)), \qquad G(x) \equiv \Delta^s(u^{[s]}(x)v^{[s]}(x)).$$

We evaluate (56b) at each point $x_i = x_0 + ih$ and sum over $0 \le i \le M$ to get

$$\begin{aligned}
(57)\qquad \frac{1}{C_s C_r}\sum_{i=0}^{M} P_s(x_i)P_r(x_i) &= \sum_{i=0}^{M} G(x_i)\Delta F(x_i)\\
&= \sum_{i=0}^{M} \Delta[F(x_i)G(x_i)] - \sum_{i=0}^{M} F(x_i)\Delta G(x_i)\\
&\quad - \sum_{i=0}^{M} [\Delta F(x_i)][\Delta G(x_i)],\\
&= F(x)G(x)\Big|_{x_0}^{x_{M+1}} - \sum_{i=0}^{M} F(x_i)\Delta G(x_i)\\
&\quad - \sum_{i=0}^{M} [\Delta F(x_i)][\Delta G(x_i)].
\end{aligned}$$

We observe that $F(x_0) = F(x_{M+1}) = 0$, whence from (53) and (57) we have converted the sum in (51) into two sums in which ΔG appears.

We may continue with this process of summation by parts to successively form sums in which higher order differences of G appear. The boundary terms vanish since, by Problem 21 and the identity (56a) for $x = x_0$ or $x = x_{M+1}$

$$\Delta^k(u^{[r]}(x)v^{[r]}(x)) = 0, \qquad k = 0, 1, \ldots, r - 1.$$

Now, from the assumption $r > s$, it follows that after $s + 1$ such applications, the term $\Delta^{2s+1}(u^{[s]}v^{[s]})$ will be a factor in all the resulting sums. But $u^{[s]}(x)v^{[s]}(x) = q_{2s}(x)$ is a polynomial of degree $2s$ and hence all the sums vanish identically since for any polynomial $p_n(x)$ of degree at most n, the difference operator reduces the degree, i.e.,

$$\Delta p_n(x) \equiv p_{n-1}(x)$$

$$\vdots$$

$$\Delta^m p_n(x) \equiv 0, \qquad \text{for } m > n.$$

The verification that (51) is valid for $r = s$ follows when we define

$$C_n = \left[\sum_{i=0}^{M} \{\Delta^n[u^{[n]}(x_i)v^{[n]}(x_i)]\}^2\right]^{-\frac{1}{2}}. \tag{58}$$

The bracket in (58) is not zero if we can show that

$$\Delta^n[u^{[n]}(x_0)v^{[n]}(x_0)] \neq 0.$$

But the polynomial

$$p_{2n}(x) \equiv u^{[n]}(x)v^{[n]}(x)$$

has only the $2n$ zeros $(x_0, x_1, \ldots, x_{n-1};\ x_{M+1}, x_{M+2}, \ldots, x_{M+n})$ and $M \geq n$. Then from (54) with $f(x) \equiv p_{2n}(x)$, only one term in the expression for $\Delta^n f(x_0)$ is non-zero, i.e.,

$$\Delta^n p_{2n}(x_0) = p_{2n}(x_n) \neq 0.$$

Hence the definition (58) is valid.

The polynomials $P_n(x)$ of (53) have been called the *Gram polynomials.*

It can be shown that the polynomials $P_n(x)/\sqrt{2/M}$ converge as $M \to \infty$ to the orthonormal polynomials defined in equation (26) (they are related to the Legendre polynomials).

Another interesting set of points for discrete least squares approximation in $[-1, 1]$ are the zeros, $(x_0, x_1, \ldots, x_M)$ of the $(M + 1)$-st Chebyshev polynomial

$$T_{M+1}(x) = 2^{-M} \cos[(M + 1)\cos^{-1} x], \qquad M = 0, 1, 2, \ldots. \tag{59}$$

In Subsection 4.2 we show that these are polynomials of the indicated degrees. Now the points $\{x_j\}$ are not uniformly spaced, but are given by

$$x_j = \cos\left[\frac{(2j+1)}{(M+1)}\frac{\pi}{2}\right], \qquad j = 0, 1, \ldots, M. \tag{60}$$

Note that the corresponding points

$$\theta_j = \cos^{-1} x_j = \frac{2j+1}{M+1}\frac{\pi}{2}$$

are uniformly spaced in $[0, \pi]$.

We say that these sets $\{x_j\}$ are interesting, because on the one hand, the discretely orthonormal polynomials are easily found in Theorem 10; and on the other hand,we prove in Subsection 5.1 that various approximation polynomials based on these points converge uniformly to any function $f(x)$ with two continuous derivatives in $[-1, 1]$.

THEOREM **10**. *For the discrete set of points* $\{x_j\}$ *defined in* (60), *the discretely orthonormal polynomials satisfying* (51) *are proportional to the Chebyshev polynomials. Specifically they are:*

$$\begin{aligned} P_0(x) &\equiv (M+1)^{-1/2}, \\ P_n(x) &\equiv 2^{1/2}(M+1)^{-1/2}\cos(n\cos^{-1} x), \qquad n = 1, 2, \ldots, M. \end{aligned} \tag{61}$$

Proof. We must verify that $P_n(x)$ defined in (61) satisfies (51). This follows directly from the discrete orthonormality property of the trigonometric functions expressed in

LEMMA **1.**

$$\sum_{j=0}^{M} \cos r\theta_j \cos s\theta_j = \begin{cases} 0 & \text{for } 0 \le r \ne s \le M; \\ \dfrac{M+1}{2} & \text{for } 0 < r = s \le M; \\ M+1 & \text{for } 0 = r = s; \end{cases} \tag{62}$$

where

$$\theta_j = \frac{2j+1}{M+1}\frac{\pi}{2}, \qquad j = 0, 1, \ldots, M. \tag{63}$$

We can most readily evaluate the sum in (62) by making use of the well-known formula,

$$e^{ix} \equiv \cos x + i \sin x, \tag{64}$$

where $i^2 = -1$ and x is a real number. Then we may write

$$\begin{aligned} \sum_{j=0}^{M} \cos r\theta_j \cos s\theta_j &= \tfrac{1}{2}\sum_{j=0}^{M} [\cos(r+s)\theta_j + \cos(r-s)\theta_j] \\ &= \tfrac{1}{2}\,\mathrm{Re}\left(\sum_{j=0}^{M} e^{i(r+s)\theta_j}\right) + \tfrac{1}{2}\,\mathrm{Re}\left(\sum_{j=0}^{M} e^{i(r-s)\theta_j}\right). \end{aligned} \tag{65}$$

But the right-hand sums may be treated as geometric series. That is, we note that $\theta_j - \theta_0 = j2\theta_0$, whence for $R = 1, 2, \ldots, 2M$,

$$\begin{aligned}
\sum_{j=0}^{M} e^{iR\theta_j} &= e^{iR\theta_0} \sum_{j=0}^{M} (e^{iR2\theta_0})^j \\
&= e^{iR\theta_0}\left(\frac{1 - e^{iR2\theta_0(M+1)}}{1 - e^{iR2\theta_0}}\right) \\
&= e^{iR\theta_0}\left(\frac{e^{-iR\theta_0(M+1)} - e^{iR\theta_0(M+1)}}{e^{-iR\theta_0} - e^{iR\theta_0}}\right)\frac{e^{iR\theta_0(M+1)}}{e^{iR\theta_0}} \\
&= \frac{\sin R(M+1)\theta_0}{\sin R\theta_0}\, e^{iR\theta_0(M+1)}.
\end{aligned}$$

By taking the real part, we find

$$\begin{aligned}
(66) \qquad \operatorname{Re}\left(\sum_{j=0}^{M} e^{iR\theta_j}\right) &= \frac{\sin R(M+1)\theta_0 \cos R(M+1)\theta_0}{\sin R\theta_0}, \\
&= \frac{\sin 2R(M+1)\theta_0}{2 \sin R\theta_0}, \\
&= \frac{\sin R\pi}{2 \sin \dfrac{R\pi}{2(M+1)}} = 0, \qquad \text{for } R = 1, 2, \ldots, 2M.
\end{aligned}$$

If we now identify $R = r \pm s$, then from (65) with $r > s$ the first part of (62) follows. The special case $r = s > 0$, of (62) results by using (65) and observing that the sum in (66), for $R = 0$ is simply

$$\operatorname{Re} \sum_{j=0}^{M} e^{i0\theta_j} = M + 1.$$

Finally, the trivial case $r = s = 0$, of (62) is directly verifiable. Thus, Lemma 1 is proven and from it follows Theorem 10. ■

We note the fact that for any set of $M + 1$ distinct points $(x_0, x_1, \ldots, x_M)$,

THEOREM **11**. *The discrete least squares approximation polynomial $Q_M(x)$ of degree at most M which minimizes*

$$\| f(x) - Q_M(x) \|_D^2 = \sum_{i=0}^{M} [f(x_i) - Q_M(x_i)]^2,$$

is the interpolation polynomial for $f(x)$ based on the distinct points $(x_0, x_1, \ldots, x_M)$.

Proof. Let $P_M(x)$ be the indicated interpolation polynomial. Then $\| f(x) - P_M(x) \|_D = 0$ and since the interpolation polynomial is unique we must have $P_M(x) \equiv Q_M(x)$. ■

We recall that with least squares approximations (discrete or not) the next higher degree approximation is obtained by simply adding a new term to the previous approximation. Theorem 11 then shows that for the discrete case, as the degree increases for a fixed set of points, the approximations "approach" the interpolation polynomial. However, for the $n + 1$ unequally spaced points (60) we show in Subsection 5.1 that while the nth degree interpolation polynomials converge like $1/\sqrt{n}$ as $n \to \infty$ (for a sufficiently smooth function), so do the discrete least square polynomials of degrees $\geq \sqrt{n}$.

We observe that one reason for working with the discrete least square method is that sums are readily computable. On the other hand, the integrals of the continuous least square method, say of the form

$$\int_a^b f(x)P_n(x)\,dx,$$

are generally only determined approximately (frequently by using quadrature formulae, i.e., sums).

The natural extension to *weighted* discrete least squares approximation is omitted. An important application of these approximation methods is to the art of fitting mathematical formulae to empirical data but we shall not treat that here.

PROBLEMS, SECTION 3

1.* A generalization of the Hilbert segments is furnished by the matrix $A = (a_{ij})$, $a_{ij} = 1/(\alpha_i + \beta_j)$, $i, j = 1, 2, \ldots, n$, where the α_i are distinct and the β_j are distinct. Show that the determinant of A is

$$\det A = \det\left|\frac{1}{\alpha_i + \beta_j}\right| = \frac{\prod_{p=1}^{n-1}\left[\prod_{q=p+1}^{n}(\alpha_p - \alpha_q)\right]\cdot\prod_{r=1}^{n-1}\left[\prod_{s=r+1}^{n}(\beta_r - \beta_s)\right]}{\prod_{i=1}^{n}\left[\prod_{j=1}^{n}(\alpha_i + \beta_j)\right]},$$

for $n \geq 2$.

[Hint: Multiply the ith row of A by $\prod_{j=1}^{n}(\alpha_i + \beta_j)$ for $i = 1, 2, \ldots, n$ and call the resulting matrix C. The elements of C are polynomials in $\{\alpha_i, \beta_j\}$, hence, det C is a polynomial $P(\{\alpha_i\}, \{\beta_j\})$ of degree at most $n(n - 1)$. Observe that P is divisible by each of the factors in the numerator of the right-hand side because a determinant vanishes if two columns or two rows are identical. Hence, P equals the numerator to within a constant factor, since the numerator has degree $n(n - 1)$. Therefore, det A equals the right-hand side to within a constant factor K_n. Determine K_n by induction.]

2.* Notice that the cofactor of any element in the above matrix, A, is the determinant of a matrix of similar form. Use the cofactor and the determinant of A to represent the elements of $A^{-1} \equiv (b_{jk})$. Express these elements in terms of the Lagrange interpolation coefficients with respect to the points α_i and β_j; the result should be

$$b_{jk} = (\alpha_k + \beta_j)A_k(-\beta_j)B_j(-\alpha_k),$$

where

$$A_k(x) = \prod_{s \neq k} \left(\frac{\alpha_s - x}{\alpha_s - \alpha_k}\right),$$

$$B_k(x) = \prod_{s \neq k} \left(\frac{\beta_s - x}{\beta_s - \beta_k}\right).$$

Verify, by using equation (2.12) that $AA^{-1} = A^{-1}A = I$.

3. Show that any polynomial of degree m which is orthogonal to the first $m + 1$ orthogonal polynomials (i.e., to all orthogonal polynomials of degree m or less) is the identically vanishing polynomial.

4. Verify that if $f(x)$ is continuous in $[a, b]$, $(x_0, x_1, \ldots, x_n)$ are distinct points in (a, b) and

$$w_M(x) \equiv \begin{cases} \dfrac{(x - x_j + \epsilon_M)}{\epsilon_M{}^2} & \text{for } x_j - \epsilon_M \leq x \leq x_j, \\ \dfrac{-(x - x_j - \epsilon_M)}{\epsilon_M{}^2} & \text{for } x_j \leq x \leq x_j + \epsilon_M, \quad j = 0, 1, \ldots, n, \\ 0 & \text{if } x \text{ is not in any of the above intervals,} \end{cases}$$

where

$$\epsilon_M = \min_{i,j} \frac{|x_i - x_j|}{M},$$

then the associated weighted least squares polynomial approximations $P_{n,M}(x)$ converge to the Lagrange interpolation polynomial as $M \to \infty$.

5. Given the linearly independent set of functions $\{g_i(x)\}$ for $i = 0, 1, 2, \ldots, n$, verify that with the definition (22), the Gram-Schmidt orthogonalization process (15)–(16) produces an orthonormal set $\{f_i(x)\}$.

6. If $w(x) \equiv 1$, $[a, b] \equiv [-1, 1]$, show that for $g_k(x) \equiv x^k$, $k = 0, 1, 2, \ldots,$ the orthonormal polynomials $P_n(x)$ resulting from (14)–(16) are

$$P_n(x) \equiv (n + \tfrac{1}{2})^{1/2} \frac{1}{n!\, 2^n} \frac{d^n}{dx^n} (x^2 - 1)^n.$$

[Hint: Verify

$$\int_{-1}^{1} P_n(x)P_m(x)\, dx = \delta_{nm}$$

by integration by parts and use uniqueness of Gram-Schmidt process.]

[The polynomials $p_n(x) \equiv P_n(x)(n + \frac{1}{2})^{-1/2}$ are called the *Legendre polynomials*, and have the properties

$$p_n(1) = 1; \qquad p_n(-1) = (-1)^n;$$

(*recurrence relation*)

$$p_{n+1}(x) = \frac{2n + 1}{n + 1} x p_n(x) - \frac{n}{n + 1} p_{n-1}(x), \qquad n \geq 1.]$$

7. Verify that the Legendre polynomials also satisfy (*differential equation*)

$$(1 - x^2)p_n'' - 2xp_n' + n(n + 1)p_n = 0.$$

[Hint: Let $u = (x^2 - 1)^n$ and apply Leibnitz' rule to get d^{n+1}/dx^{n+1} of both sides of

$$(x^2 - 1)u'(x) = 2nxu.]$$

8. Prove that for the Legendre polynomials

$$|p_n(x)| \leq 1 \qquad \text{for } |x| \leq 1.$$

[Hint: Consider

$$n(n + 1)f(x) \equiv n(n + 1)p_n^2(x) + (1 - x^2)[p_n'(x)]^2.$$

Note that $f(x) = p_n^2(x)$ if $p_n'(x) = 0$, or $x^2 = 1$. But, by using the differential equation of Problem 7, $n(n + 1)f'(x) = 2x[p_n'(x)]^2 \gtrless 0$ if $x \gtrless 0$ and hence the value of $|p_n(x)|$ at a local maximum point, $|x| < 1$ is ≤ 1.]

9. If $w(x) \equiv (1 - x^2)^{-1/2}$, $[a, b] \equiv [-1, 1]$, $g_k(x) \equiv x^k$ for $k = 0, 1, 2, \ldots$, show that the sequence defined by Problem 5 is

$$P_n(x) \equiv \sqrt{\frac{2}{\pi}} \cos(n \cos^{-1} x), \qquad n = 1, 2, \ldots,$$

$$P_0(x) \equiv \frac{1}{\sqrt{\pi}}.$$

[The polynomials

$$T_n(x) \equiv \frac{1}{2^{n-1}} \cos(n \cos^{-1} x), \qquad n = 1, 2, \ldots,$$

$$T_0(x) \equiv 2$$

are called *Chebyshev polynomials* of the first kind. They satisfy (*recurrence relation*):

$$T_{n+1}(x) = xT_n(x) - \tfrac{1}{4}T_{n-1}(x);$$

(*differential equation*):

$$(1 - x^2)T_n'' = xT_n' - n^2T_n.]$$

10. Show that $\|\cdot\|_2$ is a strict norm [see (2)].

[Hint: (a) The triangle inequality follows from the *Schwarz inequality*:

$$\int_a^b f(x)g(x)\,dx \leq FG$$

where

$$F \equiv \left\{\int_a^b [f(x)]^2\,dx\right\}^{1/2}; \qquad G \equiv \left\{\int_a^b [g(x)]^2\,dx\right\}^{1/2}.$$

Observe

$$0 \leq \int_a^b [\alpha f(x) + \beta g(x)]^2\,dx \equiv (\alpha F + \beta G)^2 + 2\alpha\beta\left[\int_a^b f(x)g(x)\,dx - FG\right].$$

For $F \neq 0 \neq G$, $\alpha F + \beta G = 0$ implies $\alpha\beta < 0$ and the inequality follows.

(b) Now to show strictness, note $||f + g|| = ||f|| + ||g||$ implies $\int_a^b f(x)g(x)\,dx = FG$, whence with non-trivial α, β,

$$0 = (\alpha F + \beta G)^2 \equiv \int_a^b [\alpha f(x) + \beta g(x)]^2\,dx.]$$

11. Given $(c_0, c_1, \ldots, c_n)$ find a polynomial $W_n(x)$ of degree at most n, such that

$$\int_a^b x^k W_n(x)\,dx = c_k \qquad \text{for } k = 0, 1, \ldots, n.$$

[Hint: Use the orthonormal polynomials $P_j(x)$, $j = 0, 1, \ldots, n$, defined by (15), (16) and (17).]

12.* Prove Theorem 8.

13.* Prove Corollary to Theorem 8.

14. Given the discrete orthonormal polynomials $\{P_n(x)\}$, for $0 \le n \le M$, on the set $\{x_j\}$ for $0 \le j \le M$ [i.e., $\sum_{j=0}^{M} P_r(x_j)P_s(x_j) = \delta_{rs}$] verify, if $n \le M$, that

$$Q_n(x) = \sum_{k=0}^{n} d_k P_k(x) \qquad \text{with } d_k = \sum_{j=0}^{M} f(x_j)P_k(x_j)$$

is the unique polynomial of degree at most n which minimizes

$$\sum_{j=0}^{M} |f(x_j) - Q_n(x_j)|^2.$$

[Hint: Show that

$$\sum_{j=0}^{M} |f(x_j) - W_n(x_j)|^2 = \sum_{j=0}^{M} |f(x_j) - Q_n(x_j)|^2 + \sum_{k=0}^{n} e_k^2,$$

where

$$W_n(x) = Q_n(x) + \sum_{k=0}^{n} e_k P_k(x).]$$

15. Show that $\|\cdot\|_D$ is a semi-norm.

[Hint: $\|f + g\|_D \le \|f\|_D + \|g\|_D$ is a consequence of the Cauchy-Schwarz inequality (see Chapter 1, Section 4)

$$\sum_{i=0}^{M} f_i g_i \le \left\{\sum_{i=0}^{M} (f_i)^2\right\}^{1/2} \left\{\sum_{i=0}^{M} (g_i)^2\right\}^{1/2}.]$$

16. $\|\cdot\|_D$ is *essentially strict* if $\|f + g\|_D = \|f\|_D + \|g\|_D$ implies there exist non-trivial α, β such that $\alpha f_i + \beta g_i = 0$ for $i = 0, 1, \ldots, M$. Prove $\|\cdot\|_D$ is essentially strict.

17. If $n \le M$, show that the polynomial $Q_n(x)$ which minimizes $\|f(x) - Q_n(x)\|_D$ is unique.

18. Verify that the orthonormal polynomials $\{P_n(x)\}$ for weight $w(x)$ and interval $[a, b]$ may be represented in the form

$$P_n(x) = c_n \frac{1}{w(x)} \frac{d^n}{dx^n} [v_n(x)],$$

with c_n a normalization constant, in the common classical cases:

(a) $[a, b] \equiv [-1, 1]$, $w(x) \equiv (1 - x)^\alpha(1 + x)^\beta$ with $\alpha > -1$, $\beta > -1$ [i.e., $v_n(x) \equiv (1 - x)^{\alpha+n}(1 + x)^{\beta+n}$] (*Jacobi polynomials*)

(b) $[a, b] \equiv [0, \infty]$, $w(x) \equiv e^{-\alpha x}$ with $\alpha > 0$ [i.e., $v_n(x) \equiv x^n e^{-\alpha x}$] (*Laguerre polynomials*)

(c) $[a, b] \equiv [-\infty, \infty]$, $w(x) \equiv e^{-\alpha^2 x^2}$ [i.e., $v_n(x) \equiv e^{-\alpha^2 x^2}$] (*Hermite polynomials*)

[Hint:

$$P_n(x) \equiv \frac{1}{w(x)} \frac{d^n}{dx^n} [v_n(x)]$$

is a polynomial of at most degree n, if $v_n(x)$ satisfies

$$\frac{d^{n+1}}{dx^{n+1}} \left[\frac{1}{w(x)} \frac{d^n}{dx^n} (v_n(x))\right] = 0;$$

furthermore, $\int_a^b P_n(x)P_m(x)w(x)\, dx = 0$ for $n > m$ is implied with the use of integration by parts from

$$\int_a^b \frac{d^n}{dx^n} [v_n(x)]P_m(x)\, dx = -\int_a^b \frac{d^{n-1}}{dx^{n-1}} [v_n(x)]P_m'(x)\, dx$$

$$\vdots$$

$$= (-1)^{m+1} \int_a^b \frac{d^{n-m-1}}{dx^{n-m-1}} [v_n(x)]P_m^{(m+1)}(x)\, dx$$

if

$$\left.\frac{d^r}{dx^r} [v_n(x)]\right|_{x=a,\,b} = 0 \qquad \text{for } r = 0, 1, \ldots, n - 1.]$$

19. By the use of Problem 18 find another representation for the Chebyshev polynomials.

20.* Prove Theorem 9 for $r > 2$.

21. Verify that with the definitions (54) and (55),

$$\Delta u^{[n]}(x) = nhu^{[n-1]}(x), \qquad n \geq 1.$$

4. POLYNOMIALS OF "BEST" APPROXIMATION

Another measure of the deviation between a function, $f(x)$, and an approximating polynomial of degree n,

$$P_n(x) = a_0 + a_1 x + \cdots + a_n x^n, \tag{1}$$

is the so-called *maximum norm*:

$$\|f(x) - P_n(x)\|_\infty \equiv \max_{a \leq x \leq b} |f(x) - P_n(x)| \equiv D(f, P_n). \tag{2}$$

A polynomial which minimizes this norm is conventionally called a polynomial of "best" approximation.

Equation (2) defines a function of the $n + 1$ coefficients $\{a_i\}$ that is not as explicit as (3.4),

$$d(a_0, a_1, \ldots, a_n) \equiv \max_{a \le x \le b} |f(x) - P_n(x)|. \tag{3}$$

A polynomial of best approximation is characterized by a point $\hat{\mathbf{a}}$ in $(n + 1)$-space at which $d(\mathbf{a})$ is a minimum. The existence of such a polynomial is shown by

THEOREM 1. *Let $f(x)$ be a given function continuous in $[a, b]$. Then for any integer n there exists a polynomial $\hat{P}_n(x)$, of degree at most n, that minimizes $\|f(x) - P_n(x)\|_\infty$.*

Proof. We shall verify the hypotheses of Theorem 0.1 to obtain existence of a minimizing polynomial $\hat{P}_n(x) \equiv \sum_{i=0}^{n} \hat{a}_i x^i$.

Clearly, $\|\cdot\|_\infty$ is a norm in the space of continuous functions on $[a, b]$. We only need to establish (0.5) for this norm. That is, we must show that on the subset of polynomials $\{P_n(x)\}$ such that

$$\sum_{j=0}^{n} a_j^2 = 1, \min \|P_n(x)\|_\infty \equiv m_n > 0.$$

By the argument in the proof of Theorem 0.1, $\|P_n(x)\|_\infty$ is a continuous function of the variables $\{a_i\}$. If $\sum$ is the closed bounded set $\sum_{j=0}^{n} a_j^2 = 1$, we may apply the Weierstrass theorem which assures us that there is a point $\{\tilde{a}_j\}$ for which

$$m_n = \min_{\Sigma} \|P_n(x)\|_\infty,$$

is attained. But at this point

$$m_n = \left\| \sum_{j=0}^{n} \tilde{a}_j x^j \right\|_\infty \neq 0,$$

since

$$\sum_{j=0}^{n} \tilde{a}_j^2 = 1$$

and any non-trivial polynomial, of degree at most n can have at most n zeros (i.e., if $\|\tilde{P}_n(x)\|_\infty = 0$ then $\tilde{P}_n(x) \equiv 0$). ∎

At this point we observe that $\|\cdot\|_\infty$ is not a strict norm, and hence we cannot use Theorem 0.2 to establish uniqueness of $\hat{P}_n(x)$. Nevertheless, the "best" approximation polynomial is unique and we will prove this fact in Theorem 3. It is of interest to note that Theorem 1 remains true if the norm $\|\cdot\|_\infty$ of (2) is replaced by the semi-norm $\|\cdot\|_{\infty,S}$ defined as

$$\|f(x)\|_{\infty,S} \equiv \sup_{x\in S} |f(x)|$$

where S is any subset of $[a, b]$ containing at least $n + 1$ points.

From the Weierstrass approximation theorem, it follows that

$$\lim_{n\to\infty} D(f, \hat{P}_n) = 0.$$

Furthermore, if $f(x)$ has r continuous derivatives in $[a, b]$, then by the convergence result for expansions in Chebyshev polynomials (see Theorem 9 in Subsection 3.4)

$$D(f, \hat{P}_n) = \mathcal{O}(n^{1-r}) \qquad \text{for } r \geq 2.$$

4.1. The Error in the Best Approximation

It is a relatively easy matter to obtain bounds on the deviation of the best approximation polynomial of degree n. Let us call this quantity

$$(4) \qquad d_n(f) \equiv \min_{\{a_0,\ldots,a_n\}} d(a_0, \ldots, a_n) \equiv \min_{\{P_n(x)\}} D(f, P_n).$$

Then for any polynomial $P_n(x)$ we have the upper bound

$$d_n(f) \leq D(f, P_n).$$

Lowers bounds can be obtained by means of

THEOREM 2 (DE LA VALLÉE-POUSSIN). *Let an nth degree polynomial $P_n(x)$ have the deviations from $f(x)$*

$$(5) \qquad f(x_j) - P_n(x_j) = (-1)^j e_j, \qquad j = 0, 1, \ldots, n + 1,$$

where $a \leq x_0 < x_1 < \cdots < x_{n+1} \leq b$ and all $e_j > 0$ or else all $e_j < 0$. Then

$$(6) \qquad \min_j |e_j| \leq d_n(f).$$

Proof. Assume that for some polynomial $Q_n(x)$,

$$(7) \qquad D(f, Q_n) < \min_j |e_j|.$$

Then the nth degree polynomial

$$Q_n(x) - P_n(x) = [f(x) - P_n(x)] - [f(x) - Q_n(x)]$$

has the same sign at the points x_j as does $f(x) - P_n(x)$. Thus, there are $n + 2$ sign changes and consequently, at least $n + 1$ zeros of this difference.

But then, this nth degree polynomial identically vanishes and so $P_n(x) \equiv Q_n(x)$, which from (7) and (2) is impossible. This contradiction arose from assuming (7); hence $D(f, Q_n) \geq \min_j |e_j|$ for every polynomial $Q_n(x)$, and (6) is established. ∎

To employ this theorem we need only construct a polynomial which oscillates about the function being approximated at least $n + 1$ times. This can usually be done by means of an interpolation polynomial of degree n.

A necessary and sufficient condition which characterizes a best approximation polynomial and establishes its uniqueness is contained in

THEOREM 3 (CHEBYSHEV). *A polynomial of degree at most n, $P_n(x)$, is a best approximation of degree at most n to $f(x)$ in $[a, b]$ if and only if $f(x) - P_n(x)$ assumes the values $\pm D(f, P_n)$, with alternate changes of sign, at least $n + 2$ times in $[a, b]$. This best approximation polynomial is unique.*

Proof. Suppose $P_n(x)$ has the indicated oscillation property. Then let x_j with $j = 0, 1, \ldots, n + 1$ be $n + 2$ points at which this maximum deviation is attained with alternate sign changes. Using these points in Theorem 2 we see that $|e_j| = D(f, P_n)$ and hence

$$d_n(f) \geq D(f, P_n).$$

From equation (4), the definition of $d_n(f)$, it follows that $D(f, P_n) = d_n(f)$ and the $P_n(x)$ in question is a best approximation polynomial. This shows the sufficiency of the *uniform oscillation property*.

To demonstrate the necessity, we will show that if $f(x) - P_n(x)$ attains the values $\pm D(f, P_n)$ with alternate sign changes at most k times where $2 \leq k \leq n + 1$, then $D(f, P_n) > d_n(f)$. Let us assume, with no loss in generality, that

$$f(x_j) - P_n(x_j) = (-1)^j D(f, P_n), \qquad j = 1, 2, \ldots, k,$$

where $a \leq x_1 < x_2 < \cdots < x_k \leq b$. Then, there exist points $\xi_1, \xi_2, \ldots, \xi_{k-1}$, separating the x_j, i.e.,

$$x_1 < \xi_1 < x_2 < \xi_2 < \cdots < \xi_{k-1} < x_k$$

and an $\epsilon > 0$ such that $|f(\xi_j) - P_n(\xi_j)| < D(f, P_n)$ and

$$-D(f, P_n) \leq f(x) - P_n(x) < D(f, P_n) - \epsilon,$$

for x in the "odd" intervals, $[a, \xi_1]$, $[\xi_2, \xi_3]$, $[\xi_4, \xi_5], \ldots$; while

$$-D(f, P_n) + \epsilon < f(x) - P_n(x) \leq D(f, P_n),$$

for x in the "even" intervals, $[\xi_1, \xi_2]$, $[\xi_3, \xi_4], \ldots$. For example, we may define $\xi_1 = \frac{1}{2}(\eta_1 + \zeta_1)$ where $\eta_1 = \text{g.l.b.}\,\{\eta\}$ for $a \leq \eta \leq x_2$ and $f(\eta) - P_n(\eta) = D(f, P_n)$; and similarly: $\zeta_1 = \text{l.u.b.}\,\{\zeta\}$ for $a \leq \zeta \leq x_2$ and $f(\zeta) - P_n(\zeta) = -D(f, P_n)$. Then $x_1 \leq \zeta_1 < \eta_1 \leq x_2$; otherwise, we may insert η_1 and ζ_1 in place of x_1 in the original sequence and find $k + 1$ alternations of sign. That is, alternately for each of the k intervals $[a, \xi_1], \ldots, [\xi_{k-1}, b]$, the deviation $f(x) - P_n(x)$ takes on only one of the extreme deviations $\pm D(f, P_n)$ and is bounded away from the extreme of opposite sign. The polynomial

$$r(x) = (x - \xi_1)(x - \xi_2)\cdots(x - \xi_{k-1})$$

has degree $k - 1$ and is of one sign throughout each of the k intervals in question. Let the maximum value of $|r(x)|$ in $[a, b]$ be M. Now define $q(x) \equiv (-1)^k r(x)/2M$ and consider the nth degree polynomial (since $k - 1 \leq n$)

$$Q_n(x) = P_n(x) + \epsilon q(x),$$

for sufficiently small positive ϵ. We claim that $D(f, Q_n) < D(f, P_n)$, and so $P_n(x)$ could not be a best approximation. Indeed, in the interior of any of the "odd" intervals $(a, \xi_1), (\xi_2, \xi_3), \ldots$, we have that $-\frac{1}{2} \leq q(x) < 0$ and conversely in the "even" intervals $(\xi_1, \xi_2), (\xi_3, \xi_4), \ldots$, we have that $0 < q(x) \leq \frac{1}{2}$. However, recalling the above inequalities,

$$-D(f, P_n) - \epsilon q(x) \leq f(x) - Q_n(x) \leq D(f, P_n) - \epsilon[1 + q(x)], \quad x \text{ in odd intervals};$$

$$-D(f, P_n) + \epsilon[1 - q(x)] \leq f(x) - Q_n(x) \leq D(f, P_n) - \epsilon q(x), \quad x \text{ in even intervals}.$$

From the signs and magnitude of $q(x)$ in each interval, it easily follows that $D(f, Q_n) < D(f, P_n)$ and the proof of necessity is completed.

To demonstrate uniqueness we assume that there are two best approximations say, $P_n(x)$ and $Q_n(x)$, both of degree at most n. Since by assumption $D(f, P_n) = D(f, Q_n) = d_n(f)$, we have in $[a, b]$,

$$\left|f(x) - \tfrac{1}{2}[P_n(x) + Q_n(x)]\right| \leq \tfrac{1}{2}|f(x) - P_n(x)| + \tfrac{1}{2}|f(x) - Q_n(x)| \leq d_n(f).$$

Thus, $\frac{1}{2}[P_n(x) + Q_n(x)]$ is another best approximation and we must have, by the first part of the theorem,

$$\left|f(x) - \tfrac{1}{2}[P_n(x) + Q_n(x)]\right| = d_n(f)$$

at $n + 2$ distinct points in $[a, b]$. From the inequality, it follows that at these points $f(x) - P_n(x) = f(x) - Q_n(x) = \pm d_n(f)$. Thus, the difference $[f(x) - P_n(x)] - [f(x) - Q_n(x)] = Q_n(x) - P_n(x)$ vanishes at $n + 2$ distinct points. Since this difference is an nth degree polynomial, it vanishes identically, i.e., $Q_n(x) \equiv P_n(x)$, and the proof is complete. ∎

This theorem can be used to recognize the best approximation polynomial. It is also the basis, along with Theorem 2, of various methods for approximating the best approximation polynomial. There is no finite procedure for constructing the best approximation polynomial for arbitrary continuous functions. However, the best approximation is known in some important special cases; see, for example, the next subsection and Problem 2.

As an obvious consequence of Theorem 3, it follows that the best approximation, $P_n(x)$, of degree at most n is *equal* to $f(x)$, the function it approximates, *at* $n + 1$ *distinct points*, say $x_0, x_1, \ldots, x_n$. Thus, $P_n(x)$ is *the* interpolation polynomial for $f(x)$ with respect to the points $\{x_i\}$ (since by Lemma 2.1 the interpolation polynomial of degree at most n is unique). Of course, for an arbitrary continuous function, $f(x)$, a corresponding set of interpolation points $\{x_i\}$ is not known a priori. Thus, this observation cannot, in general, be used to determine $P_n(x)$. However, if $f(x)$ has $n + 1$ continuous derivatives, Theorem 2.1 applies since $P_n(x)$ is an interpolation polynomial, and we have determined a form for the error in the best approximation of degree at most n. In summary, these observations can be stated as a

COROLLARY. *Let $f(x)$ have a continuous $(n + 1)$st derivative in $[a, b]$ and let $P_n(x)$ be the best polynomial approximation to $f(x)$ of degree at most n in this interval. Then, there exist $n + 1$ distinct points $x_0, x_1, \ldots, x_n$ in $a < x < b$ such that*

$$R_n(x) \equiv f(x) - P_n(x) = \frac{(x - x_0)(x - x_1)\cdots(x - x_n)}{(n + 1)!} f^{(n+1)}(\xi), \tag{8}$$

where $\xi = \xi(x)$ is in the interval:

$$\min(x, x_0, \ldots, x_n) < \xi < \max(x, x_0, \ldots, x_n).$$ ■

4.2. Chebyshev Polynomials

In the expression (8) for $R_n(x)$, the error of the best approximation, it will, in general, not be known at what point, $\xi = \xi(x)$, the derivative is to be evaluated. Hence, the value of the derivative is not known. An exception to this is the case when $f^{(n+1)}(x) =$ constant, which occurs if and only if $f(x)$ is a polynomial of degree at most $n + 1$. In this special case, the error (8) can be minimized by choosing the points $x_0, x_1, \ldots, x_n$ such that the polynomial

$$(x - x_0)(x - x_1)\cdots(x - x_n) \tag{9a}$$

has the smallest possible maximum absolute value in the interval in question (say, $a \leq x \leq b$). In the general case, the choice of these same interpolation points may be expected to yield a reasonable approximation to the best polynomial, i.e., the smaller the *variation* of $f^{(n+1)}(x)$ in $[a, b]$, the better the approximation.

We are thus led to consider the following problem: *Among all polynomials of degree $n + 1$, with leading coefficient unity, find the polynomial which deviates least from zero in the interval* $[a, b]$. In other words, we are seeking the best approximation to the function $g(x) \equiv 0$ among polynomials of the form

$$x^{n+1} - P_n(x), \tag{9b}$$

where $P_n(x)$ is a polynomial of degree at most n. Alternatively, the problem can then be formulated as: *find the best polynomial approximation of degree at most n to the function x^{n+1}.*

For this latter problem, Theorem 1 is applicable and we conclude that such a polynomial exists and it is uniquely characterized in Theorem 3. Thus, we need only construct a polynomial of the form (9) whose maximum absolute value is attained at $n + 2$ points with alternate sign changes.

Consider, for the present, the interval $[a, b] \equiv [-1, 1]$. We introduce the change of variable

$$x = \cos\theta, \tag{10}$$

which takes on each value in $[-1, 1]$ once and only once when θ is restricted to the interval $[0, \pi]$. Furthermore, the function $\cos(n + 1)\theta$ attains its maximum absolute value, unity, at $n + 2$ successive points with alternate signs for

$$\theta = j\left(\frac{\pi}{n + 1}\right), \qquad j = 0, 1, \ldots, n + 1.$$

Therefore, the function

$$T_{n+1}(x) = A_{n+1}\cos(n + 1)\theta = A_{n+1}\cos[(n + 1)\cos^{-1} x], \tag{11}$$

has the required properties as regards its extrema. To show that $T_{n+1}(x)$ is also a polynomial in x of degree $n + 1$ we consider the standard trigonometric addition formula

$$\cos(n + 1)\theta + \cos(n - 1)\theta = 2\cos\theta\cos n\theta, \qquad n = 0, 1, \ldots. \tag{12}$$

Let us define

$$t_n(x) \equiv \cos(n\cos^{-1} x), \qquad n = 0, 1, 2, \ldots, \tag{13a}$$

in terms of which (12) becomes

$$t_{n+1}(x) = 2t_1(x)t_n(x) - t_{n-1}(x), \qquad n = 1, 2, 3, \ldots. \tag{13b}$$

Clearly, from (13a), $t_0(x) = 1$, $t_1(x) = x$ and so, by induction, it follows from (13b) that $t_{n+1}(x)$ is a polynomial in x of degree $n + 1$. It also follows by induction that

$$t_{n+1}(x) = 2^n x^{n+1} + q_n(x), \qquad n = 0, 1, 2, \ldots, \tag{13c}$$

where $q_n(x)$ is a polynomial of degree at most n. Thus, with the choice $A_{n+1} = 2^{-n}$ in (11), these results imply

$$\begin{aligned} T_{n+1}(x) &= 2^{-n} \cos [(n + 1) \cos^{-1} x] = 2^{-n} t_{n+1}(x) \\ &= x^{n+1} + 2^{-n} q_n(x). \end{aligned} \tag{14}$$

At the $n + 2$ points

$$\xi_k = \cos \frac{k\pi}{n + 1}, \qquad k = 0, 1, \ldots, n + 1, \tag{15a}$$

which are in $[-1, 1]$, we have from (14)

$$T_{n+1}(\xi_k) = 2^{-n} \cos k\pi = 2^{-n}(-1)^k. \tag{15b}$$

Thus we have proven that $T_{n+1}(x)$ is the polynomial of form (9) which deviates least from zero in $[-1, 1]$; the maximum deviation is 2^{-n}.

The polynomials in (14) are called the Chebyshev polynomials (of the first kind—see Problem 9 of Section 3). If the zeros of the $(n + 1)$-st such polynomial are used to construct an interpolation polynomial of degree at most n, then for x in $[-1, 1]$ the coefficient of $f^{(n+1)}(\xi)$ in the error (8) of this approximation will have the least possible absolute maximum.

If the interval of approximation for the continuous function $g(y)$ is $a \le y \le b$, then the transformation

$$x = \frac{a - 2y + b}{a - b} \quad \text{or} \quad y = \tfrac{1}{2}[(b - a)x + (a + b)] \tag{16}$$

converts the problem of approximating $g(y)$ into that of approximating $f(x) \equiv g[y(x)]$ in the x-interval $[-1, 1]$. The zeros of $T_{n+1}(x)$ are at

$$x_k = \cos \left(\frac{2k + 1}{n + 1} \frac{\pi}{2}\right), \qquad k = 0, 1, \ldots, n; \tag{17a}$$

and the corresponding interpolation points in $[a, b]$ are then at

$$y_k = \tfrac{1}{2}[(b - a)x_k + (a + b)], \qquad k = 0, 1, \ldots, n. \tag{17b}$$

The value of the maximum deviation of $\prod_{j=0}^{n} (y - y_j)$ from zero in $[a, b]$ is then, using (16) and (17b):

$$\begin{aligned} \max_{a \le y \le b} \prod_{j=0}^{n} |y - y_j| &= \left|\frac{b - a}{2}\right|^{n+1} \cdot \max_{-1 \le x \le 1} \prod_{j=0}^{n} (x - x_k) \\ &= \frac{1}{2^n} \left|\frac{b - a}{2}\right|^{n+1}. \end{aligned} \tag{18}$$

We stress that when the points in (17b) are employed to determine an interpolation polynomial for $g(y)$ over $[a, b]$, this polynomial will not, in general, be the Chebyshev best approximation polynomial of degree n. However, these points may be used to get an estimate for the error of the best approximation polynomial. Iterative methods, based on Theorems 2 and 3, have been devised to compute with arbitrary precision the polynomial of best approximation.

PROBLEMS, SECTION 4

1. Prove that the nth Chebyshev polynomial can be expressed as:

$$T_n(x) = \tfrac{1}{2}[(x + \sqrt{x^2 - 1})^n + (x - \sqrt{x^2 - 1})^n].$$

2. Find the best approximations of degrees 0 and 1 to $f(x) \in C_2[a, b]$ provided $f_2''(x) \neq 0$ in $[a, b]$ [i.e., calculate the coefficients of these best approximation polynomials in terms of properties of $f(x)$].

5. TRIGONOMETRIC APPROXIMATION

We say that $S_n(x)$ is a *trigonometric sum of order at most n*, if

$$S_n(x) = \tfrac{1}{2}a_0 + \sum_{k=1}^{n} (a_k \cos kx + b_k \sin kx). \tag{1a}$$

The coefficients a_k and b_k are real numbers. By using the exponential function

$$e^{i\theta} \equiv \cos\theta + i \sin\theta, \qquad \cos\theta = \tfrac{1}{2}(e^{i\theta} + e^{-i\theta}), \tag{1b}$$

$$\text{and} \qquad \sin\theta = \frac{-i}{2}(e^{i\theta} - e^{-i\theta})$$

where now $i^2 = -1$, it follows that (1a) can be written in a simpler form with simpler coefficients:

$$S_n(x) = \sum_{k=-n}^{n} c_k e^{ikx}. \tag{1c}$$

Here

$$c_0 = \frac{a_0}{2}, \quad c_k = \tfrac{1}{2}(a_k - ib_k), \quad c_{-k} = \bar{c}_k = \tfrac{1}{2}(a_k + ib_k),$$
$$\text{for } k = 1, 2, \ldots, n.$$

A basic result on approximation by such trigonometric sums is again due to Weierstrass and can be stated as:

THEOREM 1 (WEIERSTRASS). *Let $f(\theta)$ be continuous on $[-\pi, \pi]$ and periodic with period 2π. Then for any $\epsilon > 0$ there exists an $n = n(\epsilon)$ and a trigonometric sum, $S_n(\theta)$, such that*

$$|f(\theta) - S_n(\theta)| < \epsilon$$

for all θ.

Proof. A proof of this theorem can be given by employing the Weierstrass polynomial approximation theorem of Section 1.

We sketch a simpler proof based on the Weierstrass polynomial approximation theorem for a continuous function $g(x, y)$ in the square, $-1 \leq x, y \leq 1$ (see Problem 2, Section 1). Define $g(x, y) \equiv \rho f(\theta)$ for $x = \rho \cos \theta$, $y = \rho \sin \theta$, $0 \leq \rho \leq 2$, $-\pi \leq \theta \leq \pi$.

Clearly, $g(x, y)$ is defined and continuous in the square. Hence, given $\epsilon > 0$, there exists a polynomial $P_n(x, y)$ such that $|g(x, y) - P_n(x, y)| \leq \epsilon$. But, then for $\rho = 1$, we have $g(x, y) \equiv f(\theta)$, $x^2 + y^2 = 1$, and therefore, $|f(\theta) - P_n(\cos \theta, \sin \theta)| \leq \epsilon$. We leave as Problem 3, the verification that $P_n(\cos \theta, \sin \theta)$ may be written as a trigonometric sum, $S_n(\theta)$. ∎

We proceed to show that the previous methods of polynomial approximation have corresponding trigonometric counterparts.

5.1. Trigonometric Interpolation

If the points of interpolation are *equally spaced*, it is relatively easy to determine a trigonometric sum which takes on specified values at the appropriate points. Let $f(x)$ be continuous and have period 2π. For this section only, we introduce the convention

$$\sum_{j=-n}^{n}{}' a_j \equiv \sum_{j=-n}^{n} a_j - \tfrac{1}{2}(a_n + a_{-n}).$$

With this notation, we define the trigonometric sum

$$U_n(x) = \sum_{j=-n}^{n}{}' c_j e^{ijx}. \tag{2}$$

On the interval $[-\pi, \pi]$ we place the $2n + 1$ equally spaced points

$$x_k = kh, \qquad k = 0, \pm 1, \pm 2, \ldots, \pm n, \quad h = \frac{\pi}{n}. \tag{3}$$

The interpolation problem is to find coefficients c_j such that

$$U_n(x_k) = f(x_k), \qquad k = 0, \pm 1, \ldots, \pm n. \tag{4a}$$

Later we consider interpolation on a different set of uniformly spaced points. Since $f(x)$ and $U_n(x)$ have period 2π, $f(x_n) = f(x_{-n})$ and $U_n(x_n) =$

$U_n(x_{-n})$, so that there are only $2n$ independent conditions in (4a) to determine the $2n + 1$ coefficients, c_k. We require that

$$c_n = c_{-n}, \tag{4b}$$

as the extra condition, and it will be shown to be consistent with the conditions (4a).

A simple calculation, based on summing a geometric series (see Subsection 3.5), reveals that

$$\sum_{k=-n}^{n}{}' e^{ijx_k}e^{-imx_k} = \begin{cases} 0 & \text{if } j \not\equiv m \pmod{2n},\dagger \\ 2n & \text{if } j \equiv m \pmod{2n}. \end{cases} \tag{5}$$

In direct analogy with orthogonal functions over an interval, we see that the quantities $\{e^{ikx_j}\}$ are orthogonal with respect to the summation $\sum_{j=-n}^{n}{}'$.
Hence, we set $x = x_k$ in (2), multiply both sides of (2) by e^{-imx_k} and sum with respect to k to find upon the use of (5):

$$\begin{aligned}
\sum_{k=-n}^{n}{}' e^{-imx_k} U_n(x_k) &= \sum_{k=-n}^{n}{}' e^{-imx_k} \sum_{j=-n}^{n}{}' c_j e^{ijx_k}, \\
&= \sum_{j=-n}^{n}{}' c_j \sum_{k=-n}^{n}{}' e^{ijx_k}e^{-imx_k} \\
&= \begin{cases} 2nc_m & \text{if } |m| < n, \\ 2n\left(\dfrac{c_n + c_{-n}}{2}\right) & \text{if } |m| = n. \end{cases}
\end{aligned}$$

By applying the conditions (4), then

$$c_j = \frac{1}{2n} \sum_{k=-n}^{n}{}' f(x_k)e^{-ijx_k}, \qquad j = 0, \pm 1, \ldots, \pm n. \tag{6}$$

It is now easy to check, by using (5), that the trigonometric sum (2) with coefficients given by (6) satisfies the conditions (4); i.e., the required interpolatory trigonometric sum is determined.

If we define new coefficients α_j and β_j by

$$\alpha_j = c_j + c_{-j}, \quad \beta_j = i(c_j - c_{-j}), \qquad j = 0, 1, 2, \ldots, n,$$

then the sum (2) becomes, upon recalling (4b) and (1b),

$$U_n(x) = \tfrac{1}{2}\alpha_0 + \sum_{j=1}^{n} \alpha_j \cos jx + \sum_{j=1}^{n-1} \beta_j \sin jx. \tag{7}$$

† If $j - m$ is an integral multiple of $2n$, we say that j and m are congruent modulo $2n$, or we write $j \equiv m \pmod{2n}$. If not, we write $j \not\equiv m \pmod{2n}$.

From (6) it follows that these coefficients are real numbers given by

$$
\text{(8)} \qquad
\begin{aligned}
\alpha_j &= \frac{1}{n} \sum_{k=-n}^{n}{}' f(x_k) \cos jx_k, \qquad j = 0, 1, \ldots, n, \\
\beta_j &= \frac{1}{n} \sum_{k=-n}^{n}{}' f(x_k) \sin jx_k, \qquad j = 1, 2, \ldots, n - 1.
\end{aligned}
$$

Equations (7) and (8) are the real form for the trigonometric interpolation sum. This form is suitable for computations without complex arithmetic.

It can be shown that the trigonometric sum (7) satisfying the conditions (4) is unique. This follows from Lemma 2 in Subsection 5.3.

We may also determine unique interpolatory trigonometric sums of order n that take on specified values at $2n + 1$ distinct points *arbitrarily spaced* in, say, $-\pi \leq x < \pi$ (not including both endpoints). The coefficients of such a sum are the solutions of a non-singular linear system. The non-singularity of this system and the interpolating trigonometric sum are treated in Problems 1 and 2.

However, another trigonometric interpolation scheme for equally spaced points can be based on the orthogonality property expressed in Lemma 1 of Subsection 3.5. That is, using θ as the independent variable, we consider the $n + 1$ points,

$$
\text{(9)} \qquad \theta_j = \theta_0 + j\Delta\theta, \quad \theta_0 \equiv \frac{\Delta\theta}{2}, \quad \Delta\theta \equiv \frac{\pi}{n + 1}, \qquad j = 0, 1, \ldots, n.
$$

These θ_j are equally spaced in $[0, \pi]$, there may be an odd or even number of them, and they do not include the endpoints [in contrast to those points in (3)]. Now we seek a special trigonometric sum of order n in the form

$$
\text{(10a)} \qquad C_n(\theta) = \tfrac{1}{2}\gamma_0 + \sum_{r=1}^{n} \gamma_r \cos r\theta,
$$

such that for some function $g(\theta)$, continuous in $[0, \pi]$,

$$
\text{(11)} \qquad C_n(\theta_j) = g(\theta_j), \qquad j = 0, 1, \ldots, n.
$$

That is, we seek to interpolate $g(\theta)$ at the points (9) with a sum of the form (10). Using the form (10) in (11) we multiply by $\cos s\theta_j$ and sum over j to get by (3.62)

$$
\text{(10b)} \qquad \gamma_s = \frac{2}{n + 1} \sum_{j=0}^{n} g(\theta_j) \cos s\theta_j, \qquad s = 0, 1, \ldots, n.
$$

Thus, the interpolation problem is solved with the coefficients (10b) in the trigonometric sum (10a). [Compare the formulae (8) and (10b).] We note that the sum (10a) is an even function of θ. Hence, if the same is true of $g(\theta)$, then $C_n(\theta)$ is the interpolation sum at $2(n + 1)$ equally spaced points in $[-\pi, \pi]$. Again, uniqueness of the nth order sum easily follows in this case from Lemma 2 of Subsection 5.3. If $g(\theta)$ is not even, then the approximation (10) should be used only over $[0, \pi]$.

An important convergence property of this approximation procedure is contained in

THEOREM 2. *Let $g(\theta)$ be an even function with period 2π and a continuous second derivative on $[-\pi, \pi]$. Then the trigonometric interpolation sums $C_n(\theta)$ given by* (10a and b), *which satisfy* (11) *on the equally spaced points* (9), *converge uniformly as $n \to \infty$ to $g(\theta)$ on $[-\pi, \pi]$. In fact,*

$$|g(\theta) - C_n(\theta)| = \mathcal{O}\left(\frac{1}{\sqrt{n}}\right), \qquad \textit{for all } |\theta| \leq \pi.$$

Proof. We first estimate the rate of decay of the coefficients γ_s for large s. Clearly from (10b), since $|\cos s\theta| \leq 1$,

$$\begin{aligned}
|\gamma_s| &\leq \frac{2}{n+1} \sum_{j=0}^{n} |g(\theta_j)|, \\
&\leq 2 \max_{0 \leq \theta \leq \pi} |g(\theta)|.
\end{aligned} \tag{12}$$

Such a bound holds for the coefficients in any sum of the form (10).

With the spacing $\Delta\theta = \pi/(n + 1)$ of (9), we define the function

$$G(\theta) \equiv \frac{g(\theta + \Delta\theta) - 2g(\theta) + g(\theta - \Delta\theta)}{\Delta\theta^2}. \tag{13}$$

This function satisfies the same smoothness and periodicity conditions as $g(\theta)$. If we set

$$B_n(\theta) \equiv \left(\frac{1}{\Delta\theta}\right)^2 [C_n(\theta + \Delta\theta) - 2C_n(\theta) + C_n(\theta - \Delta\theta)], \tag{14}$$

then from (11) and (13) it follows that

$$B_n(\theta_j) = G(\theta_j), \qquad j = 0, 1, \ldots, n,$$

and so, $B_n(\theta)$ is the unique nth order trigonometric interpolation sum for $G(\theta)$ with respect to the points θ_j in (9). If we use (10a) and the identities

$$\begin{aligned}
\cos(\phi + h) - 2\cos\phi + \cos(\phi - h) &= \cos\phi(2\cos h - 2), \\
&= -4\sin^2\frac{h}{2}\cos\phi,
\end{aligned}$$

in (14), we obtain

$$B_n(\theta) = -\left(\frac{2}{\Delta\theta}\right)^2 \sum_{r=1}^{n} \gamma_r \sin^2 \frac{r\Delta\theta}{2} \cos r\theta.$$

Thus, $B_n(\theta)$ has the form (10a) and by the remark after (12) it now follows that

$$(15) \qquad \left|\gamma_r r^2 \left(\frac{2}{r\Delta\theta} \sin \frac{r\Delta\theta}{2}\right)^2\right| \leq 2 \max_{0\leq\theta\leq\pi} |G(\theta)|, \qquad r = 1, 2, \ldots, n.$$

From Taylor's theorem

$$g(\theta \pm \Delta\theta) = g(\theta) \pm \Delta\theta g'(\theta) + \frac{(\Delta\theta)^2}{2} g''(\theta \pm \phi_\pm \Delta\theta), 0 < \phi_\pm < 1.$$

By using this and the continuity of $g''(\theta)$ in (13), we find

$$|G(\theta)| \leq | \max_{0\leq\varphi\leq\pi} |g''(\varphi)|, \qquad \theta \in [0, \pi].$$

This bound and the inequality (see Problem 5)

$$\left|\frac{h}{\sin h}\right| \leq \frac{\pi}{2}, \qquad \text{for } 0 < h \leq \frac{\pi}{2}.$$

in (15) yield finally

$$(16) \qquad |\gamma_r| \leq \frac{\pi^2}{2r^2} \max_{0\leq\theta\leq\pi} |g''(\theta)|,$$

$$= \mathcal{O}\left(\frac{1}{r^2}\right), \qquad r = 1, 2, \ldots, n.$$

This is the required estimate of the coefficients in (10).

In Subsection 5.2 we define Fourier series, and for $g(\theta)$ as above it follows from Theorem 3 that

$$(17) \qquad |g(\theta) - S_m(\theta)| = \mathcal{O}\left(\frac{1}{m}\right), \quad a_m = \mathcal{O}\left(\frac{1}{m^2}\right), \qquad m = 1, 2, \ldots$$

where the partial sums, $S_m(\theta)$, and coefficients, a_m, are defined by

$$(18a) \qquad S_m(\theta) \equiv \frac{a_0}{2} + \sum_{k=1}^{m} a_k \cos k\theta, \qquad m = 1, 2, \ldots$$

$$(18b) \qquad a_k \equiv \frac{1}{\pi} \int_{-\pi}^{\pi} g(\theta) \cos k\theta \, d\theta, \qquad k = 0, 1, 2, \ldots$$

[The sine terms are absent since $g(\theta)$ is even, and hence the $b_k \equiv 0$.] For any $m \leq n$, we define the truncated trigonometric interpolation sum

$$(19) \qquad C_{n,m}(\theta) \equiv \tfrac{1}{2}\gamma_0 + \sum_{r=1}^{m} \gamma_r \cos \theta,$$

where the γ_r are given in (10b).

We now consider

$$
\begin{aligned}
|g(\theta) - C_n(\theta)| &= |g(\theta) - S_m(\theta) + S_m(\theta) \\
&\qquad - C_{n,m}(\theta) + C_{n,m}(\theta) - C_n(\theta)|, \\
&\leq |g(\theta) - S_m(\theta)| + |S_m(\theta) - C_{n,m}(\theta)| \\
&\qquad + |C_{n,m}(\theta) - C_n(\theta)|.
\end{aligned} \tag{20}
$$

From (19), (10), and the estimate (16) we have

$$
\begin{aligned}
|C_{n,m}(\theta) - C_n(\theta)| &= |\sum_{r=m+1}^{n} \gamma_r \cos r\theta|, \\
&\leq \sum_{r=m+1}^{n} |\gamma_r|, \\
&\leq \frac{\pi^2}{2} \max |g''(\theta)| \sum_{r=m+1}^{\infty} \frac{1}{r^2}, \\
&= \mathcal{O}\left(\frac{1}{m}\right).
\end{aligned} \tag{21}
$$

To estimate the middle term in (20) we note from (9), (10b), (18b), and the evenness of $g(\theta)$ that

$$
a_k - \gamma_k = \frac{2}{\pi}\left[\int_0^\pi g(\theta)\cos k\theta \, d\theta - \sum_{j=0}^{n} g(\theta_j)\cos k\theta_j \, \Delta\theta\right], \qquad k = 0, 1, \ldots.
$$

This sum is clearly an approximation to the integral. It is, in fact, the midpoint quadrature formula (see Chapter 7) and since the integrand has a continuous second derivative it is easily shown that (see Problem 6)

$$
\begin{aligned}
|a_k - \gamma_k| &\leq \frac{\pi^2}{12(n+1)^2} \max_{0 \leq \theta \leq \pi} \left|\frac{d^2}{d\theta^2}[g(\theta)\cos k\theta]\right| \\
&= \begin{cases} \mathcal{O}(k^2/n^2), & k \geq 1, \\ \mathcal{O}(1/n^2), & k = 0. \end{cases}
\end{aligned}
$$

Using this estimate we have

$$
\begin{aligned}
|S_m(\theta) - C_{n,m}(\theta)| &= \left|\tfrac{1}{2}(a_0 - \gamma_0) + \sum_{k=1}^{m} (a_k - \gamma_k)\cos k\theta\right|, \\
&\leq \frac{1}{n^2}\left|\mathcal{O}(1) + \sum_{k=1}^{m} \mathcal{O}(k^2)\right|, \\
&= \mathcal{O}\left(\frac{m^3}{n^2}\right).
\end{aligned} \tag{22}
$$

Thus (17), (21), and (22) in (20) imply

$$|g(\theta) - C_n(\theta)| = \mathcal{O}\left(\frac{1}{m} + \frac{m^3}{n^2}\right).$$

Finally, we set $m = \sqrt{n}$. ∎

Several interesting corollaries easily follow. First, we have

COROLLARY 1. *Under the hypothesis of Theorem 2 let $\sqrt{n} \leq m \leq n$ and $C_{n,m}(\theta)$ be the mth order trigonometric sum defined by* (19) *and* (10b). *Then,*

$$|g(\theta) - C_{n,m}(\theta)| = \mathcal{O}\left(\frac{1}{\sqrt{n}}\right). \tag{23}$$

Proof. It follows as in the theorem from (17), (22), and as in (21) that $|C_{n,\sqrt{n}}(\theta) - C_{n,m}(\theta)| = \mathcal{O}(1/\sqrt{n})$. ∎

Thus various truncated trigonometric sums may furnish as good an approximation as the entire nth order sum. Next, by changing variables we obtain a result on the convergence of interpolation polynomials for special unequally spaced points. We state this as

COROLLARY 2. *Let $f(x)$ have a continuous second derivative on $[-1, 1]$. If $P_n(x)$ is the interpolation polynomial of degree at most n for $f(x)$, based on the $n + 1$ points*

$$x_j = \cos\left(\frac{2j+1}{n+1}\frac{\pi}{2}\right), \qquad j = 0, 1, 2, \ldots, n, \tag{24}$$

then $P_n(x)$ converges to $f(x)$ on $[-1, 1]$ as $n \to \infty$. In fact,

$$|f(x) - P_n(x)| = \mathcal{O}\left(\frac{1}{\sqrt{n}}\right).$$

Proof. We introduce the new variable θ in $[0, \pi]$ by

$$\theta = \cos^{-1} x,$$

and then define

$$g(\theta) = f(\cos\theta).$$

We make $g(\theta)$ even and continuous, and 2π periodic by setting $g(-\theta) = g(\theta)$. Thus, the points (24) become the points θ_j of (9) and $g(\theta)$ satisfies the hypothesis of Theorem 2. Now the nth order interpolatory trigonometric sum (10) for $g(\theta)$ becomes the interpolation polynomial of degree at most n in x (represented in terms of the Chebyshev polynomials) upon using the indicated variable change. So we have

$$|f(x) - P_n(x)| = |g(\theta) - C_n(\theta)| = \mathcal{O}\left(\frac{1}{\sqrt{n}}\right). \quad ∎$$

If we use the variable change and function $f(x)$, we find from Corollary 1 that for *some polynomials*, call them $P_{n,m}(x)$, of *degree at most m*, where $\sqrt{n} \le m \le n$,

$$|f(x) - P_{n,m}(x)| = \mathcal{O}\left(\frac{1}{\sqrt{n}}\right).$$

[Note that the $P_{n,m}(x)$ are not obtained by simply truncating the interpolation polynomial, $P_n(x)$.] It is not difficult to show however, that *$P_{n,m}(x)$ is the mth degree discrete least squares approximation to $f(x)$ with respect to the $n + 1$ points* (24), see Problem 7. Thus we have established uniform convergence of the discrete least squares polynomials when the $n + 1$ points are as in (24) and the degree is at least $\sqrt{n}$. Our present estimates of convergence indicate that no improvement occurs by interpolating at these $n + 1$ points with a polynomial of degree n. This is another indication that high order interpolation polynomials should be avoided.

5.2. Least Squares Trigonometric Approximation. Fourier Series

If $f(x)$ is periodic of period† 2π and square integrable on $[-\pi, \pi]$, we can seek a trigonometric sum of form (1a) for which

$$\|f - S_n\|_2 = \left(\int_{-\pi}^{\pi} [f(x) - S_n(x)]^2\, dx\right)^{1/2} \tag{25}$$

is a minimum with respect to all such sums. This norm now defines a quadratic function of $2n + 1$ variables

$$J(a_0, a_1, \ldots, a_n, b_1, \ldots, b_n) = \|f - S_n\|_2^2$$

which can be minimized as was (3.8). The trigonometric functions satisfy the orthogonality relations

$$\begin{aligned}
&\int_{-\pi}^{\pi} \cos jx \cos kx\, dx = \begin{cases} 0, & j \ne k, \\ \pi, & j = k \ne 0, \end{cases} \\
&\int_{-\pi}^{\pi} \sin jx \sin kx\, dx = \begin{cases} 0, & j \ne k, \\ \pi, & j = k \ne 0, \end{cases} \\
&\int_{-\pi}^{\pi} \sin jx \cos kx\, dx = 0.
\end{aligned} \tag{26}$$

By using these results in the normal system obtained by minimizing (25), we find in analogy with (3.11),

$$a_k = \frac{1}{\pi}\int_{-\pi}^{\pi} \cos kx\, f(x)\, dx, \qquad b_k = \frac{1}{\pi}\int_{-\pi}^{\pi} \sin kx\, f(x)\, dx. \tag{27}$$

† If the period of $f(x)$ is some number p, then the change of variable $\xi = 2\pi x/p$ results in a function $g(\xi) \equiv f(p\xi/2\pi)$ which has period 2π.

The trigonometric sum (1a) with coefficients given in (27) determines the best least squares approximation of order n to $f(x)$ by such sums.

We deduce as in Section 3 the corresponding Bessel's inequality [by using (26) and (27) in $\|f - S_n\|_2^2 \geq 0$]:

$$\tfrac{1}{2}a_0^2 + \sum_{k=1}^{n} (a_k^2 + b_k^2) \leq \frac{1}{\pi}\int_{-\pi}^{\pi} f^2(x)\,dx. \tag{28}$$

Since the right-hand side is independent of n, we also conclude that $\tfrac{1}{2}a_0^2 + \sum_{k=1}^{\infty} (a_k^2 + b_k^2)$ converges and that

LEMMA 1.

$$\lim_{k\to\infty} a_k = \lim_{k\to\infty} b_k = 0. \qquad \blacksquare$$

The trigonometric sum (1a) is, of course, the nth partial sum of the infinite series

$$\tfrac{1}{2}a_0 + \sum_{k=1}^{\infty} (a_k \cos kx + b_k \sin kx). \tag{29}$$

With coefficients given by (27), this is the *Fourier series* associated with the function $f(x)$.

We can now state

THEOREM 3. *Let $f(x)$ be continuous and 2π periodic. Then the partial sums $S_n(x)$ of the Fourier series, with coefficients defined in* (27), *converge in the mean to $f(x)$ and Parseval's equality holds. That is,*

$$\lim_{n\to\infty} \int_{-\pi}^{\pi} [f(x) - S_n(x)]^2\,dx = 0 \tag{30a}$$

$$\frac{a_0^2}{2} + \sum_{k=1}^{\infty} a_k^2 + b_k^2 = \frac{1}{\pi}\int_{-\pi}^{\pi} [f(x)]^2\,dx. \tag{30b}$$

Proof. Simply modify the proof of Theorem 3, Section 3. $\blacksquare$

THEOREM 4. *Let $f(x)$ have two continuous derivatives and be 2π periodic. Then*

$$|a_k| = \mathcal{O}\left(\frac{1}{k^2}\right), \qquad |b_k| = \mathcal{O}\left(\frac{1}{k^2}\right)$$

and

$$|f(x) - S_n(x)| = \mathcal{O}\left(\frac{1}{n}\right) \qquad \textit{for } -\pi \leq x \leq \pi.$$

Proof. Let a_k'', b_k'' be the Fourier coefficients corresponding to the 2π

periodic and continuous function $f''(x)$. Now, by repeated use of integration by parts,

$$a_k'' = \frac{1}{\pi}\int_{-\pi}^{\pi} \cos kx\, f''(x)\,dx = -\frac{k}{\pi}\int_{-\pi}^{\pi} \sin kx\, f'(x)\,dx$$

$$= -\frac{k^2}{\pi}\int_{-\pi}^{\pi} \cos kx\, f(x)\,dx = -k^2 a_k$$

and similarly $b_k'' = -k^2 b_k$.

But by Lemma 1, $b_k'' \to 0$, and $a_k'' \to 0$, whence

$$|a_k| = \mathcal{O}\left(\frac{1}{k^2}\right) \quad \text{and} \quad b_k = \mathcal{O}\left(\frac{1}{k^2}\right).$$

We therefore know that the Fourier series converges uniformly to a continuous function $g(x)$. Hence, we may let $n \to \infty$ under the integral in the statement (30a) of mean convergence, thus proving that $f(x) \equiv \lim_{n\to\infty} S_n(x)$. The error estimate $|f(x) - S_n(x)| = \mathcal{O}(1/n)$ follows from the boundedness of $\{\cos kx, \sin kx\}$ and the relation

$$\sum_{n}^{\infty} \frac{1}{k^2} = \mathcal{O}\left(\frac{1}{n}\right). \qquad \blacksquare$$

The theory of approximation by orthogonal functions owes much to J. Fourier, who employed trigonometric series of the form (29). In fact, least squares approximations of the form (3.7) with orthogonal polynomials or other orthogonal sets of functions are generally called Fourier series (assuming $n \to \infty$) and the coefficients given by (3.11), (3.21), or (27) are called the Fourier coefficients.

Finally, we observe a close connection between the trigonometric interpolation coefficients (8) and the Fourier coefficients (27). Recalling the definitions of $\sum'$ and the points x_k in (3) we can write (8) as

$$\alpha_j = \frac{1}{\pi}\sum_{k=1-n}^{n}\left[\frac{\cos jx_{k-1} f(x_{k-1}) + \cos jx_k f(x_k)}{2}\right](x_k - x_{k-1}),$$

$$\beta_j = \frac{1}{\pi}\sum_{k=1-n}^{n}\left[\frac{\sin jx_{k-1} f(x_{k-1}) + \sin jx_k f(x_k)}{2}\right](x_k - x_{k-1}).$$

As $n \to \infty$ we have $x_k - x_{k-1} = \pi/n \to 0$ and [say for piecewise continuous $f(x)$] these sums converge to the corresponding integrals in (27). Thus, $(\alpha_j, \beta_j) \to (a_j, b_j)$ and the trigonometric interpolating sum (7) converges, formally, to the Fourier series (29).

These sums correspond to the trapezoidal rule of numerical integration. On the other hand, the coefficients γ_j in (10b) for trigonometric interpolation with respect to the points θ_j in (9) approximate the coefficients a_j

by the midpoint rule for evaluating the integrals in (27). In the proof of Theorem 2, it is shown that for fixed j: $|\gamma_j - a_j| = \mathcal{O}(1/n^2)$. [For the case that the function $g(\theta)$ is not necessarily even, the corresponding trigonometric interpolatory coefficients are defined in Problem 4, and similarly converge to the Fourier coefficients.]

5.3. "Best" Trigonometric Approximation

If $f(x)$ is continuous on $[-\pi, \pi]$ we can seek a trigonometric sum (1), of order n, which minimizes the maximum norm

$$\|f(x) - S_n(x)\|_\infty = \max_{-\pi \leq x \leq \pi} |f(x) - S_n(x)|.$$

The existence of such a best trigonometric approximation could be demonstrated by using an analogue of Theorem 0.1. Another proof is given in Problem 9.

Results analogous to those in Theorem 4.2 and Theorem 4.3 are also valid for the best trigonometric approximation of order n. A careful glance at the proofs of these theorems reveals that the only property of polynomials employed is the fact that if a polynomial of degree n vanishes at $n + 1$ points, then it vanishes identically. Such a property is also true of trigonometric sums. In fact, best approximation by other sets of functions is possible and the property they must possess to insure a unique best approximation is called the *Haar property*, defined as follows:

A sequence of functions $\{f_0(x), f_1(x), \ldots\}$ has the Haar property if for every m the only linear combination

$$P_m(x) \equiv a_0 f_0(x) + a_1 f_1(x) + \cdots + a_m f_m(x)$$

with $m + 1$ distinct zeros† *is the identically vanishing combination* $P_m(x) \equiv 0$. It was proven by Haar that these conditions are necessary and sufficient for uniqueness. However, we shall be concerned only with the trigonometric case. Thus, we consider

LEMMA 2. *The sequence of trigonometric functions*

$$\{1, \cos x, \sin x, \cos 2x, \sin 2x, \ldots, \cos nx, \sin nx, \ldots\}$$

has the Haar property.

Proof. We need only show that every non-trivial trigonometric sum, (1), of order n has at most $2n$ roots in $-\pi \leq x < \pi$. Let us define $\xi = e^{ix}$ and note that $|\xi| = 1$. Then we have from (1c),

$$S_n(x) = \xi^{-n} \sum_{k=-n}^{n} c_k \xi^{n+k}.$$

† If the $f_k(x)$ are periodic, then the zeros must all lie in a period.

The sum on the right-hand side is a polynomial in ξ of degree at most $2n$. Thus, this sum has at most $2n$ roots if it has any non-zero coefficients. ∎

This lemma can be used to prove uniqueness of the trigonometric interpolation problems solved in Subsection 5.1 and Problem 2. We apply the lemma to prove the analog of Theorem 4.2, namely

THEOREM 5. *Let $f(x) \in C[-\pi, \pi]$ and let an nth order trigonometric sum, $S_n(x)$, have the deviations from $f(x)$*

$$f(x_j) - S_n(x_j) = (-1)^j e_j, \qquad j = 0, 1, \ldots, 2n + 1,$$

where $-\pi \leq x_0 < x_1 < \cdots < x_{2n+1} \leq \pi$ and all $e_j > 0$ or all $e_j < 0$. Then

$$\min_j |e_j| \leq \min_{\{S_n\}} \|f(x) - S_n(x)\|_\infty.$$

Proof. Assume that for some trigonometric sum of order n, say $S_n^*(x)$,

$$\|f(x) - S_n^*(x)\|_\infty < \min_j |e_j|.$$

Then, the nth order sum

$$S_n^*(x) - S_n(x) = [f(x) - S_n(x)] - [f(x) - S_n^*(x)]$$

has the same sign at the points x_j as does $f(x) - S_n(x)$. Thus there are $2n + 1$ sign changes and at least $2n + 1$ zeros of this difference in $(-\pi, \pi)$. But then, by the above lemma, this trigonometric sum vanishes identically and so $S_n(x) \equiv S_n^*(x)$ which leads to a contradiction. ∎

Continuing the analogy, we have in place of Theorem 4.3

THEOREM 6. *A trigonometric sum of order n, $S_n(x)$, is the best trigonometric approximation of order n to $f(x) \in C[-\pi, \pi]$ if and only if $f(x) - S_n(x)$ assumes the values $\pm \|f(x) - S_n(x)\|_\infty$, with alternate changes of sign, at least $2n + 2$ times in $[-\pi, \pi]$. This best approximation is unique.*

Proof. The proof is exactly analogous to that of Theorem 4.4. To show sufficiency we employ Theorem 5. To demonstrate necessity we must construct a trigonometric sum of order $2k$, say, and which has specified zeros at distinct points $\xi_1, \xi_2, \ldots, \xi_{2k}$ in $(-\pi, \pi)$. This is done by forming the determinant

$$\text{(31)} \qquad t(x) = \begin{vmatrix} 1 & \cos x & \sin x & \cdots & \cos kx & \sin kx \\ 1 & \cos \xi_1 & \sin \xi_1 & \cdots & \cos k\xi_1 & \sin k\xi_1 \\ \vdots & \vdots & \vdots & \vdots & \vdots & \vdots \\ 1 & \cos \xi_{2k} & \sin \xi_{2k} & \cdots & \cos k\xi_{2k} & \sin k\xi_{2k} \end{vmatrix}.$$

The expression $t(x)$ is used in place of the polynomial $r(x)$ to obtain a contradiction [Note the relation of $t(x)$ to the determinant in Problem 1.] The uniqueness follows by using the Haar property of the trigonometric functions. The details are left to the reader. ∎

PROBLEMS, SECTION 5

1. Let $-\pi \leq x_0 < x_1 < \cdots < x_{2n} < \pi$, and define the determinant of order $2n + 1$,

$$\Delta = \begin{vmatrix} 1 & \cos x_0 & \sin x_0 & \cdots & \cos nx_0 & \sin nx_0 \\ 1 & \cos x_1 & \sin x_1 & \cdots & \cos nx_1 & \sin nx_1 \\ \vdots & \vdots & \vdots & \vdots & \vdots & \vdots \\ 1 & \cos x_{2n} & \sin x_{2n} & \cdots & \cos nx_{2n} & \sin nx_{2n} \end{vmatrix}.$$

Show that $\Delta \neq 0$ and in fact, that

$$\Delta = (-1)^{n(n-1)/2} 2^{2n^2} \prod_{j=1}^{2n} \left[\prod_{k=0}^{j-1} \sin\left(\frac{x_j - x_k}{2}\right) \right].$$

[Hint: Express $\sin\theta$ and $\cos\theta$ in exponential form, form linear combinations of successive pairs of columns so that only one exponential appears in each element, rearrange columns so that each row forms a geometric progression. The result is a Vandermonde determinant.]

2. Show that $S_n(x)$ is a trigonometric sum which satisfies $S_n(x_j) = f(x_j)$, $j = 0, 1, \ldots, 2n$, where $\{x_k\}$ is given as in Problem 1 and

$$S_n(x) = \sum_{j=0}^{2n} f(x_j) \prod_{\substack{k=0 \\ (k \neq j)}}^{2n} \left(\frac{\sin \dfrac{x - x_k}{2}}{\sin \dfrac{x_j - x_k}{2}} \right).$$

This is the general interpolatory trigonometric sum of order n.

[Hint: Use Problem 3 and $\sin a \sin b = \frac{1}{2}[\cos(a - b) - \cos(a + b)]$.]

3. Verify that a trigonometric polynomial of degree at most n,

$$P_n(\cos\theta, \sin\theta) \equiv \sum_{i+j<n} c_{i,j} (\cos\theta)^i (\sin\theta)^j,$$

may be written as a trigonometric sum of order at most n,

$$P_n(\cos\theta, \sin\theta) \equiv S_n(\theta) \equiv \sum_{k=0}^{n} a_k \cos k\theta + b_k \sin k\theta,$$

and vice versa.

[Hint: Use (1b).]

4. Given $g(x)$ is 2π periodic, find the trigonometric sum

$$S_n(x) = \frac{a_0}{2} + \sum_{k=1}^{n-1} (a_k \cos kx + b_k \sin kx) + \frac{b_n}{2} \sin nx$$

such that

$$S_n(x_j) = g(x_j) \qquad \text{for } x_j = \frac{(2j+1)\pi}{2n}, \quad -n \le j \le n-1.$$

[Hint: Establish $\displaystyle\sum_{j=-n}^{n-1} \cos rx_j \cos sx_j = \begin{cases} 0 & r \ne s \\ n & n > r = s > 0 \\ 2n & r = s = 0 \\ 0 & r = s = n \end{cases}$

$$\sum_{j=-n}^{n-1} \cos rx_j \sin sx_j = 0$$

$$\sum_{j=-n}^{n-1} \sin rx_j \sin sx_j = \begin{cases} 0 & r \ne s \\ n & n > r = s > 0 \\ 0 & r = s = 0 \\ 2n & r = s = n, \end{cases}$$

whence

$$a_r = \frac{1}{n} \sum_{j=-n}^{n-1} g(x_j) \cos rx_j,$$

$$b_r = \frac{1}{n} \sum_{j=-n}^{n-1} g(x_j) \sin rx_j.]$$

5. Verify that

$$\left|\frac{\theta}{\sin\theta}\right| \le \frac{\pi}{2}$$

provided $|\theta| \le \pi/2$.

[Hint: Consider the chord joining $(0, 0)$ and $(\pi/2, 1)$ on the graph of $y = \sin x$.]

6. Let $f''(x)$ be continuous in $[a, b]$ and

$$x_j = x_0 + jh, \quad x_0 = a + \frac{h}{2}, \quad h = \frac{b-a}{n+1}, \qquad j = 0, 1, \ldots, n.$$

Show that

$$|E| \equiv \left|\int_a^b f(x)\,dx - \sum_{j=0}^{n} f(x_j)h\right| \le \frac{|b-a|}{24} h^2 \max_{a<x<b} |f''(x)|.$$

[Hint: Write

$$E = \sum_{j=0}^{n} \int_{x_j - h/2}^{x_j + h/2} [f(x) - f(x_j)]\,dx$$

and then use Taylor's theorem in each integrand to get

$$f(x) = f(x_j) + (x - x_j)f'(x_j) + \frac{(x - x_j)^2}{2} f''\left(x_j + \theta\frac{h}{2}\right), \qquad -1 < \theta < 1.]$$

7. Show that the mth degree discrete least squares polynomial approximation to $f(x)$ with respect to the $n + 1$ points (24) is obtained by truncating the nth order trigonometric interpolation sum (10) for $g(x)$ on the points (9) after the mth term and using the variable change $\theta = \cos^{-1} x$, $g(\theta) = f(\cos \theta)$.

[Hint: Simply change variables in the representation (3.52) by using the discrete orthonormal polynomials (3.61). The uniqueness of the discrete least square polynomials is used.]

8. With the notation of equations (2), (3), (4), and (6), verify the *discrete* analogue of the *Parseval equality*,

$$\frac{1}{2n} \sum_{k=-n}^{n}{}' |f(x_k)|^2 = \frac{1}{2n} \sum_{k=-n}^{n}{}' |U_n(x_k)|^2 = \sum_{j=-n}^{n}{}' |c_j|^2.$$

[Hint: Use equation (5).]

9. Prove the existence of a best trigonometric approximation in the form (1) for the given function $f(x)$ on the interval $[-\pi, \pi]$.

[Hint: In Problem 2, if $|f(x)| < M$ and $\{x_i\}$ are distinct, then the trigonometric sum $S_n(x)$ has bounded coefficients. For fixed n, consider the non-empty set, C, of trigonometric sums $\{S_n(x)\}$ such that

$$\|f(x) - S_n(x)\|_\infty \leq \frac{M}{2}.$$

Note that by the above remark, the coefficients of all of the sums in C are bounded. Let $S_n{}^\nu(x)$, $\nu = 1, 2, \ldots$, be a *minimizing sequence* of trigonometric sums in C, that is,

$$\|f - S_n{}^\nu\|_\infty \rightarrow \underset{C}{\text{g.l.b.}} \|f - S_n\|_\infty.$$

Pick a subsequence of $S_n{}^\nu$ such that their coefficients converge, i.e.,

$$a_k{}^{\nu_j} \rightarrow \hat{a}_k, \; b_k{}^{\nu_j} \rightarrow \hat{b}_k.$$

The sum

$$\hat{S}_n(x) \equiv \frac{\hat{a}_0}{2} + \sum_{k=1}^{n} \hat{a}_k \cos kx + \hat{b}_k \sin kx$$

is a best approximation.]

6

Differences, Interpolation Polynomials, and Approximate Differentiation

0. INTRODUCTION

Interpolation polynomials are of particular importance in numerical analysis, and so we devote special attention to them in this chapter. We are led naturally to the study of differences; both divided differences for arbitrarily spaced points and ordinary differences for equally spaced points. Not only can differences be neatly arranged in tables, for convenient hand computation (presumably an affair of the past), but they permit one to easily estimate the error in the approximation. Hence, methods based on differences are useful in this age of digital computers as they suggest very efficient computing techniques and can be used for checking the accuracy of a calculation.

We examine the error in interpolation, when the polynomial passes through equally spaced points, in some detail. This error is generally much less near the center of the interval of interpolation points and grows rapidly outside this interval, i.e., for what is termed extrapolation. Therefore, we construct special forms of the interpolation formulae which are convenient for evaluation near the center of the interpolation interval.

We use interpolation polynomials to determine formulae for the numerical approximation of derivatives of the interpolated function.

1. NEWTON'S INTERPOLATION POLYNOMIAL AND DIVIDED DIFFERENCES

We have shown in Chapter 5, Section 2, that the interpolation polynomial exists and is unique. Furthermore, for a fixed set of interpolation points, it is easily constructed using the Lagrange interpolation coefficients. The Lagrange representation has the defect that if another point of interpolation were added, then the new higher degree interpolation polynomial could not be obtained by easily modifying the previous one. (This is in contrast, say, to Taylor's series expansion or to least squares expansion in orthogonal functions where the next order approximation is obtained by simply adjoining a term to the present approximation.) We seek then a representation of the interpolation polynomial which has the property that the next higher degree interpolation polynomial is found by simply adding a new term.

Specifically, let $Q_k(x)$ be the interpolation polynomial for $f(x)$, of degree at most k, with respect to the $k + 1$ distinct points $x_0, x_1, \ldots, x_k$. We seek the successive interpolation polynomials, $\{Q_k(x)\}$, of degree at most k in the form $Q_0(x) \equiv f(x_0)$ and

$$\text{(1a)} \qquad Q_k(x) = Q_{k-1}(x) + q_k(x), \qquad \text{for } k = 1, 2, \ldots, n,$$

where $q_k(x)$ has at most degree k. Since we require

$$Q_k(x_j) = f(x_j) = Q_{k-1}(x_j), \qquad j = 0, 1, \ldots, k - 1$$

it follows that $q_k(x_j) = 0$ at these k points. Thus, we may write

$$\text{(1b)} \qquad q_k(x) = a_k \prod_{j=0}^{k-1} (x - x_j),$$

which represents the most general polynomial of degree at most k that vanishes at the indicated k points. The constant a_k remains to be determined. But, in order that $Q_k(x_k) = f(x_k)$, it follows from (1a and b) that

$$\text{(1c)} \qquad a_k = \frac{f(x_k) - Q_{k-1}(x_k)}{\prod_{j=0}^{k-1} (x_k - x_j)}, \qquad \text{for } k = 1, 2, \ldots, n.$$

The zero degree interpolation polynomial for the initial point x_0 is, trivially, $Q_0(x) \equiv f(x_0)$. Thus, with $a_0 = f(x_0)$, we obtain by using (1a and b) recursively, for $k = 1, 2, \ldots, n$,

$$\text{(2)} \qquad Q_n(x) = a_0 + (x - x_0)a_1 + \cdots + (x - x_0)\cdots(x - x_{n-1})a_n.$$

The kth coefficient is called the *kth order divided difference* and is usually expressed in the notation

$$(3) \qquad \begin{aligned} a_0 &= f[x_0], \\ a_k &= f[x_0, x_1, \ldots, x_k], \qquad k = 1, 2, \ldots. \end{aligned}$$

The values of $f(x)$ which enter into the determination of a_k are those at the arguments of $f[x_0, x_1, \ldots, x_k]$. We now obtain a representation for these coefficients which is more explicit than the recursive form given in (1). Since $Q_n(x)$ in (2) is *the unique* interpolation polynomial of degree n, we may equate the leading coefficient, a_n, in this form with that obtained by using the Lagrange form, see (2.5) in Chapter 5. That is, from

$$Q_n(x) = \sum_{j=0}^{n} f(x_j) \prod_{\substack{k=0 \\ k \neq j}}^{n} \frac{x - x_k}{x_j - x_k}$$

the coefficient of x^n is

$$(4) \qquad a_n = f[x_0, x_1, \ldots, x_n] = \sum_{j=0}^{n} \frac{f(x_j)}{\prod\limits_{\substack{k=0 \\ (k \neq j)}}^{n} (x_j - x_k)}.$$

This form could also be deduced directly from (1); see Problem 1.

From the representation (4) it follows that *the divided differences are symmetric functions of their arguments.* That is, if we adopt the additional notation

$$f_{i,j,k,\ldots} \equiv f[x_i, x_j, x_k, \ldots]$$

then this symmetry is expressed by

$$(5) \qquad f_{0,1,\ldots,n} = f_{j_0,j_1,\ldots,j_n}$$

where $(j_0, j_1, \ldots, j_n)$ is any permutation of the integers $(0, 1, \ldots, n)$.

We may derive a more convenient form than (4) for computing the divided differences by again making use of the *uniqueness* of the interpolation polynomial. That is, we may construct the polynomial $Q_n(x)$ by matching the values of $f(x_j)$ in the reverse order $j = n, n - 1, \ldots, 1, 0$. In this way we would obtain, say,

$$(2') \qquad Q_n(x) \equiv b_0 + (x - x_n)b_1 + \cdots + (x - x_n)(x - x_{n-1})\cdots(x - x_1)b_n$$

where

$$b_k = f[x_n, x_{n-1}, \ldots, x_{n-k}] \quad \text{and} \quad b_0 = f[x_n] = f(x_n).$$

But $a_n = b_n$ since they are the coefficients of x^n in the unique polynomial $Q_n(x)$ of (2) and (2').

Now if we subtract equation (2′) from equation (2) but display only the terms which contribute to the coefficients of x^n and x^{n-1} we obtain, using $a_n = b_n$,

$$0 \equiv [(x - x_0) - (x - x_n)](x - x_1)\cdots(x - x_{n-1})a_n \\ + (a_{n-1} - b_{n-1})x^{n-1} + p_{n-2}(x)$$

where $p_{n-2}(x)$ is a polynomial of degree at most $n - 2$ in x. Since this expression vanishes identically the coefficient of x^{n-1} vanishes and hence $0 = [x_n - x_0]a_n + (a_{n-1} - b_{n-1})$. Now the symmetry of the divided differences, proven above, implies that

$$b_{n-1} = f[x_n, x_{n-1}, \ldots, x_1] = f[x_1, x_2, \ldots, x_n],$$

whence from $a_n = (a_{n-1} - b_{n-1})/(x_0 - x_n)$, we have

$$\text{(6)} \qquad f[x_0, x_1, \ldots, x_n] = \frac{f[x_0, x_1, \ldots, x_{n-1}] - f[x_1, x_2, \ldots, x_n]}{x_0 - x_n}, \qquad n = 1, 2, \ldots.$$

We leave it to the reader to verify (6) directly from (4) in Problem 9. This recursion formula justifies the use of the name divided difference. Of course, we then define for completeness

$$f[x_0] = f(x_0).$$

The interpolation polynomial (2) may now be written as

$$\text{(7)} \qquad Q_n(x) = f[x_0] + (x - x_0)f[x_0, x_1] + \cdots \\ + (x - x_0)\cdots(x - x_{n-1})f[x_0, x_1, \ldots, x_n].$$

This form is known as *Newton's divided difference interpolation formula.* Note that to obtain the next higher degree such polynomial we need only add a term similar to the last term but involving a new divided difference of one higher order.

In fact, let us set $k = n + 1$ in (1b and c) and (3) and then replace x_{n+1} by x. We obtain

$$\text{(8)} \qquad f(x) - Q_n(x) = \left[\prod_{j=0}^{n} (x - x_j)\right] f[x_0, x_1, \ldots, x_n, x],$$

which for x distinct from $\{x_j\}$ defines the indicated divided difference. On the other hand, this identity gives another representation of the *error* in polynomial interpolation.

By means of formula (6) applied to $(x_j, x_{j+1}, \ldots, x_{j+n})$ we can construct a table of divided differences in a symmetric manner based on

$$\text{(6)}_j \qquad f_{j,j+1,\ldots,j+n} = \frac{f_{j+1,j+2,\ldots,j+n} - f_{j,j+1,\ldots,j+n-1}}{x_{j+n} - x_j}.$$

Table 1 *Divided Differences*

x	$f(x)$	$f[x, x]$	$f[x, x, x]$	$f[x, x, x, x]$	$\cdots$
x_0	f_0				
		$\dfrac{f_1 - f_0}{x_1 - x_0} \equiv f_{01}$			
x_1	f_1		$\dfrac{f_{12} - f_{01}}{x_2 - x_0} \equiv f_{012}$		
		$\dfrac{f_2 - f_1}{x_2 - x_1} \equiv f_{12}$		$\dfrac{f_{123} - f_{012}}{x_3 - x_0} \equiv f_{0123}$	
x_2	f_2		$\dfrac{f_{23} - f_{12}}{x_3 - x_1} \equiv f_{123}$		$\cdots$
		$\dfrac{f_3 - f_2}{x_3 - x_2} \equiv f_{23}$		$\dfrac{f_{234} - f_{123}}{x_4 - x_1} \equiv f_{1234}$	
x_3	f_3		$\dfrac{f_{34} - f_{23}}{x_4 - x_2} \equiv f_{234}$	$\vdots$	
		$\dfrac{f_4 - f_3}{x_4 - x_3} \equiv f_{34}$	$\vdots$		
x_4	f_4				
		$\vdots$			
$\vdots$	$\vdots$				

See Table 1. The divided difference required to determine $Q_{n+1}(x)$ from $Q_n(x)$ is easily obtained from Table 1 by just completing another "diagonal" line of differences. This simple property is not shared by the Lagrange form of the interpolation polynomial.

Another representation of the divided differences, which is quite useful for estimating their magnitude as well as for many theoretical purposes, is contained in

THEOREM 1. *Let $x, x_0, x_1, \ldots, x_{k-1}$ be $k + 1$ distinct points and let $f(y)$ have a continuous derivative of order k in the interval*

$$\min(x, x_0, \ldots, x_{k-1}) < y < \max(x, x_0, \ldots, x_{k-1}).$$

Then for some point $\xi = \xi(x)$ in this interval

$$f[x_0, \ldots, x_{k-1}, x] = \frac{f^{(k)}(\xi)}{(k)!}. \tag{9}$$

Proof. From equation (8) with n replaced by $k - 1$ we write

$$f(x) - Q_{k-1}(x) = (x - x_0)(x - x_1)\cdots(x - x_{k-1})f[x_0, \ldots, x_{k-1}, x].$$

However, since $Q_{k-1}(x)$ is an interpolation polynomial which is equal to

$f(x)$ at the k points $x_0, x_1, \ldots, x_{k-1}$ it follows from Theorem 2.1 of Chapter 5 that

$$f(x) - Q_{k-1}(x) \equiv R_{k-1}(x)$$
$$= (x - x_0)(x - x_1)\cdots(x - x_{k-1})\frac{f^{(k)}(\xi)}{(k)!}.$$

But by the hypothesis $(x - x_0)(x - x_1)\cdots(x - x_{k-1}) \neq 0$, and the theorem follows by equating the above right-hand sides. ■

A generalization, permitting coincident values, is established as Corollary 2 of Theorem 2 that follows.

As an immediate consequence of Theorem 1, we can obtain some information on the divided differences of polynomials. These results may be stated as the

COROLLARY. *Let*

$$P_n(x) = \alpha_0 + \alpha_1 x + \cdots + \alpha_n x^n, \qquad \alpha_n \neq 0,$$

be any polynomial of degree n and let $x_0, x_1, \ldots, x_k$ *be any* $k + 1$ *distinct points. Then*

$$P_n[x_0, x_1, \ldots, x_k] = \begin{cases} \alpha_n & \text{if } k = n, \\ 0 & \text{if } k > n. \end{cases}$$

Proof. The corollary follows from Theorem 1 since $d^n P_n(x)/dx^n = n!\,\alpha_n$; and higher derivatives vanish. ■

We shall require some continuity and differentiability properties of divided differences for our later discussion of the error in numerical differentiation and integration. Most of these results can be derived from still another representation of the divided differences which we state as

THEOREM 2. *Let* $f(x)$ *have a continuous nth derivative in the interval* $\min(x_0, x_1, \ldots, x_n) \leq x \leq \max(x_0, x_1, \ldots, x_n)$. *Then if the points* $x_0, x_1, \ldots, x_n$ *are distinct,*

$$(10)_n \qquad f[x_0, x_1, \ldots, x_n] = \int_0^1 dt_1 \int_0^{t_1} dt_2 \cdots \int_0^{t_{n-1}} dt_n$$
$$\times f^{(n)}(t_n[x_n - x_{n-1}] + \cdots + t_1[x_1 - x_0] + x_0),$$

where $n \geq 1$, $t_0 = 1$.

Proof. For an inductive proof, we first show that

$$f[x_0, x_1] = \int_0^1 dt_1\, f'(t_1[x_1 - x_0] + x_0).$$

Let a new variable of integration, ξ, be introduced by (since $x_1 \neq x_0$):

$$\xi = t_1[x_1 - x_0] + x_0, \qquad dt_1 = d\xi/[x_1 - x_0].$$

The integration limits become

$$(t_1 = 0) \to (\xi = x_0); \qquad (t_1 = 1) \to (\xi = x_1).$$

Therefore, we have

$$\int_0^1 dt_1 f'(t_1[x_1 - x_0] + x_0) = \frac{1}{x_1 - x_0} \int_{x_0}^{x_1} d\xi f'(\xi)$$
$$= \frac{f(x_1) - f(x_0)}{x_1 - x_0} = f[x_0, x_1].$$

Now we make the inductive hypothesis that

$$f[x_0, \ldots, x_{n-1}] = \int_0^1 dt_1 \int_0^{t_1} dt_2 \cdots \int_0^{t_{n-2}} dt_{n-1}$$
$$\times f^{(n-1)}(t_{n-1}[x_{n-1} - x_{n-2}] + \cdots + t_1[x_1 - x_0] + x_0).$$

In the integral in $(10)_n$ we replace the integration variable t_n by

$$\xi = t_n[x_n - x_{n-1}] + \cdots + t_1[x_1 - x_0] + x_0,$$
$$dt_n = d\xi/[x_n - x_{n-1}].$$

The corresponding limits become

$$(t_n = 0) \to (\xi = \xi_0 \equiv t_{n-1}[x_{n-1} - x_{n-2}] + \cdots + t_1[x_1 - x_0] + x_0),$$
$$(t_n = t_{n-1}) \to (\xi = \xi_1 \equiv t_{n-1}[x_n - x_{n-2}]$$
$$+ t_{n-2}[x_{n-2} - x_{n-3}] + \cdots + t_1[x_1 - x_0] + x_0).$$

Now the innermost integral in $(10)_n$ is, since $x_n \neq x_{n-1}$,

$$\int_0^{t_{n-1}} dt_n f^{(n)}(t_n[x_n - x_{n-1}] + \cdots + t_1[x_1 - x_0] + x_0)$$
$$= \int_{\xi_0}^{\xi_1} f^{(n)}(\xi) \frac{d\xi}{x_n - x_{n-1}}$$
$$= \frac{f^{(n-1)}(\xi_1) - f^{(n-1)}(\xi_0)}{x_n - x_{n-1}}.$$

However, by applying the inductive hypothesis we have

$$\int_0^1 dt_1 \int_0^{t_1} dt_2 \cdots \int_0^{t_{n-2}} dt_{n-1} \frac{f^{(n-1)}(\xi_1) - f^{(n-1)}(\xi_0)}{x_n - x_{n-1}}$$
$$= \frac{f[x_0, \ldots, x_{n-2}, x_n] - f[x_0, \ldots, x_{n-2}, x_{n-1}]}{x_n - x_{n-1}}$$
$$= f[x_0, \ldots, x_n].$$

■

Notice that the integrand on the right-hand side of (10) is a continuous function of the $n + 1$ variables $x_0, x_1, \ldots, x_n$, and hence the right-hand side is a continuous function of these variables. Thus (10) defines uniquely the continuous extension of $f[x_0, x_1, \ldots, x_n]$ when the arguments lie in any interval of continuity of the nth derivative of $f(x)$. That is, since the divided differences have only been defined for distinct sets of arguments [see (1) and the discussion leading to it], we are at liberty to define them when some of the arguments are not distinct. Naturally, we do this in a manner which maintains, if possible, the above observed continuity. If $f^{(n)}(x)$ is continuous, then Theorem 2 shows how this can be done for all differences of $f(x)$ of orders $0, 1, \ldots, n$. These remarks can be summarized as

COROLLARY 1. *Let $f^{(n)}(x)$ be continuous in $[a, b]$. For any set of points $x_0, x_1, \ldots, x_k$ in $[a, b]$ with $k \leq n$ let $f[x_0, x_1, \ldots, x_k]$ be given by* $(10)_k$. *The divided difference thus defined is a continuous function of its $k + 1$ arguments in $[a, b]$ and coincides with that defined by* (4), *or* (6), *when the arguments are distinct.* ∎

In fact, as in the proof of the First Mean Value Theorem for integrals, $(10)_n$ yields

$$m \int_0^1 dt_1 \int_0^{t_1} dt_2 \cdots \int_0^{t_{n-1}} dt_n \leq f[x_0, \cdots, x_n] \leq M \int_0^1 dt_1 \int_0^{t_1} dt_2 \cdots \int_0^{t_{n-1}} dt_n$$

where $m \equiv \min f^{(n)}(x)$ and $M \equiv \max f^{(n)}(x)$ for x in

$$\min(x_0, \ldots, x_n) \leq x \leq \max(x_0, \ldots, x_n).$$

Then by the continuity of $f^{(n)}$ there is a point μ_n in this interval such that

$$f[x_0, \ldots, x_n] = \frac{f^{(n)}(\mu_n)}{n!}.$$

Hence, we have established a generalization of Theorem 1, since the points x_i need not be distinct, in

COROLLARY 2. *If $f^{(n)}(x)$ is continuous in $[a, b]$ and $x_0, x_1, \ldots, x_n$ are in $[a, b]$, then*

$$f[x_0, x_1, \ldots, x_n] = \frac{f^{(n)}(\xi)}{n!}, \tag{11}$$

where

$$\min(x_0, x_1, \ldots, x_n) \leq \xi \leq \max(x_0, x_1, \ldots, x_n).$$ ∎

A particular case is

COROLLARY 3. *If $f^{(n)}(x)$ is continuous in a neighborhood of x, then*

$$(12) \qquad f[\underbrace{x, x, \ldots, x}_{n+1 \text{ terms}}] = \frac{f^{(n)}(x)}{n!}. \qquad \blacksquare$$

Now, we can deduce yet another representation of the divided difference, when several multiplicities† occur.

COROLLARY 4. *If $f^{(n)}(x)$ is continuous in $[a, b]$, $y_0, y_1, \ldots, y_n$ are in $[a, b]$, and x is distinct from any y_i, then*

$$(13) \qquad f[x, y_0, y_1, \ldots, y_n] = \frac{f[x, y_1, \ldots, y_n] - f[y_0, y_1, \ldots, y_n]}{x - y_0},$$

gives the unique continuous extension of the definition of divided difference.

Proof. By Theorem 2, the right-hand side of (13) is uniquely defined and continuous in $(y_0, y_1, \ldots, y_n)$ for $y_0 \neq x$. Hence, the left-hand side is the unique continuous extension of the definition of divided difference no matter what multiplicities occur in $(y_0, y_1, \ldots, y_n)$. Observe that only the continuity of the nth derivative is used. $\blacksquare$

COROLLARY 5. *If $x_i \neq y_j$, for $0 \leq i \leq p$, $0 \leq j \leq q$; $f^{(m)}(x)$ continuous in $[a, b]$; $\{x_i\}$, $\{y_i\}$ in $[a, b]$; $0 \leq p, q \leq m$, then*

$$(14) \qquad \begin{aligned} f[x_0, \ldots, x_p, y_0, \ldots, y_q] &= g[x_0, \ldots, x_p] \\ &= h[y_0, \ldots, y_q] \end{aligned}$$

where

$$g(x) \equiv f[x, y_0, \ldots, y_q], \qquad h(y) \equiv f[x_0, \ldots, x_p, y],$$

provides the unique continuous extension of definition of divided difference.

Proof. By (13), $g(x)$ has m continuous derivatives for $x \neq y_i$, $0 \leq i \leq q$. Therefore, by the theorem, $g[x_0, \ldots, x_p]$ is defined and continuous in $x_0, \ldots, x_p$ if $x_i \neq y_j$, as postulated. Furthermore, $g(x)$ is continuous as a function of the parameters $(y_0, \ldots, y_q)$ if $x \neq y_i$, by Corollary 4. Hence, the representation $(10)_p$ of $g[x_0, \ldots, x_p]$ yields the continuity of $g[x_0, \ldots, x_p]$ with respect to all variables $(x_0, \ldots, x_p, y_0, \ldots, y_q)$ provided merely that $x_i \neq y_j$.

Now the function $f[x_0, \ldots, x_p, y_0, \ldots, y_q]$ as defined in (14) is continuous (if $x_i \neq y_j$) in its variables; hence we conclude that (14) is the unique continuous extension, since (14) is valid when the arguments are all distinct. $\blacksquare$

† The conclusions of Corollaries 4–7 that follow, concern continuity properties of and representations for divided differences that are easily established when no multiplicities occur among the arguments. When multiplicities do occur, the corollaries establish the same representations under the hypothesis of minimal differentiability of $f(x)$.

COROLLARY 6. *If $f(x)$ has a continuous derivative of order m in $[a, b]$; $x_0, \ldots, x_p, y_0, \ldots, y_q, z_0, \ldots, z_r$ are in $[a, b]$; $x_i \neq y_j$, $x_i \neq z_k$, $y_j \neq z_k$ for all i, j, k; $0 \leq p, q, r \leq m$; then*

$$f[x_0, \ldots, x_p, y_0, \ldots, y_q, z_0, \ldots, z_r] = \frac{1}{p!\,q!\,r!} \frac{\partial^p}{\partial x^p} \frac{\partial^q}{\partial y^q} \frac{\partial^r}{\partial z^r} f[x, y, z]\bigg|_{(\xi, \eta, \zeta)} \tag{15}$$

where

$$\min(x_0, \ldots, x_p) \leq \xi \leq \max(x_0, \ldots, x_p),$$
$$\min(y_0, \ldots, y_q) \leq \eta \leq \max(y_0, \ldots, y_q),$$
$$\min(z_0, \ldots, z_r) \leq \zeta \leq \max(z_0, \ldots, z_r).$$

Proof. Let

$$\begin{aligned} g(x) &\equiv f[x, y_0, \ldots, y_q, z_0, \ldots, z_r] \\ h(y) &\equiv f[x, y, z_0, \ldots, z_r] \\ k(z) &\equiv f[x, y, z]. \end{aligned} \tag{16}$$

By Corollaries 2 and 5 [appropriately generalized for sets of variables $(\{x_i\}, \{y_j\}, \{z_k\})$], we have

$$f[x_0, \ldots, x_p, y_0, \ldots, y_q, z_0, \ldots, z_r] = g[x_0, \ldots, x_p] \tag{17}$$

$$g[x_0, \ldots, x_p] = \frac{1}{p!} \frac{\partial^p}{\partial x^p} g(x)\bigg|_{x=\xi} \tag{18}$$

$$g(x) = h[y_0, \ldots, y_q] = \frac{1}{q!} \frac{\partial^q}{\partial y^q} h(y)\bigg|_{y=\eta} \tag{19}$$

$$h(y) = k[z_0, \ldots, z_r] = \frac{1}{r!} \frac{\partial^r}{\partial z^r} k(z)\bigg|_{z=\zeta}. \tag{20}$$

The conclusion (15) follows from (17), (18), (19), and (20). ∎

A special case is contained in

COROLLARY 7. *If $f^{(m)}(x)$ is continuous in $[a, b]$; x, y, z are distinct points in $[a, b]$; $0 \leq p, q, r \leq m$; then*

$$f[\underbrace{x, \ldots, x}_{p+1}\, \underbrace{y, \ldots, y,}_{q+1}\, \underbrace{z, \ldots, z}_{r+1}] = \frac{1}{p!\,q!\,r!} \frac{\partial^p}{\partial x^p} \frac{\partial^q}{\partial y^q} \frac{\partial^r}{\partial z^r} f[x, y, z]. \tag{21}$$ ∎

We leave to Problems 3, 4, 5, 6, and 7, the independent proof of some simple differentiability properties of divided differences, which are needed later.

PROBLEMS, SECTION 1

1. Deduce the representation (4) for divided differences directly from the expression (1).

[Hint: Use induction and the form (2) for $Q_{k+1}(x_k)$.]

2. If $P_n(x)$ is a polynomial of degree n, show that $P_n[x_0, x]$ for $x \neq x_0$ is a polynomial of degree at most $n - 1$ in x.

3. Prove the following

LEMMA **1.** *If $x, x_0, \ldots, x_n$ are $n + 2$ distinct points then*

$$f[x_0, \ldots, x_n, x] = \sum_{j=0}^{n} \frac{f[x, x_j]}{\prod\limits_{\substack{k=0 \\ (k \neq j)}}^{n} (x_j - x_k)}.$$

[Hint: Use equation (8) and the Lagrange form of the interpolation polynomial.] This is another representation of the divided differences which is very useful in deriving their continuity and differentiability properties.

4. Prove directly the following:

THEOREM **3.** *If $f(x) \in C[a, b]$ and $f'(x_j)$ exists for some fixed $x_j \in [a, b]$, then $f[x, x_j]$ is a continuous function of x in $[a, b]$ if we assign to it the value at $x = x_j$: $f[x_j, x_j] = f'(x_j)$.*

5. Prove the following:

THEOREM **4.** *Let $f'(x) \in C[a, b]$ and $f''(x)$ be continuous in an (arbitrarily small) interval about some fixed $x_j \in [a, b]$. Then $df[x, x_j]/dx$ is a continuous function of x in $[a, b]$.*

[Hint: Form $df[x, x_j]/dx$ for $x \neq x_j$; use the Taylor expansion about x_j and take limits as $x = x_j \pm h \to x_j$.]

6. Use the results of Problems 3, 4, and 5 to state and prove, if $(x_0, x_1, \ldots, x_n)$ are distinct and in $[a, b]$,

(i) a theorem on the continuity of $f[x_0, \ldots, x_n, x]$ for $x \in [a, b]$;
(ii) a theorem on the continuity of $(d/dx)f[x_0, \ldots, x_n, x]$ for $x \in [a, b]$.

7. By using the theorem under (i) of Problem 6, note that $f[x_0, \ldots, x_n, x_n] = \lim_{h \to 0} f[x_0, \ldots, x_n, x_n + h]$. Therefore, show that

$$f[x_0, \ldots, x_n, x_n] = \frac{d}{dx_n} f[x_0, \ldots, x_n].$$

Prove that this representation is valid under the conditions: $f'(x_n)$ is defined and $x_0, x_1, \ldots, x_n$ are distinct.

[Hint: use the lemma of Problem 3 and the formula (6).]

8. Prove the symmetry of the divided difference by constructing the $Q_n(x)$ in (2) using the points $(x_0, x_1, \ldots, x_n)$ in an arbitrary permuted order $(x_{j_0}, x_{j_1}, \ldots, x_{j_n})$. [This is a generalization of what was done in deriving (2′) and proving $a_n = b_n$.]

9. Verify equation (6) (the divided difference property) directly from equation (4).

10.* (*Osculatory interpolation*). If $f(x)$ and its derivatives of order $r_0 - 1$, $r_1 - 1, \ldots, r_n - 1$ are defined respectively at the distinct points $(x_0, x_1, \ldots, x_n)$

in $[a, b]$, then there exists a unique polynomial $Q_N(x)$ of degree at most N, where $N = \sum_{k=0}^{n} r_k - 1$, such that $Q_N^{(k)}(x_j) = f^{(k)}(x_j)$ for $k = 0, 1, \ldots, r_j - 1$, $j = 0, 1, \ldots, n$. The special case $r_0 = r_1 = \cdots = r_n = 2$ has been studied in Chapter 5, Subsection 2.2.

[Hint: Show that the Newton form of the polynomial may be derived by satisfying successively all of the r_0 conditions at x_0 before proceeding to satisfy successively the r_1 conditions at x_1, etc.
Arrive at the scheme

$$Q_0(x) \equiv f(x_0)$$

$$Q_s(x) \equiv Q_{s-1}(x) + b_s(x - x_0)^s, \qquad \text{for } 1 \leq s < s_0 \text{ where } s_0 = r_0$$

$$Q_s(x) \equiv Q_{s-1}(x) + b_s\left[\prod_{k=0}^{j-1} (x - x_k)^{r_k}\right](x - x_j)^{s-s_{j-1}},$$

$$\text{for } s_{j-1} \leq s < s_j, \qquad \text{with } j = 1, 2, \ldots, n, \qquad \text{where } s_j = \sum_{k=0}^{j} r_k$$

Show that the b_s may be recursively defined. The proof of uniqueness may be based on the fact that if a polynomial has degree at most N and at least $N + 1$ zeroes (counting multiplicities), then the polynomial is identically zero.]

11.* If $f^{(p)}(x)$ is continuous in $[a, b]$, $0 \leq r_i - 1 \leq p$, $\{x_i\}$ in $[a, b]$, then show that the coefficients b_s of Problem 10 are divided differences of $f(x)$ of order s, based on the first $s + 1$ arguments in the sequence $x_0, x_0, \ldots, x_0$; $x_1, x_1, \ldots,$ where each x_i appears r_i times (the divided differences have been defined in Theorem 2, Corollary 5).

12.* Verify that the error in osculatory interpolation (see Problem 10) is

$$R_N(x) \equiv f(x) - Q_N(x) = \frac{\prod_{i=0}^{n} (x - x_i)^{r_i}}{(N + 1)!} f^{(N+1)}(\xi),$$

if f has $N + 1$ continuous derivatives in $[a, b]$, where ξ is in $[a, b]$.

13. Given the values

$$\sin(1.6) = .9995736030 \qquad \cos(1.6) = -.0291995223$$
$$\sin(1.7) = .9916648105 \qquad \cos(1.7) = -.1288444943$$

approximate $\sin(1.65)$ to seven decimal places by evaluating Taylor's series (about $x = 1.6$) including the third derivative term. Estimate the error by examining the remainder term in the formula. Calculate $\sin(1.65)$ correct to 9 decimal places and verify that the above estimate of error is correct.

14. Use the table in Problem 13 and calculate $\sin(1.65)$ by linear interpolation. Verify that the magnitude of the error is consistent with the remainder term as given by (8) and (9), or equivalently by equation (2.9) of Chapter 5.

15. Construct a table of divided differences from the values given in Problem 13 for the function $\sin x$ with the repeated arguments (1.6) (1.6) (1.7) (1.7); and find the Newton form of the osculating polynomial of degree 3. Calculate $\sin(1.65)$ by evaluating the osculating polynomial, and verify that the magnitude of the error is explained by the formula in Problem 12.

Table for Problem 15

	$f(x)$	$f[x, x]$	$f[x, x, x]$	$f[x, x, x, x]$
(1.6)	sin (1.6)			
		cos (1.6)		
(1.6)	sin (1.6)			
		10 (sin 1.7 − sin 1.6)		
(1.7)	sin (1.7)			
		cos (1.7)		
(1.7)	sin (1.7)			

16. Given the repeated arguments x_0, x_0, x_0, x_1, x_1 and the values $f^{(p)}(x_i)$ in the accompanying divided difference table. Complete the table and verify that the fourth order difference has the value given by (21) if $x_1 = x_0 + h$.

Table for Problem 16

	$f[x]$	$f[x, x]$	$f[x, x, x]$	$f[x, x, x, x]$	$f[x, x, x, x, x]$
x_0	f_0				
		f_0'			
			$\frac{f_0''}{2}$		
x_0	f_0				
		f_0'			
			$\frac{f_{01} - f_0'}{h}$		
x_0	f_0				
		f_{01}			
			$\frac{f_1' - f_{01}}{h}$		
x_1	f_1				
		f_1'			
x_1	f_1				

17. Given the $m + n + p + 3$ points $(x_0, x_1, \ldots, x_m, y_0, y_1, \ldots, y_n, z_0, z_1, \ldots, z_p)$ and $f(x)$ which has derivatives of order (m, n, p) respectively, in a neighborhood of each of the distinct points (x, y, z). Show that $f[x_0, \ldots, z_p] = g^{(m)}(x) + h^{(n)}(y) + k^{(p)}(z)$ if $x_i = x$, $y_j = y$, $z_r = z$ for all i, j, r, where

$$g(x) \equiv \frac{f(x)}{\prod_{j=0}^{n} (x - y_j) \prod_{r=0}^{p} (x - z_r)},$$

$$h(y) \equiv \frac{f(y)}{\prod_{i=0}^{m} (y - x_i) \prod_{r=0}^{p} (y - z_r)},$$

$$k(z) \equiv \frac{f(z)}{\prod_{i=0}^{m} (z - x_i) \prod_{j=0}^{n} (z - y_j)}.$$

[Hint: $f[x_0, x_1, \ldots, z_{p-1}, z_p] = g[x_0, x_1, \ldots, x_m] + h[y_0, y_1, \ldots, y_n] + k[z_0, z_1, \ldots, z_p]$ for distinct points $(x_0, \ldots, z_p)$. Therefore,

$$f[x_0, \ldots, x_m, y_0, \ldots, y_n, z_0, \ldots, z_p] = g^{(m)}(\xi) + h^{(n)}(\eta) + k^{(p)}(\zeta).$$

Now let $x_i \to x$, $y_j \to y$, $z_r \to z$, all i, j, r.]

2. ITERATIVE LINEAR INTERPOLATION

The Newton form of the interpolation polynomial permits one to increase easily the accuracy (actually the degree) of the approximating polynomial. It has many important applications and is indeed well suited for computations when the data are available in the appropriate tabular form. However, it can be viewed as one of a class of methods for generating successively higher order interpolation polynomials which we consider briefly. These procedures are iterative and can be very effectively employed on modern digital computers since they are based on successive linear interpolations.

The lemma underlying the *iterative linear interpolation* schemes can be stated as

LEMMA 1. *Let* $x_{i_1}, x_{i_2}, \ldots, x_{i_n}$ *be* n *distinct points and denote by* $P_{i_1, i_2, \ldots, i_n}(x)$ *the interpolation polynomial of degree* $n - 1$ *such that*

$$P_{i_1, i_2, \ldots, i_n}(x_{i_\nu}) = f(x_{i_\nu}), \qquad \nu = 1, 2, \ldots, n.$$

Then if x_j, x_k, *and* x_{i_ν}, $\nu = 1, 2, \ldots, m$ *are any* $m + 2$ *distinct points*

$$(1) \quad P_{i_1, i_2, \ldots, i_m, j, k}(x) \equiv \frac{(x - x_k)P_{i_1, i_2, \ldots, i_m, j}(x) - (x - x_j)P_{i_1, i_2, \ldots, i_m, k}(x)}{x_j - x_k},$$

$$\text{for } m = 0, 1, 2, \ldots.$$

Proof. We establish (1) by observing that the right-hand side defines a polynomial of degree at most $m + 1$ which takes on the values $f(x_{i_\nu})$ at x_{i_ν} for $\nu = 1, \ldots, m$, $f(x_j)$ at x_j and $f(x_k)$ at x_k. Hence, the polynomial on the right-hand side of (1) is the unique interpolation polynomial which appears on the left-hand side of (1). ∎

The variety of schemes which employ Lemma 1 to determine successively higher order interpolation polynomials differ in the order in which the pairs of values $(x_j, f(x_j))$ are used. For many applications, particularly on digital computers, the function values are generated sequentially and it may not be known in advance how many values are to be generated.

Table 1

x_0	$f(x_0)$				
x_1	$f(x_1)$	$P_{0,1}(x)$			
x_2	$f(x_2)$	$P_{1,2}(x)$	$P_{0,1,2}(x)$		
$\vdots$	$\vdots$	$\vdots$	$\vdots$	$\cdot$	
x_k	$f(x_k)$	$P_{k-1,k}(x)$	$P_{k-2,k-1,k}(x)$	$\cdots$	$P_{0,1,\ldots,k}(x)$

For such cases we may employ always the latest pair of values and compute each row sequentially to find the array in Table 1. Any $P_{\ldots}(x)$ in this scheme is obtained by using the two quantities to its left and diagonally above. Thus, to determine, say, the $(k + 1)$st row, only the kth row need be retained. Of course, as more points are generated, rows of greater length must be saved. If it is known in advance that a fixed number, say $k + 1$, of function values is to be generated, then a different order of computing is appropriate. That is, we may compute by columns and when a particular column has been evaluated the preceding column may be discarded. The schemes based on Table 1 are known as *Neville's iterated interpolation*.

Another sequence of interpolants are used in *Aitken's iterative interpolation* as is indicated in Table 2. Again, computation by columns is

Table 2

x_0	$f(x_0)$				
x_1	$f(x_1)$	$P_{0,1}(x)$			
x_2	$f(x_2)$	$P_{0,2}(x)$	$P_{0,1,2}(x)$		
$\vdots$	$\vdots$	$\vdots$	$\vdots$	$\cdot$	
x_k	$f(x_k)$	$P_{0,k}(k)$	$P_{0,1,k}(x)$	$\cdots$	$P_{0,1,\ldots,k}(x)$

appropriate for a known fixed value of k. Note that the $(k + 1)$st row can be computed if we save only the "diagonal" elements $f(x_0), P_{01}(x), \ldots, P_{0,1,\ldots,k}(x)$. In brief, the basic difference between these two procedures is that in Aitken's, the interpolants on the row with x_k use points with subscripts nearest 0, while in Neville's they use points with subscripts nearest k, as we read the entries from left to right.

A particularly important application of Neville's method is to what is called *iterative inverse iterpolation*. Given $y = f(x)$ we define the inverse function, $x = g(y)$, such that $y = f(g(y))$ and $x = g(f(x))$. Then it is desired to find a particular value of x, say $x = \bar{x}$, such that $f(\bar{x}) = \bar{y}$.

Thus, we need only find the value of $g(\bar{y})$. Given pairs of values $(x_i, f(x_i)) = (g(y_i), y_i)$, interpolation may be used to approximate $g(\bar{y})$. As additional pairs $(x_j, f(x_j))$ become available, better or rather higher order interpolation can be obtained by using Neville's scheme. Now, however, in Table 1, the first two columns are interchanged and the argument of the polynomials is f (or y). The computation should be done by rows. After several steps it is advisable to use only a few of the row elements and to discard those rows involving values x_i which were "early" iterates and thus presumably not sufficiently close to the desired value, $\bar{x}$. It should be noted that this procedure is essentially one of the generalizations of the method of false position or of Aitken's δ^2-method (see Chapter 3, Subsections 2.3 and 2.4). When the evaluation of the function $f(x)$ is not a difficult task, most workers prefer to use only a single linear interpolation at each stage, where the values $[x_k, f(x_k)]$ and $[x_{k+1}, f(x_{k+1})]$ are the most recent, i.e., regula falsi. Of course, inverse interpolation, in general, is meaningful only if $f(x)$ is defined as a single-valued function over the interval in question.

PROBLEMS, SECTION 2

1. Given the accompanying table for sin (x), interpolate for sin (1.65) by determining the value $P_3(1.65)$ of the Lagrange interpolation polynomial with the use of both the Neville and the Aitken schemes of successive linear interpolations. (Make up Tables corresponding to Tables 1 and 2.)

Table for Problem 1

sin (1.5)	= .9974949866
sin (1.6)	= .9995736030
sin (1.7)	= .9916648105
sin (1.8)	= .9738476309

2. Evaluate sin (1.65) by finding the Newton form of $P_3(1.65)$. Compare the amount of work in Problems 1 and 2, when done by hand.

3. FORWARD DIFFERENCES AND EQUALLY SPACED INTERPOLATION POINTS

Many of the results in Section 1 are simplified and additional important consequences are obtained if the points of interpolation are equally spaced. Very many, if not most, of the practical applications are with such sets of points. Thus, we take x_0 to be an arbitrary fixed point and let $h > 0$ be the spacing between adjacent points. Then the points to be considered are

$$x_j = x_0 + jh; \qquad j = 0, \pm 1, \pm 2, \ldots. \tag{1}$$

Note that $x_j - x_k = (j - k)h$ and, since j may be negative, x_0 need not be an endpoint of the interval under consideration.

Associated with equally spaced points is the (first order) *forward difference* which is defined by

$$\Delta f(x) \equiv f(x + h) - f(x). \tag{2}$$

Higner order differences are defined in the obvious way as

$$\Delta^{n+1}f(x) = \Delta^n[\Delta f(x)] = \Delta^n f(x + h) - \Delta^n f(x); \qquad n = 1, 2, \ldots. \tag{3}$$

In analogy with the divided difference table, we can easily construct the higher order forward differences for the points (1) by means of Table 1.

Table 1

x	$f(x)$	Δ	Δ^2	Δ^3	$\cdots$
x_0	$f(x_0)$				
		$f(x_1) - f(x_0) = \Delta f(x_0)$			
x_1	$f(x_1)$		$\Delta f(x_1) - \Delta f(x_0) = \Delta^2 f(x_0)$		
		$f(x_2) - f(x_1) = \Delta f(x_1)$		$\Delta^2 f(x_1) - \Delta^2 f(x_0) = \Delta^3 f(x_0)$	
x_2	$f(x_2)$		$\Delta f(x_2) - \Delta f(x_1) = \Delta^2 f(x_1)$		$\cdots$
		$f(x_3) - f(x_2) = \Delta f(x_2)$		$\Delta^2 f(x_2) - \Delta^2 f(x_1) = \Delta^3 f(x_1)$	
x_3	$f(x_3)$		$\Delta f(x_3) - \Delta f(x_2) = \Delta^2 f(x_2)$		
		$f(x_4) - f(x_3) = \Delta f(x_3)$		$\vdots$	
x_4	$f(x_4)$		$\vdots$		
$\vdots$	$\vdots$				

A relation between divided differences with equally spaced arguments and forward differences is easily obtained from the representation (1.6) and the definition (3). Thus by taking x to be any one of the points x_j of (1), say x_0, we have

$$\Delta f(x_0) = f(x_1) - f(x_0) = (x_1 - x_0)\frac{f(x_1) - f(x_0)}{x_1 - x_0} = hf[x_0, x_1].$$

Now to proceed by induction on n, the order of the difference, we assume that

$$\Delta^n f(x_0) = n!\, h^n f[x_0, x_1, \ldots, x_n], \tag{4}$$

and obtain

$$\begin{aligned}\Delta^{n+1}f(x_0) &= \Delta^n f(x_1) - \Delta^n f(x_0)\\ &= n!\,h^n f[x_1, x_2, \ldots, x_{n+1}] - n!\,h^n f[x_0, x_1, \ldots, x_n]\\ &= n!\,h^n(x_{n+1} - x_0)\frac{f[x_1, x_2, \ldots, x_{n+1}] - f[x_0, x_1, \ldots, x_n]}{(x_{n+1} - x_0)}\\ &= (n + 1)!\,h^{n+1}f[x_0, x_1, \ldots, x_{n+1}].\end{aligned}$$

Thus (4) is established for all $n \geq 1$.

Another representation of the forward differences can now be obtained by specializing the representation (1.4) to equally spaced points. We note that, by (1)

$$\begin{aligned}\prod_{\substack{j=0\\(j\neq i)}}^{n}(x_i - x_j) &= \prod_{\substack{j=0\\(j\neq i)}}^{n}(i - j)h\\ &= h^n\prod_{j=0}^{i-1}(i - j)\prod_{l=i+1}^{n}(i - l)\\ &= (-1)^{n-i}h^n(i)!\,(n - i)!\end{aligned}$$

By using this result in (1.4) we obtain

$$f[x_0, x_1, \ldots, x_n] = \frac{1}{n!\,h^n}\sum_{i=0}^{n}(-1)^{n-i}\binom{n}{i}f(x_i), \tag{5}$$

where

$$\binom{n}{i} = \frac{n!}{i!\,(n - i)!}$$

are the usual binomial coefficients and $0! = 1$. From (5) in (4) there results

$$\Delta^n f(x_0) = \sum_{i=0}^{n}(-1)^{n-i}\binom{n}{i}f(x_i). \tag{6}$$

A final expression for the forward differences is obtained by using (4) in (1.9) with $x = x_k$ to get

$$\Delta^n f(x_0) = h^n f^{(n)}(\xi); \qquad x_0 < \xi < x_n. \tag{7}$$

Of course, (7) is valid assuming that $f(x)$ has an nth derivative in the indicated interval.

It is clear, from (7), that the nth forward difference of a polynomial of degree n is a constant and that higher order differences vanish. More generally, if $f(x)$ has all derivatives bounded, say $|f^{(n)}(x)| \leq M$ for all n, then (7) implies that

$$|\Delta^n f| \leq Mh^n.$$

Thus if $h < 1$ the magnitude of nth order differences of $f(x)$ will in general decrease as n increases. On the other hand, if the nth derivative of $f(x)$ grows with n, the nth difference will decrease only if h is "sufficiently small." This may be illustrated by the function $f(x) = e^{\alpha x}$ where $\alpha > 0$. Clearly, $f^{(n)}(x) = \alpha^n e^{\alpha x}$ and using (7) $\Delta^n f(x_0) = h^n \alpha^n e^{\alpha \xi}$. If the interpolation points are to be confined to the interval $x_0 \leq x \leq x_0 + L$, then $0 \leq nh \leq L$ and the differences will generally decrease only if $h < 1/\alpha$ (we here neglect the variation in $e^{\alpha \xi}$ which may occur). Finally, if $f^{(n)}(x)$ is not bounded for all n we can expect $|\Delta^n f|$ to decrease, for sufficiently small h, only for the first several values of n. This heuristic observation is the basis of a method for detecting isolated errors in forward difference tables of supposedly smooth functions.

To describe this method we first observe that

$$\Delta^n[f(x) + g(x)] = \Delta^n f(x) + \Delta^n g(x).$$

Now suppose that $f(x)$ is a smooth function, say for simplicity with all derivatives bounded, and that in tabulating this function an error of amount δ is made in the single entry $f(x_j)$. Thus the function actually tabulated can be written as $f(x) + g(x)$ where

$$g(x_i) = \begin{cases} 0, & i \neq j; \\ \delta, & i = j. \end{cases}$$

Applying the representation (6) we see that the column of nth differences of g will contain quantities of the form

$$(-1)^{n-k}\binom{n}{k}\delta.$$

Thus in examining the higher differences of the tabulated data $f(x) + g(x)$, since $\Delta^n f$ decreases with n, an error will be apparent if the entries, from some column on, alternate in sign and the magnitudes of these varying entries are proportional to the appropriate binomial coefficients. The power of this method is illustrated in Problem 2.

This procedure will not be practical if the isolated error δ is of the same order of magnitude as the roundoff errors generally present in all of the tabular data. In fact, we shall see that if roundoff errors are present, differences of a sufficiently high order may have no significance. Let the tabular entries be, $f(x) + \rho(x)$, where $\rho(x)$ is the rounding error.

Let ϵ be a bound on the rounding error, i.e., $|\rho(x)| \leq \epsilon$. The worst possibility for the distribution of these errors is

$$\rho(x_j) \equiv (-1)^j \epsilon. \tag{8}$$

Table 2

$\rho(x)$	Δ	Δ^2	Δ^3	Δ^4
ϵ				
	-2ϵ			
$-\epsilon$		4ϵ		
	2ϵ		-8ϵ	
ϵ		-4ϵ		16ϵ
	-2ϵ		8ϵ	
$-\epsilon$		4ϵ		
	2ϵ			

This is made clear by the table of differences for such a distribution (Table 2). Any other distribution leads to some entries which would be less in absolute value than the corresponding entries above. From Table 2 we see that

$$|\Delta^n \rho(x_j)| = 2^n \epsilon.$$

[This result may be easily proved for $j = 0$ by using (8) in (6).] Thus the roundoff error present in the nth difference is at most $2^n\epsilon$. If there are s decimals retained with a roundoff error of at most one-half unit in the last place then $\epsilon = \frac{1}{2}\,10^{-s}$, and the bound on the roundoff error in the nth difference becomes $2^{n-1}\,10^{-s}$.

3.1. Interpolation Polynomials and Remainder Terms for Equidistant Points

The Lagrange and Newton forms of the interpolation polynomials become simplified when the interpolation points are equally spaced. To be consistent with the notation (1) we introduce a new independent variable, t, by setting

$$x = x_0 + th. \tag{9}$$

Thus, t measures $x - x_0$ in units of h and is an integer only at the points x_j of (1).

Now the Lagrange interpolation coefficients, (2.6) of Chapter 5, can be written as

$$\begin{aligned}
\phi_{n,j}(x) = \phi_{n,j}(x_0 + th) &= \prod_{\substack{k=0 \\ (k \neq j)}}^{n} \frac{x - x_k}{x_j - x_k} \\
&= \prod_{\substack{k=0 \\ (k \neq j)}}^{n} \frac{t - k}{j - k} \\
&= \frac{(-1)^{n-j}}{n!}\binom{n}{j} \prod_{\substack{k=0 \\ (k \neq j)}}^{n} (t - k).
\end{aligned}$$

It is convenient to introduce

$$
(10)\qquad \begin{aligned} \pi_0(t) &\equiv t, \\ \pi_n(t) &\equiv t(t-1)\cdots(t-n), \qquad n = 1, 2, \ldots. \end{aligned}
$$

The function $\pi_n(t)$ is a polynomial in t of degree $n + 1$ and is frequently called the $(n + 1)$*st factorial polynomial.* In terms of this polynomial, the Lagrange interpolation coefficients become

$$
\phi_{n,j}(x_0 + th) = \frac{\pi_n(t)}{n!}\binom{n}{j}\frac{(-1)^{n-j}}{t-j}.
$$

Using this formula in (2.5) of Chapter 5, the Lagrange form of the interpolation polynomial simplifies to

$$
(11)\qquad P_n(x_0 + th) = \frac{\pi_n(t)}{n!}\sum_{j=0}^{n}(-1)^{n-j}\binom{n}{j}\frac{f(x_j)}{t-j}.
$$

Newton's form of the interpolation polynomial is simplified by using (4), (9), and (10) in (1.7) to get

$$
(12)\qquad \begin{aligned} Q_n(x_0 + th) = f(x_0) &+ \frac{\pi_0(t)}{1!}\Delta f(x_0) + \frac{\pi_1(t)}{2!}\Delta^2 f(x_0) + \cdots \\ &+ \frac{\pi_{n-1}(t)}{n!}\Delta^n f(x_0). \end{aligned}
$$

The error in these interpolation polynomials is, by (1.8), (9), (10), and Theorem 1.1,

$$
(13)\qquad \begin{aligned} R_n(x) = R_n(x_0 + th) &= \pi_n(t)h^{n+1}f[x_0, \ldots, x_n, x] \\ &= \pi_n(t)h^{n+1}\frac{f^{(n+1)}(\xi)}{(n+1)!}. \end{aligned}
$$

$R_n(x)$ may also be called the *remainder* for the interpolation formula. Of course, the final form is valid only if $f(x)$ has $n + 1$ derivatives in the interval containing x, x_0 and $x_0 + nh$. To obtain a clear idea of the behavior of this error, we shall study some properties of the factorial polynomials $\pi_n(t)$.

From the definition (10) it is clear that $\pi_n(t)$ has $n + 1$ real roots and they are at $t = 0, 1, \ldots, n$. These polynomials have certain symmetries about the point $t = n/2$ which is the midpoint of the zeros of $\pi_n(t)$.

LEMMA 1. *For n odd,*

$$
\pi_n\left(\frac{n}{2} - \tau\right) = \pi_n\left(\frac{n}{2} + \tau\right),
$$

[*i.e.*, $\pi_n(t)$ *is symmetric about* $t = n/2$]; *for n even:*

$$\pi_n\left(\frac{n}{2} - \tau\right) = -\pi_n\left(\frac{n}{2} + \tau\right)$$

[*i.e.*, $\pi_n(t)$ *is anti-symmetric about* $t = n/2$].

Proof. Clearly, $\pi_n(n/2 - \tau)$ and $\pi_n(n/2 + \tau)$ are polynomials of degree $n + 1$ in τ and have the same $n + 1$ roots:

$$\tau = \frac{n}{2}, \frac{n}{2} - 1, \frac{n}{2} - 2, \ldots, -\frac{n}{2}.$$

Thus, these polynomials differ by at most a constant factor. Comparing coefficients of the leading terms in each then yields

$$\pi_n\left(\frac{n}{2} + \tau\right) = (-1)^{n+1}\pi_n\left(\frac{n}{2} - \tau\right). \qquad \blacksquare$$

A further result which contains a comparison of the magnitudes of $\pi_n(t)$ at various points is contained in

LEMMA 2. (a) *Let* $t + 1$ *be any non-integral point in* $0 < t + 1 \leq n/2$. *Then*

$$|\pi_n(t + 1)| < |\pi_n(t)|.$$

(b) *Let t be any non-integral point in* $n/2 \leq t < n$. *Then*

$$|\pi_n(t)| < |\pi_n(t + 1)|.$$

Proof. Since, in part (a), $t + 1$ is non-integral, $t < n$ is also non-integral and we may form

$$\left|\frac{\pi_n(t + 1)}{\pi_n(t)}\right| = \left|\frac{(t + 1)(t)(t - 1)\cdots(t - n + 1)}{(t)(t - 1)\cdots(t - n + 1)(t - n)}\right|$$

$$= \left|\frac{t + 1}{t - n}\right| = \left|\frac{t + 1}{n - t}\right| = \frac{t + 1}{(n + 1) - (t + 1)}$$

$$\leq \frac{n/2}{(n + 1) - (n/2)} = \frac{1}{1 + 2/n} < 1.$$

Thus, part (a) is proved. Part (b) follows from part (a) by using the symmetry properties of Lemma 1. $\blacksquare$

The properties of $\pi_n(t)$ proven in Lemmas 1 and 2 are illustrated in Figure 1 where $\pi_n(t)$ is plotted for $n = 5$ and $n = 6$. It easily follows from Lemma 2 that the maximum of $|\pi_n(t)|$ in $[0, n]$ occurs in the interval

(0, 1), or equivalently in $(n - 1, n)$. A *lower bound* on this magnitude is furnished by

$$\max_{0 \leq t \leq n} |\pi_n(t)| \geq |\pi_n(\tfrac{1}{2})| = \prod_{j=0}^{n} |\tfrac{1}{2} - j|$$

$$= \frac{(2n)!}{2^{2n+1}n!}.$$

Using this bound we may compare the quantity $h^{n+1}\pi_n(t)$ for equally spaced interpolation points with the corresponding *error factor*, $\prod_{j=0}^{n} (x - x_j)$, for the Chebyshev interpolation points (i.e., with that factor which deviates least from zero determined in Chapter 5, Subsection 4.2). If the interpolation is to be employed over the interval $[a, b]$, then for the Chebyshev points we have by (4.17) of Chapter 5

$$M_{\text{Ch}} \equiv \max_{a \leq x \leq b} \left| \prod_{j=0}^{n} (x - x_j) \right| = \frac{1}{2^n} \left(\frac{b - a}{2} \right)^{n+1};$$

and this value is attained at least $n + 2$ times in the interval. For equal spacing in $[a, b]$, we have $h = (b - a)/n$ and so

$$M_{\text{Eq}} \equiv \max_{0 \leq t \leq n} |h^{n+1}\pi_n(t)| > \frac{(2n)!}{2^{2n+1}n!} \left(\frac{b - a}{n} \right)^{n+1}.$$

Thus, using Stirling's formula we find that for n large

$$\frac{M_{\text{Ch}}}{M_{\text{Eq}}} < \frac{n}{\sqrt{2}} \left(\frac{e}{4} \right)^n = \frac{n}{\sqrt{2}} (0.6796\ldots)^n.$$

So if interpolation is to be employed over the entire range $[a, b]$, we find that the ratio of the maximum error factors essentially decreases, at least exponentially, for large n. Thus, the Chebyshev fit is better in the above comparison. However, we may only want to employ the interpolation polynomial near the center of the interval $[a, b]$. Specifically for n odd, say that $n = 2m + 1$, and that we are interested in the error over

$$\left[\frac{a + b}{2} - \frac{h}{2}, \frac{a + b}{2} + \frac{h}{2} \right],$$

i.e., an interval of length h centered in $[a, b]$. Now we find that

$$M_{\text{Eq}}{}^* \equiv \max_{m \leq t \leq m+1} |h^{n+1}\pi_n(t)| = h^{n+1}|\pi_n(m + \tfrac{1}{2})|$$

$$= \left(\frac{b - a}{n} \right)^{n+1} \left(\frac{n!}{2^n m!} \right)^2.$$

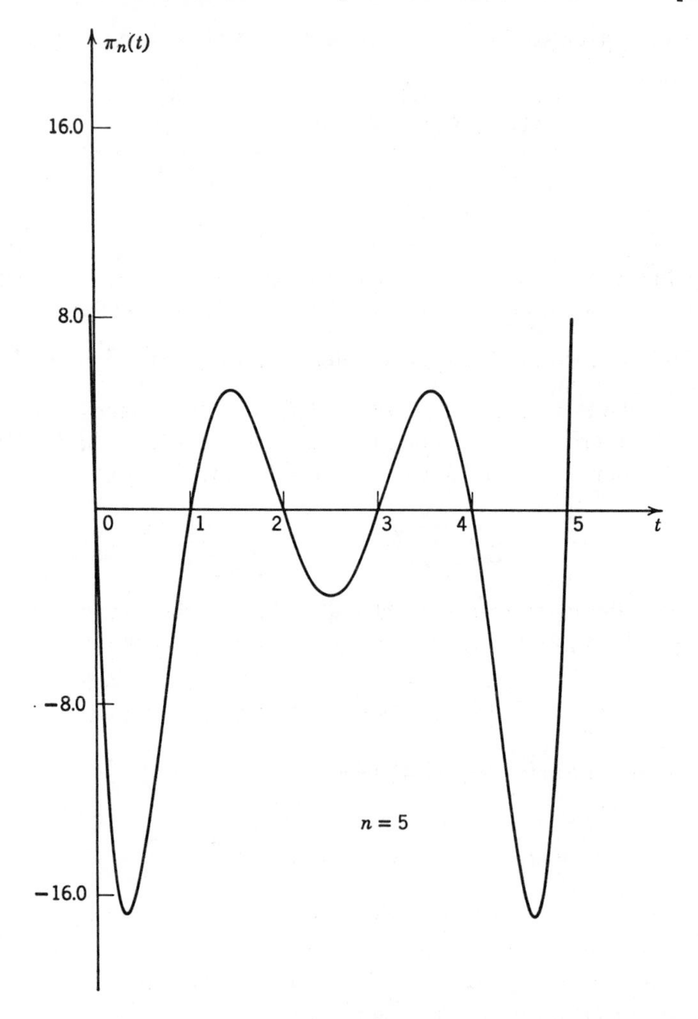

Figure 1*a*. Error factor for interpolation with 6 equidistant points.

Hence it follows that for large $n = 2m + 1$,

$$\frac{M_{\text{Ch}}}{M_{\text{Eq}}{}^*} \approx \tfrac{1}{2}\left(\frac{e}{2}\right)^n = \tfrac{1}{2}(1.3591\ldots)^n.$$

We thus find that for interpolation near the center of the interval of interpolation points the error factor for equally spaced points is exponentially smaller than the maximum for the Chebyshev fit over the entire interval.

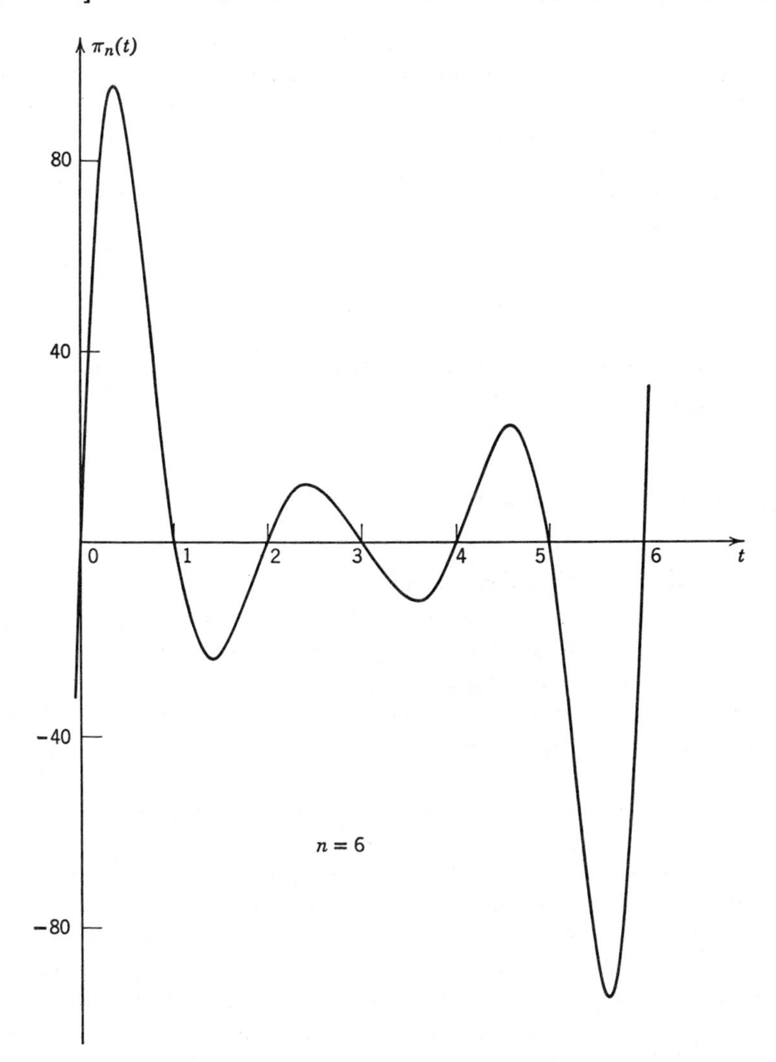

Figure 1*b*. Error factor for interpolation with 7 equidistant points.

It should also be observed (see Figure 1) that the error factor $|\pi_n(t)|$ grows rapidly for $t < 0$ and $t > n$. Thus if the "equally spaced" interpolation polynomial is used to extrapolate a function (i.e., to estimate values outside the interval of the interpolation points) we may expect the error to be much larger, in general, than it is for interpolation. Of course, the same is also true of the extrapolation error using the Chebyshev points. In fact, for any distribution of interpolation points, the growth of the magnitude of the error factor can be bounded, outside the interval of interpolation points, by the error factor with the Chebyshev points. More precisely, for any choice of interpolation points $x_0, x_1, \ldots, x_n$ in $[-1, 1]$ the error factor is, say, $\prod_{j=0}^{n} (x - x_j) \equiv p_{n+1}(x)$. If the Chebyshev points are used, then

$$\prod_{j=0}^{n} (x - x_j) = T_{n+1}(x).$$

Now if $M \equiv \max_{-1 \leq x \leq 1} |p_{n+1}(x)|$, it can be shown (see Problem 5) that for all x such that $|x| > 1$

$$|p_{n+1}(x)| \leq M2^n |T_{n+1}(x)|.$$

The equality can only hold if $p_{n+1}(x) \equiv T_{n+1}(x)$.

3.2. Centered Interpolation Formulae

Consider the error (13) in the interpolation polynomial for equal spacing. This error may be estimated if $f^{(n+1)}(\xi)$ does not vary "too much" in the interval $\min(x_0, x) < \xi < \max(x_n, x)$. [An idea of this variation may be obtained, as a result of (7), by examining the differences $\Delta^{n+1}f$.] If the variation is not too large, then as an *estimate* of the error we may use

$$R_n(x) \equiv R_n(x_0 + th) \cong \frac{\pi_n(t)}{(n+1)!} \Delta^{n+1} f(x_0).$$

An approximate *bound* on the error is obtained if $\pi_n(t)$ is replaced by its maximum absolute value in the interval in question.

Since, in general, we do not know $f^{(n+1)}(\xi)$, the best that can be done in order to obtain the smallest possible error is to use the interpolation polynomials only for that range of t where $\pi_n(t)$ has its least absolute value, i.e., by Theorem 2, for t near $n/2$ or equivalently for x near the midpoint of $[x_0, x_n]$. So if there are tabular points equally distributed about the interval of interpolation, then the interpolation polynomial to be employed should use tabular points centered as nearly as possible about the interval of interpolation. It is clear (see Figure 1) that when x is outside the interval $[x_0, x_n]$, or near the endpoints, $|\pi_n(t)|$ may be relatively large and, if so, may cause the extrapolation or interpolation error to be relatively large.

It is rather inconvenient, in general, to locate in the Difference Table 1 those differences which must be employed in (12) when t is at $n/2$. For this purpose we derive special formulae which simplify the task. Let us assume that the interval in which the interpolation is to be done is

$x_0 < x < x_1$, and that we have arbitrarily many tabular points x_j, $j = 0$, ± 1, $\pm 2, \ldots$, about this interval. Then the ordinary Newton interpolation polynomial, using successively the points $x_0, x_1, x_{-1}, x_2, x_{-2}, \ldots$, will have the desired features with regard to the interval $[x_0, x_1]$. This polynomial is of the form

$$(14) \qquad Q_n(x) = f_0 + (x - x_0)f_{0,1} + (x - x_0)(x - x_1)f_{0,1,-1} + (x - x_0)(x - x_1)(x - x_{-1})f_{0,1,-1,2} + \cdots,$$

the form of the final term depending upon the oddness or evenness of n.

However, since the divided differences are symmetric functions of their arguments we may write

$$\begin{aligned} f_{0,1,-1} &= f_{-1,0,1}, \\ f_{0,1,-1,2} &= f_{-1,0,1,2}, \\ &\vdots \\ f_{0,1,-1,\ldots,m,-m} &= f_{-m,\ldots,-1,0,1,\ldots,m}. \end{aligned}$$

Then using (4) with the appropriate shifts in the subscripts,

$$\begin{aligned} f_{-1,0,1} &= \frac{1}{2!\,h^2}\Delta^2 f_{-1}, \\ f_{-1,0,1,2} &= \frac{1}{3!\,h^3}\Delta^3 f_{-1}; \\ f_{-m,\ldots,-1,0,1,\ldots,m} &= \frac{1}{(2m)!\,h^{2m}}\Delta^{2m} f_{-m}, \\ f_{-m,\ldots,-1,0,1,\ldots,m,m+1} &= \frac{1}{(2m+1)!\,h^{2m+1}}\Delta^{2m+1} f_{-m}. \end{aligned}$$

The interpolation polynomial (14) can now be written as, for even $n = 2m$:

$$(15a) \qquad Q_n(x_0 + th) = f_0 + t\Delta f_0 + \frac{t(t-1)}{2!}\Delta^2 f_{-1} + \frac{t(t-1)(t+1)}{3!}\Delta^3 f_{-1} + \cdots + \frac{t(t-1)(t+1)\cdots(t-m)}{(2m)!}\Delta^{2m} f_{-m};$$

and for odd $n = 2m + 1$:

$$(15b) \qquad Q_n(x_0 + th) = f_0 + t\Delta f_0 + \cdots + \frac{t(t-1)(t+1)\cdots(t-m)(t+m)}{(2m+1)!}\Delta^{2m+1} f_{-m}.$$

This is the *Gaussian (forward) form* for the interpolation polynomials.

The differences used in forming these polynomials are on the line containing x_0 and the line between x_0 and x_1 (see Table 3).

Table 3 *Differences*

x	$f(x)$	Δ	Δ^2	$\cdots$	Δ^{2m}	Δ^{2m+1}
x_{-m}	f_{-m}					
		Δf_{-m}				
$\vdots$	$\vdots$	$\vdots$	$\vdots$			
x_{-1}	f_{-1}				$\vdots$	$\vdots$
		Δf_{-1}				
x_0	f_0		$\Delta^2 f_{-1}$	$\cdots$	$\Delta^{2m} f_{-m}$	
		Δf_1		$\cdots$		$\Delta^{2m+1} f_{-m}$
x_1	f_1				$\vdots$	$\vdots$
$\vdots$	$\vdots$	$\vdots$	$\vdots$			
		Δf_{m-1}				
x_m	f_m					

A more symmetric form of the interpolation polynomial can be obtained when $n = 2m + 1$ (i.e., of odd degree). For this purpose we must introduce the centered difference notation, for even order differences,

(16a)
$$\delta^2 f_r \equiv \Delta f_r - \Delta f_{r-1} = \Delta^2 f_{r-1};$$
$$\delta^{2k} f_r = \delta^2(\delta^{2(k-1)} f)_r = \Delta^{2k} f_{r-k}, \qquad k = 2, 3, \ldots.$$

The point x_r is always the midpoint about which these differences are centered. With this notation, all odd order ordinary differences, higher than the first, can be written as the difference of two centered differences:

(16b)
$$\Delta^{2m+1} f_{-m} = \delta^{2m} f_1 - \delta^{2m} f_0; \qquad m = 1, 2, \ldots.$$

Using (16) in (15b) yields:

$$Q_n(x_0 + th) = f_0 + t\Delta f_0 + \frac{t(t-1)}{2!}\Delta^2 f_{-1} + \frac{t(t-1)(t+1)}{3!}\Delta^3 f_{-1} + \cdots$$
$$+ \frac{t(t-1)(t+1)\cdots(t-m)}{(2m)!}\Delta^{2m} f_{-m}$$
$$+ \frac{t(t-1)(t+1)\cdots(t-m)(t+m)}{(2m+1)!}\Delta^{2m+1} f_{-m}$$

$$
\begin{aligned}
&= f_0 + t(f_1 - f_0) + \frac{t(t-1)}{2!}\,\delta^2 f_0 \\
&\quad + \frac{t(t-1)(t+1)}{3!}\,(\delta^2 f_1 - \delta^2 f_0) + \cdots \\
&\quad + \frac{t(t-1)(t+1)\cdots(t-m)}{(2m)!}\,\delta^{2m} f_0 \\
&\quad + \frac{t(t-1)(t+1)\cdots(t-m)(t+m)}{(2m+1)!}\,(\delta^{2m} f_1 - \delta^{2m} f_0) \\
&= tf_1 + \frac{t(t-1)(t+1)}{3!}\,\delta^2 f_1 + \cdots \\
&\quad + \frac{t(t-1)(t+1)\cdots(t-m)(t+m)}{(2m+1)!}\,\delta^{2m} f_1 \\
&\quad + (1-t)f_0 + \frac{t(t-1)}{3!}\,(2-t)\delta^2 f_0 + \cdots \\
&\quad + \frac{t(t-1)(t+1)\cdots(t-m)}{(2m+1)!}\,(m+1-t)\delta^{2m} f_0.
\end{aligned}
$$

By introducing $s \equiv 1 - t$ we may simplify the coefficients of $\delta^{2k} f_0$ and the above finally takes on the symmetric form:

$$
\begin{aligned}
(17) \qquad Q_n(x_0 + th) = \quad & sf_0 + \frac{s(s^2 - 1^2)}{3!}\,\delta^2 f_0 + \cdots \\
& + \frac{s(s^2 - 1^2)\cdots(s^2 - m^2)}{(2m+1)!}\,\delta^{2m} f_0 \\
& + tf_1 + \frac{t(t^2 - 1^2)}{3!}\,\delta^2 f_1 + \cdots \\
& + \frac{t(t^2 - 1^2)\cdots(t^2 - m^2)}{(2m+1)!}\,\delta^{2m} f_1.
\end{aligned}
$$

This is known as *Everett's form* of the interpolation polynomial.

3.3. Practical Observations on Interpolation

In this subsection, we gather and comment on some of the "rules of thumb" which are used by the practitioners of interpolation.

(a) A convenient "rule" to determine approximately the magnitude of the error in linear interpolation is

$$
|f(x_0 + th) - tf(x_0 + h) - (1 - t)f(x_0)| \le \left|\frac{\Delta^2 f(x_0)}{8}\right|.
$$

The factor $\frac{1}{8}$ is an upper bound for $t(t - 1)/2$, if $0 \leq t \leq 1$, while by (7) $h^2 f''(\xi) \cong \Delta^2 f(x_0)$ in the remainder term (13).

(b) A "rule" for estimating the magnitude of the error in general polynomial interpolation is to use the magnitude of the first neglected term in the Newton form. That is, the error in using (12), (15a), or (15b) is approximately the next term in the series. By (13) this estimate is seen to be good if the ratio $f^{(n+1)}(\xi)/f^{(n+1)}(\eta)$ is near to 1 for all ξ and η in an interval containing x and all the indicated x_i.

(c) In a table of differences, we may compute the "average value," $\tilde{\Delta}^p$, of a column of pth order differences from the definition:

$$\tilde{\Delta}^p = \frac{1}{n} \sum_{i=0}^{n-1} \Delta^p f_i.$$

It is easy to show (see Problem 7), that if an isolated error is made in f_k, for some k satisfying $p \leq k \leq n - p$, then $\tilde{\Delta}^p$ is unaffected. This observation provides a simple way of locating k and estimating the $p + 1$ errors that arise in the column of pth order differences from an isolated error in f_k; and hence yields the error in f_k approximately.

A table user could difference a printed table (whose accuracy has not been established) in order to weed out isolated typographical errors and to decide upon the order of interpolation that may be necessary.

(d) In the construction of a mathematical table, one tries to present a listing that provides a reasonable number of decimal places (or significant figures) and also permits a simple interpolation process to attain almost the full accuracy of the table, without its being too voluminous. To this end, some table makers list not only $f(x_i)$ but $\delta^2 f(x_i)$, where $\delta^2 f(x_i)$ is called a *modified second difference* (only the significant figures of δ^2 are listed). The modification is based on the use of the Everett form of the interpolation formula (17) with $m = 2$:

$$f(x_0 + th) \cong tf(x_1) + \frac{t(t^2 - 1)}{6} \left[\delta^2 f(x_1) + \frac{t^2 - 4}{20} \delta^4 f(x_1) \right]$$
$$+ sf(x_0) + \frac{s(s^2 - 1)}{6} \left[\delta^2 f(x_0) + \frac{s^2 - 4}{20} \delta^4 f(x_0) \right],$$

where $s = 1 - t$. In order to incorporate most of the "effect" of the fourth difference into the second difference, one uses an "average value" for the coefficient $(t^2 - 4)/20$. Very simply, since

$$\int_0^1 \frac{p^2 - 4}{20} \, dp = -\tfrac{11}{60} \cong -0.18333,$$

we may define the modified second difference by

$$\delta^2 f(x) = \delta^2 f(x) - \tfrac{11}{60} \delta^4 f(x),$$

and then use, for interpolation in the table,

$$f(x_0 + th) \cong tf(x_1) + \frac{t(t^2 - 1)}{6}\,\delta^2 f(x_1) + sf(x_0) + \frac{s(s^2 - 1)}{6}\,\delta^2 f(x_0).$$

Other, more sophisticated arguments have been given to justify the use of other "average values" for the coefficient $(t^2 - 4)/20$, e.g., -0.18393 (see Problem 6 for justification).

3.4. Divergence of Sequences of Interpolation Polynomials

It is not generally true that higher degree interpolation polynomials yield more accurate approximations. In fact, for equidistant points of interpolation one should use polynomials of relatively low order. We shall illustrate this by examining the interpolation error, $R_n(x)$, as a function of n and x for a particular function.

Specifically we take a function considered by Runge:

$$\text{(18a)} \qquad f(x) \equiv \frac{1}{1 + x^2},$$

and consider, in $[-5, 5]$, the equally spaced points

$$\text{(18b)} \qquad x_j = -5 + j\Delta x, \qquad j = 0, 1, 2, \ldots, n, \quad \Delta x = \frac{10}{n}.$$

For each n there is a unique polynomial $P_n(x)$ of degree at most n such that $P_n(x_j) = f(x_j)$. This is the interpolation polynomial for (18a) using the points (18b). We shall show that $|f(x) - P_n(x)|$ will become arbitrarily large at points in $[-5, 5]$ if n is sufficiently large. This occurs even though the interpolation points $\{x_j\}$ become dense in $[-5, 5]$ as $n \to \infty$.

The remainder in interpolation is, by (1.8)

$$\text{(19)} \qquad \begin{aligned} R_n(x) &= f(x) - P_n(x), \\ &= \prod_{j=0}^{n} (x - x_j) f[x_0, \ldots, x_n, x]. \end{aligned}$$

However, with the function and points in (18) we claim that

$$\text{(20)} \qquad f[x_0, \ldots, x_n, x] = f(x) \cdot \frac{(-1)^{r+1}}{\prod_{j=0}^{r} (1 + x_j^2)} \cdot \begin{cases} 1, & \text{if } n = 2r + 1; \\ x, & \text{if } n = 2r. \end{cases}$$

We first prove this for the odd case, $n = 2r + 1$, by induction on r. Note that in this case there are an even number of interpolation points in (18b) and they satisfy $x_j = -x_{n-j}$. For $r = 0$ we have $n = 1$ and $x_0 = -x_1$.

Then a direct calculation from the divided difference representation (1.4) using (18a) yields

$$f[x_0, x_1, x] = -\left(\frac{1}{1 + x^2}\right)\left(\frac{1}{1 + x_0^2}\right) = f(x)\frac{(-1)}{1 + x_0^2},$$

and the first step of the induction is established. Now assume (20) to be valid for $n = 2r + 1$, with any $x_0, \dots, x_n$ that are pairwise symmetric (i.e., $x_j = -x_{n-j}$), and let $m = n + 2 = 2(r + 1) + 1$. We define the function $g(x)$ by

$$g(x) \equiv f[x_1, x_2, \dots, x_{m-1}, x],$$

where now $x_j = -x_{m-j}$, and use (1.6) to write

$$f[x_0, x_1, \dots, x_m, x] = g[x_0, x_m, x].$$

However, by the inductive hypothesis it follows that

$$g(x) = f(x)\cdot A_r, \qquad A_r \equiv \frac{(-1)^{r+1}}{\prod_{j=1}^{r}(1 + x_j^2)}.$$

Also from (18b) we note that $x_0 = -x_m$ and hence by the previous calculation

$$g[x_0, x_m, x] = f(x)\frac{(-1)}{1 + x_0^2}A_r,$$

which upon substitution for A_r concludes the induction.

The verification of (20) for $n = 2r$ is similar to the above and is stated as Problem 8. Another proof of (20) is given as Problem 9.

Since $(x - x_j)(x - x_{n-j}) = x^2 - x_j^2$ by (18b) and recalling that $x_r = 0$ for $n = 2r$ we have

$$(21) \qquad \prod_{j=0}^{n}(x - x_j) = \prod_{j=0}^{r}(x^2 - x_j^2)\cdot\begin{cases} 1 & \text{if } n = 2r + 1; \\ 1/x & \text{if } n = 2r. \end{cases}$$

From (21) and (20) in (19) the error can be written as

$$(22) \qquad R_n(x) = (-1)^{r+1}f(x)g_n(x), \qquad g_n(x) \equiv \prod_{j=0}^{r}\frac{x^2 - x_j^2}{1 + x_j^2}.$$

Note that $\frac{1}{26} \leq f(x) \leq 1$ for x in $[-5, 5]$ and so the convergence or divergence properties are determined by $g_n(x)$. Further, since $g_n(x) = g_n(-x)$, we need only consider the interval $[0, 5]$. [It is also of interest to note that $R_n(x)$ is, in fact, an even function of x for all n.]

To examine $|g_n(x)|$ for large n, or equivalently for large r, we write

$$|g_n(x)| = [e^{\Delta x \ln |g_n(x)|}]^{1/\Delta x}, \tag{23a}$$

where from the definition in (22)

$$\Delta x \ln |g_n(x)| = \sum_{j=0}^{r} \ln \left|\frac{x^2 - x_j^2}{1 + x_j^2}\right| \Delta x. \tag{23b}$$

Now restrict x such that for some constant θ

$$1 \le x \le 5, \quad \min_{0 \le j \le r} |x + x_j| \ge \theta \frac{\Delta x}{2}, \quad 0 < \theta < 1 \tag{24}$$

For these values of x the sum in (23b) converges uniformly as $n \to \infty$; that is, explicitly,

$$\begin{aligned} \lim_{n \to \infty} \Delta x \ln |g_n(x)| &= \lim_{r \to \infty} \sum_{j=0}^{r} \ln \left|\frac{x^2 - x_j^2}{1 + x_j^2}\right| \Delta x, \\ &= \int_{-5}^{0} \ln \left|\frac{x^2 - \xi^2}{1 + \xi^2}\right| d\xi, \\ &\equiv q(x). \end{aligned} \tag{25}$$

To demonstrate this convergence, we note that

$$\ln \left|\frac{x^2 - x_j^2}{1 + x_j^2}\right| = \ln |x + x_j| + \ln |x - x_j| - \ln |1 + x_j^2|,$$

and similarly, with x_j replaced by ξ. Next we show that each of the three sums converges to the corresponding integrals. Those sums corresponding to the last two terms converge to their corresponding integrals by the definition of Riemann integrals since the corresponding integrands are continuous functions of ξ [recall that $x \ge 1$ by (24) and $x_j \le 0$ for $j \le r$ by (18b)]. Hence, we need only show that

$$\lim_{r \to \infty} \sum_{j=0}^{r} \ln |x + x_j| \Delta x = \int_{-5}^{0} \ln |x + \xi| \, d\xi, \tag{26}$$

provided x satisfies (24). Let $\delta < 1$ be an arbitrarily small fixed positive number. Then

$$\begin{aligned} &\lim_{r \to \infty} \sum_{|x + x_j| > \delta} \ln |x + x_j| \Delta x \\ &\quad = \int_{-5}^{-x-\delta} \ln |x + \xi| \, d\xi + \int_{-x+\delta}^{0} \ln |x + \xi| \, d\xi \end{aligned} \tag{27a}$$

since $\ln |x + \xi|$ is a continuous function of ξ over the indicated intervals of integration. The missing part of the integral in (26) is

$$\text{(27b)} \qquad \int_{-x-\delta}^{-x+\delta} \ln |x + \xi| \, d\xi = 2 \int_0^{\delta} \ln \eta \, d\eta = 2(\delta \ln \delta - \delta).$$

The remaining part of the sum in (26) can be bounded if we take r so large that $\Delta x < \delta$. Then recalling (24) we have

$$\begin{aligned}
\left| \sum_{|x+x_j| \leq \delta} \ln |x + x_j| \, \Delta x \right| &\leq 2\Delta x \left| \ln \frac{\theta \Delta x}{2} \right| + \left| \sum_{\Delta x \leq |x+x_j| \leq \delta} \ln |x + x_j| \, \Delta x \right|, \\
\text{(27c)} \qquad &\leq 2\Delta x \left| \ln \frac{\theta \Delta x}{2} \right| + 2 \left| \int_0^{\delta} \ln \eta \, d\eta \right|, \\
&\leq 2(\Delta x |\ln \Delta x - \ln 2 + \ln \theta| + \delta |\ln \delta - 1|) \\
&= \mathcal{O}(\delta |\ln \delta|).
\end{aligned}$$

The first term on the right in the first line is obtained from the two terms say, x_k and x_{k+1} which are nearest to $-x$. The remaining sum has been bounded by the integral by means of the monotonicity of the function $\ln x$. That is, since $\delta < 1$, we use

$$0 > \Delta x \ln |x + x_j| > \int_{|x+x_{j-1}|}^{|x+x_j|} \ln \eta \, d\eta,$$

if $|x + x_{j-1}| < |x + x_j|$ (otherwise limits of integration are $|x + x_j|$, $|x + x_{j+1}|$). Letting $r \to \infty$ in (27c) and using (27a and b) we get (26) since δ is arbitrarily small. This concludes the proof of (25).

In Problem 10 we indicate how $q(x)$ can be explicitly evaluated and it is required to show that

$$\text{(28a)} \qquad q(x) = 0 \qquad \text{at } x = 3.63\ldots;$$

$$\text{(28b)} \qquad q(x) < 0 \qquad \text{for } |x| < 3.63\ldots;$$

$$\text{(28c)} \qquad q(x) > 0 \qquad \text{for } 3.63\ldots < |x| \leq 5.$$

Now let x satisfy (24) and $x > 3.63\ldots$ as $n \to \infty$. Then by (25) and (28c) in (23a) we have, recalling that $\Delta x = 10/n$,

$$\lim_{n \to 0} |g_n(x)| = \infty.$$

That is, from (22), $|R_n(x)| \to \infty$ as $n \to \infty$ for x as above. Also, since $R_n(x) = R_n(-x)$ the points of divergence are symmetrically located on the axis.

This example illustrates part of the general convergence theory for sequences of interpolation polynomials based on uniformly spaced points in an interval $[a, b]$. According to this theory, if $f(z)$ is "analytic" in a domain of the complex plane containing $[a, b]$, then the sequence of interpolation polynomials for $f(z)$ converges inside the largest "lemniscate" whose interior is in the domain of analyticity of $f(z)$. The "lemniscate" passing through $z = \pm 3.63\ldots$ also passes through the points $z = \pm\sqrt{-1}$ at which $f(z) = 1/(1 + z^2)$ is singular.

The "lemniscates" associated with an interval $[a, b]$ are simple closed curves which are analogous to the circles that characterize the domain of convergence of a power series expansion about a given point. That is, the sequence

$$S_n(z) = \sum_{k=0}^{n} \frac{f^{(k)}(\alpha)}{k!}(x - \alpha)^k$$

converges to the function $f(z)$ for all z inside the largest circle $|z - \alpha| = r$ about α in which $f(z)$ is analytic. For $f(z) = 1/(1 + z^2)$ we obtain the sequence

$$L_{2n}(z) \equiv \sum_{k=0}^{n} (-1)^k z^{2k},$$

which converges for $|z| < 1$. This is the largest circle about the origin not containing the singular points $z = \pm\sqrt{-1}$.

PROBLEMS, SECTION 3

1. Without using $\Delta^n f(x) = h^n f^{(n)}(\xi)$, derive the value of $\Delta^n P_n(x)$ where $P_n(x) = a_0 + a_1 x + \cdots + a_n x^n$.

2. Find the errors in the following table of function values taken at evenly spaced arguments: 50173, 53503, 56837, 60197, 63522, 66871, 70226, 73566, 76950, 80320, 83695, 87084, 90459, 93849, 97244, 100634, 104049, 107460. In this example it suffices to examine the column of second differences.

[Hint: See Problem 7.]

3. Compare error factors, $\prod_{j=0}^{n} (x - x_j)$, in equally spaced and Chebyshev interpolation over $[a, b]$. (That is, use Stirling's approximation to $n!$ and verify the estimates of $M_{\text{Ch}}/M_{\text{Eq}}$ and $M_{\text{Ch}}/M_{\text{Eq}}^*$, following Lemma 2.)

4. Derive the result that $\Delta\pi_n(t) = (n + 1)\pi_{n-1}(t)$, $n = 1, 2, \ldots$; where the spacing in Δ is $h = 1$.

5.* Prove the following

THEOREM. *If $p_n(x)$ is a polynomial of degree at most n and $\max_{|x|\le 1} |p_n(x)| = M$ then for all x in $|x| > 1$,*

$$|p_n(x)| \le M2^{n-1}T_n(x) \equiv M \cos(n \cos^{-1} x).$$

[Hint: For a proof by contradiction, consider the polynomial

$$q_n(x) = p_n(\xi)\frac{T_n(x)}{T_n(\xi)} - p_n(x),$$

with ξ a point where the conclusion is invalid show $q_n(x)$ has $n + 1$ zeros.]

6.* The technique of *L. J. Comrie* for modifying the second difference uses the idea of selecting a constant, $-c$, to replace $(t^2 - 4)/20$ and $(s^2 - 4)/20$ in the Everett formula so that the maximum error in the resulting interpolation formula is a minimum. Supply the missing details in the following sketch:

Let the Everett form of the interpolation polynomial be

$$P_5(x_0 + ph) = qu_0 + pu_1 + e(q)\delta^2 u_0 + e(p)\delta^2 u_1 + d(q)\delta^4 u_0 + d(p)\delta^4 u_1,$$

where

$$q = 1 - p, \quad e(p) = \frac{p(p^2 - 1)}{6}, \quad d(p) = \frac{p(p^2 - 1)(p^2 - 4)}{120}.$$

Then

$$P_5(x_0 + ph) = qu_0 + pu_1 + e(p)(\delta^2 u_1 - c\delta^4 u_1) + e(q)(\delta^2 u_0 - c\delta^4 u_0) + R,$$

with

$$R = [d(p) + ce(p)]\delta^4 u_1 + [d(q) + ce(q)]\delta^4 u_0.$$

If we try to pick c so as to minimize $\max_{0 \le p \le 1} |R|$, we see that c must depend on u.

Hence, we simplify the problem by noting that $\delta^4 u_1 = \delta^4 u_0 + \Delta\delta^4 u_0$. Now if $\Delta\delta^4 u_0$ is much smaller than $\delta^4 u_0$ we may neglect the fifth difference and minimize

$$\max_{0 \le p \le 1} |d(p) + d(q) + c[e(p) + e(q)]|$$
$$= \max_{0 \le p \le 1} \left| \frac{p(p^2 - 1)(p - 2)}{24} + c\,\frac{p(p - 1)}{2} \right|.$$

If we let the polynomial inside the absolute value sign be $g(p, c)$, then the maximum occurs when

$$\frac{\partial g}{\partial p} \equiv \tfrac{1}{6}(p - \tfrac{1}{2})(p^2 - p - 1 + 6c) = 0$$

or $p = \frac{1}{2}, \frac{1}{2} \pm \sqrt{\frac{5}{4} - 6c}$. Since $p(p - 1)$ is of one sign, $|g(p, c)|$ should have equal values at its maxima, in order that they be minimum. Set

$$|g(\tfrac{1}{2})| = |g(\tfrac{1}{2} \pm \sqrt{\tfrac{5}{4} - 6c})|$$

which yields

$$\frac{3 - 16c}{128} = \frac{(1 - 6c)^2}{24}.$$

Only the larger one of the roots, c, is appropriate: $c \cong 0.18393$.

7. Given a table of $f(x)$ for $x_j = x_0 + jh$, $0 \le j \le n$. Show that if f_k is replaced by $f_k + \delta$, for any k with $p \le k \le n - p$ then $\sum_{j=0}^{n-q} \Delta^q f_j/(n-q+1)$ is unaffected, for $1 \le q \le p$.

8. Let $n = 2p$, $x_j = -x_{n-j}$, for $j = 0, 1, \ldots, p$, (i.e., $x_p = 0$) and $f(x) \equiv 1/(1 + x^2)$. Show that

$$f[x_0, x_1, \ldots, x_n, x] = (-1)^{p+1} \frac{x}{1 + x^2} \frac{1}{\prod_{j=0}^{p} (1 + x_j^2)}.$$

[Hint: Use mathematical induction on p.]

9. Verify that if $f(x) \equiv 1/(x + c)$

$$f[x_0, x_1, \ldots, x_n] = (-1)^n \frac{1}{\prod_{j=0}^{n} (x_j + c)}.$$

Hence, establish (20) by writing

$$\frac{1}{1 + x^2} = \frac{1}{2i}\left(\frac{1}{x - i} - \frac{1}{x + i}\right),$$

where $i^2 = -1$.

10. Verify for the function

$$q(x) \equiv \int_0^5 \ln\left|\frac{x^2 - \xi^2}{1 + \xi^2}\right| d\xi$$

that

(a) $q(x) = 0$ at $x \cong 3.63\ldots$;
(b) $q(x) < 0$ for $|x| < 3.63\ldots$;
(c) $q(x) > 0$ for $3.63\ldots < |x| \leq 5$.

[Hint: Derive for $0 \leq x \leq 5$,

$$\begin{aligned} q(x) &\equiv \int_0^x \ln(x - \xi)\, d\xi + \int_x^5 \ln(\xi - x)\, d\xi \\ &\qquad + \int_0^5 \{\ln(\xi + x) - \ln(1 + \xi^2)\}\, d\xi \\ &\equiv (5 + x)\ln(5 + x) + (5 - x)\ln(5 - x) \\ &\qquad - 5\ln 26 - 2\arctan 5.] \end{aligned}$$

4. CALCULUS OF DIFFERENCE OPERATORS

When dealing with equally spaced data there is a very useful operator method available to suggest new formulae and to aid in recalling the fundamental ones. The basic operators are:

(1)

(a) Identity $\quad If(x) \equiv f(x)$;

(b) Displacement $\quad Ef(x) \equiv f(x + h)$;

(c) Difference $\quad \Delta f(x) \equiv f(x + h) - f(x)$;

(d) Derivative $\quad Df(x) \equiv \dfrac{df(x)}{dx}.$

Note that the displacement and difference operators imply a fixed spacing, h, by which the argument is to be shifted. We assume that E and Δ use the same such value unless otherwise specified. To employ D, the function on which it operates must be differentiable. In fact, the classes of functions to which all symbolic formulae apply must generally be restricted. We shall consider only the class of polynomials (but more general extensions are possible). Two operators, A and B, are said to be equal if $Af(x) = Bf(x)$ for every function $f(x)$ of the class under consideration, i.e., for every polynomial.

From the definitions (1), it is clear that the four operators are linear, i.e., if A is any one of them then

$$A[\alpha f(x) + \beta g(x)] = \alpha Af(x) + \beta Ag(x), \tag{2}$$

for arbitrary numbers α, β and functions $f(x)$, $g(x)$. The product, AB, and the sum, $A + B$, of two operators A and B are defined by

$$(AB)f(x) \equiv A[Bf(x)], \tag{3a}$$

$$(A + B)f(x) \equiv Af(x) + Bf(x). \tag{3b}$$

From the definition (3a) the integral powers, A^n, of any operator A may be defined inductively as

$$\begin{aligned} A^0 &\equiv I, \\ A^n &\equiv AA^{n-1}, \qquad n = 1, 2, \ldots. \end{aligned} \tag{4}$$

In addition, we define non-integral powers of the displacement (or shift) operator, E^s, by

$$E^s f(x) \equiv f(x + sh), \tag{5}$$

where s is any real number, and observe that $E^s E^r = E^{s+r}$.

Using the definitions (1a), (1b), and (3b) we have

$$\begin{aligned} Ef(x) &= f(x + h) \\ &= f(x) + f(x + h) - f(x) \\ &= (I + \Delta)f(x). \end{aligned}$$

Thus, we conclude, from the definition of equality of operators, that

$$E = I + \Delta. \tag{6}$$

Equivalently, we have $\Delta = E - I$ and from the definition of powers of operators it easily follows that

$$\begin{aligned} \Delta^n &= (E - I)^n, \\ &= \sum_{i=0}^{n} (-1)^i \binom{n}{i} E^{(n-i)}. \end{aligned} \tag{7}$$

This result may be proved by induction, just as is the usual binomial expansion. However, by applying (7) to $f(x_0)$ we obtain (3.6) which has previously been derived for general functions. Thus (3.6) yields an independent proof of (7).

From the extended definition (5) we may write

$$f(x + sh) = E^s f(x).$$

On the other hand, the Newton form of the interpolation polynomial gives

$$f(x + sh) = \left[1 + \pi_0(s)\Delta + \frac{\pi_1(s)}{2!}\Delta^2 + \cdots + \frac{\pi_k(s)}{(k + 1)!}\Delta^{k+1} + \cdots\right]f(x),$$

where we note that the series terminates with Δ^p if $f(x)$ is a polynomial of degree p.

But the formal binomial series expansion of $(I + \Delta)^s$, for s arbitrary, is identical with the series on the right-hand side, and hence we adopt it as the definition for s non-integral. That is, with this convention,

$$E^s = (I + \Delta)^s \qquad \text{for } s \text{ arbitrary.} \tag{8}$$

Thus, it is clear that the steps leading to (8) are not a derivation of Newton's formula but can now be used to recall that formula when required.

Similarly, such manipulations can be employed to suggest new formulae which can then be verified independently. For example, we define the backward difference operator, ∇, by

$$\nabla f(x) \equiv f(x) - f(x - h).$$

Then as in deriving (6) we find that

$$E^{-1} = (I - \nabla),$$

and proceeding as in (8) we obtain, formally

$$f(x - sh) = \left[1 + \pi_0(-s)\nabla + \frac{\pi_1(-s)}{2!}\nabla^2 + \cdots + \frac{\pi_k(-s)}{(k + 1)!}\nabla^{k+1} + \cdots\right]f(x). \tag{9}$$

The formula suggested is, in fact, well known as *Newton's backward difference formula.* By introducing centered difference operators,

$$\delta f(x) \equiv f\left(x + \frac{h}{2}\right) - f\left(x - \frac{h}{2}\right),$$

we could derive other formulae.

To relate D to the other operators we write the formal Taylor's series expansion

$$f(x + h) = f(x) + \frac{h}{1!}f'(x) + \frac{h^2}{2!}f''(x) + \cdots + \frac{h^n}{n!}f^{(n)}(x) + \cdots$$

in the symbolic form

$$Ef(x) = \left[1 + \frac{hD}{1!} + \frac{h^2D^2}{2!} + \cdots + \frac{h^nD^n}{n!} + \cdots\right] f(x).$$

Since the series in the brackets is the expansion of e^{hD} and the equality is valid for all polynomials, we have the interesting result:

$$e^{hD} = E = I + \Delta. \tag{10}$$

[We recall that this merely says that the operator E and the operator $I + hD + \cdots + h^nD^n/n!$ are equivalent when applied to a polynomial of degree n, for all positive integers n.] Formally by taking logarithms in equation (10) we find that

$$\begin{aligned} hD &= \ln (I + \Delta) \\ &= \Delta - \tfrac{1}{2}\Delta^2 + \tfrac{1}{3}\Delta^3 - \cdots + (-1)^{n+1}\frac{1}{n}\Delta^n + \cdots. \end{aligned} \tag{11}$$

This formula suggests that hD and the first n terms on the right-hand side might be equivalent when applied to any polynomial of degree n. To verify this we use the Newton formula, (8), for any polynomial $f(x)$ of degree n:

$$f(x + sh) = \left[1 + s\Delta + \frac{s(s-1)}{2!}\Delta^2 + \cdots + \frac{s(s-1)\cdots(s-n+1)}{n!}\Delta^n\right] f(x).$$

Differentiate with respect to s and evaluate at $s = 0$ to get

$$hf'(x) = \left[\Delta - \tfrac{1}{2}\Delta^2 + \tfrac{1}{3}\Delta^3 - \cdots + (-1)^{n+1}\frac{1}{n}\Delta^n\right] f(x),$$

which was to be shown. The relation (11) may now be employed to obtain forward difference approximations to the derivative of a tabulated function. The general problem of approximating the derivatives of a function is considered in more detail in the next section.

Symbolic methods may also be employed to determine formulae for the approximate evaluation of integrals. Thus we define

$$Jf(x) \equiv \int_x^{x+h} f(\xi)\, d\xi, \tag{12}$$

and by using (5) we have, formally,

$$
\begin{aligned}
(13) \qquad Jf(x) &= h\int_0^1 E^s f(x)\,ds \\
&= h\int_0^1 E^s\,ds\,f(x) = h\int_0^1 e^{s\ln E}\,ds\,f(x) \\
&= h\frac{E-I}{\ln E}f(x) \\
&= \frac{h\Delta}{\ln(I+\Delta)}f(x).
\end{aligned}
$$

By using (11) this result implies $J = \Delta/D$ or $JD = DJ = \Delta$ which may be verified directly. However, if we write

$$\ln(I+\Delta) = \Delta(I-R),$$

where by definition

$$R = \tfrac{1}{2}\Delta - \tfrac{1}{3}\Delta^2 + \cdots + (-1)^n\frac{1}{n}\Delta^{n-1} + \cdots$$

then (13) becomes, symbolically,

$$
\begin{aligned}
(14) \qquad J &= \frac{h}{I-R} \\
&= h(I + R + R^2\cdots).
\end{aligned}
$$

Again this might be interpreted as meaning that when applied to any polynomial $f(x)$ of degree n, J is equivalent to the first $n+1$ terms on the right, or more simply just those terms involving Δ^k for $k \le n$. Usually, (14) is written in powers of Δ, i.e.,

$$(15) \qquad J = h(I + \tfrac{1}{2}\Delta - \tfrac{1}{12}\Delta^2 + \tfrac{1}{24}\Delta^3 - \tfrac{19}{720}\Delta^4 + \cdots).$$

To justify (14) we note that $Jx^n = \Delta x^{n+1}/(n+1)$. Now (11) and the definition of R give

$$
\begin{aligned}
hDx^{n+1} &= h(n+1)x^n \\
&= \Delta(I-R)x^{n+1} \\
&= \Delta x^{n+1} - R\Delta x^{n+1}.
\end{aligned}
$$

Thus, $\Delta x^{n+1} = h(n+1)x^n + R\Delta x^{n+1}$ and iterating this result yields, since $R^{n+1}\Delta x^{n+1} = 0$,

$$
\begin{aligned}
\Delta x^{n+1} &= h(n+1)x^n + R[h(n+1)x^n + R\Delta x^{n+1}] \\
&\ \ \vdots \\
&= h(n+1)(I + R + R^2 + \cdots + R^n)x^n.
\end{aligned}
$$

Thus, we have shown that

$$Jx^n = h(I + R + \cdots + R^n)x^n. \tag{16}$$

Since this holds for all integers n, the validity of (14) when applied to polynomials follows. Different expressions for J can be obtained by using other identities to eliminate E in the derivation of (13). However, Chapter 7 is devoted to the detailed study of approximation methods for the evaluation of integrals.

The symbols $1/D$ and $1/\Delta$ are not operators in the same sense as D, Δ, E, etc., since

$$\frac{1}{D}f(x) = \{F(x)\} \equiv \text{the set of polynomials } F(x) \text{ such that } F'(x) = f(x). \tag{17}$$

$$\frac{1}{\Delta}f(x) = \{G(x)\} \equiv \text{the set of polynomials } G(x) \text{ such that } G(x+h) - G(x) = f(x). \tag{18}$$

Nevertheless, if $f(x)$ is a polynomial of degree n, $\{F(x)\}$ and $\{G(x)\}$ have the same structure, that is,

$$\{F(x)\} \equiv \{P_{n+1}(x) + c\}, \qquad c \text{ any constant},$$

$$P_{n+1}(x) \text{ a fixed polynomial of degree } n+1.$$

$$\{G(x)\} \equiv \{Q_{n+1}(x) + d\}, \qquad d \text{ any constant},$$

$$Q_{n+1}(x) \text{ a fixed polynomial of degree } n+1.$$

Hence,

$$(E^p - E^q)\frac{1}{D} \qquad \text{and} \qquad (E^p - E^q)\frac{1}{\Delta}$$

are well defined operators. We leave as Problems 2 and 3 the proof that the corresponding formal power series in Δ respectively satisfy (19) and (20):

$$(E^p - E^q)\frac{1}{D}f(x) = [(I+\Delta)^p - (I+\Delta)^q]\frac{h}{\log(I+\Delta)}f(x) = \int_{x+qh}^{x+ph} f(\xi)\,d\xi, \tag{19}$$

where $f(x)$ is a polynomial;

$$(E^p - E^q)\frac{1}{\Delta}f(x) = [(I+\Delta)^p - (I+\Delta)^q]\frac{1}{\Delta}f(x) = \sum_{j=q}^{p-1} f(x+jh), \tag{20}$$

if $p > q$ are integers and $f(x)$ is a polynomial.

Equations (19) and (20) permit the development of formulae for integration and summation of polynomials [see Problems (4) and (5)]. Another kind of representation of (20), with $q = 0$, arises from replacing the term $1/\Delta$ on the far left by the formal power series in D by setting, from (10),

$$\frac{1}{\Delta} = \frac{1}{e^{hD} - I} = \frac{1}{hD + \dfrac{h^2D^2}{2!} + \dfrac{h^3D^3}{3!} + \cdots} \tag{21}$$

$$= \frac{1}{hD} - \frac{1}{2} + \frac{1}{12}hD - \frac{1}{720}h^3D^3 + \frac{1}{30{,}240}h^5D^5 - \cdots.$$

We obtain the formula,

$$\begin{aligned}\sum_{j=0}^{p-1} f(x + jh) = \frac{1}{h}\int_x^{x+ph} f(\xi)\,d\xi &- \tfrac{1}{2}[f(x + ph) - f(x)] \\ &+ \frac{h}{12}[f'(x + ph) - f'(x)] \\ &- \frac{h^3}{720}[f^{(3)}(x + ph) - f^{(3)}(x)] \\ &+ \frac{h^5}{30{,}240}[f^{(5)}(x + ph) - f^{(5)}(x)] - \cdots,\end{aligned} \tag{22}$$

called the *Euler-Maclaurin summation formula.*

PROBLEMS, SECTION 4

1. Verify that the factorial polynomials $W_0(x) \equiv 1$, $W_n(x) \equiv x(x - h)\cdots[x - (n - 1)h]$, $n = 1, 2, \ldots$, satisfy $\Delta W_0(x) = 0$, $\Delta W_{n+1}(x) = h(n + 1)W_n(x)$, and may be used as a basis for polynomials on which the calculus of difference operators is applied, e.g., with

$$P_n(x) \equiv \sum_{j=0}^{n} \alpha_j W_j(x),$$

$$\Delta P_n(x) = h \sum_{j=1}^{n} j\alpha_j W_{j-1}(x).$$

2. Verify (19). Hint: $[\{(I + \Delta)^{p-1} + (I + \Delta)^{p-2} + \ldots + (I + \Delta)^q\}\{(I + \Delta) - I\} = (I + \Delta)^p - (I + \Delta)^q.]$

3. Verify (20). [Hint: Same as in Problem 2.]

4. Use (19) with $p = 2$, $q = 0$ to find Simpson's rule valid for all polynomials of degree ≤ 3;

$$\int_x^{x+2h} f(\xi)\,d\xi = \frac{h}{3}[f(x) + 4f(x + h) + f(x + 2h)].$$

5. Use (20) with $f(x) = x^3$, $h = 1$, $q = 0$, $p = n + 1$, $x = 1$ to get a simple explicit expression for $\sum_{j=1}^{n} j^3$. (Construct a table of differences for $j = 1, 2, 3, 4$.)

6. Use (22) to derive the formula for $\sum_{j=1}^{n} j^3$.

7.* Prove the Euler-Maclaurin summation formula (22) is correct for polynomials. [Assume that you can use (21) to formally get an infinite series for $1/(e^{hD} - I)$.]

5. NUMERICAL DIFFERENTIATION

A problem of importance in many applications is to approximate the derivative of a function, being given only several values of the function. An obvious approach to this problem is to employ the derivative of an interpolation polynomial as the desired approximation to the derivative of the function. This can also be done for higher derivatives, but clearly the approximation must, in general, deteriorate as the order of the derivative increases. We have seen in Section 3 that the interpolation error factor is least near the center of the interval of interpolation (for equally spaced points), and indeed an analogous result is also true for numerical differentiation.

Denote by $P_n(x)$ the nth degree interpolation polynomial for $f(x)$ with respect to the $n + 1$ distinct points $x_0, x_1, \ldots, x_n$. Then as an approximation to

$$\frac{d^k f(x)}{dx^k} \equiv f^{(k)}(x),$$

with $k < n$, we use $P_n^{(k)}(x)$. However, to assess this approximation we require some convenient representation for the error:

$$R_n^{(k)}(x) \equiv f^{(k)}(x) - P_n^{(k)}(x). \tag{1}$$

If $f^{(n+1)}(x)$ is continuous in the interval, I_x, which includes the x_j and x, it has been shown in Theorem 2.1 of Chapter 5 that

$$R_n(x) = \prod_{j=0}^{n} (x - x_j) \frac{f^{(n+1)}(\xi)}{(n + 1)!},$$

where $\xi = \xi(x)$ is an unknown point in I_x for each x. It is tempting to differentiate this expression for $R_n(x)$ in order to obtain $R_n^{(k)}(x)$ but this is not generally legitimate. First of all, $\xi(x)$ may not be single valued, let alone differentiable k times and secondly, $f(x)$ may not be $n + 1 + k$

times differentiable. If $f(x)$ does have these differentiability properties, then another alternative is presented by recalling (1.7) in the form

$$R_n(x) = \prod_{j=0}^{n} (x - x_j) f[x_0, \ldots, x_n, x].$$

It now follows by an application of Theorem 1.2 that this representation is k times differentiable. However, the resulting expression is rather complicated and only useful in the case of first derivatives, $k = 1$, in which $x = x_i$ is one of the points of interpolation. The error becomes in this special case

$$\text{(2)} \qquad f'(x_i) - P_n'(x_i) = R_n'(x_i) = \prod_{\substack{j=0 \\ (j \neq i)}}^{n} (x_i - x_j) f[x_0, x_1, \ldots, x_n, x_i]$$

$$= \prod_{\substack{j=0 \\ (j \neq i)}}^{n} (x_i - x_j) \frac{f^{(n+1)}(\eta)}{(n+1)!}.$$

The last expression in (2) can be deduced from Theorems 1.2 and 1.1.

To obtain practical error estimates for numerical differentiation in the more general case, we return to Rolle's theorem which was the basis for the original interpolation error estimates of Theorem 2.1 in Chapter 5. The results may be stated as

THEOREM **1**. *Let the interpolation points be ordered by* $x_0 < x_1 < \cdots < x_n$. *Let* $f^{(n+1)}(x)$ *be continuous. Then for each* $k \leq n$,

$$\text{(3)} \qquad R_n^{(k)}(x) = \prod_{j=0}^{n-k} (x - \xi_j) \frac{f^{(n+1)}(\eta)}{(n+1-k)!};$$

where the $n + 1 - k$ *distinct points,* ξ_j, *are independent of* x *and lie in the intervals*

$$\text{(4)} \qquad x_j < \xi_j < x_{j+k}, \qquad j = 0, 1, \ldots, n-k;$$

and $\eta = \eta(x)$ *is some point in the interval containing* x *and the* ξ_j.

Proof. Since $R_n(x) = f(x) - P_n(x)$ has $n + 1$ continuous derivatives and vanishes at $x = x_j$, $j = 0, 1, \ldots, n$, we may apply Rolle's theorem $k \leq n$ times. In applying this theorem we can keep track of the location of the implied zeros of the higher derivatives of $R_n(x)$ by means of Table 1. The kth column lists the open intervals, (x_j, x_{j+k}), in each of which (by means of Rolle's theorem) at least one distinct root, ξ_j, of $R_n^{(k)}(x)$ must lie. Thus the points ξ_j of (4) are defined and we note that they depend only upon the function $f(x)$ and the interpolation points x_j but not upon x.

Table 1 *Zeros of Higher Derivatives of $R_n(x)$*

$R_n(x)$	$R_n^{(1)}(x)$	$R_n^{(2)}(x)$	$\cdots$	$R_n^{(k)}(x)$
x_0				
x_1	(x_0, x_1)			
x_2	(x_1, x_2)	(x_0, x_2)		
x_3	(x_2, x_3)	(x_1, x_3)		
$\vdots$	$\vdots$	$\vdots$		
x_k	(x_{k-1}, x_k)	(x_{k-2}, x_k)	$\cdots$	(x_0, x_k)
$\vdots$	$\vdots$	$\vdots$		$\vdots$
x_n	(x_{n-1}, x_n)	(x_{n-2}, x_n)		(x_{n-k}, x_n)

We now define the function

$$F(z) = R_n^{(k)}(z) - \alpha \prod_{j=0}^{n-k} (z - \xi_j),$$

and note that $F(\xi_j) = 0$ for $j = 0, 1, \ldots, n - k$. For any fixed x distinct from the ξ_j we pick $\alpha = \alpha(x)$ such that $F(x) = 0$. Then $F(z)$ has $n - k + 2$ distinct zeros and we may apply Rolle's theorem again [noting that $F(z)$ has $n - k + 1$ continuous derivatives]. We deduce that $F^{(n-k+1)}(z)$ has a zero, say at η, in the interval containing x and the ξ_j. From this result follows:

$$\begin{aligned} 0 = F^{(n-k+1)}(\eta) &= R_n^{(n+1)}(\eta) - \alpha(n - k + 1)! \\ &= f^{(n+1)}(\eta) - \alpha(n - k + 1)!, \end{aligned}$$

or

$$\alpha = \frac{f^{(n+1)}(\eta)}{(n - k + 1)!}.$$

By using this value of α in $F(x) = 0$ the result (3) follows for all x. That is, (3) holds also for $x = \xi_j$ with arbitrary η since $F(\xi_j) = 0$ for arbitrary α. ∎

The expression (3) for the error in numerical differentiation is valid for all x and so is of much more general applicability than expressions of the form (2). Using the known intervals (4) it is possible to obtain bounds on the error. For instance, if x and the x_j all lie in $[a, b]$ and in this interval $|f^{(n+1)}(x)| \leq M$ then clearly,

$$|R_n^{(k)}(x)| \leq \frac{M|b - a|^{n-k+1}}{(n - k + 1)!}.$$

Sharper estimates may be obtained by a more careful use of the inequalities (4) in bounding the error factor, $\prod_{j=0}^{n-k} (x - \xi_j)$.

There are other ways to determine numerical differentiation formulae and their errors. Suppose the value $f^{(k)}(a)$ is to be approximated by using the values $f(x_i)$, $i = 1, 2, \ldots, m$. With an $f(x)$ that has $n + 1$ continuous derivatives where $n + 1 \geq m$ we define $h_i \equiv x_i - a$ and use Taylor's theorem to write

$$\begin{aligned} f(x_i) &= f(a + h_i) \\ &= f(a) + h_i f^{(1)}(a) + \frac{h_i^2}{2!} f^{(2)}(a) + \cdots + \frac{h_i^n}{n!} f^{(n)}(a) \\ &\qquad + \frac{h_i^{n+1}}{(n+1)!} f^{(n+1)}(a + \theta_i h_i), \qquad i = 1, 2, \ldots, m. \end{aligned}$$

Here, of course, $0 < \theta_i < 1$. We now form a linear combination of these equations with weights, α_i, to be determined.

$$\begin{aligned} \sum_{i=1}^{m} \alpha_i f(x_i) &= \left(\sum_{i=1}^{m} \alpha_i\right) f(a) + \left(\sum_{i=1}^{m} \alpha_i h_i\right) f^{(1)}(a) + \cdots \\ &\quad + \left(\sum_{i=1}^{m} \alpha_i h_i^n\right) \frac{f^{(n)}(a)}{n!} + \frac{1}{(n+1)!} \sum_{i=1}^{m} \alpha_i h_i^{n+1} f^{(n+1)}(a + \theta_i h_i). \end{aligned} \tag{5}$$

We choose the α_i in order that the linear combination of the values $f(x_i)$ be the most accurate approximation to $f^{(k)}(a)$. Thus we impose the m conditions on the m unknowns α_i:

$$\sum_{i=1}^{m} \alpha_i h_i^{\nu} = \nu! \delta_{\nu k}, \qquad \nu = 0, 1, \ldots, m - 1. \tag{6}$$

It is clear that the system (6) has a unique solution since the coefficient determinant is a *Vandermonde* determinant. Thus a necessary and sufficient condition for (6) to have a non-trivial solution is that it be non-homogeneous, i.e., $m > k$. Hence, *in order to approximate a kth derivative we need more than k points.* With the solution of the system (6) in (5) we obtain, recalling that $n + 1 \geq m$

$$\begin{aligned} f^{(k)}(a) &= \sum_{i=1}^{m} \alpha_i f(x_i) - \frac{1}{m!}\left(\sum_{i=1}^{m} \alpha_i h_i^m\right) f^{(m)}(a) - \cdots \\ &\quad - \frac{1}{n!}\left(\sum_{i=1}^{m} \alpha_i h_i^n\right) f^{(n)}(a) - \frac{1}{(n+1)!} \sum_{i=1}^{m} \alpha_i h_i^{n+1} f^{(n+1)}(a + \theta_i h_i), \\ &\qquad m > k. \end{aligned} \tag{7}$$

This procedure is equivalent to what may be called the *method of undetermined coefficients*: if we are given m function values, $f(x_i)$, we seek that linear combination of the values at these points which would give the exact value of the derivative $f^{(k)}(a)$ for all polynomials of as high a degree as possible, at the fixed point a. Specifically for the first derivative, since

$$\left.\frac{dx^\nu}{dx}\right|_{x=a} = \nu a^{\nu-1},$$

we seek α_i such that

$$\sum_{i=1}^{m} \alpha_i x_i^{\nu} = \nu a^{\nu-1}, \qquad \nu = 0, 1, \ldots, m-1. \tag{8}$$

This system also has a unique solution and it is, in fact, the same as the solution of the system (6) with $k = 1$ (assuming that the quantities x_j, h_j, and a are related by $h_j = x_j - a$). This verification is posed as Problem 1. In the present derivation of the approximation formula no estimate of the error term is obtained but this could be remedied. It should also be observed that the method of undetermined coefficients can be used to determine approximations to higher derivatives.

5.1. Differentiation Using Equidistant Points

Naturally the numerical differentiation formulae are somewhat simplified when equally spaced data points are used. For instance, the operator identity (4.11) yields approximations of the form

$$f'(x) = \frac{1}{h}\left[\Delta f(x) - \tfrac{1}{2}\Delta^2 f(x) + \cdots + (-1)^{n+1}\frac{1}{n}\Delta^n f(x)\right] + R_n'(x). \tag{9a}$$

Here the data required are $f(x)$, $f(x+h), \ldots, f(x+nh)$, so that this formula only approximates the derivative at a tabular point and uses only data on one side of this point. This formula is obtained by differentiating Newton's forward difference formula (3.12) and evaluating the result at $t = 0$. Thus the error determined in (2) is applicable and becomes in this case

$$R_n'(x) = \frac{(-1)^n h^n}{n+1} f^{(n+1)}(\eta), \qquad x < \eta < x + nh. \tag{9b}$$

Another example is furnished by differentiating the Gaussian form of the interpolating polynomial, say (3.15a) with $n = 2m$, and again setting $t = 0$,

$$\begin{aligned} f'(x_0) = \frac{1}{h}\Bigg[&\Delta f(x_0) - \frac{1}{2!}\Delta^2 f(x_{-1}) - \frac{1}{3!}\Delta^3 f(x_{-1}) + \cdots \\ &+ (-1)^m \frac{m!(m-1)!}{(2m)!}\Delta^{2m} f(x_{-m})\Bigg] + R_{2m}'(x_0). \end{aligned} \tag{10a}$$

Here the tabular points involved are symmetrically placed about x_0 and again the error formula (2) is valid:

$$(10b) \qquad R_{2m}'(x_0) = (-1)^m \frac{(m!)^2}{(n+1)!} h^n f^{(n+1)}(\eta),$$

$$x_0 - mh < \eta < x_0 + mh.$$

For n and $m \gg 1$ Stirling's approximation for $n!$ implies, since $n = 2m$,

$$\frac{(m!)^2}{(2m+1)!} \approx \frac{n^{-\frac{1}{2}}\sqrt{2\pi}}{2^{n+1}}.$$

Thus a comparison of (9b) and (10b) indicates, for differentiation, the superiority of centering the data points about the point of approximation. An important special case of (10) occurs for $n = 2$; this can be written as

$$(11) \quad f'(x) = \frac{f(x+h) - f(x-h)}{2h} - \frac{h^2}{6} f^{(3)}(\eta), \qquad x - h < \eta < x + h.$$

The approximation formula in (11) is called the centered difference approximation to the first derivative.

The second derivative, or in fact, any even order derivative, can be approximated by a centered formula obtained by differentiating the other Gaussian interpolating polynomial, (15b) with $n = 2m + 1$. For example, with $n = 3$ the approximation of $f''(x_0)$ becomes on setting $x_0 = x$:

$$(12a) \qquad f''(x) = \frac{\Delta^2 f(x-h)}{h^2} + R_3^{(2)}(x)$$

$$= \frac{f(x+h) - 2f(x) + f(x-h)}{h^2} + R_3^{(2)}(x).$$

The error term can now be estimated by Theorem 1:

$$(12b) \qquad |R_3^{(2)}(x)| \le h^2 |f^{(4)}(\eta)|, \qquad x - h < \eta < x + 2h.$$

But a better bound for the error can be found by the Taylor expansion procedure indicated by equations (6) and (7). That is, if we set $k = 2$, $m = 4$, $a = x$, and $h_i = (i - 2)h$, $i = 1, 2, 3, 4$, we find that

$$\alpha_1 = \alpha_3 = \frac{1}{h^2}, \; \alpha_2 = \frac{-2}{h^2}, \; \alpha_4 = 0.$$

The error expression, given by the last terms in (7), becomes, upon setting $n = 3$,

$$R_3^{(2)}(x) = -\frac{h^2}{4!}[f^{(4)}(\xi_1) + f^{(4)}(\xi_2)], \qquad x - h < \xi_1 < x < \xi_2 < x + h.$$

But $f^{(4)}(x)$ was assumed continuous in this derivation so for some ξ in $\xi_1 < \xi < \xi_2$ we must have $\frac{1}{2}[f^{(4)}(\xi_1) + f^{(4)}(\xi_2)] = f^{(4)}(\xi)$. Thus the error is

$$R_3^{(2)}(x) = -\frac{h^2}{12}f^{(4)}(\xi), \qquad x - h < \xi < x + h. \tag{12c}$$

Note the improvement over the bound (12b) both in the factor $\frac{1}{12}$ and the decreased range of the argument of the fourth derivative. It is also of interest to observe that the same approximation (12a) and error (12c) are obtained for $m = 3$ and $n = 3$ in (7) with the above choice of h_i; that is, improving upon the accuracy by using data at one additional point is not always possible. (See Problem 2.)

PROBLEMS, SECTION 5

1. Verify that the set of coefficients $\{\alpha_i\}$ produced by solving the system (8) is the same as the solution $\{\alpha_i\}$ of system (6) for $k = 1$, $h_i \equiv x_i - a$.

2. Verify that if $\{\alpha_i\}$ produces the differentiation formula of maximum accuracy

$$f^{(k)}(x) = \sum_{i=-r}^{r} \alpha_i f(x + ih),$$

then

$$f^{(k)}(x) = \sum_{i=-r}^{r+1} \beta_i f(x + ih)$$

can be no more accurate, and is of the same accuracy only if $\beta_{r+1} = 0$, $\beta_i = \alpha_i$ for $i = -r, \ldots, +r$. Show also that the coefficients α_i satisfy $\alpha_p = \alpha_{-p}$ if k is even; $\alpha_p = -\alpha_{-p}$ if k is odd.

6. MULTIVARIATE INTERPOLATION

The problems of polynomial interpolation and approximate differentiation for functions of several independent variables are important but the methods are less well developed than in the case of functions of a single variable. An immediate indication of the difficulties inherent in the higher dimensional case can be seen in the lack of uniqueness in the general interpolation problem. That is, we ask if $p_1, p_2, \ldots, p_n$ are n distinct points, say in the x, y-plane, then is there a unique polynomial of specified degree which attains specified values, say $f(p_j)$, at these points? Clearly the answer, in general, must be *no* since if all of the points $[p_j, f(p_j)]$ lie on a straight line in x, y, z-space then there are infinitely many planes (i.e., linear polynomials) and perhaps higher degree polynomials of the form $z = P(x, y)$ containing this line. We shall not dwell on these aspects of interpolation in higher dimensions but shall show how to construct appropriate polynomials when the points of interpolation are specially chosen. It will also be found that in these special cases the interpolation polynomials are unique. For simplicity, we concentrate on functions of two variables but extension to more dimensions offers no difficulty.

Let us be given the $(m + 1)(n + 1)$ distinct points $p_{ij} \equiv (x_i, y_j)$; $i = 0, 1, \ldots, m, j = 0, 1, \ldots, n$ and corresponding function values $f(x_i, y_j)$. These points form a rectangular array which is the set of intersections of the vertical lines $x = x_i$ with the horizontal lines $y = y_j$ in the x, y-plane. We seek a polynomial, $P(x, y)$, *of degree at most m in x and at most n in y such that*

$$P(x_i, y_j) = f(x_i, y_j), \qquad i = 0, 1, \ldots, m, \qquad j = 0, 1, \ldots, n.$$

Such a polynomial must have the form

$$P(x, y) = \sum_{i=0}^{m} \sum_{j=0}^{n} a_{ij} x^i y^j, \tag{1}$$

with $(m + 1)(n + 1)$ coefficients, a_{ij}, to be determined. This problem is easily solved, due to the special form of the points p_{ij}, with the use of the Lagrange interpolation coefficients. Let us write the Lagrange coefficients for the points $\{x_i\}$ and $\{y_j\}$ as

$$\begin{aligned} X_{m,i}(x) &= \prod_{\substack{k=0 \\ (k \neq i)}}^{m} \frac{x - x_k}{x_i - x_k}, \qquad i = 0, 1, \ldots, m; \\ Y_{n,j}(y) &= \prod_{\substack{k=0 \\ (k \neq j)}}^{n} \frac{y - y_k}{y_j - y_k}, \qquad j = 0, 1, \ldots, n. \end{aligned} \tag{2}$$

Then clearly, the polynomial $X_{m,i}(x) Y_{n,j}(y)$ is of degree m in x, of degree n in y and vanishes when $(x, y) = p_{\nu\mu}$ unless $\nu = i$ and $\mu = j$ in which case it is unity. Thus the required polynomial satisfying $P(x_i, y_j) = f(x_i, y_j)$ can be written as

$$P(x, y) = \sum_{i=0}^{m} \sum_{j=0}^{n} X_{m,i}(x) Y_{n,j}(y) f(x_i, y_j). \tag{3}$$

Since the number of coefficients in the general polynomial (1) of degree m in x and n in y is equal to the number of conditions imposed we may expect that the interpolation polynomial (3) is unique. A formal proof of this fact is indicated in Problem 1. The extension to more independent variables is obvious.

Another representation of the interpolation polynomial (3) can be obtained by using Newton's divided difference formulae. With the $m + 1$ distinct points x_j we have, recalling (1.7) and (1.8),

$$(4) \qquad f(x, y) = \sum_{k=0}^{m} \omega_{k-1}(x) f[x_0, x_1, \ldots, x_k; y] + \omega_m(x) f[x_0, \ldots, x_m, x; y].$$

Here we have introduced

$$\omega_{-1}(x) \equiv 1; \quad \omega_k(x) = \omega_{k-1}(x)(x - x_k), \quad k = 0, 1, \ldots;$$

and the divided differences of a function of several variables are formed by keeping all but one variable fixed and taking the indicated differences with respect to the free variable. Hence $f[x_0, x_1, \ldots, x_k; y]$ as a function of the independent variable y has the Newton representation, using the $n + 1$ points y_j:

$$(5) \quad f[x_0, x_1, \ldots, x_k; y] = \sum_{j=0}^{n} \omega_{j-1}(y) f[x_0, x_1, \ldots, x_k; y_0, y_1, \ldots, y_j] + \omega_n(y) f[x_0, \ldots, x_k; y_0, \ldots, y_n, y]$$

We use (5) for $k = 0, 1, \ldots, m$ in (4) to obtain

$$f(x, y) = Q(x, y) + R(x, y)$$

where

$$(6) \qquad Q(x, y) \equiv \sum_{k=0}^{m} \sum_{j=0}^{n} \omega_{k-1}(x)\omega_{j-1}(y) f[x_0, \ldots, x_k; y_0, \ldots, y_j]$$

and

$$(7a) \qquad R(x, y) = \omega_n(y) \sum_{k=0}^{m} \omega_{k-1}(x) f[x_0, \ldots, x_k; y_0, \ldots, y_n, y] + \omega_m(x) f[x_0, \ldots, x_m, x; y].$$

It is clear that $R(x_i, y_j) = 0$ at the $(m + 1)(n + 1)$ points (x_i, y_j) and hence by the uniqueness proof mentioned we can conclude that $P(x, y) \equiv Q(x, y)$. The derivation of the interpolation polynomial also yields an expression for the interpolation error, $R(x, y)$. To simplify this expression we again use Newton's formula and the $m + 1$ points x_i to write

$$f[x; y_0, \ldots, y_n, y] = \sum_{i=0}^{m} \omega_{i-1}(x) f[x_0, \ldots, x_i; y_0, \ldots, y_n, y] + \omega_m(x) f[x_0, \ldots, x_m, x; y_0, \ldots, y_n, y].$$

If we multiply this identity by $\omega_n(y)$ and subtract the result from (7a) we obtain finally,

$$(7b) \quad R(x, y) = \omega_m(x) f[x_0, \ldots, x_m, x; y] + \omega_n(y) f[x; y_0, \ldots, y_n, y] - \omega_m(x)\omega_n(y) f[x_0, \ldots, x_m, x; y_0, \ldots, y_n, y].$$

If $f(x, y)$ has continuous partial derivatives of orders $m + 1$ and $n + 1$, respectively, in x and y and the appropriate mixed derivative of order $m + n + 2$ then by applying the obvious extension of Theorem 1.1 the error becomes

$$\text{(7c)} \qquad R(x, y) = \frac{\omega_m(x)}{(m+1)!} \frac{\partial^{m+1} f(\xi, y)}{\partial x^{m+1}} + \frac{\omega_n(y)}{(n+1)!} \frac{\partial^{n+1} f(x, \eta)}{\partial y^{n+1}} - \frac{\omega_m(x)\omega_n(y)}{(m+1)!\,(n+1)!} \frac{\partial^{m+n+2} f(\xi', \eta')}{\partial x^{m+1}\, \partial y^{n+1}}.$$

This error formula is not of the form of the two-dimensional Taylor series error term, as was the case in one dimension, since different orders of differentiation occur here.

By specializing the interpolation points to be equally spaced we can obtain special forms of (3) and (6). These forms may be written in terms of the difference operators of Section 4, generalized so that they operate with respect to a particular independent variable. An example of such a representation is to be found in Problem 2.

The interpolation problem solved above, by (3) or (6), does not specify the degree of the polynomial in question but rather the maximum degree in x and y, separately. If a polynomial in two variables is to have total degree at most n, say, then it must have the form

$$\text{(8)} \qquad P_n(x, y) = \sum_{k=0}^{n} \sum_{j=0}^{n-k} b_{kj} x^k y^j.$$

We note that the coefficients b_{kj} can be naturally arranged in a triangular array of $\frac{1}{2}(n + 1)(n + 2)$ numbers. [In contrast, the a_{ij} in (1) formed a rectangular array of $(m + 1)(n + 1)$ quantities.] We shall show that with an appropriate "triangular" array of points, (x_k, y_j), the interpolation problem for polynomials of the form (8) can be uniquely solved.

Let $\{x_j\}$ and $\{y_j\}$ be two sets of $n + 1$ distinct points, where $j = 0, 1, \ldots, n$. Then we consider the array of points:

$$\text{(9)} \qquad p_{kj} \equiv (x_k, y_j); \qquad j + k = 0, 1, \ldots, n.$$

[This array is actually triangular only if the values x_j and y_j are ordered by j and uniformly spaced, which we do not assume.] There are $\frac{1}{2}(n + 1)(n + 2)$ such points and with them we pose the interpolation problem: *find a polynomial in x and y of degree at most n such that $P_n(x_k, y_j) = f(x_k, y_j)$ for $0 \leq j + k \leq n$.* Newton's divided difference formulae easily yield the solution of this problem. We obtain, upon replacing m by n in (4) and n by $n - k$ in (5):

$$f(x, y) = P_n(x, y) + R_n(x, y)$$

where

$$(10)\qquad P_n(x, y) = \sum_{k=0}^{n} \sum_{j=0}^{n-k} \omega_{k-1}(x)\omega_{j-1}(y)f[x_0, \ldots, x_k; y_0, \ldots, y_j];$$

and

$$R_n(x, y) = \sum_{k=0}^{n} \omega_{k-1}(x)\omega_{n-k}(y)f[x_0, \ldots, x_k; y_0, \ldots, y_{n-k}, y] + \omega_n(x)f[x_0, \ldots, x_n, x; y],$$

$$(11)\qquad = \sum_{k=0}^{n+1} \frac{\omega_{k-1}(x)\omega_{n-k}(y)}{k!\,(n-k+1)!} \left(\frac{\partial}{\partial x}\right)^k \left(\frac{\partial}{\partial y}\right)^{n-k+1} f(\xi_k, \eta_k).$$

The polynomial (10) has degree at most n. If we assume the indicated partial derivatives of $f(x, y)$ to be continuous, then (11) yields the error which vanishes at all points in (9). The uniqueness of this polynomial follows from Problem 3. Thus the interpolation problem is solved by a polynomial of the form (8) on any set of points of the form (9). If these points are equally spaced and in monotone order, then the polynomial (10) can be simplified by introducing difference operators (see Problem 4). The remainder term in (11) is now analogous to that in Taylor's formula. In fact, if we let $x_\nu \to x_0$ and $y_\nu \to y_0$ for $\nu = 1, 2, 3, \ldots, n$ then (10) formally goes over into the Taylor expansion.

To approximate the partial derivatives of functions of several independent variables we could proceed as in Section 5 for functions of one variable. By using the error expressions of the form (7b) we could also obtain representations for the error in these numerical differentiation methods (if the function is sufficiently differentiable). However, in practice it turns out that relatively low order approximations to partial derivatives are usually all that are required. In these circumstances, it is easy to use the method of undetermined coefficients or the Taylor expansion method developed in Section 5. If no mixed derivatives occur and the points to be used are on a coordinate line in the direction of differentiation then the one-dimensional analysis is valid. For mixed derivatives the points employed in the expansion procedure must not be collinear. In Chapter 9, where partial differential equations are treated, specific applications are made in several examples.

PROBLEMS, SECTION 6

1. Show that every polynomial $Q(x, y)$ of degree m in x and n in y which vanishes at the $(m + 1)(n + 1)$ distinct points (x_i, y_j); $i = 0, 1, \ldots, m$; $j = 0, 1, \ldots, n$; vanishes identically.

[Hint: Any such polynomial has the form

$$Q(x, y) = \sum_{\nu=0}^{m} \sum_{\mu=0}^{n} a_{\nu\mu} x^{\nu} y^{\mu} = \sum_{\nu=0}^{m} b_{\nu}(y) x^{\nu}.$$

Set $y = y_j$ and then note that the polynomial $Q(x, y_j)$ of degree m in x vanishes at $m + 1$ distinct points. Thus, $b_{\nu}(y_j) = 0$ for $\nu = 0, 1, \ldots, m$. Next show that all $a_{\nu\mu} = 0$.]

2. We define the difference operators Δ_x and Δ_y by:

$$\Delta_x f(x, y) \equiv f(x + h, y) - f(x, y),$$

$$\Delta_y f(x, y) \equiv f(x, y + k) - f(x, y).$$

If $x_i = x_0 + ih$ and $y_j = y_0 + jk$ then show that the interpolation polynomial of degree m in x and n in y for $f(x, y)$ using the points (x_i, y_j); $i = 0, 1, \ldots, m$; $j = 0, 1, \ldots, n$ is:

$$P(x_0 + sh, y_0 + tk) = \sum_{\nu=0}^{m} \sum_{\mu=0}^{n} \frac{\pi_{\nu}(s)\pi_{\mu}(t)}{\nu!\,\mu!} \Delta_x{}^{\nu} \Delta_y{}^{\mu} f(x_0, y_0).$$

3. State and prove the analog of Problem 1 for polynomials in x and y of degree at most n using points of the form (9).

4. Use the difference operators of Problem 2 and special equally spaced points of the form (9) to derive from (10):

$$P_n(x_0 + sh, y_0 + tk) = \sum_{\nu=0}^{n} \sum_{\mu=0}^{n-\nu} \frac{\pi_{\nu}(s)\pi_{\mu}(t)}{\nu!\,\mu!} \Delta_x{}^{\nu} \Delta_y{}^{\mu} f(x_0, y_0).$$

Define the corresponding backward difference operators ∇_x and ∇_y; use them to write an interpolation polynomial of degree n in the plane; describe the set of interpolation points employed.

7

Numerical Integration

0. INTRODUCTION

Simple explicit formulae cannot be given for the indefinite integrals of most functions. Furthermore, in many problems the integrand, $f(x)$, is not known precisely but perhaps is given by tabular data or defined as the solution of some differential equation (which cannot be solved explicitly). Thus, we seek appropriate numerical procedures to approximate the value of the definite integral, say

$$I\{f\} \equiv \int_a^b f(x)\,dx. \tag{1}$$

Unless otherwise specified $[a, b]$ is a finite closed interval.

The types of approximation to (1) that we shall consider are all essentially of the form

$$I_n\{f\} \equiv \sum_{j=1}^{n} \alpha_j f(x_j). \tag{2}$$

When employed as an approximation to an integral, a sum of this form is called a numerical *quadrature* or numerical *integration formula.* For brevity, "numerical" is usually dropped. The n distinct points, x_j, are called the *quadrature points* or *nodes* and the quantities α_j are called the quadrature *coefficients.* The basic problems in numerical integration are concerned with choosing the nodes and coefficients so that $I_n\{f\}$ will be a "close" approximation to $I\{f\}$ for a large class of functions, $f(x)$. As with polynomial approximation we note that different criteria may be used to measure the *quadrature error,*

$$E_n\{f\} \equiv I\{f\} - I_n\{f\}$$

even though it is a scalar; these criteria suggest different types of quadrature formulae.

One particularly useful notion which measures the error of a quadrature formula is its so-called *degree of precision*; this is by definition the maximum integer m such that $E_n\{x^k\} = 0$ for $k = 0, 1, \ldots, m$ but $E_n\{x^{m+1}\} \neq 0$. Thus, if a formula has degree of precision m all polynomials of degree at most m are integrated exactly by that formula.

In fact, an expression for the error $E_n\{f\}$ of such a scheme is given in

THEOREM 1. *If* (2) *has degree of precision m and* $f(x)$ *has a continuous derivative of order* $m + 1$, *then*

$$E_n\{f\} \equiv I\{f\} - I_n\{f\} = \frac{1}{(m+1)!}\int_c^d f^{(m+1)}(\zeta)G_{n,m}(\zeta)\,d\zeta \tag{3}$$

where

$$G_{n,m}(\zeta) \equiv (m+1)[I\{(x-\zeta)_+{}^m\} - I_n\{(x-\zeta)_+{}^m\}],$$

with

$$(x-\zeta)_+{}^m \equiv \begin{cases} 0 & x \leq \zeta \\ (x-\zeta)^m & x \geq \zeta \end{cases}$$

and $[c, d]$ *is the smallest interval containing* $[a, b]$ *and all* x_j.

Proof. By Taylor's theorem (with remainder)

$$f(x) = T_m(x) + R_m(x),$$

where

$$T_m(x) \equiv \sum_{k=0}^{m} \frac{1}{k!} f^{(k)}(c)(x-c)^k,$$

$$R_m(x) \equiv \frac{1}{m!}\int_c^x f^{(m+1)}(\zeta)(x-\zeta)^m\,d\zeta.$$

Clearly,

$$R_m(x) = \frac{1}{m!}\int_c^d f^{(m+1)}(\zeta)(x-\zeta)_+{}^m\,d\zeta, \qquad c \leq x \leq d. \tag{4}$$

$I\{\ \}$ and $I_n\{\ \}$ are *linear operators*.† Hence, since $I_n\{\ \}$ has degree of precision m, $I\{T_m\} = I_n\{T_m\}$ and

$$E_n\{f\} = I\{R_m\} - I_n\{R_m\}.$$

But for $R_m(x)$ as given in (4), we find the expression (3), by interchanging the order of the operations on the variables x and ζ. ■

† $J\{\ \}$ is called a linear operator iff for all scalars a, b and functions $f(x)$, $g(x)$

$$J\{af(x) + bg(x)\} \equiv aJ\{f\} + bJ\{g\}.$$

It is left to Problem 1, to show that $G_{n,m}(c) = G_{n,m}(d) = 0$. In the following sections, simpler expressions than (3) for the error are found in special cases.

If a "close" approximation to $f(x)$ in $a \leq x \leq b$ is known, then the integral of the approximating function will be "close" to the integral of $f(x)$. That is, if

$$|f(x) - g(x)| \leq \epsilon,$$

then

$$\left| \int_a^b f(x)\, dx - \int_a^b g(x)\, dx \right| \leq |b - a|\epsilon.$$

This simple result is the motivation for developing most numerical integration methods. Of course, it is desirable that the approximating function should have a simple explicit indefinite integral. Hence polynomial approximations are naturally suggested and of these the interpolation polynomials are most frequently employed. Although there are quadrature formulae of great utility which are not necessarily motivated by the use of simple interpolation polynomials, we shall see nevertheless that *all such methods of general value are* what we will call *interpolatory*. There is considerable freedom to choose the position of the interpolation points relative to the interval of integration, so as may be expected there are a large number of numerical integration formulae. The choice of which formula to employ in a given case should depend upon its accuracy and relative ease of application.

In Sections 1 through 4, we consider *simple* quadrature formulae; in Section 5 we treat *composite* quadrature formulae. A composite formula is obtained by applying a simple formula to successive subintervals of $[a, b]$. In this fashion the problem of uniformly approximating the integrand $f(x)$ over $[a, b]$ is treated by using polynomials of a fixed "low" degree over each of the "small" subintervals into which the interval $[a, b]$ is divided.

In many integration problems the integrand cannot be accurately approximated by a polynomial. Such cases may arise, for example, if $f(x)$ is discontinuous at some points of the interval. Special considerations are required in these problems and we study some of them in Section 6.

In Section 7, we briefly treat the subject of approximating multiple integrals where the current state of the theory is not fully developed.

PROBLEMS, SECTION 0

1. Under the conditions of Theorem 1, show that

$$G_{n,m}(c) = G_{n,m}(d) = 0.$$

1. INTERPOLATORY QUADRATURE

Let $n + 1$ distinct points x_j be ordered by

$$x_0 < x_1 < \cdots < x_n.$$

With these points as interpolation points, we form the interpolation polynomial $P_n(x)$ of degree at most n [for the continuous function $f(x)$] such that $f(x_j) = P_n(x_j)$, $j = 0, 1, \ldots, n$. Then as an approximation to the integral (0.1), we set

$$I_{n+1}\{f\} \equiv \int_a^b P_n(x)\,dx. \tag{1}$$

This integral is easily evaluated. In fact, by using the Lagrange form for the interpolation polynomial

$$P_n(x) = \sum_{j=0}^{n} \phi_{n,j}(x) f(x_j) \tag{2a}$$

$$\phi_{n,j}(x) = \frac{\omega_n(x)}{(x - x_j)\omega_n'(x_j)} \qquad j = 0, 1, \ldots, n, \tag{2b}$$

where $\omega_n(x) \equiv (x - x_0)(x - x_1)\cdots(x - x_n)$, we obtain from (1) the quadrature formula

$$I_{n+1}\{f\} = \sum_{j=0}^{n} w_{n,j} f(x_j), \tag{3a}$$

with the coefficients given by

$$w_{n,j} = \int_a^b \phi_{n,j}(x)\,dx. \tag{3b}$$

It is clear that the coefficients $w_{n,j}$ are determined completely by the endpoints of the interval of integration and by the interpolation points x_j, which also are the nodes of the formula (3a); the coefficients are independent of the integrand. Any quadrature formula of the form (3a and b) is called an *interpolatory quadrature formula.*

The error in approximating the continuous function $f(x)$ by $P_n(x)$ is, by (1.8) of Chapter 6 and the definition of $\omega_n(x)$ above,

$$f(x) - P_n(x) = \omega_n(x) f[x_0, \ldots, x_n, x].$$

Integrate this equation over $[a, b]$ and use (0.1) and (1) to obtain the *interpolatory quadrature error*

$$\begin{aligned} E_{n+1}\{f\} &\equiv I\{f\} - I_{n+1}\{f\} \\ &= \int_a^b \omega_n(x) f[x_0, \ldots, x_n, x]\,dx. \end{aligned} \tag{4a}$$

If $f(x)$ is a polynomial of degree n or less then $E_{n+1}\{f\} = 0$ follows from the corollary to Theorem 1.1 of Chapter 6. Thus we have shown that *any interpolatory quadrature formula using* $n + 1$ *nodes has degree of precision at least* n. We shall, in fact, see later that even higher degrees of precision are possible if the nodes are specially placed with respect to the interval of integration. From (4a) a simple error bound is found,

$$\text{(4b)} \qquad |E_{n+1}\{f\}| \leq \max_{x \in [a,b]} |f[x_0, x_1, \ldots, x_n, x]| \int_a^b |\omega_n(x)| \, dx,$$

where without loss of generality we assume $a \leq b$.

To examine the error in more detail let us consider the *special case in which* $\omega_n(x)$ *does not vanish in the open interval* (a, b) [i.e., $\omega_n(x)$ does not change sign there]:

If $f'(x)$ is continuous on $[a, b]$, it follows that $f[x_0, \ldots, x_n, x]$ is continuous on $[a, b]$ by Theorem 1.3 of Chapter 6. Thus the mean value theorem for integrals can be employed in (4a) to yield

$$\text{(5a)} \qquad E_{n+1}\{f\} = f[x_0, \ldots, x_n, \eta] \int_a^b \omega_n(x) \, dx, \qquad a < \eta < b.$$

If, in addition, $f^{(n+1)}(x)$ is continuous on the smallest closed interval, $[c, d]$, containing $[a, b]$ and the nodes $\{x_j\}$, then by (1.9) of Chapter 6, (4a) becomes

$$\text{(5b)} \qquad E_{n+1}\{f\} = \frac{1}{(n+1)!} \int_a^b \omega_n(x) f^{(n+1)}(\xi) \, dx, \qquad \xi \equiv \xi(x) \in (c, d).$$

Now apply the same mean value theorem to (5b), where $\omega_n(x)$ is of one sign and $f^{(n+1)}(\xi(x))$ is continuous in x because

$$f^{(n+1)}(\xi(x)) \equiv (n+1)! f[x_0, x_1, \ldots, x_n, x],$$

to find

$$\text{(5c)} \qquad E_{n+1}\{f\} = \frac{f^{(n+1)}(\zeta)}{(n+1)!} \int_a^b \omega_n(x) \, dx, \qquad \zeta \in (c, d).$$

In the general case, some nodes will lie in the interval of integration and the simple error formula (5c) is not generally valid. Our aim is to find a suitable replacement for (5c).

Specifically *let there be* $r - 1 > 0$ *interpolation points or nodes in the open interval* (a, b), as in Figure 1,

$$x_0 < \cdots < x_i \leq a < x_{i+1} < \cdots < x_{i+r-1} < b \leq x_{i+r} < \cdots < x_n.$$

For convenience of notation we introduce the $r + 1$ quantities, ξ_j,

$$\xi_0 = a; \quad \xi_k = x_{i+k}, \quad k = 1, 2, \ldots, r-1; \quad \xi_r = b.$$

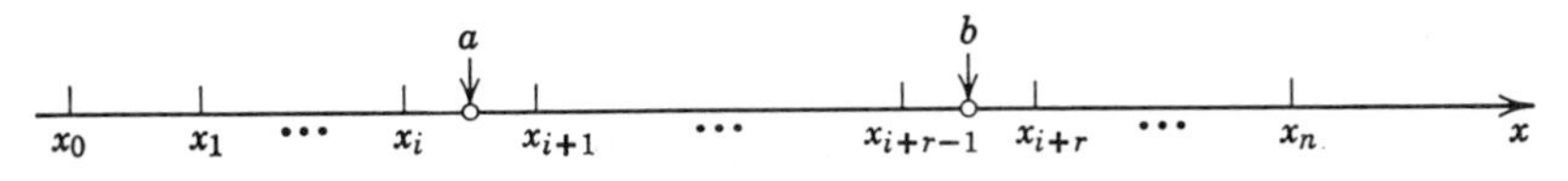

Figure 1

Now the error expression in (4a) can be written as

$$E_{n+1}\{f\} = \int_{\xi_0}^{\xi_r} \omega_n(x) f[x_0, \ldots, x_n, x]\, dx$$

$$= \sum_{k=1}^{r} \int_{\xi_{k-1}}^{\xi_k} \omega_n(x) f[x_0, \ldots, x_n, x]\, dx.$$

In each of the intervals $[\xi_{k-1}, \xi_k]$ the quantity $\omega_n(x)$ is of one sign and it changes sign at the points $\xi_1, \xi_2, \ldots, \xi_{r-1}$. Thus as in the derivation of (5a) we now conclude that

$$\text{(6a)} \qquad E_{n+1}\{f\} = \sum_{k=1}^{r} C_k f[x_0, \ldots, x_n, \eta_k], \qquad \xi_{k-1} < \eta_k < \xi_k,$$

where

$$\text{(6b)} \qquad C_k = \int_{\xi_{k-1}}^{\xi_k} \omega_n(x)\, dx, \qquad k = 1, 2, \ldots, r.$$

If $f^{(n+1)}(x)$ is continuous in $[x_0, x_n]$ then, as in (5c), we obtain

$$E_{n+1}\{f\} = \sum_{k=1}^{r} \frac{C_k}{(n+1)!} f^{(n+1)}(\zeta_k), \qquad x_0 < \zeta_k < x_n.$$

However, it is clear from (6b) that the constants C_k alternate in sign; that is,

$$\operatorname{sign} C_{k+1} = \operatorname{sign}[-C_k], \qquad k = 1, 2, \ldots, r - 1.$$

So the last form for the error can be written as two sums, each with coefficients of the same sign:

$$\text{(6c)} \qquad E_{n+1}\{f\} = \begin{cases} \dfrac{1}{(n+1)!}\{C_1 f^{(n+1)}(\zeta_1) + C_3 f^{(n+1)}(\zeta_3) + \cdots\} \\[2ex] + \dfrac{1}{(n+1)!}\{C_2 f^{(n+1)}(\zeta_2) + C_4 f^{(n+1)}(\zeta_4) + \cdots\}. \end{cases}$$

To simplify further, we require the following:

LEMMA 1. *Let $g(x)$ be a continuous function in $[a, b]$ and let $\alpha_1, \alpha_2, \ldots, \alpha_n$ be any set of non-negative numbers such that*

$$\sum_{k=1}^{n} \alpha_k = A.$$

Then for each set of n points $x_k \in [a, b]$ there exists a $\xi \in [a, b]$ such that

$$\sum_{k=1}^{n} \alpha_k g(x_k) = Ag(\xi).$$

Proof. Since $g(x)$ is continuous on a closed interval, it has there a finite maximum, M, a finite minimum, m, and actually takes on these values and all intermediate values as x ranges over $[a, b]$. Thus for each of the x_k

$$m \leq g(x_k) \leq M, \qquad k = 1, 2, \ldots, n.$$

Since the numbers α_k are non-negative, this implies

$$\alpha_k m \leq \alpha_k g(x_k) \leq \alpha_k M.$$

Sum these inequalities for $k = 1, 2, \ldots, n$, to find

$$Am \leq \sum_{k=1}^{n} \alpha_k g(x_k) \leq AM.$$

Hence the value of the sum must be equal to $Ag(\xi)$ for some $\xi \in [a, b]$. ∎

It should be observed that this lemma is analogous to the mean value theorem for integrals and that this proof copies the usual proof of that theorem.

Returning to (6c) we now have, by an obvious application of Lemma 1 to each of the sums in brackets

$$\text{(7a)} \qquad E_{n+1}\{f\} = \frac{1}{(n+1)!}\,[K_o f^{(n+1)}(\zeta_o) - K_e f^{(n+1)}(\zeta_e)],$$

where

$$\text{(7b)} \qquad K_o \equiv C_1 + C_3 + \cdots, \qquad K_e \equiv -C_2 - C_4 - \cdots,$$

$$x_0 < \zeta_e,\, \zeta_o < x_n.$$

The constants K_e and K_o *have the same sign* and so some cancellation is suggested by (7a). In fact, since

$$|K_o - K_e| = |\sum_{j=1}^{r} C_j| = |\int_a^b \omega_n(x)\,dx|,$$

the error expression (7a) formally reduces to (5c) if $\zeta_o = \zeta_e = \xi$. In general, ζ_o and ζ_e are unknown points and $f^{(n+1)}(x)$ may change sign in (x_0, x_n) so that the above reduction in the error may not occur. However, there are important special cases where, in fact, this maximum suggested cancellation does occur. We shall consider them later (see Theorem 2).

If the interpolation points or nodes are equally spaced, the above results can be modified to exhibit the dependence of the error on the spacing of the points. That is, let the interpolation points be of the form

$$x_j = x_0 + jh, \qquad j = 0, 1, \ldots, n,$$

and introduce the change of variable from x to t,

$$x = x_0 + th.$$

Now, from (3.10) of Chapter 6, we have

$$\omega_n(x) = h^{n+1}\pi_n(t),$$

and the integrals C_k in (6b) can be written as

$$C_k = h^{n+2}B_k$$

where

$$B_k = \int_{i+k-1}^{i+k} \pi_n(t)\,dt, \quad k = 2, 3, \ldots, r-1; \quad B_1 = \int_{t_a}^{i+1} \pi_n(t)\,dt;$$

$$B_r = \int_{i+r-1}^{t_b} \pi_n(t)\,dt.$$

The limits of integration t_a and t_b are given by

$$t_a = \frac{a - x_0}{h}, \qquad t_b = \frac{b - x_0}{h};$$

and lie in the interval

$$i \leq t_a < i + 1, \qquad i + r - 1 < t_b \leq i + r.$$

Since $\pi_n(t)$ is of one sign in these intervals, B_1 and B_r can be bounded by

$$|B_1| \leq \left| \int_i^{i+1} \pi_n(t)\,dt \right|, \qquad |B_r| \leq \left| \int_{i+r-1}^{i+r} \pi_n(t)\,dt \right|,$$

and these bounds are independent of h. By using these results in (7), we obtain the error representation

$$\text{(8a)} \qquad E_{n+1}\{f\} = \frac{h^{n+2}}{(n+1)!}\,[L_o f^{(n+1)}(\zeta_o) - L_e f^{(n+1)}(\zeta_e)],$$

where $x_0 < \zeta_e, \zeta_o < x_n$ and

$$\text{(8b)} \qquad L_o \equiv B_1 + B_3 + \cdots, \qquad L_e \equiv -B_2 - B_4 - \cdots.$$

The constants L_o and L_e have the same sign and, in fact,

$$|L_o - L_e| = \left| \sum_{j=1}^{r} B_j \right| = \left| \int_{t_a}^{t_b} \pi_n(t)\,dt \right| \leq \int_{t_a}^{t_b} |\pi_n(t)|\,dt.$$

We have thus shown, in general, that *if* $f^{n+1}(x)$ *is continuous in the smallest interval*, $[c, d]$, *containing all the* x_j, a *and* b, *then an interpolatory quadrature formula which uses* $n + 1$ *equally spaced nodes, of spacing* h, *has an error of the form* (8). Furthermore, by (8) or (4b):

$$|E_{n+1}\{f\}| \leq \frac{h^{n+2}}{(n+1)!} \int_{t_a}^{t_b} |\pi_n(t)| \, dt \max_{\xi \in [c,d]} |f^{(n+1)}(\xi)|.$$

This estimate is valid independent of the location of the nodes relative to the interval of integration. We give the analogue of formula (5c) which is valid in *the special case that* (a, b) *contains none of the uniformly spaced points* $\{x_i\}$:

$$\text{(8c)} \qquad E_{n+1}\{f\} = \frac{h^{n+2}}{(n+1)!} f^{(n+1)}(\zeta) \int_{t_a}^{t_b} \pi_n(t) \, dt, \qquad \zeta \in (c, d).$$

In the next subsection, we treat the Newton-Cotes formulae, and we find representations of $E_{n+1}\{f\}$ that are of the form (5c) or (8c), *even though* $\omega_n(x)$ *changes sign in* (a, b).

1.1. Newton-Cotes Formulae

Let the interpolation points, x_j, be equally spaced, say as before,

$$\text{(9a)} \qquad x_j = x_0 + jh, \qquad j = 0, 1, \ldots, n;$$

but now let the endpoints of the interval of integration be placed such that

$$\text{(9b)} \qquad x_0 = a, \quad x_n = b, \quad h = \frac{b-a}{n}.$$

With this choice of nodes the quadrature formula (3) as an approximation to the integral (0.1) is called a *closed Newton-Cotes formula.* Note that all of the nodes are in the integration interval $[a, b]$ and the word "closed" means that the endpoints a and b are the extreme nodes of the formula (3).

To examine the error, we again introduce the change of variable $x = x_0 + th$ and obtain $\omega_n(x) = h^{n+1}\pi_n(t)$. Now, however, t ranges over the interval $[0, n]$ and so we deduce properties of $\omega_n(x)$ over $[a, b]$ analogous to those of $\pi_n(t)$ developed in Lemmas 3.1 and 3.2 of Chapter 6. With the notation

$$x_{n/2} = \frac{a+b}{2} = x_0 + \frac{n}{2}h,$$

these properties are restated in

LEMMA **2.**

$$\omega_n(x_{n/2} + \xi) = (-1)^{n+1}\omega_n(x_{n/2} - \xi). \qquad \blacksquare$$

LEMMA **3.** (a) *For* $a < \xi + h \leq x_{n/2}$ *and* $\xi \neq x_j$, $j = 0, 1, \ldots, n$:

$$|\omega_n(\xi + h)| < |\omega_n(\xi)|;$$

(b) *For* $x_{n/2} \leq \xi < b$ *and* $\xi \neq x_j$, $j = 0, 1, \ldots, n$:

$$|\omega_n(\xi)| < |\omega_n(\xi + h)|. \qquad \blacksquare$$

Let us introduce the functions

$$\Omega_n(x) \equiv \int_a^x \omega_n(\xi)\, d\xi, \qquad n = 1, 2, \ldots, \tag{10}$$

which will be used to estimate the error in the closed Newton-Cotes formulae. For these functions, we have

LEMMA **4.** *For n even*

(a) $\Omega_n(a) = \Omega_n(b) = 0$;
(b) $\Omega_n(x) > 0, \qquad a < x < b.$

Proof. From the definition (10) it follows that $\Omega_n(a) = 0$. Since n is even, by Lemma 2, the integrand in $\Omega_n(b)$ is antisymmetric about the midpoint of the interval of integration and hence $\Omega_n(b) = 0$.

For part (b) we observe that $a, x_1, x_2, \ldots, x_{n-1}, b$ are the only zeros of $\omega_n(x)$, and hence $\omega_n(x) < 0$ for $x < a$ (since $\omega_n(x)$ is of odd degree). Then $\omega_n(x) > 0$ for $a < x < x_1$ and thus

$$\Omega_n(x) > 0 \qquad \text{for } a < x \leq x_1.$$

But by Lemma 3, we see that the negative contribution of $\omega_n(x)$ over $[x_1, x_2]$ to $\Omega_n(x)$ is in magnitude less than the positive contribution over $[a, x_1]$. Therefore,

$$\Omega_n(x) > 0 \qquad \text{for } a < x < x_2,$$

This argument can be repeated to cover the interval $a < x < x_{n/2}$. For $x > x_{n/2}$, Lemma 2 is employed. $\blacksquare$

Notice that these arguments can be used to yield

LEMMA **5.** *For n odd:*

(a) $\Omega_n(a) = 0, \qquad \Omega_n(b) = 2\Omega_n(x_{n/2})$;
(b) $\Omega_n(x) < 0, \qquad a < x \leq b.$ $\blacksquare$

However, we shall not require this lemma in our analysis of the error in quadrature formulae.

We are now prepared to estimate the error, E_{n+1}, given by (4a), for the closed Newton-Cotes formulae. We first treat the case of n even and assume that the integrand, $f(x)$, has $n + 2$ continuous derivatives. By using (10), integration by parts [note that the continuity of $(d/dx)f[x_0, x_1, \ldots, x_n, x]$ is assured by Problem 1.6 of Chapter 6], and Lemma 4a, the error is

$$\begin{aligned} E_{n+1}\{f\} &= \int_a^b \frac{d\Omega_n(x)}{dx} f[x_0, \ldots, x_n, x]\,dx \\ &= \Omega_n(x)f[x_0, \ldots, x_n, x]\Big|_a^b - \int_a^b \Omega_n(x)\frac{d}{dx}f[x_0, \ldots, x_n, x]\,dx \\ &= -\int_a^b \Omega_n(x)\frac{d}{dx}f[x_0, \ldots, x_n, x]\,dx. \end{aligned}$$

Hence,

$$E_{n+1}\{f\} = -\int_a^b \Omega_n(x)\frac{f^{(n+2)}(\xi(x))}{(n+2)!}\,dx$$

(from Problem 1.7 and Corollary 2 to Theorem 1.2, all of Chapter 6).

Now

$$f^{(n+2)}(\xi(x)) \equiv \frac{d}{dx}f[x_0, \ldots, x_n, x](n+2)!$$

is continuous by Problem (1.6) of Chapter 6. By Lemma 4b, $\Omega_n(x) \geq 0$. Hence, we may apply the mean value theorem for integrals in the above to get

$$E_{n+1}\{f\} = -\frac{f^{(n+2)}(\eta)}{(n+2)!}\int_a^b \Omega_n(x)\,dx, \qquad a < \eta < b.$$

In addition, integration by parts and Lemma 4 yield

$$\begin{aligned} \int_a^b \Omega_n(x)\,dx &= x\Omega_n(x)\Big|_a^b - \int_a^b x\frac{d}{dx}\Omega_n(x)\,dx \\ &= -\int_a^b x\omega_n(x)\,dx > 0. \end{aligned}$$

These results have established

THEOREM 1. *Let the points of* (9) *divide* $[a, b]$ *into an even number of equal intervals. Let* $f(x)$ *have a continuous derivative of order* $n + 2$ *on* $[a, b]$. *Then the error,* (4a), *in the closed Newton-Cotes quadrature formula,* (3), *for* n *even is*

$$E_{n+1}\{f\} = \frac{K_n}{(n+2!)}f^{(n+2)}(\eta), \qquad a < \eta < b;$$

where

$$K_n \equiv \int_a^b x\omega_n(x)\, dx < 0. \qquad \blacksquare$$

We deduce from this theorem the interesting result that the closed Newton-Cotes formula with an even number, n, of intervals has degree of precision $n + 1$ (even though the interpolation polynomial employed is of degree n).

To treat the case of odd n, we could employ Lemma 5. This would lead to an error expression containing two terms involving different order derivatives of $f(x)$. However, to obtain the simpler form of Theorem 1 we first recall that $\omega_n(x)$ does not change sign in $[b - h, b]$. Then (4a) yields by the mean value theorem for integrals and (1.9) of Chapter 6,

$$\begin{aligned} E_{n+1}\{f\} &= \int_a^{b-h} \omega_n(x)f[x_0, \ldots, x_n, x]\, dx + \int_{b-h}^b \omega_n(x)f[x_0, \ldots, x_n, x]\, dx \\ &= \int_a^{b-h} \omega_n(x)f[x_0, \ldots, x_n, x]\, dx + \frac{f^{(n+1)}(\xi')}{(n+1)!}\int_{b-h}^b \omega_n(x)\, dx, \\ &\qquad b - h < \xi' < b. \end{aligned}$$

To treat the first integral, we write

$$\omega_n(x) = \omega_{n-1}(x)(x - x_n) \quad \text{and} \quad \Omega_{n-1}(x) = \int_a^x \omega_{n-1}(\xi)\, d\xi.$$

Then the properties of divided differences given in (1.5) and (1.6) of Chapter 6 permit

$$\begin{aligned} \int_a^{b-h} \omega_n(x)f[x_0, \ldots, x_n, x]\, dx = \int_a^{b-h} \frac{d\Omega_{n-1}(x)}{dx} (f[x_0, \ldots, x_{n-1}, x] \\ - f[x_0, \ldots, x_n])\, dx \end{aligned}$$

Now $n - 1$ is even, and so $\Omega_{n-1}(a) = \Omega_{n-1}(b - h) = 0$, or

$$\int_a^{b-h} \frac{d\Omega_{n-1}(x)}{dx} dx = 0.$$

Hence we may neglect the integral involving the constant $f[x_0, \ldots, x_n]$. For the remaining integral, an integration by parts and application of the mean value theorem for integrals as before yield

$$\int_a^{b-h} \omega_n(x)f[x_0, \ldots, x_n, x]\, dx = -\frac{f^{(n+1)}(\xi'')}{(n+1)!}\int_a^{b-h} \Omega_{n-1}(x)\, dx, \qquad a < \xi'' < b - h.$$

Thus we have deduced that

$$E_{n+1}\{f\} = -[Af^{(n+1)}(\xi') + Bf^{(n+1)}(\xi'')],$$

where

$$A = -\frac{1}{(n+1)!}\int_{b-h}^{b} \omega_n(x)\,dx, \qquad B = \frac{1}{(n+1)!}\int_a^{b-h} \Omega_{n-1}(x)\,dx.$$

However, since $x = b$ is the largest zero of $\omega_n(x)$ and $\omega_n(x) > 0$ for $x > b$, it follows that $\omega_n(x) \leq 0$ in $[b - h, b]$, and so $A > 0$. That $B > 0$ follows from Lemma 4, since $n - 1$ is even. Thus, if $f^{(n+1)}(x)$ is continuous on $[a, b]$ an application of Lemma 1 implies that there exists a point ξ in $[\xi', \xi'']$ such that

$$E_{n+1}\{f\} = -(A + B)f^{(n+1)}(\xi).$$

Since

$$\omega_n(x) = \frac{d\Omega_{n-1}(x)}{dx}(x - b),$$

we have through integration by parts and Lemma 4

$$\int_a^{b-h} \omega_n(x)\,dx = \Omega_{n-1}(x)(x - b)\Big|_a^{b-h} - \int_a^{b-h} \Omega_{n-1}(x)\,dx$$

$$= -\int_a^{b-h} \Omega_{n-1}(x)\,dx.$$

Thus

$$A + B = -\frac{1}{(n+1)!}\int_a^b \omega_n(x)\,dx.$$

In summary, we have

THEOREM **2**. *If the points of* (9) *divide* $[a, b]$ *into an odd number of equal intervals and* $f(x)$ *has a continuous derivative of order* $(n + 1)$ *on* $[a, b]$ *then the error,* (4a), *in the closed Newton-Cotes quadrature formula* (3), *for n odd is*

$$E_{n+1}\{f\} = \frac{K_n}{(n+1)!}f^{(n+1)}(\xi), \qquad a < \xi < b;$$

where

$$K_n = \int_a^b \omega_n(x)\,dx < 0. \qquad \blacksquare$$

The formula covered by this theorem has degree of precision n. Note that the result in Theorem 2 is formally similar to that in (5c).

To express the dependence of the errors given in Theorems 1 and 2 on the interval size, h, we use the change of variable, $x = x_0 + th$, and find

COROLLARY. *Under the hypotheses of Theorems* 1 *and* 2, *respectively,*

$$
(11)\quad E_{n+1}\{f\} = \begin{cases} \dfrac{M_n}{(n+2)!} h^{n+3} f^{(n+2)}(\xi), & M_n \equiv \displaystyle\int_0^n t\pi_n(t)\,dt < 0, \\ & n \text{ even;} \\ \dfrac{M_n}{(n+1)!} h^{n+2} f^{(n+1)}(\xi), & M_n \equiv \displaystyle\int_0^n \pi_n(t)\,dt < 0, \\ & n \text{ odd.} \end{cases}
$$
■

Since the closed formulae are exact for polynomials of degree at most $n + 1$ when $n + 1$ is odd, and are exact for polynomials of degree at most n when $n + 1$ is even, it is generally preferable to employ the odd formulae, i.e., those with an odd number of interpolation points. Also, it clearly does not pay, in general, to add one point to a scheme with even n; rather, points should be added in pairs.

Another useful integration formula with equal intervals is found by using the points

$$(12a)\qquad x_j = x_0 + jh, \qquad j = 0, 1, \ldots, n;$$

where

$$(12b)\qquad h = \frac{b-a}{n+2}, \quad x_0 = a + h, \qquad x_n = b - h.$$

The endpoints are then labeled $x_{-1} = a$ and $x_{n+1} = b$. Since we do not employ the endpoints in formula (3), it is now called an *open Newton-Cotes formula*. In this procedure, all $n + 1$ points of interpolation are interior to the interval of integration.

To examine the error we introduce, in place of $\Omega_n(x)$, the functions $J_n(x)$ defined by

$$(13)\qquad J_n(x) = \int_a^x \omega_n(\xi)\,d\xi, \qquad n = 1, 2, \ldots.$$

These differ from the functions in (10) since now $a < x_0$ and $x_n < b$. However, as in the proof of Lemma 4, it follows that for n even

$$J_n(a) = J_n(b) = 0; \quad J_n(x) < 0, \quad a < x < b.$$

Then, exactly as in the derivation of Theorem 1, we have

THEOREM **1′**. *Replace "(9)" by "(12)" and "closed" by "open" in the statement of Theorem* 1. *Then the formula for $E_{n+1}\{f\}$ becomes*

$$E_{n+1}\{f\} = \frac{K_n'}{(n+2)!} f^{(n+2)}(\xi), \qquad a < \xi < b;\ n \text{ even},$$

where

$$K_n' = \int_a^b x\omega_n(x)\,dx > 0.$$
■

Similarly, for n odd, by procedures analogous to those for closed-type formulae, we find

THEOREM **2′**. *Use the hypothesis of Theorem* 1′, *but with n odd. Then*

$$E_{n+1}\{f\} = \frac{K_n'}{(n+1)!} f^{(n+1)}(\xi), \qquad a < \xi < b; \;\; n \text{ odd},$$

where

$$K_n' = \int_a^b \omega_n(x)\, dx > 0. \qquad \blacksquare$$

These errors for the open formula may be expressed in terms of the spacing h as

COROLLARY. *Under the hypothesis of Theorems* 1′ *and* 2′, *respectively,*

$$E_{n+1}\{f\} = \frac{M_n'}{(n+2)!} h^{n+3} f^{(n+2)}(\xi),$$

(14a)

$$M_n' = \int_{-1}^{n+1} t\pi_n(t)\, dt > 0, \qquad n \text{ even};$$

$$E_{n+1}\{f\} = \frac{M_n'}{(n+1)!} h^{n+2} f^{(n+1)}(\xi),$$

(14b)

$$M_n' = \int_{-1}^{n+1} \pi_n(t)\, dt > 0, \qquad n \text{ odd}. \qquad \blacksquare$$

Again we find that for n even, the degree of precision is $n + 1$, while for n odd it is only n. A comparison of (11) and (14) indicates that only the values of the coefficients M_n and M_n' differ in the form of the error estimates for open and closed Newton-Cotes formulae based on the same number, $n + 1$, of nodes. [However, for any fixed number of intervals in $[a, b]$ say m, the closed formulae use $m + 1$ nodes and the open formulae use $m - 1$ nodes. Hence, the closed method has a degree of precision two more than the open method on this basis, but requires two more evaluations of the function $f(x)$.]

There are useful quadrature formulae that are neither open nor closed but which have uniformly spaced nodes [e.g., see Problems 2, 3, and 5]. The formulae of Problems 2 and 3 are the basis for Adam's method for the numerical solution of ordinary differential equations (see Table 2.1 of Chapter 8).

1.2. Determination of the Coefficients

Let the interpolation points be equally spaced, say of the form $x_j = x_0 + jh$, $j = 0, 1, \ldots, n$, and let the endpoints a, b of the integration interval also be of this form, say $a = x_p = x_0 + ph$ and $b = x_q = x_0 + qh$, (but not necessarily interpolation points). Then the coefficients, $w_{n,j}$, for quadrature formulae of the form (3) can be written as, using $x = x_0 + th$,

$$w_{n,j} = \int_a^b \phi_{n,j}(x)\,dx$$

$$= h \int_p^q \prod_{\substack{k=0 \\ (k \neq j)}}^{n} \frac{t-k}{j-k}\,dt$$

$$= (-1)^{n-j} \frac{h}{n!} \binom{n}{j} \int_p^q \frac{\pi_n(t)}{t-j}\,dt, \qquad j = 0, 1, \ldots, n.$$

Thus if we define the quantities

$$\text{(15a)} \qquad A_{n,j}(p, q) = \frac{(-1)^{n-j}}{n!} \binom{n}{j} \int_p^q \frac{\pi_n(t)}{t-j}\,dt, \qquad j = 0, 1, \ldots, n;$$

the coefficients are simply

$$\text{(15b)} \qquad w_{n,j} = hA_{n,j}(p, q).$$

In the special case of the closed Newton-Cotes schemes we have $p = 0$ and $q = n$; for the open schemes of Section 1.1, $p = -1$ and $q = n + 1$. It should be noted that the $A_{n,j}(p, q)$ are independent of the spacing, h, and thus may be tabulated as functions of the parameters n, j, p, and q. Appropriate coefficients for a particular spacing, h, are then determined by using (15b). Since the integrand in (15a) is a polynomial of degree n the $A_{n,j}(p, q)$ will all be rational numbers for rational p and q. Further, we note that $A_{n,j}(0, n) = A_{n,n-j}(0, n)$ and more generally, $A_{n,j}(-r, n + r) = A_{n,n-j}(-r, n + r)$ for any r. For p and q of these important special forms, we need only tabulate the values for $j \leq n/2$. Tables 1 and 2 list the simplest closed and open Newton-Cotes formulae. The coefficients $A_{n,j}(n, n + 1)$ are to be found in Problem 6, for $n = 1, 2, 3$, and 4.

An alternative indirect procedure, called the *method of undetermined coefficients*, can also be employed to determine the coefficients for the quadrature formula (3). This method is quite practical for unequally spaced nodes as well as for fairly large values of n. In addition, it can be used to prove

Table 1 *Closed Newton-Cotes Formulae*

$$\int_{x_0}^{x_1} f(x)\,dx = \frac{h}{2}(f_0 + f_1) - \frac{h^3}{12} f^{(2)}(\xi), \qquad x_0 < \xi < x_1, \qquad \text{(trapezoidal rule)}$$

$$\int_{x_0}^{x_2} f(x)\,dx = \frac{h}{3}(f_0 + 4f_1 + f_2) - \frac{h^5}{90} f^{(4)}(\xi), \qquad x_0 < \xi < x_2, \qquad \text{(Simpson's rule)}$$

$$\int_{x_0}^{x_3} f(x)\,dx = \frac{3h}{8}(f_0 + 3f_1 + 3f_2 + f_3) - \frac{3h^5}{80} f^{(4)}(\xi), \qquad x_0 < \xi < x_3.$$

$$\int_{x_0}^{x_4} f(x)\,dx = \frac{2h}{45}(7f_0 + 32f_1 + 12f_2 + 32f_3 + 7f_4) - \frac{8h^7}{945} f^{(6)}(\xi), \qquad x_0 < \xi < x_4.$$

$$\int_{x_0}^{x_5} f(x)\,dx = \frac{5h}{288}(19f_0 + 75f_1 + 50f_2 + 50f_3 + 75f_4 + 19f_5) - \frac{275h^7}{12096} f^{(6)}(\xi), \qquad x_0 < \xi < x_5.$$

Table 2 *Open Newton-Cotes Formulae*

$$\int_{x_0}^{x_2} f(x)\,dx = 2hf_1 + \frac{h^3}{3} f^{(2)}(\xi), \qquad x_0 < \xi < x_2, \qquad \text{(midpoint rule)}$$

$$\int_{x_0}^{x_3} f(x)\,dx = \frac{3h}{2}(f_1 + f_2) + \frac{h^3}{4} f^{(2)}(\xi), \qquad x_0 < \xi < x_3.$$

$$\int_{x_0}^{x_4} f(x)\,dx = \frac{4h}{3}(2f_1 - f_2 + 2f_3) + \frac{28h^5}{90} f^{(4)}(\xi), \qquad x_0 < \xi < x_4.$$

$$\int_{x_0}^{x_5} f(x)\,dx = \frac{5h}{24}(11f_1 + f_2 + f_3 + 11f_4) + \frac{95h^5}{144} f^{(4)}(\xi), \qquad x_0 < \xi < x_5.$$

$$\int_{x_0}^{x_6} f(x)\,dx = \frac{6h}{20}(11f_1 - 14f_2 + 26f_3 - 14f_4 + 11f_5) + \frac{41h^7}{140} f^{(6)}(\xi), \qquad x_0 < \xi < x_6.$$

$$\int_{x_0}^{x_7} f(x)\,dx = \frac{7h}{1440}(611f_1 - 453f_2 + 562f_3 + 562f_4 - 453f_5 + 611f_6) + \frac{5257h^7}{8640} f^{(6)}(\xi), \qquad x_0 < \xi < x_7.$$

THEOREM 3. *A quadrature formula which uses $n + 1$ distinct nodes is an interpolatory formula iff it has degree of precision at least n.*

Proof. The necessity has been demonstrated by equation (4b). To prove sufficiency, we let the $n + 1$ distinct points x_j, $j = 0, 1, \ldots, n$, be the given nodes. If the quadrature formula, with these points and coefficients α_j, has degree of precision at least n in approximating $\int_a^b f(x)\,dx$, then we must have

$$\begin{aligned}
\sum_{j=0}^{n} \alpha_j &= (b - a) \\
\sum_{j=0}^{n} \alpha_j x_j &= \tfrac{1}{2}(b^2 - a^2) \\
&\ \vdots \\
\sum_{j=0}^{n} \alpha_j x_j^{\,n} &= \frac{1}{n+1}(b^{n+1} - a^{n+1}).
\end{aligned} \tag{16}$$

This may be considered as a system of $n + 1$ equations for the determination of the $n + 1$ coefficients, α_j. Since the coefficient matrix of the system (16) is of the Vandermonde form, and the x_j are distinct, there exists a *unique solution*. On the other hand, note that the interpolatory formula (3), with the same points x_j, is exact when applied to the powers $1, x, \ldots, x^n$. Hence, the system (16) is satisfied with the α_j replaced by the $w_{n,j}$. The fact that (16) has a unique solution shows $\alpha_j = w_{n,j}$. ∎

We shall see that most of the popular quadrature formulae are interpolatory. This follows by means of Theorem 3 and its extension, in Section 4, to weighted quadrature formulae.

The *method of undetermined coefficients* consists in solving the system (16) for the α_j. To determine the degree of precision of an interpolatory quadrature fomula, we simply form the quantities

$$E_{n+1}\{x^k\} \equiv \frac{1}{k+1}(b^{k+1} - a^{k+1}) - \sum_{j=0}^{n} \alpha_j x_j^{\,k}, \qquad k = n + 1, n + 2, \ldots;$$

and determine the least integer k such that $E_{n+1}\{x^k\} \neq 0$. The degree of precision is then $k - 1$.

If it is known that the error in the integration formula has the form $E_{n+1}\{f\} = A_m f^{(m)}(\xi)$ for some integer m, then we can determine the coefficient A_m by this method. That is, we must have $m = k$ where $k - 1$ is the determined degree of precision. Hence, use $f(x) \equiv x^k$ to get

$$E_{n+1}\{x^k\} = \frac{1}{k+1}(b^{k+1} - a^{k+1}) - \sum_{j=0}^{n} \alpha_j x_j^{\,k} = k!\, A_k, \tag{17}$$

which can be solved for A_k. For example, in (11) for the closed Newton-Cotes formulae, we may use the quantities A_{n+2} or A_{n+1} to evaluate the appropriate coefficient M_n.

As an illustration of the application of the method of undetermined coefficients, we consider the closed formula with one segment, $n = 1$, and the two nodes $x_0 = a$ and $x_1 = b$. The system (16) now becomes

$$\begin{aligned}
\alpha_0 + \alpha_1 &= (b - a); \\
a\alpha_0 + b\alpha_1 &= \tfrac{1}{2}(b^2 - a^2);
\end{aligned}$$

and the solution of this system is

$$\alpha_0 = \alpha_1 = \tfrac{1}{2}(b - a).$$

To determine the degree of precision we apply the formula to $x^2, x^3, \ldots,$ and get first

$$E_2\{x^2\} = \tfrac{1}{3}(b^3 - a^3) - \tfrac{1}{2}(b - a)(a^2 + b^2)$$
$$= -\tfrac{1}{6}(b - a)^3 \neq 0.$$

Thus the degree of precision is 1, as we knew since it is a closed Newton-Cotes formula with $n = 1$. The error can be written as

$$E_2\{f\} = A_2 f^{(2)}(\xi)$$

where $A_2 = (M_1/2!)h^3$ and $h = b - a$. By using $f(x) = x^2$ we find that

$$E_2\{x^2\} = -\tfrac{1}{6}(b - a)^3 = 2A_2.$$

Thus $A_2 = -\frac{1}{12}(b - a)^3$ and $M_1 = -\frac{1}{6}$. The formula determined above is the familiar *trapezoidal rule* which can now be written as

$$(18) \qquad \int_a^b f(x)\,dx = \frac{b - a}{2}[f(a) + f(b)] - \frac{(b - a)^3}{12} f^{(2)}(\xi), \qquad a < \xi < b.$$

PROBLEMS, SECTION 1

1. Add to the hypothesis of Lemma 1: r of the coefficients (α_k) are non-zero. What is the smallest integer r for which the stronger conclusion $\xi \in (a, b)$ is valid?

2. From equation (4.19) of Chapter 6, derive the formula

$$\int_x^{x+h} f(\xi)\,d\xi = h[d_0 f(x) + d_1 \Delta f + d_2 \Delta^2 f + \cdots + d_m \Delta^m f] + h^{m+2} d_{m+1} f^{(m+1)}(\xi)$$

where $d_0 = 1$, and recursively

$$d_k - \frac{d_{k-1}}{2} + \frac{d_{k-2}}{3} + \cdots + (-1)^k \frac{d_0}{k + 1} = 0, \qquad k = 1, 2, \ldots, m + 1;$$

$$x < \xi < x + mh; \qquad \Delta^k f \equiv \sum_{j=0}^{k} (-1)^{k-j} \binom{k}{j} f(x + jh).$$

[Hint: The coefficients $\{d_i\}$ satisfy the identity in Δ,

$$\Delta \equiv \log(I + \Delta)(d_0 I + d_1 \Delta + \cdots + d_m \Delta^m + \cdots).]$$

Check the following listing:

k	0	1	2	3	4	5
d_k	1	$\frac{1}{2}$	$-\frac{1}{12}$	$\frac{1}{24}$	$-\frac{19}{720}$	$\frac{3}{160}$

3. From equation (4.19) of Chapter 6, derive the formula

$$\int_{x-h}^{x} f(\xi)\,d\xi = h[e_0 f(x) + e_1\Delta f + e_2\Delta^2 f + \cdots + e_m\Delta^m f] + h^{m+2}e_{m+1}f^{(m+1)}(\eta), \qquad \text{where } e_0 = 1,$$

and recursively,

$$e_k - \frac{e_{k-1}}{2} + \frac{e_{k-2}}{3} + \cdots + (-1)^k\frac{e_0}{k+1} = (-1)^k, \qquad k = 1, 2, \ldots, m+1;$$

$$x - h < \eta < x + mh; \qquad \Delta \text{ as defined in Problem 2.}$$

[Hint: The coefficients $\{e_i\}$ satisfy the identity

$$\Delta - \Delta^2 + \Delta^3 + \cdots + (-1)^m\Delta^{m+1} + \cdots \equiv \left(\Delta - \frac{\Delta^2}{2} + \frac{\Delta^3}{3} + \cdots\right)(e_0 I + e_1\Delta + e_2\Delta^2 + \cdots).]$$

Check the following listing:

k	0	1	2	3	4	5
e_k	1	$-\frac{1}{2}$	$\frac{5}{12}$	$-\frac{3}{8}$	$\frac{251}{720}$	$\frac{-95}{288}$

4. With $\{d_i\}$ and $\{e_i\}$ as defined in Problems 2 and 3, and $e_{-1} = 0$, show that

$$d_m = e_m + e_{m-1}.$$

5. From Everett's form of the interpolation polynomial, (3.17) of Chapter 6, define the coefficients $\{r_i\}$ and C_m of the formula

$$\frac{1}{h}\int_{x_0}^{x_1} f(x)\,dx = r_0(f_0 + f_1) + r_1(\delta^2 f_0 + \delta^2 f_1) + \cdots + r_m(\delta^{2m}f_0 + \delta^{2m}f_1) + C_m h^{2m+2}f^{(2m+2)}(\xi),$$

where

$$x_{-m} < \xi < x_{m+1}, \qquad x_j = x_0 + jh.$$

Explicitly find r_0 and r_1.

6. Make a table listing the coefficients $A_{n,j}(n, n+1)$, for $n = 1, 2, 3$, and 4. [Hint: Use the result of Problem 3.]

7. Prove Lemma 5. [Hint: Let $n = 2m + 1$ and $0 < \epsilon \leq \frac{1}{2}$. Show that $|\pi_n(t+1)/\pi_n(t)| < 1$ for $t = m - \epsilon$.]

2. ROUNDOFF ERRORS AND UNIFORM COEFFICIENT FORMULAE

In almost all practical applications of integration formulae of the form (0. 2) the exact function values, $f(x)$, will not be available. This fact is usually due to limitations in the calculation of these function values (or in their measurement). Thus the quantities actually employed may be written as $\tilde{f}(x_j) = f(x_j) + \rho_j$ where ρ_j is the local roundoff error made in computing (or error in measuring) $f(x_j)$. The error in approximating (0.1) by (0.2) with these values is then

$$\int_a^b f(x)\,dx - \sum_{j=1}^{n} \alpha_j \tilde{f}(x_j) = E_n\{f\} - R_n\{f\},$$

where $E_n\{f\} = I\{f\} - I_n\{f\}$ is the quadrature or *truncation error* and the *accumulated roundoff error* is:

(1) $$R_n\{f\} = \sum_{j=1}^{n} \alpha_j \rho_j.$$

If, as is frequently the case, we know that the local errors are bounded, say $|\rho_j| \leq \rho$ for $j = 1, 2, \ldots, n$, then

(2) $$|R_n\{f\}| \leq \rho \sum_{j=1}^{n} |\alpha_j|.$$

Let us also assume that the formula (0.2) has degree of precision ≥ 0, i.e., that at least a constant is integrated correctly. Then with the integrand $f(x) \equiv 1$ we find

(3) $$\sum_{j=1}^{n} \alpha_j = (b - a).$$

Thus *if all the coefficients*, α_j, *are of one sign* the bound (2) becomes

(4) $$|R_n\{f\}| \leq \rho|b - a|.$$

If the coefficients are not all of one sign, then clearly,

$$\sum_{j=1}^{n} |\alpha_j| > \left|\sum_{j=1}^{n} \alpha_j\right| = |b - a|$$

and a larger maximum accumulated roundoff is possible. To attain the maximum value requires that

$$\rho_j = \rho \operatorname{sign} \alpha_j$$

which is, of course, very special. By comparing (4) and (2), we find a practical advantage in having the coefficients of a quadrature formula all of one sign, especially if the truncation error is smaller than the rounding error.

By introducing *statistical notions of roundoff* (or measurement) errors, we can, in fact, show that it is of even greater advantage to have all of the coefficients of the same value. There are several ways in which the statistical notion of "*randomness*" of the local errors can be introduced. For instance, suppose we ask for those coefficients, α_j, for which some measure of $R_n\{f\}$ is a minimum for all functions f, of some particular class, F. Since the errors $\rho_j = \rho_j\{f\}$ usually depend in an extremely complicated way on the functions f, direct attempts at such a minimization do not seem possible. However, as f varies over the class F the errors $\rho_j\{f\}$, can be expected to vary in an erratic manner. By making specific assumptions about the nature of this variation and introducing a measure of "volume" in F we can calculate various "averages" of $R_n\{f\}$ over F.

Specifically, let us consider for F a one parameter family of functions of x, say $f(x; \tau)$, where the parameter τ ranges over $0 \leq \tau \leq T$. The roundoff error in evaluating $f(x_k; \tau)$ for each τ and $k = 1, 2, \ldots, n$ will be denoted by $\rho_k(\tau)$. We assume that $|\rho_k(\tau)| \leq \rho$ and that all values in this range are equally likely for each value of τ; or in particular, we assume that the "average" roundoff over the family F vanishes; i.e., that

$$\text{(5a)} \qquad \bar{\rho}_k \equiv \frac{1}{T}\int_0^T \rho_k(\tau)\,d\tau = 0, \qquad k = 1, 2, \ldots, n.$$

Further, we assume that the roundoff errors at distinct points x_j and x_k are uncorrelated; i.e.,

$$\text{(5b)} \qquad \int_0^T \rho_j(\tau)\rho_k(\tau)\,d\tau = 0, \qquad \text{if } j \neq k.$$

In effect, this means that the error committed at x_j is independent of the error at x_k for all the functions, $f(x; \tau)$ in F. Finally, let us make the assumption that all the local errors have the same mean-square value, say,

$$\text{(5c)} \qquad \sigma^2 \equiv \frac{1}{T}\int_0^T \rho_j^2(\tau)\,d\tau, \qquad j = 1, 2, \ldots, n.$$

Note that $\sigma \leq \rho$.

We now consider some measures of the accumulated roundoff error for the family F. We define for any τ in $0 \leq \tau \leq T$:

$$R_n(\tau) = R_n\{f(x; \tau)\}$$
$$= \sum_{j=1}^{n} \alpha_j \rho_j(\tau).$$

The *mean-accumulated roundoff* for the family F is, by (5a),

$$\bar{R}_n \equiv \frac{1}{T}\int_0^T R_n(\tau)\,d\tau = \sum_{j=1}^{n} \alpha_j \bar{\rho}_j = 0.$$

Thus the coefficients, α_j, of the quadrature formula have no effect on the average accumulated roundoff. Next we compute the *mean-square roundoff error*, using (5b and c), to get

$$(6)\qquad \begin{aligned} r_n^2 &\equiv \frac{1}{T}\int_0^T R_n^2(\tau)\,d\tau \\ &= \sum_{i=1}^{n}\sum_{j=1}^{n} \alpha_i\alpha_j \frac{1}{T}\int_0^T \rho_i(\tau)\rho_j(\tau)\,d\tau \\ &= \sigma^2 \sum_{j=1}^{n} \alpha_j^2. \end{aligned}$$

The coefficients clearly have an effect on this measure of the roundoff error. Let us seek $\{\alpha_j\}$ which minimize the sum in the last line of (6) but which also satisfy (3). This problem is easily solved by using the method of Lagrange multipliers and we find for the minimizing coefficients:

$$(7)\qquad \alpha_1 = \alpha_2 = \cdots = \alpha_n = \frac{b-a}{n}.$$

Thus to reduce the *root mean-square roundoff error*, r_n, as much as possible, while retaining at least zero order precision, the coefficients should be equal. Using (7) in (6) yields for the minimum of r_n the value

$$(8)\qquad r_n = \frac{\sigma|b-a|}{\sqrt{n}}.$$

This result is somewhat surprising as it indicates that the root mean-square roundoff error actually decreases as the number of quadrature points (and hence of sources of error) increases! It should be noted that when the weights are equal the bound (4) applies. Compare the maximum bound, $\rho|b-a|$, with the statistical result in (8) to find a reduction in the dependence on n by the factor $1/\sqrt{n}$ for the statistical estimate. This is a common feature of statistical estimates of roundoff. It is frequently found in practice that the statistical estimate is a more realistic approximation of the error than is the maximum-type estimate.

These results can be interpreted in a slightly different, perhaps more familiar and intuitive way. We think of a family of calculations of the quadrature formula applied to the same function, $f(x)$. Each calculation of the family is determined by the particular rounding employed in the set of values $f(x_j)$, $j = 1, 2, \ldots, n$. That is, $\rho_j(\tau)$ is the roundoff error in the computation characterized by the parameter value τ when computing $f(x_j)$. Of course, any fixed program for an electronic computer is represented by a single value of the parameter. Thus, the intended interpretation is not repeating the same calculation, but rather, altering the computational procedure slightly each time. The above averages are then averages over an appropriate family of calculations. With this interpretation numerical experiments using "randomly" generated and independent rounding errors are easily devised.

We shall find that equal coefficient formulae of the form (0.2) cannot approximate integrals of the form (0.1), with degree of precision n for all values of n (see, however, Subsection 4.1).

2.1. Determination of Uniform Coefficient Formulae

An integration formula with uniform coefficients and n nodes has the simple form

$$I_n\{f\} \equiv \alpha_n \sum_{j=1}^{n} f(x_j). \tag{9a}$$

In order that this yield the exact result for

$$I\{f\} = \int_a^b f(x)\,dx$$

when $f(x) \equiv 1$, it is clear that

$$\alpha_n = \frac{b-a}{n}. \tag{9b}$$

We now try to determine the n nodes, x_j, such that (9) has as high a degree of precision as possible. Thus we impose the n conditions, $I_n\{x^\nu\} = I\{x^\nu\}$ for $\nu = 1, 2, \ldots, n$ to get

$$\alpha_n \sum_{j=1}^{n} x_j{}^\nu = \frac{1}{\nu+1}(b^{\nu+1} - a^{\nu+1}), \qquad \nu = 1, 2, \ldots, n. \tag{10}$$

If these equations have as solution n distinct real values, x_j, then the corresponding quadrature formula (9) has degree of precision at least n, while only n nodes are employed (as in the Newton-Cotes formulae for an even number of uniform intervals or odd number of nodes). Thus, by Theorem 1.3, we can conclude that such equal coefficient formulae are interpolatory. The error estimates of Section 1 are then applicable [say, of the form (1.4)].

Let us set $n = 1$ in (9) and (10). We find $\alpha_1 = b - a$, $x_1 = \frac{1}{2}(b + a)$, and the quadrature formula is simply the midpoint rule,

$$I_1\{f\} = (b-a)f\left(\frac{b+a}{2}\right),$$

which is clearly exact for linear functions.

For $n = 2$, the system (10) is easily reduced to a quadratic equation which yields the nodes

$$x_1 = \frac{b+a}{2} - \frac{b-a}{2\sqrt{3}}, \qquad x_2 = \frac{b+a}{2} + \frac{b-a}{2\sqrt{3}}.$$

Note that in each of these cases the nodes are symmetrically located about the center of the interval of integration.

In general, the solution of the system (10) determines the nth degree polynomial

$$P_n(x) \equiv (x - x_1)(x - x_2)\cdots(x - x_n)$$

whose roots are the required nodes. This polynomial can be written in the form

$$P_n(x) = x^n + \sigma_1 x^{n-1} + \sigma_2 x^{n-2} + \cdots + \sigma_n, \tag{11}$$

where the coefficients are the classical elementary symmetric functions of the roots, i.e.,

$$\begin{aligned} \sigma_1 &= -(x_1 + x_2 + \cdots + x_n) \\ \sigma_2 &= (x_1x_2 + x_1x_3 + \cdots + x_{n-1}x_n) \\ &\vdots \\ \sigma_n &= (-1)^n x_1x_2\cdots x_n. \end{aligned}$$

However, the values of these symmetric functions can be obtained from the sums of the powers of the roots

$$S_\nu \equiv x_1^\nu + x_2^\nu + \cdots + x_n^\nu, \qquad \nu = 1, 2, \ldots, n,$$

which are directly determined in (10). The relations between the σ_j and the S_ν are known as *Newton's identities* (see Problem 2),

$$\begin{aligned} &S_1 + \sigma_1 = 0 \\ &S_2 + S_1\sigma_1 + 2\sigma_2 = 0 \\ &\vdots \\ &S_n + S_{n-1}\sigma_1 + S_{n-2}\sigma_2 + \cdots \\ &\qquad + S_1\sigma_{n-1} + n\sigma_n = 0. \end{aligned} \tag{12}$$

Thus the determination of the nodes has been reduced to finding the roots of the polynomial (11). The coefficients of this polynomial are recursively computed from (12) by using the known values of the S_ν.

The nodes for any interval $[a, b]$ are easily obtained from the nodes for the special interval $[-1, 1]$. For this purpose we introduce the usual linear change of variable

(13) $$x = \frac{b+a}{2} + y\frac{b-a}{2},$$

and then note that

$$I\{f\} \equiv \int_a^b f(x)\,dx = \frac{b-a}{2}\int_{-1}^{1} g(y)\,dy \equiv \frac{b-a}{2} J\{g(y)\},$$

where

$$g(y) \equiv f\left(\frac{b+a}{2} + y\frac{b-a}{2}\right).$$

The n-point uniform coefficient quadrature formula which approximates $J\{g\}$ is written as

$$J_n\{g\} = \beta_n \sum_{j=1}^{n} g(y_j).$$

In order that this formula have degree of precision at least n, we must have $J\{y^\nu\} = J_n\{y^\nu\}$ for $\nu = 0, 1, \ldots, n$. Thus $\beta_n = 2/n$ and

(14) $$S_\nu \equiv \sum_{j=1}^{n} y_j^\nu = \frac{n}{2(\nu+1)}[1 + (-1)^\nu], \qquad \nu = 1, 2, \ldots, n.$$

We note that the odd order power sums now vanish, i.e., $S_1 = S_3 = \cdots = 0$. Newton's identities (12) become in this case

(15) $$\begin{aligned} \sigma_1 &= 0 \\ \frac{n}{3} + 2\sigma_2 &= 0 \\ \sigma_3 &= 0 \\ \frac{n}{5} + \frac{n}{3}\sigma_2 + 4\sigma_4 &= 0 \\ \sigma_5 &= 0 \\ &\vdots \end{aligned}$$

Thus we find that the odd order elementary symmetric functions vanish and the polynomial for the determination of the nodes, y_j, becomes

(16) $$P_n(y) = y^n + \sigma_2 y^{n-2} + \sigma_4 y^{n-4} + \cdots.$$

The roots, y_j, of $P_n(y) = 0$ are thus symmetric with respect to the origin and if n is odd, then $y = 0$ is a root. Using the transformation (13) we obtain for the nodes of the general quadrature formula (9a) the values

$$x_j = \frac{b+a}{2} + y_j\frac{b-a}{2}.$$

From the properties of the y_j it follows that the nodes x_j are symmetrically located with respect to the midpoint of the interval of integration and that for n odd the midpoint is a node. If n is an even integer then by the symmetry of the y_j we have

$$\sum_{j=1}^{n} y_j^{n+1} = 0.$$

That is, $J_n\{y^{n+1}\} = J\{y^{n+1}\}$ and the degree of precision is $n + 1$, when an even number, n, of nodes is employed. The same must, of course, be true in the general case (9) for n even.

In order to determine an n-point quadrature formula of the form (9) which has degree of precision at least n, the polynomial (16) must have n real distinct roots. However, for $n = 8$ and for all $n \geq 10$ it can be shown that a pair of complex roots occurs. For $n \leq 7$ and $n = 9$ the roots have the required properties and are known to many decimals. We list in Table 1 these roots in $0 \leq y \leq 1$; the others are obtained by symmetry.

Table 1.

	$n = 1$	$n = 2$	$n = 3$	$n = 4$
y_j	0.0	0.57735 02692	0.0	0.18759 24741
			0.70710 67812	0.79465 44723

	$n = 5$	$n = 6$	$n = 7$	$n = 9$
y_j	0.0	0.26663 54015	0.0	0.0
	0.37454 14096	0.42251 86538	0.32391 18105	0.16790 61842
	0.83249 74870	0.86624 68181	0.52965 67753	0.52876 17831
			0.88386 17008	0.60101 86554
				0.91158 93077

Quadrature formulae with uniform coefficients and any number of nodes arise in Subsection 4.1, in another setting.

PROBLEMS, SECTION 2

1. Verify that (7) characterizes the solution to the problem of minimizing $\sum_{j=1}^{n} \alpha_j^2$ subject to $\sum_{j=1}^{n} \alpha_j = b - a$.

2. Verify Newton's identities (12).

[Hint: Let $P_n(x) \equiv \prod_{j=1}^{n} (x - x_j) \equiv \sum_{k=0}^{n} \sigma_k x^{n-k}, \qquad \sigma_0 \equiv 1.$

$$P_n'(x) = \frac{d}{dx} \prod_{j=1}^{n} (x - x_j) = \sum_{j=1}^{n} \frac{P_n(x)}{x - x_j} = \sum_{j=1}^{n} \frac{P_n(x) - P_n(x_j)}{x - x_j}$$

But,

$$\begin{aligned}\frac{P_n(x) - P_n(x_j)}{x - x_j} &= \sum_{k=0}^{n} \sigma_k \left(\frac{x^{n-k} - x_j^{n-k}}{x - x_j} \right) \\ &= \sum_{k=0}^{n} \sigma_k (x^{n-k-1} + x_j x^{n-k-2} + \cdots + x_j^{n-k-2} x + x_j^{n-k-1}) \\ &= x^{n-1} + \sum_{p=1}^{n-1} (\sigma_p + \sigma_{p-1} x_j + \cdots + \sigma_1 x_j^{p-1} + x_j^p) x^{n-p-1}.\end{aligned}$$

Hence

$$P_n'(x) = nx^{n-1} + \sum_{p=1}^{n-1} (n\sigma_p + S_1 \sigma_{p-1} + \cdots + S_{p-1}\sigma_1 + S_p) x^{n-p-1}].$$

3. GAUSSIAN QUADRATURE; MAXIMUM DEGREE OF PRECISION

In Section 1 it is shown that, given $n + 1$ fixed nodes, we can determine the coefficients of a quadrature formula which has degree of precision at least n (by forming the appropriate interpolatory formula). In Subsection 2.1 we investigated the problem of determining n nodes such that all the coefficients are equal and the degree of precision is at least n. Now we allow a choice of both n nodes and n coefficients in order to determine formulae with maximum degree of precision. Of course, the degree of precision for such a formula will not be less than the corresponding degree for the interpolatory formula using the same nodes. Hence by Theorem 1.3 we conclude that *the quadrature formula with maximum degree of precision is interpolatory*.

If the formula is to have the n nodes $x_1, x_2, \ldots, x_n$, it can be written as

$$\int_a^b f(x)\, dx = \sum_{j=1}^{n} \alpha_j f(x_j) + E_n\{f\}. \tag{1}$$

However, since it must be interpolatory, the error can be written as, recalling (1.4a),

$$E_n\{f\} = \int_a^b p_n(x) f[x_1, \ldots, x_n, x]\, dx, \tag{2}$$

where we have introduced

$$p_n(x) \equiv (x - x_1)(x - x_2) \cdots (x - x_n). \tag{3}$$

Clearly, $E_n\{f\} = 0$ if $f(x)$ is a polynomial of degree $n - 1$ or less. We seek points x_j such that the error also vanishes when $f(x)$ is any polynomial of degree $n + r$ where $r = 1, 2, \ldots, m$ and m is to be as large as possible.

To determine such nodes we first recall that the nth divided difference of any polynomial of degree $n + r$ is a polynomial of degree at most r. (This follows by repeated application, n times, of the result in Problem 1.2 of Chapter 6.) Thus from (2) we conclude that the necessary and sufficient conditions for $E_n\{f\}$ to vanish for all polynomials, f, of degree $n + m$ are that

$$\text{(4)} \qquad \int_a^b p_n(x)x^r dx = 0, \qquad r = 0, 1, \ldots, m.$$

However, these are just the conditions that the polynomial $p_n(x)$ be orthogonal, over $[a, b]$, to all polynomials of degree at most m. In fact, if we take for $p_n(x)$ the nth orthogonal polynomial, then (4) is satisfied for $m = n - 1$. Further, (4) cannot be satisfied for $m = n$ or else $p_n(x)$ would have to vanish identically which is impossible. These results may be summarized as

THEOREM 1. *The quadrature formula in* (1) *can have the maximum degree of precision* $2n - 1$. *This is attained iff the n nodes,* x_j, *are the zeros of* $p_n(x)$, *the nth orthogonal polynomial over* $[a, b]$, *and the formula is interpolatory.* ∎

The formulae determined by Theorem 1 are called Gaussian quadrature formulae. From Theorem 3.4 of Chapter 5 it follows that the nodes are all interior to the interval of integration, (a, b). The coefficients in these formulae are easily obtained (once the nodes are determined) since they are interpolatory; we get as in (1.3b)

$$\text{(5)} \qquad \alpha_j = \frac{1}{p_n'(x_j)} \int_a^b \frac{p_n(x)}{x - x_j}\, dx, \qquad j = 1, 2, \ldots, n.$$

Although it is not apparent from this expression, we have

THEOREM 2. *The coefficients,* α_j, *in the Gaussian quadrature formulae are positive for all* $j = 1, 2, \ldots, n$ *and all n.*

Proof. Since the Gaussian quadrature formula with n nodes has degree of precision $2n - 1$, it yields the exact value for $\int_a^b f(x)\, dx$ when $f(x)$ is any polynomial of degree $2n - 1$ or less. In particular, then, it is exact for

$$q_j(x) \equiv \frac{p_n^2(x)}{(x - x_j)^2}, \qquad j = 1, 2, \ldots, n,$$

which are polynomials of degree $2n - 2$; i.e.,

$$\int_a^b q_j(x)\,dx = \sum_{k=1}^{n} \alpha_k q_j(x_k).$$

However, it is clear that

$$q_j(x_k) = 0, \qquad \text{for } k \neq j;$$

$$q_j(x_j) = \prod_{\substack{i=1\\(i\neq j)}}^{n} (x_j - x_i)^2 = [p_n'(x_j)]^2 > 0.$$

Thus we find

$$\alpha_j = \frac{1}{q_j(x_j)} \int_a^b q_j(x)\,dx = \frac{1}{[p_n'(x_j)]^2} \int_a^b \frac{p_n^2(x)}{(x - x_j)^2}\,dx > 0. \qquad ■$$

Note that the only property of Gaussian quadrature used in this proof is the fact that the formula with n nodes has degree of precision at least $2n - 2$. Thus we may also conclude that any quadrature formula of the form in (1) using n nodes and having degree of precision $2n - 2$ has positive coefficients. Another formula for computing the coefficients α_j is derived in the next section [see equations (4.5)].

We can obtain expressions for the error in Gaussian quadrature which are more useful than that given in (2). The first such result can be stated as

THEOREM 3. *Let $f'(x)$ be continuous in the closed interval $[a, b]$. Let $\xi_1, \xi_2, \ldots, \xi_n$ be any n distinct points in $[a, b]$ which do not coincide with the zeros, $x_1, x_2, \ldots, x_n$, of the nth orthogonal polynomial, $p_n(x)$, over $[a, b]$. Then the error in n point Gaussian quadrature applied to $\int_a^b f(x)\,dx$ is*

$$(6) \quad E_n\{f\} = \int_a^b p_n(x)(x - \xi_1)\cdots(x - \xi_n) f[x_1, \ldots, x_n, \xi_1, \ldots, \xi_n, x]\,dx.$$

Proof. Using the $2n$ distinct points x_j and ξ_j, the function $f(x)$ can be written as

$$f(x) = P_{2n-1}(x) + R_{2n-1}(x),$$

where $P_{2n-1}(x)$ is the interpolation polynomial of degree at most $2n - 1$ agreeing with $f(x)$ at the $2n$ points x_j and ξ_j, and $R_{2n-1}(x)$ is the interpolation error. With Newton's form for the remainder we write this error as

$$R_{2n-1}(x) = \prod_{j=1}^{n} [(x - x_j)(x - \xi_j)] f[x_1, \ldots, x_n, \xi_1, \ldots, \xi_n, x].$$

These expressions in (1) yield

$$(7) \qquad \int_a^b P_{2n-1}(x)\,dx + \int_a^b R_{2n-1}(x)$$
$$= \sum_{j=1}^{n} \alpha_j P_{2n-1}(x_j) + \sum_{j=1}^{n} \alpha_j R_{2n-1}(x_j) + E_n\{f\}.$$

However, since the degree of precision of the quadrature formula is $2n - 1$ we must have

$$\int_a^b P_{2n-1}(x)\,dx = \sum_{j=1}^{n} \alpha_j P_{2n-1}(x_j).$$

Also, since $f'(x)$ is continuous, it follows that $f[x_1, \ldots, x_n, \xi_1, \ldots, \xi_n, x_j]$ has a finite value for $j = 1, 2, \ldots, n$ and so $R_{2n-1}(x_j) = 0$. By using these results in (7), we obtain (6). ∎

We note that there is great freedom in the choice of the n points ξ_j in Theorem 3. Further, the conditions on $f(x)$ could be relaxed somewhat to require only continuity of $f(x)$ on $[a, b]$ and differentiability at the points x_j and ξ_j and the error representation (6) remains valid (see Problems 1.4 through 1.6 of Chapter 6). By requiring more differentiability of $f(x)$, the result in Theorem 3 can be simplified. The most common such simplification is stated as the

COROLLARY. *Let $f(x)$ have a continuous derivative of order $2n$ in $[a, b]$. Then the error in n-point Gaussian quadrature applied to $\int_a^b f(x)\,dx$ is*

$$(8) \qquad E_n\{f\} = \frac{f^{(2n)}(\xi)}{(2n)!} \int_a^b p_n^2(x)\,dx,$$

where ξ is some point in (a, b).

Proof. Under the assumed continuity conditions on $f(x)$ the integrand in (6) is a continuous function of the n points $\xi_1, \xi_2, \ldots, \xi_n$. Thus it is legitimate in this integral to let $\xi_j \to x_j$ for $j = 1, 2, \ldots, n$, and obtain, by applying the mean value theorem for integrals,

$$E_n\{f\} = \int_a^b p_n^2(x) f[x_1, \ldots, x_n, x_1, \ldots, x_n, x]\,dx$$
$$= f[x_1, \ldots, x_n, x_1, \ldots, x_n, \eta] \int_a^b p_n^2(x)\,dx, \qquad a < \eta < b.$$

The result (8) now follows from the extension of Corollary 2 of Theorem 1.2 in Chapter 6. ∎

It should be recalled in all of these results that $p_n(x)$ is *not the normalized* orthogonal polynomial of degree n over $[a, b]$, but, by (3), is the one with leading coefficient unity. So if the nth degree *orthonormal* polynomial is

$$Q_n(x) = a_n x^n + a_{n-1}x^{n-1} + \cdots + a_0,$$

then, since $p_n(x)$ and $Q_n(x)$ have the same zeros,

$$p_n(x) = \frac{1}{a_n} Q_n(x).$$

Thus we deduce that

$$\text{(9)} \qquad \int_a^b p_n^2(x)\, dx = \frac{1}{a_n^2} \int_a^b Q_n^2(x)\, dx = \frac{1}{a_n^2}.$$

For example if $a = -1$ and $b = 1$ then the Legendre polynomials, $P_n(x)$, are the relevant orthogonal polynomials. It can be shown that

$$P_n(x) = \frac{1}{2^n n!} \frac{d^n}{dx^n} (x^2 - 1)^n$$

and they are normalized by forming $\sqrt{\dfrac{2n+1}{2}}\, P_n(x)$. Thus we find in this case that

$$a_n = \frac{(2n)!}{2^n (n!)^2} \sqrt{\frac{2n+1}{2}},$$

and the error expression (8) becomes for Gaussian quadrature over $[-1, 1]$:

$$\text{(10)} \qquad E_n\{f\} = \frac{2}{(2n+1)!} \left[\frac{2^n (n!)^2}{(2n)!}\right]^2 f^{(2n)}(\xi), \qquad -1 < \xi < 1.$$

4. WEIGHTED QUADRATURE FORMULAE

It is of practical and theoretical interest to consider the approximate evaluation, for a fixed weight function $w(x)$, of integrals of the form

$$\text{(1)} \qquad W\{g\} = \int_a^b g(x) w(x)\, dx,$$

by quadrature formulae of the form

$$\text{(2)} \qquad W_n\{g\} = \sum_{j=1}^{n} \beta_j g(x_j).$$

We again call the points x_j the nodes and the β_j the coefficients of the formula. However, only the factor $g(x)$ in the integrand enters directly into the evaluation of the integration formula (2). The weight factor, $w(x)$, enters into the determination of the coefficients and nodes. Once these are determined the formula may be applied to integrals of the form (1) with different functions $g(x)$ but the same weight function. Formulae of the form (2), when applied to approximate integrals of the form (1), are called *weighted quadrature formulae*. To evaluate integrals of the form (0.1) by such formulae, write

$$\int_a^b f(x)\,dx = \int_a^b \frac{f(x)}{w(x)}\,w(x)\,dx,$$

and use (2) with $g(x) \equiv f(x)/w(x)$. As we shall see, there are frequently advantages to such procedures.

Many of the previous results are valid, with but slight changes, for the weighted quadrature formulae. Their degree of precision is defined as before; i.e., (2) has degree of precision m if

$$W_n\{x^k\} = W\{x^k\}, \qquad \text{for } k = 0, 1, \ldots, m,$$

but not for $k = m + 1$.

Given $n + 1$ distinct points $x_0, x_1, \ldots, x_n$ the *weighted interpolatory formula* with these points as nodes and weight function $w(x)$ over $[a, b]$ is, say,

$$\text{(3a)} \qquad W_{n+1}\{g\} = \sum_{j=0}^{n} w_{n,j}\,g(x_j);$$

where

$$\text{(3b)} \qquad w_{n,j} = \int_a^b \phi_{n,j}(x)w(x)\,dx, \qquad j = 0, 1, \ldots, n.$$

Here the $\phi_{n,j}(x)$ are the Lagrange interpolation coefficients for the points $x_0, x_1, \ldots, x_n$. Since $g(x) = P_n(x) + \omega_n(x)g[x_0, \ldots, x_n, x]$ where $P_n(x)$ is the Lagrange interpolation polynomial of degree n, the error in (3a and b) is

$$\begin{aligned}\text{(3c)} \qquad E_{n+1}\{g\} &= W\{g\} - W_{n+1}\{g\}\\ &= \int_a^b [g(x) - P_n(x)]w(x)\,dx\\ &= \int_a^b \omega_n(x)g[x_0, \ldots, x_n, x]w(x)\,dx.\end{aligned}$$

By assuming sufficient differentiability of $g(x)$ we can simplify this error expression. Also if $\omega_n(x)w(x)$ does not change sign in $[a, b]$, even further simplification is possible. The case of equally spaced nodes does not yield particularly simple error expressions for arbitrary weight functions, $w(x)$, and so we do not study these formulae further. It should be observed, however, that a modification of the method of undetermined coefficients still applies (i.e., the right-hand sides in (1.16) are changed from $\int_a^b x^\nu\,dx$ to $\int_a^b x^\nu w(x)\,dx$). Hence, *the result of Theorem* 1.3 *is valid* when modified to refer to weighted formulae.

If the weight function, $w(x)$, is positive in $[a, b]$ and say for simplicity continuous, then the Gaussian quadrature formulae also generalize in an obvious manner. These generalizations are best derived by seeking the n nodes, x_j, and coefficients, α_j, such that the weighted formula (2) will have the maximum degree of precision. We find now with $q_n(x) \equiv (x - x_1)\cdots(x - x_n)$ that

THEOREM 1. *The weighted quadrature formula* (2) *has degree of precision at most* $2n - 1$. *This maximum degree of precision is attained iff the n nodes,* x_j, *are the zeros of* $q_n(x)$, *the nth orthogonal polynomial with respect to the weight* $w(x)$ *over* $[a, b]$. *The formula is a weighted interpolatory one.*

Proof. The details of the proof are left as an exercise to the reader. They follow closely the proof of Theorem 3.1. ∎

The coefficients of the weighted formula of maximum precision are given by

$$(4) \qquad \beta_j = \frac{1}{q_n'(x_j)} \int_a^b \frac{q_n(x)}{x - x_j}\, w(x)\,dx, \qquad j = 1, 2, \ldots, n.$$

Exactly as in Theorem 3.2 it follows that $\beta_j > 0$. The coefficients, β_j, are called the Christoffel numbers. The formulae (2) are of the type frequently called *weighted Gaussian quadrature* with special names applied for special weight functions (see Subsection 4.1).

The coefficients β_j of the weighted Gaussian formulae can be expressed in a simpler form than that given by (4). For this purpose let $P_n(x)$ denote the nth *orthonormal* polynomial over $[a, b]$ with respect to the given weight function, $w(x)$. If the leading coefficient of $P_n(x)$ is a_n, we have $P_n(x) = a_n q_n(x)$ and hence from (4)

$$\beta_j = \frac{1}{P_n'(x_j)} \int_a^b \frac{P_n(x)}{x - x_j}\, w(x)\,dx.$$

Now set $\xi = x_k$ in the Christoffel-Darboux relation (3.25) of Chapter 5, multiply the result by $w(x)/(x - x_k)$ and integrate over $a \leq x \leq b$. This yields, since $P_n(x_k) = 0$ and

$$P_0(x) = \left[\int_a^b w(x)\,dx\right]^{-1/2};$$

$$1 = \frac{-a_n}{a_{n+1}} \int_a^b \frac{P_n(x)}{(x - x_k)} w(x)\,dx\, P_{n+1}(x_k), \qquad k = 1, 2, \ldots, n.$$

Use this in the previous expression for β_j to obtain

$$\text{(5a)} \qquad \beta_j = \frac{-a_{n+1}}{a_n P_n'(x_j) P_{n+1}(x_j)}.$$

From the three term recursion of Theorem 3.5 in Chapter 5 we find that, since x_j is a zero of $P_n(x)$,

$$P_{n+1}(x_j) = -\frac{a_{n+1}a_{n-1}}{a_n^2} P_{n-1}(x_j),$$

and (5a) becomes

$$\text{(5b)} \qquad \beta_j = \frac{a_n}{a_{n-1} P_n'(x_j) P_{n-1}(x_j)}.$$

It should be observed that the coefficients, α_j, of the ordinary Gaussian quadrature formulae are also given by the above formulae, (5), in which the $P_n(x)$ are orthonormal with respect to the uniform weight, $w(x) \equiv 1$.

The errors of the weighted Gaussian formulae are derived exactly as in Theorem 3.3 and its corollary. Thus under appropriate continuity conditions on $g(x)$ we have

$$\begin{aligned}
\text{(6)} \qquad E_n\{g\} &= W\{g\} - W_n\{g\} \\
&= \int_a^b q_n(x)(x - \xi_1)\cdots(x - \xi_n) \\
&\qquad\qquad \times g[x_1, \ldots, x_n, \xi_1, \ldots, \xi_n, x] w(x)\,dx \\
&= \int_a^b q_n^2(x) g[x_1, \ldots, x_n, x_1, \ldots, x_n, x] w(x)\,dx \\
&= g[x_1, \ldots, x_n, x_1, \ldots, x_n, \eta] \int_a^b q_n^2(x) w(x)\,dx \\
&= \frac{g^{(2n)}(\xi)}{(2n)!} \int_a^b q_n^2(x) w(x)\,dx, \qquad a < \xi < b.
\end{aligned}$$

4.1. Gauss-Chebyshev Quadrature

The polynomials orthogonal over $[-1, 1]$ with respect to the weight $w(x) = (1 - x)^{-p}(1 + x)^{-q}$, provided $p < 1$ and $q < 1$, are known as the

Jacobi polynomials. The special case, $p = q = \frac{1}{2}$, arises in the treatment of integrals of the form

$$W\{g\} \equiv \int_{-1}^{1} \frac{g(x)}{\sqrt{1 - x^2}}\, dx. \tag{7}$$

That is, consider the orthonormal polynomials over $[-1, 1]$ with respect to the weight function $(1 - x^2)^{-1/2}$, say $Q_n(x)$, such that

$$\int_{-1}^{1} Q_n(x) Q_m(x) \frac{dx}{\sqrt{1 - x^2}} = \delta_{n,m}.$$

Introduce the change of variable

$$x = \cos\theta, \qquad 0 \leq \theta \leq \pi,$$

and these integrals reduce to the form

$$\int_{0}^{\pi} Q_n(\cos\theta) Q_m(\cos\theta)\, d\theta = \delta_{n,m}.$$

In Problem 3.9 and equation (4.13) of Chapter 5 we verify that the polynomials are

$$\begin{aligned} Q_n(x) &\equiv \sqrt{\frac{2}{\pi}} \cos(n \cos^{-1} x), \qquad n = 1, 2, \ldots, \\ Q_0(x) &\equiv \frac{1}{\sqrt{\pi}}. \end{aligned} \tag{8}$$

The nodes for the n-point quadrature formula of maximum degree of precision are, by Theorem 1, the points x_j such that

$$Q_n(x_j) = 0, \qquad -1 \leq x_j \leq 1.$$

But from (8) the zeros are

$$x_j \equiv \cos\theta_j = \cos\left(\frac{2j - 1}{2n}\pi\right), \qquad j = 1, 2, \ldots, n. \tag{9}$$

The Christoffel numbers, β_j, for this best formula are most easily evaluated by using (5). That is, from (4.13) of Chapter 5 and (8)

$$Q_n(x) = 2^{n-1} \sqrt{\frac{2}{\pi}}\, x^n + \cdots, \qquad \text{for } n = 1, 2, \ldots.$$

Hence

$$\begin{aligned} a_n &= \frac{2^n}{\sqrt{2\pi}}, \qquad \text{for } n = 1, 2, \ldots, \\ a_0 &= \frac{1}{\sqrt{\pi}}. \end{aligned}$$

But by using (5b)

$$\beta_j = \frac{2}{Q_n'(x_j)Q_{n-1}(x_j)}, \qquad \text{for } n = 2, 3, \ldots;$$

therefore, from (9) with $\dfrac{d}{dx} = \dfrac{d\theta}{dx}\dfrac{d}{d\theta}$,

$$\beta_j = \frac{\pi}{n}\frac{\sin\theta_j}{\sin n\theta_j \cos(n-1)\theta_j}.$$

Now $\cos(n-1)\theta_j = \sin\theta_j \sin n\theta_j$, whence

$$\beta_j = \frac{\pi}{n}, \qquad j = 1, 2, \ldots, n. \tag{10}$$

The quadrature formulae thus derived are

$$W_n\{g\} = \frac{\pi}{n}\sum_{j=1}^{n} g(x_j), \qquad n = 1, 2, \ldots. \tag{11}$$

It is of great interest to note that *each such formula has uniform coefficients* and that the n-point *Gauss-Chebyshev* formula (11) has degree of precision $2n - 1$. Thus the problem posed in Section 2, of choosing coefficients β_j, to minimize the mean-square roundoff error in evaluating the sum (2), is solved by the same coefficients that yield maximum precision in approximating integrals of the form (7).

PROBLEM, SECTION 4

1. Carry out the proof of Theorem 1.

5. COMPOSITE QUADRATURE FORMULAE

By Theorem 1.3 (and its generalization for weighted quadrature) we see that all of the formulae considered thus far have been interpolatory. Thus, in effect, the integrand has been approximated by a single polynomial over the entire interval of integration and the integral of this polynomial is the approximation to the integral. (This is the justification of the name *simple quadrature formula.*) In order to get reasonable accuracy over a large integration interval, low degree polynomial approximations would in general not suffice. We learned in Subsection 3.4 of Chapter 6 that a high order interpolation polynomial may be a poor approximation to a smooth function in the case that the nodes are uniformly spaced. Hence we avoid using integration formulae based on interpolation polynomials of high degree and uniformly spaced nodes. On the other hand, the coefficients and nodes for formulae with maximum degree of precision are not available for large orders and are difficult to compute with great accuracy. Nevertheless, it is possible to devise quadrature formulae which have simple coefficients and nodes, and yet yield accurate approximations. These are the so-called *composite rules* which, in brief, are devised by dividing the integration interval into subintervals (usually of equal length) and then applying some formula of relatively low degree of precision over each of the subintervals. There are many composite quadrature formulae, and we only examine those most commonly used.

Let the integral to be approximated be

$$\int_a^b f(x)\,dx. \tag{1}$$

Given integers m and n, define

$$H \equiv \frac{b-a}{m}, \qquad h \equiv \frac{H}{n}, \tag{2a}$$

and divide $[a, b]$ into m subintervals, each of length H, by the points

$$y_j = a + jH, \qquad j = 0, 1, \ldots, m. \tag{2b}$$

Each of these subintervals is divided into finer subintervals of length h by the points

$$x_k = a + kh, \qquad k = 0, 1, \ldots, mn. \tag{2c}$$

Now (1) may be written as

$$\int_a^b f(x)\,dx = \sum_{j=1}^{m} \int_{y_{j-1}}^{y_j} f(x)\,dx. \tag{3}$$

By using the appropriate points x_k of (2c) each of the m integrals on the right-hand side of (3) can be approximated by a closed Newton-Cotes formula with $n + 1$ nodes. That is, by adapting the notation of Section 1,

$$\int_{y_{j-1}}^{y_j} f(x)\,dx = \sum_{k=0}^{n} w_{n,k}^{(j)} f(y_{j-1} + kh) + E_{n+1}^{(j)}\{f\}, \tag{4}$$

$$j = 1, 2, \ldots, m;$$

where $E_{n+1}^{(j)}\{f\}$ is the error in the $(n + 1)$-point formula applied to the jth integral and $w_{n,k}^{(j)}$ is the kth coefficient for the jth integral.

The coefficients, $w_{n,k}^{(j)}$, are independent of j. In fact, from (1.15), we may write

$$w_{n,k}^{(j)} = hA_{n,k}(0, n) \equiv hA_{n,k}, \tag{5}$$

where $A_{n,k}$ depends only upon the integers n and k. That is, corresponding coefficients in each subinterval are equal. With the use of (4) and (5), equation (3) becomes

$$(6)\quad \int_a^b f(x)\,dx = h\sum_{j=1}^{m}\sum_{k=0}^{n} A_{n,k} f(y_{j-1} + kh) + \sum_{j=1}^{m} E_{n+1}^{(j)}\{f\}$$
$$= h\sum_{k=0}^{n} A_{n,k}\left[\sum_{j=1}^{m} f(y_{j-1} + kh)\right] + \sum_{j=1}^{m} E_{n+1}^{(j)}\{f\}.$$

However, since $y_j = y_{j-1} + nh$ it is seen that the values of the integrand at the points y_j with $j = 1, 2, \ldots, m - 1$ appear twice in (6). We account for this repetition and rewrite the sum in the form

$$(7)\quad \int_a^b f(x)\,dx = h\left\{A_{n,0} f(y_0) + A_{n,n} f(y_m) + (A_{n,0} + A_{n,n})\sum_{j=1}^{m-1} f(y_j)\right.$$
$$\left. + \sum_{k=1}^{n-1} A_{n,k}\left[\sum_{j=1}^{m} f(y_{j-1} + kh)\right]\right\} + E_{m,n+1}\{f\}.$$

Here we have introduced the *composite error*

$$(8)\quad E_{m,n+1}\{f\} \equiv \sum_{j=1}^{m} E_{n+1}^{(j)}\{f\}.$$

Since the same closed Newton-Cotes formula has been employed over each interval $[y_{j-1}, y_j]$, we deduce from (1.11) and Lemma 1.1 applied to (8), that

$$(9a)\quad \begin{aligned} E_{m,n+1}\{f\} &= \frac{mM_n}{(n+2)!} h^{n+3} f^{(n+2)}(\xi), \\ M_n &\equiv \int_0^n t\pi_n(t)\,dt < 0, \qquad n \text{ even}; \\ E_{m,n+1}\{f\} &= \frac{mM_n}{(n+1)!} h^{n+2} f^{(n+1)}(\xi), \\ M_n &\equiv \int_0^n \pi_n(t)\,dt < 0, \qquad n \text{ odd}. \end{aligned}$$

Here $a < \xi < b$ and we have assumed that the indicated derivative of $f(x)$ is continuous on $[a, b]$. By (2a) this error can be written as

$$(9b)\quad E_{m,n+1}\{f\} = \begin{cases} \dfrac{b-a}{(n+2)!}\dfrac{M_n}{n^{n+3}} H^{n+2} f^{(n+2)}(\xi), & n \text{ even}; \\ \dfrac{b-a}{(n+1)!}\dfrac{M_n}{n^{n+2}} H^{n+1} f^{(n+1)}(\xi), & n \text{ odd}. \end{cases}$$

Thus for fixed n, the error can be made arbitrarily small by letting $H \to 0$. In this manner, we find that composite quadrature formulae may be very accurate when applied to integrals whose integrands do not possess high order derivatives.

The most common composite formulae are those with $n = 1$ (trapezoidal rule) and $n = 2$ (Simpson's rule). For the trapezoidal rule we have

$$h = H = \frac{b-a}{m}, \quad A_{1,0} = A_{1,1} = \tfrac{1}{2}, \quad M_1 = -\tfrac{1}{6};$$

and (7) and (9) yield

$$(10) \qquad \int_a^b f(x)\,dx = \frac{h}{2}\left\{f(a) + f(b) + 2\sum_{j=1}^{m-1} f(a + jh)\right\} - \frac{(b-a)}{12} h^2 f^{(2)}(\xi).$$

For Simpson's rule, with $n = 2$ in (5) and (9),

$$h = \frac{H}{2} = \frac{b-a}{2m}, \quad A_{2,0} = A_{2,2} = \tfrac{1}{3}, \quad A_{2,1} = \tfrac{4}{3}, \quad M_2 = -\tfrac{4}{15};$$

so that (7) becomes

$$(11) \qquad \int_a^b f(x)\,dx = \frac{h}{3}\left\{f(a) + f(b) + 2\sum_{j=1}^{m-1} f(a + 2jh) + 4\sum_{j=1}^{m-1} f(a + [2j-1]h)\right\} - \frac{(b-a)}{180} h^4 f^{(4)}(\xi).$$

We note that in formulae (10) and (11) the coefficients are all positive and so the roundoff errors are not generally magnified. In fact, the Newton-Cotes closed formulae have positive coefficients for $n \leq 8$.

In practice, the nodal tabulation of $f(x)$ in $[a, b]$ may not permit the use of the composite Simpson's rule because the number of net points, $N + 1$, is even. That is, the uniformly spaced points are $x_j = x_0 + jh$, such that $x_0 = a$, $x_N = b$.

In this case we could use the closed formula

$$\int_a^{a+3h} f(x)\,dx = \frac{3h}{8}[f(a) + 3f(a+h) + 3f(a+2h) + f(a+3h)] + E_4,$$

with

$$E_4 = -\frac{3h^5}{80} f^{(4)}(\xi), \qquad a < \xi < a + 3h.$$

The remaining integral

$$\int_{a+3h}^{b} f(x)\,dx$$

can then be evaluated by the composite Simpson's rule.

The error term E_4 is comparable to the error term

$$E_3 = -(h^5/90)f^{(4)}(\eta).$$

This illustrates the general principle of forming composite rules with simple formulae of comparable accuracy.

5.1. Periodic Functions and the Trapezoidal Rule

Experience has shown that if $f(x)$ is periodic, i.e., $f(x + L) = f(x)$, the formula for the integral over a period,

$$\int_0^L f(x)\,dx \cong \frac{L}{N}\sum_{j=0}^{N-1} f(x_j), \quad x_j = jh, \quad h = \frac{L}{N}, \tag{12}$$

is remarkably accurate. One possible explanation is that (12) arises from the composition of formulae having an arbitrarily high degree of precision. That is, from the Euler-Maclaurin summation formula (4.22) of Chapter 6 (also see Problem 4.7 of Chapter 6) for $p = 1$,

$$\begin{aligned}\int_{x_0}^{x_0+h} f(x)\,dx = hf(x_0) &+ \frac{h}{2}[f(x_1) - f(x_0)] \\ &- \frac{h^2}{12}[f'(x_1) - f'(x_0)] \\ &+ \frac{h^4}{720}[f^{(3)}(x_1) - f^{(3)}(x_0)] \\ &- \cdots.\end{aligned}$$

If the above is composed for all of the intervals (x_j, x_{j+1}) where $0 \leq j \leq N - 1$, we find that the terms in brackets cancel in the interior for any function $f(x)$, but also cancel at x_0 and x_N since $f(x)$ is periodic.

We have, in fact,

THEOREM **1**. *If $f(x) \in C^{2m+2}[0, L]$ and $f(x)$ is L-periodic, then the composite trapezoidal rule*, (12), *has the error*

$$e_N \equiv \int_0^L f(x)\,dx - \frac{L}{N}[\tfrac{1}{2}f(x_0) + f(x_1) + \cdots + f(x_{N-1}) + \tfrac{1}{2}f(x_N)]$$

where

$$e_N = h^{2m+2}LC_m f^{(2m+2)}(\xi),$$

with some ξ such that $0 \leq \xi \leq L$, and with a constant C_m that is independent of N and $f(x)$.

Proof. Note the central difference integration formula that is derived in Problem 1.5,

$$\int_{x_j}^{x_{j+1}} f(x)\,dx = \frac{h}{2}(f_j + f_{j+1}) + hr_1(\delta^2 f_j + \delta^2 f_{j+1}) + \cdots$$
$$+ hr_m(\delta^{2m} f_j + \delta^{2m} f_{j+1}) + C_m h^{2m+3} f^{(2m+2)}(\xi_j).$$

Add the formulae for each interval (x_j, x_{j+1}), $0 \leq j \leq N-1$. The difference terms contribute nothing to this sum. That is, with the notation

$$\psi_j \equiv \tfrac{1}{2}f_j + f_{j+1} + \cdots + f_{j+N-1} + \tfrac{1}{2}f_{j+N},$$

the integral becomes

$$\int_{x_0}^{x_N} f(x)\,dx = h[\psi_0 + 2r_1\delta^2\psi_0 + 2r_2\delta^4\psi_0 + \cdots + 2r_m\delta^{2m}\psi_0]$$
$$+ h^{2m+3}C_m \sum_{j=0}^{N-1} f^{(2m+2)}(\xi_j).$$

But it is easy to see from the periodicity of $f(x)$ that

$$\psi_j = \sum_{s=0}^{N-1} f_{j+s} = \psi_p$$

for all integers p. Hence, in particular,

$$\delta^{2k}\psi_0 = 0 \qquad \text{for all integers } k \geq 1.$$

Therefore, by Lemma 1.1 and the definition of h in (12),

$$\text{(13)} \qquad \int_{x_0}^{x_N} f(x)\,dx = h[\tfrac{1}{2}f_0 + f_1 + f_2 + \cdots + f_{N-1} + \tfrac{1}{2}f_N]$$
$$+ h^{2m+3}C_m \sum_{j=0}^{N-1} f^{(2m+2)}(\xi_j)$$
$$= h[\tfrac{1}{2}f_0 + f_1 + f_2 + \cdots + f_{N-1} + \tfrac{1}{2}f_N]$$
$$+ h^{2m+2}LC_m f^{(2m+2)}(\xi), \qquad 0 \leq \xi \leq L. \qquad \blacksquare$$

5.2. Convergence for Continuous Functions

In the event that the function $f(x)$ is merely continuous (or piecewise continuous, with jump discontinuities), we can still prove convergence of composite quadrature formulae that have non-negative coefficients and degree of precision $s > 0$, as in

THEOREM 2. *With the notation of* 2 (a, b, and c), *let*

$$\text{(14)} \qquad S_{m,n}\{f\} \equiv \sum_{j=1}^{m} S_{m,n}^{(j)}\{f\},$$

where

$$(15) \qquad S_{m,n}^{(j)}\{f\} \equiv \sum_{k=0}^{n} \alpha_{n,k}^{(j)} f(y_{j-1} + kh).$$

If $f(x)$ is continuous in $[a, b]$, $\alpha_{n,k}^{(j)} \geq 0$ and if $S_{m,n}^{(j)}$ has degree of precision s (i.e.,

$$S_{m,n}^{(j)}\{g\} = \int_{y_{j-1}}^{y_j} g(\xi)\, d\xi$$

for $g(\xi) \equiv \xi^p$, $0 \leq p \leq s$), then

$$\lim_{m\to\infty} S_{m,n}\{f\} = \int_a^b f(x)\, dx.$$

Proof. As m tends to infinity the closed intervals $[y_{j-1}, y_j]$ become arbitrarily small. Hence given any $\epsilon > 0$ there is an M such that if $m \geq M$ there exist polynomials $\{P_s^{(j)}(x)\}$ for $j = 1, 2, \ldots, m$, of degree at most s, such that

$$\max_{y_{j-1}\leq x\leq y_j} |f(x) - P_s^{(j)}(x)| \leq \epsilon \qquad \text{for } j = 1, 2, \ldots, m.$$

Now

$$(16) \quad \left| \int_{y_{j-1}}^{y_j} f(x)\, dx - S_{m,n}^{(j)}\{f\} \right| \leq \left| \int_{y_{j-1}}^{y_j} f(x)\, dx - \int_{y_{j-1}}^{y_j} P_s^{(j)}(x)\, dx \right| + \left| \int_{y_{j-1}}^{y_j} P_s^{(j)}(x)\, dx - S_{m,n}^{(j)}\{f\} \right|$$
$$\leq \epsilon|y_j - y_{j-1}| + |S_{m,n}^{(j)}\{P_s^{(j)} - f\}|$$

The fact that $S_{m,n}^{(j)}$ has degree of precision s was used to obtain the last term on the right-hand side.

The fact that $S_{m,n}^{(j)}\{g\}$ is exact for $g(\xi)$ identically constant and that $\alpha_{n,k}^{(j)} \geq 0$, implies that

$$\sum_{k=0}^{n} |\alpha_{n,k}^{(j)}| = |y_j - y_{j-1}|.$$

Hence (16) yields

$$\left| \int_{y_{j-1}}^{y_j} f(x)\, dx - S_{m,n}^{(j)}\{f\} \right| \leq 2\epsilon|y_j - y_{j-1}|.$$

Therefore,

$$\left| \int_a^b f(x)\, dx - S_{m,n}\{f\} \right| \leq 2\epsilon|b - a|. \qquad \blacksquare$$

By picking $P_s^{(j)}(x) \equiv f(\xi_j)$, where $\xi_j \equiv (y_{j-1} + y_j)/2$, we find the

COROLLARY. *Under the hypothesis of Theorem 2,*

$$\left|\int_a^b f(x)\,dx - S_{m,n}\{f\}\right| \le 2|b-a|\omega(f;\delta),$$

where ω is the modulus of continuity and $\delta \equiv \max_j |y_{j-1} - y_j|/2.$ ■

We leave to Problems 8 and 9 the formulation of generalizations of Theorem 2.

PROBLEMS, SECTION 5

1. At what interval in x and to how many decimal places must $f(x)$ be tabulated in order to evaluate $\int_0^5 f(x)\,dx$ correctly to six decimal places by using:

(a) composite trapezoidal rule,

(b) composite Simpson's rule, for $f(x) \equiv \cos x$?

2. The composite midpoint rule is based on the single node, open Newton-Cotes formula with error, E_1:

$$\int_a^{a+2h} f(x)\,dx \equiv 2hf(a+h) + E_1.$$

Find the expression for the composite formula and error term when the integral to be evaluated is

$$\int_a^{a+2mh} f(x)\,dx, \qquad m = \frac{b-a}{2h}.$$

3. Answer Problem 1 for the composite midpoint rule (see Problem 2).

4*. Use the notation of Problem 1.5 to derive the *composite trapezoidal rule with end corrections*:

$$\int_{x_0}^{x_N} f(x)\,dx = h\Big[(\tfrac{1}{2}f_0 + f_1 + \cdots + f_{N-1} + \tfrac{1}{2}f_N)$$
$$+ \sum_{k=1}^{m} r_k(d_{N,k} - d_{0,k})\Big] + C_m(x_N - x_0)h^{2m+2}f^{(2m+2)}(\xi),$$

where

$$d_{j,k} \equiv \Delta^{2k-1}f_{j-k} + \Delta^{2k-1}f_{j-k+1},$$

and

$$x_{-k} \le \xi \le x_{N+k}.$$

[Hint: Use equation (3.16a) of Chapter 6 to get

$$\delta^{2k}f_s = \Delta^{2k-1}f_{s-k+1} - \Delta^{2k-1}f_{s-k},$$

whence

$$\sum_{s=1}^{N-1} \delta^{2k}f_s = \Delta^{2k-1}f_{N-k} - \Delta^{2k-1}f_{1-k}.]$$

Since the coefficients of $\{f_j\}$ which occur in $\Delta^n f_0$ are the alternating binomial coefficients $(-1)^{n-j}\binom{n}{j}$, we see that

$$d_{j,k} = \sum_{p=-k}^{k} c_p{}^{(k)} f_{j+p},$$

where

$$c_0{}^{(k)} = 0, \quad c_p{}^{(k)} = -c_{-p}^{(k)}, \qquad p = 1, 2, \ldots, k.]$$

5.* Given the integer $N > 0$, let $h = 1/N$, $x_0 = 0$, $x_j = x_0 + jh$;

$$t_N \equiv h(\tfrac{1}{2}f_0 + f_1 + \cdots + f_{N-1} + \tfrac{1}{2}f_N).$$

(a) From Problem 4, verify that if $f(x) \in C^{2m+2}$,

$$e_N \equiv \int_0^1 f(x)\,dx - t_N = \sum_{j=1}^{m} a_j h^{2j} + \mathcal{O}(h^{2m+2}).$$

[Hint: Expand the values f_p and f_{N+p}, which appear in the end corrections, by Taylor's series about x_0 and x_N respectively. Collect terms with like powers of h.]

(b) Let

$$s_N \equiv h[f_{1/2} + f_{3/2} + \cdots + f_{(2N-1)/2}].$$

s_N is the *composite midpoint rule* for evaluating

$$\int_{x_0}^{x_N} f(x)\,dx$$

(with intervals $h/2$—see Problem 2). Verify that

$$t_{2N} = \tfrac{1}{2}(t_N + s_N),$$

and

$$e_{2N} \equiv \int_0^1 f(x)\,dx - t_{2N} = \sum_{j=1}^{m} a_j\left(\frac{h}{2}\right)^{2j} + \mathcal{O}(h^{2m+2}).$$

(c) Show that

$$\frac{4e_{2N} - e_N}{3} = \int_0^1 f(x)\,dx - \left(\frac{4t_{2N} - t_N}{3}\right)$$

$$= \sum_{j=2}^{m} b_j h^{2j} + \mathcal{O}(h^{2m+2}),$$

and $b_r = 0$ if $a_r = 0$.

Call $t_{2N}^{(1)} \equiv (4t_{2N} - t_N)/3$ the *first extrapolation* of the trapezoidal rule.

6.* *Romberg's method*: With the notation of Problem 5, consider the sequence of subdivisions obtained by halving, i.e., $N = 1, 2, 4, \cdots, 2^k, \ldots$. Define

$$t_{2^k}^{(0)} \equiv t_{2^k} \qquad k = 0, 1, 2, \ldots;$$

$$t_{2^k}^{(1)} \equiv \frac{4t_{2^k}^{(0)} - t_{2^{k-1}}^{(0)}}{3} \qquad k = 1, 2, 3, \ldots;$$

$$t_{2^k}^{(2)} \equiv \frac{4^2 t_{2^k}^{(1)} - t_{2^{k-1}}^{(1)}}{4^2 - 1} \qquad k = 2, 3, 4, \ldots;$$

$$\vdots$$

$$t_{2^k}^{(p)} \equiv \frac{4^p t_{2^k}^{(p-1)} - t_{2^{k-1}}^{(p-1)}}{4^p - 1} \qquad k = p, p + 1, \ldots.$$

$t_{2^k}^{(p)}$ is called the *pth extrapolation* of the trapezoidal rule. Show, by induction on p, that for fixed $p \geq 0$,

$$\int_0^1 f(x)\,dx - t_{2^k}^{(p)} = \mathcal{O}(2^{-k(2p+2)}) \qquad \text{if } p \leq m.$$

Romberg's method consists in successively constructing the rows of the triangular matrix R: $(r_{k,p})$, with $r_{k,p} \equiv t_{2^k}^{(p)}$. If $a_p \neq 0$, $f(x) \in C^{2m+2}[-\delta, 1+\delta]$ for some $\delta > 0$, and $0 < p \leq m$, then, by (c), the entries in the pth column converge more rapidly than those of the $(p-1)$st column to $\int_0^1 f(x)\,dx$. In Romberg's method we may achieve the degree of precision of the end correction formula and avoid the evaluation of $f(x)$ outside of the interval $[x_0, x_N]$. In practice, the rows of R may be successively computed until the elements in some column have "converged" enough.

7.* Prove that the pth extrapolation of the trapezoidal rule is a quadrature formula with non-negative weights and degree of precision at least $2p + 1$. That is, to approximate $\int_0^1 f(x)\,dx$,

$$t_{2^k}^{(p)} \equiv 2^{-k} \sum_{j=0}^{2^k} c_{k,j}^{(p)} f(j2^{-k}),$$

where

$$\sum_{j=0}^{2^k} c_{k,j}^{(p)} = 2^k, \qquad c_{k,j}^{(p)} \geq 0.$$

8.* State and prove a generalization of Theorem 2, to cover the case where $f(x)$ is piecewise continuous (with only a finite number of jump discontinuities) and where the quadrature formula is not based on uniformly spaced nodes.

9.* Under the conditions of Theorem 2, if $f(x)$ has a continuous derivative of order r, show that

(a) If $r \leq s$,

$$\left| \int_a^b f(x)\,dx - S_{m,n}\{f\} \right| \leq 2|b - a| \frac{\delta^r}{r!} \omega(f^{(r)}; \delta),$$

where $\delta = \max_j |y_{j-1} - y_j|/2$.

[Hint: Pick

$$P_s^{(j)}(x) = f(\xi_j) + f^{(1)}(\xi_j)(x - \xi_j) + \cdots + \frac{f^{(r)}(\xi_j)}{r!}(x - \xi_j)^r,$$

with $\xi_j = (y_{j-1} + y_j)/2$. Verify that

$$|f(x) - P_s^{(j)}(x)| \leq \frac{\delta^r}{r!} \omega(f^{(r)}; \delta) \qquad \text{for } y_{j-1} \leq x \leq y_j.]$$

(b) If $r > s$,

$$\left| \int_a^b f(x)\,dx - S_{m,n}\{f\} \right| \leq 2|b - a| \frac{\delta^{s+1}}{(s+1)!} K,$$

where $K \equiv \max_x |f^{(s+1)}(x)|$.

6. SINGULAR INTEGRALS; DISCONTINUOUS INTEGRANDS

In deriving most of the quadrature formulae of this chapter and the appropriate error formulae, it was either stated or implied that the integrand and various of its higher order derivatives were continuous. (An exception is found in the weighted quadrature formulae where the weight need not be continuous.) If these conditions are violated, a quadrature method may still yield a good approximation but the error will, in general, be much larger than predicted. In less favorable circumstances, of course, the approximation will be meaningless. There are a number of cases of rather frequent occurrence in which such difficulties can be anticipated and satisfactorily resolved.

6.1. Finite Jump Discontinuities

If the integrand has a finite jump discontinuity at a known point (or any finite number of them), say at $x = c$ in the interval of integration, then we write

$$\int_a^b f(x)\,dx = \int_a^c f(x)\,dx + \int_c^b f(x)\,dx. \tag{1}$$

Now if $f(z)$ has sufficiently many continuous derivatives in $[a, c]$ and $[c, b]$ the two integrals on the right-hand side may be accurately approximated by any of a variety of quadrature formulae. This simple procedure can be considered as an application of a special composite rule, not necessarily one with equal spacing.

If the integrand is continuous but has a discontinuity in some low order derivative, a similar procedure can be employed. For example, if $f(x) = |x|$ then $f'(x)$ has a finite jump at $x = 0$. In this case, a composite rule with $x = 0$ as an endpoint of a subinterval could be used.

6.2. Infinite Integrand

We consider the case in which $f(x)$ becomes infinite as $x \rightarrow a$, the lower limit of integration. The upper limit can be similarly treated and an interior discontinuity, say at $x = c$, is reduced to the endpoint cases by using (1). We assume for the present that the integral is of the form

$$I \equiv \int_a^b \frac{g(x)}{(x - a)^\theta}\,dx, \qquad 0 < \theta < 1, \tag{2}$$

where $g(x)$ has continuous derivatives in $[a, b]$ of "sufficiently high" order. The restriction on θ insures that the integral in (2) exists for rather general functions $g(x)$ [i.e., it is not required that $g(a) = 0$].

METHOD I. For any positive number ϵ in $0 < \epsilon < b - a$ we write (2) as $I = I_1 + I_2$ where

$$(3) \qquad I_1 \equiv \int_a^{a+\epsilon} \frac{g(x)}{(x-a)^\theta}\,dx, \qquad I_2 \equiv \int_{a+\epsilon}^{b} \frac{g(x)}{(x-a)^\theta}.$$

The range of integration in I_2 is now such that the integrand has derivatives of high order there. Thus, I_2 can be approximated by many of the standard procedures previously described, and in principle the error in this approximation can be estimated. It remains to approximate I_1 to within a known error.

By Taylor's theorem we have for $x \geq a$:

$$g(x) = g(a) + (x-a)g'(a) + \frac{(x-a)^2}{2!}g''(a) + \cdots + \frac{(x-a)^s}{s!}g^{(s)}(a) + \frac{(x-a)^{s+1}}{(s+1)!}g^{(s+1)}(\xi(x)).$$

Use this expansion in I_1 and perform the indicated integrations to find

$$(4) \qquad I_1 = \epsilon^{1-\theta}\left[\frac{g(a)}{1-\theta} + \frac{\epsilon}{1!}\frac{g'(a)}{2-\theta} + \frac{\epsilon^2}{2!}\frac{g''(a)}{3-\theta} + \cdots + \frac{\epsilon^s}{s!}\frac{g^{(s)}(a)}{s+1-\theta}\right] + \frac{1}{(s+1)!}\int_a^{a+\epsilon}(x-a)^{s+1-\theta}g^{(s+1)}(\xi(x))\,dx.$$

If the first (bracketed) term on the right in (4) is used as the approximation to I_1 we obtain the error bound

$$(5) \qquad |E^{(1)}| \leq \frac{\epsilon^{s+2-\theta}}{(s+1)!\,(s+2-\theta)} \max_{a \leq \xi \leq a+\epsilon} |g^{(s+1)}(\xi)|.$$

For fixed s this bound is clearly an increasing function of ϵ. Or for fixed $\epsilon < 1$, if the derivatives of $g^{(s)}(x)$ do not grow too fast with s it will also be a decreasing function of s.

If the error in evaluating I_2 is $E^{(2)}$ we must now determine conditions on ϵ, s and the quadrature formula such that $|E^{(1)}| + |E^{(2)}| < \delta$, where δ is the maximum permissible error in approximating I. Of course, the parameters should be chosen such that $|E^{(1)}| \cong |E^{(2)}|$ since then some cancellation of error may take place. For definiteness, let us assume that I_2 is approximated by a composite rule using m subintervals and a closed $(n+1)$-point Newton-Cotes quadrature formula with equal spacing over each subinterval. Furthermore, we will assume n to be even. Then from (5.9) we must have

$$E^{(2)} = \frac{mM_n}{(n+2)!}h^{n+3}\frac{d^{n+2}}{dx^{n+2}}\frac{g(x)}{(x-a)^\theta}\bigg|_{x=\xi}, \qquad a+\epsilon < \xi < b;$$

where $h = (b - a - \epsilon)/(mn)$. From this expression we obtain the bound

$$(6) \qquad |E^{(2)}| \le \frac{mM_n}{(n+2)!}\left(\frac{b-a}{mn}\right)^{n+3} \max_{a+\epsilon \le x \le b}\left|\frac{d^{n+2}}{dx^{n+2}}\left[\frac{g(x)}{(x-a)^\theta}\right]\right|,$$

in which the coefficient of the derivative term is independent of ϵ. If the derivatives entering into (5) and (6) can be estimated, say

$$|g^{(s+1)}(\xi)| \le M^{(s+1)},$$

and

$$\max_{a+\epsilon \le x \le b}\left|\frac{d^{n+2}\dfrac{g(x)}{(x-a)^\theta}}{dx^{n+2}}\right| = N^{(n+2)}(\epsilon),$$

then for fixed s, n, and δ we can find ϵ and m such that

$$\frac{\epsilon^{s+2-\theta}}{(s+1)!\,(s+2-\theta)} M^{(s+1)} = mK_n\left(\frac{b-a}{mn}\right)^{n+3} N^{(n+2)}(\epsilon) = \frac{\delta}{2}.$$

The bound $N^{(n+2)}(\epsilon)$ will, in general, become large for small ϵ and so may imply an unusually large m. Thus, we consider an alternative procedure obtained by modifying these considerations.

METHOD II. Let us rewrite the Taylor expansion of $g(x)$ as

$$(7a) \qquad g(x) = G_s(x) + \frac{(x-a)^{s+1}}{(s+1)!} g^{(s+1)}(\xi(x)),$$

where

$$(7b) \qquad G_s(x) \equiv g(a) + (x-a)g'(a) + \cdots + \frac{(x-a)^s}{s!} g^{(s)}(a).$$

Now the integral (2) is identically represented by

$$(8) \qquad I = \int_a^b \frac{g(x) - G_s(x)}{(x-a)^\theta}\,dx + \int_a^b \frac{G_s(x)}{(x-a)^\theta}\,dx \equiv I_N + I_E.$$

The second integral, I_E, can be evaluated explicitly, just as was the first part of I_1 in (4), and we have

$$(9) \qquad I_E = (b-a)^{1-\theta}\left[\frac{g(a)}{1-\theta} + \frac{(b-a)}{1!}\frac{g'(a)}{2-\theta} + \cdots + \frac{(b-a)^s}{s!}\frac{g^{(s)}(a)}{s+1-\theta}\right].$$

However, the first integral in (8), I_N, no longer has a singular integrand at $x = a$ so it can be approximated by many of the standard quadrature formulae. In fact, the first s derivatives of

$$\frac{g(x) - G_s(x)}{(x-a)^\theta}$$

are finite at $x = a$ and so the error in any closed formula with $n + 2 \leq s$ is bounded. For example, in a composite formula employing Simpson's rule to approximate I_N the error becomes, from (5.9) with $n = 2$:

$$E^{(N)} = -\frac{m}{90} h^5 \frac{d^4}{dx^4} \left[\frac{g(x) - G_s(x)}{(x - a)^\theta} \right]_{x=\xi}$$

where

$$h \equiv \frac{b - a}{2m}, \qquad \text{and} \qquad a < \xi < b.$$

In order that the indicated derivative remain bounded as $\xi \to a$, it is sufficient that $s = 4$. Of course, if $0 \leq s < 4$ the quadrature formula will still yield an accurate approximation (see Theorem 5.1) but the above form for the error cannot be used.

METHOD III. As a third alternative, which is restricted to singular integrands of the form (2), we consider the change of variable

$$(x - a) = t^\phi \qquad dx = \phi t^{\phi - 1}\, dt.$$

Then (2) becomes, if $\phi = k/(1 - \theta)$ for k any positive integer,

$$I = \phi \int_0^{(b-a)^{1/\phi}} g(a + t^\phi) t^{k-1}\, dt. \tag{10}$$

Now, since $k \geq 1$, the integrand of the above integral is continuous at $t = 0$. Thus numerical quadrature formulae may be directly applied to (10). In fact, if $\theta = q/p$, i.e., rational, and $k = p - q$, the integrand is smooth as long as g is smooth.

Methods I and II are applicable to other types of singularities. In fact, if the integral is of the form

$$\int_a^b g(x) S(x)\, dx \tag{11}$$

where $S(a)$ is infinite, Methods I and II may be applied if integrals of the form

$$\int_a^{a+\epsilon} (x - a)^k S(x)\, dx, \qquad k = 0, 1, \ldots, s, \tag{12}$$

can be explicitly evaluated. For example, if $S(x) \equiv \ln(x - a)$, we can employ these methods.

METHOD IV. In many cases of interest the singular part of the integrand, i.e., $S(x)$ in (11), is of one sign throughout $[a, b]$. Then, in principle, the weighted quadrature methods outlined in Section 4 can be employed. In particular, the weighted Gaussian quadrature formulae are frequently very effective for evaluating such singular integrals. Of course, such applications require the determination of the polynomials orthogonal over $[a, b]$ with respect to the weight $S(x)$. For many special forms of $S(x)$ these polynomials are known; for instance, in the case

$$S(x) \equiv (x - a)^{-1/2}(b - x)^{-1/2}, \tag{13}$$

the Gauss-Chebyshev formula derived in Section 4.1 is the relevant scheme. However, even if the required polynomials are not well known, it may be of value to construct them and to devise the appropriate quadrature formula. This is especially true if many integrals containing the same singular part are to be evaluated.

6.3. Infinite Integration Limits

It is clear that an integral of the form

$$I = \int_a^\infty g(x)\,dx \tag{14}$$

cannot, in general, be accurately approximated by the standard quadrature methods (which employ a finite number of finite subintervals). The usual approach to such problems is to write again $I = I_1 + I_2$ where now

$$I_1 \equiv \int_a^b g(x)\,dx, \qquad I_2 \equiv \int_b^\infty g(x)\,dx. \tag{15}$$

Then if b is "sufficiently large," it may be possible by analytical means to show that I_2 is negligible. Or alternatively, $g(x)$ may be approximated for $x > b$ by some function from which I_2 is then approximated; in this case good error estimates are usually difficult to obtain. Another procedure, too frequently disregarded, is to reduce I_2 to an integral over a finite interval.

That is, introduce the change of variable $x = 1/\xi$ and obtain

$$\begin{aligned} I_2 &= -\int_{1/b}^0 g\left(\frac{1}{\xi}\right)\xi^{-2}\,d\xi \\ &= \int_0^{1/b} f(\xi)\,d\xi. \end{aligned} \tag{16}$$

Here we have introduced the function

$$f(\xi) \equiv \frac{g(1/\xi)}{\xi^2}. \tag{17}$$

Now if $f(\xi)$ is not singular at $\xi = 0$, then I_2 may be evaluated, in the form (16), by standard quadrature methods. If $f(\xi)$ is singular at $\xi = 0$, then the evaluation of I_2 might be reduced to the previous case of Subsection 6.2.

In fact, a sufficient condition for I_2 defined in (15) to converge (absolutely) is that $g(x)$ be continuous and that

$$\lim_{x \to \infty} x^{1+\epsilon} g(x) = 0 \qquad \text{for some } \epsilon > 0. \tag{18}$$

This, by (17), is equivalent to

$$\lim_{\xi \to 0} \xi^{1-\epsilon} f(\xi) = 0.$$

Now this condition will be satisfied if $f(\xi)$ behaves at $\xi = 0$ like $\xi^{-\theta}$ where $\theta < 1 - \epsilon$. If $\epsilon < 1$, the integral in (16) may have a singularity of the form indicated in (2).

Finally, we point out that special *weighted Gaussian quadrature formulae* may be effective for various integrals over infinite intervals. In particular, the orthogonal polynomials over $[-\infty, \infty]$ with respect to the weight function e^{-x^2} are well known. They are called *Hermite polynomials*, $H_n(x)$, and they can be shown to be given by (see Problem 3.18 in Chapter 5)

$$H_n(x) = (-1)^n e^{x^2} \frac{d^n}{dx^n} (e^{-x^2}). \tag{19a}$$

It is not difficult to deduce that

$$H_{n+1}(x) = 2xH_n(x) - \frac{d}{dx} H_n(x), \tag{19b}$$

and hence by induction, since $H_0(x) = 1$, that

$$H_n(x) = 2^n x^n + \cdots. \tag{19c}$$

By repeatedly using integration by parts we can show that the normalized Hermite polynomials are

$$\frac{H_n(x)}{\sqrt{2^n n! \pi^{1/2}}}. \tag{19d}$$

The formulae based on the $H_n(x)$ are called *Gauss-Hermite quadrature formulae* and are used to approximate integrals of the general form

$$\int_{-\infty}^{\infty} e^{-x^2} g(x)\, dx.$$

For integrals over $[0, \infty]$, the *Laguerre polynomials*, $L_n(x)$, defined as

$$L_n(x) = (-1)^n e^x \frac{d^n}{dx^n} (x^n e^{-x}), \tag{20a}$$

are sometimes useful. They are orthogonal over $[0, \infty]$ with respect to the weight e^{-x}. It can be shown that

$$L_n(x) = x^n + \cdots, \tag{20b}$$

and that the normalized Laguerre polynomials are $(1/n!)L_n(x)$. The *Gauss-Laguerre quadrature formulae* are based on these polynomials and are used to approximate integrals of the form

$$\int_0^\infty e^{-x}g(x)\,dx.$$

PROBLEMS, SECTION 6

1. Evaluate

$$\int_0^1 \sqrt{1 - x^2}\,dx$$

by Method II, using the composite Simpson's rule, and obtain four-decimal-place accuracy. What is the largest interval h that is permissible?

2. Use Method III and the composite Simpson's rule to evaluate

$$\int_0^1 \sqrt{1 - x^2}\,dx$$

correctly to four decimal places. What is the largest interval h that is permissible?

3. Substitute the new variable $x = \cos\theta$ and use the composite Simpson's rule to evaluate

$$\int_0^1 \sqrt{1 - x^2}\,dx$$

correctly to four decimal places. What is the largest permissible interval $h = \Delta\theta$?

4. Verify the properties of the Hermite and Laguerre polynomials given in the text [see equations (19) and (20)].

7. MULTIPLE INTEGRALS

The problem of efficiently approximating multiple integrals numerically has not been completely solved. An obvious source of complexity is the variety of domains of integration in higher dimensions compared to just intervals in our study of one dimensional integrals. However, even if the domain is restricted, say to the unit cube, then the resulting problem is still not in a satisfactory state. A fundamental difficulty is essentially the great degree of freedom in locating the nodes or equivalently in the large number of, say, uniformly spaced nodes required to get reasonable accuracy.

One of the basic methods for approximating multiple integrals is, as in the one dimensional case, to integrate a polynomial approximation of the integrand. But since interpolation theory in higher dimensions is not well developed, we again have difficulty in devising practical schemes. Generalizations of the method of undetermined coefficients offer many possibilities, but only a few of these have been exploited for multiple integrals. Finally, we point out that the difficulties increase as the dimension of the domain of integration increases. This seems related to the fact that the ratio of the surface area to the volume, for an n-dimensional unit cube increases with n. We shall consider numerical methods for evaluating double integrals for the most part. Many of the procedures extend in an obvious way to higher dimensions, with perhaps a subsequent loss in efficiency or accuracy. Approximation methods for double integrals are frequently called *cubature formulae* since they approximate the *volume* associated with the integrand.

In general, the problem is to approximate an integral of the form

$$J\{f\} \equiv \int_D f(\mathbf{x})\,d\mathbf{x}, \tag{1}$$

where $\mathbf{x} = (x_1, \ldots, x_p)$ and $d\mathbf{x} = dx_1 \cdots dx_p$ are a point and a volume element in the p-dimensional space, respectively, and D is a domain in this space. The approximations considered are all to be of the form

$$J_N\{f\} \equiv \sum_{\nu=1}^{N} A_\nu f(\mathbf{x}_\nu). \tag{2}$$

Here the N points, $\mathbf{x}_\nu$, are the nodes of the formula and the A_ν are the coefficients. We say that formula (2) has degree of precision m as an approximation to the integral (1) if $J\{P(\mathbf{x})\} = J_N\{P(\mathbf{x})\}$ for all polynomials, $P(\mathbf{x})$, in $\mathbf{x}$ of degree† at most m but not for some polynomial of degree $m + 1$.

We cannot proceed as in Section 1 to study general interpolatory schemes since the general interpolation problem is not well posed in higher dimensions. However, if the nodes are specially chosen, say as in Section 6 of Chapter 6, then interpolation can be used and we consider such cases first.

† We say $P(\mathbf{x})$ is of degree at most m in $\mathbf{x}$, if $P(\mathbf{x})$ is a polynomial in $(x_1, \ldots, x_p)$ of the form

$$P(\mathbf{x}) \equiv \sum_{0 \le j_1 + j_2 + \cdots + j_p \le m} C_{j_1 j_2 \cdots j_p} x_1^{j_1} x_2^{j_2} \ldots x_p^{j_p}.$$

7.1. The Use of Interpolation Polynomials

Let the integral in (1) be over a plane domain, say

$$J\{f\} \equiv \iint_D f(x, y)\, dx\, dy. \tag{3}$$

Let us pick as nodes the $N = (m + 1)(n + 1)$ distinct points: (x_i, y_j), $i = 0, 1, \ldots, m$; $j = 0, 1, \ldots, n$; where the $m + 1$ distinct numbers $\{x_i\}$ and $n + 1$ distinct numbers $\{y_j\}$ are, at present, arbitrary. Then a polynomial $P(x, y)$, of degree m in x and n in y which is equal to $f(x, y)$ at these N nodes is given by (6.3) in Chapter 6. We use this polynomial to define the cubature formula

$$\begin{aligned} J_N\{f\} &\equiv \iint_D P(x, y)\, dx\, dy \\ &= \sum_{i=0}^{m} \sum_{j=0}^{n} A_{ij} f(x_i, y_j). \end{aligned} \tag{4a}$$

Here we have introduced the coefficients, A_{ij}, by the definitions

$$A_{ij} = \iint_D X_{m,i}(x) Y_{n,j}(y)\, dx\, dy, \qquad i = 0, 1, \ldots, m,\ j = 0, 1, \ldots, n; \tag{4b}$$

and the Lagrange interpolation coefficients $X_{m,i}(x)$ and $Y_{n,j}(y)$ are defined in (6.2) of Chapter 6.

While this procedure is formally valid for very general domains, D, it is only practical when the integrals in (4b) can be evaluated explicitly. A particularly simple and important special case is that of a rectangular domain,

$$D\colon \{x, y \mid a \le x \le b;\ c \le y \le d\}.$$

In this case we have

$$A_{ij} = \alpha_i \beta_j; \quad \alpha_i \equiv \int_a^b X_{m,i}(x)\, dx, \quad \beta_j \equiv \int_c^d Y_{n,j}(y)\, dy, \tag{5}$$

and the quantities α_i and β_j are just the coefficients for appropriate one dimensional interpolatory quadrature formulae. Furthermore, if the numbers x_i are equally spaced in $[a, b]$, and the y_j are equally spaced in $[c, d]$, then the α_i and β_j are the coefficients in the $(m + 1)$-point and $(n + 1)$-point Newton-Cotes quadrature formulae respectively (see Problem 1).

The error in the cubature formula (4) as an approximation to the integral

(3) is, upon recalling the expression (6.7c) of Chapter 6 for the error in the interpolation polynomial,

$$
\begin{aligned}
(6) \qquad E_N\{f\} &\equiv \iint_D [f(x, y) - P(x, y)]\, dx\, dy \\
&= \iint_D R(x, y)\, dx\, dy \\
&= \iint_D \left\{ \frac{\omega_m(x)}{(m+1)!} \left(\frac{\partial}{\partial x}\right)^{m+1} f(\xi(x), y) \right. \\
&\qquad + \frac{\omega_n(y)}{(n+1)!} \left(\frac{\partial}{\partial y}\right)^{n+1} f(x, \eta(y)) \\
&\qquad - \frac{\omega_m(x)\omega_n(y)}{(m+1)!\,(n+1)!} \left(\frac{\partial}{\partial x}\right)^{m+1} \left(\frac{\partial}{\partial y}\right)^{n+1} \\
&\qquad \left. \times f(\xi'(x), \eta'(y)) \right\} dx\, dy.
\end{aligned}
$$

We deduce from this that the formula (4) has degree of precision at least $\min(m, n)$. For instance, if as is frequently the case, we take $m = n$, then by using $N = (n + 1)^2$ nodes, a formula with degree of precision at least n is obtained.

However, a formula using only $M = \frac{1}{2}(n + 1)(n + 2)$ nodes can be devised which also has degree of precision at least n. For this purpose, we integrate the interpolation polynomial $P_n(x, y)$, given by (6.10) of Chapter 6, over D. The general result is somewhat cumbersome to write down in the form (2). First, divided differences of the type $f[x_0, \ldots, x_k; y_0, \ldots, y_j]$ must be expanded as linear combinations of the function values, $f(x_\nu, y_\mu)$, and then all terms containing such function values must be combined to determine the coefficients, $B_{\nu\mu}$. For small values of n, say $n \leq 3$, this is easily done (see Problem 2). However, for equally spaced x_k and y_j difference operators may be employed to simplify the notation and even the calculations. We indicate the general formula obtained in this manner as

$$
\begin{aligned}
(7) \qquad K_M\{f\} &= \iint_D P_n(x, y)\, dx\, dy \\
&= \sum_{k=0}^{n} \sum_{j=0}^{n-k} f[x_0, \ldots, x_k; y_0, \ldots, y_j] \\
&\qquad \times \iint_D \omega_{k-1}(x)\omega_{j-1}(y)\, dx\, dy \\
&= \sum_{\nu=0}^{n} \sum_{\mu=0}^{n-\nu} B_{\nu\mu} f(x_\nu, y_\mu).
\end{aligned}
$$

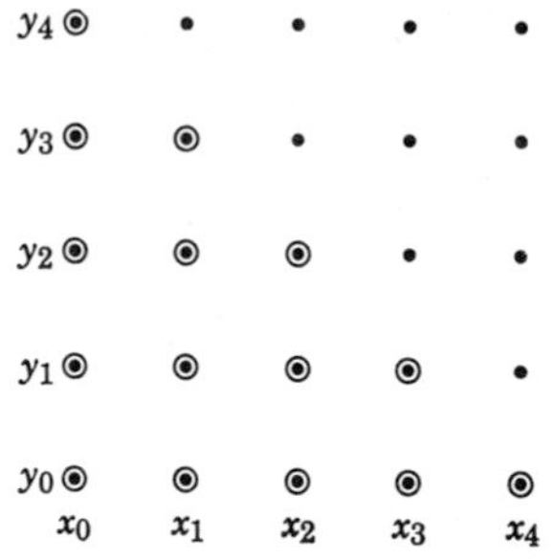

Figure 1a

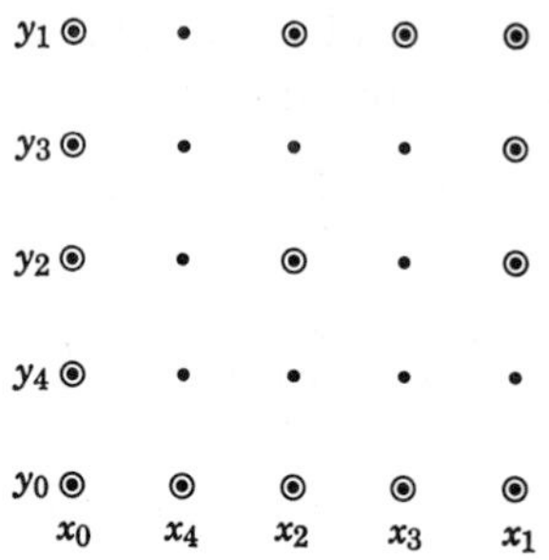

Figure 1b

The error in this formula is easily obtained from (6.11) of Chapter 6. We only wish to observe that it implies that (7) has degree of precision at least n.

The nodes that enter into this formula are a subset of the rectangular array of $(n + 1)^2$ points (x_i, y_j), $i, j = 0, 1, \ldots, n$. If the numbers x_i and y_j are monotonically ordered, say $x_0 < x_1 < x_2 < \cdots$, then the nodes in (7) are those on and below the main diagonal of a schematic $n + 1$ by $n + 1$ matrix of dots (see Figure 1a). However, any other selection of points obtained by permuting rows or columns could also be employed. This just corresponds to a renumbering of the x_i and y_j, say as in Figure 1b.

Both of the interpolatory cubature formulae (4) and (7) are easily extended to integrals in higher dimensions. As the dimension increases, there is a greater saving in number of nodes in extensions of $K_M\{f\}$ formulae compared to $J_N\{f\}$ formulae while maintaining precision of degree at least n. Thus, in the plane, $J_N\{f\}$ requires $N = (n + 1)^2$ nodes and $K_M\{f\}$ requires $M = \frac{1}{2}(n + 1)(n + 2)$ nodes for each to have degree of precision at least n. The ratio of the number of nodes required is

$$\frac{M}{N} = \frac{n + 2}{2(n + 1)} \approx \frac{1}{2}, \qquad \text{for large } n.$$

In three dimensions the ratio becomes

$$\frac{M}{N} = \frac{(n + 1)(n^2 + 5n + 6)/6}{(n + 1)^3} \approx \frac{1}{6}, \qquad \text{for large } n.$$

7.2. Undetermined Coefficients (and Nodes)

The general formula (2) can be written for double integrals as

$$J_N\{f\} = \sum_{\nu=1}^{N} A_\nu f(x_\nu, y_\nu). \tag{8}$$

We note that there are $3N$ parameters which determine such a scheme; the N coefficients, A_ν, and the $2N$ coordinates of the nodes, (x_ν, y_ν). As an approximation to the integral (3) the cubature formula (8) will have degree of precision at least n if for non-negative integers r and q:

$$\sum_{\nu=1}^{N} A_\nu x_\nu^{\ r} y_\nu^{\ q} = \iint_D x^r y^q \, dx \, dy, \qquad r + q = 0, 1, \ldots, n. \tag{9}$$

There are $\frac{1}{2}(n + 1)(n + 2)$ conditions imposed in (9). Hence, there are at least as many free parameters in (8) as there are conditions in (9) if

$$N \geq \frac{(n + 1)(n + 2)}{6} \approx \frac{n^2}{6} \tag{10a}$$

or

$$n \leq \tfrac{1}{2}(\sqrt{1 + 24N} - 3) \approx \sqrt{6N}. \tag{10b}$$

We note from (10a) that the number of nodes for which a degree of precision n might possibly be obtained is about $\frac{1}{3}$ the number used in the cubature formula (7) and about $\frac{1}{6}$ the number used in (4).

This procedure is practical only if the integrals on the right-hand side of the equations in (9) can be evaluated explicitly (or perhaps if they can be accurately approximated with ease). This is, of course, the case if D is a rectangle or, in fact, any polygonal domain. However, the resulting system of $\frac{1}{2}(n + 1)(n + 2)$ non-linear equations in $3N$ unknowns must also have a real solution with nodes in D. There are many special cases in which the procedure can be employed successfully. Let us consider, for example, the simple case of one node, $N = 1$. Then from (10) we find that $n = 1$ and there are only three equations in (9), namely,

$$A_1 = \iint_D dx \, dy, \quad A_1 x_1 = \iint_D x \, dx \, dy, \quad A_1 y_1 = \iint_D y \, dx \, dy.$$

Thus we find that the coefficient, A_1, is the area of the domain D and that the node, (x_1, y_1), is at the centroid of D. The resulting cubature formula

$$J_1\{f\} = A_1 f(x_1, y_1)$$

is exact for all linear integrands in (3). This derivation and formula trivially generalize to any number of dimensions.

The next simplest case of only two nodes, $N = 2$, yields $n = 2$ by (10), and hence a system of six non-linear algebraic equations of the form (9) must be solved. However, it is easy to show that this system does not always have a solution (see Problem 5). Thus we cannot, in general, determine a two point cubature formula with degree of precision two.

For domains which have symmetry about the x- and y-axes, the analysis of the system (9) can be simplified if the nodes are required to be symmetrically placed and have equal coefficients at corresponding locations. In this way various cubature schemes for integration over rectangles and circles may easily be derived.

When a formula of the form (8) satisfies the conditions (9), and hence has degree of precision n or more, an expression for the error can be derived by analogy with the proof of Theorem 0.1. For this purpose we require that the integrand function, $f(x, y)$, have continuous partial derivatives of all orders up to at least the $(n + 1)$st. Then we can expand the integrand about some point (x_0, y_0) into a finite Taylor's series with remainder in the form

$$f(x, y) = T_n(x, y) + R_n(x, y).$$

Here $T_n(x, y)$ is a polynomial of degree at most n and $R_n(x, y)$ is the known remainder which can be written symbolically as

$$R_n(x, y) = \frac{1}{(n + 1)!}\left[(x - x_0)\frac{\partial}{\partial x} + (y - y_0)\frac{\partial}{\partial y}\right]^{n+1} f(\xi, \eta).$$

The error in the cubature formula is then

$$\begin{aligned} E_N\{f\} &\equiv J\{f\} - J_N\{f\} \\ &= J\{T_n\} + J\{R_n\} - J_N\{T_n\} - J_N\{R_n\} \\ &= J\{R_n\} - J_N\{R_n\}. \end{aligned}$$

Here we have used the fact that since the degree of precision is n, $J\{T_n\} = J_N\{T_n\}$. In somewhat expanded form this error expression is

$$(11) \qquad \begin{aligned} E_N\{f\} = \frac{1}{(n + 1)!}\Bigg\{&\iint_D \left[(x - x_0)\frac{\partial}{\partial x} + (y - y_0)\frac{\partial}{\partial y}\right]^{n+1} \\ &\times f(\xi, \eta)\, dx\, dy - \sum_{\nu=1}^{N} A_\nu \\ &\times \left[(x_\nu - x_0)\frac{\partial}{\partial x} + (y_\nu - y_0)\frac{\partial}{\partial y}\right]^{n+1} f(\xi_\nu, \eta_\nu)\Bigg\}. \end{aligned}$$

The integrand in (11) is, of course, just symbolic since (ξ, η) depends upon (x, y) for purposes of the integration but not for the differentiations. Note that if the *maximum distance from* (x_0, y_0) *to any point in D or any node* (x_ν, y_ν) *is h* then the error satisfies $E_N\{f\} = \mathcal{O}(h^{n+1})$. In particular, *if the coefficients* A_ν *are all non-negative*, so that

$$\sum_{\nu=1}^{N} A_\nu = \iint_D dx\, dy,$$

and all $(n + 1)$st order derivatives of $f(x, y)$ are bounded by M_{n+1}, say, then with h as above we deduce from (11) that

$$|E_N\{f\}| \leq \frac{(2h)^{n+1}}{(n+1)!} 2M_{n+1} \iint_D dx\, dy. \tag{12}$$

This estimate holds for any cubature formula which has non-negative coefficients and degree of precision $n \geq 0$, provided that the integrand has the appropriate smoothness.

7.3. Separation of Variables

Perhaps the most obvious way to devise approximations for multiple integrals is by the repeated use of one dimensional quadrature formulae. The domain, D, must be somewhat special, or else it must be the union of special subdomains, in order for us to apply this method of separation of variables. In two dimensions the restriction is that vertical (or horizontal) lines have at most one segment in common with D. Integrals of the form (3) can then be written as

$$J\{f\} = \int_a^b \int_{y_1(x)}^{y_2(x)} f(x, y)\, dy\, dx, \tag{13}$$

where the segment $y_1(x) \leq y \leq y_2(x)$ is in D for all x in $[a, b]$. If we introduce for each $f(x, y)$ a function of the single variable x by the definition

$$G(f; x) = \int_{y_1(x)}^{y_2(x)} f(x, y)\, dy, \qquad a \leq x \leq b, \tag{14a}$$

then the double integral (13) becomes

$$J\{f\} = K\{G\} \equiv \int_a^b G(f; x)\, dx. \tag{14b}$$

Now let us approximate the integral $K\{G\}$ by some n-point quadrature formula with coefficients α_j and *nodes*, x_j, *which all lie in* $[a, b]$, say

$$K_n\{G\} = \sum_{j=1}^{n} \alpha_j G(f; x_j). \tag{15}$$

The numbers $G(f; x_j)$ which are required to evaluate this formula are given in (14a) as single integrals and hence can be approximated by applying other one dimensional quadrature formulae. For ease of presentation we use an m-point formula for each j and write the approximations to the $G(f; x_j)$ as

$$G_m(f; x_j) \equiv \sum_{k=1}^{m} \beta_{jk} f(x_j, y_{jk}), \qquad j = 1, 2, \ldots, n. \tag{16}$$

Here the coefficients β_{jk} and nodes y_{jk} must, in general, depend upon the value x_j since the interval of integration, $[y_1(x), y_2(x)]$ in (14a) depends upon x. By using (16) the value $K_n\{G\}$ of (15) finally yields the cubature formula

$$J_{mn}\{f\} = \sum_{j=1}^{n} \sum_{k=1}^{m} \alpha_j \beta_{jk} f(x_j, y_{jk}). \tag{17}$$

This formula employs mn nodes and is somewhat similar to that given by (4a) with coefficients (5). If the domain were a rectangle, then the same m-point formula could reasonably be used in (16) for all j. Then in (17) we could replace β_{jk} and y_{jk} by β_k and y_k, respectively, to get formal agreement with (4a) and (5).

The error in the cubature formula (17) is defined as

$$E_{mn}\{f\} = J\{f\} - J_{mn}\{f\}.$$

Let us introduce for the quadrature errors in (15) and (16) the notation:

$$\begin{aligned} e_n\{G\} &= K\{G\} - K_n\{G\}, \\ e_{mj}\{f\} &= G(f; x_j) - G_m(f; x_j), \qquad j = 1, 2, \ldots, n. \end{aligned} \tag{18}$$

Then from (14b) we have

$$\begin{aligned} E_{mn}\{f\} &= K\{G\} - K_n\{G\} + K_n\{G\} - J_{mn}\{f\} \\ &= e_n\{G\} + \sum_{j=1}^{n} \alpha_j [G(f; x_j) - G_m(f; x_j)] \\ &= e_n\{G\} + \sum_{j=1}^{n} \alpha_j e_{mj}\{f\}. \end{aligned} \tag{19}$$

It is interesting to note that when the degrees of precision of the quadrature formulae (15) and (16) are known we do not, in general, know the degree of precision of the cubature formula (17). This is because $G(f; x)$ is not generally a polynomial in x when $f(x, y)$ is a polynomial. In fact, it is easy to see that $J_{mn}\{f\}$ may not even be exact for constant integrands when the quadrature formulae employed have arbitrarily high degrees of precision.

If the bounding curves $y_1(x)$ and $y_2(x)$ of D are polynomials of degree at most $s \geq 0$, then lower bounds can be given for the degree of precision. If $f(x, y)$ is a polynomial of degree p then, by (14a), $G(f; x)$ is a polynomial of degree at most $s(p + 1)$. So if (15) has degree of precision $s(p + 1)$ and (16) has degree of precision p then $J_{mn}\{f\}$ has degree of precision at least p. Of course, if the domain is a rectangle, i.e., $s = 0$, then $J_{mn}\{f\}$

has a degree of precision which is at least the minimum of those for (15) and (16).

From this result it follows that cubature formulae of high degree of precision with relatively few nodes may be devised if the domain of integration is a rectangle. To get degree of precision n in such a case we use $[(n + 1)/2]$-point Gaussian quadrature formulae as the two relevant schemes for (15) and (16). It is here assumed that n is odd and the total number of nodes required is then only $\frac{1}{4}(n + 1)^2$. For large values of n this is about half the number of points that were required in the efficient interpolation quadrature scheme (7) with the same degree of precision (and about $\frac{2}{3}$ the number required in Subsection 7.2 by the method of undetermined coefficients). However, none of the nodes in the Gaussian scheme can be on the boundary of the domain and hence its usefulness in composite cubature formulae is reduced.

The extension to higher dimensions of the method of separation of variables is fairly clear. The restrictions on the domain are somewhat complicated, but, for instance, it is sufficient for the domain to be convex. In particular, for rectangular parallelopipeds, only a single one dimensional quadrature formula need be specified for each dimension. If the appropriate Gaussian schemes are used in this case, we obtain degree of precision n (odd) by using only $[(n + 1)/2]^p$ nodes in p dimensions.

7.4. Composite Formulae for Multiple Integrals

Just as in the case of one dimensional integrals, it may be necessary to decompose the integral (1) into a sum of integrals over smaller non-overlapping domains. That is, if $D_i \cap D_j$ has no inner points for $i \neq j$ and

$$D = D_1 \cup D_2 \cup \cdots \cup D_M,$$

then

$$J\{f\} \equiv \int_D f(\mathbf{x})\, d\mathbf{x} = \sum_{i=1}^{M} \int_{D_i} f(\mathbf{x})\, d\mathbf{x}. \tag{20}$$

If N nodes are used to calculate the integral over each of the primitive domains D_i say

$$J_N\{f, D_i\} \equiv \sum_{j=1}^{N} \alpha_{ij} f(\mathbf{x}_{ij}),$$

then *at most* MN evaluations of $f(\mathbf{x})$ are used in (20). We say *at most* because a node $\mathbf{x}$ of D_i may also be a node of D_j but $f(\mathbf{x})$ need only be found once for such a node.

If the region D is a p-dimensional rectangular parallelopiped, then a corner (or vertex) node of D_i may also occur in as many as $2^p - 1$

adjoining cells D_j. Hence the amount of work necessary to evaluate the integrand may be minimized by selecting as many nodes as possible to be vertex nodes, then edge nodes, then face nodes, etc. For example, in two dimensions, the scheme of Figure 1b is more efficient than the scheme of Figure 1a.

If the cubature formula used in D_i, $J_N\{f; D_i\}$, has degree of precision n, and has non-negative weights $\{\alpha_{ij}\}$, then just as in the derivation of (12),

$$\left|\int_{D_i} f(\mathbf{x})\, d\mathbf{x} - \sum_{j=1}^{N} \alpha_{ij} f(\mathbf{x}_{ij})\right| \leq \frac{(ph)^{n+1}}{(n+1)!} 2M_{n+1} \int_{D_i} d\mathbf{x}, \tag{21}$$

when D_i is contained in a cube of side $2h$, and

$$\left|\frac{\partial^{n+1}}{\partial x_1^{j_1} \cdots \partial x_p^{j_p}} f(\mathbf{x})\right| \leq M_{n+1}$$

for all $\mathbf{x}$ in D and all $\{j_k\}$ satisfying $\sum_{k=1}^{p} j_k = n + 1$. With these conventions, it is then a simple matter to derive the fundamental estimate of the error in the composite cubature formula,

$$\left|J\{f\} - \sum_{i=1}^{M} J_N\{f; D_i\}\right| \leq \frac{(ph)^{n+1}}{(n+1)!} 2M_{n+1} \int_D d\mathbf{x}. \tag{22}$$

PROBLEMS, SECTION 7

1. Devise the cubature schemes indicated by (4a) and (5) for equally spaced nodes when (1) $m = n = 0$; (2) $m = 0, n = 1$; (3) $m = n = 1$; (4) $m = n = 2$. Case (4) is the generalization of Simpson's rule and (3) is the generalization of the trapezoidal rule to integrals over rectangles.

2. Determine the general cubature schemes for $n = 0, 1, 2$ determined by integrating $P_n(x, y)$, given in (6.10) of Chapter 6, over an arbitrary domain D. Specialize these results for a rectangle $a \leq x \leq b$, $c \leq y \leq d$. Take uniform spacing in this rectangle in each case to simplify further. (*Note*: These schemes are not uniquely determined; see Figure 1.)

3. Compare the schemes (3) and (4) of Problem 1 to those with $n = 1$ and $n = 2$, respectively, in Problem 2 by approximating the integral

$$\int_0^1 \int_1^2 x\sqrt{9 - y^2}\, dx\, dy.$$

Try at least two of the nodal schemes for each case of the methods of Problem 2.

4. Determine the ratio M/N for the number of nodes required in four and five dimensions, to extend the formulae $J_N\{f\}$ and $K_M\{f\}$ of Subsection 7.1 which have a degree of precision at least n.

5. Consider the case $N = n = 2$ in the equations (9). Show that the resulting system does not have a solution in general by considering the special case

$$\iint_D x\,dx\,dy = \iint_D y\,dx\,dy = \iint_D xy\,dx\,dy = 0.$$

[Hint: Introduce the notation

$$A_1x_1 = \xi, \quad A_1y_1 = \eta, \quad A_1 = \zeta, \quad \iint_D dx\,dy = \zeta_0$$

and show that the system reduces to

$$\xi\eta\left(\frac{1}{\zeta} + \frac{1}{\zeta_0 - \zeta}\right) = 0, \qquad \xi^2\left(\frac{1}{\zeta} + \frac{1}{\zeta_0 - \zeta}\right) = \iint_D x^2\,dx\,dy,$$

$$\eta^2\left(\frac{1}{\zeta} + \frac{1}{\zeta_0 - \zeta}\right) = \iint_D y^2\,dx\,dy.]$$

6. Give the proof of (22) in detail.

8

Numerical Solution of Ordinary Differential Equations

0. INTRODUCTION

In order to study the effectiveness of various methods for the numerical solution of differential equation problems, we illustrate the theory for the case of the general first order ordinary differential equation

$$\text{(1a)} \qquad \frac{dy}{dx} = f(x, y),$$

subject to the initial condition

$$\text{(1b)} \qquad y(a) = y_0.$$

It is required to find a solution, $y = y(x)$, of the problem (1) in some interval, say $a \le x \le b$. Under suitable restrictions† on the function $f(x, y)$, it is well known that a unique solution exists.

The class of methods to be discussed uses a subdivision of the interval $I \equiv [a, b]$, by a finite set of distinct points

$$\text{(2)} \qquad I_\Delta\colon\ x_0 = a, \quad x_{i+1} = x_i + \Delta x_i, \quad i = 0, 1, \ldots, N.$$

Finer subdivisions also play a role and are denoted by the same generic symbol I_Δ. In the present context, the set of points defining a subdivision is frequently called a *net*, *grid*, *lattice*, or *mesh*. The quantities Δx_i are called the *net spacings* or *mesh widths*. Corresponding to each point of the net we seek a quantity, say u_i, which is to approximate $y_i \equiv y(x_i)$, the exact

† For instance, existence and uniqueness of the solution are assured if $f(x, y)$ is bounded, continuous in x, and Lipschitz continuous with respect to y in some sufficiently large rectangle R_C: $[a \le x \le b, |y - y_0| \le C]$, see equation (1.5).

solution at the corresponding net point. The set of values $\{u_i\}$ is called a *net function*. Clearly, the values $\{y_i\}$ also form a net function on I_Δ. We use the generic symbol $\{u_i\}$ to denote a net function on any subdivision.

For most of the methods treated in this chapter, the quantities $\{u_i\}$ are to be determined from a set of (usually non-algebraic) equations which in *some sense* approximate the system (1); these approximating equations are called *difference equations*. The natural requirements for the approximating difference equations are that for any function $f(x, y)$ (in some class of sufficiently often differentiable functions):

(a) They have a unique solution.
(b) Their solution, at least for "sufficiently small" net spacings, should be "close" to the exact solution of (1).
(c) Their solution should be "effectively computable."

Property (a) is trivially satisfied by many of the difference equations to be studied, the so-called *explicit schemes*. Whether or not the *implicit schemes* satisfy condition (a) is determined by a study of the roots of a sequence of equations (or systems), of the form $z = g(z)$ (see Section 2). In general, if Δx_i is small enough, the implicit equations have a unique solution.

Property (b) is related to the question of the convergence, as $\max_i \Delta x_i \to 0$, of $\{u_i\}$ to $\{y_i\}$. The study of such *convergence properties* of the difference solution shall occupy a considerable part of this chapter. In Sections 1 through 3 we examine separately the convergence of each of several special methods. In Sections 5 and 6 we give a general treatment of convergence which includes the previous cases.

The vaguely formulated property (c) involves two important considerations: (i) the number of single precision computations required; (ii) the growth of roundoff errors in the computed difference "solution." Of course, these two points are related since having to compensate for rounding errors by using more significant figures usually entails additional computations. A trivial first approximation of (i) is based on the operational count for infinite precision arithmetic. The growth of the roundoff error is related to the notion of *stability* of difference equations. The stability theory of difference equations treated in Section 5 is based on the study of difference equations with constant coefficients developed in Section 4. We establish the main general theorem of this chapter in Section 5 (i.e., stability is equivalent to convergence for consistent methods).

There are a number of systematic ways in which one can "derive" or rather generate difference equations that approximate or are consistent with (1). That is, these difference equations seem to be discrete models for the continuous problem (1). But, no matter how reasonable the derivation, the efficacy of such difference equations can only be determined by checking conditions (a)–(c). In fact, in Subsection 1.4 we derive a discrete model which seems quite reasonable, but is absolutely useless since the growth of the roundoff error cannot be controlled (i.e., it is unstable).

Later, in Section 5, a simple criterion is developed for recognizing when a finite difference scheme is stable and convergent.

It should be recalled that some of the numerical methods for approximating solutions of ordinary differential equations, and systems of them, also have important theoretical applications. In fact, one of the basic existence and uniqueness proofs uses the *Euler-Cauchy difference method* of the next section. We resist the temptation to present such a proof here. Rather, we will assume that Problem (1) is "well-posed," i.e., it has a unique solution with a certain number of continuous derivatives and furthermore, the solution depends differentiably on the initial data. As indicated in the footnote on page 364, we can guarantee the well-posedness of Problem (1) for a wide class of functions $f(x, y)$. We will be interested in showing that certain difference methods have properties (a)–(c) for such a class of functions, $f(x, y)$.

At the present time there seems to be no general way of formulating an "ideal method" for solving (1). An "ideal method" is one which requires the least amount of work (number of single precision computations) to produce an approximate solution of (1) accurate to within a given $\epsilon > 0$.

In the following sections a number of inequalities are derived with the use of two simple lemmas:

LEMMA 1. *For all real numbers z:*

$$1 + z \leq e^z, \tag{3}$$

(*where the equality holds only for $z = 0$*).

Proof. Since the function e^z has continuous derivatives of all orders we have by Taylor's theorem

$$e^z = 1 + z + \frac{z^2}{2} e^{\theta z}, \qquad 0 < \theta < 1.$$

But the last term on the right-hand side is non-negative and vanishes only when $z = 0$ and so the lemma follows. ∎

A simple corollary of this result is contained in

LEMMA 1′. *For all z such that $1 + z \geq 0$,*

$$0 \leq (1 + z)^n \leq e^{nz}, \qquad n \geq 0. \tag{4}$$

Proof. Obvious from Lemma 1. ∎

1. METHODS BASED ON APPROXIMATING THE DERIVATIVE: EULER-CAUCHY METHOD

To illustrate the basic concepts we consider the simple difference approximation to (0.1a) which results from approximating the derivative by a forward difference quotient,

$$\text{(1a)} \qquad \frac{u_{j+1} - u_j}{h} = f(x_j, u_j), \qquad j = 0, 1, \ldots, N - 1.$$

Here, for simplicity only, we have chosen a *uniform net*

$$\text{(2)} \qquad I_h\colon\ x_0 = a;\quad x_j = x_0 + jh,\quad j = 0, \ldots, N;\quad h = \frac{b - a}{N}.$$

The initial condition (0.1b) is replaced by

$$\text{(1b)} \qquad u_0 = y_0 + e_0,$$

where we have intentionally permitted the introduction of an initial error, e_0. Equations (1) are the difference equations of the Euler-Cauchy method. This method is also called the *polygon method*, where the polygon is constructed by joining successive points (x_j, u_j) with straight line segments. Each segment has the slope given by the value of f at the left endpoint.

The existence of a unique solution u_j of the difference equations follows from writing (1a) as

$$\text{(3)} \qquad u_{j+1} = u_j + hf(x_j, u_j), \qquad j = 0, 1, \ldots, N - 1.$$

Then with u_0, given in (1b), the above yields recursively $u_1, u_2, \ldots, u_N$ provided only that $f(x_j, u_j)$ is defined.

The present analysis of (3) is based on using infinite precision arithmetic. That is, the numbers that would be calculated in finite precision arithmetic satisfy $U_{j+1} = [U_j + hf(x_j, U_j)] + \rho_{j+1}$ where ρ_{j+1} is the rounding error made in evaluating the term in brackets. Later on, in Subsection 2, we study the error, $U_j - y_j$, as $h \to 0$.

We now turn to a consideration of the error $\{e_j\}$ defined by

$$\text{(4)} \qquad e_j = u_j - y_j, \qquad j = 0, 1, \ldots, N.$$

For this study we require that, in some region S to be specified later, $f(x, y)$ be continuous in x and satisfy a uniform Lipschitz condition in the variable y:

$$\text{(5)} \qquad |f(x, y) - f(x, y')| \leq K|y - y'|, \qquad \begin{array}{l}\text{for some constant } K > 0;\\ \text{all } (x, y) \text{ and } (x, y') \in S.\end{array}$$

[If $K = 0$, then $f(x, y)$ is independent of y and the problem (0.1) is a simple problem of quadrature treated in Chapter 7.] In addition, we will need a measure of the error by which the exact solution of (0.1) fails to satisfy the difference relation (1a). This error is called the *local truncation error* or discretization error and is defined by:

$$\tau_{j+1} = \frac{y_{j+1} - y_j}{h} - f(x_j, y_j), \qquad j = 0, 1, \ldots, N - 1.$$

This relation is frequently written as

$$(6) \qquad y_{j+1} = y_j + hf(x_j, y_j) + h\tau_{j+1}, \qquad j = 0, 1, \ldots, N - 1.$$

An explicit representation of the τ_j will be derived shortly [under additional conditions on $f(x, y)$]. If the τ_j vanish as $h \to 0$, we say that the difference equations are *consistent* with the differential equation. But as will be seen in Subsection 2, for another consistent scheme, the corresponding difference solution $\{u_j\}$ can diverge as $h \to 0$, from the exact solution $\{y_j\}$ even though e_0 is small. However, in the present case, we have

THEOREM 1. *Let* $\{u_j\}$ *be the solution of* (1) *and* $y(x)$ *the solution of* (0.1) *where* $f(x, y)$ *satisfies* (5) *in the strip* S: $[a \leq x \leq b, |y| < \infty]$. *Then, with the definitions* (6) *of* $\{\tau_j\}$:

$$(7) \qquad |u_j - y(x_j)| \leq e^{K(x_j - a)}\left[|e_0| + \frac{\tau}{K}\right], \qquad j = 0, 1, \ldots, N,$$

where

$$\tau \equiv \max_j |\tau_j|.$$

Proof. The subtraction of (6) from (3) yields

$$e_{j+1} = e_j + h[f(x_j, u_j) - f(x_j, y_j)] - h\tau_{j+1}, \qquad j = 0, 1, \ldots, N - 1.$$

By means of the Lipschitz condition we deduce that

$$|e_{j+1}| \leq (1 + hK)|e_j| + h\tau.$$

This inequality yields recursively

$$\begin{aligned}
|e_{j+1}| &\leq (1 + hK)^2|e_{j-1}| + [1 + (1 + hK)]h\tau, \\
&\leq (1 + hK)^3|e_{j-2}| + [1 + (1 + hK) + (1 + hK)^2]h\tau, \\
&\;\;\vdots \\
&\leq (1 + hK)^{j+1}|e_0| + \left[\frac{(1 + hK)^{j+1} - 1}{K}\right]\tau,
\end{aligned}$$

where we have summed the geometric progression. Since $K > 0$, we may apply Lemma 0.1′ in the form

$$(1 + hK)^{j+1} \leq e^{(j+1)hK} = e^{K(x_{j+1} - x_0)}.$$

Hence for all $j \leq N - 1$,

$$(8) \qquad |e_{j+1}| \leq e^{K(x_{j+1} - x_0)}\left(|e_0| + \frac{\tau}{K}\right),$$

and the theorem follows. ∎

The simple bound of Theorem 1 shows that the error at any net point, x_j, will be small if both the initial error, e_0, and the maximum local truncation error, τ, are small. Now, the value of $|e_0|$ is determined by the accuracy with which the number y_0, the initial condition, is approximated. On the other hand, we may guarantee that τ can be made arbitrarily small by picking h sufficiently small, if d^2y/dx^2 is continuous in $[a, b]$. That is, from Taylor's theorem,

$$y(x_j + h) = y(x_j) + h\frac{dy(x_j)}{dx} + \frac{h^2}{2}\frac{d^2y(x_j + \theta_j h)}{dx^2},$$

$$0 < \theta_j < 1; \qquad j = 0, 1, \ldots, N - 1.$$

However, since $y(x)$ is the solution of (0.1), $dy(x_i)/dx = f(x_i, y_i)$ and a comparison of the above with (6) yields

$$(9) \qquad \tau_{j+1} = \frac{h}{2}\frac{d^2y(x_j + \theta_j h)}{dx^2}, \qquad 0 < \theta_j < 1, \qquad j = 0, 1, \ldots, N - 1.$$

Using this representation of τ_j in Theorem 1 and the formula obtained from (0.1a) by differentiation

$$\frac{d^2y}{dx^2} = f_x(x, y) + f_y(x, y)\frac{dy}{dx},$$

we obtain a result which may be summarized as the

COROLLARY. *If, in addition to the hypothesis of Theorem 1, $f_x(x, y)$ and $f_y(x, y)$ are continuous in S, then*

$$|e_j| \leq e^{K(x_j - a)}\left(|e_0| + h\frac{M_2}{2K}\right) \leq e^{K(b - a)}\left(|e_0| + h\frac{M_2}{2K}\right)$$

where†

$$M_2 = \max_{a \leq x \leq b}\left|\frac{d^2y(x)}{dx^2}\right|.$$ ∎

If $e_0 = 0$ or $|e_0| \leq \alpha h$ for some constant α, then as a consequence of the corollary, $\lim_{h \to 0} e_j = 0$, or more precisely, the maximum norm of the error $\{e_j\}$ is at least $\mathcal{O}(h)$ and converges uniformly to zero since the rightmost

† If in S, $|f_x| \leq P$, $|f_y| \leq Q$, and $|f| \leq R$, then $\left|\frac{d^2y}{dx^2}\right| \leq P + QR$. Hence for such a class of functions $f(x, y)$, we find the a priori bound $M_2 \leq P + QR$.

bound is independent of j. Note that since f_y is assumed continuous in the corollary, the condition (5) need not be postulated but can, in fact, be deduced if $K \equiv \sup_S |f_y|$ is finite.

In general, the bounds on the error are usually tremendous overestimates. It is possible, however, to obtain more precise expressions for the error, essentially under the conditions of the corollary. These expressions are in turn not practical since they cannot be evaluated explicitly. But since they do have analytical significance we present

THEOREM 2. *If* $\{e_j\}$ *and* $\{\tau_j\}$ *are defined by* (4) *and* (6) *respectively, and* $f_y(x, y)$ *is continuous in S, then there exist numbers* ϕ_i *in* $0 < \phi_i < 1$ *such that*

$$e_i = A_{i,0}e_0 - h \sum_{j=1}^{i} A_{i,j}\tau_j, \qquad i = 1, 2, \ldots, N; \tag{10}$$

where $A_{0,0} \equiv 1$ *and*

$$A_{i,j} \equiv \begin{cases} 0 & j \geq i + 1 \\ 1 & j = i, \\ \alpha_{i-1}A_{i-1,j} & j < i, \ \alpha_i = 1 + hf_y(x_i, y_i + \phi_i e_i). \end{cases} \tag{11}$$

Proof. The proof is similar to that of Theorem 1 but now the mean value theorem is used in place of the Lipschitz condition. Thus, from (6) and (3),

$$\begin{aligned} e_{i+1} &= e_i + h[f(x_i, u_i) - f(x_i, y_i)] - h\tau_{i+1} \\ &= \alpha_i e_i - h\tau_{i+1}. \end{aligned}$$

To show that the algebraic manipulations, in the recursive application of the above, yield quantities of the form (10) and (11), we proceed by induction. Then with $i = 0$ in the above

$$\begin{aligned} e_1 &= \alpha_0 e_0 - h\tau_1 \\ &= A_{1,0}e_0 - hA_{1,1}\tau_1. \end{aligned}$$

Now we assume (10) to be valid and use (11) to obtain

$$\begin{aligned} e_{i+1} &= \alpha_i A_{i,0}e_0 - h \sum_{j=1}^{i} \alpha_i A_{i,j}\tau_j - h\tau_{i+1} \\ &= A_{i+1,0}e_0 - h \sum_{j=1}^{i} A_{i+1,j}\tau_j - hA_{i+1,i+1}\tau_{i+1} \\ &= A_{i+1,0}e_0 - h \sum_{j=1}^{i+1} A_{i+1,j}\tau_j. \end{aligned}$$

The induction is thus complete and the theorem follows. ■

By restricting h to be sufficiently small the exact error expression (10) can be reduced to a simple form which has practical significance. We state this result as

COROLLARY 1. *Under the hypothesis of Theorem 2 let*

$$d \equiv \inf_S f_y(x, y), \qquad D \equiv \sup_S f_y(x, y)$$

be finite and restrict h so that

$$1 + hd \geq 0. \tag{12}$$

Then for each $i = 1, 2, \ldots, N$, *there exist three numbers* p_i, q_i, *and* t_i *in the intervals*

$$d \leq p_i \leq D, \quad d \leq q_i \leq D, \quad \min_{1 \leq j \leq i} \tau_j \leq t_i \leq \max_{1 \leq j \leq i} \tau_j, \tag{13}$$

such that

$$e_i = (1 + hp_i)^i e_0 - \left[\frac{(1 + hq_i)^i - 1}{q_i}\right] t_i; \qquad i = 1, 2, \ldots, N. \tag{14}$$

Proof. We note that from (11) and (12) it follows that

$$0 \leq 1 + hd \leq \alpha_j \leq 1 + hD, \qquad j = 0, 1, \ldots, N - 1.$$

Then

$$(1 + hd)^i \leq A_{i,0} = \alpha_0 \alpha_1 \cdots \alpha_{i-1} \leq (1 + hD)^i,$$

and hence there is a number p_i in the interval $[d, D]$ such that

$$A_{i,0} = (1 + hp_i)^i.$$

Now define the quantities

$$S_i = \sum_{j=1}^{i} A_{i,j}, \qquad t_i = \sum_{j=1}^{i} \left(\frac{A_{i,j}}{S_i}\right) \tau_j. \tag{15}$$

The $A_{i,j}$ are non-negative as a result of condition (12). Hence t_i, which is an average with non-negative weights of the τ_j, must satisfy condition (13) (see Lemma 1.1 of Chapter 7 which can be used to prove this assertion). We also note, using (11), that

$$0 \leq (1 + hd)^{i-j} \leq A_{i,j} = \alpha_j \alpha_{j+1} \cdots \alpha_{i-1} \leq (1 + hD)^{i-j}, \qquad j < i;$$

$$A_{i,i} = 1.$$

Then from the definition of S_i

$$\begin{aligned} 1 + (1 + hd) + \cdots + (1 + hd)^{i-1} &\leq S_i \\ &\leq 1 + (1 + hD) + \cdots + (1 + hD)^{i-1}, \end{aligned}$$

or by summing the progressions

$$\frac{(1 + hd)^i - 1}{hd} \leq S_i \leq \frac{(1 + hD)^i - 1}{hD}; \qquad i = 1, 2, \ldots, N.$$

But the functions $[(1 + z)^i - 1]/z$ are continuous functions of z and hence there exist numbers q_i, in the interval $[d, D]$, such that

$$S_i = \frac{(1 + hq_i)^i - 1}{hq_i}; \qquad i = 1, 2, \ldots, N.$$

The corollary now follows by using the expressions for $A_{i,0}$, S_i, and t_i in (10). ■

The form of the error given in (14) can be used to derive practical information in many cases, Clearly, if d and D are known, or can be estimated, we can obtain upper and *lower* bounds on the factors which multiply the two error terms (e_0 and t_i). A more striking application occurs, however, when $f(x, y)$ is such that $D < 0$ (i.e., $f_y < 0$). Now clearly, $p_i < 0$, $q_i < 0$, and by the condition (12) imposed on h:

$$0 \leq (1 + hp_i) < 1, \qquad 0 \leq (1 + hq_i) < 1.$$

Then by Lemma 0.1′

$$(1 + hp_i)^i < e^{ihp_i} = e^{(x_i - a)p_i},$$

or since $p_i < 0$, this may be written as

$$(1 + hp_i)^i < e^{-|p_i(x_i - a)|} < 1.$$

Similarly, $0 \leq (1 + hq_i)^i < 1$, and by taking absolute values in (14), we find

COROLLARY 2. *If $f_y < 0$, then the hypotheses of Theorem 2 and its Corollary 1 imply*

$$|e_i| \leq e^{-|p_i(x_i - a)|}|e_0| + \left|\frac{t_i}{D}\right|, \qquad i = 1, 2, \ldots, N. \tag{16}$$ ■

This result shows that the initial error cannot grow if $f_y < 0$ and further, that the local truncation errors in this case contribute at most an amount $\tau/|D|$.

1.1. Improving the Accuracy of the Numerical Solution

We now improve upon the corollary to Theorem 1 by characterizing the $\mathcal{O}(h)$ term in the error $\{e_j\}$.

THEOREM 3. *Let the solution $y = y(x)$ of* (0.1) *have three continuous and bounded derivatives: let $f_{yy}(x, y)$ be continuous and bounded, and let the initial error e_0 in the difference solution $\{u_i\}$ of* (1) *be*

$$e_0 = \xi_0 h,$$

where ξ_0 is independent of h. Then

$$e_j = h\xi(x_j) + \mathcal{O}(h^2), \qquad j = 0, 1, \ldots, N,$$

where $\xi(x)$ is the solution† *of the linear problem*

$$\frac{d\xi}{dx} = f_y(x, y(x))\xi - \tfrac{1}{2}y''(x),$$

$$\xi(a) = \xi_0.$$

Proof. As in the proof of Theorem 2,

$$e_{i+1} = \alpha_i e_i - h\tau_{i+1}.$$

But now in (9) and (11) we use the extra differentiability properties to obtain

$$\alpha_i = 1 + hf_y(x_i, y_i) + hf_{yy}(x_i, y_i + \phi_i' e_i)\phi_i e_i$$

$$\tau_{i+1} = \frac{h}{2}y''(x_i) + \frac{h}{2}y'''(x_i + \theta_i' h)\theta_i h, \qquad 0 < \phi_i, \phi_i', \theta_i, \theta_i' < 1.$$

Then from this,

$$e_{i+1} = [1 + hf_y(x_i, y_i)]e_i - \frac{h^2}{2}y''(x_i) + h[\mathcal{O}(e_i^2) + \mathcal{O}(h^2)].$$

By using the differential equation which defines $\xi(x)$, Taylor's expansion yields

$$\xi(x_{i+1}) = \xi(x_i) + h\xi'(x_i) + \frac{h^2}{2}\xi''(x_i + \psi_i h)$$

$$= [1 + hf_y(x_i, y_i)]\xi(x_i) - \frac{h}{2}y''(x_i) + \mathcal{O}(h^2).$$

We now form the quantities

$$\delta_j \equiv e_j - h\xi_j$$

and find

$$\delta_{i+1} = [1 + hf_y(x_i, y_i)]\delta_i + h[\mathcal{O}(e_i^2) + \mathcal{O}(h^2)], \qquad i = 1, 2, \ldots.$$

† Under the hypothesis, $\xi(x)$ exists and has a continuous second derivative. In fact, $\xi(x)$ can be explicitly represented by quadratures.

Now we observe, as remarked after the corollary to Theorem 1, that $|e_i| = \mathcal{O}(h)$. Hence we may delete the term $\mathcal{O}(e_i^2)$. But, from the specific initial conditions chosen for $\xi(x)$ we have $\delta_0 = 0$, and hence a recursive application of the formulae for δ_i yields, as in the derivation of (7),

$$|\delta_j| \leq e^{K(b-a)} \frac{[\mathcal{O}(h^2)]}{K} = \mathcal{O}(h^2)$$

and the theorem follows. ■

To apply Theorem 3, we introduce the notation $u_j(h)$ to indicate the dependence of the numerical solution on the net spacing. Then the theorem states that with, say $x_j = z$,

$$u_j(h) = y(z) + h\xi(z) + \mathcal{O}(h^2).$$

Similarly, with the net spacing $h/2$ and $x_{2j} = z$, we have

$$u_{2j}\left(\frac{h}{2}\right) = y(z) + \frac{h}{2}\,\xi(z) + \mathcal{O}(h^2).$$

Then

$$2u_{2j}\left(\frac{h}{2}\right) - u_j(h) = y(z) + \mathcal{O}(h^2),$$

and an extra order of magnitude in accuracy is obtained if we use as the difference approximation, at any point $x_j = z$ of the net with spacing h, the quantity

$$\bar{u}_j = 2u_{2j}\left(\frac{h}{2}\right) - u_j(h).$$

This requires computations with two nets of spacings h and $h/2$ respectively. It should be observed that the formula for $\bar{u}_j$ is similar to the formula which arises in Aitken's δ^2-process in the iterative solution of arbitrary equations (see Subsection 2.4 of Chapter 3). In the present context this procedure is called *Richardson's deferred approach to the limit*, or *extrapolation to zero mesh width*. This extrapolation may be applied, in an appropriately modified form, to many of the numerical methods to be considered here.

1.2. Roundoff Errors

In actually performing the calculations required to evaluate (1), roundoff errors will, in general, be introduced. Thus the numbers actually obtained will not be the set $\{u_i\}$ but, say, some quantities $\{U_i\}$. These numbers satisfy equations of the form

$$\text{(17a)} \qquad U_{i+1} = U_i + hf(x_i, U_i) + \rho_{i+1}, \qquad i = 0, 1, \ldots, N-1;$$

where ρ_{i+1} represents the error introduced by inexact evaluation of the quantity $U_i + hf(x_i, U_i)$. The ρ_i are called the *local roundoff errors*. If we Let ρ_0 be the initial roundoff error committed in evaluating y_0 then the initial condition becomes

$$U_0 = y_0 + \rho_0. \tag{17b}$$

Let the errors between the U_i, the actual numbers obtained in the computation, and the u_i, the exact solution of the difference equations, be denoted by

$$\epsilon_i \equiv U_i - u_i, \qquad i = 0, 1, \ldots, N, \tag{18}$$

Then from (1) and (17) we obtain

$$\begin{aligned} &\epsilon_0 = \rho_0 - e_0; \\ &\epsilon_{i+1} = \epsilon_i + h[f(x_i, U_i) - f(x_i, u_i)] + \rho_{i+1}, \quad i = 0, 1, \ldots, N-1. \end{aligned} \tag{19}$$

It is clear that these equations for ϵ_i are formally similar to those which determine the quantities e_i. In fact, the previous theorems and corollaries can be restated in an obvious way to give bounds and representations of the errors ϵ_i. We shall return to the study of the growth of the ϵ_i in Section 5. But, we now consider the more important total errors

$$E_i \equiv U_i - y(x_i) = e_i + \epsilon_i, \qquad i = 0, 1, \ldots, N, \tag{20}$$

between the actual numerical solution, U_i, and the exact solution of the differential equation, $y(x_i)$. In an obvious manner, we find that

$$\begin{aligned} &E_0 = \rho_0; \\ &E_{i+1} = E_i + h[f(x_i, U_i) - f(x_i, y(x_i))] - (h\tau_{i+1} - \rho_{i+1}), \\ &\qquad\qquad i = 0, 1, \ldots, N-1. \end{aligned} \tag{21}$$

Again we may prove the analogs of the previous results.

THEOREM **4**. *Under the conditions of the corollary to Theorem* 1, *we find that the error* (20) *satisfies*

$$|E_j| \leq e^{K(b-a)}\left[|\rho_0| + \frac{1}{K}\left(\frac{hM_2}{2} + \frac{\rho}{h}\right)\right], \qquad \text{for } j = 0, 1, \ldots, N, \tag{22a}$$

where the roundoff errors ρ_i *are defined in* (17) *and*

$$\rho = \max_{1 \leq i \leq N} |\rho_i|, \qquad M_2 = \max_{a \leq x \leq b} \left|\frac{d^2y(x)}{dx^2}\right|. \tag{22b}$$

■

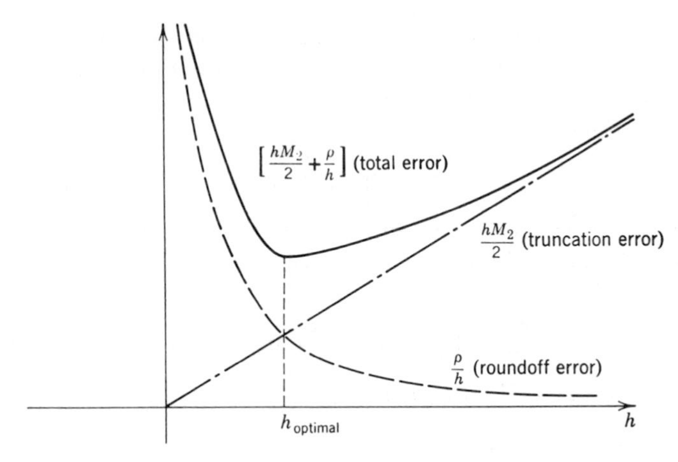

Figure 1. Comparison of truncation and roundoff error bounds as functions of h.

The dependence of this error bound on the net spacing, h, is illustrated in Figure 1.

Clearly, the choice of h for which the bound in (22) is a minimum is obtained when

$$\frac{hM_2}{2} = \frac{\rho}{h}, \quad \text{or} \quad h = \sqrt{2\rho/M_2}. \tag{23}$$

For this optimal value of h,

$$\frac{hM_2}{2} + \frac{\rho}{h} = \sqrt{2M_2\rho}.$$

In many calculations performed on electronic computers $\rho \ll M_2$, and so the "optimal" value for h will be unnecessarily small and need not be employed. Furthermore, the bound (22) indicates that for fixed h no greater accuracy is obtained by reducing the roundoff error so that

$$\rho < \frac{h^2 M_2}{2};$$

in fact, any extra labor required for such computational precision is essentially wasted. If the relation (23) is approximately satisfied there might be some fortuitous cancellation of local roundoff and truncation errors.

On the other hand, we remark that (22) establishes the *convergence* as $h \to 0$ of the Euler-Cauchy method (17) if the rounding error satisfies

$$|\rho_i| \leq \rho = \mathcal{O}(h^2), \qquad \text{for } i = 1, 2, \ldots, N$$

while initially

$$|E_0| = |\rho_0| = \xi_0 h.$$

In fact, under these circumstances

$$|E_j| = \mathcal{O}(h); \qquad j = 0, 1, \ldots, N.$$

We leave to Problem 2 the proof of the validity of the Richardson extrapolation to zero mesh width provided $\rho = \mathcal{O}(h^3)$.

1.3. Centered Difference Method

To obtain greater accuracy with a fixed mesh size we seek difference approximations with smaller local truncation errors. One such modification of (1a) is suggested by attempting to approximate the derivative at x_i by a more accurate expression than the forward difference quotient. We shall briefly examine here the use of the *centered* formula

$$\text{(24a)} \qquad \frac{u_{i+1} - u_{i-1}}{2h} = f(x_i, u_i), \qquad i = 1, 2, \ldots, N - 1.$$

However, in order to use these difference equations to compute $\{u_j\}$, two starting values are required, say

$$\text{(24b)} \qquad u_0 = y_0 + e_0, \qquad u_1 = y_1 + e_1.$$

The first value u_0 is again the approximation of the exact initial data, while u_1 should be determined such that $|e_1|$ is "small." This could be done by employing the Euler-Cauchy method in the interval $0 \leq x \leq h$ with some smaller spacing, say $h' = h/N'$ with $N' \geq 1$, or by developing the Taylor's series

$$u_1 = y_0 + hy_0' + \frac{h^2}{2} y_0'',$$

with

$$y_0' = f(x_0, y_0), \qquad y_0'' = f_x(x_0, y_0) + y_0' f_y(x_0, y_0).$$

However, this problem or similar ones will occur again and shall be discussed in more detail later.

The truncation error in (24a) is now defined by

$$\text{(25)} \qquad y_{i+1} = y_{i-1} + 2hf(x_i, y_i) + 2h\tau_{i+1}, \qquad i = 1, 2, \ldots, N - 1.$$

Let us make sure that $y(x)$ has three continuous derivatives by assuming that $f(x, y)$ has continuous second derivatives. Then by Taylor's theorem

$$y_{i+1} = y(x_i + h) = y_i + hy_i' + \frac{h^2}{2} y_i'' + \frac{h^3}{3!} y'''(\xi_{i+1}), \quad x_i < \xi_{i+1} < x_{i+1};$$

$$y_{i-1} = y(x_i - h) = y_i - hy_i' + \frac{h^2}{2} y_i'' - \frac{h^3}{3!} y'''(\xi_{i-1}), \quad x_{i-1} < \xi_{i-1} < x_i.$$

These equations imply

$$y_{i+1} = y_{i-1} + 2hy_i' + \frac{h^3}{3!}[y'''(\xi_{i+1}) + y'''(\xi_{i-1})],$$

and since $y_i' = f(x_i, y_i)$, a comparison with (25) yields

$$(26)\qquad \tau_{i+1} = \frac{h^2}{6}\frac{y'''(\xi_{i+1}) + y'''(\xi_{i-1})}{2},$$

$$= \frac{h^2}{6}y'''(\xi_i), \qquad i = 1, 2, \ldots, N-1.$$

Here we have used the continuity of $y'''(x)$ to replace the average of third derivatives by an intermediate value at ξ_i. Hence the local truncation error of the centered scheme is smaller than the truncation error (9) of the Euler-Cauchy method as $h \to 0$. The present τ_i vanish to second order in h and we thus call the centered scheme (24a) a second order method.

To show that these effects are indeed relevant for the convergence of the finite difference solution, we again consider the errors $e_i \equiv u_i - y(x_i)$ and find from (24a) and (25)

$$e_{i+1} = e_{i-1} + 2h[f(x_i, u_i) - f(x_i, y(x_i))] - 2h\tau_{i+1}$$

$$= e_{i-1} + 2h\frac{\partial f(x_i, y(x_i) + \theta_i e_i)}{\partial y}e_i - 2h\tau_{i+1}, \qquad i = 1, 2, \ldots, N-1.$$

To demonstrate convergence, let us introduce the bounds

$$(27)\qquad K \geq 2\left|\frac{\partial f}{\partial y}\right|, \quad M_3 \geq |y'''(x)|, \quad \tau = \frac{h^2}{3}M_3 \geq 2|\tau_i|.$$

Hence by taking absolute values

$$(28)\qquad |e_{i+1}| \leq hK|e_i| + |e_{i-1}| + h\tau; \qquad i = 1, 2, \ldots, N-1.$$

To obtain bounds on the $|e_i|$ we introduce a comparison or majorizing set of quantities $\{a_i\}$ defined by

$$(29)\qquad \begin{aligned} a_0 &\equiv \max(|e_0|, |e_1|), \\ a_{i+1} &= (1 + hK)a_i + h\tau, \qquad i = 0, 1, \ldots, N-1. \end{aligned}$$

From the definition it is clear that $a_0 \geq |e_0|$. We will show by induction that $a_j \geq |e_j|$, $j = 0, 1, \ldots, N$. Assume that $a_j \geq |e_j|$, $j = 0, 1, \ldots, i$. Equation (28) yields

$$\begin{aligned} |e_{i+1}| &\leq hKa_i + a_{i-1} + h\tau \\ &\leq (1 + hK)a_i + h\tau \\ &= a_{i+1}. \end{aligned}$$

Here we have employed (29) and the obvious relation $a_i \geq a_{i-1}$. Hence the induction proof is complete and $a_j \geq |e_j|$ for $j = 0, 1, \ldots, N$. However, by the usual recursive application of (29) we now obtain

$$(30) \qquad \begin{aligned} |e_N| &\leq a_N \\ &= (1 + hK)^N a_0 + \frac{(1 + hK)^N - 1}{K} \tau \\ &\leq (1 + hK)^N \left(a_0 + \frac{\tau}{K}\right) \\ &\leq e^{K(b-a)}\left[\max(|e_0|, |e_1|) + h^2 \frac{M_3}{3K}\right]. \end{aligned}$$

By comparing (30) with the result in the corollary to Theorem 1, we now find that the error is of order h^2 if the initial errors, e_0 and e_1, are proportional to h^2. So in the present centered scheme, the higher order local truncation error (26) is reflected in faster convergence of $\{u_i\}$ to $\{y(x_i)\}$ as $h \to 0$. We might naturally expect this to be the case in general, and so seek difference equations with local truncation errors of arbitrarily high order in h. However, as is demonstrated in the next subsection, this expectation is not always realized.

It should be mentioned that roundoff effects can also be included in the study of the present centered scheme.

THEOREM 5. *Let the roundoffs* ρ_i *satisfy*

$$U_0 = y_0 + \rho_0, \qquad U_1 = y_1 + \rho_1,$$

where

$$U_{i+1} = U_{i-1} + 2hf(x_i, U_i) + \rho_{i+1}, \qquad \text{for } i = 1, 2, \ldots, N - 1,$$

$$\max_{2 \leq i \leq N} |\rho_i| = \rho.$$

Then $E_i = U_i - y_i$ *can be bounded by*

$$|E_i| \leq e^{K(b-a)}\left\{\max(|\rho_0|, |\rho_1|) + \left[h^2 \frac{M_3}{3K} + \frac{\rho}{hK}\right]\right\}, \qquad i = 0, 1, \ldots, N,$$

provided (27) *holds.* ∎

Now the error is at least of order h^2, if the maximum roundoff error satisfies $\rho = \mathcal{O}(h^3)$ as $h \to 0$, while $\rho_0 = \mathcal{O}(h^2)$, $\rho_1 = \mathcal{O}(h^2)$.

1.4. A Divergent Method with Higher Order Truncation Error

To demonstrate the care which must be taken in generating difference schemes, we consider a case with third order local truncation error but which is completely unsuitable for computation. The basis for this scheme is an attempt to approximate better the derivative at x_i, and thus to obtain a local truncation error which is higher order in h. For this purpose we consider a difference equation of the form

$$\text{(31)} \qquad a_1u_{i+1} + a_2u_i + a_3u_{i-1} + a_4u_{i-2} = f(x_i, y_i), \qquad i = 2, 3, \ldots, N - 1;$$

and seek coefficients $a_1, \ldots, a_4$ such that the local truncation error is of as high an order, in h, as possible. This is essentially the *method of undetermined coefficients* applied to the problem of approximating derivatives (see Subsection 5.1 of Chapter 6). That is, recalling $y'(x_i) = f(x_i, y(x_i))$, we define τ_i by

$$\text{(32)} \qquad a_1y(x_{i+1}) + a_2y(x_i) + a_3y(x_{i-1}) + a_4y(x_{i-2}) - y'(x_i) = \tau_{i+1}.$$

Then if $y(x)$ has four continuous derivatives, Taylor's theorem yields

$$\text{(33)} \qquad \begin{aligned} y_{i+1} &= y(x_i + h) = y_i + hy'(x_i) + h^2y''(x_i)/2 + h^3y'''(x_i)/3! \\ &\qquad + h^4y^{\text{iv}}(\xi_{i+1})/4! \\ y_{i-1} &= y(x_i - h) = y_i - hy'(x_i) + h^2y''(x_i)/2 - h^3y'''(x_i)/3! \\ &\qquad + h^4y^{\text{iv}}(\xi_{i-1})/4! \\ y_{i-2} &= y(x_i - 2h) = y_i - 2hy'(x_i) + 4h^2y''(x_i)/2 - 8h^3y'''(x_i)/3! \\ &\qquad + 16h^4y^{\text{iv}}(\xi_{i-2})/4! \end{aligned}$$

Forming the sum indicated in (32) and requiring as many terms as possible to vanish, we find

$$\begin{aligned} a_1 + a_2 + a_3 + a_4 \quad &= 0 \\ (a_1 + 0 - a_3 - 2a_4)h &= 1 \\ (a_1 + 0 + a_3 + 4a_4)h^2 &= 0 \\ (a_1 + 0 - a_3 - 8a_4)h^3 &= 0. \end{aligned}$$

This system of linear equations has the unique solution

$$\text{(34)} \qquad a_1 = \frac{1}{3h}, \quad a_2 = \frac{1}{2h}, \quad a_3 = -\frac{1}{h}, \quad a_4 = \frac{1}{6h}.$$

Now (32) can be written as

$$\text{(35)} \qquad y(x_{i+1}) = -\tfrac{3}{2}y(x_i) + 3y(x_{i-1}) - \tfrac{1}{2}y(x_{i-2}) + 3hf(x_i, y(x_i)) + 3h\tau_{i+1}$$

where the local truncation error is from (33) and (34) in (32)

$$\text{(36)} \qquad \tau_{i+1} = \frac{h^3}{4!}\left[\tfrac{1}{3}y^{\text{iv}}(\xi_{i+1}) - y^{\text{iv}}(\xi_{i-1}) + \tfrac{8}{3}y^{\text{iv}}(\xi_{i-2})\right].$$

The difference equations (31) become

(37a) $$u_{i+1} = -\tfrac{3}{2}u_i + 3u_{i-1} - \tfrac{1}{2}u_{i-2} + 3hf(x_i, u_i),$$
$$i = 2, 3, \ldots, N-1;$$

and to these must be adjoined starting values, say,

(37b) $$u_0 = y_0 + e_0, \quad u_1 = y_1 + e_1, \quad u_2 = y_2 + e_2.$$

As before, the "extra" values u_1 and u_2 would have to be obtained by some other procedure. However, we proceed to show that this scheme, which has third order truncation errors, does not converge in general.

Let us first consider a case in which f_y is continuous and

$$\left|\frac{\partial f}{\partial y}\right| \leq K, \qquad \tau = h^3 M_4 \geq \max_i |\tau_i|.$$

Then with $e_i = u_i - y(x_i)$ we obtain, from (35) and (37), in the usual way

$$|e_{i+1}| \leq (\tfrac{3}{2} + 3hK)|e_i| + 3|e_{i-1}| + \tfrac{1}{2}|e_{i-2}| + 3h\tau.$$

If we introduce a majorizing set $\{a_i\}$, analogous to (29), we find that

$$\begin{aligned}|e_N| &\leq (5 + 3hK)^N\left[\max(|e_0|, |e_1|, |e_2|) + \frac{3h^4M_4}{4 + 3hK}\right],\\ &\leq 5^N e^{3K(b-a)/5}\left[\max(|e_0|, |e_1|, |e_2|) + \frac{3h^4M_4}{4 + 3hK}\right],\\ &= 5^{(b-a)/h}e^{3K(b-a)/5}\left[\max(|e_0|, |e_1|, |e_2|) + \frac{3h^4M_4}{4 + 3hK}\right].\end{aligned}$$

However, as $h \to 0$ this bound becomes infinite. While this does not prove divergence, we strongly suspect it.

To actually show that the third order scheme (37) cannot converge in general, we apply it to the special case where $f(x, y) \equiv -y$; i.e., to the equation

$$\frac{dy}{dx} = -y,$$

whose solution $y = y_0 e^{-(x-a)}$ satisfies $y(a) = y_0$.

Now (37a) can be written as

(38) $$u_{i+1} + (\tfrac{3}{2} + 3h)u_i - 3u_{i-1} + \tfrac{1}{2}u_{i-2} = 0, \qquad i = 2, 3, \ldots.$$

This is a linear difference equation with constant coefficients (see Section 4) and it can be solved exactly. That is, we seek a solution of the form

$$u_j = \alpha^j, \qquad j = 0, 1, \ldots.$$

But then from (38),

$$[\alpha^3 + (\tfrac{3}{2} + 3h)\alpha^2 - 3\alpha + \tfrac{1}{2}]\alpha^{i-2} = 0, \qquad i = 2, 3, \ldots.$$

Thus, in addition to the trivial solution, $u_j \equiv 0$, we have three solutions of the form $u_j = \alpha_\nu{}^j$, where α_ν for $\nu = 1, 2, 3$ are the roots of

$$\alpha^3 + (\tfrac{3}{2} + 3h)\alpha^2 - 3\alpha + \tfrac{1}{2} = 0. \tag{39}$$

It is easily verified that, for h sufficiently small, these three roots are distinct.

It is easy to check that a linear combination

$$u_j = A_1\alpha_1{}^j + A_2\alpha_2{}^j + A_3\alpha_3{}^j, \tag{40}$$

is also a solution. The coefficients A_ν are determined from the assumed known data for $j = 0, 1, 2$, by satisfying

$$\begin{aligned} A_1 \quad + A_2 \quad + A_3 \quad &= u_0 \\ A_1\alpha_1 + A_2\alpha_2 + A_3\alpha_3 &= u_1 \\ A_1\alpha_1{}^2 + A_2\alpha_2{}^2 + A_3\alpha_3{}^2 &= u_2. \end{aligned}$$

Since the coefficient determinant is a Vandermonde determinant and the α_ν are distinct, the A_ν are uniquely determined. Then the u_j for $j \geq 3$ are also uniquely determined by (40).

Let us write

$$p(\alpha, h) \equiv \alpha^3 + (\tfrac{3}{2} + 3h)\alpha^2 - 3\alpha + \tfrac{1}{2},$$

and denote the roots of $p(\alpha, h) = 0$ by $\alpha_\nu(h)$. Since

$$p(\alpha, 0) \equiv (\alpha - 1)(\alpha^2 + \tfrac{5}{2}\alpha - \tfrac{1}{2})$$

we have, with the ordering $\alpha_1 \leq \alpha_2 \leq \alpha_3$,

$$\alpha_1(0) = -\frac{5 + \sqrt{29}}{4}, \quad \alpha_2(0) = \frac{\sqrt{29} - 5}{4}, \quad \alpha_3(0) = 1.$$

For h sufficiently small, $|\alpha_\nu(h) - \alpha_\nu(0)|$ can be made arbitrarily small and $\alpha_1(h) < -2$. So the solution (40), for large j, behaves like

$$u_j \approx A_1[\alpha_1(h)]^j$$

or in particular for $x_N = b$,

$$u_N \approx A_1[\alpha_1(h)]^{(b-a)/h}.$$

Thus as $h \to 0$ the difference solution becomes exponentially unbounded *at any point* $x_N = b > a$. Furthermore, notice that $\alpha_1(h)$ is negative, and hence u_j oscillates. This behavior is typical of "*unstable*" *schemes* (see

Section 5). Of course, we have assumed here that the initial data is such that $A_1 \neq 0$. In fact,

$$A_1 = \frac{\det \begin{vmatrix} u_0 & 1 & 1 \\ u_1 & \alpha_2 & \alpha_3 \\ u_2 & {\alpha_2}^2 & {\alpha_3}^2 \end{vmatrix}}{\det \begin{vmatrix} 1 & 1 & 1 \\ \alpha_1 & \alpha_2 & \alpha_3 \\ {\alpha_1}^2 & {\alpha_2}^2 & {\alpha_3}^2 \end{vmatrix}}.$$

Hence if

$$u_1 = u_0 + ch, \quad u_2 = u_0 + dh,$$

then in general, it follows that

$$A_1 = h\beta + \mathcal{O}(h^2), \qquad \beta \neq 0.$$

This is based on the fact that $\alpha_i(h)$ can be developed in the form

$$\alpha_i(h) = \alpha_i(0) + h\alpha_i'(0) + \mathcal{O}(h^2).$$

(For example, in the exceptional case

$$u_1 = \alpha_3 u_0 \quad \text{and} \quad u_2 = {\alpha_3}^2 u_0,$$

the quantity $A_1 = 0$.)

For the actual calculations of the quantities U_j, the local roundoff error at any net point x_j will set off such an exponentially growing term. Hence this method is divergent!

PROBLEMS, SECTION 1

1. If $f(x, y)$ is independent of y, i.e., the Lipschitz constant $K = 0$, show that the error estimates of Theorem 1 and its corollary are respectively

(a) $$|e_j| \leq |e_0| + |x_j - x_0|\tau, \qquad j \geq 0;$$

(b) $$|e_j| \leq |e_0| + h|x_j - x_0|M_2/2, \qquad j \geq 0.$$

2. Carry out the proof of the validity of the Richardson extrapolation to zero mesh width for the Euler-Cauchy method defined in (17) with rounding errors $\rho_0 = \xi_0 h$ and for $i = 1, 2, \ldots, |\rho_i| \leq \rho = \mathcal{O}(h^3)$.

3. Find the coefficient $\alpha_i'(0)$ in the expansion

$$\alpha_i(h) = \alpha_i(0) + h\alpha_i'(0) + \mathcal{O}(h^2)$$

by formally substituting this expression into (39) and setting the coefficient of h equal to zero.

2. MULTISTEP METHODS BASED ON QUADRATURE FORMULAE

The study of the divergent scheme introduced in (1.37) shows that more accurate approximations of the derivative do not lead to more accurate numerical methods. But a way to determine convergent schemes with an arbitrarily high order of accuracy is *suggested* by converting the original differential equation into an equivalent integral equation. Thus by integrating (0.1a) over the interval $[a, x]$ and using (0.1b) we obtain

$$y(x) = y_0 + \int_a^x f(\xi, y(\xi))\, d\xi. \tag{1}$$

Clearly, any solution of (0.1) satisfies this integral equation and, by differentiation, we find that any solution† of (1) also satisfies (0.1a) and (0.1b). With the subscript notation, $y_i \equiv y(x_i)$, the solution of (0.1) or (1) also satisfies

$$y_{i+1} = y_{i-p} + \int_{x_{i-p}}^{x_{i+1}} f(x, y(x))\, dx. \tag{2}$$

This is obtained by integrating (0.1a) over $[x_{i-p}, x_{i+1}]$ for any $i = 0, 1, \ldots$ and any $p = 0, 1, \ldots, i$. For a given choice of p a variety of approximations are suggested by applying quadrature formulae to evaluate the integral in (2). The number of schemes suggested in this manner is great, but in practice only relatively few of them are ever used.

We shall limit our study to the case of uniformly spaced net points and interpolatory quadrature formulae. In order to classify these methods in a fairly general manner we distinguish two types of quadrature formulae:

TYPE A. *Closed on the right*, i.e., with $n + 1$ nodes

$$x_{i+1}, x_i, x_{i-1}, \ldots, x_{i+1-n};$$

or else

TYPE B. *Open on the right*, i.e., with $n + 1$ nodes

$$x_i, x_{i-1}, \ldots, x_{i-n}.$$

The difference equations suggested by these two classes of methods can be written as

$$u_{i+1} = u_{i-p} + h \sum_{j=0}^{n} \alpha_j f(x_{i+1-j}, u_{i+1-j}); \tag{3a}$$

$$u_{i+1} = u_{i-p} + h \sum_{j=0}^{n} \beta_j f(x_{i-j}, u_{i-j}). \tag{3b}$$

† It is easy to see that any continuous solution $y(x)$ of (1) is differentiable. This follows since the right-hand side of (1) is differentiable, if $f(x, y)$ and $y(x)$ are continuous.

We have denoted the coefficients of the quadrature formulae by $h\alpha_j$ and $h\beta_j$ and note that they are independent of i. If the net spacing were not uniform, the coefficients would depend on i, in general, and the schemes would require the storage of more than $n + 1$ coefficients. This is one of the main reasons for the choice of uniform spacing in methods based on quadrature formulae.

When the integers p and n are specified, the coefficients are determined as in Subsection 1.2 of Chapter 7. We see from (1.15) of Chapter 7 that the quantities α_j and β_j are independent of h, the net spacing. It also follows from Theorem 1.3 of Chapter 7 that the interpolatory quadrature formulae in (3) have the maximum degree of precision possible with the specified nodes; this is reflected in their having the smallest "truncation error," defined later in equation (8a and b).

In order to compute with a method of type A (closed on the right) we must have available the quantities $u_i, \ldots, u_{i+1-n}$ and u_{i-p}. Thus the points x_{i+1} for which (3a) may be used satisfy

$$\text{(4a)} \qquad i \geq \max(n - 1, p) = t,$$

provided that $u_0, u_1, \ldots, u_t$ are given. Similarly, method B requires $u_i, \ldots, u_{i-n}$ and u_{i-p}, so that the points x_{i+1} for which (3b) can be used satisfy

$$\text{(4b)} \qquad i \geq \max(n, p) = r,$$

provided that $u_0, u_1, \ldots, u_r$ are given. Special procedures are required to obtain these starting values in either case (see Section 3).

The fundamental difference between the open and closed methods is the ease with which the difference equations can be solved. There is no difficulty in solving the equations based on the open formulae. In fact, since formula (3b) is an explicit expression for u_{i+1} these are called *explicit methods*. But, the closed formulae define *implicit methods* since the equation for the determination of u_{i+1} is implicit.† That is, (3a) is of the form

$$\text{(5a)} \qquad u_{i+1} = g_i(u_{i+1}),$$

where

$$g_i(z) \equiv c_i + h\alpha_0 f(x_{i+1}, z),$$

with

$$c_i \equiv u_{i-p} + h \sum_{j=1}^{n} \alpha_j f(x_{i+1-j}, u_{i+1-j})$$

† In the special case that $f(x, y)$ is linear in y, i.e., $f(x, y) \equiv a(x)y + b(x)$, the implicit equation (3a) is easily solved explicitly for u_{i+1} if $1 - h\alpha_0 a(x_{i+1}) \neq 0$.

and so

$$\text{(5b)} \qquad \frac{dg_i(z)}{dz} = h\alpha_0 \frac{\partial f(x_{i+1}, z)}{\partial z}.$$

Clearly the method of functional iteration is natural for solving (5a). By Theorem 1.2 of Chapter 3 we know that, if a "sufficiently close" initial estimate $u_{i+1}^{(0)}$ of the root is given, the iterations

$$u_{i+1}^{(\nu+1)} = g_i(u_{i+1}^{(\nu)}), \qquad \nu = 0, 1, \ldots,$$

will converge provided h is sufficiently small, e.g.,

$$\text{(6)} \qquad h < \frac{1}{|\alpha_0 K|}, \qquad K \equiv \max \left|\frac{\partial f(x, y)}{\partial y}\right|.$$

On the other hand, we may apply Theorem 1.1 of Chapter 3 to show existence of a unique root in the interval $[c_i - \rho, c_i + \rho]$. That is, by selecting $u_{i+1}^{(0)} = c_i$, we have

$$|u_{i+1}^{(1)} - u_{i+1}^{(0)}| \leq h\alpha_0 M$$

where $M \equiv \max |f(x, y)|$. Hence for any

$$\rho \geq \frac{h\alpha_0 M}{1 - h\alpha_0 K}$$

we have an interval in which a unique root of (5) exists. We shall see that it is not necessary to find the root of (5) in order to preserve the high accuracy of the method. In fact, the *predictor-corrector* technique, which we next study, uses only one iteration of (5) without a loss in accuracy.

The *predictor-corrector method* is defined by

$$\text{(7a)} \qquad u_{i+1}^* = u_{i-q} + h \sum_{j=0}^{m} \beta_j f(x_{i-j}, u_{i-j});$$

$$\text{(7b)} \qquad u_{i+1} = u_{i-p} + h \sum_{k=1}^{n} \alpha_k f(x_{i+1-k}, u_{i+1-k}) + h\alpha_0 f(x_{i+1}, u_{i+1}^*);$$

$$\text{(7c)} \qquad i \geq s \equiv \max\,[p, q, m, n - 1].$$

Here, an $m + 1$ point quadrature formula open on the right has been used [in the *predictor* (7a)] to approximate an integral over $[x_{i-q}, x_{i+1}]$. The closed formula (7b) (called the *corrector*) is similar to that of (3a) but u_{i+1}^* has been used in place of u_{i+1} in the right-hand side. Thus as previously indicated, the corrector is the first iteration step in solving the implicit equation (5) with the initial guess furnished by the predictor. Hence *only two evaluations* of the function $f(x, y)$ are required for each step of the predictor-corrector method; i.e., $f(x_i, u_i)$ and $f(x_{i+1}, u_{i+1}^*)$.

The procedure (7) can only be employed after the values $u_0, u_1, \ldots, u_s$ have been determined. Here s, defined by (7c), is determined from the open and closed formulae (7a and b). To compute the u_i, $i < s$, we refer to the procedures of Section 3.

It is clear, from the analysis given in the remainder of this section, that the predictor-corrector method (7) has the same order of accuracy as the implicit method of type A defined by the corrector (3a), provided that the explicit predictor is sufficiently accurate. In other words, we avoid the necessity of repeatedly iterating the corrector, as in (5), by using a good initial approximation. We shall first develop estimates of the error in solving (1) by the predictor-corrector method (7). Next, we shall show how to modify the error estimates to cover the case of finite precision arithmetic, i.e., with rounding errors. Finally, we indicate briefly how the error estimates may be derived for the methods (3a) or (3b).

The predictor-corrector method has the advantage of permitting the detection of an isolated numerical error through the comparison of u_{i+1}^* with u_{i+1} (or U_{i+1}^* with U_{i+1} defined later).

The truncation error of (7) is obtained as follows. We define σ_{i+1}^* and σ_{i+1} in terms of the exact solution $y = y(x)$ by

(8a)
$$y_{i+1} = y_{i-q} + h \sum_{j=0}^{m} \beta_j f(x_{i-j}, y_{i-j}) + h\sigma_{i+1}^*;$$

(8b)
$$y_{i+1} = y_{i-p} + h \sum_{k=0}^{n} \alpha_k f(x_{i+1-k}, y_{i+1-k}) + h\sigma_{i+1}.$$

Then with the definition

(8c)
$$y_{i+1}^* \equiv y_{i+1} - h\sigma_{i+1}^*,$$

the *local truncation error*, τ_{i+1}, of the predictor-corrector method (7) is defined by

(9)
$$y_{i+1} = y_{i-p} + h \sum_{k=1}^{n} \alpha_k f(x_{i+1-k}, y_{i+1-k}) + h\alpha_0 f(x_{i+1}, y_{i+1}^*) + h\tau_{i+1}, \qquad i \geq s.$$

To obtain a more explicit expression for this error we subtract (9) from (8b) and use (8c) to get

(10)
$$\begin{aligned}\tau_{i+1} &= \sigma_{i+1} + \alpha_0[f(x_{i+1}, y_{i+1}) - f(x_{i+1}, y_{i+1} - h\sigma_{i+1}^*)] \\ &= \sigma_{i+1} + h\sigma_{i+1}^*\alpha_0 \overline{\frac{\partial f}{\partial y}}.\end{aligned}$$

Here we have assumed f_y to be continuous in y and used the mean value theorem; $\bar{f}_y$ is a value of f_y at some intermediate point in the obvious interval. The quantities $h\sigma^*_{i+1}$ and $h\sigma_{i+1}$ are the errors in the $(m + 1)$- and $(n + 1)$-point quadrature formulae, (8a and b). Explicit expressions for these quadrature errors have been obtained for various cases in Section 1 of Chapter 7. It should be noted from (10) that the order as $h \to 0$ of the truncation error in the predictor-corrector method is the same as the order for the corresponding closed formula used alone, provided that the order of $h\sigma^*$ is not less than the order of σ. Table 1 has a brief listing of commonly used predictor-corrector schemes.

Table 1

Table 1 *Some Common Predictor-Corrector Methods*

Associated Name	$m, q; n, p$		Weights $j = 0$	1	2	3	4	σ^*	σ
Modified Euler	$0, 0; 1, 0$	$\beta_j \equiv$	1					$hy^{(2)}$	$-\frac{1}{12}h^2y^{(3)}$
		$\alpha_j \equiv$	$\frac{1}{2}$	$\frac{1}{2}$					
Milne's Method (3 points)	$2, 3; 2, 1$	$\beta_j \equiv$	$\frac{8}{3}$	$-\frac{4}{3}$	$\frac{8}{3}$			$\frac{28}{90}h^4y^{(5)}$	$-\frac{1}{90}h^4y^{(5)}$
		$\alpha_j \equiv$	$\frac{1}{3}$	$\frac{4}{3}$	$\frac{1}{3}$				
Improved Adams, or Moulton's Method (4 points)	$3, 0; 3, 0$	$\beta_j \equiv$	$\frac{55}{24}$	$-\frac{59}{24}$	$\frac{37}{24}$	$-\frac{9}{24}$		$\frac{251}{720}h^4y^{(5)}$	$-\frac{19}{720}h^4y^{(5)}$
		$\alpha_j \equiv$	$\frac{9}{24}$	$\frac{19}{24}$	$-\frac{5}{24}$	$\frac{1}{24}$			
Milne's Method (5 points)	$4, 5; 4, 3$	$\beta_j \equiv$	$\frac{33}{10}$	$-\frac{21}{5}$	$\frac{39}{5}$	$-\frac{21}{5}$	$\frac{33}{10}$	$\frac{41}{140}h^6y^{(7)}$	$-\frac{8}{945}h^6y^{(7)}$
		$\alpha_j \equiv$	$\frac{14}{45}$	$\frac{64}{45}$	$\frac{24}{45}$	$\frac{64}{45}$	$\frac{14}{45}$		

2.1. Error Estimates in Predictor-Corrector Methods

To examine convergence of the scheme (7) we introduce the *errors*

$$e_i \equiv u_i - y_i, \qquad e_i^* \equiv u_i^* - y_i^*. \tag{11}$$

Then subtraction of (9) from (7) yields, if f_y is continuous,

$$e_{i+1} = e_{i-p} + h \sum_{k=1}^{n} \alpha_k g_{i+1-k} e_{i+1-k} + h\alpha_0 g_{i+1} e^*_{i+1} - h\tau_{i+1}.$$

Here we have used the mean value theorem to introduce

$$g_j \equiv \frac{\partial f}{\partial y}(x_j, \bar{y}_j), \quad \bar{y}_j \in (y_j, u_j) \quad \text{for } j \neq i + 1, \quad \text{while}$$

$$\bar{y}_{i+1} \in (y^*_{i+1}, u^*_{i+1}).$$

However, from (7a) and (8a and c) we obtain

$$e^*_{i+1} = e_{i-q} + h \sum_{j=0}^{m} \beta_j g_{i-j} e_{i-j},$$

and using this in the above implies finally

$$(12) \qquad e_{i+1} = e_{i-p} + h \sum_{k=1}^{n} \alpha_k g_{i+1-k} e_{i+1-k} + h\alpha_0 g_{i+1} e_{i-q}$$
$$+ h^2 \alpha_0 g_{i+1} \sum_{j=0}^{m} \beta_j g_{i-j} e_{i-j} - h\tau_{i+1}, \qquad i \geq s.$$

To estimate these errors we introduce

$$(13) \qquad K \equiv \max \left|\frac{\partial f}{\partial y}\right|; \qquad \tau \equiv \max |\tau_j|;$$
$$A \equiv \sum_{k=0}^{n} |\alpha_k|; \qquad B \equiv \sum_{j=0}^{m} |\beta_j|.$$

Then by taking the absolute value of both sides of (12), we have

$$(14) \qquad |e_{i+1}| \leq |e_{i-p}| + hK(|\alpha_0|\,|e_{i-q}| + \sum_{k=1}^{n} |\alpha_k|\,|e_{i+1-k}|)$$
$$+ h^2K^2|\alpha_0| \sum_{j=0}^{m} |\beta_j|\,|e_{i-j}| + h|\tau_{i+1}|; \qquad i \geq s.$$

We again introduce a comparison or majorizing set, $\{a_i\}$, defined by

$$(15a) \qquad a_0 \equiv \max (|e_0|, |e_1|, \ldots, |e_s|),$$

$$(15b) \qquad a_{i+1} = (1 + hKA + h^2K^2|\alpha_0|B)a_i + h\tau;$$

and claim that

$$|e_j| \leq a_j; \qquad j = 0, 1, \ldots, N.$$

The proof of this inequality is easily given by induction. From (15), $\{a_i\}$ is a non-decreasing sequence. Therefore, (15a) establishes the inequality for $j \leq s$. Now assume the inequality holds for all $j \leq i$ where $i \geq s$. Then (14) implies

$$
\begin{aligned}
|e_{i+1}| &\leq a_{i-p} + hK\left(|\alpha_0|a_{i-q} + \sum_{k=1}^{n} |\alpha_k|a_{i+1-k}\right) \\
&\qquad + h^2K^2|\alpha_0| \sum_{j=0}^{m} |\beta_j|a_{i-j} + h\tau, \\
&\leq \left[1 + hK\left(|\alpha_0| + \sum_{k=1}^{n} |\alpha_k|\right) + h^2K^2|\alpha_0| \sum_{j=0}^{m} |\beta_j|\right]a_i + h\tau \\
&= (1 + hKA + h^2K^2|\alpha_0|B)a_i + h\tau \\
&= a_{i+1}.
\end{aligned}
$$

Note that from the recursive expression for e^*_{i+1}, we have the single estimate $|e^*_{i+1}| \leq (1 + hBK)|a_i|$. The application of (15b) recursively, in the by now familiar manner, yields the final result which can be summarized as

THEOREM 1. *Let the predictor-corrector method* (7) *be applied to solve* (0.1) *or* (1) *in* $a \leq x \leq b$ *with "initial" values,* u_i*, satisfying*

$$
\text{(16a)} \qquad |u_i - y(x_i)| \leq a_0, \qquad i = 0, 1, \ldots, s.
$$

Let $f_y(x, y)$ *be continuous and bounded in* S: $\{(x, y) \mid a \leq x \leq b; |y| < \infty\}$. *Then with the definitions* (9) *and* (13) *the errors in the numerical solution satisfy, for* $a \leq x_j \leq b$,

$$
\text{(16b)} \qquad |u_j - y(x_j)| \leq \left[a_0 + \frac{\tau}{K(A + hK|\alpha_0|B)}\right] \times \exp\,[(x_j - a)K(A + hK|\alpha_0|B)],
$$

$$
\text{(16c)} \qquad |u_j^* - y^*(x_j)| \leq \left[a_0 + \frac{\tau}{K(A + hK|\alpha_0|B)}\right] \times \exp\,[(x_j - a)K(A + hK|\alpha_0|B) + hBK]. \qquad ■
$$

From this theorem it follows that u_j converges to $y(x_j)$ as $h \to 0$ if $a_0 \to 0$ and $\tau \to 0$. The order in h of the estimate (16) is the minimum of the orders in h of a_0 and τ. We say that the methods for selecting the initial data and method (7) are *balanced* if a_0 and τ vanish to the same order in h. If they do not, then some "extra accuracy" has been wasted.

If the exact solution, $y(x)$, has sufficiently many continuous derivatives,† then the local truncation error, τ, can be simply expressed by using the

† If all partial derivatives of order p of $f(x, y)$ are continuous, then $y(x)$ has a continuous derivative of order $p + 1$. This results by differentiating (1) often enough.

methods of Section 1, Chapter 7. For instance, from the crude estimates of the form (1.8) of Chapter 7 and (10) we find

$$\tau = \mathcal{O}(h^{m+2}) + \mathcal{O}(h^{n+1}).$$

From these estimates, we see that the predictor and corrector formulae are balanced if $m + 1 = n$. The method is said to be of order $t \equiv \min(m + 2, n + 1)$. [In the special case $p = n - 1$, $q = m + 1$, with m and n both even;

$$\tau = \mathcal{O}(h^{m+3}) + \mathcal{O}(h^{n+2}),$$

which results from the estimate of error in the Newton-Cotes formulae (1.11) of Chapter 7. Parity makes $m = n$ optimal.]

Of course, roundoff errors are committed when the formulae in (7) are evaluated. If we call U_j and U_j^* the numbers actually obtained in these evaluations, then we can write

$$\text{(17a)} \qquad U_{i+1}^* = U_{i-q} + h \sum_{j=0}^{m} \beta_j f(x_{i-j}, U_{i-j}) + \rho_{i+1}^*,$$

$$\text{(17b)} \qquad U_{i+1} = U_{i-p} + h \sum_{k=1}^{n} \alpha_k f(x_{i+1-k}, U_{i+1-k}) + h\alpha_0 f(x_{i+1}, U_{i+1}^*) + \rho_{i+1}, \qquad i \geq s.$$

Here ρ_j^* and ρ_j are the roundoff errors introduced into each of the indicated computations. Now we define the errors

$$E_j \equiv U_j - y(x_j), \qquad E_j^* \equiv U_j^* - y_j^*;$$

and obtain from (8), (9), and (17), as in the derivation of (12)

$$\text{(18)} \qquad E_{i+1} = E_{i-p} + h \sum_{k=1}^{n} \alpha_k g_{i+1-k} E_{i+1-k} + h\alpha_0 g_{i+1} E_{i-q} + h^2 \alpha_0 g_{i+1} \sum_{j=0}^{m} \beta_j g_{i-j} E_{i-j} - h\tau_{i+1} + \rho_{i+1} + h\alpha_0 g_{i+1} \rho_{i+1}^*, \qquad i \geq s,$$

with

$$g_j = f_y(x_j, \bar{y}_j) \quad \text{and} \quad \bar{y}_j \in (y_j, U_j) \quad \text{for } j \neq i + 1,$$

while

$$\bar{y}_{i+1} \in (y_{i+1}^*, U_{i+1}^*).$$

By applying the previous method of analysis to this system we find the total error bound in

THEOREM 2. *Under the hypothesis of Theorem* 1, *with the notation* (13) *and* (17),

$$(19) \qquad |U_j - y(x_j)| \leq \left[b_0 + \frac{\tau + |\alpha_0|K\rho^* + (1/h)\rho}{K(A + hK|\alpha_0|B)}\right] \times \exp\left[(x_j - a)K(A + hK|\alpha_0|B)\right], \qquad a \leq x_j \leq b,$$

where

$$(20) \qquad \rho = \max_{s<j<N} |\rho_j|, \qquad \rho^* = \max_{s<j<N} |\rho_j^*|, \qquad b_0 = \max_{0<j<s} |U_j - y(x_j)|. \qquad ■$$

It is of interest to note that in the bound (19) the corrector roundoff enters in the form ρ/h while that from the predictor has a coefficient independent of h. However, it is unlikely that any special measures in actual computation could be adopted to balance these different orders in h of the roundoff. If in Theorem 2 we know that

$$b_0 = \mathcal{O}(h), \quad \rho^* = \mathcal{O}(h), \quad \tau = \mathcal{O}(h), \quad \text{and} \quad \rho = \mathcal{O}(h^2),$$

then (19) yields $|E_j| \leq Vh$ for $a \leq x_j \leq b$ for a constant V independent of h.

It should be observed that if $\alpha_0 \equiv 0$ in (7), the predictor is never used. The corrector in this case is an open formula, and the above error analysis then applies to the method based on the use of a *single open formula*. The corresponding result for the method based on the use of a *single closed formula* (i.e., the *implicit method*) is obtained by a slight modification of the above technique (see Problem 1). Now if, in the predictor-corrector method, more than one iteration is employed, the estimates (16) and (19) no longer apply. But a comparison of the error bounds of the predictor-corrector method and the corresponding implicit corrector method shows that there is no great gain to be expected in using the corrector more than once, provided $h\sigma_{i+1}^*$ and σ_{i+1} are of the same order in h.

We can remark further that in Theorem 2 the requirement that $f(x, y)$ and $f_y(x, y)$ be bounded and continuous for $|y| < \infty$ can be replaced by the milder restriction that $f(x, y)$ and $f_y(x, y)$ be bounded and continuous in the strip

$$S': \{(x, y) \mid a \leq x \leq b, |y - y(x)| \leq d\} \qquad \text{for some } d > 0,$$

provided that:

(a) h is sufficiently small,
(b) $b_0 = \mathcal{O}(h)$, $\rho^* = \mathcal{O}(h)$, $\tau = \mathcal{O}(h)$ and,
(c) $\rho = \mathcal{O}(h^2)$.

To show that the estimate (19) holds, we now could show inductively that the values U^*_{i+1} and U_{i+1} exist and are in the strip S', and therefore (19) is satisfied for $j = i + 1$. The constant $K \equiv \max_{S'} |f_y|$ replaces the previous definition of K in (19).

2.2. Change of Net Spacing

During the course of a computation based on a predictor-corrector method, we should keep track of the "measure of error,"

$$U_{i+1} - U^*_{i+1} \equiv \eta_{i+1}.$$

That is,

$$(20) \qquad |\eta_{i+1}| = |U_{i+1} - U^*_{i+1}| = |E_{i+1} - E^*_{i+1} + h\sigma^*_{i+1}|$$
$$\leq |E_{i+1}| + |E^*_{i+1}| + h|\sigma^*_{i+1}|.$$

Hence if $|\eta_{i+1}|$ is large, we know that the actual error is probably large.

An isolated mistake in computation may be responsible for a large $|\eta_{i+1}|$. But, if the computation is correct, then the obvious way to reduce η_{i+1} is to reduce the interval size h. In practice, this is usually done by successively halving h. Alternatively if the estimate (20) becomes very small, h may be increased, say, by doubling it.

Doubling the interval size offers no difficulty if at least $2s$ points have been computed with the net spacing h. We merely discard the data at every other net point, replace h by $2h$, and continue the calculations.

On the other hand, in order to halve h, we require data at $s/2$ new intermediate points, say

$$x_i - h/2, x_{i-1} - h/2, \ldots, \qquad x_{i+1-(s/2)} - h/2.$$

These values can be determined by the application of an interpolation procedure which uses the known data at appropriate net points $x_j = x_0 + jh$. However, the accuracy of the interpolation formula must be consistent with that of the predictor-corrector formula being used. That is, *the interpolation error must be at least of the same order in h as the truncation error*, τ, given in (10). Otherwise, as Theorem 1 indicates, the accuracy of the numerical solution will not be greater than that determined by the interpolation. The caution required in order to reduce the net spacing without sacrificing accuracy is one of the disadvantages of predictor-corrector methods when compared to single-step methods of the next section. Sometimes it is feasible to actually restart the integration at x_i, by using the method employed at x_0, but with the net size $h/2$.

PROBLEMS, SECTION 2

1. Verify the entries in Table I labeled Modified Euler and Milne's method (3 points).

2. Verify the entries in Table I labeled Improved Adams and Milne's Method (5 points).

3. For each of the methods of Table I, with what interval size h, and how many decimal places should the equation

$$y' = y \qquad y(0) = 1$$

be solved in $0 \leq x \leq 5$ in order that the error satisfy

$$|E_i| \equiv |U_i - y_i| \leq 10^{-4}?$$

4. Assume that $f(x, y)$ and $f_y(x, y)$ are bounded and continuous in S:

$$\{(x, y) \mid a \leq x \leq b, |y| < \infty\}.$$

Then if $h < 1/|\alpha_0 K|$ where $K \equiv \max |f_y|$, the implicit scheme (3a) can be used to find the $\{u_i\}$, given $u_0, u_1, \ldots, u_r$, with $r = \max(n, p)$. [See discussion after equation (6).] Estimate the total error, $E_i \equiv U_i - y_i$, in solving (1) by the implicit scheme based on (3a).

[Hint: If we stop the iterative process described in (5), when the equation (3a) is satisfied with the error ρ_{i+1}, we will obtain a sequence $\{U_i\}$ with $U_0 = u_0$, $U_1 = u_1, \ldots, U_r = u_r$, that satisfies

$$U_{i+1} = U_{i-p} + h \sum_{j=0}^{n} \alpha_j f(x_{i+1-j}, U_{i+1-j}) + \rho_{i+1}.$$

Equation (8b) defines the corresponding truncation error σ_{i+1}. Show that E_i satisfies

$$E_{i+1} = E_{i-p} + h \sum_{j=1}^{n} \alpha_{i+1-j} g_{i+1-j} E_{i+1-j} + h\alpha_0 g_{i+1} E_{i+1} + \rho_{j+1} - h\sigma_{i+1},$$

where $g_j = f_y(x_j, \bar{y}_j)$ at some suitable point $\bar{y}_j$. Hence show that $|E_i| \leq a_i$, where the sequence $\{a_i\}$ is defined by

$$a_0 = \max(|E_0|, \ldots, |E_r|)$$

$$a_{i+1} = a_i\left(\frac{1 + hAK}{1 - h|\alpha_0|K}\right) + \left(\frac{\rho + h\sigma}{1 - h|\alpha_0|K}\right)$$

where $\rho = \max |\rho_i|$, $\sigma = \max |\sigma_i|$. Then show that

$$a_{i+1} = a_i(1 + hQ) + R$$

where

$$Q = \frac{K(A + |\alpha_0|)}{1 - h|\alpha_0|K}$$

$$R = \frac{\rho + h\sigma}{1 - h|\alpha_0|K}.]$$

3. ONE-STEP METHODS

The higher accuracy predictor-corrector methods of Section 2 all require special procedures for starting the calculations. That is, some approximate solution, u_j, must first be computed for $j = 0, 1, \ldots, s$. In addition, if the interval size, h, is reduced during the course of the calculation, care must be taken to preserve the accuracy of the method. The single-step methods, which we now consider, require none of these special measures. In fact, they can be used to determine the starting values and to change the net spacing in other methods. The price paid for these advantages is, in general, the requirement of a greater number of evaluations of the function $f(x, y)$ (or functions related to it) for each step in the solution.

Again we consider the initial value problem

$$\frac{dy}{dx} = f(x, y), \qquad y(a) = y_0. \tag{1}$$

By *single-step* we mean that only data at $x = x_0$ are to be employed in obtaining the approximation to $y(x)$ at $x = x_1$. Obviously such a procedure could then be employed at x_1, and so forth, to extend the solution with arbitrary step sizes. However, for convenience in exposition we shall consider calculations on a uniform net

$$x_j = a + jh, \qquad h = \frac{b - a}{N}.$$

Any single-step method for approximating the solution of (1) in $[a, b]$ can be indicated by the general form

$$u_0 = y_0 + e_0, \tag{2a}$$

$$u_{j+1} = u_j + hF\{h, x_j, u_j; f\}, \qquad j = 0, 1, \ldots, N - 1. \tag{2b}$$

Here we denote by $F\{h, x_j, u_j; f\}$ some quantity whose value is uniquely determined by the value of (h, x_j, u_j) and the function f. For example, the Euler-Cauchy scheme in (1.1) is a single-step method in which $F\{h, x, u; f\} \equiv f(x, u)$. We shall see that a variety of different choices for F is determined by using Taylor's theorem or quadrature formulae.

It is a simple matter to obtain estimates for the error in a very general class of single-step methods. To do this we first define the *local truncation errors*, τ_{j+1}, by writing

$$y(x_{j+1}) = y(x_j) + hF\{h, x_j, y(x_j); f\} + h\tau_{j+1}, \qquad j = 0, 1, \ldots, N - 1; \tag{3}$$

where $y(x)$ is the solution of (1). The largest integer p such that $|\tau_j| = \mathcal{O}(h^p)$ is called the order† of the method. As usual, the errors in the numerical solution are defined by

$$e_j \equiv u_j - y(x_j), \qquad j = 0, 1, \ldots, N. \tag{4}$$

Now in analogy with Theorem 1.1 we have

THEOREM 1. *Let u_j be the numerical solution defined in* (2) *where $F\{h, x, u; f\}$ satisfies*

$$|F\{h, x, u; f\} - F\{h, x, v; f\}| \leq K|u - v| \tag{5}$$

for all (x, u) *and* (x, v) *in the strip* S: $\{(x, y) \mid a \leq x \leq b, |y| < \infty\}$. *Then if* $y(x)$ *is the solution of* (1), *and* $\{\tau_j\}$ *is defined by* (3),

$$|u_j - y(x_j)| \leq e^{K(x_j - a)}\left(|e_0| + \frac{\tau}{K}\right), \qquad j = 0, 1, \ldots, N; \tag{6}$$

where

$$\tau \equiv \max_j |\tau_j|.$$

Proof. Subtract (3) from (2b) and use (5) to find

$$\begin{aligned} |e_{j+1}| &= |e_j + h[F\{h, x_j, u_j; f\} - F\{h, x_j, y(x_j); f\}] - h\tau_{j+1}| \\ &\leq (1 + hK)|e_j| + h\tau, \qquad j = 0, 1, \ldots, N - 1. \end{aligned}$$

The remainder of the proof follows exactly as in Theorem 1.1. ∎

If the function $f(x, y)$ and the scheme, determined by $F\{h, x, u; f\}$, have special smoothness properties it may be possible to replace (5) by a type of mean value *equality*, say,

$$F\{h, x, u; f\} - F\{h, x, v; f\} = G\{h, x, u, v; f\}(u - v). \tag{7}$$

Here G is determined by the value of (h, x, u, v) and the function f. Again in the Euler-Cauchy scheme if $\partial f/\partial y$ is continuous, then (7) holds with

$$G\{h, x, u, v; f\} = \frac{\partial f(x, \theta u + (1 - \theta)v)}{\partial y}, \qquad \text{for some } \theta \text{ in } 0 < \theta < 1.$$

When this mean value property is satisfied, we can prove exact analogs of Theorem 1.2 and its corollaries.

The roundoff errors in single-step methods can be trea[illegible] ry much as

† We assume that $f(x, y)$ has enough continuous derivatives s[illegible] may be determined by a Taylor's series expansion of $y(x_j + h) - y(x_j)$ [illegible], $x_j, y(x_j); f\}$ in powers of h.

in Subsection 1.2. Thus the numbers actually obtained, say $\{U_j\}$, in trying to evaluate the set $\{u_j\}$ from (2) will, in general, have errors due to the finite precision arithmetic. These numbers will satisfy equations of the form

$$U_0 = y_0 + \rho_0 \tag{8a}$$

$$U_{j+1} = U_j + hF\{h, x_j, U_j; f\} + \rho_{j+1}, \qquad j = 0, 1, \ldots, N - 1. \tag{8b}$$

Then, if (5) is satisfied, we deduce in an obvious manner

$$|U_j - y(x_j)| \le e^{K(x_j - a)}\left(|\rho_0| + \frac{\tau + \dfrac{\rho}{h}}{K}\right) \qquad j = 0, 1, \ldots, N, \tag{9}$$

where $\rho = \max\limits_{1 \le j} |\rho_j|$. Again we see that as $h \to 0$, while $x_j - a = jh \equiv c$ is fixed, the roundoff error may become arbitrarily large if the computing accuracy remains unchanged. This effect is due to the fact that infinitely many computations are required to get to the finite point $x = c$ as $h \to 0$. If the single-step scheme is of order p, then τ can be bounded by a term of the form Mh^p. For numerical balance then, $|\rho_0| = \mathcal{O}(h^p)$ and $\rho = \mathcal{O}(h^{p+1})$ are reasonable requirements for the magnitude of the rounding error.

3.1. Finite Taylor's Series

If the solution $y(x)$, of (1) has continuous derivatives of order $r + 1$ in $[a, b]$, then by Taylor's theorem:

$$\begin{aligned} y(x_{j+1}) = y(x_j) &+ hy^{(1)}(x_j) + \cdots + \frac{h^r}{r!} y^{(r)}(x_j) \\ &+ \frac{h^{r+1}}{(r+1)!} y^{(r+1)}(x_j + \theta_j h), \\ &\qquad 0 < \theta_j < 1; \quad j = 0, 1, \ldots, N - 1. \end{aligned} \tag{10a}$$

From the differential equation it follows that the higher order derivatives of $y(x)$ can be expressed as

$$\begin{aligned} y^{(1)}(x) &= f(x, y), \\ y^{(2)}(x) &= f_x(x, y) + f_y(x, y) y^{(1)}(x), \\ y^{(3)}(x) &= f_{xx}(x, y) + 2f_{xy}(x, y) y^{(1)}(x) \\ &\qquad + f_{yy}(x, y)[y^{(1)}(x)]^2 + f_y(x, y) y^{(2)}(x), \\ &\vdots \end{aligned} \tag{10b}$$

or in general,

$$y^{(\nu)}(x) = \frac{d^{\nu-1}}{dx^{\nu-1}} f(x, y(x)); \qquad \nu = 1, 2, \ldots. \tag{10b'}$$

Thus given the value of $y(x)$ at a point, we may determine its derivatives if we can evaluate the partial derivatives of $f(x, y)$. We use these observations in the *finite Taylor's series method* for approximating solutions of (1).

Equation (10a) suggests the scheme

$$u_{j+1} = u_j + hu_i^{(1)} + \cdots + \frac{h^r}{r!} u_j^{(r)}, \qquad j = 0, 1, \ldots, N-1, \tag{11a}$$

where in analogy with (10b) we have defined, for given (x_j, u_j),

$$\begin{aligned} u_j^{(1)} &= f(x_j, u_j), \\ u_j^{(2)} &= f_x(x_j, u_j) + f_y(x_j, u_j)u_j^{(1)}, \\ u_j^{(3)} &= f_{xx}(x_j, u_j) + 2f_{xy}(x_j, u_j)u_j^{(1)} \\ &\quad + f_{yy}(x_j, u_j)(u_j^{(1)})^2 + f_y(x_j, u_j)u_j^{(2)}, \\ &\vdots \end{aligned} \tag{11b}$$

These formulae are easily deduced from the compact symbolic formula obtained from (10b′)

$$u_j^{(\nu)} = \left[\left(\frac{\partial}{\partial x} + f(x, y)\frac{\partial}{\partial y}\right)^{\nu-1} f(x, y)\right]\Bigg|_{x = x_j,\ y = u_j}. \tag{11b′}$$

The initial value is, allowing for an error in obtaining y_0,

$$u_0 = y_0 + e_0. \tag{11c}$$

The formulation of the method is complete and the approximation $\{u_i\}$ can be computed by recursive application of (11a) through (11c).

To write the Taylor's series method in the form (2) we need only define the operator

$$F\{h, x_j, u_j; f\} \equiv u_j^{(1)} + \frac{h}{2!} u_j^{(2)} + \cdots + \frac{h^{r-1}}{r!} u_j^{(r)} \tag{12}$$

where the $u_j^{(\nu)}$ are defined in (11). Then from the expansion (10a) and the definition (3) of the truncation error for a one-step method, we obtain

$$\begin{aligned} \tau_{j+1} &= \frac{h^r}{(r+1)!} y^{(r+1)}(x_j + \theta_j h), \\ &\qquad\qquad 0 < \theta_j < 1, \quad j = 0, 1, \ldots, N-1. \end{aligned} \tag{13}$$

The method is thus of order r when the first $r + 1$ terms in the Taylor expansion are used. To verify condition (5) we use Taylor's theorem in (11b), after eliminating all $u^{(\nu)}$ and $v^{(\nu)}$, to get

$$\begin{aligned} u^{(1)} - v^{(1)} &= (u - v)[f_y] \\ u^{(2)} - v^{(2)} &= (u - v)[f_{xy} + f_y^2 + f_{yy}f] \\ &\vdots \end{aligned}$$

The arguments of $f(x, y)$ and its derivatives which occur in the brackets are all of the form $(x, \theta u + (1 - \theta)v)$ with different values of θ, in $0 < \theta < 1$, in different brackets. From the representation (11b′) we find in a straightforward manner that

$$(14)\quad u^{(\nu)} - v^{(\nu)} = (u - v)\left[\frac{\partial}{\partial y}\left\{\left(\frac{\partial}{\partial x} + f(x, y)\frac{\partial}{\partial y}\right)^{\nu-1} f(x, y)\right\}\right]\Bigg|_{y=\theta_\nu u + (1-\theta_\nu)v}$$

$$\nu = 1, 2, \ldots.$$

Hence we can conclude that if $f(x, y)$ has sufficiently many continuous and bounded partial derivatives for $(x, y) \in S$ then $F\{h, x, u; f\}$ defined in (12) satisfies (7). Thus (5) is also satisfied, say, for all $h \leq h_0$, where h_0 is some fixed spacing. The constant K entering into (5) can be written in the form

$$(15)\qquad K = M_1 + \frac{h_0}{2!} M_2 + \cdots + \frac{h_0^{r-1}}{r!} M_r,$$

where M_k is a bound on the appropriate bracket in the kth equation in (14); i.e.,

$$M_k \equiv \sup_S \left|\frac{\partial}{\partial y}\left(\frac{\partial}{\partial x} + f(x, y)\frac{\partial}{\partial y}\right)^{k-1} f(x, y)\right|, \qquad k = 1, 2, \ldots, r.$$

By applying Theorem 1 and equation (13), we find that for all $h \leq h_0$

$$(16)\quad |u_j - y(x_j)| \leq e^{K(x_j - a)}\left[|e_0| + \frac{M_{r+1}h^r}{(r + 1)!\,K}\right], \qquad j = 0, 1, \ldots, N,$$

where

$$M_{r+1} \equiv \sup_{[a, b]} |y^{(r+1)}(x)|.$$

If we neglect the initial error, i.e., set $e_0 = 0$, then the error is at most $\mathcal{O}(h^r)$. Thus the Taylor series method can be used to generate starting data which is consistent with any order predictor-corrector method provided only that $f(x, y)$ is sufficiently smooth. However, many different function evaluations, as in (11b), are required and so this method is not very efficient.

Let us assume that

$$-C^2 \leq f_y(x, y) \leq -B^2 < 0 \qquad \text{for all } x \in [a, b] \text{ and } |y| < \infty.$$

We show that in this case the bound (16) can be improved upon. Pick h_0 such that

$$K' = -B^2 + \frac{h_0}{2!} M_2 + \cdots + \frac{h_0^{r-1}}{r!} M_r < 0 \qquad \text{and} \qquad 1 + h_0 K'' > 0,$$

where

$$K'' = -C^2 - B^2 - K'.$$

Now we find, by retracing the proof of Theorem 1 with a little care that for all $h \leq h_0$

$$(17) \qquad |u_j - y(x_j)| \leq e^{K'(x_j - a)}\left[|e_0| + \frac{M_{r+1}h^r}{(r+1)!\,|K'|}\right], \qquad j = 0, 1, \ldots, N.$$

We note that the exponential here is a decreasing function of x_j.

3.2. One-Step Methods Based on Quadrature Formulae

By integrating the differential equation (1) over $[x_j, x_{j+1}]$ we get

$$(18a) \qquad y(x_{j+1}) = y(x_j) + \int_{x_j}^{x_j+h} f(x, y(x))\, dx.$$

Hence, we see that various forms for $hF\{h, x, u; f\}$ are naturally suggested by quadrature formulae. However, as we are considering one-step methods the appropriate quadrature formulae should only employ nodes in $[x_j, x_{j+1}]$, say, for example, the $n + 1$ points ξ_ν satisfying

$$x_j \leq \xi_0 < \xi_1 < \cdots < \xi_n \leq x_{j+1}.$$

But the integrand or an approximation to it must be known at these nodes and so we require approximations to $y(\xi_\nu)$, $\nu = 0, 1, \ldots, n$. We have for the exact solution at these points,

$$(18b) \qquad y(\xi_\nu) = y(x_j) + \int_{x_j}^{\xi_\nu} f(x, y(x))\, dx, \qquad \nu = 0, 1, \ldots, n.$$

Thus we could use a sequence of quadrature formulae to estimate successively the values $y(\xi_\nu)$ and ultimately $y(x_{j+1})$.

A general class of one-step methods based upon these observations is given by using (2) with

$$(19a) \qquad hF\{h, x_j, u_j; f\} = h \sum_{\mu=0}^{n} \alpha_\mu f(\xi_\mu, \eta_\mu);$$

$$(19b) \qquad \xi_0 = x_j; \qquad \xi_\nu = \xi_0 + \theta_\nu h,$$

$$\nu = 1, 2, \ldots, n.$$

$$(19c) \qquad \eta_0 = u_j; \qquad \eta_\nu = \eta_0 + h \sum_{k=0}^{\nu-1} \alpha_{\nu k} f(\xi_k, \eta_k)$$

If in (19) we regard u_ν as an approximation to $y(\xi_\nu)$, then the sums in (19a) and (19c) can be regarded as approximations to the integrals, respectively, in (18a) and (18b). In fact, these considerations *suggest* that we require

$$(20a) \qquad 0 = \theta_0 \leq \theta_1 \leq \theta_2 \leq \cdots \leq \theta_n \leq 1;$$

$$(20b) \qquad \sum_{\nu=0}^{n} \alpha_\nu = 1;$$

$$(20c) \qquad \sum_{k=0}^{\nu-1} \alpha_{\nu k} = \theta_\nu, \qquad \nu = 1, 2, \ldots, n.$$

Condition (20a) requires that all the nodes, ξ_ν, lie in $[x_j, x_{j+1}]$ and form a non-decreasing set; condition (20b) implies that the sum in (19a) has degree of precision at least 0 as a quadrature formula over $[\xi_0, \xi_n]$; conditions (20c) imply similar results for the sums in (19c) as quadrature formulae over $[\xi_0, \xi_\nu]$. If, in addition to (20b), we require that

$$\text{(21)} \qquad \sum_{\nu=0}^{n} \alpha_\nu(\theta_\nu)^p = \frac{1}{p+1}, \qquad p = 1, 2, \ldots, m;$$

then the basic quadrature scheme in (19a) has degree of precision at least m.

These considerations suggest many choices for the parameters in (19), some of which have been examined in the literature (i.e., Gaussian quadrature, equal coefficient formulae, etc.). In fact, in practice, the parameters are determined by the reasonable requirement that the local truncation error for a fixed choice of n, be of as high an order in h as possible. From (3), we determine the local truncation error, τ_{j+1}, for the one-step method defined by (19) from

$$\text{(22a)} \qquad y(x_j + h) = y(x_j) + h \sum_{\nu=0}^{n} \alpha_\nu f(\xi_\nu, y_{\nu j}) + h\tau_{j+1},$$

where

$$\text{(22b)} \qquad y_{0j} = y(x_j);\; y_{\nu j} = y_{0j} + h \sum_{k=0}^{\nu-1} \alpha_{\nu k} f(\xi_k, y_{kj}), \qquad \nu = 1, 2, \ldots, n.$$

If the parameters are given and $f(x, y)$ has sufficiently many continuous derivatives, then $y(x_j + h)$ and the $f(\xi_k, y_{kj})$ can be expanded in powers of h [about x_j, $y(x_j)$]. Equation (22) then yields, upon equating coefficients of like powers of h in (22a), an expression for τ_{j+1}. Obviously, this procedure can be used to determine the parameters in (19) such that τ_{j+1} has the highest possible order in h. The use of Taylor's theorem here is similar to its use in Section 5 of Chapter 5 to determine high order approximations to derivatives, but is now much more complicated. We do not repeat here any of these lengthy calculations, but present in Table 2 some sets of parameter values for one-step methods of indicated order. It is found, in fact, that for $n = 0$ and 1 the maximum orders are 1 and 2, respectively, and the conditions imposed are just those in (20) and (21) with $m = 1$ or 2. For $n = 2$ an order of 3 can be obtained, if, in addition to (20) and (21) with $m = 3$ one additional relation is satisfied; namely,

$$\sum_{\nu=0}^{n} \alpha_\nu \sum_{k=0}^{\nu-1} \alpha_{\nu k} \theta_k = \tfrac{1}{6}.$$

(This relation can be explained as the result of requiring the coefficients α_ν and $\alpha_{\nu k}$ to come from quadrature formulae with respective degrees of precision 2 and 1, at least.)

Table 2 *Some Standard Single-Step Difference Methods*

Associated Name	n	Coefficients and Nodes ν or j	0	1	2	3	Order in h of τ
Modified Euler	1	$\alpha_\nu =$	$\frac{1}{2}$	$\frac{1}{2}$			$\mathcal{O}(h^2)$
		$\theta_j =$	0	1			
		$\alpha_{1j} =$	1				
Heun	2	$\alpha_\nu =$	$\frac{1}{4}$	0	$\frac{3}{4}$		$\mathcal{O}(h^2)$
		$\theta_j =$	0	$\frac{1}{3}$	$\frac{2}{3}$		
		$\alpha_{1j} =$	$\frac{1}{3}$				
		$\alpha_{2j} =$	0	$\frac{2}{3}$			
Kutta	2	$\alpha_\nu =$	$\frac{1}{6}$	$\frac{2}{3}$	$\frac{1}{6}$		$\mathcal{O}(h^3)$
		$\theta_j =$	0	$\frac{1}{2}$	1		
		$\alpha_{1j} =$	$\frac{1}{2}$				
		$\alpha_{2j} =$	-1	2			
Runge-Kutta	3	$\alpha_\nu =$	$\frac{1}{6}$	$\frac{1}{3}$	$\frac{1}{3}$	$\frac{1}{6}$	$\mathcal{O}(h^4)$
		$\theta_j =$	0	$\frac{1}{2}$	$\frac{1}{2}$	1	
		$\alpha_{1j} =$	$\frac{1}{2}$				
		$\alpha_{2j} =$	0	$\frac{1}{2}$			
		$\alpha_{3j} =$	0	0	1		
Runge-Kutta	3	$\alpha_\nu =$	$\frac{1}{8}$	$\frac{3}{8}$	$\frac{3}{8}$	$\frac{1}{8}$	$\mathcal{O}(h^4)$
		$\theta_j =$	0	$\frac{1}{3}$	$\frac{2}{3}$	1	
		$\alpha_{1j} =$	$\frac{1}{3}$				
		$\alpha_{2j} =$	$-\frac{1}{3}$	1			
		$\alpha_{3j} =$	1	-1	1		

We shall now show that all the schemes included in (19) satisfy the mean value property (7), if $f_y(x, y)$ is continuous in S. Let the quantities ζ_ν be defined as the η_ν are in (19) but with u_j replaced by v_j. Now we introduce the notation $g(x, y) \equiv f_y(x, y)$ and use the continuity of g to deduce

$$\text{(23a)} \qquad (\eta_\nu - \zeta_\nu) = (\eta_0 - \zeta_0) + h \sum_{k=0}^{\nu-1} \alpha_{\nu k}[f(\xi_k, \eta_k) - f(\xi_k, \zeta_k)]$$

$$= (\eta_0 - \zeta_0) + h \sum_{k=0}^{\nu-1} \alpha_{\nu k} g_k(\eta_k - \zeta_k); \quad \nu = 1, 2, \ldots, n.$$

Here we have used

$$\text{(23b)} \qquad g_k = g(\xi_k, \phi_k\eta_k + (1 - \phi_k)\zeta_k), \quad 0 < \phi_k < 1, \quad k = 0, 1, \ldots, n.$$

By applying the equations in (23) recursively, we can determine expressions for the $(\eta_\nu - \zeta_\nu)$ in terms of $(\eta_0 - \zeta_0)$. However, this procedure is rather complicated and so we will just present the result and verify it by induction.

Let us define the quantities $B_{\nu j}$ as follows: $B_{0j} \equiv 1$;

$$
(24)\qquad
\begin{aligned}
B_{\nu j} &= 1 & j &= 0 \\
B_{\nu j} &= \sum_{k=j-1}^{\nu-1} \alpha_{\nu k} g_k B_{k,j-1} & j &= 1, 2, \ldots, \nu \qquad \nu = 1, 2, \ldots, n. \\
B_{\nu j} &= 0 & j &\geq \nu + 1
\end{aligned}
$$

Then we have

$$
(25)\qquad (\eta_\nu - \zeta_\nu) = (\eta_0 - \zeta_0)(B_{\nu 0} + hB_{\nu 1} + h^2 B_{\nu 2} + \ldots + h^\nu B_{\nu\nu}), \qquad \nu = 1, 2, \ldots, n.
$$

To verify (25) by induction, we note that from (23) and (24) with $\nu = 1$,

$$
(\eta_1 - \zeta_1) = (\eta_0 - \zeta_0)(1 + h\alpha_{10} g_0) = (\eta_0 - \zeta_0)(B_{10} + hB_{11}).
$$

Thus (25) is valid for $\nu = 1$. We now assume (25) to be valid up to $\nu - 1$ and use it in (23a) to obtain

$$
\begin{aligned}
(\eta_\nu - \zeta_\nu) &= (\eta_0 - \zeta_0)\left(1 + h \sum_{k=0}^{\nu-1} \alpha_{\nu k} g_k \sum_{m=0}^{k} h^m B_{km}\right), \\
&= (\eta_0 - \zeta_0)\left(1 + \sum_{m=0}^{\nu-1} h^{m+1} \sum_{k=m}^{\nu-1} \alpha_{\nu k} g_k B_{km}\right), \\
&= (\eta_0 - \zeta_0)\left(1 + \sum_{m=0}^{\nu-1} h^{m+1} B_{\nu,m+1}\right).
\end{aligned}
$$

The induction is thus concluded and (25) is established.

We now obtain, from the mean value theorem and (25),

$$
\begin{aligned}
F\{h, x_j, u_j; f\} - F\{h, x_j, v_j; f\} &= \sum_{\nu=0}^{n} \alpha_\nu g_\nu (\eta_\nu - \zeta_\nu) \\
&= (u_j - v_j) \sum_{\nu=0}^{n} \alpha_\nu g_\nu \sum_{k=0}^{\nu} h^k B_{\nu k},
\end{aligned}
$$

since $(\eta_0 - \zeta_0) = (u_j - v_j)$. That is, we have established

LEMMA 1. *Under the assumption that $f_y(x, y)$ is continuous every one-step scheme defined by* (19) *satisfies the generalized mean value property* (7) *with*

$$
(26)\qquad G\{h, x, u, v; f\} \equiv \sum_{\nu=0}^{n} \left(\alpha_\nu g_\nu \sum_{k=0}^{\nu} h^k B_{\nu k}\right). \qquad \blacksquare
$$

Here the $B_{\nu k}$ are defined in (24) and the g_ν and g_k are values of $g(x, y) = f_y(x, y)$ at appropriate points of the form $(x, \phi u + (1 - \phi)v)$, $0 < \phi < 1$. Again it should be observed that if $f_y < 0$ in S, then by taking h so small that $\alpha_\nu \geq 0$, (26) implies $G < 0$. Hence in such cases, the initial error in one-step methods will decay exponentially with distance.

Of course, the indicated class of single-step methods satisfies a Lipschitz condition of the form (5). To obtain a suitable constant K_0 we could use, from (26),

$$K_0 = \sup_{S,\, h \leq h_0} \left| \sum_{\nu=0}^{n} \alpha_\nu g_\nu \sum_{k=0}^{\nu} h^k B_{\nu k} \right|.$$

However, this is not readily calculable and so we shall determine an upper bound for it in terms of the parameters of the scheme and

$$M \equiv \sup_S \left| \frac{\partial f(x, y)}{\partial y} \right|.$$

We note that $|g_\nu| \leq M$ for all ν. Now define

$$\Phi \equiv \max_\nu \left(\sum_{k=0}^{\nu-1} |\alpha_{\nu k}| \right) \qquad \text{and} \qquad \beta_j \equiv \max_\nu |B_{\nu j}|,$$

and from (24) with j in $1 \leq j \leq \nu$

$$\begin{aligned} |B_{\nu j}| &\leq M \sum_{k=j-1}^{\nu-1} |\alpha_{\nu k}| \cdot |B_{k, j-1}| \\ &\leq M \cdot \max_{j \leq i \leq \nu} |B_{i-1, j-1}| \cdot \sum_{k=j-1}^{\nu-1} |\alpha_{\nu k}| \\ &\leq M \Phi \beta_{j-1}. \end{aligned}$$

Since the right-hand side is independent of ν we conclude that

$$\beta_j \leq M \Phi \beta_{j-1},$$

and by recursion using $\beta_0 = 1$

$$\beta_j \leq (M\Phi)^j, \qquad j = 0, 1, \ldots, \nu.$$

Since $|B_{\nu j}| \leq \beta_j$ for all ν we have

$$\begin{aligned} K_0 &\leq \sum_{\nu=0}^{n} |\alpha_\nu| M \left[1 + \sum_{k=1}^{\nu} h^k (\Phi M)^k \right] \\ &\leq M \sum_{\nu=0}^{n} |\alpha_\nu| \frac{1 - (h\Phi M)^{\nu+1}}{1 - (h\Phi M)}. \end{aligned} \tag{27}$$

If we require that the coefficients α_ν be non-negative and satisfy at least (20b), then

$$\sum_{\nu=0}^{n} |\alpha_\nu| = \sum_{\nu=0}^{n} \alpha_\nu = 1.$$

If further, h is chosen such that $h \leq h_0$, where $h_0 \Phi M < 1$, then the above bound simplifies to

$$K_0 \leq \frac{M}{1 - h\Phi M}. \tag{28}$$

We also note that if the $\alpha_{\nu k}$ are non-negative for all ν and all $k = 0, 1, \ldots, \nu - 1$ and if (20) is satisfied, then $\Phi = \theta_n$ and hence $0 \leq \Phi \leq 1$. *For sufficiently small h, in any event, the above bound can be made as close as we please to M which serves as the Lipschitz constant in the simple Euler method treated in Section* 1.

PROBLEMS, SECTION 3

1. Verify the entries in Table 2 under the name *Modified Euler.*

2. Verify the entries in Table 2 under the names *Heun* and *Kutta* with $n = 2$.

3. Verify the entries in Table 2 for both schemes under the name *Runge-Kutta* with $n = 3$.

4. LINEAR DIFFERENCE EQUATIONS

We recall that linear difference equations with constant coefficients have appeared previously in our study, for example, in Section 4 of Chapter 3 and in Subsection 1.4 of the present chapter. The theory of such difference equations will be sketched here because it will be used in the general treatment of difference methods given in the next section. The general linear difference equation with constant coefficients is a relation of the form

$$L(u_j) \equiv \sum_{s=0}^{n} a_s u_{j+s} = c_{j+n}, \qquad j = j_0, j_0 + 1, \ldots. \tag{1}$$

Here the quantities a_s are the coefficients, the c_{j+n} are the inhomogeneous terms, and the sequence $\{u_j\}$ is to be determined subject in general to additional conditions. Usually the sequence is desired starting from some initial index, say as indicated in (1) for $j \geq j_0$. The difference equation in (1) is said to be of order n, if $a_n a_0 \neq 0$, since then the indices on the u_i vary over $n + 1$ consecutive integers. We shall see that a solution of an nth order linear difference equation is, in general, determined by specifying n

initial conditions. That is, if j_0 is the initial index then we adjoin to (1) the conditions

$$u_{j_0} = v_0, \quad u_{j_0+1} = v_1, \ldots, u_{j_0+n-1} = v_{n-1}. \tag{2}$$

We now have

THEOREM 1. *If the difference equation* (1) *is of nth order then there is one and only one solution* $\{u_i\}$ *satisfying the initial conditions* (2).

Proof. The existence of the solution follows trivially since $a_n \neq 0$ implies from (1) that

$$u_{j+n} = -\frac{1}{a_n}\sum_{s=0}^{n-1}(a_s u_{j+s}) + \frac{c_{j+n}}{a_n}, \qquad j = j_0, j_0 + 1, \ldots. \tag{3}$$

For uniqueness let there be two solutions, $\{u_i'\}$ and $\{u_i''\}$. Then their difference $\{u_i\} \equiv \{u_i' - u_i''\}$ satisfies (2) with $v_0 = v_1 = \cdots = v_{n-1} = 0$ and (3) with $c_{j+n} \equiv 0$. Thus we find that $\{u_i\} \equiv \{0\}$ and the proof is complete. ∎

We consider the nth order *homogeneous difference equations* corresponding to (1), namely:

$$L(u_j) = 0; \qquad j = j_0, j_0 + 1, \ldots. \tag{4}$$

If the sequences $\{u_i\}$ and $\{v_i\}$ are solutions of (4) then, by the linearity of these equations, the sequence $\{\alpha u_i + \beta v_i\}$ is also a solution. Here α and β are arbitrary numbers. Thus we easily find that the set of all solutions of (4) forms a linear vector space. A set of solutions, say r of them

$$\{u_i^{(1)}\}, \{u_i^{(2)}\}, \ldots, \{u_i^{(r)}\},$$

are *linearly independent* if only the *trivial* combination of the $\{u_i^{(\nu)}\}$ vanishes identically; that is, if

$$\alpha_1 u_i^{(1)} + \alpha_2 u_i^{(2)} + \cdots + \alpha_r u_i^{(r)} = 0, \qquad \text{for } i = j_0, j_0 + 1, \ldots,$$

implies that $\alpha_1 = \alpha_2 = \cdots = \alpha_r = 0$. This is essentially the same notion as the linear independence of vectors. A set of n independent solutions of the nth order equations (4) is called a *fundamental set of solutions.*

A basic result now can be stated as

THEOREM 2. *Let* $\{u_i^{(\nu)}\}$, $\nu = 1, 2, \ldots, n$, *be a fundamental set of solutions of the homogeneous difference equations* (4). *Then any solution,* $\{v_i\}$, *of these equations can be expressed uniquely in the form*

$$\{v_i\} = \left\{\sum_{\nu=1}^{n} \alpha_\nu u_i^{(\nu)}\right\}.$$

Proof. Since the set $\{u_i^{(\nu)}\}$ is independent, Theorem 1 implies that the n vectors $\mathbf{u}^{(\nu)}$, where $\mathbf{u}^{(\nu)} \equiv (u_i^{(\nu)})$ for $i = j_0, j_0 + 1, \ldots, j_0 + n - 1$, are linearly independent. That is, according to Theorem 1, if the $\mathbf{u}^{(\nu)}$ were linearly dependent, then the corresponding infinite sequences $\{u_i^{(\nu)}\}$ would be linearly dependent. Hence the nth order matrix

$$A \equiv \begin{bmatrix} u_{j_0}^{(1)} & u_{j_0}^{(2)} & \cdots & u_{j_0}^{(n)} \\ u_{j_0+1}^{(1)} & u_{j_0+1}^{(2)} & \cdots & u_{j_0+1}^{(n)} \\ \vdots & \vdots & & \vdots \\ u_{j_0+n-1}^{(1)} & u_{j_0+n-1}^{(2)} & \cdots & u_{j_0+n-1}^{(n)} \end{bmatrix}$$

is non-singular. Thus given any n components, say, $\mathbf{v} \equiv \{v_i\}$ for $i = j_0, j_0 + 1, \ldots, j_0 + n - 1$, we can uniquely solve the nth order system

$$A\boldsymbol{\alpha} = \mathbf{v}.$$

The components α_ν of $\boldsymbol{\alpha}$ are the coefficients to be used in the theorem. Since the first n components of any solution $\{v_i\}$ can be expressed as a linear combination of the first n components of the fundamental set the theorem now follows by an application of Theorem 1. ∎

We can, furthermore, find a fundamental set of solutions of (4). We try as a solution the powers of some scalar, say

$$u_i = \alpha x^i; \qquad i = j_0, j_0 + 1, \ldots.$$

Then (4) yields

$$(a_n x^n + a_{n-1}x^{n-1} + \cdots + a_0)(\alpha x^j) = 0.$$

If $\alpha x^j = 0$ the corresponding solution is trivial and does not lead to a fundamental set. Hence, we only consider the roots of

$$\text{(5)} \qquad p_n(x) \equiv a_n x^n + a_{n-1}x^{n-1} + \cdots + a_0 = 0.$$

The nth degree polynomial $p_n(x)$ is called the *characteristic polynomial* of the difference equation (4). We easily find that if x is a root of (5) then $\{u_i\} = \{x^i\}$ is a solution of the homogeneous difference equations. *If the roots of the characteristic equation are distinct, say* $x_1, x_2, \ldots, x_n$, *then a fundamental set of solutions is given by* $\{u_i^{(\nu)}\} = \{x_\nu{}^i\}$, $\nu = 1, 2, \ldots, n$. Since $a_n a_0 \neq 0$, there is no zero root and the independence of the $\{u_i^{(\nu)}\}$ follows from the independence of the first n components. That is, let us define the matrix

$$U \equiv (\mathbf{u}^{(1)}, \quad \mathbf{u}^{(2)}, \ldots, \mathbf{u}^{(n)}),$$

whose columns are vectors obtained from the first n components of the set of solutions defined above. Then clearly, U is non-singular if the x_i are distinct, since

$$U = \begin{bmatrix} 1 & 1 & \cdots & 1 \\ x_1 & x_2 & \cdots & x_n \\ \vdots & \vdots & & \vdots \\ x_1^{n-1} & x_2^{n-1} & \cdots & x_n^{n-1} \end{bmatrix} \begin{bmatrix} x_1^{j_0} & & & 0 \\ & x_2^{j_0} & & \\ & & \ddots & \\ 0 & & & x_n^{j_0} \end{bmatrix}.$$

If the roots x_i are not distinct, we can still define a fundamental set of solutions. Let x_1 be a root of multiplicity $m_1 > 1$ of $p_n(x) = 0$. Then we use the powers to generate one solution and successive derivatives† with respect to x_1, up to order $m_1 - 1$ to generate $m_1 - 1$ additional solutions. Specifically, let $u_j^{(1)} = x_1^j$. Now try

$$\{u_j^{(2)}\} = \frac{d}{dx_1}\{u_j^{(1)}\}, \ldots, \{u_j^{(m_1)}\} = \frac{d}{dx_1}\{u_j^{(m_1-1)}\}.$$

However, since any solution can be multiplied by a non-zero constant we multiply the resulting $\{u_j^{(\nu)}\}$ by $x_1^{\nu-1}$, to retain the original powers of x_1 in corresponding terms, that is, we introduce

$$\{v_j^{(\nu)}\} = x_1^{\nu-1}\{u_j^{(\nu)}\}, \qquad \nu = 1, 2, \ldots, m_1.$$

The elements of these sequences are found to be

$$\begin{aligned} v_j^{(1)} &= x_1^j, \\ v_j^{(2)} &= jx_1^j, \\ v_j^{(3)} &= j(j-1)x_1^j, \\ &\vdots \\ v_j^{(m_1)} &= j(j-1)\cdots(j-m_1+2)x_1^j, \qquad \text{for } j = j_0, j_0+1, \ldots. \end{aligned}$$

We leave the verification that these form m_1 solutions as Problem 1.

By forming linear combinations of the solutions $\{v_j^{(\nu)}\}$ corresponding to a root, x_1, of multiplicity m_1, we find the simpler sequences $\{w_j^{(\nu)}\}$

$$\begin{aligned} w_j^{(1)} &= x_1^j, \\ w_j^{(2)} &= jx_1^j, \\ w_j^{(3)} &= j^2x_1^j, \\ &\vdots \\ w_j^{(m_1)} &= j^{m_1-1}x_1^j, \qquad j = j_0, j_0+1, \ldots. \end{aligned} \tag{6}$$

† To motivate this procedure, observe that if x_1 and $x_2 = x_1 + h$ are two roots of (5), then

$$u_i = h^{-1}[(x_1 + h)^i - x_1^i], \qquad i = j_0, j_0 + 1, \ldots,$$

is a solution of (4). But then

$$\lim_{h\to 0} u_i = ix_1^{i-1}$$

is also a solution.

These sequences are obviously linearly independent and are solutions because they can be obtained as linear combinations of the $\{v_j^{(\nu)}\}$.

Let us now consider the inhomogeneous nth order linear difference equations (1). If $\{v_j\}$ is any particular solution of equation (1) and $\{u_j\}$ is a solution of the homogeneous system (4), then $\{u_j + v_j\}$ is a solution of (1). This solution of (1) can be made to satisfy any particular initial conditions by adjusting the $\{u_j\}$. We now develop a discrete analog of what is known as *Duhamel's principle* in the theory of differential equations (where integral representations of solutions are obtained).

THEOREM 3. *Let $\{u_j^{(\nu)}\}$ be the fundamental set of solutions of the nth order homogeneous difference equation* (4) *which satisfy the initial conditions*

$$u_i^{(\nu)} = \delta_{i\nu}, \quad i = 0, 1, \ldots, n-1; \quad \nu = 0, 1, \ldots, n-1. \tag{7}$$

Then the solution of (1) *subject to the initial conditions* (2) *with* $j_0 \equiv 0$, *is given by*

$$u_j = \sum_{\nu=0}^{n-1} v_\nu u_j^{(\nu)} + \frac{1}{a_n}\sum_{k=0}^{j-n} c_{k+n}u_{j-k-1}^{(n-1)}, \qquad j = 0, 1, \ldots. \tag{8}$$

(Here we define $u_i^{(n-1)} \equiv 0$ for all $i < 0$ and $c_j \equiv 0$ for all $j < n$.)

Proof. The first sum in (8) satisfies the initial conditions (2) and the homogeneous difference equations. Thus we need only show that the second sum in (8) satisfies homogeneous initial conditions and the inhomogeneous difference equations (1), with $j_0 = 0$. Let us define

$$w_j \equiv \frac{1}{a_n}\sum_{k=0}^{j-n} c_{k+n}u_{j-k-1}^{(n-1)}; \qquad j = 0, 1, \ldots.$$

Then we have $w_j = 0$ for $j = 0, 1, \ldots, n-1$ by recalling that $u_i^{(n-1)} = 0$ for $i \leq n-2$ and $c_j = 0$ for $j < n$. In fact, for the same reason, we may write

$$w_j = \frac{1}{a_n}\sum_{k=-\infty}^{\infty} c_{k+n}u_{j-k-1}^{(n-1)},$$

since the additional terms vanish. Then

$$L(w_j) = \sum_{s=0}^{n} a_s w_{j+s} = \frac{1}{a_n}\sum_{s=0}^{n} a_s \sum_{k=-\infty}^{\infty} c_{k+n}u_{j+s-k-1}^{(n-1)}.$$

$$= \frac{1}{a_n}\sum_{s=0}^{n} a_s \sum_{k=0}^{j} c_{k+n}u_{j+s-k-1}^{(n-1)},$$

since the terms corresponding to other values of k vanish. Hence,

$$L(w_j) = \frac{1}{a_n}\sum_{k=0}^{j} c_{k+n}L(u_{j-k-1}^{(n-1)}).$$

However, it is easily verified that

$$L(u_{j-k-1}^{(n-1)}) = a_n \delta_{jk}$$

and so finally,

$$L(w_j) = c_{j+n}. \qquad \blacksquare$$

PROBLEMS, SECTION 4

1. Verify that equation (6) defines m_1 linearly independent solutions of (4).

2. Verify that a fundamental set of solutions of (4) is obtained from (6), if m_1 is replaced by m_i and x_1 by x_i for each root x_i of (5) of multiplicity m_i.

5. CONSISTENCY, CONVERGENCE, AND STABILITY OF DIFFERENCE METHODS

The numerical procedures which we have introduced in Sections 1 through 3 in order to approximate the solutions of differential equations may be called difference methods. In this section we study the convergence of a more general class of difference schemes. The analysis constitutes a uniform development for all of the commonly used methods that were treated separately in Sections 1 through 3.

The solution of the difference equations is what we try to compute, and this may have to be done for very fine meshes, i.e., for many net points. Thus, as a practical matter, it is important that these solutions should not be too sensitive to small errors in the computations (for example, roundoff errors). This sensitivity to errors is related to what is called the *stability of the difference equations*. We have already investigated such matters but without the introduction of this terminology. We shall see that for *consistent* methods, stability of the difference equations is equivalent to convergence of the difference equation solution to the solution of the differential equation problem.

As usual, we consider methods for approximating the solution, $y(x)$, of the initial value problem

$$(1) \qquad \begin{aligned} y' &= f(x, y), \qquad a \le x \le b, \\ y(a) &= y_0. \end{aligned}$$

We assume that $f(x, y)$ is in the class $\mathscr{F}$ of functions such that $f_y(x, y)$, $f_x(x, y)$, and all partial derivatives of f of some finite order $q \ge 1$ are continuous and uniformly bounded in S: $\{(x, y) \mid a \le x \le b;\ |y| < \infty\}$. For any fixed net spacing $h = (b - a)/N$, we use a uniform net $x_j = a + jh$, $j = 0, 1, \ldots, N$, and seek approximations u_j to $y(x_j)$ on this net.

The approximations are defined as the solution of some difference problem, say

(2a) $$a_n u_{j+n} + \cdots + a_0 u_j = hF\{h, x_j; u_{j-m}, \ldots, u_{j+n}; f\} + h\rho_{j+n},$$
$$j = m, m+1, \ldots, N-n;$$

where $\{a_i\}$ are real constants independent of h satisfying $a_n a_0 \neq 0$, and ρ_{j+n} is the *local rounding error* subject to

$$|\rho_{j+n}| \leq \rho(h), \qquad \text{for } j \geq m.$$

The initial data are specified as, say

(2b) $$u_0 = y_0 + \rho_0, u_1 = y_1 + \rho_1, \ldots, u_{m+n-1} = y_{m+n-1} + \rho_{m+n-1}$$

where

$$|\rho_k| \leq r(h), \text{ for } 0 \leq k \leq m+n-1.$$

We shall later require that $\rho(h) \to 0$ and $r(h) \to 0$ as $h \to 0$.

By suitably defining $F\{h, x_j; u_{j-m}, \ldots, u_{j+n}; f\}$, we may incorporate in (2) all of the schemes treated in the previous sections. On the other hand, the only properties that we need postulate for F, in order to make this general study of convergence, are easily seen to hold for all of the commonly used difference methods. That is, we require

(3a) $$F\{h, x_j; u_{j-m}, \ldots, u_{j+n}; 0\} \equiv 0;$$

(3b) $$|F\{h, x_j; v_{j-m}, \ldots, v_{j+n}; f\} - F\{h, x_j; u_{j-m}, \ldots, u_{j+n}; f\}|$$
$$\leq C \sum_{k=-m}^{n} |v_{j+k} - u_{j+k}|,$$

where the constant C depends only on the bounds of f and a finite number of its partial derivatives in S. The *local truncation error*, τ_{j+n}, is defined by

(4a) $$\sum_{k=0}^{n} a_k y_{j+k} - hF\{h, x_j; y_{j-m}, \ldots, y_{j+n}; f\} = h\tau_{j+n},$$

where $y(x)$ is a solution of (1). We further require that

(4b) $$|\tau_{j+n}| \leq \tau(h) \qquad \text{for } m \leq j \leq N-n$$

and

(4c) $$\lim_{h \to 0} \tau(h) = 0,$$

i.e., the truncation error tends to zero. Condition (4c) implies that the difference equation (2) is an "approximation" to (1), rather than some other equation. Strictly speaking, we say that (2) is *consistent* with (1) if $r(h) \to 0$ and $\tau(h) \to 0$ as $h \to 0$. [For example, let $m = 0$, $n = 1$; $a_0 = -1$; and

$$F\{h, x_j; u_{j-m}, \ldots, u_{j+n}; f\} \equiv f(x_{j+1}, u_{j+1}) + f(x_j, u_j).$$

Then (2) is not consistent with the equation (1). That is, by using Taylor series in (4a), for small h, $\tau_{j+1} \cong f(x_j, y_j)$. Hence τ_j does not approach zero. In fact, this scheme is consistent with the equation $y' = 2f(x, y)$.] If, for all $h \leq h_0$,

$$\tau_{j+n} \leq \tau(h) \equiv Mh^p, \qquad m \leq j \leq N - n, \tag{5}$$

where M depends only on the bounds of f and a finite number of its derivatives in S, we say that the truncation error of the difference method is of *order* p.

We, of course, are interested in characterizing the convergent schemes and in obtaining an estimate of the error. For a fixed mesh width, h, we define the *pointwise error*

$$e_j \equiv u_j - y_j, \qquad \text{where } y_j = y(x_j).$$

The method (2) is *convergent* if, for any $f(x, y)$ in $\mathscr{F}$, $\max\limits_{0 \leq j \leq N} |e_j| \to 0$ as $h \to 0$, provided that the rounding errors $\rho(h)$ tend to zero.

If scheme (2) is convergent for all f in $\mathscr{F}$, then it is convergent for the problem (1) with $f(x, y) \equiv 0$, $y_0 = 0$. From this simple observation, we note that if (2) is convergent and F satisfies (3a), then the solution $\{u_j\}$ of

$$\begin{aligned} &a_n u_{j+n} + a_{n-1} u_{j+n-1} + \cdots + a_0 u_j = 0 \\ &u_0 = \rho_0,\ u_1 = \rho_1, \ldots, u_{n-1} = \rho_{n-1}, \end{aligned} \tag{6}$$

must tend to the solution $y(x) \equiv 0$, for any set of initial errors $\{\rho_k\}$ such that $\max\limits_k |\rho_k| \to 0$ as $h \to 0$.

We say that the difference method (2) satisfies the *root condition* if

$$P(\zeta) \equiv a_n \zeta^n + a_{n-1}\zeta^{n-1} + \cdots + a_1\zeta + a_0, \tag{7a}$$

has only zeros ζ_i such that

$$|\zeta_i| \leq 1, \tag{7b}$$

and the multiplicities, r_i, of the ζ_i are such that if

$$|\zeta_k| = 1, \qquad \text{then } r_k = 1. \tag{7c}$$

In other words, scheme (2) satisfies the root condition, if the zeros of $P(\zeta)$ lie in the unit circle and only simple zeros may lie on the boundary of the unit circle. We can now establish a necessary condition that (2) be convergent; i.e., for the solution of (6) to tend to zero.

THEOREM 1. *If* (2) *is convergent and F satisfies* (3a), *then* (2) *satisfies the root condition* (7).

Proof. We show by contradiction that the root condition is necessary. That is, if $|\zeta_i| > 1$ and ζ_i is a complex root of $P(\zeta)$, define

(8a) $$u_j \equiv h(\zeta_i^j + \bar{\zeta}_i^j), \qquad j = 0, 1, \ldots;$$

if ζ_i is a real root set

(8b) $$u_j \equiv h_i \zeta^j.$$

Clearly, (8) is a solution of (6) with $\rho_k = u_k$ for $k = 0, 1, \ldots, n-1$, and $\max |\rho_k| \to 0$ as $h \to 0$. On the other hand, for any c in $a < c < b$ set $j = [c/h]$. But then $|u_{[c/h]}| \to \infty$ as $h \to 0$. Hence such a scheme is not convergent.

If on the other hand, $|\zeta_i| = 1$ and ζ_i is a multiple root and complex set

(9a) $$u_j = hj(\zeta_i^j + \bar{\zeta}_i^j), \qquad j = 0, 1, \ldots;$$

while if ζ_i is real, set

(9b) $$u_j = hj\zeta_i^j.$$

Now if $j = [c/h]$, $|u_{[c/h]}|$ does not approach zero as $h \to 0$. Hence such a scheme is not convergent. ∎

The requirement that the solution $\{u_i\}$ of (2) depend Lipschitz continuously on $\{\rho_k\}$ is the definition of *stability*. That is, we say that (2) is *stable* if for any f in $\mathscr{F}$, there is an h_0 and an M, such that for all $0 \leq h \leq h_0$, and $N \equiv N(h) = (b - a)/h$

(10a) $$|u_i - v_i| \leq M\epsilon, \qquad \text{for } 0 \leq i \leq N$$

whenever $\{v_i\}$ satisfies

(10b) $$a_n v_{j+n} + \cdots + a_0 v_j = hF\{h, x_j; v_{j-m}; \ldots, v_{j+n}; f\} + h\sigma_{j+n}, \qquad j \geq m,$$

(10c) $$v_0 = y_0 + \sigma_0, \ldots, v_{m+n-1} = y_{m+n-1} + \sigma_{m+n-1}$$

where

$$|\rho_k - \sigma_k| \leq \epsilon \qquad \text{for } 0 \leq k \leq N.$$

It is then easy to show

THEOREM 2. *If the scheme* (2) *is stable and F satisfies* (3a) *then the root condition* (7) *is satisfied.*

Proof. The proof follows by contradiction as did the previous theorem. Merely verify that in the case $f \equiv 0$ and $\rho_{j+n} = 0$ for $j \geq 0$, the definitions

(8) or (9) with h replaced by δ define a solution $\{u_i\}$ of (2). On the other hand, set $\sigma_k = 0$ and $v_i = 0$. Then it follows that

$$|\rho_k - \sigma_k| \leq \epsilon \qquad \text{for } 0 \leq k \leq n-1,$$

where ϵ is proportional to δ. But now, (10a) cannot be satisfied for any fixed M as $h \to 0$. ■

Next we have

THEOREM 3. *If* (2) *is consistent with* (1), i.e. *satisfies* (4) *and* $r(h) \to 0$ *as* $h \to 0$, *and* F *satisfies* (3), *then* (2) *is a convergent scheme if and only if the root condition* (7) *is satisfied.*

Proof. In Theorem 1 we have shown that the root condition is necessary for convergence. We now assume that the root condition holds and prove that (2) is convergent. By subtracting equation (4a) from equation (2a), we obtain a difference equation satisfied by the pointwise error $e_j = u_j - y_j$,

$$\text{(11)} \qquad \sum_{s=0}^{n} a_s e_{j+s} = c_{j+n}, \qquad \text{for } m \leq j \leq N-n,$$

where

$$c_{j+n} \equiv h[F\{h, x_j; u_{j-m}, \ldots, u_{j+n}; f\} - F\{h, x_j; y_{j-m}, \ldots, y_{j+n}; f\}] + h\rho_{j+n} - h\tau_{j+n}.$$

We may solve the inhomogeneous difference equation (11), by using Theorem 4.3, in the form

$$\text{(12)} \qquad e_{j+m} = \sum_{k=0}^{n-1} e_{m+k} u_j^{(k)} + \frac{1}{a_n} \sum_{k=0}^{j-n} c_{k+m+n} u_{j-k-1}^{(n-1)}, \qquad \text{for } j = 0, 1, \ldots, N-m.$$

[That is, define

$$E_j \equiv e_{j+m}, \quad C_j \equiv c_{j+m}, \qquad \text{for } j = 0, 1, \ldots.$$

Then (11) holds, i.e.,

$$\sum_{s=0}^{n} a_s E_{j+s} = C_{j+n} \qquad \text{for } 0 \leq j \leq N-m-n.$$

Hence equation (4.8) gives a representation for E_j, which reduces to (12).]

But because the root condition is satisfied, we know from Theorem 4.2 and equation (4.6) that the solutions $\{u_j^{(k)}\}$ satisfy

$$\text{(13)} \qquad |u_j^{(k)}| \leq Q$$

for some constant Q independent of k and j.

Furthermore, from the definitions of c_{j+n}, $\rho(h)$, $\tau(h)$ that appear after equations (11), (2a), and (4a) respectively, and from (3b), we find

$$(14) \qquad |c_{k+m+n}| \leq h\left[C \sum_{r=0}^{m+n} |e_{k+r}| + \rho(h) + \tau(h)\right].$$

If we use the estimates (13) and (14) in (12), we have

$$|e_{j+m}| \leq Qn \max_{0\leq k\leq n-1} |e_{m+k}| + \frac{Q}{a_n}(j - n + 1)h[(m + n + 1)C \max_{0\leq r\leq j+m} |e_r| + \rho(h) + \tau(h)]$$

This inequality simplifies, if we introduce

$$\omega_j = \max_{0\leq k\leq j} |e_k|,$$

to read

$$(15) \qquad |e_{j+m}| \leq Qn\omega_{m+n-1} + \frac{Q(j - n + 1)h}{a_n} \times [(m + n + 1)C\omega_{j+m} + \rho(h) + \tau(h)], \qquad \text{for } j = 0, 1, \ldots, N - m.$$

Since ω_{j+m} is equal to $|e_k|$ for some index $k \leq j + m$ and since $n \geq 1$, we find from (15) that a fortiori

$$(16) \qquad \omega_{j+m} \leq \kappa jh\omega_{j+m} + Qn\omega_{m+n-1} + \frac{Qjh}{a_n}[\rho(h) + \tau(h)],$$

where

$$\kappa = \frac{QC(m + n + 1)}{a_n}, \qquad \text{for } j = 0, 1, \ldots, N - m.$$

If we limit the range of j, as h tends to zero, so that

$$(17) \qquad jh\kappa \leq \tfrac{1}{2},$$

then (16) yields

$$(18) \qquad \omega_{j+m} \leq 2Q\left[n\omega_{m+n-1} + \frac{\rho(h) + \tau(h)}{2\kappa a_n}\right] \qquad \text{for } 0 \leq j \leq \frac{1}{2\kappa h}.$$

If we now employ the definition of ω_j, (18) yields, using $r(h)$ defined after (2b),

$$(19) \qquad |e_{j+m}| \leq 2Q\left[nr(h) + \frac{\rho(h) + \tau(h)}{2\kappa a_n}\right] \qquad \text{for } 0 \leq jh \leq \frac{1}{2\kappa}.$$

Equation (19) bounds the pointwise error, for a finite interval $(a, a + 1/(2\kappa))$, in terms of the bound for the initial error, ω_{m+n-1}, and the bounds $\rho(h)$ and $\tau(h)$ of the rounding and truncation errors. The length of the interval of convergence, $1/(2\kappa)$, is independent of h and is defined after (16).

Hence we may repeat this argument by beginning with the $m + n$ errors bounded by (19).

$$e_{[1/(2\kappa h)]-m-n+1},\ e_{[1/(2\kappa h)]-m-n+2},\ \ldots,\ e_{[1/(2\kappa h)]}.$$

In this way, we may successively establish pointwise convergence as $h \to 0$, in the finite number of intervals

$$\left(a + \frac{1}{2\kappa}, a + \frac{2}{2\kappa}\right), \left(a + \frac{2}{2\kappa}, a + \frac{3}{2\kappa}\right), \ldots, \left(a + \frac{R}{2\kappa}, a + (b - a)\right),$$

where $R = [(b - a)/(2\kappa)]$. The error estimates for successive intervals can then be seen to satisfy, in analogy with (19),

$$\text{(20)} \qquad I_{p+1} \le 2Q\left[nI_p + \frac{\rho(h) + \tau(h)}{2\kappa a_n}\right], \qquad \text{for } p = 1, 2, \ldots, R,$$

where I_p is the pointwise error bound for the interval

$$\left(a + \frac{p - 1}{2\kappa}, a + \frac{p}{2\kappa}\right).$$

From (20) it is then possible to recursively bound I_{R+1} and hence to bound $|e_j|$ for $0 \le j \le N$ by

$$\text{(21)} \qquad |e_j| \le (2Qn)^{R+1}r(h) + \frac{(2Qn)^{R+1} - 1}{2Qn - 1}\,\frac{Q(\rho(h) + \tau(h))}{\kappa a_n}, \qquad \text{if } 2Qn \ne 1;$$

$$\le (rh) + \frac{(R + 1)Q(\rho(h) + \tau(h))}{\kappa a_n}, \qquad \text{if } 2Qn = 1.$$

Formula (21) not only establishes convergence of the finite difference scheme (2), but gives an upper bound for the error e_j in terms of the initial error, the rounding error and the truncation error. This bound is of the same general character as were the bounds that we derived earlier for the special methods treated in Sections 1 through 3. ∎

By essentially the same arguments we could prove:

THEOREM **4**. *If the F in* (2) *satisfies* (3) *then* (2) *is a stable scheme iff the root condition* (7) *is satisfied.*

We have therefore established the important consequence of Theorems 3 and 4;

THEOREM 5. *If the scheme* (2) *is consistent with* (1) *and F satisfies condition* (3), *then the necessary and sufficient condition that* (2) *be convergent is that it be stable.* ■

It is possible to strengthen Theorem 3 by noting that F need only satisfy the Lipschitz condition (3b) in a narrow strip about the solution $y(x)$ given by S_d: $\{(x, y) \mid a \leq x \leq b;\ |y - y(x)| \leq d\}$ for any fixed constant $d > 0$. That is, for h sufficiently small, the error estimate (21) shows that if the solution of the difference equation starts in the strip $S_{d/2}$ then it remains in the strip S_d.

The special case with $m = 0$ and

$$F\{h, x_j; u_{j-m}, \ldots, u_{j+n}; f\} \equiv \sum_{s=0}^{n} b_s f(x_{j+s}, u_{j+s}) \tag{22}$$

has been treated by Dahlquist. He found the surprising result that although by proper choice of the $2n + 1$ independent parameters $\{a_s/a_n\}$, $\{b_s/a_n\}$, it is possible to construct a scheme having a truncation error of order $2n$; *only schemes with a truncation error of order at most* $n + 2$ *may be convergent.* (In fact, if n is odd then only schemes for which the truncation error is of order at most $n + 1$ may be convergent.) The implicit scheme of equation (2.3a) with $p = n - 1$, based on the Newton-Cotes quadrature formulae applied to (2.2) with $p = n - 1$, then has the maximum possible order of truncation error for convergent schemes of form (2) with F given by (22). Dahlquist's work finds other schemes having a truncation error of the same order, but shows that schemes which are both convergent and of greater accuracy do not exist.

PROBLEMS, SECTION 5

1. Define F for the following schemes treated in Sections 1 through 3 given by

(a) equation (1.1a)
(b) equation (1.24a)
(c) equation (1.37a)
(d) equation (2.3a)
(e) equation (2.3b)
(f) equation (2.7a and b)
(g) equation (3.11a and b)
(h) equation (3.8) and (3.19a, b, and c)

and verify that conditions (3a and b) are satisfied. Which of these schemes do not satisfy the root condition?

2. If (2) is convergent, show that $P(1) = \sum_{s=0}^{n} a_s = 0$.

[Hint: Let $f(x, y) \equiv 0$; $y(a) = y_0 \neq 0$; $\rho_k = 0$.]

3. In the scheme (2) with F given by (22), show that with $T(\zeta) \equiv \sum_{s=0}^{n} b_s\zeta^s$, $P'(1) = T(1)$ implies that the truncation error is of order $p \geq 1$.

[Hint: Expand the left side of equation (3c) about (x_j, y_j) in powers of h. Observe that

$$P'(1) = \sum_{s=0}^{n} sa_s, \qquad T(1) = \sum_{s=0}^{n} b_s.]$$

6. HIGHER ORDER EQUATIONS AND SYSTEMS

Any rth order ordinary differential equation,

$$\frac{d^r z}{dx^r} = g\left(x, z, \frac{dz}{dx}, \ldots, \frac{d^{r-1}z}{dx^{r-1}}\right),$$

can be replaced by an equivalent system of first order equations. There are a variety of ways in which this reduction can be performed; the most straightforward introduces the variables

$$y^{(1)}(x) \equiv z(x), y^{(2)}(x) \equiv \frac{dy^{(1)}(x)}{dx}, \ldots, y^{(r)}(x) \equiv \frac{dy^{(r-1)}(x)}{dx}.$$

Then the differential equation can be written as

$$\frac{d}{dx} y^{(1)} = y^{(2)},$$

$$\vdots$$

$$\frac{d}{dx} y^{(r-1)} = y^{(r)}$$

$$\frac{d}{dx} y^{(r)} = g(x, y^{(1)}, y^{(2)}, \ldots, y^{(r)}).$$

This is, of course, a special case of the general system

(1a) $$\frac{d\mathbf{y}}{dx} = \mathbf{f}(x; \mathbf{y}).$$

Here we have introduced the r-dimensional column vectors $\mathbf{y}$ and $\mathbf{f}$ with components

$$y^{(\nu)}(x), f^{(\nu)}(x; \mathbf{y}) = f^{(\nu)}(x, y^{(1)}, y^{(2)}, \ldots, y^{(r)}), \qquad \nu = 1, 2, \ldots, r.$$

We will study the difference methods appropriate for solving a system (1a). The initial data for such a system are assumed given in the form

(1b) $$\mathbf{y}(a) = \mathbf{y}_0,$$

where we seek a solution of (1) in the interval $a \leq x \leq b$.

All of the difference methods previously proposed for a single first order equation have their direct analogs for the system (1). With (1a) in component form,

$$\frac{dy^{(\nu)}}{dx} = f^{(\nu)}(x, y^{(1)}(x), \ldots, y^{(r)}(x)), \qquad \nu = 1, 2, \ldots, r;$$

it does not require much insight to write down the corresponding difference methods based on quadrature formulae or even the single-step methods. In fact, the general predictor-corrector becomes, in vector form,

$$\text{(2a)} \qquad \mathbf{u}^*_{i+1} = \mathbf{u}_{i-q} + h \sum_{j=0}^{m} \beta_j \mathbf{f}(x_{i-j}; \mathbf{u}_{i-j});$$

$$\text{(2b)} \qquad \mathbf{u}_{i+1} = \mathbf{u}_{i-p} + h \sum_{k=1}^{n} \alpha_k \mathbf{f}(x_{i+1-k}; \mathbf{u}_{i+1-k}) + h\alpha_0 \mathbf{f}(x_{i+1}; \mathbf{u}^*_{i+1}).$$

Similarly, the general one-step difference methods for the system (1) can be written as

$$\text{(3)} \qquad \mathbf{u}_{j+1} = \mathbf{u}_j + h\mathbf{F}\{h, x_j; \mathbf{u}_j; \mathbf{f}\}.$$

For the class of methods defined in Subsection 3.2 we take (for systems)

$$\text{(4a)} \qquad h\mathbf{F}\{h, x_j; \mathbf{u}_j; \mathbf{f}\} = h \sum_{\mu=0}^{n} \alpha_\mu \mathbf{f}(\xi_\mu; \boldsymbol{\eta}_\mu);$$

$$\text{(4b)} \qquad \xi_0 = x_j; \qquad \xi_\nu \equiv \xi_0 + \theta_\nu h;$$

$$\text{(4c)} \qquad \boldsymbol{\eta}_0 = \mathbf{u}_j; \qquad \boldsymbol{\eta}_\nu \equiv \boldsymbol{\eta}_0 + h \sum_{k=0}^{\nu-1} \alpha_{\nu k} \mathbf{f}(\xi_k; \boldsymbol{\eta}_k), \qquad \nu = 1, 2, \ldots, n.$$

Here the quantities α_μ, θ_ν, and $\alpha_{\nu k}$ are defined as in (3.20) and (3.21).

As in Section 2 we define the truncation error in predictor-corrector methods applied to a system. That is, an r-dimensional vector, $\boldsymbol{\tau}_{i+1}$, defined by

$$\text{(5a)} \qquad \mathbf{y}_{i+1} = \mathbf{y}_{i-p} + h \sum_{k=1}^{n} \alpha_k \mathbf{f}(x_{i+1-k}; \mathbf{y}_{i+1-k}) + h\alpha_0 \mathbf{f}(x_{i+1}; \mathbf{y}^*_{i+1}) + h\boldsymbol{\tau}_{i+1}.$$

Here $\mathbf{y}^*_{i+1} \equiv \mathbf{y}_{i+1} - h\boldsymbol{\sigma}^*_{i+1}$ defines $\boldsymbol{\sigma}^*_{i+1}$, and $\mathbf{y}^*_{i+1}$ is defined by the right side of (2a) with $\mathbf{u}$ replaced by $\mathbf{y}$. We find that

$$\text{(5b)} \qquad \boldsymbol{\tau} = \boldsymbol{\sigma} + h\alpha_0 \bar{J} \boldsymbol{\sigma}^*,$$

where $\boldsymbol{\sigma}^*$ and $\boldsymbol{\sigma}$ have components $\sigma^{(\nu)*}$ and $\sigma^{(\nu)}$ which are respectively the errors in applying the appropriate quadrature formulae to the integration of $f^{(\nu)}(x, y^{(1)}(x), \ldots, y^{(r)}(x))$. The elements of the matrix $\bar{J}$ are found by evaluating the corresponding elements of the *Jacobian matrix*

$$J \equiv (a_{\nu\mu}) = \left(\frac{\partial f^{(\nu)}}{\partial y^{(\mu)}}\right) \tag{5c}$$

at appropriate intermediate points. The detailed derivation of (5) is left as an exercise. By using the matrix (5c) we find that the error vectors

$$\mathbf{e}_j \equiv \mathbf{u}_j - \mathbf{y}_j, \qquad \mathbf{e}_j^* \equiv \mathbf{u}_j^* - \mathbf{y}_j^* \tag{6}$$

satisfy the systems

$$\mathbf{e}_{i+1}^* = \mathbf{e}_{i-q} + h \sum_{j=0}^{m} \beta_j J_{i-j} \mathbf{e}_{i-j}; \tag{7a}$$

$$\mathbf{e}_{i+1} = \mathbf{e}_{i-p} + h \sum_{j=1}^{n} \alpha_j J_{i+1-j} \mathbf{e}_{i+1-j} + h^2 \alpha_0 J_{i+1} \sum_{j=0}^{m} \beta_j J_{i-j} \mathbf{e}_{i-j} - h \boldsymbol{\tau}_{i+1}. \tag{7b}$$

Here again the matrices J_i have as elements the $a_{\nu\mu}$ of J evaluated at appropriate intermediate points (the elements in each row of J_j can be shown to be evaluated at the same point). A convergence proof can now be given exactly as in Subsection 2.1 (see Theorem 2.1) if we employ appropriate vector and matrix norms.

In fact, if the root condition (5.7) is satisfied, we may copy the proof of convergence and the error estimates given in Theorem 5.3 by replacing

$$y,\ f,\ u,\ v,\ e,\ F,\ \rho_k,\ \tau_k,\ c_k,\ E_k,\ C_k,\ |\ |$$

by the corresponding vector quantities

$$\mathbf{y},\ \mathbf{f},\ \mathbf{u},\ \mathbf{v},\ \mathbf{e},\ \mathbf{F},\ \boldsymbol{\rho}_k,\ \boldsymbol{\tau}_k,\ \mathbf{c}_k,\ \mathbf{E}_k,\ \mathbf{C}_k,\ \|\ \|_\infty$$

(i.e., absolute value is replaced by maximum absolute component), for the scheme

$$\sum_{j=0}^{n} a_s \mathbf{u}_{j+s} = h\mathbf{F}\{h, x_j; \mathbf{u}_{j-m}, \ldots, \mathbf{u}_{j+n}; \mathbf{f}\} + h\boldsymbol{\rho}_{j+n}, \tag{8a}$$

where $m \le j \le N - n$, with

$$\mathbf{u}_0 = \mathbf{y}_0 + \boldsymbol{\rho}_0, \qquad \mathbf{u}_1 = \mathbf{y}_1 + \boldsymbol{\rho}_1, \ldots, \mathbf{u}_{m+n-1} = \mathbf{y}_{m+n-1} + \boldsymbol{\rho}_{m+n-1}. \tag{8b}$$

PROBLEMS, SECTION 6

1. Verify the error estimate corresponding to equation (5.19), as indicated in the last sentence of Section 6, for the scheme defined by (8). That is, show that

$$\|\mathbf{e}_{j+m}\|_\infty \le 2Q\left[n \max_{0 \le k \le m+n-1} \|\mathbf{e}_k\|_\infty + \frac{\rho(h) + \tau(h)}{2\kappa a_n}\right] \qquad \text{for } 0 \le jh \le \frac{1}{2\kappa},$$

where

$$\kappa = \frac{QC(m + n + 1)}{a_n}$$

and C, which appears in the vector analog of (5.3b), is a bound for the vector norm $\| \ \|_\infty$ of **f** and of all its partial derivatives with respect to $x, y^{(1)}, y^{(2)}, \ldots, y^{(r)}$ of some finite order, in the domain S: $\{(x; \mathbf{y}) \mid a \le x \le b, \|\mathbf{y}\|_\infty < \infty\}$.

2. Verify that if $\mathbf{u} \equiv (u^{(k)})$, $\mathbf{v} \equiv (v^{(k)})$, and $f(x; \mathbf{y})$ has a continuous derivative with respect to all variables, then

$$f(x; \mathbf{u}) - f(x; \mathbf{v}) = \sum_k (u^{(k)} - v^{(k)}) \frac{\partial f}{\partial y^{(k)}} (x; \mathbf{v} + \theta(\mathbf{u} - \mathbf{v}))$$

for some θ such that $0 < \theta < 1$.

Hence, if f is replaced by a vector valued function **f**, each component $f^{(j)}$ of **f** may have its own θ_j satisfying $0 < \theta_j < 1$.

[Hint: Study $g(t) \equiv f(x;\ \mathbf{v} + t(\mathbf{u} - \mathbf{v}))$. Note that $g(t) - g(0) = tg'(\theta t)$ for some θ in $0 < \theta < 1$. Then evaluate

$$\frac{d}{dt} g(t) = \frac{d}{dt} f(x, \mathbf{v} + t(\mathbf{u} - \mathbf{v}))$$

and set $t = 1$.] This justifies the definition of J_k in (7) and $\bar{J}$ in (5b).

7. BOUNDARY VALUE AND EIGENVALUE PROBLEMS

A boundary value problem for an ordinary differential equation (or system) is one in which the dependent variable is required to satisfy specified conditions at more than one point. Since an equation of nth order has a general solution depending upon n parameters, the total number of boundary conditions required to determine a unique solution is, in general, n. However, when the total of n boundary conditions is given at *more than one point*, it is possible for more than one solution to exist or for no solution to exist. Of course, if more than n conditions are imposed, even for the initial value problem, there will, in general, be no solution. A detailed study of the existence and uniqueness theory is beyond the scope of our book. However, for linear problems, the theory is well known and we shall indicate here the elements of this theory which may be applicable to non-linear problems and to the analysis of numerical procedures used to solve such boundary value problems.

The simplest linear boundary value problem is one in which the solution of a second order equation, say

$$y'' - p(x)y' - q(x)y = 0, \tag{1a}$$

is specified at two distinct points, say

$$y(a) = \alpha, \qquad y(b) = \beta. \tag{1b}$$

The solution $y(x)$, is sought in the interval $a \leq x \leq b$. A formal approach to the exact solution of the boundary value problem is obtained by considering the related *initial value* problem,

$$Y'' - p(x)Y' - q(x)Y = 0, \tag{2a}$$

$$Y(a) = \alpha, \qquad Y'(a) = s. \tag{2b}$$

The theory of solutions of such initial value problems is well known and if, for example, the functions $p(x)$ and $q(x)$ are continuous on $[a, b]$, the existence of a unique solution of (2) in $[a, b]$ is assured. Let us denote this solution by

$$Y = Y(s; x),$$

and recall that *every* solution of (1a) or (2a) is a linear combination of two particular "independent" solutions of (1a), $y^{(1)}(x)$ and $y^{(2)}(x)$, which satisfy, say,

$$y^{(1)}(a) = 1, \qquad y^{(1)\prime}(a) = 0; \tag{3a}$$

$$y^{(2)}(a) = 0, \qquad y^{(2)\prime}(a) = 1. \tag{3b}$$

Then the unique solution of (2a) which satisfies (2b) is

$$Y(s; x) = ay^{(1)}(x) + sy^{(2)}(x). \tag{4}$$

Now if we take s such that

$$Y(s;b) \equiv ay^{(1)}(b) + sy^{(2)}(b) = \beta, \tag{5}$$

then $y(x) \equiv Y(s;x)$ is a solution of the boundary value problem (1). Clearly, there is at most one root of equation (5),

$$s = \frac{\beta - ay^{(1)}(b)}{y^{(2)}(b)},$$

provided that $y^{(2)}(b) \neq 0$. If, on the other hand, $y^{(2)}(b) = 0$ there *may not* be a solution of the boundary value problem (1). A solution would exist in this case only if $\beta = ay^{(1)}(b)$, but it would not be unique since then $Y(s; x)$ of (4) is a solution for arbitrary s.

Thus there are two mutually exclusive cases for the linear boundary value problem, the so-called *alternative principle*: *either a unique solution exists or else the homogeneous problem* (i.e., $y(a) = y(b) = 0$) *has a nontrivial solution* (which is $sy^{(2)}(x)$ in this example).

These observations permit us to study the solution of the inhomogeneous equation

$$y'' - p(x)y' - q(x)y = r(x), \tag{6}$$

subject to the boundary conditions (1b). This problem can be reduced to the previous case if a particular solution of (6), say $y^{(p)}(x)$, can be found. Then we define

$$w(x) \equiv y(x) - y^{(p)}(x), \tag{7}$$

and find that $w(x)$ must satisfy the homogeneous equation (1a). The boundary conditions for $w(x)$ become, from (7) and (1b)

$$w(a) = \alpha - y^{(p)}(a) \equiv \alpha',$$
$$w(b) = \beta - y^{(p)}(b) \equiv \beta'.$$

Thus, we can find the solution of (6) and (1b) by solving (1) with (α, β) replaced by (α', β'). A definite problem for the determination of $y^{(p)}(x)$ is obtained by specifying particular initial conditions, say

$$y^{(p)}(a) = y^{(p)\prime}(a) = 0, \tag{8}$$

which provides a standard type of initial value problem for the equation (6). Again, the alternative principle holds; see Problem 1.

The formulation of boundary value problems for linear second order equations can be easily extended to more general nth order equations or equivalently to nth order systems of first order equations (not necessarily linear). For example, in the latter case we may consider the system

$$\mathbf{y}' = \mathbf{f}(x; \mathbf{y}), \tag{9a}$$

where we use the row vectors $\mathbf{y} = (y_1, y_2, \ldots, y_n)$, $\mathbf{f} = (f_1, f_2, \ldots, f_n)$ and the functions $f_k \equiv f_k(x; \mathbf{y}) = f_k(x; y_1, \ldots, y_n)$ are functions of $n + 1$ variables. The n boundary conditions may be, say,

$$\begin{aligned} &y_1(a) = \alpha_1, \; y_2(a) = \alpha_2, \ldots, y_{m_1}(a) = \alpha_{m_1}, \\ &y_{m_1+1}(b) = \beta_1, \quad y_{m_1+2}(b) = \beta_2, \ldots, y_n(b) = \beta_{m_2}, \\ &\qquad m_1 > 0, \; m_1 + m_2 = n, \; m_2 > 0. \end{aligned} \tag{9b}$$

Thus, we specify m_1 quantities at $x = a$ and the remaining $n - m_1 = m_2$ quantities at $x = b$.

In analogy with (2) we consider the related initial value problem:

$$\mathbf{Y}' = \mathbf{f}(x, \mathbf{Y}); \tag{10a}$$

$$\begin{aligned} Y_i(a) &= \alpha_i, \qquad i = 1, 2, \ldots, m_1 \\ Y_{m_1+j}(a) &= s_j, \qquad j = 1, 2, \ldots, m_2. \end{aligned} \tag{10b}$$

We indicate the dependence on the m_2 arbitrary parameters s_j by writing

$$Y_k = Y_k(s_1, s_2, \ldots, s_{m_2}; x), \qquad k = 1, 2, \ldots, n.$$

These parameters are to be determined such that

$$Y_{m_1+j}(s_1, s_2, \ldots, s_{m_2}; b) = \beta_j, \qquad j = 1, 2, \ldots, m_2. \tag{11}$$

This represents a system of m_2 equations in the m_2 unknowns s_j. In the corresponding linear case (i.e., in which each f_k is linear in all the y_k) the system (11) becomes a linear system and its solvability is thus reduced to a study of the non-singularity of a matrix of order m_2.

Note that the alternative principle is again valid. In the general case, however, the roots of a transcendental system (11) are required and the existence and uniqueness theory is more complicated (and in fact, is not as completely developed as it is for the linear case).

We shall examine two different types of numerical methods for approximating the solutions of boundary value problems, in Subsections 7.1 and 7.2.

7.1. Initial Value or "Shooting" Methods

The initial value or "shooting" methods attempt to carry out numerically the procedure indicated in equations (2) through (5). That is, roughly, the initial data are adjusted so that the solution of an initial value problem satisfies the required boundary condition at some distant (boundary) point.

We take, for definiteness, a uniform net

$$x_0 = a, \quad x_j = x_0 + jh, \quad j = 0, 1, \ldots, N, \quad h = \frac{b-a}{N}; \tag{12}$$

and shall try to approximate thereon the solution of the linear equation (6) subject to (1b). We first approximate the solutions $y^{(1)}(x)$ and $y^{(2)}(x)$ of the initial value problems (1a) and (3). This can be done, for example, by replacing (1a) by an equivalent first order system and then using a predictor-corrector or one-step method as indicated in Section 6. In the same manner, we can approximate the particular solution $y^{(p)}(x)$ of (6) and (8). The respective numerical solutions are denoted at each point x_j of (12) by

$$u_j^{(1)}, \quad u_j^{(2)}, \quad u_j^{(p)}, \qquad j = 0, 1, \ldots, N. \tag{13a}$$

These solutions satisfy, at $x_0 = a$, the conditions

$$u_0^{(1)} = 1, \quad u_0^{(2)} = 0, \quad u_0^{(p)} = 0. \tag{13b}$$

Assume that the same numerical procedure has been used to compute each of these solutions and that we have

$$e_j^{(1)} \equiv u_j^{(1)} - y^{(1)}(x_j) = \mathcal{O}(h^r), \tag{14a}$$

$$e_j^{(2)} \equiv u_j^{(2)} - y^{(2)}(x_j) = \mathcal{O}(h^r), \tag{14b}$$

$$e_j^{(p)} \equiv u_j^{(p)} - y^{(p)}(x_j) = \mathcal{O}(h^r). \tag{14c}$$

[That is, the truncation error of the integration scheme is $\mathcal{O}(h^r)$, and the rounding errors are at most $\mathcal{O}(h^{r+1})$ so that the estimates (14) apply.]

The exact solution of (6) and (1b) is, by the previous analysis, given by

$$y(x) = y^{(p)}(x) + \alpha y^{(1)}(x) + s y^{(2)}(x), \qquad a \le x \le b; \tag{15a}$$

$$s = \frac{\beta - y^{(p)}(b) - \alpha y^{(1)}(b)}{y^{(2)}(b)}. \tag{15b}$$

Of course, we assume that $y^{(2)}(b) \neq 0$. Otherwise the homogeneous problem has a non-trivial solution and then, in general, the boundary value problem has no solution. With the use of (13), we take for the approximate solution the obvious combination

$$U_j = u_j^{(p)} + \alpha u_j^{(1)} + s_h u_j^{(2)}, \qquad j = 0, 1, \ldots, N; \tag{16a}$$

where

$$s_h = \frac{\beta - u_N^{(p)} - \alpha u_N^{(1)}}{u_N^{(2)}}. \tag{16b}$$

From (13b) and (16b) it clearly follows that, as required,

$$U_0 = \alpha, \qquad U_N = \beta,$$

where we have neglected possible roundoff errors in forming U_j and s_h. Thus, in principle, U_j is an approximate solution of the boundary value problem (6) and (1b). In practice, we need only calculate the solution of two initial value problems to evaluate U_j. That is, $y^{(p)}(x) + \alpha y^{(1)}(x)$ satisfies (6) and conditions (3a) so that $u_j^{(p)} + \alpha u_j^{(1)}$ can be computed as the solution of a single initial value problem.

Upon recalling (14), we are led to the obvious, and in fact, correct conclusion that

$$e_j \equiv U_j - y(x_j) = \mathcal{O}(h^r).$$

However, as we now show, there may still be *practical* difficulties in obtaining an accurate approximation.

Upon subtracting (15a) with $x = x_j$ from (16a) and using the definitions (14), we find

$$e_j = (e_j^{(p)} + \alpha e_j^{(1)} + s e_j^{(2)}) + (s_h - s)u^{(2)}(x_j), \qquad j = 0, 1, \ldots, N. \tag{17a}$$

Since $b = x_N$, (15b) and (16b) imply $e_N = 0$ and

$$(s_h - s) = -\frac{e_N^{(p)} + \alpha e_N^{(1)} + s e_N^{(2)}}{u_N^{(2)}}. \tag{17b}$$

Use (17b) in (17a) to find

$$e_j = (e_j^{(p)} + \alpha e_j^{(1)} + s e_j^{(2)}) - (e_N^{(p)} + \alpha e_N^{(1)} + s e_N^{(2)}) \frac{u_j^{(2)}}{u_N^{(2)}}. \tag{18}$$

From this expression for the error we see that $e_0 = e_N = 0$ and thus the error is, in general, small near the endpoints of the interval. However, whenever $|u_j^{(2)}/u_N^{(2)}|$ becomes large we may expect relatively large errors. This ratio can be computed and thus a practical assessment of the accuracy in the present method is possible. In particular, note that $u_N^{(2)} = y^{(2)}(b) + e_N^{(2)}$. Thus, whenever the fixed number $y^{(2)}(b)$ is small and opposite in sign to the error $e_N^{(2)}$, which depends upon h, we may find a magnification of the intermediate errors, e_j.

The effect of roundoff errors in the present method can be very pronounced. While performing the calculations (16), significance is frequently lost when large almost equal quantities are subtracted from each other. This may be due to the occurrence of a small value of $u_N^{(2)}$, or to rapidly growing solutions $y^{(1)}(x)$, $y^{(2)}(x)$ and/or $y^{(p)}(x)$.

By using the estimates (14) in (18) we obtain the error bound

$$(19) \qquad |U_j - y(x_j)| = |e_j| \leq Mh^r\left(1 + \left|\frac{u_j^{(2)}}{u_N^{(2)}}\right|\right), \qquad j = 1, 2, \ldots, N - 1.$$

Thus (for sufficiently small net spacing) the error behaves as theoretically expected. In practice, however, it may frequently be necessary to use many significant figures in the calculations to realize these error estimates.

The method (and its attendant difficulties) treated in this subsection is easily extended to more general *linear* boundary value problems.

We can alter the procedure slightly and, with considerably more computing effort, solve non-linear boundary value problems. For example if, in place of (1), the problem is

$$(20) \qquad y'' = f(x, y, y'); \qquad y(a) = \alpha, \quad y(b) = \beta;$$

we consider the initial value problem [in place of (2)]

$$(21) \qquad Y'' = f(x, Y, Y'); \qquad Y(a) = \alpha, \quad Y'(a) = s.$$

If $Y(s; x)$ is the solution of (21) and s^* is such that

$$(22) \qquad Y(s^*; b) = \beta,$$

then $y(x) = Y(s^*; x)$ is a solution of (20). The equation (22) is, in general, transcendental, whereas in the linear case the corresponding equation, (5), is linear in s.

The problem of solving (20) is reduced to the determination of the root (or roots) of (22). The root s^* could be found by applying the iterative methods of Chapter 3. Of course, in each step of such iteration schemes at least one evaluation of $Y(s; b)$ is required for some value of s. This may be found only approximately by integrating (21) numerically on some net (12). That is, the net function $U_j(s)$, $j = 0, 1, \ldots, N$, may be constructed by some method described in earlier sections. Then $U_j(s)$ is an approximation to $Y(s; x_j)$. If the overall error of the integration scheme is $\mathcal{O}(h^r)$ and the function $f(x, y, z)$ is sufficiently smooth, then for each s we will determine, in fact,

$$U_N(s) = Y(s; b) + \mathcal{O}(h^r).$$

If the solutions of (21) are such that (22) has a simple root, s^*, and

$$0 < \left|\frac{\partial Y(s; b)}{\partial s}\right| \le K,$$

for $|s - s^*| \le \rho$, then we can show that a functional iteration procedure will, if s_0 is close enough to s^*, produce a sequence $s_0, s_1, \ldots$, such that for some k

$$U_N(s_k) - \beta = Y(s_k; b) - \beta + \mathcal{O}(h^r) = \mathcal{O}(h^r).$$

Hence

$$|s_k - s^*| = \mathcal{O}(h^r).$$

By using sufficiently many iterations, we can thus get within $\mathcal{O}(h^r)$ of a root of (22) and hence compute a solution of (20) to within an error bounded by Mh^r. In Problems 4 and 5, we indicate some of the details of these results.

It is convenient for the application of Newton's iterative method in solving (22) to approximate $\partial Y(s; b)/\partial s$. By differentiating (21) with respect to s, we can formally find the differential equation, called the *variational equation*, i.e., satisfied by the function $W(s; x) \equiv \partial Y(s; x)/\partial s$:

$$W'' = \frac{\partial f(x, Y, Y')}{\partial y} W + \frac{\partial f(x, Y, Y')}{\partial z} W';$$

$$W(s; a) = 0, \quad W(s; b) = 1.$$

A numerical approximation to the solution of the variational equation may be computed stepwise along with the evaluation of $U_j(s)$. Hence for $j = N$ we would have an approximation for both $Y(s; b)$ and $\partial Y(s; b)/\partial s$.

7.2. Finite Difference Methods

We consider here finite difference methods which are not based on solving the initial value problem. These are called *direct methods*. The truncation error of the particular difference method we use is $\mathcal{O}(h^2)$ and the labor required for a given accuracy is comparable to that for the initial value method of some low order.

Let the boundary value problem be (6) and (1b) which we write as

(23a) $$L\{y\} \equiv y'' - p(x)y' - q(x)y = r(x);$$

(23b) $$y(a) = \alpha, \quad y(b) = \beta.$$

We impose here the restriction that

(24) $$q(x) \ge Q_* > 0, \quad a \le x \le b.$$

The most simple existence and uniqueness proofs for solutions of boundary value problems of the form (23) require such a condition but with $Q_* \geq 0$. We assume a unique solution of (23) to exist with four continuous derivatives in $a \leq x \leq b$. A uniform net will be used with $h = (b - a)/(N + 1)$.

Now rather than seek high order accuracy in a difference approximation of (23a) we use the simple difference equations

$$\text{(25a)} \qquad L_h\{u_j\} \equiv \frac{u_{j-1} - 2u_j + u_{j+1}}{h^2} - p(x_j)\frac{u_{j+1} - u_{j-1}}{2h} - q(x_j)u_j = r(x_j), \qquad j = 1, 2, \ldots, N.$$

The boundary conditions are replaced by

$$\text{(25b)} \qquad u_0 = \alpha, \qquad u_{N+1} = \beta.$$

Multiply (25a) by $-h^2/2$ to obtain

$$-\frac{h^2}{2} L_h\{u_j\} = -b_j u_{j-1} + a_j u_j - c_j u_{j+1} = -\frac{h^2}{2} r(x_j), \qquad j = 1, 2, \ldots, N,$$

where

$$\text{(26)} \qquad a_j \equiv 1 + \frac{h^2}{2} q(x_j), \qquad b_j \equiv \frac{1}{2}\left[1 + \frac{h}{2} p(x_j)\right], \qquad c_j \equiv \frac{1}{2}\left[1 - \frac{h}{2} p(x_j)\right].$$

Using this notation the system of difference equations (25a) and boundary conditions (25b) can be written in the vector form

$$\text{(27a)} \qquad A\mathbf{u} = \mathbf{r},$$

where

$$\text{(27b)} \qquad \mathbf{u} \equiv \begin{pmatrix} u_1 \\ u_2 \\ \vdots \\ u_N \end{pmatrix}, \qquad \mathbf{r} \equiv -\frac{h^2}{2}\begin{pmatrix} r_1 \\ r_2 \\ \vdots \\ r_N \end{pmatrix} + \begin{pmatrix} b_1\alpha \\ 0 \\ \vdots \\ 0 \\ c_N\beta \end{pmatrix},$$

$$A \equiv \begin{pmatrix} a_1 & -c_1 & & & 0 \\ -b_2 & a_2 & -c_2 & & \\ & \cdot & \cdot & \cdot & \\ & & \cdot & \cdot & \cdot \\ & & \cdot & \cdot & \cdot \\ & & -b_{J-1} & a_{J-1} & -c_{N-1} \\ 0 & & & -b_N & a_N \end{pmatrix}.$$

Thus to solve the difference problem (25) we must, in fact, solve the Nth order linear system (27a) with tridiagonal coefficient matrix, A, given in (27b).

Let us require that the net spacing h be so small that

$$\text{(28)} \qquad \frac{h}{2}|p(x_j)| \leq 1, \qquad j = 1, 2, \ldots, N.$$

Then from (26) it follows that

$$|b_j| + |c_j| = b_j + c_j = 1,$$

while (24) implies $a_j > 1$. So we deduce that

$$|a_1| > |c_1|;$$

$$|a_j| > |b_j| + |c_j|, \qquad 2 \leq j \leq N - 1;$$

$$|a_N| > |b_N|;$$

and hence Theorem 3.5 of Chapter 2 applies. The solution of (27a) can thus be computed by the simple direct factorization of A described in Section 3 of Chapter 2. Of course, this furnishes a proof of the existence of a unique solution of the difference equations (27) provided (28) is satisfied.

Let us now estimate the error in the numerical approximation defined above. The local truncation errors, τ_j, are defined by:

$$\text{(29)} \qquad L_h\{y(x_j)\} = r(x_j) + \tau_j; \qquad j = 1, 2, \ldots, N.$$

Since $y(x)$ is a solution of (1a) we have, assuming $y^{\text{iv}}(x)$ to be continuous,

$$\text{(30)} \qquad \tau_j = L_h\{y(x_j)\} - L\{y(x_j)\}$$

$$= \left[\frac{y(x_j - h) - 2y(x_j) + y(x_j + h)}{h^2} - y''(x_j)\right]$$

$$- p(x_j)\left[\frac{y(x_j + h) - y(x_j - h)}{2h} - y'(x_j)\right]$$

$$= \frac{h^2}{12}[y^{\text{iv}}(\xi_j) - 2p(x_j)y'''(\eta_j)]; \qquad j = 1, 2, \ldots, N.$$

Here ξ_j and η_j are in $[x_{j-1}, x_{j+1}]$ and we have used Taylor's theorem.

The basic *error estimate* can now be stated as

THEOREM **1.** *If the net spacing, h, satisfies* (28) *then*

$$\text{(31a)} \qquad |u_j - y(x_j)| \leq h^2\left(\frac{M_4 + 2P^*M_3}{12Q_*}\right), \qquad j = 0, 1, \ldots, N + 1;$$

where $y(x)$ is the solution of (6) *and* (1*b*), $\{u_j\}$ *is the solution of* (25) *and*

$$P^* \equiv \max_{[a,\,b]} |p(x)|, \qquad M_3 \equiv \max_{[a,\,b]} |y'''(x)|,$$

(31b)

$$M_4 \equiv \max_{[a,\,b]} |y^{\text{iv}}(x)|.$$

Proof. Let us define

$$e_j = u_j - y(x_j), \qquad j = 0, 1, \ldots, N + 1.$$

Then subtracting (29) from (25a) yields, with the aid of (26),

$$\text{(32)} \qquad a_j e_j = b_j e_{j-1} + c_j e_{j+1} + \frac{h^2}{2} \tau_j, \qquad j = 1, 2, \ldots, N.$$

Now with the norms

$$e \equiv \max_{0 \le j \le N+1} |e_j|, \qquad \tau \equiv \max_{1 \le j \le N} |\tau_j|,$$

we obtain by taking absolute values in (32) and using the equation after (28)

$$|a_j e_j| \le e + \frac{h^2}{2} \tau, \qquad j = 1, 2, \ldots, N.$$

However, by condition (24), $|a_j| = a_j \ge 1 + (h^2/2)Q_*$ and so the above implies that

$$\left(1 + \frac{h^2}{2} Q_*\right)|e_j| \le e + \frac{h^2}{2} \tau, \qquad j = 1, 2, \ldots, N.$$

From (23b) and (25b) we have $e_0 = e_{N+1} = 0$ and so the above inequality is valid for all j in $0 \le j \le N + 1$. Thus we conclude that

$$e \le \frac{1}{Q_*} \tau.$$

Finally, by using the quantities (31b) in (30) we find that

$$\tau \le \frac{h^2}{12} (M_4 + 2P^* M_3)$$

and the theorem follows. ∎

From Theorem 1 we see that the difference solution converges to the exact solution as $h \to 0$ and, in fact, that the error is at most $\mathcal{O}(h^2)$. For equations in which $p(x) \equiv 0$, error bounds that are $\mathcal{O}(h^4)$ are easily obtained by using a slight modification of (25a) (see Problem 6). Boundary conditions more general than those in (23b) can be treated with no essential change in these results (see Problem 7). The condition (24) can be relaxed to $q(x) \ge 0$ and a somewhat more involved argument yields a result analogous to that of Theorem 1. These arguments are based on a so-called *maximum principle* (see Problem 9). The use of this maximum principle is demonstrated in Problem 10.

The effects of roundoff in computing the solution of (25) can be estimated. In fact, let U_j be the computed quantities which, in place of (25), satisfy

$$\text{(33a)} \qquad -\frac{h^2}{2} L_h\{U_j\} = -\frac{h^2}{2} r(x_j) + \rho_j, \qquad j = 1, 2, \ldots, N;$$

and the boundary conditions

$$\text{(33b)} \qquad U_0 = \alpha + \rho_0, \qquad U_{N+1} = \beta + \rho_{N+1}.$$

The quantities ρ_j represent the local roundoff errors committed in each of the indicated computations. Now we define

$$E_j \equiv U_j - y(x_j), \qquad j = 0, 1, \ldots, N + 1$$

and exactly as in the proof of Theorem 1 we deduce that

$$\left(1 + \frac{h^2}{2} Q_*\right)|E_j| \leq E + \frac{h^2}{2} \tau + \rho, \qquad j = 1, 2, \ldots, N.$$

Here

$$\rho \equiv \max_{0 \leq j \leq N+1} |\rho_j| \text{ and } |E_0| = |\rho_0|, \ |E_{N+1}| = |\rho_{N+1}|,$$

$$E \equiv \max_{0 < j \leq N+1} |E_j|.$$

If, in addition to (28), we require that $h^2 Q_*/2 \leq 1$, then this inequality is also valid for $j = 0$, and $j = N + 1$ so we finally obtain

$$E \leq \frac{1}{Q_*}\left(\tau + 2\frac{\rho}{h^2}\right).$$

Thus for sufficiently small net spacing, h, we have

$$\text{(34)} \qquad |U_j - y(x_j)| \leq h^2\left(\frac{M_4 + 2P^* M_3}{12 Q_*}\right) + \frac{1}{h^2}\left(\frac{2\rho}{Q_*}\right),$$

$$j = 0, 1, \ldots, N + 1.$$

The roundoff affects this estimate somewhat differently than it did the corresponding estimates in Subsections 1.2 and 1.3, etc. Now to have an error bound which is $\mathcal{O}(h^2)$ we must limit the roundoff by $\rho = \mathcal{O}(h^4)$ as $h \to 0$. That is, two orders in h improvement over the local truncation error are required. Previously, only one additional order in h was required, since our difference equations were then approximations to first order differential equations (or systems of equations).

Difference methods can also be applied to fairly general non-linear second order boundary value problems. While such methods accurate to

$\mathcal{O}(h^2)$ can be determined, the difference equations are now no longer linear. Hence iterations are employed to solve these equations. It should be observed that the iterations are not employed in order to satisfy the correct boundary conditions, as would be the case in the initial value methods. The construction of iteration procedures for solving the difference equations is quite simple.

The non-linear boundary value problems we consider are of the form (20) where the function $f(x, y, z)$ is assumed to satisfy the conditions

$$(35) \qquad 0 < Q_* \leq \frac{\partial f(x, y, z)}{\partial y} \leq Q^*, \qquad \left|\frac{\partial f(x, y, z)}{\partial z}\right| \leq P^*;$$

in some sufficiently large region. Furthermore, these partial derivatives and $y^{\text{iv}}(x)$ are assumed to be continuous.

Again we use a uniform net. On this net the difference approximation of (20) is taken to be

$$(36a) \qquad \frac{u_{j-1} - 2u_j + u_{j+1}}{h^2} = f\left(x_j, u_j, \frac{u_{j+1} - u_{j-1}}{2h}\right), \qquad j = 1, 2, \ldots, N;$$

$$(36b) \qquad u_0 = \alpha, \qquad u_{N+1} = \beta.$$

The local truncation error, τ_j, of this method is defined in the usual manner by

$$(37) \qquad \frac{y_{j-1} - 2y_j + y_{j+1}}{h^2} = f\left(x_j, y_j, \frac{y_{j+1} - y_{j-1}}{2h}\right) + \tau_j, \qquad j = 1, 2, \ldots, N.$$

From the assumed continuity properties of $\partial f/\partial z$ and $y^{\text{iv}}(x)$ it follows that

$$(38) \qquad \tau_j = \frac{h^2}{12}\left[y^{\text{iv}}(\xi_j) - 2\frac{\partial f(x_j, y_j, y'(\zeta_j))}{\partial z} y'''(\eta_j)\right], \qquad j = 1, 2, \ldots, N.$$

Here, ξ_j and η_j in $[x_{j-1}, x_{j+1}]$ are the appropriate mean values used in Taylor's theorem.

To examine the convergence of this procedure we introduce $e_j \equiv u_j - y(x_j)$ and, for the further applications of Taylor's theorem,

$$(39) \qquad \begin{aligned} p_j &\equiv \frac{\partial f}{\partial z}\left(x_j, y_j + \theta_j e_j, \frac{y(x_{j+1}) - y(x_{j-1})}{2h} + \theta_j \frac{e_{j+1} - e_{j-1}}{2h}\right); \\ q_j &\equiv \frac{\partial f}{\partial y}\left(x_j, y_j + \theta_j e_j, \frac{y(x_{j+1}) - y(x_{j-1})}{2h} + \theta_j \frac{e_{j+1} - e_{j-1}}{2h}\right); \end{aligned}$$

$$0 < \theta_j < 1.$$

Then subtracting (37) from (36a) we get, with the above notation and appropriate values for the θ_j,

$$(40)\qquad \left(1 + \frac{h^2}{2} q_j\right)e_j = \frac{1}{2}\left(1 + \frac{h}{2} p_j\right)e_{j-1} + \frac{1}{2}\left(1 - \frac{h}{2} p_j\right)e_{j+1} + \frac{h^2}{2}\tau_j,$$
$$j = 1, 2, \ldots, N.$$

This system of equations is formally identical to that in (32), with the same boundary conditions, $e_0 = e_{N+1} = 0$. So we may conclude, by using (35) in place of (24), exactly as in the proof of Theorem 1 that

$$(41)\qquad |e_j| \leq h^2 \frac{M_4 + 2P^*M_3}{12Q_*}.$$

Here P^* is defined in (35) and M_3 and M_4 are the appropriate bounds on the derivatives of the solution of (20). Thus the order of convergence for the non-linear problem is the same as that for the linear case; the constants in (41) have only slightly different meanings from those in (31). The non-linear cases for which the difference method is applicable can be generalized as are the linear cases in Problems 7 and 8.

If $f(x, y, z)$ is not a linear function of y and z, then the difference equations (36) constitute a non-linear system of equations. The general methods of Chapter 3 could be applied in order to solve such systems. In particular, Newton's method is frequently well suited for this purpose, and in special cases the convergence proof given in Subsection 3.2 of Chapter 3 can be applied. However, due to the special structure of this system some other iteration schemes are naturally suggested, and we shall consider one of them here. All of these methods proceed from an initial estimate of the solution, say

$$u_j^{(0)}, \quad j = 1, 2, \ldots, N; \quad u_0^{(0)} = \alpha, \quad u_{N+1}^{(0)} = \beta.$$

A particularly simple iteration scheme for solving (36) is defined by:

$$(42a)\qquad (1 + \omega)u_i^{(\nu+1)} = \tfrac{1}{2}(u_{j-1}^{(\nu)} + u_{j+1}^{(\nu)}) + \omega u_j^{(\nu)}$$
$$- \frac{h^2}{2} f\left(x_j, u_j^{(\nu)}, \frac{u_{j+1}^{(\nu)} - u_{j-1}^{(\nu)}}{2h}\right),$$
$$j = 1, 2, \ldots, N;$$

$$(42b)\qquad u_0^{(\nu+1)} = \alpha, \qquad u_{N+1}^{(\nu+1)} = \beta.$$

Here ω is a parameter to be determined so that the iterates converge. In fact, we can show, see Problem (11), that if ω satisfies

$$(42c)\qquad \omega \geq \frac{h^2}{2} Q^*$$

then the iterates satisfy

$$|u_j^{(\nu+1)} - u_j^{(\nu)}| \leq \left(1 - \frac{h^2 Q_*}{2(1 + \omega)}\right)^\nu \max_k |u_k^{(1)} - u_k^{(0)}|; \quad j = 1, 2, \ldots, N. \tag{43}$$

From this result we see that the iterates form a Cauchy sequence. Thus not only do they converge but by the assumed continuity of $f(x, y, z)$ we can show, exactly as in the proof of Theorem 1.1 in Chapter 3, that a *unique* solution of the difference equations (36) exists.

7.3. Eigenvalue Problems

We have shown previously that a linear boundary value problem may have non-unique solutions. In fact, this occurs if and only if the corresponding homogeneous boundary value problem has a non-trivial solution. If the coefficients of the homogeneous equation depend upon some parameter it is frequently of interest to determine the values of the parameter for which such non-trivial solutions exist. These special parameter values are called *eigenvalues* and the corresponding non-trivial solutions are called *eigenfunctions*. The simplest example is furnished by the homogeneous problem

$$y'' + \lambda y = 0; \qquad y(a) = y(b) = 0.$$

For each of the parameter values

$$\lambda = \lambda_n \equiv \left[\frac{n\pi}{b - a}\right]^2, \qquad n = 1, 2, \ldots;$$

there exists a non-trivial solution

$$y(x) = y_n(x) \equiv \sin \lambda_n^{1/2}(x - a), \qquad n = 1, 2, \ldots.$$

A fairly general class of eigen-problems, which includes many of the cases that occur in applied mathematics, are the Sturm-Liouville problems,

$$L\{y\} + \lambda r(x)y \equiv [p(x)y']' - q(x)y + \lambda r(x)y = 0, \tag{44a}$$

$$\alpha_0 y'(a) - \alpha_1 y(a) = 0, \qquad \beta_0 y'(b) + \beta_1 y(b) = 0. \tag{44b}$$

Here $p(x) > 0$, $r(x) > 0$, and $q(x) \geq 0$; $p'(x)$, $q(x)$, and $r(x)$ are continuous on $[a, b]$; and the constants α_ν and β_ν are non-negative and at least one of each pair does not vanish. It is known that for such problems there exists an infinite sequence of non-negative eigenvalues

$$0 \leq \lambda_1 < \lambda_2 < \lambda_3 \cdots. \tag{45}$$

In addition, there exist corresponding eigenfunctions, $y_n(x)$, which satisfy the orthogonality relations

$$\int_a^b y_n(x)y_m(x)r(x)\,dx = \delta_{nm},$$

and the *nth eigenfunction has* $n - 1$ *distinct zeros in* $a < x < b$.

We may again relate the solution of (44) to an initial value problem. For any fixed λ we consider

$$L\{Y\} + \lambda r(x)Y = 0; \tag{46a}$$

$$\alpha_0 Y'(a) - \alpha_1 Y(a) = 0, \qquad \gamma_0 Y'(a) - \gamma_1 Y(a) = 1. \tag{46b}$$

Here γ_0 and γ_1 are any constants such that $(\alpha_1\gamma_0 - \alpha_0\gamma_1) \neq 0$. Then the two initial conditions in (46b) are linearly independent and a unique non-trivial solution of the initial value problem (46) exists. We denote this solution by $Y(\lambda; x)$. Now we consider the equation

$$\Phi(\lambda) \equiv \beta_0 Y'(\lambda; b) + \beta_1 Y(\lambda; b) = 0. \tag{47}$$

Clearly, each eigenvalue λ_n in (45) must satisfy this equation. Also every zero, λ^*, of $\Phi(\lambda)$ is an eigenvalue of (44) and the corresponding solution $Y(\lambda^*; x)$ of (46) is a corresponding eigenfunction of (44). Note that the present analysis differs from the corresponding discussion at the beginning of Section 7. Here, a parameter in the equation must be adjusted while the adjoined initial condition remains fixed, which reverses the previous situation. Of course, the present considerations apply to eigenvalue problems more general than those in (44); say for instance to problems in which the eigenvalue parameter λ enters into all of the coefficients of the equation and the boundary conditions. Extensions to homogeneous systems, of, say m second-order equations with m parameters are also clearly suggested. The initial value procedure can actually be used to prove the existence of the eigenvalues (45) and the oscillation properties of the eigenfunctions.

To approximate the eigenvalues and eigenfunctions for problems of the form (44), and various generalizations of these problems, we may apply numerical methods which are exactly analogous to those used in subsections 7.1 and 7.2. However, the proofs of convergence and estimates of the errors are now not always as easy to obtain as they were for those boundary value problems.

Some approximation methods for eigenvalue problems are based on *variational principles*. These have led to the construction of useful numerical methods. However, we do not treat them here, but refer the reader to the brief discussion of variational principles in Subsection 1.2 of Chapter 9.

A simple application of the basic error estimate for an eigenvalue of a symmetric matrix, Theorem 1.5 of Chapter 4, can be used to give an error estimate for the eigenvalue of a differential equation that is approximated by a difference method (e.g., the method in Subsection 7.2). Consider the eigenvalue problem

$$L\{y\} = \lambda y; \qquad y(a) = y(b) = 0, \tag{48}$$

where $L\{\cdot\}$ is defined in (44a).

Assume that λ is an eigenvalue and $y(x)$ a corresponding eigenfunction, with a continuous fourth derivative. Let

$$L_h\{u\} = \Lambda u; \qquad u(a) = u(b) = 0, \tag{49}$$

be a finite difference approximation to (48), on the net (12). Assume that the matrix form of (49), analogous to (27), is

$$A\mathbf{u} = -\frac{h^2}{2}\Lambda\mathbf{u}, \tag{50}$$

where A is a *symmetric* matrix. Then the truncation error, $\boldsymbol{\tau}$, of the eigensolution is defined by

$$A\mathbf{y} + \frac{h^2}{2}\lambda\mathbf{y} \equiv -\frac{h^2}{2}\boldsymbol{\tau}.$$

If $\|\boldsymbol{\tau}\|_\infty \leq Mh^2$, when $\|\mathbf{y}\|_\infty = 1$, then $\|\boldsymbol{\tau}\|_2 \leq MN^{1/2}h^2$. Furthermore, Theorem 1.5 of Chapter 4 implies

$$\min_{1\leq j\leq N} \frac{h^2}{2}|\lambda - \Lambda_j| \leq \frac{h^4}{2}MN^{1/2},$$

whence we have shown,

THEOREM **2.**

$$\min_{1\leq j\leq N} |\lambda - \Lambda_j| \leq h^2 MN^{1/2} = \mathcal{O}(h^{3/2}). \qquad \blacksquare$$

Theorem 2 states that some eigenvalue, Λ_j, of the discrete problem (49) is a good approximation to a given eigenvalue λ of (48). But as $h \to 0$, the theorem fails to identify which eigenvalue Λ_j is the closest approximation. In Problem 14, we verify that, in a special case, the smallest eigenvalues Λ_j approximate respectively the lowest eigenvalues λ_j.

PROBLEMS, SECTION 7

1. Establish the *alternative principle*. Either the equations (6) and (1b) have a unique solution or else the homogeneous problem [i.e., $r(x) \equiv 0, \alpha = \beta = 0$] has a non-trivial solution.

2. Solve by the initial value method

$$y'' = -100y; \quad y(0) = 1, \quad y(2\pi + \epsilon) = 1.$$

Use

$$y^{(1)}(x) \equiv \cos 10x, \quad y^{(2)}(x) = \frac{\sin 10x}{10}.$$

For small ϵ, show that $s = 5\epsilon + \mathcal{O}(\epsilon^3)$.Explain why the computational scheme corresponding to the initial value method would be difficult to apply for small ϵ.

3. Solve by the initial value method

$$y'' = 100y; \quad y(0) = 1, \quad y(3) = e^{-30}.$$

Use

$$y^{(1)}(x) = \frac{e^{10x} + e^{-10x}}{2},$$

$$y^{(2)}(x) = \frac{e^{10x} - e^{-10x}}{20}.$$

Explain why the computational scheme of the initial value method would have to be applied with great care.

4. The chord method for approximating the root s^* of (22) is based on the iteration scheme

$$s_{k+1} = g(s_k),$$

where

$$g(s) \equiv s - m[Y(s; b) - \beta].$$

Show that if for some $\rho > 0$,

$$0 < L \leq \left|\frac{\partial Y}{\partial s}(s; b)\right| \leq K, \qquad \text{for } |s - s^*| \leq \rho,$$

then with

$$m = \frac{2}{L + K} \operatorname{sign}\left(\frac{\partial Y(s; b)}{\partial s}\right),$$

$$|g'(s)| \leq \frac{K - L}{K + L} < 1.$$

5. Let the approximate solution of (21) be $U_j(s)$, $0 \leq j \leq N$; and assume that, in the notation of Problem 4,

$$m|U_N(s) - Y(s; b)| \leq \delta = \mathcal{O}(h^r) \qquad \text{for } |s - s^*| \leq \rho.$$

Define $\lambda \equiv (K - L)/(K + L)$ and let h be small enough so that $\delta \leq (1 - \lambda)\rho/2$. Use Theorem 1.3 of Chapter 3 and Problem 4 to show that, with $\sigma_{k+1} = \sigma_k - m[U_N(\sigma_k) - \beta]$, then

$$|\sigma_k - s^*| \leq \frac{\delta}{1 - \lambda} + \lambda^k\left(\rho - \frac{2\delta}{1 - \lambda}\right),$$

if

$$|\sigma_0 - s^*| \leq \rho - \frac{\delta}{1 - \lambda}.$$

6. For the boundary value problem: $y'' - q(x)y = r(x)$; $y(a) = y(b) = 0$ use a difference scheme of the form:

$$[u_{j+1} - 2u_j + u_{j-1}]/h^2 - [\alpha_1 q(x_{j+1})u_{j+1} + \alpha_0 q(x_j)u_j + \alpha_{-1} q(x_{j-1})u_{j-1}] = [\alpha_1 r(x_{j+1}) + \alpha_0 r(x_j) + \alpha_{-1} r(x_{j-1})],$$

for $j = 1, 2, \ldots, N$, with $u_0 = u_{N+1} = 0$ (as usual $h = (b - a)/(N + 1)$).

(a) Determine $\alpha_0, \alpha_1, \alpha_{-1}$ such that the truncation error is $\mathcal{O}(h^4)$. We assume here that y^{iv}, q^{iv}, and r^{iv} are continuous. Note that for the solution $y(x)$ we have $y^{\text{iv}} - [q(x)y]'' = r''(x)$.

(b) If $q(x) \geq Q_* > 0$, then show that for sufficiently small h:

$$|u_j - y(x_j)| \leq \frac{h^4}{720} \frac{2M_6 + 5N_4 + 5R_4}{Q_*}$$

where $M_6 \equiv \max_{[a,b]} |y^{\text{vi}}(x)|$, $N_4 \equiv \max_{[a,b]} |[q(x)y(x)]^{\text{iv}}|$, $R_4 \equiv \max_{[a,b]} |r^{\text{iv}}(x)|$. The proof is just as in Theorem 1.

7. Consider the boundary value problem

$$y'' - p(x)y' - q(x)y = r(x); \quad \alpha_0 y'(a) - \alpha_1 y(a) = \alpha,$$
$$\beta_0 y'(b) + \beta_1 y(b) = \beta$$

where α_0, β_0, α_1 and β_1 are all positive. Use the difference equations

$$\frac{u_{j+1} - 2u_j + u_{j-1}}{h^2} - p(x_j)\frac{u_{j+1} - u_{j-1}}{2h} - q(x_j)u_j = r(x_j)$$

for $j = 0, 1, \ldots, N + 1$

and the "boundary" conditions

$$\alpha_0\left(\frac{u_1 - u_{-1}}{2h}\right) - \alpha_1 u_0 = \alpha, \qquad \beta_0\left(\frac{u_{N+2} - u_N}{2h}\right) - \beta_1 u_{N+1} = \beta.$$

[Note: Values at $x_{-1} = a - h$ and $x_{N+2} = b + h$ have been introduced and the difference approximations of the differential equation have been employed at $x_0 = a$ and at $x_{N+1} = b$. Hence the values x_{-1} and x_{N+2} can be eliminated from the above difference equations.]

(a) Write these difference equations as a system of order $N + 2$ in the form (27). If the tridiagonal coefficient matrix is

$$A \equiv \begin{pmatrix} \ddots & \ddots & \ddots \\ -A_j & B_j & -C_j \\ \ddots & \ddots & \ddots \end{pmatrix}$$

with $j = 0, 1, \ldots, N + 1$, show that from (26),

$$A_j = a_j, \quad B_j = b_j, \quad C_j = c_j \qquad \text{for } j = 1, 2, \ldots, N$$

and that

$$A_0 = \left(a_0 + 2h\frac{\alpha_0}{\alpha_1} b_0\right), \quad C_0 = (c_0 + b_0).$$

Find similar expressions for A_{N+1} and B_{N+1}.

(b) If $q(x) \geq Q_* > 0$ and the solution $y(x)$ is sufficiently smooth in an open interval containing $[a, b]$ show that for sufficiently small h

$$|u_j - y(x_j)| \leq \mathcal{O}(h^2), \qquad j = 0, 1, \ldots, N + 1.$$

8.* Consider the boundary value problem:

$$[a(x)y']' - p(x)y' - q(x)y = r(x); \qquad y(a) = y(b) = 0$$

and the corresponding difference problem:

$$\left[a\left(x_j + \frac{h}{2}\right)\left(\frac{u_{j+1} - u_j}{h^2}\right) - a\left(x_j - \frac{h}{2}\right)\left(\frac{u_j - u_{j-1}}{h^2}\right)\right]$$

$$- p(x_j)\left(\frac{u_{j+1} - u_{j-1}}{2h}\right) - q(x_j)u_j = r(x_j),$$

$$j = 1, 2, \ldots, N; \quad u_0 = u_{N+1} = 0.$$

(a) If y^{iv} and a''' are continuous, show that the truncation error in this scheme is $\mathcal{O}(h^2)$.

(b) If $q(x) \geq Q_* > 0$ and $A^* \geq a(x) \geq A_* > 0$, show that

$$|u_j - y(x_j)| \leq \frac{A^*}{A_* Q_*} \tau$$

provided $A_* - (h/2)q(x_j) \geq 0$ for $j = 1, 2, \ldots, N$.

[Hint: Proceed as in the proof of Theorem 1 but now divide by $|b_j| + |c_j| = b_j + c_j \geq 2A_*$ before bounding the coefficients.]

9. We define the difference operator T by

$$Tu_j \equiv a_j u_j - b_j u_{j-1} - c_j u_{j+1}, \qquad j = 1, 2, \ldots, N,$$

where:

$$b_j > 0, \quad c_j > 0, \quad a_j \geq b_j + c_j.$$

Prove the

MAXIMUM PRINCIPLE: *Let the net function $\{V_j\}$ satisfy $TV_j \leq 0, j = 1, 2, \ldots, N$. Then*

$$\max_{0 \leq j \leq N+1} V_j = \max\{V_0, V_{N+1}\}.$$

Conversely if $TV_j \geq 0, j = 1, 2, \ldots, N$; then

$$\min_{0 \leq j \leq N+1} V_j = \min\{V_0, V_{N+1}\}.$$

[Hint: Use contradiction; assume $\max V_j \equiv M$ is at V_k for some k in $1 \leq k \leq N$ but that $V_0 \neq M$ and $V_{N+1} \neq M$. Then conclude that $V_j = M$ for all j which is a contradiction. The minimum result follows by changing sign.]

[Note: The conditions on the coefficients in T are satisfied by the quantities in (26) provided (28) is satisfied *even if we allow* $q(x) = 0$ (i.e., if condition (24) is weakened)].

10. Let T be as in Problem 9 and $\{e_j\}$ satisfy

$$Te_j = \sigma_j, \qquad j = 1, 2, \ldots, N.$$

Suppose $\{g_j\}$ satisfies $g_j > 0$ and

$$Tg_j \geq 1.$$

Then prove that

$$|e_j| \leq \max_{\nu=0,\,N+1} (|e_\nu| + \sigma g_\nu) + \sigma \cdot g,$$

where $\sigma = \max |\sigma_j|$, $g = \max |g_j|$.

[Hint: Form $\omega_j \equiv e_j \pm \sigma g_j$ and apply the maximum principle.]

11.* Prove that (43) follows from (42).

[Hint: Subtract (42a) from the corresponding equation with $\nu + 1$ replaced by ν; use Taylor's theorem and proceed as in the derivation of (41).]

12.* Consider, in place of (36), the difference equations $u_0 = u_{N+1} = 0$;

$$\frac{u_{j+1} - 2u_j + u_{j-1}}{h^2} = f\left(x_j, \frac{u_{j+1} + u_{j-1}}{2}, \frac{u_{j+1} - u_{j-1}}{2h}\right), \qquad j = 1, 2, \ldots, N.$$

(a) Show that $|u_j - y(x_j)| = \mathcal{O}(h^2)$ where $y(x)$ is the solution of (20); (28) and (35) hold; and $y^{\text{iv}}(x)$, $\partial f/\partial y$ and $\partial f/\partial y'$ are continuous.

(b) Under the above assumptions and $h(P^* + hQ^*)/2 \leq 1$, prove convergence of the iterations:

$$u_0^{(\nu+1)} = u_{N+1}^{(\nu+1)} = 0,$$

$$u_j^{(\nu+1)} = \tfrac{1}{2}[u_{j+1}^{(\nu)} + u_{j-1}^{(\nu)}] - \frac{h^2}{2} f\left(x_j, \frac{u_{j+1}^{(\nu)} + u_{j-1}^{(\nu)}}{2}, \frac{u_{j+1}^{(\nu)} - u_{j-1}^{(\nu)}}{2h}\right);$$

$$j = 1, 2, \ldots, N.$$

Note that the parameter ω is not required here, as it was in (42); i.e., we could employ the value $\omega = 0$.

13. Solve for the eigenvalues and eigenvectors of the problem $y'' + \lambda y = 0$, $y'(a) = y'(b) = 0$, by using the initial value technique. For example, use the initial values $y'(a) = 0$, $y(a) = \text{constant} \neq 0$.

14. Find the eigenvalues and eigenvectors of the scheme

$$\frac{u_{j-1} - 2u_j + u_{j+1}}{h^2} = \Lambda u_j, \qquad 1 \leq j \leq N,$$

$$u_0 = u_{N+1} = 0, \qquad h = \frac{\pi}{N+1}.$$

Compare them with the eigenvalues and eigenvectors of

$$y'' = \lambda y, \qquad y(0) = y(\pi) = 0.$$

[Hint: Solve the difference equation in the form $u_j = \alpha^j$, for an appropriate α. Show that the eigenfunctions are

$$u_j^{(k)} = A_k \sin j\left(\frac{\pi k}{N+1}\right), \qquad 0 \leq j \leq N+1,$$

for $k = 1, 2, \ldots, N$.]

15. (a) Use the results of Problems 6 and 14 to devise a difference scheme for $y'' + \lambda y = 0$, $y(0) = y(1) = 0$ which yields $\mathcal{O}(h^4)$ approximations to the eigenvalues.

(b) Find the eigenvalues of the difference scheme and verify directly, with a comparison to $\lambda_n = n^2\pi^2$, that they are actually $\mathcal{O}(h^4)$. How accurate are the eigenvectors?

16.* Derive a *variational differential equation* that is satisfied by $\partial Y(\lambda; x)/\partial\lambda$, where Y is a solution of (46). Describe how Newton's iterative method might be formulated to solve for an eigenvalue λ from (47).

9

Difference Methods for Partial Differential Equations

0. INTRODUCTION

Although considerable study has been made of partial differential equation problems, the mathematical theory—existence, uniqueness, or well-posedness—is not nearly as complete as it is in the case of ordinary differential equations. Furthermore, except for some problems that are solved by explicit formulae, the analytical methods developed for the treatment of partial differential equations are, in general, not suited for the efficient numerical evaluation of solutions. Hence, as may be expected, the theory of numerical methods for partial differential equations is somewhat fragmented. Where the theory of the differential equations is well developed there has been a corresponding development of numerical methods. But the difference methods found thus far usually do not permit the construction of schemes of an arbitrarily high order of accuracy. For certain systems of partial differential equations convergent numerical methods of arbitrarily high order of accuracy have been devised (for instance, linear first order hyperbolic systems in two unknowns); while for others (say the simple case of the Laplace equation on a square) only relatively low order methods have been proved to converge. Furthermore, in contrast to the case of the numerical solution of ordinary differential equation problems, the facility with which one may use difference methods on modern electronic computers to solve problems involving partial differential equations is severely limited by (a) size of the high speed memory, (b) speed of the arithmetic unit, and (c) difficulty of programming a problem for and communicating with the computer.

In view of the limitations of the scope of this book and of the incompleteness of the theory of difference methods, we shall illuminate some of the highlights of this theory through the treatment of problems for the

$$\text{(1a)} \qquad \frac{\partial^2 u}{\partial x^2} + \frac{\partial^2 u}{\partial y^2} = 0 \qquad \textit{Laplace equation;}$$

$$\text{(1b)} \qquad \frac{\partial^2 u}{\partial t^2} - c^2 \frac{\partial^2 u}{\partial x^2} = 0 \qquad \textit{Wave equation;}$$

$$\text{(1c)} \qquad \frac{\partial u}{\partial t} - a^2 \frac{\partial^2 u}{\partial x^2} = 0 \qquad \textit{Diffusion or heat conduction equation.}$$

The applications of these equations are so varied and well known that we do not make specific mention of particular cases. Of course, in applied mathematics other partial differential equations occur; most of these are non-linear and not covered by a complete mathematical theory of existence, uniqueness, or well-posedness.

To each of the equations in (1) we must adjoin appropriate subsidiary relations, called *boundary* and/or *initial conditions*, which serve to complete the formulation of a "meaningful problem." These conditions are related to the domain, say D, in which the equation (1) is to be solved. When the problem arises from a physical application it is usually clear (to anyone understanding the phenomenon) what these relations must be. Some familiar examples are, for the respective equations (1a, b, and c);

$$\text{(2a, i)} \qquad u = f(x, y), \qquad \text{for } (x, y) \text{ on the boundary of } D,$$

or with $\partial/\partial n$ representing the normal derivative,

$$\text{(2a, ii)} \qquad \alpha u + \beta \frac{\partial u}{\partial n} = f(x, y), \qquad \text{for } (x, y) \text{ on the boundary of } D;$$

$$\text{(2b, i)} \qquad u(0, x) = f(x), \ \frac{\partial u(0, x)}{\partial t} = g(x), \quad -\infty < x < \infty,$$

where $D \equiv \{(t, x) \mid t \geq 0, \ -\infty < x < \infty\}$, i.e., $D \equiv$ half plane, or

$$\text{(2b, ii)} \qquad \begin{cases} u(0, x) = f(x), \ \dfrac{\partial u(0, x)}{\partial t} = g(x), \\ u(t, a) = \alpha(t), \ u(t, b) = \beta(t), \ t > 0, \end{cases}$$

where $D \equiv \{(t, x) \mid t \geq 0, \ a \leq x \leq b\}$, i.e., $D \equiv$ half strip;

$$\text{(2c, i)} \qquad u(0, x) = f(x), \quad -\infty < x < \infty,$$

where D is the half plane, or

$$\text{(2c, ii)} \qquad \begin{cases} u(0, x) = f(x), \; a \leq x \leq b, \\ u(t, a) = \alpha(t), \; u(t, b) = \beta(t), \end{cases}$$

where D is the half strip.

If the functions introduced in (2a, b, and c) satisfy appropriate smoothness conditions, then each set of relations (i) or (ii) adjoined to the corresponding equation in (1) yields a problem which has been termed *well-posed* or *properly-posed* by Hadamard. This implies that each such problem has a bounded solution, that the solution is unique, and that it depends continuously on the data (i.e., a "small" change in f, g, α, or β produces a "correspondingly small" change in the solution). There are many other combinations of boundary and/or initial conditions which together with the equations in (1) (or more general equations) constitute properly posed problems. It is such problems for which there is a reasonably developed theory of difference approximations. We shall examine this theory briefly in Section 5, after first studying some special cases. However, as we shall see in Section 5, the theory serves mainly to determine whether a given method yields approximations of reasonable accuracy; but the theory does not directly suggest how to construct numerical schemes.

0.1. Conventions of Notation

For simplicity, let the domain D have boundary C and lie in the three dimensional space of variables (x, y, t). Cover this space by a *net*, *grid*, *mesh*, or *lattice* of discrete points, with coordinates (x_i, y_j, t_k) given by

$$x_i = x_0 + i\delta x, \quad y_j = y_0 + j\delta y, \quad t_k = t_0 + k\delta t;$$
$$i, j, k = 0, \pm 1, \pm 2, \ldots.$$

Here, we have taken the *net spacings* δx, δy, and δt to be uniform. The lattice points may be divided into three disjoint sets: D_δ, the *interior net points*; C_δ, the *boundary net points*; and the remainder which are *external points*. Here we assume, again for simplicity, that C is composed of sections of coordinate surfaces. The specific rules for assigning lattice points to a particular set will be clarified in the subsequent examples and discussion.

At the points of $D_\delta + C_\delta$ the function $u(x, y, t)$ is to be approximated by a net function, $U(x_i, y_j, t_k)$. It is convenient to denote the components of net functions by appropriate subscripts and/or superscripts. For instance, we may use

$$U(x_i, y_j) \equiv U_{i,j}; \quad U(x_i, y_j, t_k) \equiv U^k_{i,j}, \quad \text{etc.}$$

This notation is frequently cumbersome and at times difficult (if not unpleasant) to read. Thus, while we shall have occasion to use it, we prefer another notation, more in keeping with the usual functional notation. If U has been defined to be a net function, then we may write $U(x, y, t)$ and understand the argument point (x, y, t) to be some general point of the net on which U is defined. Furthermore, if we simply write U then the argument is understood to be a general point (x, y, t) of the appropriate net.

We shall make frequent use of various difference quotients of net functions (of course, in order to approximate partial derivatives). For this purpose we introduce a *subscript notation for difference quotients* of net functions

$$\text{(3a)} \qquad U_x(x, y, t) \equiv \frac{U(x + \delta x, y, t) - U(x, y, t)}{\delta x}$$

$$\text{(3b)} \qquad U_{\bar{x}}(x, y, t) \equiv \frac{U(x, y, t) - U(x - \delta x, y, t)}{\delta x}$$

$$\text{(3c)} \qquad U_{\hat{x}}(x, y, t) \equiv \tfrac{1}{2}[U_x(x, y, t) + U_{\bar{x}}(x, y, t)].$$

Clearly, (3a, b, and c) are just the *forward, backward,* and *centered difference quotients* with respect to x. By our previous convention we might have written the left-hand sides of (3) as just U_x, $U_{\bar{x}}$, and $U_{\hat{x}}$. This convenient notation was introduced by Courant, Friedrichs, and Lewy in a fundamental paper on difference methods for partial differential equations. The difference quotients with respect to other discrete variables are defined in analogy with (3), say U_y, $U_{\bar{t}}$, etc. It is a simple matter to verify that these difference operator commute; i.e.,

$$U_{xy} = U_{yx}, \quad U_{\hat{x}\bar{t}} = U_{\bar{t}\hat{x}}, \quad \text{etc.}$$

A particularly important case is the centered second difference quotient which can be written as

$$\text{(4)} \qquad U_{y\bar{y}} = U_{\bar{y}y} = \frac{1}{(\delta y)^2}[U(x, y + \delta y, t) - 2U + U(x, y - \delta y, t)].$$

1. LAPLACE EQUATION IN A RECTANGLE

A standard type of problem which employs the Laplace operator or Laplacian,

$$\Delta \equiv \frac{\partial^2}{\partial x^2} + \frac{\partial^2}{\partial y^2},$$

is to determine a function, $u(x, y)$, such that

$$\text{(1a)} \qquad -\Delta u(x, y) = f(x, y), \qquad (x, y) \in D;$$

$$\text{(1b)} \qquad u(x, y) = g(x, y), \qquad (x, y) \in C.$$

Here D is some domain in the x, y-plane and C is its boundary. If the boundary C and inhomogeneous terms $f(x, y)$ and $g(x, y)$ satisfy mild regularity conditions, it is well known that the problem (1) is well-posed. If $f \equiv 0$ this is called a *Dirichlet problem* for Laplace's equation while in its present form (1a) is called the *Poisson equation*. For simplicity of presentation, we take D to be a rectangle

$$D \equiv \{(x, y) \mid 0 < x < a, 0 < y < b\}; \tag{2a}$$

whose boundary C is composed of four line segments

$$C \equiv \{(x, y) \mid x = 0, a, 0 \le y \le b; y = 0, b, 0 \le x \le a\}. \tag{2b}$$

To "solve" this problem numerically we introduce the net spacings $\delta x = a/(J + 1)$, $\delta y = b/(K + 1)$, and the uniformly spaced net points

$$x_j = j\delta x, \quad y_k = k\delta y; \quad j, k = 0, \pm 1, \pm 2, \ldots.$$

Those net points interior to D we call D_δ, i.e.,

$$D_\delta \equiv \{(x_j, y_k) \mid 1 \le j \le J; 1 \le k \le K\}. \tag{3a}$$

The net points on C, with the exception of the four corners of C, we call C_δ, i.e.,

$$C_\delta \equiv \{(x_j, y_k) \mid j = (0, J + 1), 1 \le k \le K; \quad k = (0, K + 1), 1 \le j \le J\}. \tag{3b}$$

At the net points $D_\delta + C_\delta$ we seek quantities $U(x_j, y_k)$ which are to approximate the solution $u(x_j, y_k)$ of (1). The net function will, of course, be defined as the solution of a system of difference equations that replaces the partial differential equation (1a) and boundary conditions (1b) on the net.

An obvious approximation to the Laplacian is obtained by replacing each second derivative by a centered second difference quotient. Thus at each point $(x, y) \in D_\delta$ we define

$$\Delta_\delta U(x, y) \equiv U_{x\bar{x}}(x, y) + U_{y\bar{y}}(x, y). \tag{4a}$$

In the subscript notation, we could also write for each $(x_j, y_k) \in D_\delta$

$$\Delta_\delta U_{j,k} \equiv \frac{U_{j+1,k} - 2U_{j,k} + U_{j-1,k}}{(\delta x)^2} + \frac{U_{j,k+1} - 2U_{j,k} + U_{j,k-1}}{(\delta y)^2}. \tag{4b}$$

It is frequently convenient with either of these notations to indicate the net points involved in the definition of $\Delta_\delta U$ by means of a diagram as in Figure 1. The set of points marked with crosses is called the *star* or *stencil* associated with the difference operator Δ_δ.

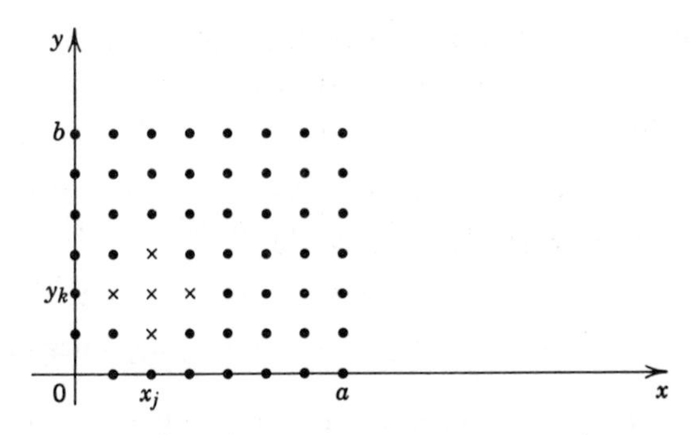

Figure 1. Net points for Laplace difference operator Δ_δ.

With the notation (4a), we write the difference problem as

$$-\Delta_\delta U(x, y) = f(x, y), \qquad (x, y) \in D_\delta; \tag{5a}$$

$$U(x, y) = g(x, y), \qquad (x, y) \in C_\delta. \tag{5b}$$

From (3), (4), and (5) we find that the values U at $JK + 2(J + K)$ net points in $D_\delta + C_\delta$ satisfy $JK + 2(J + K)$ linear equations. Hence, we may hope to solve (5) for the unknowns $U(x, y)$ in $D_\delta + C_\delta$. The $2(J + K)$ values of U on C_δ are specified in (5b) and so the JK equations of (5a) must determine the remaining JK unknowns. We shall first show that this system has a unique solution and then we will estimate the error in the approximation. Finally, we shall consider practical methods for solving the linear system (5).

To demonstrate that the difference equations have a unique solution we shall prove that the corresponding homogeneous system has only the trivial solution. For this purpose and for the error estimates to be obtained, we first prove a *maximum principle* for the operator Δ_δ.

THEOREM 1.

(a) *If* $V(x, y)$ *is a net function defined on* $D_\delta + C_\delta$ *and satisfies*

$$\Delta_\delta V(x, y) \geq 0 \qquad \textit{for all } (x, y) \in D_\delta,$$

then

$$\max_{D_\delta} V(x, y) \leq \max_{C_\delta} V(x, y).$$

(b) *Alternatively, if* V *satisfies*

$$\Delta_\delta V(x, y) \leq 0 \qquad \textit{for all } (x, y) \in D_\delta,$$

then

$$\min_{D_\delta} V(x, y) \geq \min_{C_\delta} V(x, y).$$

Proof. We prove part (a) by contradiction. Assume that at some point $P_0 \equiv (x^*, y^*)$ of D_δ, we have $V(P_0) = M$ where

$$M \geq V(P) \quad \text{for all } P \in D_\delta \qquad \text{and } M > V(P) \quad \text{for all } P \in C_\delta.$$

Let us introduce the notation $P_1 \equiv (x^* + \delta x, y^*)$, $P_2 \equiv (x^* - \delta x, y^*)$, $P_3 \equiv (x^*, y^* + \delta y)$, $P_4 \equiv (x^*, y^* - \delta y)$ and then use (4) to write

$$\Delta_\delta V(P_0) \equiv \theta_x[V(P_1) + V(P_2)] + \theta_y[V(P_3) + V(P_4)] - 2(\theta_x + \theta_y)V(P_0)$$

where $\theta_x \equiv 1/(\delta x)^2$ and $\theta_y \equiv 1/(\delta y)^2$. However, by hypothesis $\Delta_\delta V(P_0) \geq 0$, so we have

$$M = V(P_0) \leq \frac{1}{\theta_x + \theta_y}\left[\theta_x \frac{V(P_1) + V(P_2)}{2} + \theta_y \frac{V(P_3) + V(P_4)}{2}\right].$$

But $M \geq V$ implies that $V(P_\nu) = M$ for $\nu = 1, 2, 3, 4$. We now repeat this argument for each interior point P_ν instead of the point P_0. By repetition, each point of D_δ and C_δ appears as one of the P_ν for some corresponding P_0. Thus, we conclude that

$$V(P) = M \quad \text{for all } P \quad \text{in } D_\delta + C_\delta,$$

which contradicts the assumption that $V < M$ on C_δ. Part (a) of the theorem follows.†

To prove part (b), we could repeat an argument similar to the above. However, it is simpler to recall that

$$\max\,[-V(x, y)] = -\min V(x, y); \qquad \Delta_\delta(-V) = -\Delta_\delta(V).$$

Hence, if V satisfies the hypothesis of part (b), then $-V$ satisfies the hypothesis of part (a). But the conclusion of part (a) for $-V$ is identical to the conclusion of part (b) for V.† ∎

Let us now consider the homogeneous system corresponding to (5); i.e., $f \equiv g \equiv 0$. From Theorem 1, it follows that the *max* and *min* of the solution of this homogeneous system vanish; hence, the only solution is the trivial one. Thus it follows by the alternative principle for linear systems that (5) has a unique solution for arbitrary $f(x, y)$ and $g(x, y)$.

A bound for the solution of the difference equation (5) can also be obtained by an appropriate application of the *maximum principle*. The result, called an *a priori estimate*, may be stated as

THEOREM 2. *Let $V(x, y)$ be any net function defined on the sets D_δ and C_δ defined by* (3). *Then*

$$\max_{D_\delta} |V| \leq \max_{C_\delta} |V| + \frac{a^2}{2} \max_{D_\delta} |\Delta_\delta V|. \tag{6}$$

† We have in fact proved more; namely, that if the maximum, in case (a), or the minimum, in case (b), of $V(x, y)$ occurs in D_δ, then $V(x, y)$ is constant on $D_\delta + C_\delta$.

Proof. We introduce the function

$$\phi(x, y) \equiv \tfrac{1}{2}x^2$$

and observe that for all $(x, y) \in D_\delta + C_\delta$,

$$0 \leq \phi(x, y) \leq \frac{a^2}{2}; \qquad \Delta_\delta \phi(x, y) = 1.$$

Now define the two net functions $V_+(x, y)$ and $V_-(x, y)$ by

$$V_\pm(x, y) \equiv \pm V(x, y) + N\phi(x, y),$$

where

$$N \equiv \max_{D_\delta} |\Delta_\delta V|.$$

Clearly for all $(x, y) \in D_\delta$, it follows that

$$\Delta_\delta V_\pm(x, y) = \pm \Delta_\delta V(x, y) + N \geq 0.$$

Thus we may apply the maximum principle, part (a) of Theorem 1, to each of $V_\pm(x, y)$ to obtain for all $(x, y) \in D_\delta$,

$$\begin{aligned} V_\pm(x, y) &\leq \max_{C_\delta} V_\pm(x, y) \\ &= \max_{C_\delta} [\pm V(x, y) + N\phi] \leq \max_{C_\delta} [\pm V(x, y)] + N\frac{a^2}{2}. \end{aligned}$$

But from the definition of $V_\pm$ and the fact that $\phi \geq 0$,

$$\pm V(x, y) \leq V_\pm(x, y).$$

Hence,

$$\begin{aligned} \pm V(x, y) &\leq \max_{C_\delta} [\pm V(x, y)] + N\frac{a^2}{2}, \\ &\leq \max_{C_\delta} |V| + \frac{a^2}{2} N. \end{aligned}$$

Since the right-hand side in the final inequality is independent of (x, y) in D_δ the theorem follows. ∎

Note that we could readily replace $a^2/2$ in (6) by $b^2/2$ since the function $\psi(x, y) = y^2/2$ can be used in place of $\phi(x, y)$ in the proof of the theorem.

It is now a simple matter to estimate the error $U - u$. We introduce the *local truncation error*, $\tau\{\Phi\}$, for the difference operator Δ_δ on D_δ by

$$\tau\{\Phi(x, y)\} \equiv \Delta_\delta \Phi(x, y) - \Delta\Phi(x, y), \qquad (x, y) \text{ in } D_\delta, \tag{7}$$

where $\Phi(x, y)$ is any sufficiently smooth function defined on D. Now if $u(x, y)$ is the solution of the boundary value problem (1) we have from (1a) at the points of D_δ

$$-\Delta_\delta u(x, y) = f(x, y) - \tau\{u(x, y)\}.$$

Subtracting this from (5a) at each point of D_δ yields

$$\text{(8a)} \qquad -\Delta_\delta[U(x, y) - u(x, y)] = \tau\{u(x, y)\}, \qquad (x, y) \text{ in } D_\delta.$$

Also from (1b) and (5b) we obtain

$$\text{(8b)} \qquad U(x, y) - u(x, y) = 0, \qquad (x, y) \text{ in } C_\delta.$$

Now apply Theorem 2 to the net function $U(x, y) - u(x, y)$ and we get by (8)

$$\max_{D_\delta} |U(x, y) - u(x, y)| \leq \frac{a^2}{2} \max_{D_\delta} |\tau\{u\}|.$$

Upon introducing the *maximum norm* defined for any net function $W(x, y)$ by $\|W\| = \max_{D_\delta} |W|$, we have the

COROLLARY. *With u, U, and τ defined respectively by* (1), (5), *and* (7), *we have*

$$\text{(9)} \qquad \|U(x, y) - u(x, y)\| \leq \frac{a^2}{2} \|\tau\{u\}\|. \qquad \blacksquare$$

Note that the error bound is proportional to the truncation error!

It is easy to estimate $\|\tau\|$. If the solution $u(x, y)$ of (1) has *continuous and bounded fourth order partial derivatives* in D, then

$$\text{(10)} \qquad u(x \pm \delta x, y) = u(x, y) \pm \delta x \frac{\partial u(x, y)}{\partial x} + \frac{(\delta x)^2}{2!} \frac{\partial^2 u(x, y)}{\partial x^2}$$
$$\pm \frac{(\delta x)^3}{3!} \frac{\partial^3 u(x, y)}{\partial x^3} + \frac{(\delta x)^4}{4!} \frac{\partial^4 u(x + \theta_\pm \delta x, y)}{\partial x^4};$$
$$|\theta_\pm| < 1.$$

Thus we find, as in Chapter 6, that

$$u_{x\bar{x}}(x, y) - \frac{\partial^2 u(x, y)}{\partial x^2} = \frac{(\delta x)^2}{12} \frac{\partial^4 u(x + \theta \delta x, y)}{\partial x^4}, \qquad |\theta| < 1,$$

with a similar result for the y derivatives. Hence,

$$\tau\{u(x, y)\} = \Delta_\delta u(x, y) - \Delta u(x, y)$$
$$= \frac{1}{12} \left[(\delta x)^2 \frac{\partial^4 u(x + \theta \delta x, y)}{\partial x^4} + (\delta y)^2 \frac{\partial^4 u(x, y + \psi \delta x)}{\partial y^4}\right].$$

If we denote the bounds of the respective fourth order derivatives by $M_x^{(4)}$ and $M_y^{(4)}$, then

$$\text{(11a)} \qquad \|\tau\{u\}\| \leq \tfrac{1}{12}[(\delta x)^2 M_x^{(4)} + (\delta y)^2 M_y^{(4)}].$$

If $u(x, y)$ has only *continuous third order derivatives*, we terminate the expansions in (10) one term earlier and get

$$u_{x\bar{x}}(x, y) - \frac{\partial^2 u(x, y)}{\partial x^2} = \frac{\delta x}{3!}\left[\frac{\partial^3 u(x + \theta_+ \delta x, y)}{\partial x^3} - \frac{\partial^3 u(x + \theta_- \delta x, y)}{\partial x^3}\right].$$

If the moduli of continuity in D of the third derivatives $\partial^3 u/\partial x^3$ and $\partial^3 u/\partial y^3$ are denoted by $\omega_x^{(3)}(\delta)$ and $\omega_y^{(3)}(\delta)$, respectively, we have

$$\|\tau\{u\}\| \leq \tfrac{1}{6}[\delta x \omega_x^{(3)}(2\delta x) + \delta y \omega_y^{(3)}(2\delta y)]. \tag{11b}$$

Clearly by these procedures, we find that if $u(x, y)$ has only continuous second derivatives with moduli of continuity $\omega_x^{(2)}(\delta)$ and $\omega_y^{(2)}(\delta)$, then

$$\|\tau\{u\}\| \leq \omega_x^{(2)}(\delta x) + \omega_y^{(2)}(\delta y). \tag{11c}$$

With the aid of any of the estimates (11a, b, or c) that may be appropriate, the corollary establishes convergence of the approximate solution to the exact solution *as* $\delta x \to 0$ *and* $\delta y \to 0$ *in any manner*. We see that the convergence rate is generally faster for "smoother" solutions $u(x, y)$. For solutions which have more than four continuous derivatives, we cannot deduce better truncation error estimates than that given by (11a). It is possible to construct more accurate difference approximations to the Laplacian, which then have solutions U of greater accuracy than $\mathcal{O}[(\delta x)^2 + (\delta y)^2]$. But there is no general way of constructing convergence proofs for similar schemes of *arbitrarily* high order truncation error. In fact, it is unlikely that such schemes, which are of maximum order of accuracy, converge in general.

The effects of roundoff can also be estimated by means of Theorem 2. Let the numbers actually computed be denoted by $\bar{U}(x, y)$. Then we can write

$$-\Delta_\delta \bar{U}(x, y) = f(x, y) + \frac{1}{\delta x\, \delta y}\rho(x, y), \qquad (x, y) \in D_\delta; \tag{12a}$$

$$\bar{U}(x, y) = g(x, y) + \rho'(x, y), \qquad (x, y) \in C_\delta. \tag{12b}$$

Here $\rho'(x, y)$ is the roundoff error in approximating the boundary data. After noting that the coefficients in Δ_δ are proportional to $1/(\delta x)^2$ and $1/(\delta y)^2$, we have defined the roundoff errors, $\rho(x, y)$, in the computations (12a) to be proportional to a similar factor. This corresponds to the fact that the actual computations are done with the form of (4) which results after multiplication by the factor $\delta x \delta y$. We now obtain from (1), (7) and (12)

$$-\Delta_\delta[\bar{U}(x, y) - u(x, y)] = \tau\{u(x, y)\} + \frac{\rho(x, y)}{\delta x\, \delta y}, \qquad (x, y) \in D_\delta;$$

$$\bar{U}(x, y) - u(x, y) = \rho'(x, y), \qquad (x, y) \in C_\delta.$$

Thus for the net function $\bar{U}(x, y) - u(x, y)$, Theorem 2 implies

THEOREM 3. *With* u, $\bar{U}$, *and* τ *defined by* (1), (12), *and* (7) *respectively, we have*

$$\|\bar{U}(x, y) - u(x, y)\| \leq \|\rho'\| + \frac{a^2}{2}\left[\|\tau\{u\}\| + \frac{1}{\delta x\, \delta y}\|\rho\|\right]. \tag{13}$$

Here

$$\|\rho'\| = \max_{C_\delta} |\rho'(x, y)| \quad \textit{and} \quad \|\rho\| = \max_{D_\delta} |\rho(x, y)|.$$ ∎

Thus we find that the boundary roundoff error and the interior roundoff error have quite different effects on the accuracy as δx and $\delta y \to 0$. In fact, to be consistent with the truncation error, the interior roundoff error, ρ, should be of the same order as $\delta x\, \delta y \tau\{u\}$ and the boundary roundoff error, ρ', should be of the same order as τ when δx and $\delta y \to 0$. This result for ρ is analogous to that in (7.34), of Chapter 8 where simple difference approximations of an ordinary boundary value problem were considered.

The maximum principle and its applications given here can be generalized in various ways (see Problems 1–4). Extensions to rectangular domains in higher dimensions are straightforward, and non-rectangular domains may also be treated (with suitable modifications of the difference equations near the boundary surface).

1.1. Matrix Formulation

The system of linear equations (5) can be written in matrix-vector notation in various ways. For this purpose, we use the subscript notation for any net function $V(x, y)$ defined on $D_\delta + C_\delta$

$$V(x_j, y_k) \equiv V_{jk}; \qquad 0 \leq j \leq J + 1,\ 0 \leq k \leq K + 1.$$

From the values of such a net function, construct the J-dimensional vector

$$\mathbf{V}_k \equiv \begin{pmatrix} V_{1k} \\ V_{2k} \\ \vdots \\ V_{Jk} \end{pmatrix}, \qquad k = 0, 1, 2, \ldots, K + 1. \tag{14a}$$

Each vector $\mathbf{V}_k$ consists of the elements of the net function $V(x_j, y_k)$ on the coordinate line segment $y = y_k$, $x_1 \leq x \leq x_J$. (We note that the elements on the line segments $y_1 \leq y \leq y_K$, $x = x_0$, and $x = x_{J+1}$ are not included.) We also introduce the Jth order square matrices

$$I_J \equiv \begin{bmatrix} 1 & & & & \\ & \cdot & & & \\ & & \cdot & & \\ & & & \cdot & \\ & & & & 1 \end{bmatrix} \equiv (\delta_{ij}),$$

(14b)

$$L_J \equiv \begin{bmatrix} 0 & & & & & & \\ 1 & 0 & & & & & \\ & \cdot & \cdot & & & & \\ & & \cdot & \cdot & & & \\ & & & \cdot & \cdot & & \\ & & & & \cdot & \cdot & \\ & & & & & 1 & 0 \end{bmatrix} \equiv (a_{ij}), \qquad a_{ij} = \delta_{i-1,j},$$

and the quantities

(14c)
$$\delta^2 \equiv \frac{(\delta x)^2(\delta y)^2}{2[(\delta x)^2 + (\delta y)^2]}, \qquad \theta_x \equiv \frac{(\delta y)^2}{2[(\delta x)^2 + (\delta y)^2]},$$

$$\theta_y \equiv \frac{(\delta x)^2}{2[(\delta x)^2 + (\delta y)^2]}.$$

Upon multiplying (5a) by δ^2, we can write the result for $(x, y) = (x_j, y_k)$, in subscript notation as

(15)
$$U_{jk} - \theta_x(U_{j-1,k} + U_{j+1,k}) - \theta_y(U_{j,k-1} + U_{j,k+1}) = \delta^2 f_{jk},$$
$$1 \leq j \leq J,\ 1 \leq k \leq K.$$

Or with the vector and matrix notation of (14) this system becomes

(16a)
$$\begin{aligned} [I_J - \theta_x(L_J + L_J^T)]\mathbf{U}_1 - \theta_y\mathbf{U}_2 &= \delta^2\mathbf{F}_1, \\ -\theta_y\mathbf{U}_{k-1} + [I_J - \theta_x(L_J + L_J^T)]\mathbf{U}_k - \theta_y\mathbf{U}_{k+1} &= \delta^2\mathbf{F}_k, \\ &\qquad 2 \leq k \leq K-1; \\ -\theta_y\mathbf{U}_{K-1} + [I_J - \theta_x(L_J + L_J^T)]\mathbf{U}_K &= \delta^2\mathbf{F}_K. \end{aligned}$$

Here we have introduced

(16b)
$$\begin{aligned} \mathbf{F}_1 &= \mathbf{f}_1 + \frac{1}{(\delta x)^2}\mathbf{w}_1 + \frac{1}{(\delta y)^2}\mathbf{U}_0, \\ \mathbf{F}_k &= \mathbf{f}_k + \frac{1}{(\delta x)^2}\mathbf{w}_k, \qquad 2 \leq k \leq K-1, \\ \mathbf{F}_K &= \mathbf{f}_K + \frac{1}{(\delta x)^2}\mathbf{w}_K + \frac{1}{(\delta y)^2}\mathbf{U}_{K+1}, \end{aligned}$$

where

$$\mathbf{w}_k \equiv (w_{ik}) \equiv \begin{bmatrix} U_{0k} \\ 0 \\ \vdots \\ 0 \\ U_{J+1,k} \end{bmatrix},$$

i.e., $w_{ik} = 0$ for $2 \leq i \leq J - 1$; $w_{1k} = U_{0k}$; $w_{Jk} = U_{J+1,k}$. Of course, all of the U_{jk} which enter into (16b) are known quantities given in (5b).

Further simplification is obtained by introducing JK-dimensional vectors or K-dimensional *compound vectors* (i.e., vectors whose components are J-dimensional vectors)

(17a)
$$\mathbf{U} \equiv \begin{pmatrix} \mathbf{U}_1 \\ \mathbf{U}_2 \\ \vdots \\ \mathbf{U}_K \end{pmatrix} \equiv \begin{pmatrix} U_{11} \\ U_{21} \\ \vdots \\ U_{JK} \end{pmatrix}, \qquad \mathbf{F} \equiv \begin{pmatrix} \mathbf{F}_1 \\ \mathbf{F}_2 \\ \vdots \\ \mathbf{F}_K \end{pmatrix};$$

and the square matrices of order JK

$$I \equiv (\delta_{ij}),$$

(17b)
$$L \equiv \begin{bmatrix} L_J & & & \\ & L_J & & \\ & & \ddots & \\ & & & L_J \end{bmatrix}, \qquad B \equiv \begin{bmatrix} 0 & & & & \\ I_J & 0 & & & \\ & \ddots & \ddots & & \\ & & \ddots & \ddots & \\ & & & I_J & 0 \end{bmatrix},$$

$$H \equiv \theta_x(L + L^T), \quad V \equiv \theta_y(B + B^T), \quad A \equiv I - H - V.$$

Now the system (16), or equivalently (5), can be written as

(18)
$$A\mathbf{U} = \delta^2\mathbf{F}.$$

The vectors in (17a) associate a component with each net point (x, y) of D_δ. In the indicated vectors of dimension $N = JK$, the rth component is the value of the net function at the point (x_j, y_k) such that $r = j + (k - 1)J$. If the assignment of integers, r, to net points of D_δ is done in some other order, then the vectors and matrices are changed by some permutation. (Another ordering of interest would be to list the elements on lines x = constant of D_δ). The previous proof that the system (5) has a unique solution now implies that A is non-singular. We shall prove this fact directly by showing that the eigenvalues of A are positive.

Let us consider again the problem of obtaining error estimates for the approximate solution. Multiply (8a) by δ^2 and employ the present notation to obtain in place of the system (8)

$$A(\mathbf{U} - \mathbf{u}) = \delta^2 \boldsymbol{\tau}. \tag{19a}$$

Here $\mathbf{U}$ is as before, $\mathbf{u}$ is the vector of the exact solution on D_δ, and $\boldsymbol{\tau}$ is the vector of local truncation errors $\tau\{u(x, y)\}$ on D_δ with no adjustments now required as in (16b) since $U - u = 0$ on C_δ. Then as A is non-singular, we have

$$\mathbf{U} - \mathbf{u} = \delta^2 A^{-1} \boldsymbol{\tau}, \tag{19b}$$

which is an exact representation for the error. By using any vector norm and the corresponding natural matrix norm, we have from (19b),

$$\|\mathbf{U} - \mathbf{u}\| \leq \delta^2 \|A^{-1}\| \cdot \|\boldsymbol{\tau}\|. \tag{20}$$

We note from (17b), that $A = A^T$ and thus A^{-1} is symmetric. If we use the Euclidean norm in (20), i.e., for any vector $\mathbf{v}$,

$$\|\mathbf{v}\|_2 = \sqrt{\sum_{\nu=1}^{JK} v_\nu^2},$$

then

$$\|A^{-1}\|_2 = \max_{1 \leq \nu \leq JK} (1/|\Lambda_\nu|) = 1/\min_{1 \leq \nu \leq JK} |\Lambda_\nu|$$

where the Λ_ν are the eigenvalues of A. The eigenvalues of A satisfy

$$A\mathbf{W} = \Lambda \mathbf{W}. \tag{21}$$

However, we see that this is equivalent to the *finite difference eigenvalue problem*

$$-\Delta_\delta W(x, y) = \frac{\Lambda}{\delta^2} W(x, y), \qquad (x, y) \text{ in } D_\delta \tag{22a}$$

$$W(x, y) = 0, \qquad (x, y) \text{ in } C_\delta, \tag{22b}$$

since multiplication of (22a) by δ^2 yields (21).

We determine the eigenvalues of problem (21) by using the technique called *separation of variables* for (22). Let us try to find a solution of the form $W(x, y) = \phi(x)\psi(y)$ of (22a), i.e.,

$$-\Delta_\delta \phi(x)\psi(y) = -\phi_{x\bar{x}}\psi(y) - \phi(x)\psi_{y\bar{y}} = \frac{\Lambda}{\delta^2} \phi(x)\psi(y).$$

Now divide by $W(x, y)$ to get

$$-\frac{\phi_{x\bar{x}}}{\phi(x)} - \frac{\psi_{y\bar{y}}}{\psi(y)} = \frac{\Lambda}{\delta^2}, \qquad (x, y) \text{ in } D_\delta.$$

But the only way that the sum of a function of x and a function of y can be constant is for each function to be a constant. Hence we may write $\Lambda = \xi + \eta$ and have the two sets of equations

$$\text{(23a)} \qquad -\phi_{x\bar{x}}(x) = \frac{\xi}{\delta^2}\phi(x)$$

$$(x, y) \text{ in } D_\delta$$

$$\text{(23b)} \qquad -\psi_{y\bar{y}}(y) = \frac{\eta}{\delta^2}\psi(y)$$

If ξ and η are known, (23) would be ordinary difference equations of second order with constant coefficients. We solve them as we did the difference equations in Section 4 of Chapter 8. Thus, let us use the form $\phi(x) = \alpha^x$ in (23a) to get by using (14c),

$$\frac{\alpha^x}{(\delta x)^2}\left[-\alpha^{-\delta x} + \left(2 - \frac{\xi}{\theta_x}\right) - \alpha^{\delta x}\right] = 0, \qquad \delta x \le x \le a - \delta x.$$

If we set $\omega = \alpha^{\delta x}$, then these equations are satisfied provided

$$\frac{\xi}{\theta_x} = 2 - \omega - \omega^{-1}.$$

Furthermore, it is clear that $\phi(x) = \alpha^{-x}$ yields the same condition, and hence the general solution of (23a) is of the form

$$\phi(x_j) = c\alpha^{x_j} + d\alpha^{-x_j} = c\omega^j + d\omega^{-j}.$$

To satisfy the boundary conditions (22b), we have $\phi(x)\psi(y) = 0$ for (x, y) in C_δ. This implies that

$$\text{(24)} \qquad \phi(0) = \phi(x_0) = 0; \qquad \phi(a) = \phi(x_{J+1}) = 0.$$

From the condition (24) at $j = 0$, we have $c = -d$; hence at $j = J + 1$,

$$\omega^{2(J+1)} = 1.$$

The $2(J + 1)$ roots of this equation are the roots of unity

$$\omega_p = e^{i[p\pi/(J+1)]}, \qquad p = 1, 2, \ldots, 2(J + 1).$$

However, if we replace ω by ω^{-1}, the solution $\phi(x)$ of the difference equation becomes $-\phi(x)$. Hence, we need consider only the first $J + 1$ such roots. But the $(J + 1)$st root is $\omega = -1$ which leads to the trivial solution, $\phi \equiv 0$. Thus we have found J non-trivial solutions of (23a) which satisfy (24) and they are

$$\text{(25a)} \qquad \phi_p(x_j) = c(\omega_p{}^j - \omega_p{}^{-j}) = \sin\left(j\frac{p\pi}{J + 1}\right)$$

$$p = 1, 2, \ldots, J.$$

$$\text{(25b)} \qquad \xi_p = 2\theta_x\left(1 - \cos\frac{p\pi}{J + 1}\right) = 4\theta_x \sin^2\left(\frac{\pi}{2}\frac{p}{J + 1}\right)$$

Here we have chosen the arbitrary normalization constant of $\phi(x)$ to be $c = -i/2$. In an exactly analogous manner, we find K non-trivial solutions of (23b) which satisfy $\psi(0) = \psi(b) = 0$;

$$\psi_q(y_k) = \sin\left(k\frac{q\pi}{K+1}\right) \tag{26a}$$

$$q = 1, 2, \ldots, K.$$

$$\eta_q = 4\theta_y \sin^2\left(\frac{\pi}{2}\frac{q}{K+1}\right) \tag{26b}$$

By combining these results, we find the solutions of the eigenvalue problem (22)

$$\begin{aligned} W_{p,q}(x, y) &= \phi_p(x)\psi_q(y) \\ \Lambda_{p,q} &= \xi_p + \eta_q \end{aligned} \qquad 1 \le p \le J,\ 1 \le q \le K. \tag{27}$$

We have thus found JK different eigenfunctions $W_{p,q}(x, y)$, with corresponding eigenvalues $\Lambda_{p,q}$ (which may not all be distinct). In the vector representation of the net functions $W_{p,q}(x, y)$, we have JK distinct eigenvectors, $\mathbf{W}_{p,q}$, of the matrix A in (21). [In fact, it can be shown that the JK eigenvectors in (27) are orthogonal.] Hence, all of the eigenvalues of A are in the set $\Lambda_{p,q}$. We observe that all eigenvalues of A are positive and A is not only non-singular, but is also *positive definite*.

The norm of A^{-1} is now found to be

$$\begin{aligned} \|A^{-1}\|_2 &= [\min_{p,q}(\xi_p + \eta_q)]^{-1} = \frac{1}{\xi_1 + \eta_1} \\ &= \left[4\theta_x \sin^2\left(\frac{\pi}{2a}\delta x\right) + 4\theta_y \sin^2\left(\frac{\pi}{2b}\delta y\right)\right]^{-1} \\ &= \frac{1}{\delta^2\pi^2}\left(\frac{a^2b^2}{a^2 + b^2}\right)\{1 + \mathcal{O}[(\delta x)^2 + (\delta y)^2]\}. \end{aligned}$$

Thus the error estimate in (20) becomes in this norm,

$$\|\mathbf{U} - \mathbf{u}\|_2 \le \frac{a^2b^2}{\pi^2(a^2 + b^2)}\|\boldsymbol{\tau}\|_2 \cdot \{1 + \mathcal{O}[(\delta x)^2 + (\delta y)^2]\}. \tag{28}$$

This bound is similar to that in (9) but it must be recalled that the norms are different. We have presented here a convergence proof which is independent of the maximum principle. There are still other proofs that could have been given. In particular, if we were to consider the problem (1) with $f(x, y) \equiv 0$ on D, then the solution could easily be written in terms of Fourier series [assuming $g(x, y)$ to have piecewise continuous derivatives on C]. The solution of the corresponding difference problem

(5), with $f(x, y) \equiv 0$, can also be given in terms of (finite) Fourier series. A comparison of these explicit solutions would then show convergence as δx and δy vanish, and the rate of convergence would depend upon the smoothness properties of the boundary data, $g(x, y)$. Of course, the determination of the explicit solutions used in the calculation of $\|A^{-1}\|_2$ cannot be made in most of the applications of the present difference method. In particular, if the domain is not composed of coordinate lines and/or if the equation is replaced by one with variable coefficients, then these special methods must be modified to give analogous results. However, the maximum principle is readily extended to include many such applications. Often it may be possible to obtain a *bound* on $\|A^{-1}\|$ (in some norm) without having to determine the eigenvalues of A.

1.2. An Eigenvalue Problem for the Laplacian Operator

In view of the development of the previous subsection, we can readily find approximations to the *eigenfunctions*, $u(x, y)$, and *eigenvalues*, λ, of the Laplacian operator for a rectangular region. The eigenfunction is not identically zero, i.e., $u \not\equiv 0$, and for some constant, λ, (the eigenvalue) satisfies

$$\text{(29a)} \qquad -\Delta u = \lambda u, \qquad (x, y) \text{ in } D,$$

$$\text{(29b)} \qquad u = 0, \qquad (x, y) \text{ in } C.$$

We can solve this continuous problem by the *separation of variables* technique. Thus we set

$$u = f(x)g(y),$$

whence from (29a)

$$-\frac{f''}{f} - \frac{g''}{g} = \lambda,$$

while from (29b)

$$f(0) = f(a) = g(0) = g(b) = 0.$$

But since λ is a constant, we find that

$$-\frac{f''}{f} = \text{constant}, \qquad 0 \leq x \leq a,$$

$$-\frac{g''}{g} = \text{constant}, \qquad 0 \leq y \leq b.$$

The only possible *non-trivial solutions* of these differential equations and boundary conditions are proportional to

$$\text{(30a)} \qquad f_m(x) = \sin\left(m\pi\frac{x}{a}\right)$$

$$\text{(30b)} \qquad g_n(y) = \sin\left(n\pi\frac{y}{b}\right).$$

Hence the eigenfunctions and eigenvalues of (29) are

(31a) $$u_{m,n}(x, y) = \sin\left(m\pi\frac{x}{a}\right)\sin\left(n\pi\frac{y}{b}\right),$$

(31b) $$\lambda_{m,n} = \pi^2\left(\frac{m^2}{a^2} + \frac{n^2}{b^2}\right); \qquad m, n = 1, 2, \ldots.$$

[It can be shown that these are all of the independent eigenfunctions and eigenvalues of (29).] We now exhibit the eigenfunctions U and eigenvalues μ of the approximating difference equations defined by $U \not\equiv 0$,

(32a) $$-\Delta_\delta U = \mu U, \qquad (x, y) \text{ in } D_\delta$$

(32b) $$U = 0, \qquad (x, y) \text{ in } C_\delta.$$

That is, from (27), (26), and (25), with $\mu = \Lambda/\delta^2$,

(33a) $$U_{p,q}(x_j, y_k) = \sin\left(j\frac{p\pi}{J+1}\right)\sin\left(k\frac{q\pi}{K+1}\right),$$

(33b) $$\mu_{p,q} = 4\left[\frac{(J+1)^2\sin^2\left(\frac{\pi}{2}\frac{p}{J+1}\right)}{a^2} + \frac{(K+1)^2\sin^2\left(\frac{\pi}{2}\frac{q}{K+1}\right)}{b^2}\right],$$

$$1 \le p \le J, \; 1 \le q \le K.$$

From (31a) and (33a), we note that

(34a) $$u_{p,q}(x_j, y_k) = U_{p,q}(x_j, y_k).$$

This is an exceptional coincidence! On the other hand, if we use (31b) and expand (33b) for fixed (p, q) and large (J, K), we find

(34b) $$\mu_{p,q} - \lambda_{p,q} = \mathcal{O}[p^4(\delta x)^2 + q^4(\delta y)^2].$$

Equation (34b) expresses the fact, also valid in more general problems, that the lowest eigenvalues of the difference operator approximate the respective lowest eigenvalues of the differential operator with an error proportional to the square of the mesh width. Frequently the error in the approximation of the corresponding eigenfunctions is also proportional to the square of the mesh width.

In most cases, where the eigenvalues of the differential operator obey a *variational principle*, the practical problem of determining the eigenvalues of the difference operator is made simpler by characterizing them as the *stationary values* of some functional.

For example, in the case of (29) the eigenvalues are the stationary values, $\lambda = G[u]$, of

$$G[u] \equiv \frac{\int_D \int \left[\left(\frac{\partial u}{\partial x} \right)^2 + \left(\frac{\partial u}{\partial y} \right)^2 \right] dx\, dy}{\int_D \int u^2\, dx\, dy}, \tag{35}$$

where $u(x, y)$ ranges over the class $\mathscr{G}$ of non-trivial functions with continuous first derivatives and such that $u \equiv 0$ on C. We say $G[u]$ is *stationary* at u, if

$$\frac{d}{d\epsilon} G[u + \epsilon v] = 0 \qquad \text{at } \epsilon = 0,$$

for all functions v in $\mathscr{G}$. It can be shown that if $\lambda = G[u]$ is stationary at u, then u has continuous second derivatives and satisfies (29). On the other hand, the corresponding functional that characterizes the eigenvalues of the difference operator in (32) is

$$H[U] \equiv \frac{Q[U]}{L[U]} \equiv \frac{\frac{1}{2} \sum [(U_x)^2 + (U_{\bar{x}})^2 + (U_y)^2 + (U_{\bar{y}})^2]}{\sum U^2}. \tag{36}$$

The sums in (36) are taken over all net points of the infinite lattice that covers the plane and U is in the class $\mathscr{F}$ of non-trivial net functions which satisfy

$$U(x, y) = 0 \quad \text{for} \quad (x, y) \text{ not in } D_\delta.$$

THEOREM 4. *$\mu = H[U]$ is stationary at U iff μ and U are an eigenvalue and eigenfunction that satisfy* (32).

Proof. Let

$$\begin{aligned} V(x_0, y_0) &= 1, \\ V(x, y) &= 0 \qquad \text{if } (x, y) \neq (x_0, y_0). \end{aligned} \tag{37}$$

It is easy to calculate

$$\left. \frac{d}{d\epsilon} H[U + \epsilon V] \right|_{\epsilon = 0}$$

by expanding the numerator and denominator of H to first order in ϵ. That is,

$$\begin{aligned} Q[U + \epsilon V] &\cong Q[U] + \epsilon \sum U_x V_x + U_{\bar{x}} V_{\bar{x}} + U_y V_y + U_{\bar{y}} V_{\bar{y}} \\ &= Q[U] - 2\epsilon \Delta_\delta U(x_0, y_0); \\ L[U + \epsilon V] &\cong L[U] + 2\epsilon \sum UV \\ &= L[U] + 2\epsilon U(x_0, y_0). \end{aligned}$$

Hence,

$$(38) \qquad H[U + \epsilon V] \cong H[U] - \frac{2\epsilon}{L[U]} [\Delta_\delta U(x_0, y_0) + \mu U(x_0, y_0)].$$

Therefore, if $H[U]$ is stationary, (32) holds, since we may pick (x_0, y_0) to be any point in D_δ. On the other hand, in Problem 5 it is shown that (32) implies $H[U]$ is stationary. ∎

We remark that the variational principles can be used as a basis for constructing methods to determine the eigenfunctions and eigenvalues as in the Rayleigh-Ritz methods, which we do not treat. Another application of the related functionals (e.g., $H[u]$ and $G[u]$) is to determine estimates for $\mu_{p,q} - \lambda_{p,q}$ in more general cases. But we cannot pursue this topic further.

The eigenvalue problem (32) corresponds to the matrix eigenvalue problem, similar to (21),

$$A\mathbf{U} = \mu\delta^2\mathbf{U},$$

where A is symmetric. Hence we may use the argument of Theorem 7.2 of Chapter 8 to prove a result analogous to (7.51) of Chapter 8.

PROBLEMS, SECTION 1

In Problems 1, 2, and 3 we indicate how to generalize Theorems 1, 2, and 3 for a non-rectangular region. For example, let

$$D \equiv \{(x, y) \mid x^2 + y^2 < a^2\}, \qquad C \equiv \{(x, y) \mid x^2 + y^2 = a^2\};$$

in the notation of Theorem 1, let P be any lattice point and define

$$D_\delta \equiv \{P \mid P, P_1, P_2, P_3, P_4 \in D\}.$$

Now, if $P \in D$ but $P \notin D_\delta$, we set $P \in C_\delta$ and note that at least one pair of its opposite neighbors is separated by C, say

$$P_1 \notin D, \qquad P_2 \in D.$$

Let $P_C \in C$ be on the line segment PP_1; let $\theta \equiv$ distance PP_C, therefore, $0 < \theta \leq \delta x$. Define $U(P_C) \equiv u(P_C)$ for any point P_C on C.

1. *Maximum Principle*: In the above notation, for $P \in D$ but $P \notin D_\delta$, define the *linear interpolation operator*

$$B_\delta U(P) \equiv \frac{\theta U(P_2) + \delta x U(P_C)}{\delta x + \theta} - U(P).$$

Show that

(a) If

$$\Delta_\delta U(P) \geq 0, \qquad \text{for } P \in D_\delta,$$

and

$$B_\delta U(P) \geq 0, \qquad \text{for } P \in C_\delta,$$

then

$$\max_{P \in D} U(P) \leq \max_{P_C \in C} U(P_C).$$

(b) If

$$\Delta_\delta U(P) \le 0, \qquad \text{for } P \in D_\delta;$$
$$B_\delta U(P) \le 0, \qquad \text{for } P \in C_\delta,$$

then

$$\min_{P \in D} U(P) \ge \min_{P_C \in C} U(P_C).$$

(c) The equations

$$-\Delta_\delta U(P) = f(P), \qquad P \in D_\delta,$$
$$B_\delta U(P) = g(P), \qquad P \in C_\delta,$$

have a unique solution.

2. With the *linear interpolation operator*, (i.e., $B_\delta U(P)$), prove the *a priori estimate*, for lattice points in D, and any lattice function $U(P)$,

$$\max_{P \in D_\delta} |U(P)| \le \max_{P_C \in C} |U(P_C)| + \frac{a^2}{2} K,$$

where

$$K = \max \left[\max_{P \in D_\delta} |\Delta_\delta U(P)|, \qquad \max_{P \in C_\delta} |B_\delta U(P)| \right].$$

3. Derive a bound for the error, $E = U - u$, when U is found with rounding errors ρ, ρ_1 that satisfy

$$-\Delta_\delta U(P) = f(P) + \frac{\rho}{\delta x\, \delta y}, \qquad P \in D_\delta$$
$$B_\delta U(P) = \rho_1, \qquad P \in C_\delta.$$

If u has continuous derivatives of fourth order and $\delta x = \mathcal{O}(h)$, $\delta y = \mathcal{O}(h)$, show that

$$\max_{P \in D} |E(P)| = \mathcal{O}(h^2),$$

for sufficiently small ρ, ρ_1.

[Hint: Define the *truncation error* as in (7) for $P \in D_\delta$. Otherwise, set

$$\tau\{u(P)\} \equiv B_\delta u(P), \qquad P \in C_\delta.$$

Apply the a priori estimate to $E(P)$, for $P \in D$.]

4. Show how the statements of the maximum principle, the a priori estimate, and the error bound must be modified for a more general bounded domain D.

5.* With U and V in the class $\mathcal{F}$ of Theorem 4, show that

$$\sum U_x V_x + U_{\bar{x}} V_{\bar{x}} + U_y V_y + U_y V_y = -2 \sum V \Delta_\delta U.$$

Hence

$$H[U + \epsilon V] \cong H[U] - \frac{2\epsilon}{L[U]} \sum V(\Delta_\delta U + \mu U).$$

Therefore if U satisfies (32), $H[U]$ is stationary.

[Hint: Use summation by parts to remove the difference quotients of V. For example, in $\sum U_x V_x$, the value $V(P)$ for a fixed $P \in D_\delta$ occurs only in $V_x(P_2)$ and $V_x(P)$. Thus in this sum the coefficient of $V(P)$ is found to be:

$$-[U(P_2) - 2U(P) + U(P_1)]/\delta x^2.]$$

2. SOLUTION OF LAPLACE DIFFERENCE EQUATIONS

The linear algebraic system of equations determined by the difference scheme (1.5) is, for the rectangular region, of order JK. For small net spacings δx and δy, this may be extremely large since $JK = \text{constant}/(\delta x\, \delta y)$. (In practice, $JK > 2500$ is not at all unusual.) Thus the standard elimination procedures for the equivalent system (1.18) of order JK require on the order of $(JK)^3$ operations for solution and are too inefficient. Now from the definition (1.17b) of A, we see that many of its elements are zero and in fact, that it is block tridiagonal. The Gaussian elimination procedures which take account of large blocks of zero elements (in particular, the methods of Subsection 3.3 in Chapter 2) are then naturally suggested. This block elimination method requires at most on the order of J^3K operations (for rectangular regions) and is efficiently carried out on modern digital computers. (The storage requirements are for $2K - 1$ matrices of order J and one vector of order JK. But this data is used only in dealing with systems of order J and hence is not all required at the same time.) In fact, since only tridiagonal systems need to be solved, efficient organization requires only $\mathcal{O}(J^2K)$ operations!

Nevertheless, *iterative methods* seem to be the ones most often employed to solve the Laplace difference equations. Again, the large number of zero elements in the coefficient matrix greatly reduces the computational effort required in each iteration. However, some care must be taken to insure that sufficient accuracy will be obtained in a "reasonable" number of iterations. We consider such methods for the rectangular region.

The simplest iteration method begins with an initial estimate of the solution, say $U^{(0)}$, and then defines the sequence of net functions $U^{(\nu)}$ by

$$\text{(1a)} \qquad U^{(\nu+1)}(x, y) = U^{(\nu)}(x, y) + \delta^2\Delta_\delta U^{(\nu)}(x, y) + \delta^2 f(x, y), \qquad (x, y) \text{ in } D_\delta,$$

$$\text{(1b)} \qquad U^{(\nu+1)}(x, y) = g(x, y), \qquad (x, y) \text{ in } C_\delta, \quad \nu = 0, 1, \ldots.$$

Here δ^2 is defined in (1.14c) and the boundary condition (1.1b) is to be satisfied by $U^{(0)}$. From the other definitions in (1.14c) and (1.4), we find that (1a) is, in subscript notation,

$$\text{(2)} \qquad U_{j,k}^{(\nu+1)} = \theta_x(U_{j-1,k}^{(\nu)} + U_{j+1,k}^{(\nu)}) + \theta_y(U_{j,k-1}^{(\nu)} + U_{j,k+1}^{(\nu)}) + \delta^2 f_{j,k}, \qquad 1 \le j \le J,\ 1 \le k \le K.$$

[Note the relation between (2) and (1.15).] The calculations required in (1) or equivalently (2) can be carried out in any order on the net.

This iteration scheme is easily written in matrix form by using the notation of the previous subsection. We get that

$$\text{(3a)} \qquad \mathbf{U}^{(\nu+1)} = (H + V)\mathbf{U}^{(\nu)} + \delta^2\mathbf{F}, \qquad \nu = 0, 1, \ldots.$$

Here the JK-order matrices are defined by (1.14b) and (1.17b), $\mathbf{F}$ is defined in (1.16b) and (1.17a), and $\mathbf{U}^{(\nu)}$ is the vector with components $U_{j,k}^{(\nu)}$ ordered as in (1.17a). From the definition of A in (1.17b), we see that (3a) can be written as

$$\text{(3b)} \qquad \mathbf{U}^{(\nu+1)} = (I - A)\mathbf{U}^{(\nu)} + \delta^2\mathbf{F},$$

and thus, this scheme applied to (1.18) is a special case of the general iterative methods studied in Section 4 of Chapter 2. In fact, this is just the *Jacobi or simultaneous iteration scheme* of Subsection 4.1 in Chapter 2 applied to the system (1.18).

From the general theory of iterative methods, we know that the necessary and sufficient condition for the convergence of the sequence $\{\mathbf{U}^{(\nu)}\}$ to the solution $\mathbf{U}$ for an arbitrary initial guess $\mathbf{U}^{(0)}$ is that all of the eigenvalues of $(H + V)$ are in magnitude less than unity (see Theorem 4.1 of Chapter 2). The eigenvalues of this matrix are the roots of the characteristic polynomial

$$\text{(4)} \qquad \Psi(\eta) \equiv \det|\eta I - (H + V)| = \det|\eta I - (I - A)|.$$

However, we have determined the eigenvalues of A in the previous subsection; they are given in (1.27). Thus the eigenvalues of $(I - A)$, and hence the roots of $\Psi(\eta) = 0$, are

$$\text{(5a)} \qquad \eta = \eta_{p,q} \equiv 1 - \lambda_{p,q}$$
$$= 1 - 4\theta_x \sin^2\left(\frac{\pi}{2}\frac{p}{J+1}\right) - 4\theta_y \sin^2\left(\frac{\pi}{2}\frac{q}{K+1}\right),$$
$$1 \le p \le J, \; 1 \le q \le K.$$

Now we easily find that $0 < \eta < 1$ and for small δx and δy

$$\text{(5b)} \qquad \rho(H + V) \equiv \max_{p,q} |\eta_{p,q}| = \eta_{1,1} = 1 - \lambda_{1,1}$$
$$= 1 - \delta^2\pi^2\left(\frac{1}{a^2} + \frac{1}{b^2}\right) + \mathcal{O}(\delta^4).$$

Since $0 < \lambda_{1,1} < 1$, the method clearly converges and the rate of convergence is by (4.11) of Chapter 2

$$\text{(6)} \qquad R_J = -\log(1 - \lambda_{1,1})$$
$$= \delta^2\pi^2\left(\frac{1}{a^2} + \frac{1}{b^2}\right) + \mathcal{O}(\delta^4).$$

We find that the rate of convergence decreases with

$$\delta^2 = \frac{\delta x^2 \delta y^2}{2[(\delta x)^2 + (\delta y)^2]},$$

and thus for difference equations with a small net spacing we may expect very slow convergence.

The *Gauss-Seidel or successive iteration* method for the Laplace difference equations can be written as

$$U_{j,k}^{(\nu+1)} = \theta_x(U_{j-1,k}^{(\nu+1)} + U_{j+1,k}^{(\nu)}) + \theta_y(U_{j,k-1}^{(\nu+1)} + U_{j,k+1}^{(\nu)}) + \delta^2 f_{j,k}, \quad 1 \le j \le J,\ 1 \le k \le K. \tag{7a}$$

In matrix form this becomes, with the use of (1.17b),

$$[I - (\theta_x L + \theta_y B)]\mathbf{U}^{(\nu+1)} = (\theta_x L + \theta_y B)^T\mathbf{U}^{(\nu)} + \delta^2\mathbf{F}. \tag{7b}$$

In the present application this iteration scheme is frequently called the *Liebmann method.* The new iterates cannot be evaluated in a completely arbitrary order in this method. We first compute $U_{1,1}^{(\nu+1)}$ and then, in order, the other elements on the coordinate lines with $j = 1$ and $k = 1$. Next $U_{2,2}^{(\nu+1)}$ is determined, etc. By slight changes in the scheme we could start the calculations at either of the other three "corners" in D_δ. However, as we shall see, all of these methods have the same rate of convergence. This successive scheme is easier to employ on a digital computer than the simultaneous scheme since now each new component can immediately replace the previous value in storage. In addition, we shall find that the Gauss-Seidel method converges exactly twice as fast as the Jacobi method (when they are used on the same problem) and thus one should *never use the Jacobi method* on such difference problems.

The convergence of the iteration method (7) is determined by the magnitude of the eigenvalues of the matrix

$$S_1 \equiv [I - (\theta_x L + \theta_y B)]^{-1}(\theta_x L + \theta_y B)^T. \tag{8}$$

The indicated inverse exists since $\theta_x L + \theta_y B$ is a strictly lower triangular matrix. Thus the eigenvalues, ξ, of the matrix S_1 are the roots of the characteristic polynomial

$$\begin{aligned}\Phi_1(\xi) &\equiv \det |\xi[I - (\theta_x L + \theta_y B)] - (\theta_x L + \theta_y B)^T| \\ &= \det |\xi I - \theta_x(\xi L + L^T) - \theta_y(\xi B + B^T)|.\end{aligned} \tag{9}$$

To examine the roots of this polynomial we shall use the following

THEOREM **1.** *Let the matrices L, B, and A be defined as in* (1.14b) *and* (1.17b). *Then for any non-zero scalars* α *and* β

$$\det |A| = \det |I - \theta_x(\alpha L + \alpha^{-1}L^T) - \theta_y(\beta B + \beta^{-1}B^T)|. \tag{10}$$

Proof. Let the elements of A be $a_{r,s}$ where $r, s = 1, 2, \ldots, N = JK$. Then each term in the formal expansion of det $|A|$ is given by a product of the form

$$\pm a_{1,\pi(1)}a_{2,\pi(2)}\cdots a_{r,\pi(r)}\cdots a_{N,\pi(N)}.$$

Here π is one of the $N!$ permutations of the first N integers. Let each point (x_j, y_k) of the net D_δ be identified with a unique integer (see Figure 1)

$$(x_j, y_k) \leftrightarrow r \equiv j + (k - 1)J.$$

Then any given permutation π can be represented by N vectors on D_δ by drawing lines from r to $\pi(r)$ for $1 \le r \le N$ [i.e., from the point corresponding to r to the point corresponding to $\pi(r)$]. Now by the definition of the matrix A it follows that $a_{r,\pi(r)} \neq 0$ only if $r = \pi(r)$ or the point corresponding to $\pi(r)$ is one of the four neighboring net points, $(x \pm \delta x, y)$ or $(x, y \pm \delta y)$, in the star about the point (x, y) corresponding to r. Thus, the only terms in the expansion of det $|A|$ which may not vanish correspond to permutations whose geometric representation is composed entirely of unit vectors in the $(\pm x)$ and $(\pm y)$ directions and null vectors.

Now every permutation is a product of disjoint cycles and in the above representation a cycle is a closed path of vectors on D_δ (see Figure 1). Thus for any cycle corresponding to a non-vanishing product of elements, there must be the same number of unit vectors in the $(+x)$ direction as in the $(-x)$ direction and similarly for the $(\pm y)$ directions. Now we recall that $a_{r,\pi(r)}$ is an element of L if $\pi(r) = r - 1$ and is an element of L^T if $\pi(r) = r + 1$. Thus there are as many factors from L as from L^T in any non-

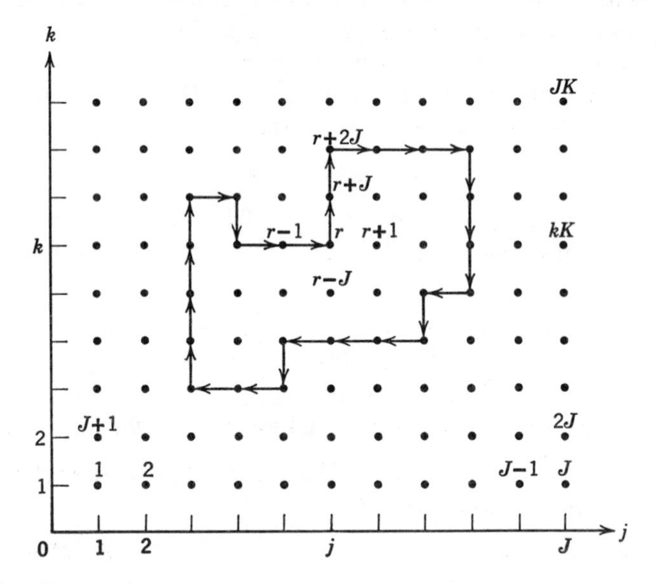

Figure 1. Geometric representation of a non-vanishing cycle. In the permutation $n \to \pi(n)$, of which this cycle is a factor, $r = j + (k - 1)J$ and the cycle is given by:

$$\pi(r) = r + J, \pi(r + J) = r + 2J, \ldots, \pi(r - 1) = r.$$

vanishing term in the expansion. Similarly, $a_{r,\pi(r)}$ is an element of B if $\pi(r) = r - J$ and of B^T if $\pi(r) = r + J$. Thus factors from B and B^T also enter pairwise in any non-vanishing term.

These results hold for the expansion of the right-hand determinant in (10) if elements from L, L^T, B, and B^T are replaced by those of αL, $\alpha^{-1}L^T$, βB, and $\beta^{-1}B^T$ respectively. Thus in any non-vanishing term the scalars α and β do not appear and the proof is concluded. ■

We note that the proof of Theorem 1 depends *only upon the location of the zero elements in the matrix A*. Hence if the non-zero elements of A are changed in any manner we have the

COROLLARY. *If μ, ξ, η, ζ, and ν are matrices with zero components wherever I, L, L^T, B, and B^T respectively have zero components, then*

$$\det|\mu - (\alpha\xi + \alpha^{-1}\eta) - (\beta\zeta + \beta^{-1}\nu)| = \det|\mu - (\xi + \eta + \zeta + \nu)|. \quad ■$$

In particular, we now consider the determinant $\Phi(\xi)$ in (9). The identity I and the matrices L and B have been multiplied by a scalar ξ, so no zero elements of A have been altered. Thus we may apply the corollary to get

$$\Phi_1(\xi) = \det|\xi I - \theta_x(\alpha\xi L + \alpha^{-1}L^T) - \theta_y(\beta\xi B + \beta^{-1}B^T)|.$$

Take $\alpha = \beta = \xi^{-1/2}$ and recall (4) to find that

$$\begin{aligned}\Phi_1(\xi) &= \det|\xi^{1/2}I|\cdot\det|\xi^{1/2}I - (H + V)| \\ &= \xi^{JK/2}\Psi(\xi^{1/2}).\end{aligned}$$

Thus every non-zero root ξ of $\Phi_1(\xi) = 0$ satisfies $\Psi(\xi^{1/2}) = 0$ and every root η of $\Psi(\eta) = 0$ satisfies $\Phi_1(\eta^2) = 0$. So all non-zero eigenvalues of the matrix S_1 in (8) are given by

$$\xi = \eta^2 = (1 - \lambda_{p,q})^2, \qquad 1 \le p \le J,\ 1 \le q \le K,$$

and from (5) we find that the maximum eigenvalue of S_1 is

$$\begin{aligned}\rho(S_1) &= [\rho(H + V)]^2 = (1 - \lambda_{1,1})^2 \\ &= 1 - 2\delta^2\pi^2\left(\frac{1}{a^2} + \frac{1}{b^2}\right) + \mathcal{O}(\delta^4).\end{aligned}$$

The *rate of convergence for the Gauss-Seidel scheme is thus*

$$R_{GS} = 2\delta^2\pi^2\left(\frac{1}{a^2} + \frac{1}{b^2}\right) + \mathcal{O}(\delta^4), \tag{12}$$

or *twice that for the Jacobi scheme.*

The convergence rate of the Gauss-Seidel method (7) may be improved by introducing an appropriate *acceleration parameter*, as discussed in Section 5 of Chapter 2. That is, set

(13a) $$V_{j,k}^{(\nu+1)} = \theta_x(U_{j-1,k}^{(\nu+1)} + U_{j+1,k}^{(\nu)}) + \theta_y(U_{j,k-1}^{(\nu+1)} + U_{j,k+1}^{(\nu)}) + \delta^2 f_{j,k}$$

and then, at the point (x_j, y_k), take

(13b) $$\begin{aligned} U_{j,k}^{(\nu+1)} &= \omega V_{j,k}^{(\nu+1)} + (1 - \omega)U_{j,k}^{(\nu)} \\ &= U_{j,k}^{(\nu)} + \omega(V_{j,k}^{(\nu+1)} - U_{j,k}^{(\nu)}). \end{aligned}$$

Here ω is the acceleration parameter to be determined. We note that for $\omega = 1$ this scheme reduces to that in (7a), i.e., to the ordinary Gauss-Seidel method. The order in which the components of the new iterates are to be computed is just as in the previous successive scheme.

To examine the convergence of the accelerated Gauss-Seidel method we first write it in matrix form. Obviously (13a) implies

(14a) $$\mathbf{V}^{(\nu+1)} = (\theta_x L + \theta_y B)\mathbf{U}^{(\nu+1)} + (\theta_x L + \theta_y B)^T\mathbf{U}^{(\nu)} + \delta^2\mathbf{F},$$

and (13b) implies

(14b) $$\mathbf{U}^{(\nu+1)} = \omega\mathbf{V}^{(\nu+1)} + (1 - \omega)\mathbf{U}^{(\nu)}.$$

Upon eliminating $\mathbf{V}^{(\nu+1)}$, we obtain

(15) $$\begin{aligned} &[I - \omega(\theta_x L + \theta_y B)]\mathbf{U}^{(\nu+1)} \\ &\qquad = [(1 - \omega)I + \omega(\theta_x L + \theta_y B)^T]\mathbf{U}^{(\nu)} + \omega\delta^2\mathbf{F}. \end{aligned}$$

The convergence of these iterations is thus determined by the magnitude of the eigenvalues of the matrix

(16) $$S_\omega \equiv [I - \omega(\theta_x L + \theta_y B)]^{-1}[(1 - \omega)I + \omega(\theta_x L + \theta_y B)^T].$$

Note that for $\omega = 1$ the above matrix reduces to the S_1 defined in (8) for the ordinary unaccelerated successive iterations. The eigenvalues of S_ω are the roots ζ of the characteristic polynomial

(17) $$\begin{aligned} \Phi_\omega(\zeta) &\equiv \det |[I - \omega(\theta_x L + \theta_y B)]\zeta \\ &\qquad - [(1 - \omega)I + \omega(\theta_x L + \theta_y B)^T]| \\ &= \det |(\zeta + \omega - 1)I - \omega\zeta(\theta_x L + \theta_y B) \\ &\qquad - \omega(\theta_x L + \theta_y B)^T|. \end{aligned}$$

The matrix in (17) has zero elements wherever the matrix A has them and so the corollary to Theorem 1 is applicable. If we use the scalars $\alpha = \beta = \zeta^{-1/2}$ then we obtain from (17) and (4)

$$\begin{aligned} \Phi_\omega(\zeta) &= \det |(\zeta + \omega - 1)I - \omega\zeta^{1/2}(\theta_x L + \theta_y B) - \omega\zeta^{1/2}(\theta_x L + \theta_y B)^T| \\ &= \det |\omega\zeta^{1/2}I| \cdot \det \left|\frac{(\zeta + \omega - 1)}{\omega\zeta^{1/2}} I - (H + V)\right| \\ &= (\omega\zeta^{1/2})^{JK}\Psi\left(\frac{\zeta + \omega - 1}{\omega\zeta^{1/2}}\right). \end{aligned}$$

From this result we conclude, for each $\omega \neq 0$, that any non-zero root ζ of $\Phi_\omega(\zeta) = 0$ satisfies $\Psi(\eta) = 0$, and that every root η of $\Psi(\eta) = 0$ satisfies $\Phi_\omega(\zeta) = 0$ provided $(\zeta + \omega - 1)/(\omega\zeta^{1/2}) = \eta$. Thus the non-zero eigenvalues of the matrix S_ω are the roots ζ of

$$\zeta + \omega - 1 = \omega\zeta^{1/2}\eta \tag{18a}$$

where η ranges over the roots of $\Psi(\eta) = 0$ [i.e., the eigenvalues of $I - A$ given in (5a)]. Since (18a) is quadratic in $\zeta^{1/2}$, we find that all ζ which satisfy this equation are given by

$$\zeta = \zeta_\pm = \left[\left(\frac{\omega\eta}{2}\right) \pm \sqrt{\left(\frac{\omega\eta}{2}\right)^2 + (1 - \omega)}\,\right]^2. \tag{18b}$$

We may now determine ω such that the iteration scheme (15) converges. First observe that since η is real it follows from (18b) that $|\zeta_-| \geq 1$ for $\omega \leq 0$. Thus an eigenvalue of S_ω will have magnitude larger than unity and we conclude that the accelerated Gauss-Seidel method is not convergent for any non-positive ω. For fixed $\omega > 0$ we see that some eigenvalues may be complex (only if $\omega > 1$) but then their magnitude is

$$|\zeta| = \omega - 1. \tag{19a}$$

For the real eigenvalues it follows from (18b) with $\omega > 0$ and $\eta > 0$ that ζ_+ is an increasing function of η and that $|\zeta_+| > |\zeta_-|$. Thus the largest real eigenvalue of S_ω is, since $\eta \leq \eta_{1,1}$,

$$\zeta = \zeta_1(\omega) \equiv \left[\frac{\omega\eta_{1,1}}{2} + \sqrt{\left(\frac{\omega\eta_{1,1}}{2}\right)^2 + (1 - \omega)}\,\right]^2. \tag{19b}$$

From (19) we obtain for $\omega > 0$

$$\rho(S_\omega) = \max\,[\omega - 1,\ \zeta_1(\omega)].$$

As $0 < \eta_{1,1} < 1$ it follows that $\rho(S_\omega) < 1$ if $0 < \omega < 2$ since in this interval, when ζ_1 is real,

$$\begin{aligned}\zeta_1(\omega) &= \left[\left(\frac{\omega\eta_{1,1}}{2}\right) + \sqrt{\left(1 - \frac{\omega}{2}\right)^2 - \left(\frac{\omega}{2}\right)^2(1 - \eta_{1,1}^2)}\,\right]^2, \\ &< \left[\left(\frac{\omega}{2}\right) + \sqrt{\left(1 - \frac{\omega}{2}\right)^2}\,\right]^2 = 1.\end{aligned}$$

On the other hand, if $\omega \geq 2$ then $\zeta_1(\omega)$ is complex, and by (19a) some eigenvalue has modulus not less than unity. Thus we have

THEOREM 2. *The accelerated Gauss-Seidel iterations converge iff the acceleration parameter ω lies in the interval* $0 < \omega < 2$. ∎

The optimal value for the acceleration parameter is that value $\omega = \omega^*$, in $0 < \omega < 2$, for which

$$\rho(S_{\omega*}) = \min_{0<\omega<2} \rho(S_\omega) = \min_{0<\omega<2} \{\max [\omega - 1, \zeta_1(\omega)]\}.$$

[We know that the indicated minimum exists since $\rho(S_\omega)$ is continuous in $0 \leq \omega \leq 2$ and satisfies $\rho(S_0) = \rho(S_2) = 1$, $\rho(S_\omega) < 1$ in $0 < \omega < 2$.] It is clear that the expression in the radical of (19b) is a decreasing function of ω for $0 < \omega < 2$. Thus $\zeta_1(\omega)$ becomes complex when this expression vanishes, i.e., for

$$\omega = \omega_b = \frac{2}{1 + \sqrt{1 - \eta_{1,1}^2}}.$$

For $\omega \geq \omega_b$ we have now $\rho(S_\omega) = \omega - 1$ and

$$\min_{\omega_b \leq \omega < 2} \rho(S_\omega) = \omega_b - 1.$$

For $0 < \omega \leq \omega_b$, since $\zeta_1'(\omega) < 0$, $\zeta_1(\omega)$ is a decreasing function of ω. Hence $\rho(S_{\omega*})$ occurs for $\omega^* = \omega_b$ at which $\omega_b - 1 = \zeta_1(\omega_b) = \rho(S_{\omega*})$. Thus, in summary, we have for the *optimal application of the accelerated Gauss-Seidel method*

(20a)
$$\omega^* = \frac{2}{1 + \sqrt{1 - \eta_{1,1}^2}},$$

(20b)
$$\rho(S_{\omega*}) = \omega^* - 1 = \frac{1 - \sqrt{1 - \eta_{1,1}^2}}{1 + \sqrt{1 - \eta_{1,1}^2}}.$$

From (5), we have

$$\rho(S_{\omega*}) = 1 - 2\delta\pi \sqrt{2\left(\frac{1}{a^2} + \frac{1}{b^2}\right)} + \mathcal{O}(\delta^2)$$

and so the rate of convergence is now

(21)
$$R_{\text{AGS}} = 2\delta\pi \sqrt{2\left(\frac{1}{a^2} + \frac{1}{b^2}\right)} + \mathcal{O}(\delta^2).$$

By comparing (21) with (6) and (12), we see that the power of δ in the rate of convergence for the optimal accelerated Gauss-Seidel method is lower than the power of δ appearing in the ordinary Gauss-Seidel or Jacobi methods. The same result is obtained if the iterations were to proceed in one of the other orders indicated after (7b). This is suggested by the form of our results in which the coordinate directions and related dimensions enter symmetrically (see, however, the discussion at the end of the next subsection).

2.1. Line or Block Iterations

Since the linear system (1.18), which we are solving iteratively, has the simple block structure indicated in (1.16a) it is rather natural to consider corresponding block iterations (i.e., Subsection 4.3 of Chapter 2). In the present application, these are more properly called "line" methods since the net function is altered by changing the data on a complete coordinate line of net points in D_δ simultaneously. A particularly simple line iteration for the system in (1.16) is

$$
\begin{aligned}
&[I_J - \theta_x(L_J + L_J^T)]\mathbf{U}_1^{(\nu+1)} - \theta_y\mathbf{U}_2^{(\nu)} = \delta^2\mathbf{F}_1, \\
\text{(22a)} \qquad & -\theta_y\mathbf{U}_{k-1}^{(\nu)} + [I_J - \theta_x(L_J + L_J^T)]\mathbf{U}_k^{(\nu+1)} - \theta_y\mathbf{U}_{k+1}^{(\nu)} = \delta^2\mathbf{F}_k, \\
&\qquad\qquad 2 \le k \le K-1, \\
& -\theta_y\mathbf{U}_{K-1}^{(\nu)} + [I_J - \theta_x(L_J + L_J^T)]\mathbf{U}_K^{(\nu+1)} = \delta^2\mathbf{F}_K.
\end{aligned}
$$

The K systems for the $\mathbf{U}_k^{(\nu+1)}$ can be solved in any order. At each of the K steps in one of these iterations, a linear system of order J must be solved with the coefficient matrix $I_J - \theta_x(L_J + L_J^T)$. However, this matrix is tridiagonal and can easily be factored by the method of Subsection 3.2 in Chapter 2. This is done only once and then each linear system in the succeeding iterations is solved by evaluating two simple recursions of the forms (3.12) and (3.13) of Chapter 2. The present scheme is frequently called a *line Jacobi method.*

By using the matrices and vectors in (1.17) we can write the iterative scheme (22a) as [compare with (3a)]

$$\text{(22b)} \qquad (I - H)\mathbf{U}^{(\nu+1)} = V\mathbf{U}^{(\nu)} + \delta^2\mathbf{F}.$$

The convergence is thus determined by the matrix

$$\text{(23a)} \qquad (I - H)^{-1}V$$

whose eigenvalues, ρ, are the roots of the characteristic polynomial

$$\text{(23b)} \qquad P(\rho) \equiv \det|\rho I - \rho H - V|.$$

It is not difficult to show that the matrices H and V have common eigenvectors (since they are symmetric and commute). In fact, the eigenvalues and eigenvectors of these matrices are easily computed. Just as the eigenvalue problems (1.21) and (1.22) are equivalent, it follows that the following pairs of eigenvalue problems are also equivalent

$$\text{(24a)} \qquad (2\theta_x I - H)\mathbf{W} = \xi\mathbf{W}, \qquad \begin{cases} -W_{x\bar{x}} = (\xi/\delta^2)W & \text{on } D_\delta, \\ W = 0 & \text{on } C_\delta; \end{cases}$$

$$\text{(24b)} \qquad (2\theta_y I - V)\mathbf{W} = \eta\mathbf{W}, \qquad \begin{cases} -W_{y\bar{y}} = (\eta/\delta^2)W & \text{on } D_\delta, \\ W = 0 & \text{on } C_\delta. \end{cases}$$

[In fact, if we set $\Lambda = \xi + \eta$ and add corresponding equations we obtain (1.21) and (1.22), by using $\theta_x + \theta_y = \frac{1}{2}$.] The problems in (24) may be solved by separating variables and recalling (1.23)–(1.27). We find that these problems have common eigenvectors $\mathbf{W}_{p,q}$ with the components

$$(25) \qquad W_{p,q}(x_j, y_k) = \sin\left(j\frac{p\pi}{J+1}\right)\sin\left(k\frac{q\pi}{K+1}\right), \qquad 1 \le p \le J,\ 1 \le q \le K,$$

and the eigenvalues

$$(26a) \qquad \xi = \xi_p = 4\theta_x \sin^2\left(\frac{\pi}{2}\frac{p}{J+1}\right) = 2\theta_x\left[1 - \cos\left(\pi\frac{p}{J+1}\right)\right], \qquad 1 \le p \le J;$$

$$(26b) \qquad \eta = \eta_q = 4\theta_y \sin^2\left(\frac{\pi}{2}\frac{q}{K+1}\right) = 2\theta_y\left[1 - \cos\left(\pi\frac{q}{K+1}\right)\right], \qquad 1 \le q \le K.$$

Each eigenvalue ξ_p of the problem (24a) has multiplicity K and each eigenvalue η_q of (24b) has multiplicity J. The eigenvalues of H and V are easily obtained from the above and are

$$2\theta_x \cos\left(\pi\frac{p}{J+1}\right) \quad \text{and} \quad 2\theta_y \cos\left(\pi\frac{q}{K+1}\right),$$

respectively.

The vectors $\mathbf{W}_{p,q}$ are also eigenvectors of $(I - H)^{-1}V$ and multiplication by this matrix yields the eigenvalues, which are also the roots of $P(\rho) = 0$

$$(27) \qquad \rho_{p,q} = \frac{2\theta_y \cos\dfrac{q\pi}{K+1}}{1 - 2\theta_x \cos\dfrac{p\pi}{J+1}}, \qquad 1 \le p \le J,\ 1 \le q \le K.$$

The maximum magnitude of the eigenvalues is found by the usual expansions and some simplification to be

$$(28a) \qquad \max_{p,q} |\rho_{p,q}| = \rho_{1,1} = 1 - \frac{\delta y^2}{2}\pi^2\left(\frac{1}{a^2} + \frac{1}{b^2}\right) + \mathcal{O}(\delta^4).$$

Hence the rate of convergence for this *line Jacobi scheme* is

$$(28b) \qquad R_{LJ} = \frac{\delta y^2}{2}\pi^2\left(\frac{1}{a^2} + \frac{1}{b^2}\right) + \mathcal{O}(\delta^4).$$

Note the similarity between this result and that in (6) for the (point) Jacobi iterations and that in (12) for the successive iterations. If $\delta x = \delta y$, then $\delta y^2 = 4\delta^2$ and the above rate is essentially that of the Gauss-Seidel method given in (12).

Of course, an analog of the method of successive iterations is also possible for the line methods. We need only use the latest improved data as soon as it is obtained. Thus in (22) we replace $\mathbf{U}_{k-1}^{(\nu)}$ by $\mathbf{U}_{k-1}^{(\nu+1)}$ for $k = 2, 3, \ldots, K$, to obtain a *line Gauss-Seidel scheme*. In matrix form this successive-line method is written as

$$(29) \qquad (I - H - \theta_y B)\mathbf{U}^{(\nu+1)} = \theta_y B^T \mathbf{U}^{(\nu)} + \delta^2 \mathbf{F}.$$

However, an accelerated version of these iterations is of interest, and we directly consider this more general procedure. As before, an intermediate iterate $\mathbf{V}^{(\nu+1)}$ is defined by

$$(30a) \qquad (I - H)\mathbf{V}^{(\nu+1)} = \theta_y B \mathbf{U}^{(\nu+1)} + \theta_y B^T \mathbf{U}^{(\nu)} + \delta^2 \mathbf{F}.$$

Then with an arbitrary parameter ω we set

$$(30b) \qquad \mathbf{U}^{(\nu+1)} = \omega \mathbf{V}^{(\nu+1)} + (1 - \omega)\mathbf{U}^{(\nu)}.$$

The calculations are performed a line at a time, as in the line Jacobi method, to determine the $\mathbf{V}_k^{(\nu+1)}$ and then the $\mathbf{U}_k^{(\nu+1)}$ before going on to $k + 1$. However, now they must be done in a fixed order (say increasing or decreasing k). For $\omega = 1$ this scheme reduces to that in (29).

The *accelerated successive-line method* becomes upon the elimination of $\mathbf{V}^{(\nu+1)}$ in (30)

$$\begin{aligned}(I - H - \omega\theta_y B)&\mathbf{U}^{(\nu+1)} \\ &= [(1 - \omega)I - (1 - \omega)H + \omega\theta_y B^T]\mathbf{U}^{(\nu)} + \omega\delta^2\mathbf{F}.\end{aligned}$$

To examine the convergence of the scheme, we must determine the eigenvalues of the matrix

$$(31a) \quad T_\omega \equiv (I - H - \omega\theta_y B)^{-1}[(1 - \omega)I - (1 - \omega)H + \omega\theta_y B^T],$$

which is determined from the roots τ of the characteristic polynomial

$$(31b) \quad Q_\omega(\tau) \equiv \det |(\tau + \omega - 1)(I - H) - \tau\omega\theta_y B - \omega\theta_y B^T|.$$

We see that the matrix in (31b) has zero elements wherever the matrix A has them and so just as in (17), we can apply the corollary to Theorem 1. With the scalars $\alpha = 1$ and $\beta = \tau^{-1/2}$ we then get from (31b) and (23b)

$$\begin{aligned} Q_\omega(\tau) &= \det |(\tau + \omega - 1)(I - H) - \omega\tau^{1/2} V| \\ &= \det |\omega\tau^{1/2} I| \cdot \det \left|\frac{(\tau + \omega - 1)}{\omega\tau^{1/2}}(I - H) - V\right| \\ &= (\omega\tau^{1/2})^{JK} P\left(\frac{\tau + \omega - 1}{\omega\tau^{1/2}}\right). \end{aligned}$$

It follows that τ and the roots ρ of $P(\rho) = 0$ are related by

$$\tau + \omega - 1 = \omega\tau^{1/2}\rho, \tag{32}$$

by the reasoning that led to (18a). For $\omega = 1$, the iterations reduce to the ordinary successive-line iterations and the non-zero roots τ are given by $\tau = \rho^2$. Thus this method converges twice as fast as the line Jacobi method. Finally, since the eigenvalues ρ lie in $0 < \rho < 1$, the arguments used in (18)–(21) can be applied to the roots $\tau(\omega)$ and the acceleration parameter ω, which satisfy (32). Now the optimal parameter value ω_* and minimum value $\rho(T_{\omega_*})$ of $\rho(T_\omega)$ become, where $\rho(T)$ denotes the spectral radius of T,

$$\omega_* = \frac{2}{1 + \sqrt{1 - \rho_{1,1}^2}}, \tag{33a}$$

$$\rho(T_{\omega_*}) = \omega_* - 1 = \frac{1 - \sqrt{1 - \rho_{1,1}^2}}{1 + \sqrt{1 - \rho_{1,1}^2}}. \tag{33b}$$

By using (28a), we find

$$\rho(T_{\omega_*}) = 1 - 2\delta y\pi\sqrt{\left(\frac{1}{a^2} + \frac{1}{b^2}\right)} + \mathcal{O}(\delta^2),$$

and hence the rate of convergence of the optimum *accelerated line Gauss-Seidel* method is

$$R_{\text{ALGS}} = 2\delta y\pi\sqrt{\left(\frac{1}{a^2} + \frac{1}{b^2}\right)} + \mathcal{O}(\delta^2). \tag{34}$$

To compare rates of convergence we note, using (21), that

$$\begin{aligned} \frac{R_{\text{ALGS}}}{R_{\text{AGS}}} &= \frac{\delta y}{\sqrt{2}\,\delta} + \mathcal{O}(\delta^2) \\ &= \left[1 + \left(\frac{\delta y}{\delta x}\right)^2\right]^{1/2} + \mathcal{O}(\delta^2). \end{aligned} \tag{35}$$

Thus it follows that *for any mesh ratio*, $\delta y/\delta x$, the optimum accelerated successive-line method has a larger rate of convergence than the corresponding optimum accelerated successive (point) iterations. For equal net spacing in the x- and y-directions the factor of improvement is, asymptotically, $\sqrt{2}$. However, if $\delta y > \delta x$, even greater improvement results. We observe here that the net lines along which the new data are obtained at each step should be in the direction of the smallest mesh width; i.e., the "closest" neighbors are grouped together on a line and improved as a group. All of the above could be repeated with H and V

interchanged which corresponds to taking lines in the y-direction. The only change in (35) that would result is the interchange of δx and δy. A decision as to whether the ALGS scheme is more efficient than the AGS scheme must depend upon the size of

$$\frac{\#\,\mathrm{ops}_{\mathrm{ALGS}}}{\#\,\mathrm{ops}_{\mathrm{AGS}}} \equiv \alpha.$$

α measures the ratio of the amounts of work involved in one iteration step for each of the two methods. If

$$\alpha \frac{R_{\mathrm{ALGS}}}{R_{\mathrm{AGS}}} < 1,$$

then the ALGS scheme is more efficient; otherwise, the AGS scheme is more efficient.

2.2. Alternating Direction Iterations

One of the most effective iteration schemes for solving the system (1.16) or (1.18) employs a combination of horizontal and vertical line iterations. In terms of an acceleration parameter ω, and recalling that $2\theta_x + 2\theta_y = 1$, such a scheme due to Peaceman and Rachford can be defined as follows:

$$\text{(36a)} \qquad [(\omega + 2\theta_x)I - H]\mathbf{U}^{\nu+1/2} = [(\omega - 2\theta_y)I + V]\mathbf{U}^{\nu} + \delta^2\mathbf{F},$$

$$\text{(36b)} \qquad [(\omega + 2\theta_y)I - V]\mathbf{U}^{\nu+1} = [(\omega - 2\theta_x)I + H]\mathbf{U}^{\nu+1/2} + \delta^2\mathbf{F}.$$

The vector $\mathbf{U}^{\nu+1/2}$ is an intermediate quantity used to define the scheme and of course it is actually computed in carrying out the procedure. The first step, (36a), is just a horizontal line scheme, similar to line-Jacobi. (In fact, with $\omega = 2\theta_y$ in (36a), we obtain (22b) with $\mathbf{U}^{\nu+1}$ replaced by $\mathbf{U}^{\nu+1/2}$.) Clearly then (36b) is essentially a vertical line-Jacobi iteration. The vector to be found in each of the stages (36a and b) is easily evaluated by solving a tridiagonal system.

To study the convergence of this scheme, we eliminate $\mathbf{U}^{\nu+1/2}$ in (36) and obtain, assuming for the moment that the required inverses exist,

$$\mathbf{U}^{\nu+1} = Q_\omega\mathbf{U}^{\nu} + \mathbf{f}_\omega,$$

where

$$\text{(37)} \qquad Q_\omega \equiv [(\omega + 2\theta_y)I - V]^{-1}[(\omega - 2\theta_x)I + H] \times [(\omega + 2\theta_x)I - H]^{-1}[(\omega - 2\theta_y)I + V],$$

and

$$\mathbf{f}_\omega \equiv [(\omega + 2\theta_y)I - V]^{-1} \times \{[(\omega - 2\theta_x)I + H][(\omega + 2\theta_x)I - H]^{-1} + I\}\delta^2\mathbf{F}.$$

The eigenvalues of Q_ω are easily obtained since the matrices $(2\theta_x I - H)$

and $(2\theta_y I - V)$ have common eigenvectors given in (25). We obtain using (24) and the eigenvalues given in (26)

$$Q_\omega \mathbf{W}_{p,q} = \frac{(\omega - \xi_p)(\omega - \eta_q)}{(\omega + \xi_p)(\omega + \eta_q)} \mathbf{W}_{p,q}.$$

Thus the eigenvalues, say $\lambda(\omega)$, of Q_ω are

$$(38) \qquad \lambda_{p,q}(\omega) \equiv \frac{(\omega - \xi_p)(\omega - \eta_q)}{(\omega + \xi_p)(\omega + \eta_q)}, \qquad \begin{cases} p = 1, 2, \dots, J, \\ q = 1, 2, \dots, K. \end{cases}$$

Since $\xi_p > 0$ and $\eta_q > 0$ for all p and q, it follows that the *alternating direction scheme* (36) *converges for any choice of* $\omega > 0$. We also note that all relevant inverses exist for positive ω.

The trick in the proper use of the alternating direction type schemes is *not* to use a single acceleration parameter ω as above but rather to use a sequence of them, say $\omega_1, \omega_2, \dots, \omega_m$ applied periodically (or cyclically). That is, the calculations in (36) are to be carried out m times (using each ω_i for a complete double sweep of the net) in order to compute $\mathbf{U}^{\nu+1}$ from $\mathbf{U}^\nu$. To actually write this scheme out we should introduce $2m - 1$ intermediate quantities $\mathbf{U}^{\nu+1/(2m)}, \mathbf{U}^{\nu+2/(2m)}, \dots, U^{\nu+1-1/(2m)}$ and successively use (36a and b) for the pairs $U^{\nu+(2j-1)/(2m)}, U^{\nu+2j/(2m)}$. As before, we find that the eigenvalues which determine convergence are now

$$(39) \qquad \lambda_{p,q}(\omega_1, \omega_2, \dots, \omega_m) \equiv \prod_{i=1}^{m} \frac{(\omega_i - \xi_p)(\omega_i - \eta_q)}{(\omega_i + \xi_p)(\omega_i + \eta_q)},$$

$$\begin{cases} p = 1, 2, \dots, J, \\ q = 1, 2, \dots, K. \end{cases}$$

If we take $m = J$ and choose $\omega_j = \xi_j$ for $j = 1, 2, \dots, J$, then it clearly follows from (39) that

$$\lambda_{p,q}(\omega_1, \omega_2, \dots, \omega_J) = 0$$

for all p and q. In this case the exact solution is obtained in a finite number of steps. Of course we could also employ $\omega_i = \eta_i$ with $m = K$ to get similar results. However, both J and K are extremely large in general and we desired to obtain an accurate approximation in only ν iterations where $m\nu \ll J$ and $m\nu \ll K$. Thus we consider the problem, with fixed small m, to find ω_i such that

$$\max_{p,q} |\lambda_{p,q}(\omega_1, \omega_2, \dots, \omega_m)|$$

is minimized with respect to all possible choices of the acceleration parameters ω_i.

This problem is related to the subject of best approximations, Section 4 of Chapter 5. Specifically let us define the function

$$F(z) \equiv \prod_{i=1}^{m} \frac{(\omega_i - z)}{(\omega_i + z)}. \tag{40}$$

Then from (39) we have, recalling (26),

$$\max_{p,q} |\lambda_{p,q}(\omega_1, \omega_2, \ldots, \omega_m)| \leq \max_{\substack{\xi_1 \leq x \leq \xi_J \\ \eta_1 \leq y \leq \eta_K}} |F(x)F(y)|.$$

Thus we seek ω_i such that the rational function $F(x)F(y)$ is the best (uniform) approximation to zero on the rectangle $\xi_1 \leq x \leq \xi_J$, $\eta_1 \leq y \leq \eta_K$. The optimization problem is further simplified by noting that for all x, y on this rectangle

$$|F(x)F(y)| \leq \max_{\alpha \leq z \leq \beta} F^2(z) \equiv \|F(z)\|_\infty^2$$

where $\alpha \equiv \min(\xi_1, \eta_1)$ and $\beta \equiv \max(\xi_J, \eta_K)$. Thus our problem is reduced to *finding the best approximation to zero of the form* (40) *on an interval* $0 < \alpha \leq z \leq \beta$. The existence and uniqueness of such a best rational approximation can be proved in a manner analogous to the treatment in Section 4 of Chapter 5 of best polynomial approximations. We shall not present the analysis here of how to determine the optimum parameters ω_i. Rather, we show how to find a set of parameters ω_i, for which we can estimate $\|F\|_\infty$ in order to compare the rate of convergence of the cyclic alternating direction method with the previously studied iterative methods.

In Problem 1, we verify that for $m = 1$ the choice $\omega_1 = \sqrt{\alpha\beta}$ minimizes $\|F\|_\infty$, and

$$\|F\|_\infty = \frac{1 - \sqrt{\alpha/\beta}}{1 + \sqrt{\alpha/\beta}}.$$

Hence we divide the interval $[\alpha, \beta]$ by points $0 < \alpha_0 = \alpha < \alpha_1 < \cdots < \alpha_m = \beta$, such that

$$\frac{\alpha_0}{\alpha_1} = \frac{\alpha_1}{\alpha_2} = \cdots = \frac{\alpha_{m-1}}{\alpha_m}.$$

The values α_j which have this property are

$$\alpha_j = \alpha\left(\frac{\beta}{\alpha}\right)^{j/m}, \qquad j = 0, 1, \ldots, m. \tag{41a}$$

We now set

$$\omega_j = \sqrt{\alpha_{j-1}\alpha_j} \tag{41b}$$

and find that, since the magnitude of each factor of $F(z)$ is bounded by unity,

$$\max_{\alpha_{i-1}\leq z\leq\alpha_i} |F(z)| \leq \max_{\alpha_{i-1}\leq z\leq\alpha_i} \left|\frac{\omega_i - z}{\omega_i + z}\right|, \qquad i = 1, 2, \ldots, m.$$

On the other hand, with the choice (41), the result in Problem 1 implies that

$$\max_{\alpha_{i-1}\leq z\leq\alpha_i} \left|\frac{\omega_i - z}{\omega_i + z}\right| = \frac{1 - \sqrt{\dfrac{\alpha_{i-1}}{\alpha_i}}}{1 + \sqrt{\dfrac{\alpha_{i-1}}{\alpha_i}}} = \frac{1 - \left(\dfrac{\alpha}{\beta}\right)^{1/2m}}{1 + \left(\dfrac{\alpha}{\beta}\right)^{1/2m}}$$

for all i. Hence

$$\|F\|_\infty \leq \frac{1 - \left(\dfrac{\alpha}{\beta}\right)^{1/(2m)}}{1 + \left(\dfrac{\alpha}{\beta}\right)^{1/(2m)}}.$$

From the definitions of α and β, and the results (1.25) and (1.26), we note that $\alpha = \mathcal{O}(\delta^2)$, $\beta \cong 1$. Therefore,

$$\|F\|_\infty \leq 1 - \mathcal{O}(\delta^{1/m}).$$

Hence we have shown that the rate of convergence of the cyclic alternating direction method is less than $\mathcal{O}(\delta^{1/m})$, for a complete cycle. But the amount of work required to compute m sweeps as in (36a and b) is equivalent to the work required for about $2m$ applications of the line accelerated schemes. Now, the convergence rate of $2m$ applications of a line accelerated scheme is $\mathcal{O}(2m\delta)$. This is much smaller than $\mathcal{O}(\delta^{1/m})$, the convergence rate for one cycle of the alternating direction method for small m. We have thus shown that the alternating direction method is more efficient than any of the other iterative schemes, even when parameters that are not necessarily optimal are employed. For detailed comparisons we refer to the book of Varga. In practice it is wise to start each cycle with the largest parameter value, ω_m, and then successively to use the smaller values.

PROBLEM, SECTION 2

1. Given $0 < \alpha < \beta$, show that

$$\min_{0\leq\omega}\left\{\max_{\alpha\leq z\leq\beta}\left|\frac{\omega - z}{\omega + z}\right|\right\} = \frac{1 - \sqrt{\alpha/\beta}}{1 + \sqrt{\alpha/\beta}},$$

and that the minimum value is attained for $\omega = \omega^* \equiv \sqrt{\alpha\beta}$.

[Hint: The function $(\omega - z)/(\omega + z)$ is a monotonic function of z for any fixed ω. Hence it attains its extreme values at $z = \alpha$ and $z = \beta$. Equal extreme values are attained for $\omega = \omega^*$.]

3. WAVE EQUATION AND AN EQUIVALENT SYSTEM

We consider the *initial value* or *Cauchy problem* for the *wave equation*: Find a function $u(x, t)$ continuous in the half plane

$$D \equiv \{x, t \mid t \geq 0, -\infty < x < \infty\}$$

which satisfies, for $t > 0$,

$$\frac{\partial^2 u}{\partial t^2} - c^2 \frac{\partial^2 u}{\partial x^2} = 0; \tag{1}$$

and for $t = 0$,

$$u(x, 0) = f(x), \tag{2a}$$

$$\frac{\partial u(x, 0)}{\partial t} = g(x). \tag{2b}$$

This problem may be solved explicitly in terms of quadratures. That is, by using the change of variables

$$\xi = x + ct, \quad \eta = x - ct, \quad \phi(\xi, \eta) \equiv u(x, y),$$

we find

$$\frac{\partial}{\partial x} \equiv \frac{\partial}{\partial \xi} + \frac{\partial}{\partial \eta} \quad \text{and} \quad \frac{\partial}{\partial t} \equiv c\left(\frac{\partial}{\partial \xi} - \frac{\partial}{\partial \eta}\right),$$

whence equation (1) reduces to

$$4c^2 \frac{\partial^2 \phi}{\partial \xi \, \partial \eta} = 0.$$

The general solution of this equation is found, by two integrations, to be of the form

$$\phi(\xi, \eta) = P(\xi) + Q(\eta).$$

Thus the general solution of (1) is

$$u(x, t) = P(x + ct) + Q(x - ct), \tag{3}$$

where P and Q are arbitrary (twice differentiable) functions. Since $P(x + ct)$ is constant along lines $x + ct =$ constant, this part of the solution can be considered as a *signal* or *wave* which propagates to the left with speed $c > 0$ as time increases. Similarly, $Q(x - ct)$ represents a wave moving to the right with speed c. The lines in the x, t-plane along which the signals travel,

$$x \pm ct = \text{constant},$$

are called the *characteristics* of equation (1).

The initial conditions (2), when applied to (3), yield

$$P(x) + Q(x) = f(x),$$

$$P'(x) - Q'(x) = \frac{1}{c} g(x).$$

Thus by integrating the second relation over $[0, x]$ and using the first, we may solve the pair of equations for $P(x)$ and $Q(x)$. That is, set $K = \frac{1}{2}[P(0) - Q(0)]$ and find

$$P(x) = \tfrac{1}{2}f(x) + \frac{1}{2c}\int_0^x g(\zeta)\,d\zeta + K,$$

$$Q(x) = \tfrac{1}{2}f(x) - \frac{1}{2c}\int_0^x g(\zeta)\,d\zeta - K.$$

If we replace x in $P(x)$ by $x + ct$ and in $Q(x)$ by $x - ct$, we get from (3) the solution of the initial value problem

$$u(x, t) = \tfrac{1}{2}[f(x + ct) + f(x - ct)] + \frac{1}{2c}\int_{x-ct}^{x+ct} g(\zeta)\,d\zeta. \tag{4}$$

Clearly, the solution at any point (x^*, t^*) depends upon the initial data only in the interval $[x^* - ct^*, x^* + ct^*]$ on the initial line, $t = 0$. This interval is cut out by the two characteristics passing through (x^*, t^*) shown in Figure 1. The shaded triangle in this figure is called the *domain of dependence* of the point (x^*, t^*) and its base is the *interval of dependence.*

The Cauchy problem, (1) and (2), can also be formulated as an initial value problem for a first order system of partial differential equations. In particular, introduce the function $v(x, t)$ and consider

$$\frac{\partial u}{\partial t} = c\frac{\partial v}{\partial x},$$

$$\frac{\partial v}{\partial t} = c\frac{\partial u}{\partial x}, \tag{5}$$

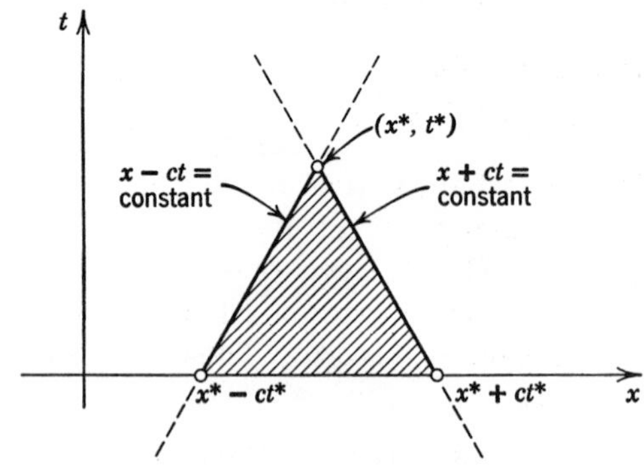

Figure 1. Characteristics and domain of dependence for the wave equation.

subject to the initial values

$$u(x, 0) = f(x), \tag{6a}$$

$$v(x, 0) = G(x). \tag{6b}$$

Equation (1) results after the elimination of $v(x, t)$ from equations (5). With the same change of variables used before, we find that the general solution of (5) has the form

$$u(x, t) = p(x + ct) + q(x - ct),$$

$$v(x, t) = p(x + ct) - q(x - ct),$$

where p and q are again arbitrary functions but only required to possess one derivative. To satisfy the initial conditions (6), we must have

$$p(x) = \tfrac{1}{2}f(x) + \tfrac{1}{2}G(x),$$

$$q(x) = \tfrac{1}{2}f(x) - \tfrac{1}{2}G(x),$$

and hence the solution of the initial value problem (5) and (6) is

$$\begin{aligned} u(x, t) &= \tfrac{1}{2}[f(x + ct) + f(x - ct)] + \tfrac{1}{2}[G(x + ct) - G(x - ct)], \\ v(x, t) &= \tfrac{1}{2}[f(x + ct) - f(x - ct)] + \tfrac{1}{2}[G(x + ct) + G(x - ct)]. \end{aligned} \tag{7}$$

A comparison of the solutions u given in (7) and (4) shows that the two Cauchy problems are equivalent if in (6b) we take

$$G(x) = \frac{1}{c}\int_0^x g(\zeta)\, d\zeta + \text{constant}.$$

This relation could have been derived directly by satisfying the first equation of (5) at $t = 0$ and using (2b).

For the system (5), the lines $x \pm ct = \text{constant}$ are again the characteristics, and the domain of dependence is still as in Figure 1. Of course, as is clear from (7), the solution at any point (x^*, t^*) is now determined by the values of the initial data (6) at the points where the characteristics through (x^*, t^*) intersect the initial line, $t = 0$. These properties of the system (5) become particularly transparent if we first add and then subtract the equations in this system to get

$$\left(\frac{\partial}{\partial t} - c\frac{\partial}{\partial x}\right)(u + v) = 0,$$

$$\left(\frac{\partial}{\partial t} + c\frac{\partial}{\partial x}\right)(u - v) = 0.$$

This is called the *characteristic form* of the system (5) and the combinations $u \pm v$ are the *characteristic* (dependent) *variables*. In the (x, t)-plane the operators $\left(\frac{\partial}{\partial t} \pm c\frac{\partial}{\partial x}\right)$ represent differentiation in two specific directions, the *characteristic directions*, and each characteristic variable is differentiated in an appropriate one of these directions. The notions of characteristic direction, characteristic variable and domain of dependence are useful in the treatment of more general equations of *hyperbolic*† *type* in two independent variables (x, t). For example, consider the simplest case of a linear system of n first order differential equations for the n functions $\{u_i\}$ that are the components of $\mathbf{u}$,

$$\frac{\partial}{\partial t}\mathbf{u} + A\frac{\partial}{\partial x}\mathbf{u} = \mathbf{b}.$$

Suppose that the square matrix $A = A(x, t)$ has n real eigenvalues (α_i) and a complete set of eigenvectors. Let P be the matrix whose columns are the eigenvectors of A. Then define $\mathbf{v}$ by $\mathbf{u} = P\mathbf{v}$ and insert in the above system. This yields

$$\frac{\partial}{\partial t}(P\mathbf{v}) + A\frac{\partial}{\partial x}(P\mathbf{v}) = \mathbf{b},$$

or, by differentiation

$$P\frac{\partial}{\partial t}(\mathbf{v}) + AP\frac{\partial}{\partial x}(\mathbf{v}) = \mathbf{b} - \left(\frac{\partial}{\partial t}P\right)\mathbf{v} - A\left(\frac{\partial}{\partial x}P\right)\mathbf{v}.$$

If we multiply both sides on the left by P^{-1}, we find

$$\left(I\frac{\partial}{\partial t} + P^{-1}AP\frac{\partial}{\partial x}\right)\mathbf{v} = P^{-1}\left[\mathbf{b} - \left(\frac{\partial}{\partial t}P\right)\mathbf{v} - A\left(\frac{\partial}{\partial x}P\right)\mathbf{v}\right].$$

This system is in the simple *characteristic form*. That is, differentiation in only a single (*characteristic*) *direction*,

$$\frac{dx}{dt} = \alpha_j,$$

occurs in each equation, since $P^{-1}AP$ is a diagonal matrix with the (α_j) on the diagonal. The components of $\mathbf{v} = P^{-1}\mathbf{u}$ are the *characteristic variables*.

We refrain from giving the definition of characteristic surface, which plays a vital role in the theory of partial differential equations in more dimensions. It is sufficient to say that the notion of domain of dependence

† A system of partial differential equations is said to be of hyperbolic type if the Cauchy initial value problem is well posed for this system. For a linear system of equations, simple algebraic properties of the coefficients have been shown to imply the hyperbolicity of the system, e.g., the conditions on the matrix A above.

is important for *hyperbolic* equations in higher dimensions but the notion of characteristic form of the system does not generalize.

The *homogeneous characteristic equation* has the form

(8a)
$$\frac{\partial w}{\partial t} + \alpha \frac{\partial w}{\partial x} = 0.$$

Now on any curve $x = x(t)$ in the (x, t)-plane, w is a function of t given by $w(x(t), t)$ and has the total derivative

$$\frac{dw}{dt} = \frac{\partial w}{\partial t} + \frac{\partial w}{\partial x}\frac{dx}{dt}.$$

Thus if the curve is chosen such that

(8b)
$$\frac{dx}{dt} = \alpha,$$

then any solution w of (8) satisfies $dw/dt = 0$ and hence, is constant on such a curve. The curves (8b) are the characteristics and if α is not a constant, they are not straight lines.

The Cauchy problems previously formulated and solved could have been solved by the *method of separation of variables*. Instead, we shall now apply this method to a special *mixed initial-boundary value problem* for the wave equation. The problem of interest is to solve the wave equation

$$\frac{\partial^2 u}{\partial t^2} - c^2 \frac{\partial^2 u}{\partial x^2} = 0$$

subject to the *initial conditions*

(9a)
$$u(x, 0) = f(x),$$

(9b)
$$\frac{\partial u}{\partial t}(x, 0) = 0, \qquad 0 < x < L,$$

and the *boundary conditions*

(10a)
$$u(0, t) = 0,$$

(10b)
$$u(L, t) = 0, \qquad t > 0.$$

The solution is to be determined in the strip $R \equiv \{x, t \mid 0 \leq x \leq L, t \geq 0\}$. For convenience we have used a homogeneous initial condition in (9b).

Let us seek solutions of the wave equation in the form

$$u(x, t) = \phi(x)\psi(t)$$

which satisfy the boundary conditions (10). Then we must have

$$\frac{\phi''(x)}{\phi(x)} = \frac{1}{c^2}\frac{\ddot{\psi}(t)}{\psi(t)} = k^2 \equiv \text{constant.}$$

and

$$\phi(0) = \phi(L) = 0,$$

where two primes indicate d^2/dx^2 and two dots indicate d^2/dt^2. The general solutions of the differential equations which result from this separation are

$$\phi(x) = \alpha e^{kx} + \beta e^{-kx},$$

$$\psi(t) = ae^{ckt} + be^{-ckt}.$$

From the boundary condition $\phi(0) = 0$, we get $\alpha = -\beta$; while $\phi(L) = 0$ implies that

$$e^{kL} = e^{-kL} \qquad \text{or} \qquad e^{2kL} = 1.$$

Thus the boundary conditions can only be satisfied if k has pure imaginary values such that $2kL = 2n\pi i$, or

$$k = i\frac{n\pi}{L}, \qquad n = 1, 2, \ldots.$$

We omit $n = 0$ since it leads to the trivial result $\phi(x) \equiv 0$. With $\alpha = 1$ and any coefficients a and b, we have shown that $\phi(x)\psi(t)$ is a solution of (1) satisfying (10), if

$$\phi(x) = \phi_n(x) = \sin\frac{n\pi x}{L}$$

$$\psi(t) = \psi_n(t) = a_n \sin c\frac{n\pi t}{L} + b_n \cos c\frac{n\pi t}{L}, \qquad n = 1, 2, \ldots.$$

Thus, formally, a solution of the wave equation which satisfies (10) is given by

$$u(x, t) = \sum_{n=1}^{\infty} \phi_n(x)\psi_n(t).$$

To satisfy the initial conditions (9) with this solution, we require that

$$\text{(11a)} \qquad \sum_{n=1}^{\infty} b_n \sin\frac{n\pi x}{L} = f(x), \qquad \sum_{n=1}^{\infty} na_n \sin\frac{n\pi x}{L} = 0.$$

Multiply each of these relations by $\sin(m\pi x/L)$ and integrate over $[0, L]$, if the series converge uniformly, to find, since

$$\int_0^{\pi} \sin n\theta \sin m\theta \, d\theta = \frac{\pi}{2}\delta_{mn},$$

$$\text{(11b)} \qquad b_n = \frac{2}{L}\int_0^L f(x)\sin\frac{n\pi x}{L}\,dx, \qquad a_n = 0, \; n = 1, 2, \ldots.$$

The coefficients b_n are just the Fourier coefficients for the expansion of $f(x)$ in a sine series. If $f'(x)$ is piecewise continuous, this series converges uniformly to $f(x)$. The solution of the mixed problem (1), (9), and (10) is given by

$$\begin{aligned} u(x, t) &= \sum_{n=1}^{\infty} b_n \sin \frac{n\pi x}{L} \cos c \frac{n\pi t}{L} \\ &= \sum_{n=1}^{\infty} \frac{b_n}{2} \left[\sin \frac{n\pi}{L}(x + ct) + \sin \frac{n\pi}{L}(x - ct) \right]. \end{aligned} \tag{11c}$$

If the function $f(x)$ has a piecewise continuous third derivative, the series in (11) defines a function $u(x, t)$ with continuous second derivatives which can be evaluated by differentiating (11c) termwise. Hence, $u(x, t)$ defined by (11) is a solution of the mixed problem. Equation (11c) again shows that $u(x, t)$ is the sum of functions of the two variables $x \pm ct$. In fact, in this special case,

$$u(x, t) = P(x + ct) + P(x - ct),$$

where

$$P(x) \equiv \sum_{n=1}^{\infty} \frac{b_n}{2} \sin \frac{n\pi x}{L} = \frac{f(x)}{2}.$$

3.1. Difference Approximations and Domains of Dependence

On the half space $t \geq 0$, $|x| \leq \infty$, we introduce the uniformly spaced net points

$$x_j = j\Delta x, \quad t_n = n\Delta t; \quad |j|,n = 0, 1, 2, \ldots.$$

The set of net points D_Δ is defined by

$$D_\Delta \equiv \{x_j, t_n \mid j = 0, \pm 1, \pm 2, \ldots; n = 1, 2, \ldots\}.$$

A direct approximation of the wave equation (1) is obtained by using centered difference quotients, as in Section 1, to replace derivatives. Thus if $U(x, t)$ is a net function, we consider the difference equations

$$U_{t\bar{t}}(x, t) - c^2 U_{x\bar{x}}(x, t) = 0, \qquad (x, t) \in D_\Delta. \tag{12a}$$

If we take the point $(x, t) = (x_j, t_n)$ and use the subscript notation $U(x_j, t_n) = U_{j,n}$, then (12a) can be multiplied by Δt^2 and the result rewritten as

$$\begin{aligned} U_{j,n+1} = 2\left[1 - \left(c\frac{\Delta t}{\Delta x}\right)^2\right] U_{j,n} \\ + \left(c\frac{\Delta t}{\Delta x}\right)^2 (U_{j+1,n} + U_{j-1,n}) - U_{j,n-1}, \\ n \geq 1, \ |j| = 0, 1, \ldots. \end{aligned} \tag{12b}$$

The star of net points entering into (12) is the same as that in Figure 1 of Section 1 (with y replaced by t). To calculate U on any time line $t = t_{n+1}$, say, the values of U must be known on the two preceding time lines. Thus in order to start the computations indicated in (12), we require data on the two initial time lines $t = 0$ and $t = \Delta t$. This is consistent with the form of initial data given for the wave equation in (2). A simple adaptation of these conditions is

$$\text{(13a)} \qquad U(x, 0) = f(x),$$

$$\text{(13b)} \qquad U_t(x, 0) \equiv U_{\bar{t}}(x, \Delta t) = g(x).$$

From (13b), we have

$$\text{(13c)} \qquad \begin{aligned} U(x, \Delta t) &\equiv U(x, 0) + \Delta t\, U_t(x, 0) \\ &= f(x) + \Delta t g(x). \end{aligned}$$

More accurate approximations to $u(x, \Delta t)$ can be obtained if we assume that f and g are sufficiently differentiable and that the wave equation (1) is satisfied on the initial line, $t = 0$. That is, by Taylor's theorem

$$u(x, \Delta t) = u(x, 0) + \Delta t \frac{\partial u(x, 0)}{\partial t} + \frac{\Delta t^2}{2!} \frac{\partial^2 u(x, 0)}{\partial t^2} + \mathcal{O}(\Delta t^3).$$

But since $u(x, t)$ satisfies (1) and (2),

$$\frac{\partial^2 u(x, 0)}{\partial t^2} = c^2 \frac{\partial^2 u(x, 0)}{\partial x^2} = c^2 f''(x),$$

hence

$$u(x, \Delta t) = f(x) + \Delta t g(x) + \frac{\Delta t^2}{2} c^2 f''(x) + \mathcal{O}(\Delta t^3).$$

This suggests replacing (13c) by the formula

$$\text{(14a)} \qquad U(x, \Delta t) = f(x) + \Delta t g(x) + \frac{\Delta t^2}{2} c^2 f_{x\bar{x}}(x),$$

or equivalently the replacement of (13b) by

$$\text{(14b)} \qquad U_t(x, 0) \equiv U_{\bar{t}}(x, \Delta t) = g(x) + \frac{\Delta t}{2} c^2 U_{x\bar{x}}(x, 0).$$

Even more accurate approximations than (14a) can be derived by continuing this procedure. For instance, the next term would involve

$$\frac{\partial^3 u(x, 0)}{\partial t^3} = c^2 \frac{\partial^3 u(x, 0)}{\partial x^2\, \partial t} = c^2 g''(x).$$

The difference problem posed by (12b) and (13a and c) [or (12b), (13a), and (14a)] is called *explicit* since it is in a form in which the solution is obtained recursively by evaluating the given formulae. (This was not the case for the elliptic difference equations of Section 1 and 2, where a major part of the task was to solve the difference equations efficiently.) A glance at (12), (14), and the star in Figure 1 on p. 447 indicates that the solution at any fixed net point, (x^*, t^*), depends only on the values of U at the net points in the triangle formed by the initial line and the two lines with slopes $\pm\Delta t/\Delta x$, say $x \pm \frac{\Delta x}{\Delta t} t =$ constant, which pass through (x^*, t^*). This region is shown in Figure 2 and it may be called the *numerical domain of dependence* for the difference equations (12).

Clearly, the numerical domain of dependence will be greater than or equal to the domain of dependence of the wave equation, for the same point (x^*, t^*), iff

$$\frac{\Delta t}{\Delta x} \leq \frac{1}{c}.$$

We refer to $1/c$ as the *characteristic slope* and to $\Delta t/\Delta x$ as the *net slope.* Therefore, if the characteristic slope is greater than or equal to the net slope, then the numerical domain of dependence includes the domain of dependence of the wave equation. We introduce the ratio of these slopes as

$$\lambda \equiv \frac{\text{net slope}}{\text{characteristic slope}} \equiv \frac{c\Delta t}{\Delta x} \tag{15}$$

and then the above condition becomes $\lambda \leq 1$. Note that since c is the speed of propagation of a signal or wave for the wave equation, λ is the

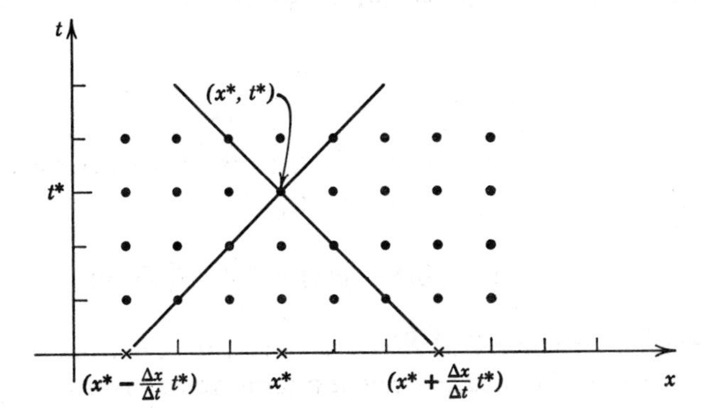

Figure 2. Net points and numerical domain of dependence for difference scheme (12).

ratio of the distance such a signal travels in one time step to the length of a spacial step of the net. Thus if such signals cannot move more than the distance Δx in the time Δt, then the numerical domain contains the analytical domain of dependence.

To understand the significance, for difference schemes, of these domains of dependence, we consider two Cauchy problems for the wave equation. The first is that posed by (1) and (2) with the solution $u(x, t)$ given in (4). In the second problem we retain (2a) and replace $g(x)$ in (2b) by

$$g^*(x) = g(x) + \begin{cases} 0 & x \leq a, \\ 4cm(x - a) & x \geq a; \end{cases}$$

where $x = a$ is an arbitrary fixed point. By using (4) with the new initial data, the solution, $u^*(x, t)$, of the altered problem is found to be

$$\text{(16a)} \quad \begin{aligned} u^*(x, t) &= u(x, t) + \begin{cases} 0 & x + ct \leq a, \\ \dfrac{1}{2c}\displaystyle\int_{\max(a,\, x - ct)}^{x+ct} 4cm(\zeta - a)\, d\zeta & x + ct \geq a, \end{cases} \\ &= u(x, t) + \begin{cases} 0 & x + ct \leq a, \\ m(x + ct - a)^2 & x + ct \geq a \geq x - ct, \\ 4cmt(x - a) & x - ct \geq a. \end{cases} \end{aligned}$$

Now for each of these problems let us consider the corresponding difference problem (12) and (13) for a net, chosen such that $x = a$ is a net point on the initial line. If the difference solutions are denoted by U and U^* respectively, then, since they have identical initial data on $x \leq a$, it follows from a consideration of the numerical domains of dependence that

$$\text{(16b)} \qquad U^*(x, t) = U(x, t), \qquad \text{if } x + \frac{\Delta x}{\Delta t}\, t \leq a.$$

If the net spacing is such that $\lambda > 1$, then there are net points (x, t) which satisfy

$$x + \frac{\Delta x}{\Delta t}\, t = a \quad \text{and} \quad x + ct > a \geq x - ct.$$

At such points we have from (16)

$$\left.\begin{aligned} u^*(x, t) - u(x, t) &= m(x + ct - a)^2 \\ U^*(x, t) - U(x, t) &= 0 \end{aligned}\right\} \qquad a - ct < x = a - \frac{\Delta x}{\Delta t}\, t.$$

If we let $\Delta x \to 0$ and $\Delta t \to 0$ while λ = constant > 1 and $x = a$ remains a net point on $t = 0$, then clearly $U(x, t)$ and $U^*(x, t)$ cannot both converge to the corresponding solutions $u(x, t)$ and $u^*(x, t)$. Thus we deduce

THEOREM 1. *In general, the difference solution of* (12) *and* (13) *cannot converge to the exact solution of* (1) *and* (2) *as* $\Delta x \to 0$ *and* $\Delta t \to 0$ *for constant* $\lambda = c\Delta t/\Delta x > 1$. ■

The requirement $\lambda \leq 1$, which by the above observations is seen to be a necessary condition for convergence in general (i.e., for "all" initial value problems) is called the *Courant-Friedrichs-Lewy condition* (or sometimes just the *Courant condition* for brevity). In other words, *the numerical domain of dependence of a difference scheme should include the domain of dependence of the differential equation or else convergence is not always possible.* We also call this the *domain of dependence condition.*

The relationship between the notion of a domain of dependence and the convergence of a difference method is easily studied for the initial value problem of the single characteristic equation (8a), in which $\alpha > 0$ is a constant. Then the characteristic curves, determined by (8b), are the lines

$$x = \alpha t + \text{constant}.$$

The solution of (8a) which satisfies the initial condition

$$w(x, 0) = f(x),$$

is thus

$$w(x, t) = f(x - \alpha t).$$

The domain of dependence of the point (x^*, t^*) is the set of points (x, t) on the characteristic, $x = \alpha t + x^* - \alpha t^*$.

With the uniform net of spacing Δx and Δt we consider the difference equations, for a net function $W(x, t)$,

$$W_t(x, t) + \alpha W_x(x, t) = 0. \tag{17a}$$

In subscript notation, with $(x, t) = (x_j, t_n)$ and $\lambda \equiv \alpha\Delta t/\Delta x$, this becomes

$$W_{j,n+1} = (1 + \lambda)W_{j,n} - \lambda W_{j+1,n}.$$

For this scheme, $W(x, t + \Delta t)$ is determined by data to the right of x or directly below it. However, since $\alpha > 0$, the exact solution depends upon data to the left of x. Thus for any $\lambda > 0$, the numerical domain of dependence cannot contain that of the differential equation. Clearly then, this procedure is not convergent, in general, as $\Delta x, \Delta t \to 0$.

Now, let us replace the forward x-difference by a backward difference to get

(17b) $$W_t(x, t) + \alpha W_{\bar{x}}(x, t) = 0,$$

or equivalently

$$W_{j,n+1} = (1 - \lambda)W_{j,n} + \lambda W_{j-1,n}.$$

Clearly, the actual domain of dependence is contained within the numerical domain if $\lambda \le 1$. If the exact solution is sufficiently smooth (i.e., say $f''(x)$ is continuous), then we find that the *local truncation error*, τ, is

$$\tau(x, t) \equiv w_t(x, t) + \alpha w_{\bar{x}}(x, t) = \mathcal{O}(\Delta t + \Delta x).$$

With the definition

$$e(x, t) = W(x, t) - w(x, t)$$

we obtain

$$e_{j,n+1} = (1 - \lambda)e_{j,n} + \lambda e_{j-1,n} - \Delta t \tau_{j,n}.$$

Now let $E_n \equiv \underset{j}{\text{l.u.b.}}\, |e_{j,n}|$ and take the absolute value of both sides to get, since $\lambda \le 1$,

$$\begin{aligned} |e_{j,n+1}| &\le (1 - \lambda)|e_{j,n}| + \lambda|e_{j-1,n}| + \Delta t|\tau_{j,n}| \\ &\le (1 - \lambda)E_n + \lambda E_n + \Delta t\mathcal{O}(\Delta t + \Delta x) \\ &\le E_n + \Delta t\mathcal{O}(\Delta t + \Delta x). \end{aligned}$$

Thus

$$E_{n+1} \le E_n + \Delta t\mathcal{O}(\Delta t + \Delta x)$$

and a simple recursion yields,

$$E_{n+1} \le E_0 + t_{n+1}\mathcal{O}(\Delta t + \Delta x),$$

or

$$|W(x, t) - w(x, t)| \le \|W(x, 0) - f(x)\|_\infty + t\mathcal{O}(\Delta t + \Delta x).$$

Convergence now follows as Δt and Δx vanish while $\lambda \le 1$, provided the initial data $W(x, 0)$ approaches $f(x)$.

The second scheme converges for special choices of the *mesh ratio*, $\Delta t/\Delta x$, and is therefore said to be *conditionally convergent*. Of course, the first scheme never satisfies the domain of dependence condition while the second converges when it does satisfy this condition. *However, there are schemes which satisfy the domain of dependence condition, are reasonable approximations to the differential equation but still do not converge for any value of the mesh ratio.* Consider, for example, the scheme

(17c) $$W_t(x, t) + \alpha W_{\hat{x}}(x, t) = 0,$$

which uses the centered x-difference quotient. The truncation error is now $\tau = \mathcal{O}(\Delta t + \Delta x^2)$, and is at least as good as in the previous case. If $\lambda \leq 1$, the domain of dependence condition is satisfied but this scheme does not converge, in general, for any mesh ratio (see Problem 1). Thus to determine convergent schemes it is not sufficient to examine domains of dependence and truncation error alone. Now the difference scheme (17c) can be modified in a simple way to yield a convergent scheme. We use

$$\frac{1}{\Delta t}\{W(x, t + \Delta t) - \tfrac{1}{2}[W(x + \Delta x, t) + W(x - \Delta x, t)]\} \\ + \alpha W_{\hat{x}}(x, t) = 0,$$

in which $W(x, t)$ in the forward difference $W_t(x, t)$ has been replaced by an average of two adjacent values. In subscript notation this becomes

$$W_{j,n+1} = \tfrac{1}{2}(1 - \lambda)W_{j+1,n} + \tfrac{1}{2}(1 + \lambda)W_{j-1,n},$$

and the truncation error is again $\mathcal{O}(\Delta t + \Delta x^2)$. If $\lambda \leq 1$, the numerical domain of dependence includes that of the differential equation (8a); the coefficients $1 \pm \lambda$ are non-negative with sum unity; and convergence can be proved as above [see also the convergence proof in (31)–(35)]. It should be noted that this difference scheme can be written as

$$W_t(x, t) + \alpha W_{\hat{x}}(x, t) - \Delta x \frac{\alpha}{2\lambda} W_{x\bar{x}}(x, t) = 0.$$

3.2. Convergence of Difference Solutions

The difference solution determined by (12) and (13) converges to the solution of the initial value problem (1) and (2), provided $\Delta x \to 0$ and $\Delta t \to 0$ while $\lambda \leq 1$. The proof of this fact is somewhat complicated for $\lambda < 1$ but is much simpler if the special mesh ratio condition $\lambda = 1$ holds. Hence we first consider this case in which the characteristic slope and net slope are equal. It follows that, with the definition

$$D_{i,j} \equiv U_{i,j} - U_{i-1,j-1},$$

the difference equations (12) can be written, when $\lambda = 1$, as

$$D_{j,n+1} = D_{j+1,n}.$$

Note that the value $U_{j,n}$ does not enter into the above difference equation. In fact, the net points may be divided into two groups, corresponding to the red and black squares on a checkerboard, and the difference equations do not couple net points of different groups. Thus we need consider only one such group of net points. Application of the above form of the difference equation recursively yields

$$\begin{aligned} D_{j,n+1} &= D_{j+1,n} \\ &= D_{j+2,n-1} \\ &\vdots \\ &= D_{j+n,1} \end{aligned}$$

These relations are equivalent to the fact that the $D_{i,m}$ are constant on the diagonal $x + ct =$ constant through (x_j, t_{n+1}). We also observe by summing the $D_{i,m}$ along the diagonal $x - ct =$ constant through (x_j, t_{n+1}) that

$$U_{j,n+1} - U_{j-n-1,0} = \sum_{\nu=0}^{n} D_{j-\nu,n-\nu+1}.$$

By combining the last two results and recalling the initial conditions (13a and c), we get

$$\begin{aligned} (18) \qquad U_{j,n+1} &= U_{j-n-1,0} + \sum_{\nu=0}^{n} D_{j+n-2\nu,1} \\ &= f_{j-n-1} + \sum_{\nu=0}^{n} \Delta t g_{j+n-2\nu} + \sum_{\nu=0}^{n} (f_{j+n-2\nu} - f_{j+n-2\nu-1}). \end{aligned}$$

This is an explicit representation of the solution of the difference problem (12)–(13). To examine convergence we shall let $\Delta x = c\Delta t \to 0$. Since $t_n = n\Delta t$ and $x_n = n\Delta x = ct_n$, it follows that $F_{j+n} \equiv F(x_{j+n}) = F(x_j + ct_n)$ for any function $F(x)$. Then if $f(x)$ has a continuous first derivative

$$\begin{aligned} f_{j+n-2\nu} - f_{j+n-2\nu-1} &= f(x_{j-2\nu} + ct_n) - f(x_{j-2\nu} + ct_n - \Delta x) \\ &= \Delta x f'(x_{j-2\nu} + ct_n + \theta_\nu \Delta x), \qquad 0 > \theta_\nu > -1. \end{aligned}$$

Now take the limit as $\Delta x \to 0$ and $n, j \to \infty$ in (18), while $t_{n+1} = t$ and $x_j = x$, for any *fixed* (x, t), to get

$$\begin{aligned} U(x, t) \\ &= f(x - ct) + \lim_{\Delta x \to 0} \frac{1}{2c} \sum_{\nu=0}^{n} g(x + ct - [2\nu + 1]\Delta x) 2\Delta x \\ &\quad + \lim_{\Delta x \to 0} \frac{1}{2} \sum_{\nu=0}^{n} f'(x + ct + \theta_\nu \Delta x - [2\nu + 1]\Delta x) 2\Delta x \\ &= f(x - ct) + \frac{1}{2c} \int_0^{2ct} g(x + ct - \xi)\, d\xi + \frac{1}{2} \int_0^{2ct} f'(x + ct - \xi)\, d\xi \\ &= \frac{1}{2} [f(x + ct) + f(x - ct)] + \frac{1}{2c} \int_{x-ct}^{x+ct} g(\eta)\, d\eta \\ &= u(x, t). \end{aligned}$$

Thus the proof of convergence, when $\lambda = 1$, has been completed by using the representation of $u(x, t)$ given in (4).

We study the case of $\lambda < 1$, first for the mixed initial-boundary value problem formulated in (1), (9), and (10) whose solution is given in (11). The difference problem can be formulated as the difference equations (12) where now D_Δ are the net points in $0 < x < L$, $t > 0$, and the mesh is such that, say,

$$(J + 1)\Delta x = L.$$

The initial conditions (13) or (14) are to be satisfied for $0 < x < L$, where $g(x) \equiv 0$, and the boundary conditions for the difference problem are

$$(19) \qquad U(0, t_n) \equiv U_{0,n} = 0, \quad U(L, t_n) \equiv U_{J+1,n} = 0, \quad n = 1, 2, \ldots.$$

We now seek solutions of the difference problem by the method of separation of variables. In fact, from the experience gained in Subsection 1.1, we try the forms

$$U(x, t) = \Psi^{(p)}(t) \sin\left(p\frac{\pi x}{L}\right), \qquad p = 1, 2, \ldots.$$

From (12)

$$\begin{aligned}\Psi^{(p)}_{t\bar{t}} \sin\left(p\frac{\pi x}{L}\right) &= c^2\Psi^{(p)}(t) \sin_{x\bar{x}}\left(p\frac{\pi x}{L}\right)\\ &= -\frac{c^2\zeta_p{}^2}{\Delta x^2}\Psi^{(p)} \sin\left(p\frac{\pi x}{L}\right)\end{aligned}$$

where $\zeta_p = 2\sin\left(\frac{\pi}{2}\frac{p}{J+1}\right)$ and we have employed the trigonometric identities

$$\sin\left[p\frac{\pi(x+\Delta x)}{L}\right] + \sin\left[p\frac{\pi(x-\Delta x)}{L}\right] = 2\cos\left(p\frac{\pi\Delta x}{L}\right)\sin\left(p\frac{\pi x}{L}\right),$$

$$1 - \cos\left(p\frac{\pi\Delta x}{L}\right) = 2\sin^2\left(p\frac{\pi\Delta x}{2L}\right).$$

It follows that $\Psi^{(p)}(t)$ must satisfy

$$\Delta t^2\Psi^{(p)}_{t\bar{t}}(t) = -\lambda^2\zeta_p{}^2\Psi^{(p)}(t),$$

and we try the quantities

$$\Psi^{(p)}(t) = \sin\mu_p t.$$

By using the same trigonometric identities, we get

$$4\sin^2\frac{\mu_p\Delta t}{2}\sin\mu_p t = \lambda^2\zeta_p{}^2\sin\mu_p t.$$

A similar result is obtained if we use $\Psi^{(p)}(t) = \cos \mu_p t$. Thus the terms

$$(20) \qquad (A_p \sin \mu_p t + B_p \cos \mu_p t) \sin\left(p\frac{\pi x}{L}\right), \qquad p = 1, 2, \ldots,$$

will satisfy the difference equations (12) if μ_p is such that

$$4 \sin^2 \frac{\mu_p \Delta t}{2} = \lambda^2 \zeta_p{}^2,$$

or

$$(21) \qquad \sin \mu_p \frac{\Delta t}{2} = \pm \lambda \sin\left(\frac{\pi}{2}\frac{p}{J+1}\right), \qquad p = 1, 2, \ldots.$$

These transcendental equations have real roots, μ_p, for all p, iff $\lambda = c\Delta t/\Delta x \leq 1$.

A linear combination of the solutions in (20) yields

$$(22) \qquad U(x, t) = \sum_{p=1}^{\infty} (A_p \sin \mu_p t + B_p \cos \mu_p t) \sin\left(p\frac{\pi x}{L}\right).$$

If this series converges, it is a solution of (12) and satisfies the boundary conditions (19). To satisfy the initial condition (13a) we must have

$$(23a) \qquad \sum_{p=1}^{\infty} B_p \sin\left(p\frac{\pi x}{L}\right) = f(x);$$

while condition (14a) requires

$$(23b) \qquad \sum_{p=1}^{\infty} (A_p \sin \mu_p \Delta t + B_p \cos \mu_p \Delta t) \sin\left(p\frac{\pi x}{L}\right)$$
$$= f(x) + \frac{\Delta t^2}{2} c^2 f_{x\bar{x}}(x)$$
$$= (1 - \lambda^2) f(x) + \lambda^2 \frac{f(x + \Delta x) + f(x - \Delta x)}{2}.$$

From (11a) we see that (23a) is satisfied if $B_p = b_p$, $p = 1, 2, \ldots$, where the b_p are defined in (11b). From the identity $1 - 2\sin^2(\theta/2) = \cos\theta$ and (21), we have

$$\cos(\mu_p \Delta t) \sin\left(p\frac{\pi x}{L}\right) = (1 - \lambda^2) \sin\left(p\frac{\pi x}{L}\right)$$
$$+ \frac{\lambda^2}{2}\left\{\sin\left[p\frac{\pi(x + \Delta x)}{L}\right] + \sin\left[p\frac{\pi(x - \Delta x)}{L}\right]\right\}.$$

hence (23b) and (23a) yield

$$\sum_{p=1}^{\infty} A_p \sin \mu_p \Delta t \sin\left(p\frac{\pi x}{L}\right) = 0.$$

Clearly, this is satisfied by the choice $A_p = 0$. Thus (22) is the solution of the difference problem (12), (13a), (14), and (19), if $A_p = 0$ and $B_p = b_p$. The series (22) converges if (11a) converges absolutely since the μ_p are real (e.g., if $f'(x)$ is continuous).

With the exact solution given by (11c), we obtain

$$(24)\quad |U(x, t) - u(x, t)| \le \left|\sum_{p=1}^{N} b_p\left[\cos \mu_p t - \cos\left(\frac{cp\pi}{L} t\right)\right] \sin\left(p\frac{\pi x}{L}\right)\right| + \left|\sum_{p=N+1}^{\infty} b_p \cos \mu_p t \sin\left(p\frac{\pi x}{L}\right)\right| + \left|\sum_{p=N+1}^{\infty} b_p \cos\left(\frac{cp\pi}{L} t\right) \sin\left(p\frac{\pi x}{L}\right)\right|.$$

By taking N sufficiently large, the last two sums can be made arbitrarily small for all Δx, Δt [since the corresponding series converge absolutely if $f'(x)$ is continuous]. If $\Delta x \to 0$ and $\Delta t \to 0$ while $\lambda = c\Delta t/\Delta x \le 1$ and $(J + 1)\Delta x = L$, we have from (21) for $1 \le p \le N$,

$$\mu_P \to \frac{cp\pi}{L}.$$

Thus $|U(x, t) - u(x, t)|$ can be made arbitrarily small. This proves convergence of the difference scheme, if $\lambda \le 1$, for the mixed initial-boundary value problem, when $f(x)$ has two continuous derivatives and $g(x) \equiv 0$. Convergence for the case $g(x) \not\equiv 0$ can be shown in a similar way.

Now if $\lambda \le 1$, convergence can be proved for the pure initial value problem, by making use of the notion of domain of dependence. That is, given an interval $[a, b]$ and time T, convergence in S: $\{(x, t) \mid a \le x \le b, 0 \le t \le T\}$, can be shown by modifying the initial data only for $x < a - (\Delta x/\Delta t)T$, and $x > b + (\Delta x/\Delta t)T$. For this modified problem, the solutions of the differential equation and of the difference equations are unchanged in S. In fact, if the initial data have been modified so as to be periodic and odd about $[a - (\Delta x/\Delta t)T - \delta, b + (\Delta x/\Delta t)T + \delta]$, for some $\delta > 0$, then the above proof establishes convergence.

The proof of convergence does not generalize to equations with variable coefficients; furthermore, it does not provide an estimate of the error in terms of the interval size Δx or Δt; neither does it provide a treatment of the effect of rounding errors. These defects are avoided in the analysis of the next subsection for the case of a first order system of equations.

3.3. Difference Methods for a First Order Hyperbolic System

We have seen in equations (5) through (7) that the initial value problem for the wave equation can be replaced by an equivalent first order system.

In fact, such systems arise more naturally in physical theories and applied mathematics than does the second order wave equation. We shall treat the simple system (5) which can be written in vector form as

$$\frac{\partial \mathbf{u}}{\partial t} = cA\frac{\partial \mathbf{u}}{\partial x}, \tag{25a}$$

where

$$\mathbf{u} \equiv \begin{pmatrix} u \\ v \end{pmatrix}, \qquad A \equiv \begin{pmatrix} 0 & 1 \\ 1 & 0 \end{pmatrix}. \tag{25b}$$

Most of our methods and results are also applicable to more general *hyperbolic systems* of first order partial differential equations in two independent variables. [For example, such systems may be formulated as in (25a) with the square matrix A of order n having real eigenvalues and simple elementary divisors, i.e., A is *diagonalizable*.] The initial conditions to be imposed can be written as

$$\mathbf{u}(x, 0) = \mathbf{u}_0(x), \tag{26a}$$

where for the problem posed in (5) and (6) we take

$$\mathbf{u}_0(x) \equiv \begin{pmatrix} f(x) \\ G(x) \end{pmatrix}. \tag{26b}$$

The solution of (25) subject to (26) is given in (7).

On the uniform net with spacing Δx and Δt we introduce the net functions $U(x, t)$ and $V(x, t)$, or in vector form the vector net function

$$\mathbf{U}(x, t) \equiv \begin{pmatrix} U(x, t) \\ V(x, t) \end{pmatrix}.$$

Then as an approximation to the system (25) with $\lambda = c\Delta t/\Delta x$, we consider

$$\begin{aligned} \mathbf{U}(x, t + \Delta t) = {} & \tfrac{1}{2}[\mathbf{U}(x + \Delta x, t) + \mathbf{U}(x - \Delta x, t)] \\ & + \frac{\lambda}{2}A[\mathbf{U}(x + \Delta x, t) - \mathbf{U}(x - \Delta x, t)]. \end{aligned} \tag{27a}$$

[It would be tempting to replace the first bracketed term on the right-hand side above by $\mathbf{U}(x, t)$, but as shown in Problem 1 that scheme is divergent.] If we subtract $\mathbf{U}(x, t)$ from each side and divide by Δt, we can write (27) in the difference quotient notation

$$\mathbf{U}_t(x, t) = cA\mathbf{U}_{\hat{x}}(x, t) + \frac{c}{2\lambda}\Delta x \mathbf{U}_{x\bar{x}}(x, t). \tag{27b}$$

Thus our difference equations are obtained by adding a term of order Δx to the divergent approximation of (25). The discussion following equation (17c) may be considered a motivation for (27a).

An immediate advantage of the scheme in (27) can be seen by considering the case $\lambda = 1$ in which the net diagonals and characteristics have the same slopes and so the numerical and analytical domains of dependence coincide. By writing the system in component form and using $\Delta x = c\Delta t$, we get from (27a)

$$\text{(28)}\qquad \begin{aligned} U(x, t + \Delta t) &= \tfrac{1}{2}[U(x + c\Delta t, t) + U(x - c\Delta t, t)] \\ &\qquad + \tfrac{1}{2}[V(x + c\Delta t, t) - V(x - c\Delta t, t)] \\ V(x, t + \Delta t) &= \tfrac{1}{2}[U(x + c\Delta t, t) - U(x - c\Delta t, t)] \\ &\qquad + \tfrac{1}{2}[V(x + c\Delta t, t) + V(x - c\Delta t, t)] \end{aligned}$$

For the initial conditions

$$\text{(29)}\qquad U(x, 0) = f(x), \qquad V(x, 0) = G(x);$$

a comparison of (28) when $t = 0$ with (7) when $t = \Delta t$ shows that the numerical solution and the exact solution are *identical* on net points for which $t = \Delta t$. By considering the exact solution at this first time step as initial data, we find that the solution is also exact for net points with $t = 2\Delta t$. By induction, we can show that *the difference scheme* (27) *with* $\lambda = 1$ *subject to the initial data* (29) *has a solution which is equal to the exact solution* (7) *of* (25) *and* (26) *at the points of the net.* (However, for higher order systems and variable coefficients, we do not get the exact solution for any fixed choice of λ. These results suggest the use of the largest value of λ for which the domain of dependence condition is satisfied.)

Let us consider the scheme (27) with λ arbitrary. From the considerations of domains of dependence we know that for $\lambda > 1$, the difference solution cannot generally converge to the exact solution. Therefore, we restrict the mesh ratio by $0 < \lambda \leq 1$ and proceed to show that the approximate solution then converges to the exact solution (which is assumed sufficiently smooth). By using the exact solution $\mathbf{u}(x, t)$ of (25) at the points of the net, we define the local truncation error $\boldsymbol{\tau}(x, t)$,

$$\text{(30)}\qquad \begin{aligned} \boldsymbol{\tau}(x, t) &\equiv \mathbf{u}_t(x, t) - cA\mathbf{u}_{\hat{x}}(x, t) - \frac{c}{2\lambda}\Delta x \mathbf{u}_{x\bar{x}}(x, t) \\ &= \left(\mathbf{u}_t - \frac{\partial \mathbf{u}}{\partial t}\right) - cA\left(\mathbf{u}_{\hat{x}} - \frac{\partial \mathbf{u}}{\partial x}\right) \\ &\qquad - \frac{c}{2\lambda}\Delta x\left(\mathbf{u}_{x\bar{x}} - \frac{\partial^2 \mathbf{u}}{\partial x^2}\right) - \frac{c}{2\lambda}\Delta x \frac{\partial^2 \mathbf{u}}{\partial x^2} \\ &= \mathcal{O}(\Delta t + \Delta x). \end{aligned}$$

If we denote the error in the difference solution by

$$\mathbf{e}(x, t) = \mathbf{U}(x, t) - \mathbf{u}(x, t),$$

then from (27b) and (30) it follows that

$$\mathbf{e}_t(x, t) = cA\mathbf{e}_{\hat{x}}(x, t) + \frac{c}{2\lambda}\Delta x\mathbf{e}_{x\bar{x}}(x, t) - \boldsymbol{\tau}(x, t),$$

or as in (27a) this is equivalent to

$$\text{(31)}\qquad \mathbf{e}(x, t + \Delta t) = \tfrac{1}{2}(I + \lambda A)\mathbf{e}(x + \Delta x, t) + \tfrac{1}{2}(I - \lambda A)\mathbf{e}(x - \Delta x, t) - \Delta t\boldsymbol{\tau}(x, t).$$

Here we have introduced the identity matrix I.

Since A is symmetric, it can be diagonalized by an orthogonal matrix. We have, in fact,

$$\text{(32)}\qquad PAP^* = \begin{pmatrix} 1 & 0 \\ 0 & -1 \end{pmatrix}, \qquad P^{-1} = P^* = P \equiv \frac{1}{\sqrt{2}}\begin{pmatrix} 1 & 1 \\ 1 & -1 \end{pmatrix}.$$

Now let us introduce the vector net function

$$\text{(33)}\qquad \boldsymbol{\epsilon}(x, t) \equiv \begin{pmatrix} \xi(x, t) \\ \eta(x, t) \end{pmatrix} = P\mathbf{e}(x, t)$$

and multiply (31) by P on the left to get

$$\boldsymbol{\epsilon}(x, t + \Delta t) = \tfrac{1}{2}(I + \lambda PAP^*)\boldsymbol{\epsilon}(x + \Delta x, t) + \tfrac{1}{2}(I - \lambda PAP^*)\boldsymbol{\epsilon}(x - \Delta x, t) - \Delta tP\boldsymbol{\tau}(x, t).$$

By taking absolute values and using (32), we find in component form

$$|\xi(x, t + \Delta t)| \le \tfrac{1}{2}|1 + \lambda|\cdot|\xi(x + \Delta x, t)| + \tfrac{1}{2}|1 - \lambda|\cdot|\xi(x - \Delta x, t)| + \frac{\Delta t}{\sqrt{2}}|\tau_1(x, t) + \tau_2(x, t)|$$

$$|\eta(x, t + \Delta t)| \le \tfrac{1}{2}|1 - \lambda|\cdot|\eta(x + \Delta x, t)| + \tfrac{1}{2}|1 + \lambda|\cdot|\eta(x - \Delta x, t)| + \frac{\Delta t}{\sqrt{2}}|\tau_1(x, t) - \tau_2(x, t)|.$$

Since $0 < \lambda \le 1$, the absolute value signs can be removed from the factors $|1 \pm \lambda|$. Then with the definitions

$$\text{(34)}\qquad E(t) \equiv \sup_x \|\boldsymbol{\epsilon}(x, t)\|, \qquad \sigma(t) \equiv \sup_x \|\boldsymbol{\tau}(x, t)\|,$$

where the norm of a vector is the maximum absolute component, we deduce

$$E(t + \Delta t) \le E(t) + \sqrt{2}\,\Delta t\sigma(t).$$

A recursive application of this inequality yields

$$E(t) \leq E(0) + \sqrt{2}\,\Delta t \sum_{\nu=1}^{t/\Delta t} \sigma(t - \nu\,\Delta t) \tag{35}$$

$$\leq E(0) + \sqrt{2}\,t\,\|\sigma(t)\|$$

where

$$\|\sigma(t)\| = \sup_{t' \leq t} \sigma(t') = \sup_{\substack{x \\ t' \leq t}} \|\boldsymbol{\tau}(x, t')\|.$$

Now we recall that, from (32) and (33), $\mathbf{e}(x, t) = P\boldsymbol{\epsilon}(x, t)$ and so

$$\begin{aligned} \|\mathbf{e}(x, t)\| &\leq \|P\| \cdot \|\boldsymbol{\epsilon}(x, t)\| \\ &\leq \sqrt{2}\,\|\boldsymbol{\epsilon}(x, t)\| \\ &\leq \sqrt{2}\,E(t). \end{aligned}$$

Thus it follows from (35) and the definitions of **e** and **E** that

$$\|\mathbf{U}(x, t) - \mathbf{u}(x, t)\| \leq \sqrt{2} \sup_x \|\mathbf{U}(x, 0) - \mathbf{u}(x, 0)\| + 2t\,\|\sigma(t)\|. \tag{36}$$

Note that the suprema on the right side need be taken only over points in the domain of dependence of the point (x, t). By using the initial data (29) and the estimate (30) of the local truncation error, the above implies

$$\|\mathbf{U}(x, t) - \mathbf{u}(x, t)\| \leq t\,\mathcal{O}(\Delta t + \Delta x). \tag{37}$$

Thus, as was to be shown, *the difference solution of* (27) *and* (29) *converges to the exact solution of* (25) *and* (26) *as* $\Delta t \to 0$ *and* $\Delta x \to 0$ *for* $\lambda = c\Delta t/\Delta x \leq 1$. The convergence here is at least first order in Δt or Δx. In Problem 2, the numerical scheme (27c) is shown to be convergent if the rounding error is of the same order as the truncation error.

We remark that the scheme (27) is convergent for hyperbolic systems of order n in the form (25a), where A is diagonalizable (see Problems 3 and 4).

PROBLEMS, SECTION 3

1. Show that the difference scheme $W_t = W_{\hat{x}}$, with constant $\lambda = \Delta t/\Delta x$, is divergent as an approximation to $\partial w/\partial t = \partial w/\partial x$.

[Hint: Show that $e^{\beta t}e^{i\alpha x}$ is a solution of the difference equation, if $i^2 = -1$ and

$$e^{\beta\Delta t} = 1 + i\frac{\Delta t}{\Delta x}\sin\alpha\,\Delta x.$$

Consider now the initial data

$$W(x, 0) = \sum_{r=0}^{\infty} 2^{-2r} \cos \frac{\pi}{2} 2^r x.$$

Show that $W(0, t) \to \infty$ if $\Delta x = 2^{-n}$ and $n \to \infty$. That is, set $\alpha_r = \pi 2^{r-1}$ and

$$W(x, t) \equiv \text{Re} \sum_{r=0}^{\infty} 2^{-2r} e^{\beta_r t} e^{i\alpha_r x}.$$

Show that the term $r = n$ dominates the sum of all of the other terms as $n \to \infty$, in

$$W(0, t) \geq -\sum_{r=n+1}^{\infty} 2^{-2r} + \text{Re} \sum_{r=0}^{n} 2^{-2r}\left(1 + i\lambda \sin \frac{\pi}{2} 2^{r-n}\right)^{t/\Delta t}.]$$

2. Show that the difference scheme

$$\mathbf{U}_t = cA\mathbf{U}_{\hat{x}} + \frac{c}{2\lambda} \Delta x \mathbf{U}_{x\bar{x}} + \boldsymbol{\rho}(x, t), \tag{27c}$$

$$\mathbf{U}(x, 0) = \mathbf{u}(x, 0) + \boldsymbol{\rho}(x)$$

converges with error

$$\|\mathbf{U}(x, t) - \mathbf{u}(x, t)\| \leq t\mathcal{O}(\Delta t + \Delta x),$$

if the rounding errors $\boldsymbol{\rho}(x, t)$ and $\boldsymbol{\rho}(x)$ are at most of magnitude $\mathcal{O}(\Delta t + \Delta x)$.

3. Carry out the proof of convergence of scheme (27) for the case of a system of n equations (25a), where A is a constant matrix having a complete set of eigenvectors.

4. If $A \equiv A(x, t)$ has a uniformly bounded matrix of real eigenvectors $P(x, t)$, with uniformly bounded inverse $P^{-1}(x, t)$, then show that (27) is a convergent scheme for (25a).

5. Given the difference scheme $W_t = W_{\hat{x}}$ (shown in Problem 1 to be divergent if $\lambda = \Delta t/\Delta x$ is constant), prove convergence if $\lambda = \mu \Delta x$ with some constant μ, for the periodic initial value problems

$$W(x, 0) = f(x) \equiv \sum_{n=-\infty}^{\infty} a_n e^{inx}$$

such that $\sum_{n=-\infty}^{\infty} n|a_n| < \infty$.

[Hint: Verify that the function

$$W(x, t) \equiv \sum_{n=-\infty}^{\infty} a_n \left(1 + i\frac{\Delta t}{\Delta x} \sin n\Delta x\right)^{t/\Delta t} e^{inx}$$

is defined by the series, satisfies the difference equation, and converges to $f(x + t)$ as $\Delta x \to 0$, if $\Delta t/\Delta x = \mu \Delta x$. That is, show by using Lemma 0.1′ of Chapter 8,

$$\begin{aligned}
\left|1 + i\frac{\Delta t}{\Delta x} \sin n\Delta x\right|^{t/\Delta t} &\leq (1 + \mu^2 \Delta x^2 \sin^2 n\Delta x)^{t/(2\Delta t)} \\
&\leq e^{(\mu t/2) \sin^2 n\Delta x} \\
&\leq e^{\mu t/2}.]
\end{aligned}$$

Such a difference scheme is rather inefficient, since too many time steps are required.

4. HEAT EQUATION

The initial value problem for the heat equation is: Find a continuous function $u(x, t)$ that satisfies

(1a) $$\frac{\partial u}{\partial t} - \frac{\partial^2 u}{\partial x^2} = 0 \qquad t > 0;$$

(1b) $$u(x, 0) = f(x) \qquad -\infty < x < \infty.$$

The solution of this problem is found to be

(2) $$u(x, t) = \int_{-\infty}^{\infty} \frac{e^{-(\xi - x)^2/4t}}{\sqrt{4\pi t}} f(\xi)\, d\xi.$$

Here we assume $f(x)$ to be bounded and continuous, and then direct differentiation under the integral sign shows that (1a) is satisfied. Since

$$\int_{-\infty}^{\infty} e^{-y^2}\, dy = \sqrt{\pi}$$

we may write (2) as

$$u(x, t) = f(x) + \int_{-\infty}^{\infty} \frac{e^{-(\xi - x)^2/4t}}{\sqrt{4\pi t}} [f(\xi) - f(x)]\, d\xi,$$

and now let $t \to 0$ from above. For all $\xi \neq x$ we have

$$\lim_{t \to 0} \frac{e^{-(\xi - x)^2/4t}}{\sqrt{4\pi t}} = 0$$

and for $\xi = x$, the remaining factor in the integrand vanishes. Thus it is plausible that we could prove that the function given by (2) is continuous and satisfies the initial condition (1b).

Now from (2), we see that if $f(x) > 0$ in an open interval (a, b) and $f(x) \equiv 0$ outside (a, b), then $u(x, t) > 0$ for all x when $t > 0$. Thus we may say that *signals propagate with infinite speed* for the heat equation. Clearly, the form of the solution in (2) shows that *the domain of dependence* of a point (x, t) with $t > 0$ *is the entire* x-axis (*or initial line*).

With the uniform net spacings Δx, Δt, and net points D_Δ in the half-space $t > 0$ (see Subsection 3.1), we consider the difference equations

(3a) $$U_t(x, t) - U_{x\bar{x}}(x, t) = 0, \qquad (x, t) \in D_\Delta.$$

In subscript notation with $(x, t) = (x_j, t_n)$, this can be written in the form

(3b) $$U_{j,n+1} = (1 - 2\lambda)U_{j,n} + \lambda(U_{j+1,n} + U_{j-1,n}), \qquad n = 0, 1, \ldots.$$

Here we have introduced the *mesh ratio*

(3c) $$\lambda \equiv \frac{\Delta t}{\Delta x^2}.$$

The net function $U(x, t)$ is also subject to initial conditions which, from (1b), we take as

$$U(x, 0) = f(x). \tag{4}$$

The net points used in the difference equation (3) have the *star* or *stencil* of Figure 1. The solution is easily evaluated by means of (3b) and we see that *the numerical interval of dependence* of a point (x, t) in the net *is the initial line segment* $[x - (\Delta x/\Delta t)t, x + (\Delta x/\Delta t)t]$. Thus in order to satisfy the domain of dependence condition of Subsection 3.1, which is again valid, we must have that $\Delta t/\Delta x \to 0$ as $\Delta t \to 0$ and $\Delta x \to 0$. Otherwise, the numerical interval of dependence of the difference equation (3) would not become arbitrarily large, and hence convergence could not occur in all cases.

If, as the net spacing goes to zero, the mesh ratio λ defined in (3c) is constant, then $\Delta t/\Delta x = \lambda\,\Delta x \to 0$ and the domain of dependence condition is satisfied. We shall show, in fact, that if $0 < \lambda \leq \frac{1}{2}$, then the difference scheme (3) and (4) is convergent; but *if* $\lambda > \frac{1}{2}$ *the difference solution does not generally converge to the exact solution.* As usual, the truncation error $\tau(x, t)$ on D_Δ is defined by writing for the exact solution $u(x, t)$ of (1)

$$u_t(x, t) - u_{x\bar{x}}(x, t) \equiv \tau(x, t), \qquad (x, t) \in D_\Delta. \tag{5a}$$

By Taylor's theorem the truncation error can be expressed, assuming u to be sufficiently smooth, as

$$\tau(x, t) = \frac{\Delta t}{2}\frac{\partial^2 \bar{u}}{\partial t^2} - \frac{\Delta x^2}{12}\frac{\partial^4 \bar{\bar{u}}}{\partial x^4}. \tag{5b}$$

With the definition

$$e(x, t) = U(x, t) - u(x, t)$$

we get from (5) and (3)

$$e_{j,n+1} = (1 - 2\lambda)e_{j,n} + \lambda(e_{j+1,n} + e_{j-1,n}) - \Delta t \tau_{j,n}.$$

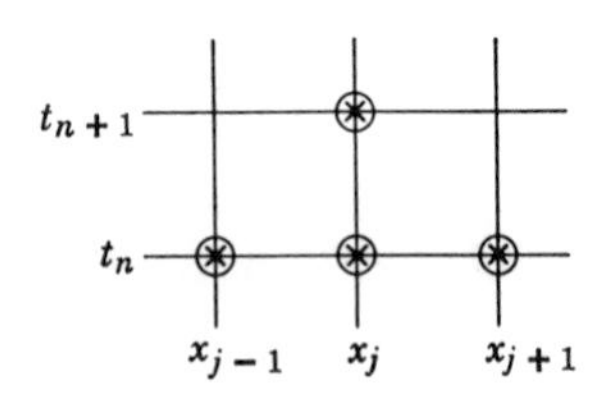

Figure 1. Net points of star for the explicit difference scheme (3).

If $0 < \lambda \le \frac{1}{2}$, then $(1 - 2\lambda) \ge 0$ and with the definitions

$$E_n \equiv \sup_j |e_{j,n}|, \qquad \tau \equiv \sup_{j,n} |\tau_{j,n}|,$$

the above yields upon taking absolute values

$$\begin{aligned}|e_{j,n+1}| &\le (1 - 2\lambda)|e_{j,n}| + \lambda(|e_{j+1,n}| + |e_{j-1,n}|) + \Delta t|\tau_{j,n}| \\ &\le E_n + \Delta t \tau.\end{aligned}$$

Or since the right-hand side is now independent of j,

$$E_{n+1} \le E_n + \Delta t \tau.$$

Hence, by a recursive application

$$E_n \le E_0 + n\Delta t \tau = E_0 + t_n \tau.$$

Thus we have deduced that

$$|u(x, t) - U(x, t)| \le \sup_x |u(x, 0) - U(x, 0)| + t\mathcal{O}(\Delta t + \Delta x^2). \tag{6}$$

Therefore, by recalling (4), $|u(x, t) - U(x, t)| \to 0$ as $\Delta t \to 0$ and $\Delta x \to 0$, if $\lambda = \Delta t/\Delta x^2 \le \frac{1}{2}$. The convergence demonstrated here is of order $\mathcal{O}(\Delta x^2)$ since $\Delta t = \lambda \Delta x^2$.

To demonstrate the divergence of the difference scheme (3) when $\lambda > \frac{1}{2}$, we first construct explicit solutions of the difference equations. We try net functions of the exponential form

$$V^{(\alpha)}(x, t) = \text{Re}\,(e^{i\alpha x - \omega t}).$$

Then

$$\begin{aligned}V_t^{(\alpha)} - V_{x\bar{x}}^{(\alpha)} &= \left[V^{(\alpha)}(x, t)\left(\frac{e^{-\omega\Delta t} - 1}{\Delta t} - \frac{e^{i\alpha\Delta x} - 2 + e^{-i\alpha\Delta x}}{\Delta x^2}\right)\right] \\ &= \left\{V^{(\alpha)}(x, t)\frac{1}{\Delta t}\{e^{-\omega\Delta t} - [(1 - 2\lambda) + \lambda e^{i\alpha\Delta x} + \lambda e^{-i\alpha\Delta x}]\}\right\} \\ &= V^{(\alpha)}(x, t)\frac{1}{\Delta t}\left[e^{-\omega\Delta t} - \left(1 - 4\lambda\sin^2\frac{\alpha\Delta x}{2}\right)\right].\end{aligned}$$

Now $V^{(\alpha)}$ is a solution of the difference equations provided that ω and α satisfy

$$e^{-\omega\Delta t} = 1 - 4\lambda\sin^2\frac{\alpha\Delta x}{2}.$$

The initial conditions satisfied by $V^{(\alpha)}$ are

$$V^{(\alpha)}(x, 0) = \text{Re}\,e^{i\alpha x} = \cos\alpha x \tag{7a}$$

and the solution can be written, since $e^{-\omega t} = (e^{-\omega\Delta t})^{t/\Delta t}$,

$$V^{(\alpha)}(x, t) = \cos\alpha x\left(1 - 4\lambda\sin^2\frac{\alpha\Delta x}{2}\right)^{t/\Delta t}. \tag{7b}$$

Clearly, for all Δx and Δt such that $\lambda \leq \frac{1}{2}$ and real α, it follows that the solution (7) satisfies $|V^{(\alpha)}(x, t)| \leq 1$. However, if $\lambda > \frac{1}{2}$, then for some α and Δx we have $|1 - 4\lambda \sin^2 (\alpha\Delta x/2)| > 1$ and so $|V^{(\alpha)}(0, t)|$ becomes arbitrarily large for sufficiently large $t/\Delta t$. We will capitalize on this *instability* (see next section) of the difference scheme (3), if $\lambda > \frac{1}{2}$, to construct a smooth initial condition for which divergence is easily demonstrated. Since the difference equations are linear and homogeneous, we may superpose solutions of the form (7) to get other solutions. With $\alpha = \alpha_\nu = 2^\nu\pi$ and coefficients $\beta_\nu > 0$, we form

$$\text{(8a)} \qquad \begin{aligned} V(x, t) &= \sum_{\nu=0}^{\infty} \beta_\nu V^{(\alpha_\nu)}(x, t) \\ &= \sum_{\nu=0}^{\infty} \beta_\nu \cos (2^\nu\pi x)\left(1 - 4\lambda \sin^2 \frac{2^\nu\pi\Delta x}{2}\right)^{t/\Delta t}. \end{aligned}$$

The corresponding initial function

$$\text{(8b)} \qquad V(x, 0) = f(x) = \sum_{\nu=0}^{\infty} \beta_\nu \cos (2^\nu\pi x),$$

has as many derivatives as we wish provided that $\beta_\nu \to 0$ sufficiently fast. Now let $\Delta x = 2^{-m}$ and $\Delta t = \lambda 4^{-m}$ so that (8a) yields

$$\begin{aligned} V(0, t) &= \sum_{\nu=0}^{\infty} \beta_\nu\left[1 - 4\lambda \sin^2 \left(2^{\nu-m}\frac{\pi}{2}\right)\right]^{t/\Delta t} \\ &= \sum_{\nu=0}^{m} \beta_\nu\left[1 - 4\lambda \sin^2 \left(2^{\nu-m}\frac{\pi}{2}\right)\right]^{t/\Delta t} + \sum_{\nu=m+1}^{\infty} \beta_\nu. \end{aligned}$$

But

$$\sin^2 \left(2^{\nu-m}\frac{\pi}{2}\right) \leq \tfrac{1}{2} \qquad \text{for } \nu = 0, 1, \ldots, m - 1,$$

and so the above yields, for $\frac{1}{2} < \lambda \leq 1$, with $\beta_\nu > 0$,

$$\begin{aligned} |V(0, t)| &\geq -\sum_{\nu=0}^{\infty} \beta_\nu + \beta_m(4\lambda - 1)^{t/\Delta t} \\ &= -f(0) + \beta_m(4\lambda - 1)^{t4^m/\lambda}. \end{aligned}$$

Now if the β_ν are chosen as $\beta_\nu = e^{-2^\nu}$, then the initial function $f(x)$ is a smooth (analytic) function and the estimate yields, for $\frac{1}{2} < \lambda \leq 1$,

$$\text{(8c)} \qquad |V(0, t)| \geq -V(0, 0) + e^{2^m[(t/\lambda)2^m \ln (4\lambda-1)-1]}.$$

Thus, as $m \to \infty$, it follows that $|V(0, t)|$ becomes unbounded, for any finite $t > 0$, since $4\lambda - 1 > 1$. Hence this difference solution cannot converge to the solution of the corresponding smooth problem with initial data given by (8b). Thus, as was to be shown, the scheme (3) and (4) does not generally converge when $\lambda > \frac{1}{2}$. We say that the difference scheme (3) is *conditionally convergent* which means that the scheme is convergent only if λ satisfies some condition, i.e., $\lambda \leq \frac{1}{2}$. In the next subsection, we will see that it is possible to construct *unconditionally convergent* schemes for the mixed initial-boundary value problem.

4.1. Implicit Methods

To demonstrate implicit difference schemes, we consider mixed initial-boundary value problems for the inhomogeneous heat equation. That is,

$$\text{(9a)} \qquad \frac{\partial u}{\partial t} - \frac{\partial^2 u}{\partial x^2} = s(x, t) \qquad 0 < x < L, \qquad t > 0;$$

$$\text{(9b)} \qquad u(x, 0) = f(x) \qquad 0 \leq x \leq L;$$

$$\text{(9c)} \qquad u(0, t) = g(t), \qquad u(L, t) = h(t), \qquad t > 0$$

The net spacing is now chosen such that

$$\Delta x = \frac{L}{J + 1}$$

and the net points in the interior of the half strip

$$D \equiv \{x, t \mid 0 \leq x \leq L, \ t \geq 0\}$$

we denote by D_Δ; i.e.,

$$D_\Delta \equiv \{x, t \mid x = j\Delta x, \ 1 \leq j \leq J; \ t = n\Delta t, \ n = 1, 2, \ldots\}.$$

For a net function $U(x, t)$, we define the *implicit* difference equations

$$\text{(10a)} \qquad U_{\bar{t}}(x, t) - U_{x\bar{x}}(x, t) = s(x, t), \qquad (x, t) \in D_\Delta.$$

In subscript notation, again with $(x, t) = (x_j, t_n)$, these equations can be written as

$$\text{(10b)} \qquad (1 + 2\lambda)U_{j,n} = U_{j,n-1} + \lambda(U_{j+1,n} + U_{j-1,n}) + \Delta t s_{j,n};$$
$$n = 1, 2, \ldots, \ 1 \leq j \leq J.$$

The only difference between (3a) and (10a) is the time difference quotient, which is forward in (3) and backward in (10). The star associated with (10) is shown in Figure 2. The initial and boundary data are specified in the obvious way

$$\text{(11a)} \qquad U(x_j, 0) \equiv U_{j,0} = f(x_j), \qquad 0 \leq x_j \leq L;$$

$$\text{(11b)} \qquad U(0, t_n) \equiv U_{0,n} = g(t_n), \qquad U(L, t_n) \equiv U_{J+1,n} = h(t_n), \qquad t_n > 0.$$

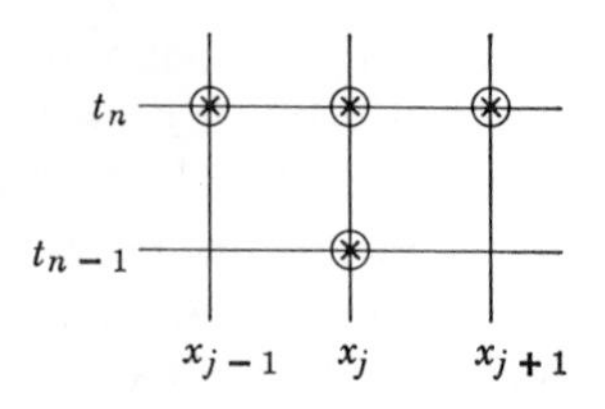

Figure 2. Net points of star for implicit difference scheme (10).

For each $t = t_n$, the equations in (10) and (11b) form a system of $J + 2$ linear equations in the unknowns $U_{j,n}$, $0 \leq j \leq J + 1$. However, since $U_{0,n}$ and $U_{J+1,n}$ are specified in (11b), it can be reduced to a system of order J. In fact, with the coefficient matrix A of order J defined by

$$(12) \qquad A \equiv \mathrm{I} + \lambda B \qquad B \equiv \begin{pmatrix} 2 & -1 & & & & 0 \\ -1 & 2 & -1 & & & \\ & \cdot & \cdot & \cdot & & \\ & & \cdot & \cdot & \cdot & \\ & & & \cdot & \cdot & -1 \\ 0 & & & & -1 & 2 \end{pmatrix}$$

and the J-dimensional vectors $\mathbf{U}_n$, $\mathbf{b}_n$, $\mathbf{s}_n$, and $\mathbf{f}$ defined by

$$(13) \qquad \mathbf{U}_n \equiv \begin{pmatrix} U_{1,n} \\ U_{2,n} \\ \vdots \\ U_{J,n} \end{pmatrix}, \qquad \mathbf{b}_n \equiv \begin{pmatrix} U_{0,n} \\ 0 \\ \vdots \\ 0 \\ U_{J+1,n} \end{pmatrix} = \begin{pmatrix} g_n \\ 0 \\ \vdots \\ 0 \\ h_n \end{pmatrix},$$

$$\mathbf{s}_n \equiv \begin{pmatrix} s_{1,n} \\ s_{2,n} \\ \vdots \\ s_{J,n} \end{pmatrix}, \qquad \mathbf{f} \equiv \begin{pmatrix} f_1 \\ f_2 \\ \vdots \\ f_J \end{pmatrix}$$

the systems (10) and (11) can be written as

$$(14) \qquad A\mathbf{U}_n = \mathbf{U}_{n-1} + \lambda \mathbf{b}_n + \Delta t \mathbf{s}_n, \qquad n = 1, 2, \ldots; \ \mathbf{U}_0 = \mathbf{f}.$$

For each n, a system with the same tridiagonal coefficient matrix A must be solved. Since $\lambda > 0$, it follows that the lemma of Subsection 3.2 in Chapter 2 applies. Thus, not only is A non-singular, but the solution of each system is easily obtained by evaluating two simple two term recursions. Of course, the factorization $A = LU$ need only be done initially. [See equations (3.10) through (3.13) of Chapter 2.]

It is clear from (14) that the difference solution $\mathbf{U}_n$ at any time t_n depends upon all components of the initial data, $\mathbf{U}_0$, for any value of λ. This is also clear from the form of the star corresponding to the difference equations (10). Thus *for any value of the mesh slope*, $\Delta t/\Delta x$, the numerical domain of dependence is the entire initial line segment and hence *the domain of dependence condition is automatically satisfied* by the implicit difference scheme. We shall show that, in addition, *the implicit difference solution converges for all values of* λ to the exact solution. In other words, the scheme is *unconditionally convergent*.

The truncation error $\tau(x, t)$ of the solution $u(x, t)$ of (9) is defined for the difference scheme (10) as

$$\tau(x, t) \equiv u_{\bar{t}}(x, t) - u_{\bar{x}\bar{x}}(x, t) - s(x, t), \qquad (x, t) \in D_\Delta. \tag{15a}$$

Since $u(x, t)$ satisfies (9a), we obtain by the usual Taylor's series expansions, assuming sufficient differentiability of the solution,

$$\tau(x, t) = \mathcal{O}(\Delta t + \Delta x^2). \tag{15b}$$

Now from (10), (11), and (15) we get for the error,

$$e(x, t) \equiv U(x, t) - u(x, t),$$

the difference problem

$$e_{\bar{t}}(x, t) - e_{x\bar{x}}(x, t) = -\tau(x, t), \qquad (x, t) \in D_\Delta; \tag{16a}$$

$$e(x, 0) = 0; \tag{16b}$$

$$e(0, t) = 0, \qquad e(L, t) = 0. \tag{16c}$$

In subscript notation (16a) yields

$$(1 + 2\lambda)e_{j,n} = e_{j,n-1} + \lambda(e_{j+1,n} + e_{j-1,n}) - \Delta t \tau_{j,n},$$
$$n = 1, 2, \ldots, \quad 1 \leq j \leq J.$$

By taking absolute values and using $E_n \equiv \max_j |e_{j,n}|$, $\tau \equiv \sup_{\substack{j \\ n' \leq N}} |\tau_{j,n'}|$, and $\lambda > 0$, we get

$$(1 + 2\lambda)|e_{j,n}| \leq E_{n-1} + 2\lambda E_n + \Delta t \tau, \qquad 1 \leq j \leq J,\ n \leq N.$$

Since the right-hand side is independent of j and $e_{0,n} = e_{J+1,n} = 0$, we may replace $|e_{j,n}|$ by E_n to get

$$E_n \leq E_{n-1} + \Delta t \tau.$$

Thus by the usual recursion technique

$$E_n \leq E_0 + t_n \tau,$$

or

$$(17) \qquad \begin{aligned} |u(x, t) - U(x, t)| &\leq \max_x |u(x, 0) - U(x, 0)| + t\tau \\ &\leq t\mathcal{O}(\Delta t + \Delta x^2), \qquad t \leq N\Delta t \equiv T. \end{aligned}$$

From this result we deduce unconditional convergence as $\Delta t \rightarrow 0$ and $\Delta x \rightarrow 0$, i.e., λ is arbitrary.

There are other implicit schemes which converge for arbitrary λ and one of them in particular has a local truncation error which is $\mathcal{O}(\Delta t^2 + \Delta x^2)$. We examine the family of schemes defined by

$$(18a) \qquad \begin{aligned} U_{\bar{t}}(x, t) - [\theta U_{x\bar{x}}(x, t) + (1 - \theta)U_{x\bar{x}}(x, t - \Delta t)] \\ = \theta s(x, t) + (1 - \theta)s(x, t - \Delta t), \qquad (x, t) \in D_\Delta. \end{aligned}$$

Here θ is a real parameter such that $0 \leq \theta \leq 1$. For $\theta = 1$, (18a) reduces to (10a); while for $\theta = 0$, (18a) is equivalent to (3). For any $\theta \neq 0$, the difference equations (18) are implicit. The boundary and initial data are as specified in (11). In subscript notation (18a) takes the form

$$(18b) \qquad \begin{aligned} &(1 + 2\theta\lambda)U_{j,n} - \theta\lambda(U_{j+1,n} + U_{j-1,n}) \\ &\quad = [1 - 2(1 - \theta)\lambda]U_{j,n-1} + (1 - \theta)\lambda(U_{j+1,n-1} + U_{j-1,n-1}) \\ &\qquad + \Delta t[\theta s_{j,n} + (1 - \theta)s_{j,n-1}], \\ &\qquad\qquad n = 1, 2, \ldots, \quad 1 \leq j \leq J. \end{aligned}$$

By using the matrices and vectors in (12) and (13), the system (18) and (11) can be written as

$$\mathbf{U}_0 = \mathbf{f},$$

$$(19) \qquad \begin{aligned} (I + \theta\lambda B)\mathbf{U}_n = [I - (1 - \theta)\lambda B]\mathbf{U}_{n-1} + \lambda[\theta \mathbf{b}_n + (1 - \theta)\mathbf{b}_{n-1}] \\ + \Delta t[\theta \mathbf{s}_n + (1 - \theta)\mathbf{s}_{n-1}], \qquad n = 1, 2, \ldots. \end{aligned}$$

These systems can be solved by factoring the tridiagonal matrix $I + \theta\lambda B$. Clearly, for $\theta \neq 0$, the domain of dependence condition is satisfied for arbitrary $\Delta t/\Delta x$.

The truncation error $\tau(x, t)$ now depends upon the parameter θ. The usual Taylor's series expansion about (x, t) yields, if the solution $u(x, t)$ of (9) has enough derivatives and we make use of (9a) to simplify,

$$\begin{aligned}(20)\qquad \tau(x, t) &= u_{\bar{t}}(x, t) - [\theta u_{x\bar{x}}(x, t) + (1-\theta)u_{x\bar{x}}(x, t-\Delta t)] \\ &\qquad - [\theta s(x, t) + (1-\theta)s(x, t-\Delta t)] \\ &= \frac{\partial u}{\partial t} - \frac{\partial^2 u}{\partial x^2} - s(x, t) \\ &\qquad - \frac{\Delta t}{2}\left[\frac{\partial^2 u}{\partial t^2} - 2(1-\theta)\left(\frac{\partial^3 u}{\partial t\,\partial x^2} + \frac{\partial s}{\partial t}\right)\right] \\ &\qquad + \mathcal{O}[(\Delta t)^2 + (\Delta x)^2] \\ &= \frac{(1-2\theta)\Delta t}{2}\frac{\partial^2 u}{\partial t^2} + \mathcal{O}(\Delta t^2 + \Delta x^2).\end{aligned}$$

Thus for the special case $\theta = \frac{1}{2}$, the truncation error is $\mathcal{O}(\Delta t^2 + \Delta x^2)$. [In this case all the difference quotients in (18) are centered about $(x, t + \Delta t/2)$ and the difference method is called the *Crank-Nicolson* scheme.] For arbitrary θ, the truncation error is $\mathcal{O}(\Delta t + \Delta x^2)$, as in the explicit and purely implicit cases. With the notation $e \equiv U - u$ we obtain from (20), (18), (11), and (9b and c)

$$(21a)\qquad e_{\bar{t}}(x, t) - [\theta e_{x\bar{x}}(x, t) + (1-\theta)e_{x\bar{x}}(x, t-\Delta t)] = -\tau(x, t), \quad (x, t) \in D_\Delta;$$

$$(21b)\qquad e(x, 0) = 0;$$

$$(21c)\qquad e(0, t) = 0, \qquad e(L, t) = 0.$$

Let us write (21) in vector form by using the matrix B of (12) and the vectors

$$\mathbf{e}_n \equiv \begin{pmatrix} e_{1,n} \\ e_{2,n} \\ \vdots \\ e_{J,n} \end{pmatrix}, \qquad \boldsymbol{\tau}_n \equiv \begin{pmatrix} \tau_{1,n} \\ \tau_{2,n} \\ \vdots \\ \tau_{J,n} \end{pmatrix},$$

to get $\mathbf{e}_0 = 0$ and

$$(22)\qquad (I + \theta\lambda B)\mathbf{e}_n = [I - (1-\theta)\lambda B]\mathbf{e}_{n-1} - \Delta t\boldsymbol{\tau}_n, \qquad n = 1, 2, \ldots.$$

Since $\lambda > 0$ and $I + \theta\lambda B$ is non-singular, we may multiply by the inverse of this matrix to get

$$(23a)\qquad \mathbf{e}_n = C\mathbf{e}_{n-1} + \Delta t\boldsymbol{\sigma}_n, \qquad n = 1, 2, \ldots,$$

where we have introduced

$$(23b)\qquad C \equiv I - (I + \theta\lambda B)^{-1}\lambda B, \qquad \boldsymbol{\sigma}_n \equiv -(I + \theta\lambda B)^{-1}\boldsymbol{\tau}_n.$$

A recursive application of (23a) yields

$$\mathbf{e}_n = C^n\mathbf{e}_0 + \Delta t \sum_{\nu=1}^{n} C^{\nu-1}\boldsymbol{\sigma}_\nu. \tag{24}$$

Upon taking norms of this representation of the error, we get

$$\begin{aligned} \|\mathbf{e}_n\| &\le \|C\|^n \cdot \|\mathbf{e}_0\| + \Delta t \sum_{\nu=1}^{n} \|C\|^{\nu-1} \cdot \|\boldsymbol{\sigma}_\nu\| \\ &\le \|C\|^n \cdot \|\mathbf{e}_0\| + \Delta t \frac{1 - \|C\|^n}{1 - \|C\|} \cdot \max_{1 \le \nu \le n} \|\boldsymbol{\sigma}_\nu\|. \end{aligned} \tag{25}$$

[This could have been deduced directly from (23a) by first taking the norm and then applying the recursion.]

Let us use a special norm in (25) for which we are able to compute $\|C\|$, i.e., $\|\mathbf{x}\| = \left(\sum_{i=1}^{n} |x_i|^2\right)^{1/2}$. Then, since B is symmetric, it follows that C in (23b) is symmetric and by (1.11) of Chapter 1,

$$\|C\| = \rho(C) \equiv \max_j |\gamma_j(C)|,$$

where $\gamma_j(C)$ is an eigenvalue of C. That is, the spectral radius of C is the corresponding natural norm. The eigenvalues of B are easily obtained. We note that the matrix B is related to the matrix H in (1.17b). Using L_J of (1.14b) we have $B = 2I - (L_J + L_J^T)$ and the calculations in (1.23)–(1.25) are applicable. Specifically the eigenvalues β_j of B are found to be

$$\beta_j = 4\sin^2\left(\frac{\pi}{2}\frac{j}{J+1}\right), \qquad j = 1, 2, \ldots, J; \tag{26a}$$

corresponding to the eigenvectors

$$\mathbf{y}^{(j)} = \begin{pmatrix} \sin\,[1j\pi/(J+1)] \\ \sin\,[2j\pi/(J+1)] \\ \vdots \\ \sin\,Jj\pi/(J+1)] \end{pmatrix}, \qquad j = 1, 2, \ldots, J. \tag{26b}$$

Thus the eigenvalues, γ_j, of C defined in (23b) are

$$\gamma_j = 1 - \frac{\lambda\beta_j}{1 + \theta\lambda\beta_j}. \tag{27}$$

In order that

$$\rho(C) \equiv \max_j |\gamma_j| < 1,$$

we must have $-1 < \gamma_j < 1$. Since $\beta_j > 0$, it follows that $\gamma_j < 1$ for all $\lambda > 0$ and $\theta \geq 0$. Now $\gamma_j > -1$ is equivalent to

$$(1 - 2\theta)\lambda\beta_j < 2,$$

and this is satisfied for all $\lambda > 0$ if $\theta \geq \frac{1}{2}$. Thus we have shown that

$$\|C\| < 1 \quad \textit{for all} \quad \lambda > 0 \ \textit{if} \ \theta \geq \tfrac{1}{2}. \tag{28a}$$

On the other hand, for $0 \leq \theta < \frac{1}{2}$, we must have $\lambda < 2/[(1 - 2\theta)\beta_j]$. Or since $0 < \beta_j < 4$ for $j = 1, 2, \ldots, J$, this implies

$$\|C\| < 1 \quad \textit{for} \quad \lambda \leq \frac{1}{2(1 - 2\theta)} \ \textit{if} \ 0 \leq \theta < \tfrac{1}{2}. \tag{28b}$$

Under either of the conditions (28), we obtain from (25) and (23b), since

$$\|\boldsymbol{\sigma}_\nu\| \leq \|(I + \theta\lambda B)^{-1}\| \cdot \|\boldsymbol{\tau}_\nu\| \leq \frac{1}{1 + \theta\lambda\beta_1}\|\boldsymbol{\tau}_\nu\| \leq \|\boldsymbol{\tau}_\nu\|,$$

that

$$\begin{aligned} \|\mathbf{e}_n\| &\leq \|\mathbf{e}_0\| + \frac{\Delta t}{1 - \|C\|} \max_{\nu \leq n} \|\boldsymbol{\tau}_\nu\| \\ &= \|\mathbf{e}_0\| + \frac{\Delta t}{1 - \|C\|} \mathcal{O}[(\theta - \tfrac{1}{2})\Delta t + \Delta t^2 + \Delta x^2]. \end{aligned} \tag{29}$$

To examine the convergence properties as $\Delta x \to 0$ and $\Delta t \to 0$, we note that for small Δx

$$\beta_1 = \left(\frac{\pi}{L}\Delta x\right)^2 + \mathcal{O}(\Delta x^4), \qquad \beta_J = 4 - \left(\frac{\pi}{L}\Delta x\right)^2 + \mathcal{O}(\Delta x^4).$$

It is easily established that $1 - [x/(1 + \theta x)]$ is a decreasing function of x and an increasing function of θ for $x > 0$ and $\theta \geq 0$. Thus

$$\|C\| = \max(\gamma_1, |\gamma_J|), \tag{30}$$

and for small Δx

$$\begin{aligned} \gamma_1 &= 1 - \lambda\left(\frac{\pi}{L}\Delta x\right)^2 + \mathcal{O}(\lambda\Delta x^4) \\ &= 1 - \Delta t\frac{\pi^2}{L^2} + \mathcal{O}(\lambda\Delta x^4). \end{aligned} \tag{31}$$

In case (28b), for any θ in $0 \leq \theta < \frac{1}{2}$ and $\lambda \leq 1/[2(1 - 2\theta)]$, we get an upper bound for $|\gamma_J|$ by picking the largest value of λ,

$$\begin{aligned} |\gamma_J| &\leq \left|1 - \frac{\beta_J}{2(1 - 2\theta) + \theta\beta_J}\right| \\ &\leq 1 - (\tfrac{1}{2} - \theta)\left(\frac{\pi}{L}\Delta x\right)^2 + \mathcal{O}(\Delta x^4). \end{aligned} \tag{32}$$

Hence, in case (28b) we find from (30), (31), and (32) that

$$1 - \|C\| \geq \begin{cases} \mathcal{O}(\Delta t), \text{ or} \\ \mathcal{O}(\Delta x^2). \end{cases}$$

Therefore (29) yields for $0 \leq \theta < \frac{1}{2}$, $0 < \lambda \leq 1/2(1 - 2\theta)$,

$$\|e_n\| \leq \|\mathbf{e}_0\| + \max(1, \lambda)\mathcal{O}[(\theta - \tfrac{1}{2})\Delta t + \Delta t^2 + \Delta x^2]. \tag{33}$$

Finally, in case (28a), $\theta \geq \frac{1}{2}$ and λ arbitrary, we get an upper bound for $|\gamma_J|$ by picking the smallest value of θ,

$$\begin{aligned} |\gamma_J| &\leq \left|1 - \frac{\lambda\beta_J}{1 + \frac{1}{2}\lambda\beta_J}\right| \\ &\leq \left|1 - \frac{4\lambda}{1 + 2\lambda} + \frac{\lambda}{(1 + 2\lambda)^2}\left(\frac{\pi}{L}\Delta x\right)^2 + \mathcal{O}(\lambda\Delta x^4)\right|. \end{aligned} \tag{34}$$

The last inequality is most useful in the case of very large λ, i.e.,

$$\begin{aligned} |\gamma_J| &\leq \frac{4\lambda}{1 + 2\lambda} - 1 - \frac{\lambda}{(1 + 2\lambda)^2}\left(\frac{\pi}{L}\Delta x\right)^2 + \mathcal{O}(\lambda\Delta x^4) \\ &\leq 1 - \frac{1}{\lambda} + o\left(\frac{1}{\lambda}\right), \qquad \lambda \gg 1. \end{aligned}$$

Therefore,

$$1 - \|C\| \geq \begin{cases} \mathcal{O}(\Delta t), \\ \mathcal{O}(1/\lambda), \qquad \theta \geq \frac{1}{2}. \end{cases}$$

Hence (29) becomes for $\theta \geq \frac{1}{2}$, λ arbitrary,

$$\|\mathbf{e}_n\| \leq \|\mathbf{e}_0\| + \max(1, \lambda\Delta t)\mathcal{O}[(\theta - \tfrac{1}{2})\Delta t + \Delta t^2 + \Delta x^2]. \tag{35}$$

Inequality (35) indicates that with the choice $\theta = \frac{1}{2}$, $\lambda\Delta t =$ constant, the error is bounded by

$$\|\mathbf{e}_n\| \leq \|\mathbf{e}_0\| + \mathcal{O}(\Delta x^2). \tag{36}$$

Of course, this error bound is of the same magnitude as the error estimate for the explicit scheme, but for a much larger time step. That is, even though the number of operations required to solve the implicit equations for one time step is of the order of twice the number of operations required for one time step of the explicit scheme, there is a tremendous saving in labor when we choose $\lambda\Delta t =$ constant for the Crank-Nicolson scheme.

PROBLEMS, SECTION 4

1. Given $u(x, t)$ continuous in

$$\bar{D} \equiv D \cup C_1 \cup C_2,$$

where

$$D \equiv \{x, t \mid 0 < x < L, 0 < t < T\},$$

$$C_1 \equiv \{x, t \mid 0 < x < L, \quad t = T\},$$

$$C_2 \equiv \{x, t \mid 0 \le x \le L, \quad t = 0;$$

$$0 = x, \quad 0 < t \le T;$$

$$x = L, 0 < t \le T\}.$$

If, in $D \cup C_1$, $u(x, t)$ has a continuous second derivative with respect to x and a continuous first derivative with respect to t, such that

$$\frac{\partial u}{\partial t} = \frac{\partial^2 u}{\partial x^2},$$

then

$$\max_{x \in \bar{D}} u(x, t)$$

is attained at some point of C_2. (This is a weak form of the *maximum principle* satisfied by the solutions of the heat equation.)

[Hint: For any $\epsilon > 0$, set

$$v(x, t) \equiv u(x, t) - \epsilon t.$$

Clearly, $v(x, t)$ cannot attain its maximum at any point P of $D \cup C_1$, since otherwise

$$\frac{\partial v(P)}{\partial t} \ge 0,$$

hence

$$\frac{\partial^2 v}{\partial x^2} \equiv \frac{\partial v}{\partial t} + \epsilon,$$

would be positive at P. Therefore, $v(x, t)$ attains its maximum *only* on C_2. But since $\epsilon > 0$ could be arbitrarily small, this implies that $u(x, t)$ must attain its maximum on C_2.]

(Note: By considering

$$W(x, t) \equiv -u(x, t),$$

we can establish the *minimum principle*,

$$\min_{x \in \bar{D}} u(x, t)$$

is attained at some point in C_2.)

2. Verify that the solution $U_{j,n}$ of (10a and b) and (11a and b) satisfies the *maximum principle*

$$V_n \le \max\{V_0, n\Delta t S_n, G_n, H_n\}$$

where

$$V_n \equiv \max_j |U_{j,n}|, \qquad n = 0, 1, 2, \ldots,$$

$$S_n \equiv \max_{\substack{0 \le t \le n\Delta t \\ 0 \le x \le L}} |s(x, t)|,$$

$$G_n \equiv \max_{0 \le t \le n\Delta t} |g(t)|,$$

$$H_n \equiv \max_{0 \le t \le n\Delta t} |h(t)|.$$

[Hint: Modify the argument that estimates E_n, after equations (16a, b, and c).]

3. Formulate and prove a maximum principle for the explicit scheme

$$\text{(10a}') \qquad \begin{aligned} &U_t(x, t) - U_{x\bar{x}}(x, t) = s(x, t), \\ &(x, t) \in D_\Delta, \qquad \text{with } \lambda \le \tfrac{1}{2}. \end{aligned}$$

Use the auxiliary conditions (11a and b).

4. Prove convergence of the finite difference solutions $U(x, t)$ to the solution of the mixed initial-boundary value problem in (9a, b, and c) with $s(x, t) \equiv 0$, under the *weak compatibility condition*

$$f(0) = g(0); \; f(L) = h(0),$$

and with $f(x)$, $g(t)$ and $h(t)$ that are continuous.

[Hint: Uniformly approximate $f(x)$, $g(t)$, and $h(t)$ by polynomials $f_m(x)$, $g_m(t)$, and $h_m(t)$ satisfying the *strong compatibility condition*,

$$f_m''(0) = g_m'(0), \qquad f_m^{(\text{iv})}(0) = g_m''(0);$$

$$f_m''(L) = h_m'(0), \qquad f_m^{(\text{iv})}(L) = h_m''(0).$$

That is, assume there exist corresponding smooth solutions $u_m(x, t)$ and, for any given $(\Delta x, \Delta t)$, difference solutions $U_m(x, t)$ as well as the continuous solution $u(x, t)$ and the difference solution $U(x, t)$. Then estimate

$$\begin{aligned} u(x, t) - U(x, t) = {} & [u(x, t) - u_m(x, t)] \\ & + [u_m(x, t) - U_m(x, t)] \\ & + [U_m(x, t) - U(x, t)], \end{aligned}$$

by using the maximum principles for the outermost bracketed terms to fix an m for which they contribute at most ϵ for all $(\Delta x, \Delta t)$. Next, pick $(\Delta x, \Delta t)$ sufficiently small so that the middle braketed term is at most ϵ.]

5. GENERAL THEORY: CONSISTENCY, CONVERGENCE, AND STABILITY

The apparently scattered results of the preceding sections can be related by a simple general theory. A more complete and rigorous development could be given with the aid of the simplest notions of functional analysis, which we forego.

A partial differential equation can be represented symbolically as

$$L(u) = f(P), \qquad P \in D; \tag{1a}$$

with the convention that only terms involving the dependent variable u are included on the left-hand side of (1) and that all inhomogeneous terms are included in f [i.e., f is a function only of the independent variables, $P \equiv (t, x, y, \ldots)$]. The domain in which (1a) is to be satisfied is denoted by D. The set of points on which boundary and/or initial data are prescribed is denoted by C. The conditions to be satisfied by u on C can be represented as

$$B(u) = g(P), \qquad P \in C. \tag{1b}$$

Here B may not be a differential operator, but (1b) merely represents the conditions imposed on various parts of C. For example, conditions (4.9b) and (4.9c) would both be incorporated in (1b) for the mixed initial-boundary value problem of (4.9). We shall only consider problems (1) for which a unique and smooth solution u exists for any data in some class of smooth functions $\{f, g\}$ (smooth means "sufficiently" differentiable).

Let us consider a net for the independent variables of the problem (1) with spacing: $\Delta t, \Delta x, \Delta y, \ldots$. Certain of these net points, say those *interior* to D will be denoted by the set D_Δ. Similarly, *boundary* net points, C_Δ, will also be defined. There are various ways in which this can be done, depending upon the difference method employed. Obviously net points lying on C may be included in C_Δ, but frequently we may also wish to include the points of intersection of C with the net lines. (In fact, for some problems, net points outside of C are included in C_Δ and points outside D are included in D_Δ, but we shall not dwell on these possibilities in the present discussion.)

At the points of $D_\Delta + C_\Delta$ a difference approximation U is defined as the solution of some set of difference equations. These may be indicated symbolically as

$$L_\Delta(U) = f(P), \qquad P \in D_\Delta, \tag{2a}$$

which is to approximate (1a); and the boundary difference approximations are indicated by

$$B_\Delta(U) = g(P), \qquad P \in C_\Delta. \tag{2b}$$

Again the notation may imply different relations over different parts of C_Δ, say as in (4.11). Of course, it is desired that the difference solution U of (2) should be a close approximation to the solution u of (1) at corresponding points of $D_\Delta + C_\Delta$ for all data that are sufficiently smooth.

Furthermore, the difference solution should be uniquely defined by (2) and its numerical evaluation should be possible without significant loss of accuracy due, say, to roundoff errors. To study these questions the three notions of *consistency*, *convergence*, and *stability* of the difference schemes are introduced. It is then easily shown that *for consistent schemes, stability implies convergence*. We begin with the definition of

CONSISTENCY. *Let $\phi(t, x, y, \ldots)$ be any function with "sufficiently many" continuous partial derivatives in $D + C$. For each such function and every point $P \in D_\Delta$, let*

$$\tau\{\phi(P)\} \equiv L(\phi(P)) - L_\Delta(\phi(P)); \tag{3a}$$

and for each point $P \in C_\Delta$ let

$$\beta\{\phi(P)\} \equiv B(\phi(P)) - B_\Delta(\phi(P)). \tag{3b}$$

Then the difference problem (2) *is consistent with problem* (1) *if*

$$\|\tau\{\phi\}\| \to 0, \qquad \|\beta\{\phi\}\| \to 0, \tag{3c}$$

when $\Delta t \to 0$, $\Delta x \to 0$, $\Delta y \to 0, \ldots,$ in some manner, and $\|\ \|$ represents norms in the appropriate sets D_Δ and C_Δ. We call $\tau\{\phi\}$ and $\beta\{\phi\}$ the local truncation errors.

If (3c) is satisfied only when some particular relationship between the $\Delta t, \Delta x, \Delta y, \ldots$ is maintained (i.e., say provided that $\Delta t/\Delta x \to 0$ as $\Delta t \to 0$ and $\Delta x \to 0$), then we say that the difference formulation is *conditionally consistent*. For example, with the heat equation operator:

$$L(u) \equiv \frac{\partial u}{\partial t} - \frac{\partial^2 u}{\partial x^2}$$

the ordinary explicit scheme of (4.3a) can be written in terms of the difference operator

$$L_\Delta(U) \equiv U_t - U_{x\bar{x}}$$

which is "unconditionally" consistent with $L(u)$. However, the *Dufort-Frankel* explicit scheme employs the difference operator

$$L_\Delta'(U) \equiv \tfrac{1}{2}(U_t + U_{\bar{t}}) - \left(U_{x\bar{x}} - \frac{\Delta t^2}{\Delta x^2}\, U_{t\bar{t}}\right),$$

which is consistent with $L(u)$ only if $\Delta t/\Delta x \to 0$ with the net spacing. In fact, if $\Delta t/\Delta x \equiv c =$ a fixed constant, then the difference operator $L_\Delta'(U)$ is consistent with the hyperbolic operator

$$L'(u) \equiv c^2 \frac{\partial^2 u}{\partial t^2} + \frac{\partial u}{\partial t} - \frac{\partial^2 u}{\partial x^2}.$$

These latter results follow from the simple calculation

$$\tau\{\phi\} \equiv L(\phi) - L_\Delta'(\phi)$$

$$= \left[\frac{\partial \phi}{\partial t} - \frac{1}{2}(\phi_t + \phi_{\bar{t}})\right] - \left(\frac{\partial^2 \phi}{\partial x^2} - \phi_{x\bar{x}}\right)$$

$$- \frac{\Delta t^2}{\Delta x^2}\left(\phi_{t\bar{t}} - \frac{\partial^2 \phi}{\partial t^2}\right) - \frac{\Delta t^2}{\Delta x^2}\frac{\partial^2 \phi}{\partial t^2}$$

$$= \left(\frac{\Delta t^2}{6}\frac{\partial^3 \phi'}{\partial t^3}\right) + \left(\frac{\Delta x^2}{12}\frac{\partial^4 \phi''}{\partial x^4}\right) - \left(\frac{\Delta t^4}{12\Delta x^2}\frac{\partial^4 \phi'''}{\partial t^4}\right) - \frac{\Delta t^2}{\Delta x^2}\frac{\partial^2 \phi}{\partial t^2}.$$

In the above consideration, we have neglected to mention that initial data must also be prescribed at $t = \Delta t$ for L_Δ' to become an explicit scheme. In many cases, say the mixed problem (4.9) where (1b) represents (4.9b and c) and (2b) represents (4.11a and b), we have $\beta\{\phi\} \equiv 0$ and so only the difference approximation to the differential equation determines consistency. On the other hand, if (1b) represents initial conditions like (3.2) for the wave equation, then (2b) represents some approximation like (3.13) or alternatively like (3.13a) and (3.14b). In the first case, we obtain $\beta\{\phi\} = \mathcal{O}(\Delta t)$ and in the second case

$$\beta\{\phi\} = \mathcal{O}\left[\Delta t^2 + \Delta t \Delta x^2 + \Delta t\left(\frac{\partial^2 \phi}{\partial t^2} - c^2\frac{\partial^2 \phi}{\partial x^2}\right)\right].$$

Here we have an example in which the order of the local truncation error is increased for special functions (i.e., solutions of the wave equation).

In practice, we will not work with the exact solution of (2a and b), because of rounding operations. Hence we will consider the solutions W defined on $D_\Delta + C_\Delta$ which satisfy the modified equations

(2c) $$L_\Delta(W) = f(P) + \rho(P) \qquad P \in D_\Delta,$$

(2d) $$B_\Delta(W) = g(P) + \sigma(P) \qquad P \in C_\Delta.$$

The functions $\rho(P)$ and $\sigma(P)$ represent the error introduced in solving (2a and b) approximately. We will refer to $\rho(P)$ and $\sigma(P)$ as *rounding errors.*

We turn now to the definition of

CONVERGENCE. *Let u be the solution of problem* (1), *and let U be the difference solution of problem* (2). *The difference solution is convergent to the exact solution iff*

$$\|u(P) - U(P)\| \to 0$$

for all $P \in D_\Delta + C_\Delta$ when $\Delta t \to 0$, $\Delta x \to 0$, $\Delta y \to 0, \ldots,$ in some manner, and $\|\cdot\|$ represents a norm in $D_\Delta + C_\Delta$.

If the difference solution is convergent for all data in some wide class of smooth functions $\{f, g\}$, we call the corresponding *difference scheme convergent*. Notice, however, that this notion is quite distinct from that of consistency and, in fact, a scheme may readily be consistent but not convergent. The schemes (3.17a and c) which are consistent approximations to (3.8a) furnish two such examples.

Care must be taken in observing that a difference scheme may be convergent for a class, $\mathscr{F}$, of smooth functions $\{f, g\}$, but not convergent for a larger class $\mathscr{G} \supset \mathscr{F}$. For example, consider the Cauchy problem given by

$$\text{(4a)} \qquad \begin{aligned} L(u) &\equiv \frac{\partial u}{\partial t} - \frac{\partial u}{\partial x} = 0, \qquad t > 0, \\ B(u) &\equiv u(x, 0) = e^{i\alpha x}; \end{aligned}$$

and the corresponding difference problem

$$\text{(4b)} \qquad \begin{aligned} L_\Delta(U) &\equiv U_t - U_{\hat{x}} = 0 \\ B_\Delta(U) &\equiv U(x, 0) = e^{i\alpha x}. \end{aligned}$$

We easily verify that for any real α, the solutions u and U of (4a) and (4b), respectively, are

$$\text{(5a)} \qquad u(x, t) = e^{i\alpha(t+x)},$$

and

$$\text{(5b)} \qquad U(x, t) = \left(1 + i\frac{\Delta t}{\Delta x}\sin \alpha\Delta x\right)^{t/\Delta t} e^{i\alpha x}.$$

But if $|\alpha| \leq M$, we have, for $0 \leq t \leq T$ and all x, the uniform convergence

$$\text{(6)} \qquad \lim_{\Delta x, \Delta t \to 0} U(x, t) = e^{i(\alpha t + \alpha x)} = u(x, t).$$

In other words, the scheme $L_\Delta(U) = 0$ which we have shown to be divergent, in general, in Problem 3.1, is convergent when the initial data are chosen from the class of finite trigonometric sums, i.e., for any initial data of the form

$$u(x, 0) = \sum_{j=1}^{N} \beta_j e^{i\alpha_j x}.$$

The reader should observe that even if the domain of dependence condition is violated by the difference scheme (4b) (i.e., $\lambda = \Delta t/\Delta x > 1$), the solution (5b) will still converge to that in (5a). Thus for this *special class of trigonometric data* the *difference scheme* (4b) *is unconditionally convergent.*

We recall that the actual evaluation of a numerical scheme produces approximate solutions, say W, which satisfy slight modifications of (2a and b), say (2c and d). Hence, in practice, we are interested primarily in schemes for which $\|W - u\| \to 0$, when $\|\rho\| \to 0$ and $\|\sigma\| \to 0$, as $\Delta t \to 0$, $\Delta x \to 0, \ldots$. (The norms $\| \;\|$ are defined respectively for the sets $D_\Delta + C_\Delta$, D_Δ and C_Δ, and the data $\{f, g\}$ are to belong to some wide class of functions.) We say that such schemes have *convergent approximate solutions.*

When a convergent scheme is known, we are assured that difference solutions of arbitrarily good accuracy *exist*. When a scheme with convergent approximate solutions is known we are assured that difference solutions of good accuracy can be *computed*! But in either case, it is important that an a priori estimate of the error can be evaluated, preferably in terms of the data and the mesh spacing. For this and other purposes we introduce the concept of

STABILITY. *A difference scheme determined by linear difference operators $L_\Delta(\cdot)$ and $B_\Delta(\cdot)$ is stable if there exists a finite positive quantity K, independent of the net spacing, such that*

$$\|U\| \leq K(\|L_\Delta(U)\| + \|B_\Delta(U)\|) \tag{7}$$

for all net functions U defined on $D_\Delta + C_\Delta$. (The norms $\|\cdot\|$ are, as usual, defined for net functions on $D_\Delta + C_\Delta$, D_Δ and C_Δ respectively.) If (7) *is valid for all net spacings, then the linear difference scheme $\{L_\Delta, B_\Delta\}$ is unconditionally stable; if* (7) *holds for some restricted family of net spacings in which $\Delta t, \Delta x, \Delta y, \ldots,$ may all be made arbitrarily small, then $\{L_\Delta, B_\Delta\}$ is conditionally stable.*

Clearly by this definition, stability of a difference scheme is a property independent of any differential equation problem. We have restricted this definition to linear difference schemes as they are the only ones treated in this chapter. However, a more general definition can be given which reduces to the above for linear problems. (This is an obvious restatement of the definition given in Section 5 of Chapter 8 for ordinary differential equations.) Briefly, if L_Δ and B_Δ are the difference operators in question, *they are stable if for every pair of net functions U and V defined on $D_\Delta + C_\Delta$, there is a $K > 0$ independent of the net spacing such that*

$$\|U - V\| \leq K(\|L_\Delta(U) - L_\Delta(V)\| + \|B_\Delta(U) - B_\Delta(V)\|).$$

If L_Δ and B_Δ are linear, then this reduces to the previous definition applied to $(U - V)$.

The factor K in the definitions of stability may depend upon the dimensions of the domain D containing D_Δ. We have already proved the stability of various difference schemes in the previous sections. For example, with the Laplace difference approximation (1.4), consider the difference equations (1.5). The corresponding difference operators are $L_\Delta \equiv -\Delta_\delta$ and $B_\Delta U \equiv U$. By applying Theorem 1.2 we deduce that, for arbitrary Δx and Δy,

$$\|U\| \equiv \max_{D_\delta + C_\delta} |U| \le K(\max_{C_\delta} |U| + \max_{D_\delta} |\Delta_\delta U|)$$

$$= K(\|B_\Delta U\| + \|L_\Delta U\|),$$

where $K \equiv \max(1, a^2/2)$. Thus the difference scheme used in (1.5) is unconditionally stable.

Next, consider the (hyperbolic) system of difference equations defined in (3.27). We define L_Δ by

$$L_\Delta(\mathbf{U}(x, t)) \equiv \mathbf{U}_t(x, t) - cA\mathbf{U}_{\hat{x}}(x, t) - \frac{c}{2\lambda}\Delta x \mathbf{U}_{x\bar{x}}(x, t),$$

and the initial data are to be given by specifying

$$B_\Delta(\mathbf{U}(x, 0)) \equiv \mathbf{U}(x, 0).$$

(The generalization to vectors $\mathbf{U}$, $\mathbf{f}$, and $\mathbf{g}$ is taken for granted.) By using these definitions in (2) we have the difference problem

$$L_\Delta(\mathbf{U}(x, t)) = \mathbf{f}(x, t), \qquad (x, t) \text{ in } D_\Delta; \qquad \mathbf{U}(x, 0) = \mathbf{g}(x).$$

However, this is just the problem posed in (3.31) for $\mathbf{e}(x, t)$ where $\boldsymbol{\tau}$ replaces $\mathbf{f}$ and $\mathbf{e}(x, 0)$ replaces $\mathbf{g}(x)$. Thus, as in the derivation of (3.35) and (3.36), we deduce for the above difference problem that if $\lambda \equiv c\Delta t/\Delta x \le 1$, then with the maximum norms over the appropriate sets

$$\|\mathbf{U}(x, t)\| \le K(\|\mathbf{g}\| + \|\mathbf{f}\|) = K(\|B_\Delta \mathbf{U}\| + \|L_\Delta \mathbf{U}\|),$$

where $K \equiv \max(\sqrt{2}, 2t)$. Hence, conditional stability is established (i.e., for $c\Delta t \le \Delta x$) and we note that the constant K grows with the time interval included in D.

Finally, consider the explicit difference equations for the heat equation, which we write as,

$$L_\Delta U(x, t) \equiv U_t(x, t) - U_{x\bar{x}}(x, t) = f(x, t).$$

If, initially, we take

$$B_\Delta U(x, 0) \equiv U(x, 0) = g(x),$$

then exactly as in the derivation of (4.6) from (4.3) and (4.4) we get, provided $\lambda = \Delta t/\Delta x^2 \le \frac{1}{2}$,

$$\|U(x, t)\| \equiv \max_x |U(x, t)| \le K(\|f(x)\| + \max_{t' \le t} \|s(x, t')\|),$$

where $K \equiv \max(1, t)$. Again we have conditional stability, for $\Delta t \leq \Delta x^2/2$, and the constant K grows with the time. The special choice $s(x, t) \equiv 0$ and $f(x) \equiv V(x, 0)$ given by (4.8b), for the above difference problem shows that *this explicit difference scheme is unstable for fixed* $\lambda > \frac{1}{2}$. The purely implicit difference equations given in (4.10) with initial and boundary data specified as in (4.11) are easily shown, by the methods used in (4.15) through (4.17), to form an unconditionally stable difference scheme.

The basic result connecting the three concepts which we have introduced in this section may be stated as

THEOREM 1. *Let L_Δ and B_Δ be linear difference operators which are stable and consistent with L and B on some family of nets in which $\Delta t, \Delta x, \Delta y, \ldots,$ may be made arbitrarily small. Then the difference solution U of* (2) *is convergent to the solution u of* (1).

Proof. For each point $P \in D_\Delta$ and for any of the above family of nets, we obtain by subtracting (1a) from (2a)

$$\begin{aligned} 0 &= L_\Delta(U(P)) - L(u(P)) \\ &= [L_\Delta(U(P)) - L_\Delta(u(P))] + [L_\Delta(u(P)) - L(u(P))]. \end{aligned}$$

From the assumed linearity of L_Δ and the definition of the local truncation error, $\tau\{\phi\}$, we then have

$$\text{(8a)} \qquad L_\Delta(U - u) = \tau\{u(P)\}, \qquad P \in D_\Delta.$$

In an analogous manner (1b) and (2b) imply

$$\text{(8b)} \qquad B_\Delta(U - u) = \beta\{u(P)\}, \qquad P \in C_\Delta.$$

However, the difference operations in (8) have been assumed stable on the family of nets employed here. Thus it follows that for the net function $(U - u)$,

$$\text{(9)} \qquad \|U - u\| \leq K(\|\tau\{u\}\| + \|\beta\{u\}\|).$$

Now by the assumed consistency we may let $\Delta t \to 0, \Delta x \to 0, \Delta y \to 0, \ldots,$ in such a manner that $\|\tau\| \to 0$ and $\|\beta\| \to 0$. Then, obviously, $\|U - u\| \to 0$ and convergence is demonstrated. ∎

It should be recalled, in the above proof, that the solution u of (1) is to have as many continuous derivatives as are required for the derivation of consistency. We then see from (6) that the error in the difference solution is estimated in terms of the local truncation errors. With little change in the proof, Theorem 1 is applicable if L_Δ and B_Δ are nonlinear stable difference operators. (It should also be observed that the linearity of L and B have not been assumed in the above proof. Their linearity follows from the required consistency with linear difference operators.)

5.1. Further Consequences of Stability

The stability of a linear difference scheme comes very close to insuring that computations with the proposed scheme are practical. More precisely, what we are assured is that stable linear difference equations *have a unique solution* and that, at least in principle, the growth of roundoff errors is bounded. In this case, the scheme has convergent approximate solutions.

Let the difference problem (2a and b) be linear. Then the difference equations form a linear system whose order is equal to the number of net points in $D_\Delta + C_\Delta$. Now we *assume* that the number of unknowns and equations are equal. Having made this important assumption (which in particular cases is easily verified) we may, in order to show that (2a and b) has a solution, either show that some coefficient matrix is non-singular or, equivalently, show that the corresponding homogeneous problem has only the trivial solution. However, from the assumed stability of L_Δ and B_Δ we get from (2a and b)

$$\|U\| \leq K(\|f\| + \|g\|).$$

It follows that the system has only the trivial solution if $f \equiv g \equiv 0$. Thus the unique solvability of the linear difference problem is a simple consequence of stability.

The consideration of the effect of roundoff errors is also quite simple. If by $W(P)$ we represent the numbers actually obtained in numerically approximating the solution of (2a and b), then

$$L_\Delta W = f(P) + \rho(P) \qquad P \in D_\Delta; \tag{10a}$$

$$B_\Delta W = g(P) + \sigma(P) \qquad P \in C_\Delta. \tag{10b}$$

Here $\rho(P)$ and $\sigma(P)$ represent the effects of rounding, which cause W to be in error, and hence not quite satisfy the system (2a and b). As in the proof of Theorem 1, we now derive by means of the linearity of L_Δ and B_Δ

$$L_\Delta(W - u) = \tau\{u(P)\} + \rho(P) \qquad P \in D_\Delta;$$

$$B_\Delta(W - u) = \beta\{u(P)\} + \sigma(P) \qquad P \in C_\Delta.$$

From the assumed stability of the difference problem it now follows that

$$\|W - u\| \leq K(\|\tau\| + \|\beta\| + \|\rho\| + \|\sigma\|). \tag{11}$$

Thus we have shown that *a stable and consistent linear scheme has convergent approximate solutions.*

To maintain accuracy consistent with the local truncation errors, the local roundoffs ρ and σ should be of the same order in the net spacing. The quantities ρ and σ introduced in (10) are usually the actual rounding errors committed in computing, divided by some multiple of the net spacing. This is because one does not compute with the equations in the form (2), but rather with multiples of these equations in which the coefficients are bounded as the net spacing vanishes. So the actual rounding errors should be reduced like the truncation error times a power of the net spacing.

The definition of stability that we have given is considerably more restrictive than is required to prove convergence in many cases. In fact, the "stability constant" K may be allowed to depend upon the net spacing and be unbounded as, say, $\Delta t \to 0$. But if in this case

$$\lim_{\Delta t \to 0} (\Delta t)^p K = 0 \tag{12}$$

for some $p > 0$, then convergence still follows if $\|\tau\| + \|\beta\| = \mathcal{O}(\Delta t^p)$. Convergent approximate solutions are also obtained if the norms of the rounding errors, $\|\rho\|$ and $\|\sigma\|$, are required to be at least $\mathcal{O}(\Delta t^p)$. [When a condition of the form (12) holds for all $p \geq p_0 > 0$ but not for $p < p_0$, the scheme is frequently said to be *weakly stable*.] In addition, as we have seen in the examples, many proofs of stability yield inequalities of the form

$$\|U\| \leq K_0\|L_\Delta U\| + K_1\|B_\Delta U\|, \tag{13}$$

which then yield stability with the constant $K = \max(K_0, K_1)$. However, if for example $K_1 \to 0$ as the net spacing vanishes, then (13) would imply convergence if only L_Δ were consistent with L but not necessarily B_Δ with B. Thus *it is possible to have convergence without consistency* provided a stronger form of stability holds. In the case of such stronger stability it is clear that the error bound (11) can be replaced by

$$\|W - u\| \leq K_0(\|\tau\| + \|\rho\|) + K_1(\|\beta\| + \|\sigma\|). \tag{14}$$

Thus a poorer approximation of the boundary conditions need not affect the overall accuracy if, as the net spacing vanishes,

$$K_1(\|\beta\| + \|\sigma\|) \approx K_0(\|\tau\| + \|\rho\|).$$

5.2. The von Neumann Stability Test

There is an important special class of difference schemes for which a simple algebraic criterion always furnishes a necessary and sometimes even a sufficient condition for stability. Simply stated the difference schemes in question should have constant coefficients and correspond to pure initial value problems with periodic initial data.

We shall illustrate this theory first by considering rather general *explicit* difference schemes of the form

$$\text{(15)} \qquad \mathbf{U}(t + \Delta t, x) = \sum_{j=-m}^{m} C_j \mathbf{U}(t, x + j\Delta x); \qquad 0 \le t \le T - \Delta t,\ 0 \le x \le 2\pi.$$

Here $\mathbf{U}(t, x)$ is, say, a p-dimensional vector to be determined on some net with spacing $\Delta x, \Delta t$ and the $C_j \equiv C_j(\Delta x, \Delta t)$ are matrices of order p which are independent of x, t, but, in general, depend upon Δx and Δt. Initial data are also specified by, say,

$$\text{(16)} \qquad \mathbf{U}(0, x) = \mathbf{g}(x),$$

where $\mathbf{g}(x + 2\pi) = \mathbf{g}(x)$. The difference equations (15) employ data at $2m + 1$ net points on level t in order to compute the new vector at one point on level $t + \Delta t$. We use the assumed periodicity of $\mathbf{U}(t, x)$ in order to evaluate (15) for x near 0 or 2π. With no loss in generality then, assume that $m\Delta x < 2\pi$, and $\mathbf{U}(t, x)$ is defined for all x in $0 \le x \le 2\pi$, and $t = 0, \Delta t, 2\Delta t, \ldots$ (see Problem 2).

Since $\mathbf{U}(t, x)$ is to be periodic in x and the C_j are constants, we can formally construct Fourier series solutions of (15). That is, $\mathbf{U}(t, x)$ is of the form

$$\text{(17)} \qquad \mathbf{U}(t, x) = \sum_{k=-\infty}^{\infty} \mathbf{V}(t, k) e^{ikx}.$$

Upon recalling the orthogonality over $[0, 2\pi]$ of e^{ikx} and e^{iqx} for $k \ne q$, we find that this series satisfies (15) iff

$$\text{(18a)} \qquad \mathbf{V}(t + \Delta t, k) = G(k, \Delta x, \Delta t)\mathbf{V}(t, k)$$

where

$$\text{(18b)} \qquad G(k, \Delta x, \Delta t) \equiv \sum_{j=-m}^{m} C_j(\Delta x, \Delta t) e^{ijk\Delta x}, \qquad |k| = 0, 1, 2, \ldots.$$

From (16) and (17), it follows that the $\mathbf{V}(0, k)$ are just the Fourier coefficients of the initial data, $\mathbf{g}(x)$; i.e.,

$$\mathbf{V}(0, k) = \frac{1}{2\pi} \int_0^{2\pi} \mathbf{g}(x) e^{-ikx}\, dx.$$

Repeated application of (18a) now yields

$$\text{(19)} \qquad \mathbf{V}(t, k) = G^n(k, \Delta x, \Delta t)\mathbf{V}(0, k), \qquad n = t/\Delta t.$$

The matrices $G(k, \Delta x, \Delta t)$, of order p as defined in (18b) are called the *amplification matrices* of the scheme (15), since they determine the growth of the Fourier coefficients of the solutions of (15).

Because $\mathbf{U}(t, x)$ is defined for $0 \leq x \leq 2\pi$, we may introduce as a norm

$$(20) \qquad \|\mathbf{U}(t)\| \equiv \left\{\frac{1}{2\pi}\int_0^{2\pi} |\mathbf{U}(t, x)|^2\, dx\right\}^{1/2}, \qquad t = 0, \Delta t, 2\Delta t, \ldots,$$

which for each t is called the $\mathscr{L}^2$-norm of $\mathbf{U}(t, x)$. (By $|\mathbf{U}|$, we denote the Euclidean norm of the vector $\mathbf{U}$.) However, by the Parseval equality (see Theorem 5.3 of Chapter 5), or directly by using (17) in (20), we have

$$(21) \qquad \|\mathbf{U}(t)\| = \left\{\sum_{k=-\infty}^{\infty} |\mathbf{V}(t, k)|^2\right\}^{1/2}.$$

One of the main reasons for using the $\mathscr{L}^2$-norm is that it can be simply related, as above, to the sum of the squares of the Fourier coefficients. Then from (19) and (21), we conclude that

$$(22) \qquad \|\mathbf{U}(t)\| \leq \max_k \|G^n(k, \Delta x, \Delta t)\| \left\{\sum_{k=-\infty}^{\infty} |\mathbf{V}(0, k)|^2\right\}^{1/2}$$
$$= (\max_k \|G^n(k, \Delta x, \Delta t)\|)\|\mathbf{U}(0)\|, \quad t = n\Delta t,\ n = 1, 2, \ldots.$$

The matrix norm $\|G^n\|$ to be used in (22) is, of course, any norm compatible with the Euclidean vector norm $|\mathbf{V}|$. As previously observed in Section 1 of Chapter 1 the natural norm induced by this vector norm is the smallest such compatible matrix norm and so we shall employ it here. Thus with no loss in generality let $\|G^n\|$ be the spectral norm of G^n [see the definition in (1.11) of Chapter 1 and the discussion preceding Lemma 1.2 thereof].

Let us say, for the present, that *the difference scheme* (15) *and* (16) *is stable* (*in the* $\mathscr{L}^2$-*norm*) *iff there exists a constant K, independent of the net spacing, such that*

$$(23) \qquad \|\mathbf{U}(t)\| \leq K\|\mathbf{U}(0)\|, \qquad 0 \leq t \leq T,$$

for all solutions $\mathbf{U}(t, x)$ *of* (15) *and* (16) *for all* $\mathbf{g}(x)$ *with finite* $\mathscr{L}^2$-*norm.* But then we see from (22) that stability is a consequence of the uniform boundedness of the powers of the amplification matrices. To be more precise we introduce the definition: the family of matrices

$$G^n(k, \Delta x, \Delta t) \qquad \begin{cases} m\Delta x \leq 2\pi \\ 0 < n\Delta t \leq T \\ k = 0, \pm 1, \pm 2, \ldots \end{cases}$$

is *uniformly bounded* if there exists a constant K independent of k, Δx and Δt such that

$$\|G^n(k, \Delta x, \Delta t)\| \leq K \qquad \begin{cases} \text{for } m\Delta x \leq 2\pi \\ 0 < n\Delta t \leq T \\ |k| = 0, 1, 2, \ldots. \end{cases} \tag{24}$$

Now we have the simple

THEOREM 2. *The difference equations* (15) *are stable in the* $\mathscr{L}^2$*-norm iff the amplification matrices* (18b) *satisfy the uniform boundedness condition* (24).

Proof. As indicated above, (24) and (22) imply (23) thus showing the sufficiency. On the other hand, if (24) is not satisfied for any finite K, then (23) cannot be satisfied for any finite K, for all $\mathbf{U}(0, x)$. That is, given any K, a single Fourier term of (17) will be a solution of (15), if (18) is satisfied, and can be chosen so that (23) is violated. ∎

Having established the importance of uniform boundedness we now state a simple necessary condition for stability due to *von Neumann*.

THEOREM 3. *If the scheme* (15) *is stable* (*in the* $\mathscr{L}^2$*-norm*) *then there exists a constant* M, *independent of the net spacing, such that*

$$\rho(G(k, \Delta x, \Delta t)) \leq 1 + M\Delta t, \qquad \textit{for } k = 0, \pm 1, \pm 2, \ldots. \tag{25}$$

Proof. If (15) is stable, then by Theorem 2 the uniform boundedness condition (24) holds. But upon recalling Lemma 1.2 of Chapter 1, we have

$$\begin{aligned} \rho^n(G(k, \Delta x, \Delta t)) &= \rho(G^n(k, \Delta x, \Delta t)) \\ &\leq \|G^n(k, \Delta x, \Delta t)\| \\ &\leq K, \qquad |k| = 0, 1, 2, \ldots, \ 0 < n\Delta t \leq T, \ m\Delta x \leq 2\pi. \end{aligned}$$

With no loss in generality we may take $K \geq 1$ and thus for $T = n\Delta t$,

$$\rho(G(k, \Delta x, \Delta t)) \leq K^{1/n} = K^{\Delta t/T} \leq 1 + K\frac{\Delta t}{T}, \qquad \Delta t \leq T.$$

The last inequality is established in Problem 3 and with $M \equiv K/T$ the result (25) follows. ∎

We call (25) the *von Neumann condition* and Theorem 3 shows that it is necessary for stability in the $\mathscr{L}^2$-norm. In some cases it may also be a sufficient condition. For instance, if the amplification matrices $G(k, \Delta x, \Delta t)$ are Hermitian, then $\|G^n\| = \|G\|^n = \rho^n(G)$ and Theorem 2 implies that the von Neumann condition is sufficient for stability. In any event, one should always try to compute the eigenvalues of $G(k, \Delta x, \Delta t)$ to rule out possibly unstable schemes. If (25) is valid only for special nets, say with $\Delta x = \phi(\Delta t)$, then the scheme may only be conditionally stable (see Problems 4 and 5).

Difference schemes of the form (15) can be applied to quite general classes of partial differential equation problems. For example, systems of the form (3.25a) can be treated provided the matrix cA is constant and a uniformly spaced net is used. More generally, we can treat systems of the form

$$L\mathbf{u}(t, x) \equiv \frac{\partial \mathbf{u}(t, x)}{\partial t} - \mathscr{A}\mathbf{u}(t, x) = 0 \tag{26}$$

where $\mathscr{A}$ is some differential operator with respect to the x variable and has constant coefficients. Thus the heat equation (4.1a) and other higher order systems are included.

The previous analysis is easily extended to more general difference schemes. The most obvious such generalization is to implicit schemes which may be written as

$$\sum_{|j| \leq m} B_j \mathbf{U}(t + \Delta t, x + j\Delta x) = \sum_{|j| \leq m} C_j \mathbf{U}(t, x + j\Delta x), \qquad 0 \leq t \leq T - \Delta t. \tag{27}$$

Here the B_j are matrices of order p, independent of x and t, but dependent, in general, on Δx and Δt. We now need only change the definition of the amplification matrices from (18b) to

$$G(k, \Delta x, \Delta t) \equiv \left(\sum_{|j| \leq m} B_j e^{ijk\Delta x}\right)^{-1} \left(\sum_{|j| \leq m} C_j e^{ijk\Delta x}\right), \qquad |k| = 0, 1, 2, \ldots \tag{28}$$

and of course require that the indicated inverse exists (see Problem 4). We can treat difference schemes with more than two time levels but then the amplification matrices are of higher order, say of order pq for $q + 1$ time levels. Extensions to more independent variables are not difficult.

We recall that the stability definition (23) which has been used in this subsection is not the same as that in (7), which is employed in the basic Theorem 1. However, for the class of equations, say of the form (26), for which difference schemes of the form (15) or (27) are appropriate, we can show that (23) [or (24)] is equivalent to (7). Specifically, we have

THEOREM **4**. *Let $B_\Delta \equiv I$, the identity operator, and let L_Δ be defined by*

$$\Delta t L_\Delta \mathbf{U}(t, x) \equiv \mathbf{U}(t + \Delta t, x) - \sum_{|j| \leq m} C_j \mathbf{U}(t, x + j\Delta x). \tag{29}$$

Then for difference problems of the form (15) *and* (16) *the definition* (23) *of $\mathscr{L}^2$-stability is equivalent to that in* (7) *with appropriate norms.*

Proof. The appropriate norms to employ in (7) are, in terms of the $\mathscr{L}^2$-norm (20),

$$\|\mathbf{U}\| \equiv \sup_{0 \le t_n \le T} \|\mathbf{U}(t_n)\|, \qquad \|L_\Delta \mathbf{U}\| \equiv \sup_{0 < t_n \le T} \|L_\Delta \mathbf{U}(t_n)\|,$$

$$\|B_\Delta \mathbf{U}\| \equiv \|\mathbf{U}(0)\|.$$

Then trivially it follows that (7) implies (23) for all solutions $\mathbf{U}$ of (15) and (16); that is, for all $\mathbf{U}(t, x)$ satisfying

$$L_\Delta \mathbf{U} = 0 \text{ on } D_\Delta, \qquad \text{i.e., } 0 < t \le T;$$

$$B_\Delta \mathbf{U} \equiv \mathbf{U}(0, x) = \mathbf{g}(x) \text{ on } C_\Delta, \qquad \text{i.e., } t = 0.$$

Now let $\mathbf{U}(t, x)$ be *any* function with convergent Fourier series (17), *not necessarily a solution of* (15). Insert the series on the right-hand side of (29), multiply the result by $(1/2\pi)e^{-ikx}$, and integrate over $[0, 2\pi]$. We obtain with the definition (18b),

$$\mathbf{V}(t + \Delta t, k) = G(k, \Delta x, \Delta t)\mathbf{V}(t, k) + \frac{\Delta t}{2\pi} \int_0^{2\pi} e^{-ikx} L_\Delta \mathbf{U}(t, x)\, dx.$$

By applying this result recursively in t, with the notation $G \equiv G(k, \Delta x, \Delta t)$ and $n\Delta t = t$, it follows that

$$\mathbf{V}(t, k) = G^n \mathbf{V}(0, k) + \frac{\Delta t}{2\pi} \sum_{\nu=0}^{n-1} G^\nu \int_0^{2\pi} e^{-ikx} L_\Delta \mathbf{U}(t - \nu\Delta t, x)\, dx,$$

$$0 < n\Delta t \le T, \quad |k| = 0, 1, 2, \ldots.$$

Upon taking the Euclidean norm of this vector equation, we have

$$|\mathbf{V}(t, k)| \le \max_{\nu \le n} \|G^\nu(k, \Delta x, \Delta t)\|$$

$$\times \left\{ |\mathbf{V}(0, k)| + \Delta t \sum_{\nu=0}^{n-1} \left| \frac{1}{2\pi} \int_0^{2\pi} e^{-ikx} L_\Delta \mathbf{U}(t - \nu\Delta t, x)\, dx \right| \right\}.$$

Finally, by using this result in (21) (see Problem 6), we get with the aid of the Schwarz inequality, the inequalities

$$2ab \le a^2 + b^2 \quad \text{and} \quad a + b \ge \sqrt{a^2 + b^2},$$

the estimate

$$\text{(30)} \qquad \|\mathbf{U}(t)\| \le \sqrt{2} \sup_{\substack{\nu \le n \\ |k| < \infty}} \|G^\nu(k, \Delta x, \Delta t)\| \{ \|\mathbf{U}(0)\| + t \max_{\tau \le t} \|L_\Delta \mathbf{U}(\tau)\| \},$$

$$\text{for } n\Delta t \le T, \quad |k| = 0, 1, 2, \ldots.$$

Here we have introduced the $\mathscr{L}^2$-norm of $L_\Delta \mathbf{U}(t, x)$, as in (20), and used the Parseval equality to deduce that

$$\|L_\Delta \mathbf{U}(\tau)\|^2 = \frac{1}{2\pi} \sum_{k=-\infty}^{\infty} \left| \int_0^{2\pi} e^{-ikx} L_\Delta \mathbf{U}(\tau, x)\, dx \right|^2.$$

Now let the scheme (15) be stable in the sense of (23). Then by Theorem 2 the amplification matrices $G^{\nu}(k, \Delta x, \Delta t)$ are uniformly bounded, as in (24), and we find from (30) that

$$\|\mathbf{U}\| \leq K'\{\|\mathbf{U}(0)\| + \|L_{\Delta}\mathbf{U}\|\}$$

where $K' \equiv \sqrt{2}\, K \max(1, T)$. But this holds for all (sufficiently smooth) functions $\mathbf{U}(t, x)$ in $0 \leq x \leq 2\pi$, $0 \leq t \leq T$, and so the stability defined in (7) holds for the difference operator L_{Δ} of (29). ∎

We conclude with the remark that the condition of stability thus far has only been shown to be sufficient for convergence, But, by using elementary ideas of functional analysis, Lax has proved that for linear well-posed initial value problems, the *consistent difference schemes* (15) or (27) *are stable iff the schemes are convergent*!

PROBLEMS, SECTION 5

1.* With the notation $\sum_{j=-n}^{n}{}'$ and x_k of Subsection 5.1 of Chapter 5, set $h = \Delta x$ and verify that the 2π periodic solution $W(x, t)$ of $W_t = W_{\hat{x}}$, with $W(x_k, 0) = f(x_k)$ for $k = 0, \pm 1, \ldots, \pm n$, satisfies

$$\frac{1}{2n}\sum_{k=-n}^{n}{}' |W(x_k, t)|^2 = \sum_{j=-n}^{n}{}' \left|b_j\left(1 + i\frac{\Delta t}{\Delta x}\sin j\Delta x\right)^{t/\Delta t}\right|^2$$

$$\leq e^{\mu t}\sum_{j=-n}^{n}{}' |b_j|^2 = \frac{e^{\mu t}}{2n}\sum_{k=-n}^{n}{}' |W(x_k, 0)|^2,$$

where

$$b_j = \frac{\Delta x}{2\pi}\sum_{k=-n}^{n}{}' f(x_k)e^{-ijx_k} \quad \text{and} \quad \Delta t = \mu\Delta x^2.$$

That is, if we define $\| \ \|$ by

$$\|W(t)\|^2 = \frac{1}{2n}\sum_{k=-n}^{n}{}' |W(x_k, t)|^2,$$

we have shown that

$$\|W(t)\| \leq e^{\mu t/2}\|W(0)\|$$

and the difference scheme is stable for $\Delta t = \mu\Delta x^2$ for any constant μ (see Problem 3.5).

2. Explain how if $\mathbf{U}(0, x)$ is given, (15) may be used to define $\mathbf{U}(\Delta t, x)$, for $0 \leq x \leq 2\pi$—even though (15) is a difference equation.

3. Verify the inequality, with $K \geq 1$,

$$K^x \leq 1 + Kx, \qquad \text{for } 0 \leq x \leq 1.$$

[Hint: Study $f(x) \equiv e^{x \ln K}$, show that $f'(x) > 0, f''(x) > 0$.]

4. (a) Show that the amplification matrices for the Crank-Nicolson like scheme (4.18) are the scalars

$$G(k, \Delta x, \Delta t) = \frac{1 - (1 - \theta)2\lambda(1 - \cos k\Delta x)}{1 + \theta 2\lambda(1 - \cos k\Delta x)}, \qquad \lambda = \frac{\Delta t}{\Delta x^2},$$

(b) Since this is a case where G is Hermitian, determine by means of the von Neumann condition the restrictions on λ for stability for any θ in $0 \leq \theta \leq 1$. Compare your results with (4.28).

5. (a) Find the amplification matrices of the scheme (3.27). Verify that the von Neumann condition is satisfied when the Courant condition, $\lambda \equiv c\Delta t/\Delta x < 1$, is satisfied.

(b) Apply the von Neumann test to the divergent scheme

$$\mathbf{U}(x, t + \Delta t) = \mathbf{U}(x, t) + \frac{\lambda}{2} A[\mathbf{U}(x + \Delta x, t) - \mathbf{U}(x - \Delta x, t)],$$

$$A = \begin{pmatrix} 0 & 1 \\ 1 & 0 \end{pmatrix}.$$

6. If $a(k, n) = b(k) + c \sum_{\nu=0}^{n-1} d(k, \nu)$, for real numbers a, b, c, and d, show that

$$[a(k, n)]^2 \leq 2\left\{[b(k)]^2 + c^2\left[\sum_{\nu=0}^{n-1} d(k, \nu)\right]^2\right\}.$$

Hence, show that

$$[a(k, n)]^2 \leq 2\left\{[b(k)]^2 + nc^2 \sum_{\nu=0}^{n-1} [d(k, \nu)]^2\right\}.$$

[Hint: Apply Schwarz' inequality: $(\sum 1 \cdot d)^2 \leq (\sum 1)(\sum d^2)$.]

Bibliography

GENERAL TEXTS

Hildebrand, F. B., *Introduction to Numerical Analysis*, McGraw-Hill Book Co., New York, 1956.

Householder, Alston S., *Principles of Numerical Analysis*, McGraw-Hill Book Co., New York, 1953.

John, F., *Advanced Numerical Methods* (lecture notes), New York University, Courant Institute of Mathematical Sciences, New York, 1956. Revised as *Lectures on Numerical Analysis*, Gordon and Breach, New York, to be published.

Lanczos, Cornelius, *Applied Analysis*, Prentice-Hall, Englewood Cliffs, N.J., 1956.

Milne, W. E., *Numerical Calculus*, Princeton University Press, Princeton, N.J., 1949.

Todd, John (editor), *Survey of Numerical Analysis*, McGraw-Hill Book Co., New York, 1962.

SPECIAL REFERENCES FOR CHAPTERS 1, 2, 3, 4

Faddeev, D. K., and V. N. Faddeeva, *Computational Methods of Linear Algebra* (translated by R. C. Williams), W. H. Freeman and Co., San Francisco, 1963.

Forsythe, G. E., and P. Henrici, "The cyclic Jacobi method for computing the principal vectors of a complex matrix," *Trans. Amer. Math. Soc.*, **94**, 100–115 (1958).

Francis, J. G. F., "The *QR* transformation—a unitary analogue to the *LR* transformation," *Computer Journal*, **4**, 265–271, 332–345 (1961/62).

Givens, W., "A method of computing eigenvalues and eigenvectors suggested by classical results on symmetric matrices," *Appl. Math. Ser. Natl. Bur. Stand.*, **29**, 117–122 (1953).

———, "Numerical computation of the characteristic values of a real symmetric matrix," *Oak Ridge Natl. Lab. Report:* ORNL 1574, 1954.

Householder, Alston, *The Theory of Matrices in Numerical Analysis*, Blaisdell Publishing Co., New York, 1964.

Ostrowski, A. M., *Solution of Equations and Systems of Equations*, Academic Press, New York, 1960.

Ostrowski, A. M., "On the convergence of the Rayleigh quotient iteration for the computation of the characteristic roots and vectors, I–VI," *Arch. Rational Mech. Anal.*, **1**, 233–241 (1958); **2**, 423–428 (1959); **3**, 325–340, 341–367, 472–481 (1959); **4**, 153–165 (1960).

Partlett, B., "The development and use of methods of *LR* type," *SIAM Review*, **6**, 275–295 (1964).

Rutishauser, H., "Solution of eigenvalue problems with the *LR*-transformation," *Appl. Math. Ser. Natl. Bur. Stand.*, **49**, 47–81 (1958).

Wilkinson, J. H., *Rounding Errors in Algebraic Processes*, Prentice-Hall, Englewood Cliffs, N.J., 1963.

———, *The Algebraic Eigenvalue Problem*, Oxford University Press, London, 1965.

SPECIAL REFERENCES FOR CHAPTERS 5, 6, 7

Achieser, N. I., *Theory of Approximation* (translated by C. J. Hyman), Frederick Ungar Publishing Co., New York, 1956.

Davis, P. J., *Interpolation and Approximation*, Blaisdell Publishing Co., New York, 1965.

Krylov, V. I., *Approximate Calculation of Integrals* (translation by A. H. Stroud), The Macmillan Co., New York, 1962.

Langer, R. E. (editor), *On Numerical Approximation*, University of Wisconsin Press, Madison, Wisc., 1959.

Lorentz, G. G., *Bernstein Polynomials*, University of Toronto Press, Toronto, 1953.

Milne-Thomson, L. M., *Calculus of Finite Differences*, Macmillan and Co., London, 1933.

Steffensen, J. F., *Interpolation*, Chelsea Publishing Co., New York, 1927.

Stiefel, E., "Note on Jordan elimination, linear programming and Tchebycheff approximation," *Numer. Math.*, **2**, 1–17 (1960).

Stroud, A. H., "A bibliography on approximate integration," *Math. Comp.*, **15**, 52–80 (1961).

Szego, G., *Orthogonal Polynomials*, American Mathematical Society, New York, 1959.

SPECIAL REFERENCES FOR CHAPTER 8

Dahlquist, G., "Convergence and stability in the numerical integration of ordinary differential equations," *Math. Scand.*, **4**, 33–53 (1956).

Fox, L., *Numerical Solution of Two-Point Boundary Problems*, Clarendon Press, Oxford, 1957.

Henrici, P., *Discrete Variable Methods in Ordinary Differential Equations*, John Wiley and Sons, New York, 1962.

———, *Error Propagation for Difference Methods*, John Wiley and Sons, New York, 1963.

Milne, W. E., *Numerical Solution of Differential Equations*, John Wiley and Sons, New York, 1953.

SPECIAL REFERENCES FOR CHAPTER 9

Collatz, L., *The Numerical Treatment of Differential Equations*, 3rd ed., Springer-Verlag, Berlin, 1960.

Courant, R., K. O. Friedrichs, and H. Lewy, "Uber die partiellen Differenzengleichungen der mathematischen Physik," *Math. Ann.*, **100**, 32–74 (1928). (Translation by P. Fox: "On the partial difference equations of mathematical physics," New York University, Courant Institute of Mathematical Sciences, Report NYO-7689, 1956.)

Douglas, J., Jr., "On the relation between stability and convergence in the numerical solution of linear parabolic and hyperbolic differential equations," *SIAM Journal*, **4**, 20–37 (1956).

———, "Survey of numerical methods for parabolic differential equations," *Advances in Computers*, **2**, Academic Press, New York, 1961.

Forsythe, G. E., and W. R. Wasow, *Finite-Difference Methods for Partial Differential Equations*, John Wiley and Sons, New York, 1960.

Kreiss, H. O., "On difference approximations of the dissipative type for hyperbolic differential equations," *Comm. Pure Appl. Math.*, **17**, 335–353 (1964).

Lax, P., "Numerical solution of partial differential equations," *Amer. Math. Monthly*, **72**, No. 2, part II, 74–84 (1965).

Peaceman, D. W. and H. H. Rachford, Jr., "The numerical solution of parabolic and elliptic differential equations," *SIAM Journal*, **3**, 28–41 (1955).

Richtmyer, R. D., *Difference Methods for Initial Value Problems*, Interscience, New York, 1957.

Varga, R. S., *Matrix Iterative Analysis*, Prentice-Hall, Englewood Cliffs, N.J., 1962.

Index

ring tone warbling in his ear for several long seconds before a man's voice finally answered.

"Donald?"

Wolfe spoke slowly and clearly, aware that the line was most likely protected by levels of encryption far more advanced than even his own at Fort Detrick: Bilderberg's most powerful attendees took no chances with their anonymity.

"It is time," he said. "Oppenheimer is in position and ready to strike, as are my men. I only need you to give me the go-ahead and assurance of my security."

The voice replied, calm and in control. "Everything is in place, Donald. As soon as you order your men in, your role in this will be unidentifiable. We will contact you directly at the next Bilderberg meeting once everything has been achieved and the dust has settled. By then, everyone will have forgotten about Jeb Oppenheimer and his crusade."

The line went dead in Wolfe's ear. He immediately punched in a second number and waited for the line to pick up.

"Hoffman."

Red Hoffman was breathing heavily, as though he were slogging his way up a hill.

"What's your status?" Wolfe asked without preamble.

"We're within two miles of them," Hoffman said under his breath. Wolfe could hear other footfalls around him, the sound of troops marching. "We'll have everything under control within the hour."

Wolfe breathed a sigh of relief.

"As soon as you do," he said with finality, "obtain a live subject and leave the area. There must be absolutely no witnesses. Do you understand?"

Hoffman's reply was brisk and uncompromising.

"Understood, sir."

"Bring the subject back to me as soon as you have him, in person."

"Will do. Hoffman out."

Jarvis nodded, turning slowly as he did so, and then his eyes settled on the United Nations headquarters, the flags of its 192 member states arranged in alphabetical order in front of the building, fluttering on their high poles.

And a sudden, terrible realization shot through him.

"I've got to go."

Jarvis clicked off the phone as he struck out across Forty-sixth Street toward the UN buildings, glancing at his watch and hoping against hope that he was wrong. He dialed another number, this time getting Butch Cutler on the other end, sounding as though he was traveling in a vehicle.

"Doug? What's the story?"

"Get the Santa Fe County Sheriff's Office and get them into the SkinGen building as fast as you can. We've got the evidence you need, but there's no time to collate it all and present it to the attorneys. Just go in and find out what the hell they've been up to in there."

"Any idea what they'll be looking for?" Cutler asked.

"Tissues belonging to a victim of the 1918 Spanish Flu pandemic," Jarvis said as he jogged down the street. "I think Donald Wolfe's planning to infect the United Nations General Assembly during his speech there. I need to know how he might do that."

Butch Cutler didn't reply for what felt like a long time as Jarvis jogged toward the vast edifice of the United Nations General Assembly, wishing with every step that he exercised more regularly.

When the reply came, it was tinged with horror.

"There're only two ways he could do it," Cutler replied. "It's either going to be in the air or it's going to be in the water. My guess is he'll infect the water that they're drinking, either through the water supply or directly into their glasses somehow. Viruses don't survive long in the open air."

"Got it."

Jarvis shut off his phone and broke into a run toward the north entrance of the complex that opened onto a landscaped plaza, where the curved facade of the General Assembly Building

The line went dead. Wolfe shut the phone off, unclipped its rear panel, and slipped the SIM card out. He tossed it onto the floor and smashed one heel down on the delicate card. Then he slipped the untraceable cell phone into his pocket and turned, striding across the chamber toward the exit.

He was almost there when he saw two security guards flanking the doors, talking to an old man in a smart blue suit. Wolfe froze on the spot as he recognized Douglas Jarvis, gasping for breath, his face flushed with urgency, gesturing wildly at the two guards.

Wolfe turned and hurried away to take a different exit from the hall, walking across the connection between the General Assembly Hall and the Conference Building, cantilevered over Franklin D. Roosevelt Drive. He then took an elevator to the fourth floor. He weaved his way to the Delegates' Dining Room and just beyond it to a kitchen that served VIPs in the dining room and was often used as a shortcut by delegates between the Conference Building's dining rooms and the General Assembly Hall. Wolfe strode into the kitchens, one hand in his pocket as several members of the staff within glanced at him.

"Can I help you, sir?"

Wolfe smiled at the head chef, gesturing to the kitchens beyond.

"Just a routine search of the premises before the assembly convenes," he said, flashing his USAMRIID identity card. "Ask your staff to give me the room, if you will. I'll walk through here, then exit the same way I came in and report to you there. Is there anything out of order today in here that I need to know about?"

"No, sir." The man shook his head and headed for the kitchen door, instantly recognizing Wolfe's identity and rank. "All's running smoothly."

Donald Wolfe waited until the staff had all left the kitchen, keeping his head up and his eyes alert as he turned to stride between the endless worktops, steel vats, pans, and ovens. As he walked he saw two rows of approximately fifty large glass jugs filled with water. Each would be taken down to the Assembly Hall and used to fill the glasses of hundreds of world leaders as they sat listening to the lectures and speeches that were part of the General Assembly's convention.

Wolfe lifted his hand from his pocket, a large plastic syringe concealed beneath his sleeve. As he walked casually past the huge jugs, he lifted his hand and squirted brief jets of clear fluid, one for each jug, one after another. He then turned at the end of the row and repeated the action down the second row on the other side, releasing jets of infected water into the jugs until his syringe was empty. Wolfe slipped the syringe back into his pocket and walked toward the exit, leaving the kitchens and nodding to the chef by the doors as he departed.

By his best estimate, given the travels of world leaders, the handshakes, the hordes of staff, the telephones and faxes and interviews, cars, aircraft, and beds, from the United Nation's General Assembly Building to a world pandemic would take less than two weeks.

And the best of it was, nobody would show symptoms for at least four days after infection. Within ten, they would be dead.

Doug Jarvis ran across the connection between the General Assembly Hall and the Conference Building, staggering into the elevator and punching the button for the fourth floor. He sucked in air with ragged gasps as he leaned on the aluminum walls and watched the digital floor counter change agonizingly slowly. The elevator alarm pinged, and the doors slid open.

Jarvis took a step out and stared straight into the eyes of Donald Wolfe.

The colonel filled the corridor before him, resplendent in his uniform. Before Jarvis could react, Wolfe rushed forward and slammed his shoulder into Jarvis's chest, plunging them both back into the elevator with a crash of bodies against metal. Wolfe thumped down on top of Jarvis and the impact forced the air from the older man's lungs. Jarvis saw him punch a fist out at the buttons beside them and the elevator doors closed before he reached down and drew a small ceremonial silver pistol from a holster at his waist.

Jarvis struggled against Wolfe's iron grip, but the younger man was too strong for him.

"You're finished, Colonel," Jarvis growled up at him. "Doesn't matter what happens here now. We know everything: Brevig Mission, the flu corpse, SkinGen's involvement. It's over."

Wolfe nodded, jabbing the pistol against Jarvis's cheek.

"Yes, it is indeed over. Or at least it is for the majority of the world's population. It doesn't matter what happens to me now, Mr. Jarvis. This is more important than my survival, or yours. This is about the survival of our species. One way or another, by the end of today one hundred ninety-two world leaders and their staff will walk out of United Nations Plaza carrying the most virulent influenza virus ever to have existed. They will contaminate one another and pass the infection on at a trimetric rate throughout the global population. Hundreds of millions will meet an early grave, for the benefit of those remaining." Wolfe grinned a hawkish smile. "Cruel to be kind, as they say."

Jarvis shook his head.

"You'll never get that far," he said. "SkinGen's already being raided as we speak, and Jeb Oppenheimer's little experiments in New Mexico have unraveled already. The police will be here at any moment."

Wolfe chuckled as he glared down at Jarvis with his piercing gray eyes.

"Not soon enough," Wolfe said. "I'm due onstage in ten minutes. This will all be over by then, if not before. As for you, your time's already done."

Wolfe raised the pistol above his head and brought it crashing down on Jarvis's temple with a sickening crack. Jarvis felt an instant of skull-piercing pain, and then everything turned black.

61

Misery Hole
New Mexico
7:48 a.m.

Ethan jogged forward in a low crouch, dodging left and right between bushes of thorn scrub as he tracked north along a dry riverbed weaving through a rugged valley. The tops of the hills were now bathed in brilliant sunlight that flared brightly off the rocks. The flutter of bats' wings whispered through the air above him as the tiny mammals raced away from the spreading dawn. Behind him Lopez followed his every footstep, whispering as she did so.

"You sure you know where you're going?"

Ethan, his pistol held before him in both hands, nodded.

"Damn straight I do. Be quiet, we're nearly there."

Ethan saw that the valley ahead became steeper, and to his right a narrow track heavily lined with trees and scrub climbed the side of the hills before vanishing entirely on a ridge above them. Somewhere within, he now knew, was an entrance to the caverns concealed from humanity by Ellison Thorne and his men more than 150 years before.

Ethan crept up to the ridge and looked over the edge as Lopez joined him.

Below them, the angular, stacked rocks of the hillside vanished into a yawning chasm perhaps thirty feet across, surrounded by trees and the lechuguilla bushes that had given the mysterious cave somewhere within its name.

"Misery Hole," Lopez said. "Looks deep."

Ethan peered over the edge into the depths and felt his guts convulse as vertigo scrambled his senses.

"Maybe a hundred feet," he whispered, sweating from more now than just the heat. "I can't see the bottom, too dark."

"We've got to get down there," Lopez said. "Fast."

Ethan nodded as he stepped back from the edge and took a deep breath, staring into the distance. Lopez looked at him for a moment and then chuckled.

"Oh, you're kidding me," she said in delight. "The rough and tough marine's scared of heights?"

"Why do you think I joined the marines and not an airborne unit?" Ethan muttered.

"Oh, come on." Lopez punched his shoulder. "Surely you must have jumped out of airplanes or something?"

"Sure we did." Ethan nodded. "Never had a problem with that, due to having a goddamned parachute on my back. But this . . ." He gestured to Misery Hole. "This is different."

Ethan spotted a thick rope ladder bound to some trees fifty yards away. He reluctantly pointed to it, and Lopez followed him around the ragged rocks and through thick scrub until they reached the ladder. Ethan glanced at her.

"You want me to go first?" Lopez offered in an ingratiating, motherly voice.

"Like hell."

Ethan took hold of the ladder, easing himself onto it and descending out over the edge. Don't look down. Just face the rocks. *Semper fi.*

The ladder hung vertically over the abyss, vanishing into the plunging depths below as he clambered one careful step after another. The shadows of the cave engulfed him as he descended, the ladder swinging and twisting precariously as Lopez swayed on the rungs above him.

"Jesus! You wanna keep still?" he snapped, looking up at her and wincing as the whirling vertigo spun his senses again.

"Sorry, honey," Lopez called back with a smile. "Baby steps now, okay?"

A strange odor assaulted Ethan's senses as they clambered down, a damp and almost metallic reek of ancient soil that offered a welcome diversion of his attention. A breeze drifted up past them, cool and moist compared to the swiftly warming desert air above. Rock walls engulfed them as they climbed down, and the pure blue sky above them grew smaller. As his eyes adjusted to the gloom, Ethan was relieved to glimpse below him a dimly illuminated floor strewn with the detritus of rock falls and the desiccated timbers of long-fallen trees, a maze of obstacles crisscrossing the sandy floor.

He climbed down the last few feet of the ladder and dropped onto the dust with a heavy sigh of relief near the entrance to Lechuguilla Cave, a broad and forbidding maw hewn into the side of the rock walls that exhaled the metallic air of the earth itself. He had the brief impression of being in a giant dirty fishbowl, the light streaming down from above in shimmering shafts and the earth littered with debris as though a tornado had blown through.

Lopez dropped down alongside him and looked into the depths of the cave.

"Now what?" she asked.

Ethan was about to reply when the mouth of the cave spoke for him.

"Satan's teeth, you just don't know when to quit, do you, boy?"

From the inky blackness within Lechuguilla Cave, five bayonets glinted in the light beaming down from above. Ethan watched as Ellison Thorne led his companions out into the light, their rifles pointed unwaveringly at him.

Ethan raised his hands, his pistol still in his jeans.

"We're not here to fight."

"How the hell did you find us?" Ellison Thorne growled at him down the barrel of the rifle.

Ethan walked forward to stand within a few yards of Thorne and his men, keeping his hands raised as he spoke.

"We followed the trail," he said.

"We didn't leave a trail," Copthorne shot back.

"I didn't say we followed *your* trail," Ethan replied, and pointed up to the entrance high above. "We followed theirs."

He glanced up into the distant disk of blue sky far above. The four soldiers followed his gaze to see small bats fluttering past against the heavens, some of them diving down toward Lechuguilla and flashing past them into the darkness beyond. Ellison Thorne looked at Ethan, his eyes narrowing.

"Those critters roost in any one o' a thousand caves around these parts," he said.

Ethan slowly lowered his hands, letting his arms hang loosely by his sides as he explained.

"We had a rough idea of where you'd be," he said. "Saffron Oppenheimer's friends have seen you out this way before now. You said you don't know what's happened to you, that Hiram Conley shouldn't have gone to Tyler Willis for help when he started becoming sick. Well, if he hadn't, I wouldn't have been able to find you now. You need to listen to what I have to say."

Copthorne, McQuire, Wren, and Cochrane all looked at Ellison Thorne.

"We ain't got nothing to lose," Kip Wren said. "Let 'im say his piece, then we'll be gone."

Ellison Thorne sighed and lowered his rifle.

"Speak your mind, boy," he rumbled, "but be quick about it."

Ethan nodded.

"Tyler Willis told us that Hiram Conley wasn't blessed with some kind of miracle gene that stopped him from aging. He told us that he was infected with bacteria that must have somehow been able to reside in his body, and in doing so been able to repair cell damage and therefore prevent aging."

Lopez lowered her hands as she spoke. "The reason Conley became sick is that for some reason the population of bacteria in his body slowly decreased with time as they died off, and he began to age rapidly."

"You sayin' there's something living inside us?" Nathaniel asked in horror.

"Everything that's alive has bacteria inside," Ethan explained quickly. "That's just normal. What's unusual is that the bacteria inside your bodies keeps you from aging, and for that to happen it must have evolved over extremely long periods of time to be able to exist symbiotically with human beings."

"Symbi-whatally?" Copthorne muttered in confusion.

"Symbiotically," Lopez said. "It means that you need *it* to be alive, but it also needs *you*. All the bacteria in our bodies are as dependent upon us for survival as we are upon them."

Ethan gestured up to the bats fluttering through the dawn sky above them.

"The only mammalian species that lives in these caves are Mexican free-tailed bats," he said. "They've lived here for millennia, and over that time the bacteria that also live in the caves must have somehow evolved to exist symbiotically with the bats, which are known to have extremely long life spans for their size."

Thorne frowned in confusion.

"But we didn't eat any bats when we came here."

"You didn't have to," Ethan said. "The bacteria would have found a way inside you somewhere in these caves, maybe through cuts or in the water that you drank here. Humans are mammals, just like bats, and the bacteria would have been able to survive inside you in the same way. Once inside, the bacteria would have multiplied, requiring some form of energy from your bodies in return for their existing and altering your cells to prevent aging."

"Iron," Ellison Thorne said. "We needed extra iron in our diets, else we became anemic. These bacteria must have used it as a fuel."

"And many mammals have been known to get minerals from the walls of caves," Lopez said, "which would replace the bat's own iron deficiencies after infection."

"All this is fascinating," Ellison Thorne said, "but right now all I give a damn about is whether we have ourselves a cure here."

"That I can't tell you," Ethan said. "The only thing we can do is get you all out of here and somewhere safe before Jeb Oppenheimer and his crew arrive."

"There isn't anywhere safe for us but Misery Hole," Ellison Thorne snapped. "We're not going anywhere with you or anyone else, y'hear?"

"We don't have much time," Lopez said urgently. "Oppenheimer's men could be here any minute."

"We'll take our chances," Ellison replied.

Ethan was about to speak when a gunshot shattered the silence and Kip Wren let out a cry of agony as his right leg buckled and his rifle spun from his grasp.

Ethan hurled himself to his right as the old soldiers scattered for cover. As he rolled through the dust and pulled his pistol from his jeans, he saw dozens of figures dressed in black lining the edge of Misery Hole far above them. A hail of tracer fire zipped down into the darkness amid the clatter of automatic weapons.

"Oppenheimer!" he shouted. "How the hell did he find us so quickly?"

Lopez, crouched low behind a scattering of rocks, shouted back, "I don't know! We need cover right now!"

Ethan looked at the swarms of mercenaries now running toward the ladder, and knew that there was only one direction left for them to take.

"Fall back!" he shouted. "Into the cave!"

62

Ethan ducked down low and dashed toward the dark maw of the cave behind them as bullets whipped the dust at his feet and sprayed wood chips from fallen branches around him.

He followed Ellison Thorne as the big man hefted Kip Wren onto his shoulder and loped into the darkness of Lechuguilla Cave. Almost immediately they were engulfed in a billowing cloud of cordite smoke as Copthorne, McQuire, and Cochrane opened fire with their Springfield rifles straight up at the ladder above. As Ethan crouched he saw the shots hit their marks to the sound of dull thumps, the leaden Minié balls slamming mercilessly through clothes and flesh. Three of the attackers screamed and fell from the ladder to plunge with their limbs clawing the air.

Ethan and Lopez leaped aside as the bodies thumped down onto the floor of the cave in puffs of dust and the crackle of splintering bones. Ethan aimed his pistol at one of them as he lay screaming in agony and fired a rapid double-tap. The first round hit the man in his belly, the second high on his temple, silencing him instantly.

"Keep them off that ladder!" Ethan shouted.

Realizing that their quarry was armed and now veiled in wreaths of thick white smoke, the attackers suddenly scattered for cover, throwing themselves down behind rocks on the lip of Misery Hole seventy feet above them.

"Three down!" Cochrane shouted. "I'm reckonin' about ninety-six to go!"

Ethan flinched as wild shots smacked into the solid bedrock nearby, stone chips and a fine powder of shocked rock spilling into the air around them. Lopez crawled to Ethan's side.

"We're outnumbered here," she shouted above the crackle of gunfire. "There's no way out but up."

Edward Copthorne reloaded his rifle and glanced across at her.

"There's no way for us but down!" he pointed out. "We've got thousands of passages behind us. We go inside, they'll never find us."

"We'll stand a better chance against those guys inside the caves," Ethan agreed. "Out here, they could rush us at once and it'll all be over and we can't hold them off that ladder forever. Besides, they look like amateurs, missing more than they hit."

Lopez didn't reply, but she got to her feet and dashed off into the darkness behind them. Ethan looked at the old soldiers alongside him, and gestured with his pistol.

"Let another volley off, then go. I'll cover you with the pistol."

Copthorne nodded, and along with the others he aimed his rifle straight up at the lip of Misery Hole.

"Fire!"

A blast of gunfire smacked Ethan's eardrums as thick smoke drifted across his field of view, stinging his eyes and burning in his throat. Copthorne, McQuire, and Cochrane leaped to their feet and dashed for the cave entrance, using the drifting smoke to conceal their flight. Ethan aimed at several black shapes moving among the bushes and rocks high above, firing at them randomly. He was about to turn and flee when a flash of white caught his eye, a man moving from left to right. Ethan squinted through the smoke, and saw the man's awkward gait as he struggled from cover to cover, dragging someone else along behind him. Ethan recognized Jeb Oppenheimer immediately. At that moment, four small, round objects thumped down to land on the cavern floor around him. Ethan needed only one glance to see what they were.

"Grenades!" he shouted.

He fired off another two rounds and then fled into the darkness of the cave. He hurled himself down behind some rocks as four deafening blasts rocked the cavern behind him.

"Y' git any of 'em?" Copthorne asked from somewhere in the darkness.

"Just the one," Ethan said, and reminded himself that he'd fired seven shots, which left eight in the pistol. "Mostly just kept their heads down. How's Kip?"

Ellison Thorne's voice reached them from somewhere deeper in the cave.

"He ain't right. Bullet's in his leg and it ain't goin' nowhere."

Damn. Right now they needed all the firepower and hands

they could muster while they tried to figure out a way of escaping the assault. Ethan looked at the entrance to the cave, the low crevice appearing brightly lit now compared to the depths in which they huddled. The cave behind descended gently, ragged walls and a low ceiling vanishing into impenetrable blackness from which drifted a gentle breeze, as though the earth itself were breathing. There were potentially hundreds of miles of unexplored caverns within, plunging to depths of more than two thousand feet belowground, some flooded, others prone to collapse. Ethan was suddenly acutely aware that they lacked even the most basic expeditionary gear required for a descent into such a complex underground maze. Thousands of people over the years had ventured into such places never to return, hopelessly lost and doomed to starvation and a cold, lonely death far belowground in absolute blackness.

"Heads up!"

John Cochrane's voice startled Ethan as it echoed through the cavern, and he turned to see the shapes of men drop into view from the ladder and rush toward the cave entrance. Instantly, three Springfield rifles blasted out a volley of shots that cut down three of the men. Ethan fired on a fourth, another double-tap that sliced through the attacker with a visible fine spray of blood that spilled onto rocks at the cave entrance. The man fell to his knees before toppling over sideways.

"They're trying to overwhelm us!" McQuire shouted as he struggled to reload his weapon in the eerie half-light.

John Cochrane and McQuire fired at the same time, killing two more men who materialized at the entrance to the cave as they searched for cover from which to continue their assault. The Minié balls slammed into them as they ran and the two men cartwheeled into the dust.

Ethan's ears began aching and ringing from the infernal blasts of the rifles, the terrible noise amplified in the narrow confines of the cave as several more mercenaries tumbled into view. From behind Ethan, Ellison Thorne loomed, a pistol in each hand that spurted smoke and flame as he fired both weapons at once, cutting an attacker down in mid-stride to fall face-first onto the rocks.

"Stand tall, lads!" Ellison bellowed, his huge chest and the narrow cave amplifying his voice to thunder out above the gunfire. "Up an' at 'em!"

With a sudden and unexpected war cry, Ellison led his men at a run toward the cave entrance, their bayonets and knives flashing, and for an instant Ethan's vision blurred as he no longer saw the mercenaries plunging toward them, or the antiquated rifles or the pistol in his hand. A flash vision of other caves from years before filled his mind, the bitter peaks of Afghanistan's Hindu Kush mountains, the screams of the Taliban and United States Marines locked in hand-to-hand combat in the deadly warrens of the Tora Bora cave complex. Young men, far from home, flashing long knives and bayonets, the primeval grunts and cries of mortal combat as crazed terrorists with no fear of death fought idealistic soldiers struggling to preserve their way of life.

Semper fi.

Ethan leaped up on impulse and rushed forward as Ellison's men collided with six soldiers who had managed to breach the entrance to the cave. He saw Ellison Thorne grab the barrel of one man's M16 and twist it aside before stepping in and head-butting him with a sickening crunch, the mercenary's legs crumpling as he sagged onto the floor of the cave. Screams soared through the darkness as bayonets plunged deep into flesh, stubby machine guns batted aside by the long-barreled Springfields before they could be brought to bear.

Ethan charged at the nearest attacker, glimpsing a wiry-looking man with scars that spread from the corners of his mouth back toward his ears, the victim of a mugging or maybe some kind of gang ritual. Either way he didn't look like a soldier, more like one of the heartless bastards Ethan had fought in those terrible caves in Afghanistan, the kind of men who stoned women to death, the kind of men who might have abducted his fiancée, Joanna, in Gaza. The ugly barrel of the soldier's M16 swiveled toward Ethan, who rushed in with almost suicidal rage and grabbed the gun's stock and smashed it aside before driving his thumb in behind the trigger. A jab of pain lanced his thumb as the man squeezed the trigger desperately, only for nothing to happen. In the light of the cave entrance Ethan saw the soldier's features dissolve into panic, just as Ethan hopped violently and drove his right knee up into the man's groin with as much force as he could muster.

The mercenary gagged, his tongue quivering in his gaping mouth and his eyes wide and sightless as exquisite pain racked his body. Ethan yanked the M16 from his grasp and stepped one pace back, spinning the weapon into his grip and then whipping the butt up to crack the man under his jaw. The mercenary's teeth cracked together and crunched through his tongue, the useless muscle spilling sideways out of his mouth in a torrent of thick blood as he collapsed to his knees in agony. Ethan aimed at the kneeling man's head and fired a single shot. The round hit the man square in his right temple and neatly blew off the left side of his face in a spray of blood and bone chips. Ethan blinked as the body wavered for a moment and then seemed to collapse in slow motion onto its side.

A deep silence filled the cave, Ethan's world suddenly filled with a ringing in his ears and the deep, rapid rhythm of his own breathing. His heart thundered in his chest and his eyelids fluttered erratically as adrenaline surged through his synapses.

"Sweet mother of Jesus," McQuire's voice uttered in the humming silence as he looked at Ethan's victim. "You sure know how to pain a man."

Ethan blinked as his hearing returned and he saw half a dozen bodies lying in the entrance to the cave, some groaning, some weeping and clasping gunshot wounds and savagely mauled stomachs, others staring sightlessly at the ceiling of the cave. He looked down at the ruined body of the man he had slain and felt a sudden chill of shame.

"This ain't over," Ellison Thorne said, wiping the bloodied

blade of his knife on his pants. "Let's fall back afore they try again."

Ethan was about to follow him when from somewhere outside a voice shouted down to them.

"Give up the cave, or this woman dies!"

Ethan edged forward and squinted up into the bright light to see two men standing on either side of Jeb Oppenheimer far above, whose hands were cupped around his mouth.

Kneeling beside them was Lillian Cruz. As he watched, Oppenheimer's mercenaries began descending the rope ladder, prodding and pushing Lillian before them as a human shield.

63

United Nations General Assembly Hall
New York City

Donald Wolfe stood to one side of the speaker's podium and stared out across the huge blue, green, and gold General Assembly Hall beneath the towering, domed, seventy-five-foot ceiling. Representatives of member states sat behind tables that faced the raised speaker's rostrum. At the podium sat the president of the General Assembly, with the secretary-general of the United Nations to his right and the undersecretary-general for General Assembly Affairs and Conference Services to his left.

The secretary-general stood at the rostrum as he addressed the leaders of the majority of the world's nations arrayed before him in the enormous amphitheater. Behind him, two massive screens flanked an eight-foot-diameter United Nations emblem, each screen showing his face as he spoke. Heads of state and their translators occupied hundreds of chairs rising up and away before him, listening intently to the man often considered to be the true leader of the free world.

"Our purpose, today, is to deal with the debate on our population: its growth, health, and future over the next fifty to one hundred years. There have been a number of studies conducted by official bodies both within and outside of government, each attempting to provide a solution to the growing problem of supply and demand, in real terms, across the globe and, now, the situation is reaching a crisis point."

Donald Wolfe sat among the audience, trying to ignore the prickly heat irritating the collar of his shirt. He was nervous, he realized, more so than he had expected to be as he listened to the secretary-general.

"Ladies and gentlemen, to lead the discussion on this most serious of topics and the implications for the spread of infectious disease, please welcome Colonel Donald Wolfe of the United States Army Medical Research Institute for Infectious Diseases."

There was a polite ripple of applause as Wolfe got to his feet and walked across to the dais, shaking the secretary-general's hand before clearing his throat and looking up at the massed ranks of world leaders.

"Ladies and gentlemen, let me begin by warning you that what I have to say to you today will not be palatable. It will not be considered politically correct, perhaps not even morally correct, but in the light of the threat we face from an unspoken truth that is becoming more dangerous to human survival than climate change and nuclear proliferation I believe that somebody must speak out, and I take this opportunity to do so."

Three hundred faces were fixed upon his, an absolute silence as every man and woman listened to the sound of his voice. Perhaps for the first time in his life, Wolfe realized why people fought their way to the top of the political sphere: for the power. The knowledge that what you were saying could change the lives of millions, perhaps even billions, was intoxicating. He took a breath before continuing.

"Today, the population of our planet stands at approximately seven billion. Of those, just over one billion live in the developed Western world. We are the consumers, the guzzlers of gas. But daily we are being joined by millions more from the developing world, as emerging powerhouse economies like China, India, and a resurgent Russia seek to emulate our success, our prowess, our prodigious consumption of energy and resources." Wolfe paused for a moment before hitting them with a big one. "It has been estimated that by 2006, we were consuming resources forty percent faster than the earth could replace them. When those resources are gone, we will see the utter collapse of modern society into something approximating medieval Europe."

A ripple of whispers fluttered through the audience, and Wolfe capitalized on the energy among them as he continued.

"It has been calculated by scientists that due to technologies improving such things as farming, humans are now ten thousand times more populous on our planet than nature would normally allow any predator to be. Your own United Nations Environment Program, involving fourteen hundred scientists over five years, concluded that human consumption now far outstrips available resources." Wolfe smiled, and gestured outside. "New York City consumes as much electricity in one day as the island of Crete consumes in one year."

Wolfe scanned his audience with eagle eyes, saw that they were waiting for his next revelation.

"Our planet has what's known as a carrying capacity: an ability to support a finite volume of flora and fauna. Since the middle of the last century, our bloated consumption of the most crucial substance to the survival of life—freshwater—has caused a chain reaction that now threatens to be the catalyst for the fall of mankind. Without sufficient water, agriculture, the very thing that has allowed our species to spread across the globe, is being undermined. A global decline in world agricultural capacity began in the 1990s and continues to this day, growing exponentially. Groundwater aquifers around the world are failing, producing a grain deficit. Despite all our efforts and all our charities, the number of people malnourished in the world has grown by one hundred million per decade, not because we haven't exported grain to those countries who need it, but because their populations have swollen to such gargantuan proportions as to be unsustainable even with our agricultural technology."

Wolfe looked around at the various individual countries' heads of state.

"Algeria, Egypt, Iran, Mexico, and Pakistan are all importing grain due to internal deficits or are on the verge of doing so. Northern China's food production is forecast to decline by almost forty percent by the end of this century. Here in America, the Los Angeles basin, an area reckoned to be naturally able to support a maximum of one million people, is home to a megacity of thirty million people, and our population is the only one in the developed world still growing, predicted to rise to four hundred thirty million by the year 2030. By that time, we may no longer have enough resources to export the grain that keeps so many other countries from starving, because we'll barely have enough water to produce enough grain to feed ourselves."

Wolfe subtly changed his tone to a more hopeful oratory.

"We have talked, endlessly, of reducing our carbon footprints, of recycling and carbon capture, of alternative energy sources and radical economic strategies to reduce our consumption of both the electrical energy and resources available to us on our planet. We have passed laws to drill into pristine environments in the pursuit of oil, gone as far as the Antarctic to assuage our thirst for energy, and yet all this time we have ignored the fact that as fast as we seek solutions to our crises, they are made irrelevant by the increased demands of our growing population. It is estimated that global human population could exceed twelve billion people in the next fifty years." Wolfe paused. "Just think about that number for a moment. Twelve billion people, almost twice as many as are alive today."

Wolfe shook his head.

"We are fighting a losing battle, when within our reach all

along has been the solution to our problem. Population itself. It has been estimated by the National Research Institute on Food and Nutrition that world population needs to be reduced by two-thirds in order to achieve a sustainable use of natural resources." Wolfe let the weight of his words sink in. "If all nations were to follow this advice, resources would no longer be an issue for humanity, nor would pollution or the specter of anthropogenic climate change."

Wolfe looked about at the hundreds of faces watching him.

"It is time to act before the escalation of resource wastage becomes irreversible and our security upon this planet unsupportable. If we are not willing to do it for ourselves then at least we should consider doing it for our children, for it is they who will shoulder the burden of our inaction. But it will not be conflict through inadequate resources that will bring about their downfall. There is a far greater disaster awaiting them if we do not act."

Silence reigned throughout the chamber as they listened.

"Forty-five percent of humanity live in cities, often in conditions of squalor and disease. It is only a matter of time before a major pandemic, one that festers and flourishes in these hotbeds of sickness and is able to spread rapidly, begins the next major disaster in human history. We think of pandemics as explosions of some new and exotic disease that appear without warning, but in fact they fester for years, decades even, before becoming established. The HIV-1 virus entered the United States many times before taking hold in 1969, more than a decade before it was identified. Modern flu viruses like HN51 appear and disappear, struggling to take a permanent hold in human populations. Soon they too will emerge as true killers, and they will decimate the populations they infect."

Wolfe looked around the amphitheater at the people of power arrayed before him.

"I have provided you with press packs that cover these facts in more detail. Pass them on," Wolfe said, "to your staff, to your ministers, to anyone with the will and the means to influence what we do next on the global stage to prevent the eradication of our species by disease. We have a choice, ladies and gentlemen: either we control population for ourselves, without unnecessary loss of life, or I predict that within the next five decades nature will control population for us, and millions, perhaps billions, will pay the price. Our sons and daughters will be among them."

Donald Wolfe reached into his pocket and produced a bottle of water that he had purchased a week before—one of several thousand he now stored in his home to guard against the coming apocalypse—and lifted it to show the entire audience before him.

"Water. Something so simple that we take it for granted—we even bottle and sell it when we can so easily access it from our taps. Yet one billion people worldwide cannot perform this simple act, of accessing clean water. Ladies and gentlemen, I'd like you to raise your glasses and show your support for action against the destruction of our world, of our species, which has begun already in the draining of the world's freshwater supply. We must act now, before it's too late to act at all." Wolfe paused. "Is there anyone in the hall who is not willing to act?"

Wolfe watched as every single man and woman sitting before him lifted their glasses.

"To humanity," Wolfe toasted. "And our role in saving it."

64

United Nations General Assembly
New York City

Doug Jarvis awoke as pain ripped across his skull. Bright light seared his eyeballs as a voice spoke in his ear.

"Just lie still, sir, you're going to be fine."

Jarvis struggled to focus against the pain, and saw a man leaning over him, smiling encouragingly.

"Where am I?" Jarvis rasped.

"You're going to be fine, just relax. You've had quite a bash."

Jarvis saw that he was in a small room, a first-aid box hanging from the wall nearby. A sudden flurry of memories bolted through his brain and he fought to sit upright. The sound of metal on metal clanked in his ears, and he realized that he was cuffed to a gurney.

"What the hell?"

"Please stay still, sir," the man beside him said.

"Why the hell am I cuffed? Where's Donald Wolfe? He's trying to poison the General Assembly and—"

"There's no need to worry about that," said the doctor softly.

Jarvis peered at him and saw the kindly smile on his face, the look of a man who was tending to a child.

"Have you got my identification?" Jarvis demanded.

"You didn't have any," the doctor said, dabbing gently with a crimson-stained cloth at Jarvis's head. "You were found slumped in an elevator. You must have taken quite a fall. It's a wonder how you got into the Conference Building at all."

Jarvis felt a sudden impending doom descend upon him as he looked up at the doctor.

"Sir, I've been stripped of my identification by Colonel Donald Wolfe. The man is armed. He intends to infect the water within the General Assembly Hall with a strain of Spanish Flu that is highly lethal. You know what Spanish Flu is, I take it?"

The doctor looked down at Jarvis.

"Yes, I do. But I can assure you that there is no danger. Colonel Wolfe was the man who found you and called the medical team. He seemed quite concerned."

"He was the one who hit me!" Jarvis shouted, pain barraging his weary brain with salvos of agony. "Get on the phone to the Defense Intelligence Agency, Lieutenant-General Abraham Mitchell. He'll confirm who I am!"

"I'm sure he will," the doctor nodded patronizingly, and continued to dab at the head wound.

Jarvis mustered every last ounce of his patience.

"Doctor, if you don't make that call and get me off this goddamned gurney, a lot of people are going to die. Do you understand? If I'm lying, you'll have wasted two minutes on a call. If I'm not, you'll have wasted millions of lives. What's it going to be? A hundred-second phone call or a hundred million lives?" He leveled the doctor with a steady gaze. "Your call."

Lakewood
New Mexico

"All vehicles: suspects armed and considered dangerous, proceed with caution."

Lieutenant Zamora keyed off his microphone and glanced in his rearview mirror at the line of four patrol cars following him in convoy down I-285, their lights flashing.

Alongside him sat Butch Cutler, while in the rear seat of his car, cuffed and silent, sat Saffron Oppenheimer.

"I'm taking a hell of a goddamned risk bringing you out here," he said, keeping one eye on the road ahead. "You try to make a break for it, I'll hand your case over to U.S. Marshals and have them hunt you down day and night."

"Story of my life," Saffron replied without concern. "How much farther?"

"We're ten minutes away from the caverns."

"Stay north of the hills," Saffron said. "I've never seen those old men head south of them."

Lieutenant Zamora shook his head in confusion.

"What the hell is it with those people? If they even exist, why the hell would they stay in the state at all? If SkinGen and people like Jeb Oppenheimer are hunting for them, surely they'd be better hidden in rural Wyoming or something?"

"I don't know," Saffron replied. "Maybe they have people here in New Mexico who help them, keep them supplied with food and water and such like. Whatever the reason, it's enough to keep them here. We wouldn't have seen them in the deserts as much as we did if they were roaming the entire continent."

Cutler stared thoughtfully at the road ahead for a moment before speaking. "What makes you so sure that Nicola Lopez will betray Ethan Warner? Near as I could see, they're pretty tight together."

"I know what I saw," Saffron insisted. "Lopez met with my grandfather and there's only one reason that he would do that—to buy her off. Did you get the warrant to access her bank accounts?"

Cutler looked at Zamora, who said nothing. Saffron smiled coldly.

"So you did get them," she purred. "Incriminating, were they?"

"Money is money," Zamora said. "There was an anomalous sum deposited in one of her accounts yesterday, yes, but we haven't traced its source yet. It could be anything."

"Yeah," Saffron uttered, "course it could."

65

United Nations General Assembly Hall
New York City

Donald Wolfe lifted his bottle of water, watching as the hundred-strong audience did the same. Just as the liquid touched his lips, the ceremonial entrance smashed open with a crack that echoed through the hall as hundreds of presidents, prime ministers, and their staff turned away from their glasses at the noise.

Doug Jarvis burst into the hall, shrieking at the top of his lungs. "Don't drink the water!"

Donald Wolfe stared wide-eyed at Jarvis, who staggered unsteadily into the chamber, his head wrapped in a bloodied bandage and two security guards flanking him.

"What the hell is this?" he uttered, the microphone amplifying his voice.

Jarvis shouted as loud as he could in the hall.

"The water is infected!"

A rush of horrified gasps echoed through the hall as dignitaries put the glasses down as though they were filled with venom. Jarvis pointed up at Donald Wolfe.

"He's organized a pandemic, starting from this hall!"

Wolfe stammered his response even as hundreds of heads turned to look at him.

"That's ridiculous!"

"He's armed!" Jarvis shouted, and the two security guards flanking him placed their hands on their weapons. "We know everything," Jarvis said. "It's time to come clean about New Mexico and Alaska, Donald."

Wolfe felt a tingling sensation creep uncomfortably down his spine as he felt the eyes of the entire amphitheater watching him.

"What in the name of God are you talking about?"

"The death of Tyler Willis," Jarvis replied with an impassive expression. "The abduction of Lillian Cruz. The men you have dispatched to New Mexico to assist Jeb Oppenheimer of SkinGen Corp in abducting men for biological experiments into longevity, and the murder of a scientist in Brevig Mission, Alaska."

Wolfe opened his mouth to reply but Jarvis cut across him, turning to look up at the surrounding world leaders.

"Whatever you do, do not drink the water in your glasses. It's infected with a strain of 1918 Spanish Flu, obtained by Donald Wolfe from a mass grave in Alaska and brought here to New York."

A rush of gasps crashed across the delegation.

"That's preposterous!" Wolfe snapped in alarm. "Why on earth would I do something like that?"

"In the next half hour, more people could die," Jarvis shot back at him. "We have tracked your movements ever since you left SkinGen in Santa Fe two days ago, and the FBI are already at Fort Detrick and searching your office."

Wolfe was flustered behind the dais, looking this way and that for an escape.

"SkinGen is a private corporation! I have no connection with them, and the FBI can't just walk in and—"

"Yes, they can," Jarvis replied. "Jeb Oppenheimer is on the run, Donald. It's over. Your little scheme to reduce the world's population through disease is finished, and your partner in crime is wanted for the murder of Tyler Willis. This is your chance to ensure that more innocent people don't die. Tell us exactly where your men are and what they're doing, before this becomes a multiple homicide investigation."

Wolfe gaped at Jarvis and tried to force his brain to feed words to his unwilling vocal chords. "I really don't know what you're talking about. I'm in the middle of an address and you're—"

"The hell with your address," Jarvis shot back. "I've had men watching you for forty-eight hours. You dispatched more than one hundred mercenaries into New Mexico yesterday under cover of a training mission to kill American citizens!"

More gasps filled the auditorium as Wolfe felt dread plunge through him. His throat dried out as he struggled to speak.

"This is preposterous!" he shouted, and turned to the secretary-general. "I demand that this man be removed from the chamber!"

The secretary-general glanced at Jarvis and then at Wolfe.

"Denied. Explain yourself."

Wolfe was about to speak but Jarvis cut across him again.

"You sent biological cleanup teams into Santa Fe to investigate claims that you made of infected blood or agents being spilled in precise locations within the city," he said. "We have this on record as your own agenda for sending the teams in the first place. However, we also know that only two people knew that there was infected blood in those locations, and both of them are involved in the murders. My problem, Colonel Wolfe, is that unless you'd been informed by those individuals of the presence of infected blood, you could not have known that it was present at the scene."

A silence descended over the auditorium. Wolfe felt the weight of the world's political machine bearing down upon him. Jarvis took another step closer to the dais.

"In addition, you flew from Santa Fe to New York City yesterday, to attend a function last night and this United Nations meeting. However, your flight took some eight hours longer than it should have. I have proof that you traveled to Brevig Mission in Alaska, and that since your arrival there a scientist working on the glacier has been found buried in the grave of a victim of Spanish Flu." A flurry of horrified whispers filled the hall as Jarvis went on. "Tissue samples from the infected corpse were found in the laboratories of SkinGen Corp just moments ago, sir. Can you explain how you came to acquire them while in Alaska, or how they ended up in the hands of Jeb Oppenheimer?"

Donald Wolfe tried to answer but his jaw ached as he gaped and he couldn't think of a single thing to say. Jarvis gave him no quarter, speaking before he could muster a reply.

"You're not here to convince the nations of this world to reduce their populations. You're here to spread a lethal virus across the globe while trying to ensure the longevity of select groups of businessmen. You're a eugenicist, using disease and genetics to shape the human population just how you think it should be, trying to remove those you deem to somehow be less worthy of life." Jarvis tapped his own forehead. "If I'm lying, hand your gun to these police officers and let them run tests on the blood residue that I know is on the handle. It will match mine!"

Wolfe stared with wild eyes into the middle distance, as though aware of his exposure and yet unable to bring himself to focus upon his inevitable demise. Jarvis took another couple paces toward the dais, his voice carrying across the entire amphitheater.

"Sir, I do not care if you are innocent, framed, incompetent, or just plain guilty. Right now, all that concerns me is that if you do not inform us of where those men are, right now, several citizens of this country that you *swore* to protect will die. Do you understand?"

Wolfe swallowed thickly and then nodded once. As he did so, he saw one of the dignitaries watching the exchange discreetly reach into his pocket and retrieve a cell phone upon which he began typing.

"Where are they?" Jarvis repeated, getting Wolfe's attention once again. "Where are Jeb Oppenheimer and the soldiers you sent into the desert?"

The words fell from Wolfe's mouth as if of their own accord.

"Near Rattlesnake Canyon, Carlsbad," he uttered. "East of the Guadelupe Mountains."

A deep silence filled the hall as Wolfe's guilt was laid bare for all to see.

Jarvis turned and called out to the police behind him, who immediately began running for the exits with radios to their mouths. Wolfe glanced up at the dignitary with the phone, a Bilderberg member, who was looking down at him with a disapproving gaze. Wolfe realized in that moment that it was all over for him.

Jarvis turned toward him.

"Sir, if you would accompany the police to—"

Wolfe stepped down off the dais in one fluid motion and reached down, slipping the ceremonial pistol from its holster. He barely heard the cries of alarm from the chamber around him as he cocked the weapon, scarcely saw the security guards draw their own weapons with amazing speed to point at him. Wolfe turned to face the secretary-general, saluted once with his free hand, and then put the pistol's barrel to his head.

He saw Doug Jarvis rush toward him, his mouth open and his eyes wide.

"Donald, no!"

A deafening blast filled the room and Wolfe's world vanished into blackness.

66

Lechuguilla Cave
New Mexico

Ellison Thorne peered up from the shelter of the cave, and Ethan watched as his craggy face screwed up in fury, his big hands clenching and unclenching around the handles of his pistols as he bellowed back in reply.

"You're sure one cowardly tyke, Mr. Oppenheimer."

A distant laugh rippled down toward them from far above, the old man obscured now by the mercenaries working their way down the ladder.

"A determined one," Oppenheimer replied. "You're the ones cowering in a cave."

"Why don't you come down here and join us?" Ellison Thorne thundered back. "If'n you're hankerin' so bad to be here?"

No reply came, and Ethan turned to Ellison Thorne.

"He's already murdered one man," he said. "He'll have no problem killing Lillian Cruz."

Ellison squinted up at their attackers for a long moment and then made a decision.

"We don't take orders from you, Jeb!" he boomed. "We'll decide how this goes down!"

Oppenheimer's voice echoed down at them in reply. "You've got five minutes before we finish this for good!"

Ethan and the rest of the soldiers withdrew into the cave to see Lopez hurrying to join them.

"Kip's not good," she said. "He's bleeding out and there's nothing I can do about it."

Copthorne, McQuire, and Cochrane all looked to Ellison Thorne, who in turn looked at Ethan.

"If'n that man gets hold of what's in these caves, you say he'll destroy the human race?"

Ethan nodded.

"He'll let only a certain few people have access to the drugs he's going to develop using the bacteria in your bodies. Those people will become biologically immortal, just as you are, while the rest of the world will be prevented from raising families. Oppenheimer, and people like him, will rule without end over the population."

"But we're dying," McQuire said. "Whatever this thing is, it doesn't last forever."

"It will, once Oppenheimer's had his chance to genetically mess with it," Lopez said. "Imagine a man like Oppenheimer being able to rejuvenate himself: young, fit, immortal, and in control of a drug that every human being on the planet would kill to acquire. The world will be ruled by a dictator class with him at its head."

Ellison Thorne turned away and rubbed his temples.

"If'n we get deep enough into the caves they won't be able to follow," he said quickly. "We can get ourselves sorted and come back out fighting."

"You'll never be able to do it quickly enough to prevent Jeb from killing Lillian Cruz," Ethan said. "She's innocent in all this."

"So are we!" Ellison shouted. "We didn't ask for this, but how can we now surrender it just because of one man's greed?!"

Ethan spoke quietly.

"Because this isn't about just you, or us, or Lillian Cruz. The simple truth is that people like Jeb Oppenheimer think they can control everything and they can't. Sooner or later, the science of this bacteria will be lost, or stolen, or leaked. No matter how hard they try they won't be able to prevent it from reaching the public domain, through a disgruntled employee or maybe one of those heroic whistle-blowers who leak these things to the public for no financial gain, or maybe even through a population-wide revolution. It happened in the Middle East, the people toppling their dictators one after the other."

Lopez stepped forward.

"When that happens," she said, "and it will, then we'll have an entire population of people who will never die. Can you imagine what will happen when the population rises so fast that the resources dry up, when there's no more water, no more food, no more fuel? When people will kill one another for a morsel of food or a sip of water? One day, this will not result in a human population that lives forever. It will end our species, completely. We will all die."

Ellison Thorne stared at Ethan and Lopez for a long moment, and then looked at the other soldiers.

"What the hell do you think?"

McQuire, Copthorne, and Cochrane exchanged glances, and then they nodded slowly. As they did, a voice came from behind them.

"I agree too."

Ethan turned in surprise as Kip Wren limped toward them, using his rifle as a cane. His right thigh was bandaged with a makeshift tourniquet.

"You're in no condition to be standin'," Ellison said.

"Well, I am standin'," Kip Wren replied. "But I'm all played out, and I don't intend to spend my last sweet seconds sitting on my ass back there. You ever remember that old sayin' we used to live by during the war when we din' know if we'd ever see our homes and families agin?"

When none of the soldiers replied, Kip Wren spoke for them.

"The greatest act of courage a man can make is the surrenderin' of his own life for others, when nobody will ever know of his sacrifice."

Ethan felt a rush of emotion as he heard the words, thoughts of countless comrades in the marines and in other units who had served and died to protect democracy and the way of life that he and so many millions of others cherished. Through the world wars and countless other conflicts, souls far too young to die had sacrificed themselves upon the altar of freedom for the benefit of generations they would never meet. Ethan wondered briefly just how many heroes had anonymously given their lives throughout the history of mankind, their courage lost in time.

A sound from Misery Hole caught his attention, and he saw dozens of black-suited soldiers drop onto the floor of the chasm and dash for cover behind rocks and fallen branches around the mouth of Lechuguilla Cave. Ethan and his companions crouched down behind cover as they watched Lillian Cruz being manhandled by the soldiers nearby, and the white-suited Jeb Oppenheimer appear at the bottom of the ladder, protected by an honor guard of his men. Ellison Thorne watched the gathering troops for a long moment, struggling between the desire to live and the altruism that Ethan suspected ran strongly through his veins. The big man finally sighed and looked at Kip Wren.

"What d'you propose?" he asked. "That we bring the whole damned cavern down on top of ourselves?"

Kip Wren grinned against the pain racking his body, and from his pocket produced a stick of dynamite that looked almost as old as he did.

"I've got me an idea, and I think you'll like it."

67

"We want to make an exchange!"

Jeb Oppenheimer heard the sound of Ellison Thorne's voice boom out of Lechuguilla Cave as though a god of the underworld were addressing him. He peered down at the darkened, ragged cavern entrance, unable to see where the big man was hiding. The entrance was littered with the bodies of eleven men, some writhing in delirium or agony, others motionless.

"Of what?" Oppenheimer called.

"One of my men for your hostage!"

Oppenheimer looked at Hoffman, who frowned.

"One of them was injured," he said. "I'm damned sure I got him in the leg before they ran into that cave."

Oppenheimer cleared his throat and shouted his reply. "I don't want tainted goods!"

Ellison Thorne's reply thundered back from within the cave.

"Your men can't shoot for shit, Oppenheimer! He's got a minor wound to his leg, nothing more, but he needs attention. We'll send him up to you for treatment, you send us the woman. Everybody wins."

Oppenheimer heard Hoffman speaking from behind him.

"Can't hurt, and it gets us what we want real easy."

Oppenheimer peered at Hoffman.

"If something looks too good to be true," he growled, "then it probably is. Their man is most likely mortally wounded and they're looking to buy us off long enough to escape."

Hoffman appeared to consider this for a moment and then chuckled.

"Well, we can't storm their little hidey-hole there without incurring too many casualties, so I say we let them send their man up. We send this woman of yours down, and then besiege them just the same. We win, both ways."

Oppenheimer nodded.

"For once, I'm in complete agreement with you," he said, looking back at Lillian Cruz where she knelt just out of sight behind them. "Bind and gag her so she can't warn them, then arrange the exchange. We'll take her back as soon as we're ready."

"Looks like they're goin' forrit," Nathaniel McQuire said.

Ellison Thorne snorted in the half-darkness.

"The hell they are," he rumbled. "That Oppenheimer's as slippery as a rattlesnake and no less deadly. He'll mean to kill us all, one way or another."

Ethan watched as Kip Wren concealed his dynamite stick up his sleeve and took a cigarette lighter from Nathaniel McQuire.

"You sure about this?" he asked the soldier, whose breath was labored and his brow sprinkled with sweat.

"I am," Kip replied. "I can't lose: either I die here or I die up there. But be quick about it. I ain't certain how long I'll be able to stand."

Ellison Thorne looked down at Kip.

"Once you're up there, light it fast and get a hold of Oppenheimer just as tight as you can," he advised him. "This thing'll smoke like all hell once it's lit, and they'll try to get it off you."

Lopez frowned.

"What if they shoot him before it goes off?" she asked.

"Then Oppenheimer will definitely die," Ethan said. "Kip's body will seize and they'll never get him off the old bastard in time."

Kip nodded, his teeth gritted in an awkward smile.

"Damn straight," he said. "Then this will all be over."

A voice drifted down to them from beyond the cave entrance.

"We're ready, bring him out!"

Ethan watched as, with surprising care, Ellison Thorne mopped the sweat from Kip Wren's face and brow. Kip drank from Copthorne's water bottle before turning to Ellison. The two men stared at each other for a long beat, and Ethan realized that they were parting company for the first time in a hundred forty years.

Kip Wren, leaning on his rifle, held out his hand to Ellison. The big man took it, and then stepped in and briefly embraced his old comrade for a few moments before releasing him.

"As you were, Corporal," he said finally, his gravelly voice taut.

Kip Wren saluted once and then began limping away toward the light.

Ethan and Lopez followed, seeing Oppenheimer's soldiers arranged behind rocks near the entrance, their rifles aimed at the old soldier limping toward them. Kip Wren, mastering his pain, strode out into the cavern.

Ethan watched as Lillian Cruz, bound and gagged, was prodded toward the cave by two men carrying assault rifles. Carefully she descended toward them, her eyes fixed on the cave ahead and the old man making his way up toward her. The two glanced at each other as they passed, before Lillian reached the mouth of the cave, hesitated, and looked back over her shoulder. Kip Wren strode toward their enemy, visibly favoring his left leg but valiantly reaching the spot where Oppenheimer crouched.

"Do it," Copthorne whispered from behind gritted teeth, "do it now, Kip."

Ethan shifted his position and edged forward as Lillian Cruz stepped cautiously into the depths of the cavern.

"I've waited a very long time for this," Oppenheimer said, looking at Kip Wren as though he were studying a work of art or similar.

"Not as long as I have," Wren replied, one hand closed around his cigarette lighter, the other holding the dynamite stick out of sight up his sleeve. Shafts of light shimmered down through the veils of cordite smoke swirling above their heads, the smell of blood and of scorched flesh staining the air.

"I assure you that you will not be harmed," Oppenheimer promised. "I've already realized that you're no good to me dead."

Kip Wren peered at Oppenheimer.

"What are your intentions?"

"To study you," Oppenheimer said. "To discover what makes you biologically immortal, and use it to enhance the quality of the human race, the better for all mankind."

"The better for you and your ilk, Oppenheimer," Kip corrected him. "You're in this for yourself, not for anybody else, and always have been."

Oppenheimer gave a brief shrug.

"The world's full of winners and losers, the haves and the have-nots," he replied with a cold smile. "I'm just going to ensure that the losers are bred out of existence, leaving only the haves and the have-mores."

Kip Wren felt a surge of hatred mix violently with the pain seething through his body, and with a cry of vengeance that echoed through the chamber he lunged forward and grabbed Oppenheimer with both arms, whirling him around as he pulled the old man against his chest. His hands met across Oppenheimer's waist, the dynamite stick and the cigarette lighter clear for the mercenaries to see.

"Stand down!" Kip Wren shouted. "Or I'll blow this bastard sky-high!"

Oppenheimer squirmed in panic as he struggled to free himself from Kip Wren's grip, his rheumy eyes fixed on the explosive held against his belly. Five soldiers rushed in to try to pry him off even as Oppenheimer shouted at them.

"Get back!"

Hoffman waved frantically at his soldiers.

"Don't shoot! We need him alive!"

Kip watched as the mercenaries fell back, then turned to look over his shoulder. Lillian Cruz was already in the shadows of Lechuguilla Cave, almost out of sight.

"Come on out, Ellison!" Kip yelled with a maniacal chuckle. "I got 'em surrounded!"

He was about to manhandle Jeb Oppenheimer toward the ladder when a voice yelled out at him.

"No!"

68

Saffron Oppenheimer leaped out of Lieutenant Zamora's patrol car before it had even stopped rolling, sprinting away up the hillside track toward Misery Hole at a terrific pace. Butch Cutler and Enrico Zamora labored after her in pursuit, but she gave them no quarter as she raced ever upward and then came to a sliding halt at the dizzying edge of Misery Hole.

She spotted the rope ladders hastily rigged at the edge of the shaft nearby, and rushed around the edge before clambering onto the nearest one and hurriedly climbing down into the shadows. From her vantage point high above the floor of Misery Hole, she could see what was happening. Some kind of gunfight had filled the floor of the cavern with thick smoke that boiled up around her, stinging her eyes and choking her throat, but it also concealed her presence as she slid the last few feet to the ground behind the massed ranks of Oppenheimer's mercenaries, all aiming assault weapons at a low cave slicing across one wall of Misery Hole.

Saffron didn't wait for Lieutenant Zamora or Butch Cutler to reach her. Instead she charged at full sprint past the crouching ranks of soldiers toward where her grandfather writhed and twisted in the hands of an old man with a thick mustache, a stick of dynamite grasped in one bony-looking hand.

"No!" Saffron shouted, her arms outstretched as she ran. "Don't kill him! There has to be a better way!"

Jeb Oppenheimer's crinkled jaw fell open as he stared in amazement at Saffron.

Kip Wren yelled, "Stay aback, ma'am! I ain't bluffin'!"

"What the hell are you doing down here?" Oppenheimer said. "This isn't your business, Saffy, get out of the way."

"It's my goddamned business now, Grandpa," Saffron shot back, "ever since I went to the police. They know everything."

Oppenheimer stared at her for a moment longer.

"You wouldn't." He smirked. "You wouldn't risk the jail time."

The voice that answered came from behind them all. "Yes, she would." Oppenheimer, Hoffman, and the mercenaries turned to see Zamora standing behind them in full uniform, his pistol aimed at Oppenheimer. "It's over, Jeb, no matter what happens. The entire New Mexico State Police Department is on its way here right now."

Butch Cutler stood alongside Zamora, clearly not intimidated by the mercenaries as he aimed a large Colt revolver at them.

"So is USAMRIID," he reported. "This whole thing's been blown sky-high."

Oppenheimer appeared completely overwhelmed, unable to speak. Saffron stared at him in horror.

"How could you do it?" she asked. "All those millions of people? You're planning to kill them all."

Oppenheimer's features hardened again as he took his eyes off the dynamite at his belly.

"It will happen sooner or later all on its own," he spat back at her. "I'm just controlling the situation."

"You can't control nature!" Saffron wailed. "It's not possible!"

"Either way," Butch Cutler said, "it's not going to happen. No pandemic, no population control, no nothing. It's over. The police will be here at any moment."

Hoffman peered suspiciously at Cutler.

"If so," he asked, "then where the hell are they?"

Neither Cutler nor Zamora replied.

Saffron looked at the old soldier holding her grandfather hostage. His face was twisted in upon itself in agony, sweat thick on his brow and his legs trembling with the effort of staying upright. She took a careful pace forward.

"I can help you," she said. "Let him go and we can get you to a hospital."

Kip Wren glared at her, but his strength was failing and she saw in his expression the realization that he was out of time. He smiled at her, almost regretfully.

"Apologies, ma'am," he said through gritted teeth, "but I can't do that."

The old soldier's hands twisted, and she saw him light the dynamite stick's fuse in a flash of sparks and smoke.

Saffron hurled herself at Kip Wren and threw a fast right jab that drove her fingers into Wren's eyes. The old soldier shrieked and jerked backward, and as he did so Saffron smashed her hands down across the smoldering dynamite stick and yanked it from Kip Wren's hands. Instantly, half a dozen soldiers plunged down onto Kip in a frenzied tangle of limbs and shouts, binding his arms and legs.

Saffron hurled the hissing dynamite stick into the depths of Lechuguilla Cave and whirled toward her grandfather.

"Get down!"

To her surprise, the mercenaries and Jeb Oppenheimer ignored her and fled to the entrance to Lechuguilla Cave in a chaotic tumble. Confused, Saffron turned and saw the dynamite stick come whirling back out of the darkness to land at her feet.

"Oh shit."

Saffron turned and ran hard, hurling herself flat to the ground behind scattered rocks as the dynamite exploded with a deafening boom and hurled the severed body parts of mercenaries in all directions to land with soggy thumps around her. She staggered to her feet, her ears ringing and pink blood spots staining her white T-shirt. She heard a laugh echo around the chamber as Jeb Oppenheimer struggled to his feet from behind a boulder and slapped his spindly thigh with one hand as he looked down at the cave entrance.

"We've got our man!" he shouted, and turned to Hoffman. "Kill them all."

Hoffman, his M16 rifle once again cradled in his grip, grinned and strode toward Jeb Oppenheimer. One huge hand reached out and gripped the old man by the throat with enough force to bulge his eyes. Oppenheimer gagged in shock as he was lifted onto his toes, his cane still dangling from his wrist.

"Your time, old man," Hoffman hissed, "is over."

With a heave of effort, Hoffman turned and hurled Oppenheimer toward Lechuguilla Cave, the old man's limbs flailing as he tumbled through the desert dust to land in a heap near the entrance. A ripple of grim laughs from Hoffman's men followed him.

Saffron stared in disbelief at Hoffman. "What the hell are you doing?"

Hoffman ignored her as he turned to Cutler and Zamora.

"Drop your weapons and get in that cave or we'll blow you away right here and now."

Half the mercenaries whirled and trained their assault rifles on Cutler and Zamora. Cutler looked at them and then glanced across at Zamora.

"Fancy going down in a blaze of glory?"

"Not so much."

Cutler slowly lowered his revolver as he called out to Hoffman.

"Whatever Wolfe is paying you, it won't be nearly enough to get you out of this."

Hoffman appeared unimpressed. "I don't really care," he said with a grim smile that conveyed an utter ruthlessness. "Because neither of you will be around to report anything, and we'll be gone within minutes."

Hoffman gestured to his men, and they disarmed Cutler and Zamora and shoved them down toward the depths of the cave, stumbling in the darkness as they struggled to see their way. Saffron hurried down with them and took her grandfather's arm. Oppenheimer struggled to his feet, steadied himself with his cane, and turned to glare up at Hoffman.

"I'll pay you double," he shouted, "triple, whatever you want!"

Hoffman shook his head.

"You're a damned fool, old man," he shouted. "You think that you're powerful because you're rich, but you're a small fish in a very big pond and I work for the sharks. You're nothing, Oppenheimer, a nobody compared to who I work for!"

Hoffman turned away and looked at one of the soldiers next to him.

"Don't shoot any of them. We need this to look like an accident. Get the explosives out and blow the cave. It'll hide any evidence that they were here at all, and if we need anything in the future we can come back when they've all rotted to hell."

69

Ethan, along with Lopez, Ellison Thorne, and the remaining soldiers, had watched the entire exchange in amazement, from Saffron's unexpected wrecking of their plan to Oppenheimer's being hurled toward them to land at the mouth of the cave.

"What the hell?" McQuire uttered in disbelief.

Ethan replied grimly, "Looks like Oppenheimer's finally getting what he deserves."

Beside them, Lillian Cruz scowled at the mercenaries.

"They've got one of your men now," she said. "He's still alive. Whatever you all planned hasn't worked."

Ethan looked across at Ellison Thorne. "You sure Kip's not going to make it?"

Ellison Thorne shook his head.

"Can't see how, even with all those clever sawbones and newfangled contraptions they've got today in hospitals. He'll have mustered out by noon."

"None of it matters a damn," Saffron snapped, and pointed

at Lopez. "You were all doomed when she cut herself a deal with my grandfather."

A silence descended within the cave, even the breeze from the depths seeming to hang listless as Ethan stared at Saffron for a long and disbelieving moment before he turned to look at Lopez. Everyone in the cave was staring at her, and Ethan watched her glaring back at them defiantly like a cornered wildcat.

"What the hell, Nicola?" Ethan asked, staring at her.

Lopez glared back at Saffron.

"And we had a ticket out of here until you dropped in and screwed everything up saving that worthless old bastard," she shot back, pointing at Jeb. "Nice work."

Ellison Thorne glowered at Lopez from the shadows.

"What trickery has she gone an' done on us?" he growled. "We wondered how the old man found us so swiftly."

"She betrayed all of you," Saffron said, and looked at Jeb Oppenheimer. "How much did it cost you, Grandpa, for another traitor to join your ranks?"

Oppenheimer was not looking at them, staring instead with interest into the dark depths of the cave, but he answered her question.

"A quarter million bucks," he muttered as he hobbled off into the darkness of the caves behind them. "Cheap at twice the price."

Ethan stared at Lopez in horror. To his surprise, Lopez smirked as she leveled Saffron with a cold gaze.

"Predictable, to the last," she said coldly. "You're damned right I took his bribe, and you're damned right he paid me to guide his mercenaries to this cave. Gave me a GPS tracker to reveal the location of Lechuguilla Cave. If you'd thought about it, you'd have also guessed that I'd tossed the GPS tracker long before we got here."

"Not long enough to avoid a bloodbath!" Saffron shouted. "People are dying because of what you did!"

Nathaniel McQuire turned his rifle to point at Lopez. "Well, if that don't beat the Dutch! You sold out on us!"

Ethan leaped forward between Lopez and the weapon. He looked at her and saw a glimmer of shame in her dark eyes.

"Why?" he finally managed to ask.

"We needed a way to get support here quickly, if we needed it," she snapped back at him. "I figured if I dropped the tracker somewhere close by, Jarvis would be able to locate it if he had the IP address of the device."

Ethan eyed her uncertainly.

"That's a hell of a risk, Nicola," he said. "It could have backfired on us."

"You're goddamned right there!" Saffron hissed, reaching out for Lopez.

"Cut it out!" Ethan snapped, blocking Saffron's way. "We can sort this out another time; right now we need to get the hell out of here."

"There ain't no way out," Ellison Thorne said. "We searched plenty."

Ethan turned to Butch Cutler and Zamora.

"Some rescue you've pulled off here. Any other great ideas?"

"We were too late," Cutler said. "Couldn't get in front of them in time."

"Where's the cavalry?" Lopez asked them.

McQuire laughed at her. "You can't get horses down here, ma'am."

Cutler looked at McQuire oddly before replying to Lopez.

"They can't move unless your man, Jarvis, can prove Donald Wolfe's involved. I haven't heard from him and our damned cell phones won't work down here, so now all we've got is the tracker you've supposedly dumped somewhere in the deserts."

Lopez smiled coldly.

"They got here real fast, so they almost certainly found it," she said. "Which means that Oppenheimer most likely picked it up, being the scrooge that he is." Lopez looked across at the old man. "He won't have left it behind."

Oppenheimer stared at Lopez in horror and then reached

into his jacket pocket. He pulled out a small black device that lay in the palm of his hand, his face twisted with rage at Lopez's betrayal. The old man hurled the device into the darkness and turned away, hobbling deeper into the cave.

Ellison Thorne looked at Lopez for a long moment as he realized what she had done.

"I'll be damned," he murmured, his eyes twinkling with suppressed admiration.

Saffron shook her head.

"You've risked all our lives on a gamble, and made a profit to boot!" Ethan was about to reply when he saw movement at the mouth of the caves. Mercenaries were sneaking forward carrying boxes of what looked suspiciously like explosives and were mounting them near the cave mouth.

"Damn," he said. "They've got Kip Wren, so now they're going to blow the cave mouth in and seal it."

Ellison Thorne nodded.

"That's what we did, more than a hundred years ago now," he said. "Blew the entrance so that nobody would find it. Was a bunch of goddamned scientists that decided to dig through the rubble and found these caves back in 1986."

Ethan turned as he heard Oppenheimer arguing with Saffron as they backed into the cave. The old man hurried away from her with his awkward gait, his cane clicking in the darkness until they could no longer hear it.

"Now where the hell is he goin'?" asked John Cochrane.

"He still wants the bacteria," Ethan said. "Even now it's all that he's interested in."

"To hell with him," John Cochrane said.

"I can still change him," Saffron protested. "He's not completely destroyed yet."

"Leave him," Ethan said to her, grabbing her arm. "He doesn't care about you or anybody else. All he gives a damn about is his own immortality."

Saffron shook her head as she leaped to her feet. "I can't leave him down here!"

Ethan watched her dash away before he turned to Ellison Thorne. The big man sighed heavily.

"We don't have time for this." He looked around the cave at them all, and then his features creased with concern. "Where's Lillian?"

Ethan scanned around them, but could see no sign of the medical examiner.

"Oh, you're kidding," Lopez uttered, looking behind her into the depths of the cave.

"She's going after the bacteria too?" McQuire said in disbelief.

Ellison loaded his rifle as he gestured to the cave entrance and looked at Ethan.

"You've got to get out of here," he said. "All you need to do is take Saffron and Lillian with you. We'll hold out until your reinforcements arrive, if they ever do. We can blow this cave in from the inside once you're clear, which will prevent Oppenheimer's men from getting to us. We've got enough dynamite left to do it, and Kip won't survive beyond noon. We win."

Ethan glanced at the mercenaries outside the cave, and saw in the background two of them trying to stabilize Kip Wren, an intravenous line now in his arm.

"Don't worry," Ethan replied, "Doug won't let us down. Shall we do this the old way, as one?"

Ellison Thorne nodded, and as he did so he and his comrades began loading their rifles once more. Behind him, Lopez picked up Kip Wren's rifle and began stuffing a Minié ball down the muzzle, following it with wadding. Ethan checked his pistol—six shots remaining, and he had a spare magazine. Twenty-one shots in total, plus five rifles: twenty-six shots, against maybe seventy or eighty heavily armed men.

"It's suicidal," Lopez pointed out. "Even if we do get clear, we'll still be outnumbered and we can't climb back up and out of Misery Hole without getting shot."

"Ain't got no choice," John Cochrane said. "I'd like to have had all eight of us here and an army behind us, but those days are long past."

Ethan stared at Cochrane, and then suddenly something smashed through his thoughts like a freight train as he squatted in the half darkness, watching Ellison and his men loading up.

"Eight of you?" Ethan said.

Ellison Thorne glanced at him but said nothing as he finished loading his rifle and aimed it at the cave entrance. Ethan slapped his head in disbelief.

"My God, I've been such an idiot!"

Lopez smiled in the darkness.

"Nothing new," she chortled, and looked at Ellison Thorne. "Something you need to tell us?"

Ellison shook his head, but Ethan spoke for him.

"The supplies," he said. "The medicines and clothes and everything that you would have needed to survive this long. Paperwork, documents, evidence that you weren't a hundred fifty years old. Damn it, you *do* have someone on the inside protecting you."

"That ain't no concern of yours," Ellison warned him with a pointed finger.

Ethan wasn't about to be intimidated, and on an impulse he fished the old photograph from his pocket and looked at it.

"You all said the same thing," he pointed out. "This photograph was taken after the Battle of Glorieta Pass, 1862. I didn't realize it until we got here this morning, but now I get it. This photograph was taken *after* you'd escaped from the Confederate retreat into Texas. It was taken *after* you'd sheltered in these caves."

Lopez frowned curiously.

"So what? There's seven of them in the picture," she said.

"Sure there are." Ethan nodded. "But who was holding the camera?"

Lopez stared at him for a moment as she realized his point. Ellison Thorne was about to answer when Edward Copthorne shouted out a warning.

"Enemy to the front!"

Ethan whirled to see four mercenaries plunge into the entrance and open fire randomly into the darkness, the staccato clatter of their assault rifles deafening in the confines of the cave. Behind them, Ethan glimpsed a half dozen more men carrying what might have been plastic explosives, hugging the walls of the cave as they prepared to blow the entrance.

"Take out the shooters first!" Ethan hollered as Ellison's men took aim.

Ethan ducked his head away from the noise and the smoke as all five rifles fired at once, the barrel of Lopez's weapon barely a foot from his head as she blasted one of the attackers deep in the belly, the soldier folding over the round and tumbling to his knees.

Ethan aimed at one of the men carrying the explosives and fired, catching him cleanly in the chest. The soldier crumpled and dropped his explosives as Ethan leaped up from behind cover and charged forward through the thick veils of smoke, aware of the bayonets glinting alongside him as Ellison, Copthorne, McQuire, and Cochrane all dashed out toward the entrance with a volley of war cries.

As they crashed into another wave of attackers rushing down toward them, Ellison Thorne shouted to Ethan above the din.

"Find Lillian and Saffron! I don't want any o' these bastards slipping past and killing them, and you'll need to get them out before we can blow the entrance!"

Ethan whirled to see Lopez ramming the wicked bayonet of her rifle straight into the chest of a screaming soldier. He shouted her name and tossed his pistol in a graceful arc toward her. Lopez whirled and caught the pistol before diving for cover behind scattered rocks and opening fire on their attackers.

Ethan dashed back into the cave, struggling to see as he plunged into the darkness.

70

Jeb Oppenheimer cackled to himself as he clambered awkwardly over endless jagged rocks in the darkness, his way lit only by the solid-gold lighter he held like a lantern in front of him, a white handkerchief wrapped around it to protect his fingers from the heat. The interior of the cave was low, forcing him to stoop in order to move forward. But he could smell a breeze that drifted into his face from somewhere ahead in the impenetrable blackness, cool air touched with the scent of damp but also of something else, an almost clinical smell that he could not identify but which seemed somehow familiar.

The noise of fighting behind him had faded, the complex turns and twists in the cave deadening all sound. Drops of water plopped in fat drips into puddles on the ground, seeping through the bedrock from hilltops hundreds of feet above his head. The thought of millions of tons of solid rock bearing down upon the chamber from above sent a wriggle of fear twisting through his gullet but he pushed on, driven by the knowledge of what resided somewhere deep within these prehistoric caves.

Ahead, the weakly flickering flame of his lighter reflected off something embedded in the rocks that glittered like pearls. Oppenheimer slowed as the low ceiling of the tunnel rose and he squeezed through a narrow vertical cleft in the rocks into a chamber filled with a shimmering pool of crystalline water so clear that the light of his flame illuminated the floor perhaps twenty feet beneath the surface.

But that was not what drew his eyes and caught the breath in his throat.

Above his head, immense crystals like giant geometric tree trunks were lodged at angles to span the width of the chamber above the shimmering water. Like giant causeways made of translucent glass, they crisscrossed above the water and sparkled in the weak light of the flame as though encrusted with jewels.

"Gypsum," Oppenheimer gasped, recognizing the immaculate nature of what was otherwise a nondescript mineral.

But here it possessed a purity the likes of which he'd never seen. He began easing his way into the cave, staring in awe at the crystals and the flickering water. The strange scent he'd detected earlier tainted the air around him, and he recognized it as ammonia. A flickering motion on the cavernous ceiling caught his eye, and he looked up to see bats roosting in their thousands above him, their wings fluttering as they clung to their rocky domain. As they did so, he saw an occasional droplet of fluid fall from the heights, dropping into the water with a tiny splash and ripple, the cause of the endless shimmering of the surface.

Slowly, placing his feet near the edge of the pool, Oppenheimer peered over the edge. There, deep below the surface, he watched the tiny droplets fall through the beautifully clear water to join a bizarrely colored deposit deep beneath the surface, a kaleidoscopic multitude of fungi and mosses. Oppenheimer guessed that the droppings in the water must clear overnight when the bats were out hunting, settling on the bottom of the pool. In the reflection from the surface of the water that illuminated his wrinkled face, Oppenheimer saw his own smile beaming back at him like a shimmering ghost as a disembodied voice echoed through the cave around him.

"It's guano."

He whirled to see Lillian Cruz watching him from the entrance to the chamber. Oppenheimer regarded her for a moment and then decided that she was no threat to him as he turned back to the water.

"The guano has ammonia in it," he said, almost to himself.

Lillian stepped into the chamber, gesturing to the water. "It also has high levels of phosphorus and nitrogen," she said. "Along with ammonia it contains uric, oxalic, phosphoric, and carbonic acids, various earth salts, and nitrates."

Jeb Oppenheimer's mind was working overtime as he nodded to himself, gesturing to the giant gypsum crystals soaring above the chamber.

"The gypsum and sulfur crystals mean speleogenesis: cave forming by sulfuric acid dissolution," Oppenheimer said. "The limestone cavern would have formed from the bottom up, in contrast to the normal top-down carbonic acid dissolution mechanism of cave formation. Sulfuric acid, derived from hydrogen sulfide, would have migrated from nearby oil deposits."

"The cave then floods over time with water draining through fissures from the ground above, creating these pools," Lillian added.

Oppenheimer nodded eagerly, gesturing up at the crystals with his cane. "The waterfalls," he said, "hitting the crystals and sometimes taking with it bacteria that was encased within the crystals when they formed millions of years ago, bacteria like *Bacillus permians.*"

Lillian nodded.

"The bacteria fall into the water and mix with the guano at the bottom. Phosphorus in guano is an essential plant macronutrient," she said. "That's why it's used so heavily in fertilizers. The guano, laden with the bacteria, are kept in solution by the water in the pool. The bacteria, provided with a nutrient source by the guano, are reanimated and come into contact with all manner of mosses, fungi, and bottom-feeding invertebrates."

Oppenheimer's laugh rattled out in the chamber, echoing back and forth around them as he spoke.

"Some insect and invertebrate species are semiaquatic, and others live on the surface. They consume the bacteria-laden guano, and are likewise consumed by the bats that hunt them!" Slowly he turned to face Lillian, his wrinkled features alive now as finally, after so many years, he realized that he had found something that had existed in folklore for millennia. "The bats carry the bacteria, giving them their unusually long life spans. They also process the bacteria through their guts and excrete many of them back into the water, or spill blood through injury into the pool."

Lillian nodded, and despite the fact that he knew she hated him, she smiled.

"Which over time ladens the water with the very fluids the bats have ingested, alive with a form of *Bacillus permians* that has evolved within these caves to live symbiotically within mammals."

"But what was the fuel?" Oppenheimer struggled to understand. "What metabolism was required to sustain them for such long periods inside human beings?"

Lillian no longer held the truth back from Oppenheimer. In fact, she appeared to enjoy revealing to him what she had learned. "Iron, from the hemoglobin in blood," she replied. "Anyone who carries the infection will suffer from anemia if iron supplements are not provided in their diets."

Oppenheimer looked at her pleadingly, like a child who has misbehaved yet yearns desperately for one last chance.

"But how could it have made the transition to humans through a single encounter?"

Lillian regarded the old man for a long moment before replying.

"Cross-species communication is possible in bacteria through something known as quorum sensing. The bacteria use it to coordinate gene expression via the density of their population. If there're enough of them present in a biological species, the genes are activated and any infection shows symptoms."

"My God," Oppenheimer exclaimed. "Like the bioluminescent luciferase in fish that glow underwater, produced by *Vibrio fischeri*. It would not be visible if it were produced by a single cell, only when the population is large enough does the production of luciferase begin."

"The bacteria's ability to express the gene is activated only when enough are consumed by the host species," Lillian confirmed.

Oppenheimer gasped, touching his head with one hand.

"The only people who have ever been down here long enough to consume enough of the bacteria to activate them were those Civil War soldiers. Which means that they must have gotten their infection from . . ."

Oppenheimer stared at the beautiful waters at his feet as Lillian took a few paces forward to join him. Her voice, soft as it was, carried throughout the cavern and into Oppenheimer's ears with the words he had once believed he would never hear.

"This is the water," she said quietly. "This is the elixir, the *real* fountain of youth. Ellison Thorne and his men drank the water here while they waited for the Confederate army to pass them by in 1862. They did not age from that day onward."

Oppenheimer, his eyes alight with joy, let his cane fall onto the rocks beside him as he got down on his knees, tears dripping from his face to ripple into the water.

"And this is our ticket out of here," he whispered to his own reflection. "They dare not shoot us, if we're uninjured and already carrying the infection."

Slowly, he lowered his lips and they finally touched the surface. It was icy cold, clear, finer than the most expensive wine he had ever tasted. It surged through him as though he were forcing ice cubes down his throat, filled him with a tingling sensation as though his very nerve endings were sparking electricity onto the charged air in the chamber.

Finally, Oppenheimer stopped drinking and turned as he knelt beside the water, looking up at Lillian Cruz. He smiled broadly, just in time to see Lillian's features melt into an expression of pure hatred as she lunged down and grabbed the back of his head and plunged it beneath the surface. As the freezing water swallowed his head, Oppenheimer heard Lillian's voice shouting at him above the bubbles and splashes as he fought for his life.

"You wanted to be here so much? Now you can damned well stay here!"

Oppenheimer's ruined lungs ached, his aged heart thumped in his emaciated chest, and his eyes bulged as he fought the urge to breathe. The clear view of the bottom of the pool swirled and starred as his vision faded. He was losing consciousness when he saw Lillian's hand plunge into the water beside his head, holding a small plastic container that held what looked like a ball of iron surrounded by flesh. Water from the pool filled the container, and then it vanished again as a black cloud descended over his vision. He heard a faint voice from somewhere on the periphery of his consciousness.

"Get off him!"

Suddenly the immoveable weight of Lillian's body vanished, and Oppenheimer lurched upright and out of the water. He sucked a huge volume of air into his lungs. His vision returned as he sagged backward onto the damp rocks just in time to see Saffron hurl Lillian Cruz to one side.

Ethan burst into the chamber just in time to see Lillian Cruz staggering to her feet, water pouring from her arms. Oppenheimer sat in a drenched huddle beside the pool, Saffron standing protectively over him.

"We've got to get out of here!" Ethan yelled, grabbing Lillian and propelling her out of the chamber. "Get to the surface!"

Lillian glared at Jeb Oppenheimer and Saffron, but she obeyed and dashed out of the chamber. Ethan turned to Saffron.

"It's time to go," he said.

"I'm not leaving him here," Saffron shot back.

"Fine!" Ethan shouted, losing patience. "Let's just get out of here!"

Jeb Oppenheimer struggled to his feet.

"I'm not leaving without samples," he insisted, gesturing at the huge crystals with his cane. "Help me get them before we leave."

Ethan almost laughed.

"Like hell," he said, and grabbed Saffron's arm. "Come on, let's go."

Ethan had almost turned his back when he heard the sound of a gun's mechanism being cocked. He turned to see the old man holding a small, snub-nosed pistol in his right hand. Ethan froze as Oppenheimer smiled grimly.

"I never leave home without one," he said. "Now get up there and get me some of those crystals or I'll put a bullet in you."

Ethan stared in disbelief as Oppenheimer walked away from the pool and positioned himself between the chamber exit and Ethan and Saffron.

"Are you really that insane?" Ethan demanded. "We could be buried alive in here at any moment."

"Best hurry then!" Oppenheimer cackled, gesturing with the pistol. "Move!"

Ethan shook his head.

"No. You'll never be able to climb up there on your own, so without me you're screwed."

Oppenheimer's wrinkled face screwed up on itself in furious defiance.

"Not quite."

Oppenheimer shifted his aim and before Ethan could even register what he was about to do, he fired a single shot that rang out deafeningly loud in the chamber. Saffron cried out as the bullet thumped into her belly and out through her side, ricocheting off nearby rocks and zipping away into the chamber.

71

Ethan lunged forward and caught Saffron as she toppled sideways from the impact, her eyes wide with shock and disbelief and her face suddenly pale. He lowered her down as gently as he could on the rocks, pushing fist-size rocks out of her way before setting her down. He let his hand fall on one of the rocks as he spoke to her.

"Don't panic," he said desperately. "Keep your heartbeat as slow as you can so you don't bleed too quickly."

"It's a stomach wound," Oppenheimer cackled, moving closer. "She'll leak the contents of her gut into her bloodstream and die from blood poisoning. You've got about five minutes before it'll be too late to save her."

Ethan let go of Saffron's body and turned to glare at Oppenheimer. For a moment he considered simply doing what the old man said, but suddenly he was overcome with an intense desire to deny the old bastard what he wanted once and for all.

"Go to hell," he uttered. "I'd sooner die."

Oppenheimer glared at Ethan.

"Get up there and collect those crystals or I swear I'll shoot you where you stand."

"Do it," Ethan said. "You'll never get them, not from me and not from Saffron. You're finished, Jeb, totally finished. At least let Saffron live even if you kill me."

Oppenheimer turned slightly and pointed the pistol at Saffron again.

"Do it, you cretin, or I'll shave another few minutes off her life."

Ethan looked down at Saffron, who stared up at him through

her pain and shook her head vigorously. Ethan looked back up at the crystals and then turned, swinging his arm up to hurl the rock in his hand up into the cavernous vault of the chamber. The rock smashed into the ceiling above and instantaneously the giant flock of bats screeched in unison, spilling from the chamber's roof in a screaming black avalanche of wings and teeth. Ethan ducked down as the bats raced past in a thick black fog and blasted into Jeb Oppenheimer as they raced for the cavern exit. Ethan heard the old man curse and drop his cane as the bats slammed into him, the pistol firing a wild shot in Ethan's direction.

Ethan launched himself forward and crashed into Oppenheimer, pinning the pistol between them as they smashed into the rocks. Oppenheimer gagged in agony as sharp stones stabbed through his dirtied suit and punctured his skin, flecks of saliva and mucus spilling from his mouth as he cursed and scratched at Ethan's face with his nails. Ethan pulled away from the attack, keeping hold of Oppenheimer's gun and twisting it from his grasp. The old man cried out in fury, reaching down to one foot with his free hand. Ethan glimpsed a small knife that Oppenheimer grabbed from a sheath at his ankle and whipped around toward Ethan's flank. Ethan reached out for the blade but he couldn't move quickly enough to block the blow. Something grated against his ribs and vibrated through his flesh as the blade plunged hilt deep into his side with a dull thud. He jerked away from the blade and rolled off the old man as he grabbed the blade's handle. Oppenheimer scrambled to his feet, the pistol still in his grasp as he aimed it between Ethan's eyes and glowered down at him as his chest heaved for breath.

"Nice knowing you," Oppenheimer cackled, and squeezed the trigger.

"Grandpa!"

Ethan glanced behind Oppenheimer, to see Saffron on her feet and looming behind Jeb. The old man whirled in surprise, just in time for Saffron to catch his gun wrist in her left hand and twist it violently upon itself. Saffron's shrill scream of anguish echoed through the chamber as she yanked Oppenheimer's arm down and brought her knee up into his elbow. A crack like the snapping of a twig echoed through the cavern as Oppenheimer's arm broke mid-joint, the pistol clattering to the rocks at his feet.

Ethan, pain searing his body, watched as Saffron glared down at her grandfather as she held him by his broken arm.

"My grandmother would have hated what you've become," she hissed at him.

Oppenheimer, his voice tight with agony and fear, pleaded with her.

"This is worth it, Saffy," he croaked. "Every step of this journey, it's worth it."

Saffron twisted his arm harder, provoking a squeal of pain.

"Not for me."

Saffron drove her knee into Oppenheimer's gut. The old man's eyes bulged as a blast of foul air spilled from him lips. Saffron spun on her heel and pulled hard, hauling his wiry body over her shoulder. Jeb Oppenheimer screamed in pain as he was flipped over her body and plunged backward into the pool, Saffron's weakened legs buckling as she fell on top of him and clasped her hands around his throat.

Ethan dragged himself up onto his elbow to see the old man thrashing hopelessly in the water, his white suit weighing him down and his broken arm useless by his side as he sank below the surface in Saffron's furious grasp. Above the thrashing water of her grandfather's desperate death throes he could hear Saffron crying, and then the water fell still beneath her, expanding ripples drifting out toward the distant reaches of the chamber. Saffron stared down at the roiling surface of the water and cradled her bleeding stomach. Ethan struggled over the pain in his side and clambered up the rock face beside him to regain his feet and limp across to her. He looked down into the pool, where Jeb Oppenheimer sank slowly away to sprawl motionless on the bottom, staring up with wide, lifeless eyes.

"I'd say this pool's been contaminated," he said.

Saffron nodded and then fell sideways. Ethan managed to catch her in one arm, his other hand still on the handle of the knife wedged in his side.

"We need to get out of here," he said quickly, and turned them both toward the cavern exit.

Saffron was breathing heavily as she struggled to keep moving through the low, awkward passage, and Ethan could feel thick blood flowing warmly from around the knife still wedged in his side. He tried not to think about what damage the steel blade might be causing inside him, forcing himself onward one step at a time toward the dim light ahead from the cave entrance.

Another blast of noise from explosives thundered down the passage, a shock wave hitting Ethan and a fine mist of particles stinging his face. The cavern ahead was filled with thick smoke, and he could hear the occasional crack of a rifle as he staggered toward the light.

He instantly saw John Cochrane lying halfway between the cave entrance and the last line of defense, sprawled on his back and staring wide-eyed at the ceiling. Half of his forehead was missing, his ear hanging from his scalp in a web of tattered fronds of skin and bone. The cave entrance was littered with bodies and fallen chunks of rock.

Lopez saw him coming over her shoulder as she lay with her pistol propped up on the rocks.

"We're pinned down!" she shouted. Then she saw their wounds. "Are you okay?"

Ethan nodded and managed to struggle up alongside her before both he and Saffron collapsed, breathless and sweating, in the darkness. Lillian Cruz dashed across to them, her eyes wide as she took in their injuries. Lopez saw the knife in Ethan's side, and her skin turned pale as she looked at him.

"Jesus, Ethan, we need to get you to a hospital."

"We need to get out of here first," Ethan said, his own voice a raspy whisper in his ears.

"They're still trying to blow the entrance to the caves," Lillian shouted above the crackle of gunfire. "We can't hold them off much longer."

Ethan looked across at Ellison Thorne. His shoulder was stained with crimson blood where a bullet had hit him, his features tight with terminal defiance. Nathaniel McQuire appeared uninjured, but he was clearly exhausted, his face smeared with soot and scratches from the blasts of the explosives. Edward Copthorne was lying unconscious alongside them, several patches of blood on his clothes betraying multiple bullet wounds.

"How much ammunition have you got left?" Ethan asked.

"Plenty," Ellison replied. "We've made our own ammunition for years using makeshift forges, just like we did back in the Civil War. But I wasn't intending to keep shooting." He gestured across to the cave entrance. "It's time we end this, once and for all."

Ethan nodded, one hand clutching the blade impaled in his side as he looked at the cave entrance. There, he could see sticks of dynamite piled loosely around the sides of the cavern, connected by a fuse wire that ran to a simple detonator lying beside Edward Copthorne's inert body.

Ethan crawled to his knees and picked up his pistol from where it lay beside Lopez.

"We're ready," he said.

Lillian Cruz helped Saffron to her feet and Lopez stood up with her rifle held at port arms, bayonet fixed and ready. Ethan looked down at Ellison Thorne one last time as the big man picked up a handful of dynamite sticks and cradled them in his arms, a cigarette lighter in his big fist.

"Who was holding the camera, Ellison?" he asked. "They'll still be in danger, they'll need help."

Ellison Thorne smiled. "Not now they won't," he said.

Before Ethan could challenge him again, Ellison lit the dynamite in his grasp in a fizzing cloud of sparks and blue smoke. Lillian prodded them all toward the edge of the cave entrance as Ellison staggered to his feet, holding his lethally blazing explosives and nodding once at Ethan. Ethan turned and hurried toward the light, squinting through the smoke and aiming his pistol ahead as they burst out into the vertical shafts of sunlight beaming down to the bottom of Misery Hole.

A single soldier was squatting amid boulders right in front of Ethan, fiddling with a pack of C4. Ethan aimed and fired a single shot, the round snapping the man's head sideways as though he'd been clubbed with a baseball bat. He saw Lopez fire a round at a man farther up the slope, hitting him squarely in the chest and taking him clean off his feet to land with a thud on his back.

Ethan split to the right of the cave entrance with Saffron and Cutler as Lillian and Zamora hurled themselves in the opposite direction.

Ethan shouted out as he crouched down with one arm across Saffron's shoulders and Cutler shielding the pair of them with his body. "Get down, now!"

Ethan glimpsed dozens of mercenaries all aiming assault rifles at them, when the cave entrance exploded as Ellison Thorne's dynamite ignited.

72 |

A superheated blast wave of shattered rocks and flame roared from the mouth of the cave, spitting a cloud of supersonic debris that radiated out across Misery Hole. The mercenaries besieging the cave hurled themselves aside as they were hammered by the deadly shrapnel, shielding their faces as they hit the ground. The infernal blast rolled and echoed through the cavern as a tumbling shower of dust, stone chips, and shattered vegetation fell from around them. Ethan took his hands from over his head, wiping dust from his eyes as he looked back to see the mouth of Lechuguilla Cave now obscured by a mountain of shattered rocks still tumbling down to thump onto the ground nearby.

Lopez and Lillian Cruz lifted their faces from the dust, but Saffron remained motionless. Ethan struggled across to her side and gently rolled her over. Her face was pale and her eyes rolled up in their sockets, but she was conscious, groaning with the pain that was now beginning to seep like acid through her body as the contents of her stomach leaked into her bloodstream.

"She needs a hospital now," Cutler said, "or she won't make it."

Ethan was about to reply when some of the mercenaries began to recover themselves and leaped from cover, M16s at the ready as they surrounded them, a tall man with pallid, pockmarked skin at their head. His red hair glistened in the sunlight above his paunchy face as he looked down at them.

"At last," he said, lifting his rifle to point at Ethan, "it's time to bring this to a close."

Zamora, on his knees and with a rifle pointing at him, shook his head.

"Don't do this."

The man ignored him and aimed his assault rifle, the ugly black barrel pointing straight between Ethan's eyes, and then a deafening gunshot burst the air around Ethan's head. He heard himself cry out and grab his face in shock, only to see the assault rifle aimed at him spin wildly away through the air. The pale-skinned man doubled over in shock as he stared at the ragged, bloody hole torn through his hand.

A barrage of voices shouted out, laden with fearsome rage as they raced closer, and as Ethan looked up through the shimmering beams of light sweeping down through the clouds of dust and smoke, he saw dozens of figures rappelling down toward him.

"Put your weapons down! *Down down down!*"

Ethan, his eyes itching with grit, watched as United States Marines plunged into the cavern around them. Snipers enshrouded in bush camouflage popped up from the edge of the ridge a hundred feet above, machine guns pointed unwaveringly at Hoffman and his men. The sudden sound of distant thunder became a deafening, hammering chorus as a pair of Bell Boeing V-22 Osprey aircraft roared past against the disk of blue sky above, escorted by two Apache AH-64 gunships, the sound of their amassed rotors thundering down through the cavern like the footfalls of a charging army of giants.

"That'll be the cavalry," Lopez said as she lowered her pistol and spat dust from her mouth. She crawled to Ethan's side and gently pressed her hand around the blade lodged in his side, trying to stem the bleeding.

Ethan saw Hoffman being forced to his knees by a marine, blubbering like a child as he was cuffed and forced flat onto his stomach. Other marines thundered down to surround them as Lopez called out.

"Medic!"

Within moments a pair of medics were alongside them, dropping their Bergens and fishing out IV lines and saline bags with practiced efficiency. Ethan lay back, ignoring the line being inserted into his arm and watching as the other medic worked on Saffron Oppenheimer nearby. The man turned his head, holding a small microphone attached to his helmet as he spoke into it.

"Delta-Four, CASEVAC, repeat, casualty evacuation immediate, standby."

Ethan heard a warbled response through the medic's earpiece, and then the noise from the approaching Ospreys became overpowering.

Cutler and Zamora appeared beside him, concerned looks on their faces, and then they were pushed aside by four marines who surrounded him.

He lay back and looked up past the towering rocky walls of Misery Hole, saw swaths of golden desert dust sweeping across the sky as the two huge aircraft landed somewhere nearby. The four marines lifted him onto a makeshift field stretcher, the medic carrying his saline bag as he was lofted upon their shoulders and attached to a winch lowered hastily by troops far above.

Ethan saw Lopez watching him with a furtive expression, her long black hair caked with dust, and then he was hoisted up and away from the floor of the cavern, spiraling slowly as he was pulled all the way to the lip of Misery Hole and then lifted clear of the chasm.

One of the Ospreys had landed "hot" on a plateau a hundred yards away, the huge pivoting engines directing their immense thrust downward for vertical landing. The blades were kicking up a fearsome dust storm around the aircraft, but the medic carrying the saline bag shielded Ethan's face with one hand as he was rushed aboard the aircraft. The marines lay him down on a canvas bed that had been folded down from one side of the fuselage wall.

Ethan lay back, watching as Saffron was lifted aboard moments later and carried with reverential care by the battle-hardened marines to lie nearby on another fold-down bed. The marines rushed off the Osprey and moments later Ethan felt his stomach plunge as the aircraft's engines roared even louder and it lifted off and accelerated into forward flight.

Medics fussed over Saffron's inert form nearby, and Ethan watched with concern until a soldier carrying what looked like a laptop computer squatted beside him and rested the computer on his chest. A flickering image of Doug Jarvis offered him a brief smile from beneath a hastily bandaged and bloodied forehead.

"Good morning."

Ethan closed his eyes briefly before replying.

"I wouldn't call it good," he said, glancing again at Saffron. "Saffron's been hit bad."

Jarvis nodded.

"Don't worry, she's in good hands now and will be in the best of care once you get back to Holloman. I've alerted their best people to take care of her once you arrive. I take it your meager flesh wounds won't prevent you from attending a detailed debrief once we're there?"

Ethan sighed.

"Sure. What's a few inches of Toledo steel between friends? Where the hell are you? And what happened to your face?"

Jarvis grimaced.

"New York, and if you think I look bad, you should see Donald Wolfe. Get some down time, Ethan. I've already got the DIA on the case."

"How did you know where to find us?" he asked.

"Lopez sent me a text," Jarvis said. "An IP address for a GPS tracker. I'd only just gotten out of the UN building, but we managed to figure out what she'd done. I dispatched the marines right after."

"She lied to me, Doug," Ethan said with a sigh.

"She got the job done," Jarvis countered. "Anyway, don't worry about the details—my teams are already on their way to clear up, and by lunchtime none of this will ever have happened."

Ethan rested his head against his pillow, staring up at the ceiling of the Osprey.

"I doubt that," he said quietly.

"Why?" Jarvis asked. "Jeb Oppenheimer suffered a tragic accident this morning and will be buried in a private ceremony—he didn't have any friends to speak of, so no need for a public funeral. His mercenaries will be tried by military court and convicted of conspiracy to murder DIA agents. Donald Wolfe effectively confessed to conspiracy to murder Tyler Willis and for the homicide of an unknown male in Alaska in front of the entire United Nations before he shot himself. Several UN representatives are in the hospital after drinking the water that Wolfe infected with Spanish Flu, but thankfully most did not touch their drinks. I take it that Lillian Cruz has been liberated and will no doubt testify as and when required regarding her abduction. And as for the soldiers you located out here . . ."

Ethan nodded slowly.

"They're all dead," he guessed. "Have been for decades."

"Longer," Jarvis said, and his voice became somber. "The marines found one of them in the hands of Oppenheimer's mercenaries. He wasn't alive."

"Kip Wren," Ethan identified him. "A very brave man. They all were."

Jarvis nodded.

"Their remains will be allowed to decay as appears to be the

norm and then be retrieved from the caves for further study. Butch Cutler and Officer Zamora will be signing nondisclosure forms within the hour. The chamber within Lechuguilla Cave will be sealed and its location removed from any public forum. Best that we keep whatever's in there under lock and key, for obvious reasons."

"What about Saffron?" Ethan asked.

"Tricky," Jarvis said. "For whatever reason, she developed a conscience and handed herself in to state police, told them everything. I'd have sent people in to retrieve the incriminating evidence on her behalf, but the police are already there and apparently have recovered video tapes that implicate her in an unsolved homicide some four years ago."

Ethan sighed heavily.

"The circumstances were different from how they appear on the images, apparently," Ethan said. "After all this, they'll use everything they can to send her down for life. You need to do something to help her, Doug, she doesn't deserve this."

On-screen, Jarvis shrugged.

"What can I do? This case is resolved and everything the DIA needs is now in their hands, including Kip Wren's body and the site itself. They're not going to let me interfere in a civil case."

Ethan looked at him seriously.

"They might have to," he said. "Because it's not over."

"What do you mean?" Jarvis asked. "There's nobody left."

"Yes, there is," Ethan explained. "There were eight of them, not seven, because someone was holding the camera when that photograph was taken in 1862."

Jarvis's features creased as a realization dawned in his eyes.

"The subject is still alive?" he asked.

"Very much so," Ethan said. "Protect Saffron and I'll help figure out who it is. Let her go down, and the DIA can go to hell for its information."

73

CHRISTUS St. Vincent Regional Medical Center
Santa Fe
May 19
3:47 p.m.

Ethan packed the handful of dust-covered clothes that he'd worn in Misery Hole, pausing to glance at the thick bloodstains on his shirt before stuffing it into a bag and dropping it into the trash can in his hospital room. The movement caused a painful twinge low on his flank where Oppenheimer's blade had sunk into his flesh, the scar tissue only half healed beneath the surgical dressing wrapped around his waist. He slipped into the fresh clothes sent to him by Doug Jarvis and turned his back on the room, walking down the corridor outside to where he knew Saffron Oppenheimer was staying, easily identifiable by the two state troopers guarding the door. Ethan introduced himself to them, and they stood aside to let him in.

The room was pleasantly spacious, and Saffron lay with her eyes closed at a comfortable angle on her bed with only a saline drip to show any evidence of her trauma. Ethan slipped quietly into the room and closed the door behind him.

"About time you showed up, tough guy," Saffron said, opening one eye to peer at him. "Last I heard you were at death's door with a little splinter in your side."

Ethan grinned.

"Go to hell," he said, crossing the room to her bedside. "The doctors have told me I'm not to exert myself for five days. It could lead to complications."

"Sure," Saffron said, but her eyes were dancing with humor.

"How're you doing?" he asked.

"I'll be fine," she said, shifting her position slightly and propping herself more upright on her pillows, "now that I don't have my breakfast running through my bloodstream. Took a while for them to stitch me up but I'll make it."

"You did good," Ethan said. "If you hadn't walked into that police station, Lopez and I would have ended up beneath a few tons of New Mexico rubble and Jeb Oppenheimer would have been planning who lived and who died in his new world order."

Saffron smiled faintly.

"Didn't do me much good though," she said. "Guards at my hospital door and a lengthy prison sentence to look forward to. You should have left me in that cavern with Jeb."

Ethan sighed and sat on the edge of the bed.

"I'm doing what I can," he said. "I've got the DIA over a barrel about what happened in Lechuguilla Cave, both last week and in 1862. They're looking at options for you."

"Why doesn't that phrase fill me with confidence?"

"It's not easy," Ethan admitted. "You've become a high-profile victim of these events, but it's still going to be hard to keep you out of jail."

Saffron became melancholy, staring at her bedsheets.

"I killed him," she said finally. "I killed my own grandfather."

"You defended yourself," Ethan replied. "He shot you first, remember?"

Saffron didn't appear to hear his last words as she reached out and gripped his hand.

"Promise me you'll call your parents," she said.

"Why?"

"Because you still can," Saffron replied. "Don't leave it another day, okay? Just do it, before you wake up one morning and realize you no longer can."

Ethan held her gaze for a long moment and saw the determination in Saffron's eyes. He nodded and flipped her a mock salute.

"Okay, I'll do it. I promise. Scout's honor. But you've got to do something for me."

Saffron raised a questioning eyebrow as Ethan gestured out the window.

"You're the heir to SkinGen Corp," he said.

"I don't want any of it," she said. "The whole goddamned company can go to hell and—"

"That's what I mean," Ethan cut across her. "You want to change things for the better? Selling the company off for nothing will achieve exactly that: nothing. Take the helm, organize something, even if it's just distributing drugs to countries and people who can't afford them. SkinGen is not a legacy to be avoided, it's an opportunity. Use it."

Saffron leaned back against her pillows with a sigh and released his hand.

"I wouldn't know where to start," she said.

"Anywhere will do," Ethan said. "Just start. Jeb Oppenheimer was a monster but much of what he believed made sense, especially to ordinary people. Find those things, act on them. Don't leave SkinGen tainted with the memory of Jeb: reinvent it in the image of Saffron."

Saffron laughed briefly, but she saw Ethan's expression and the laugh faded.

"You think it's worth it?"

"It's a no-brainer," Ethan replied.

"Fine, I'll do it," Saffron said. "Now get out of here, I've got a genuine injury to recover from."

Ethan grinned and walked to the door. He was about to open it when she called over to him.

"I won't be seeing you again, will I?"

Ethan hesitated at the door.

"If the DIA gets you out of this, you'll be hidden away for quite a while."

"Maybe I'd have looked you up."

"Maybe you still should."

Saffron shook her head and smiled. "I don't have a chance while your Mexican spitfire's watching your back."

"Lopez?" Ethan said. "We're just partners and—"

"Like hell," Saffron cut across him. "She likes you, I can tell it from a mile away. You just watch your step with her though."

Ethan sighed softly. "You mean because she sold out to Jeb Oppenheimer? You know she tossed the tracking device before we found the caves, right after she'd sent the DIA the IP address. She had a plan."

"She looks out for you all right, but she's a firecracker, hard to handle."

Ethan wasn't sure how to react. "What do you mean?"

"Put it this way," Saffron said. "If you're walking about with a grenade in your pocket, you step lightly because it could kill you just as sure as it could save your life. She's a wildcard, Ethan. Nobody knows what she might do next."

Ethan opened the door, thoughts of Lopez clouding his mind.

"I'll bear that in mind."

74

The New Mexico sun blazed brightly in a flawless blue sky as Ethan stepped out of the hospital foyer, shielding his eyes. Across the street stood Nicola Lopez, leaning against the trunk of a Ford Taurus and jangling the keys in her hand. Ethan stepped across to her, the hot wind rippling his shirt and lifting one corner to reveal the thick dressings. He touched them self-consciously as he walked.

"You're not going to start showing that thing off to people, are you?" Lopez asked.

"Looks good, doesn't it?" he said. "Another war wound to add to the collection."

Lopez smiled and shook her head.

"People will just assume you've had your appendix out or something. Jesus, you were in the hospital for two whole days. Anyone would think you got your head blown off."

Ethan, affronted, covered the wound back up.

"C'mon, I suffered for this one. Doctor said a half inch higher or lower and I'd have been in real trouble."

"Yeah," Lopez agreed. "My point exactly."

Ethan shook his head in dismay.

"Just because you can walk out of a gunfight looking like you've been modeling swimwear doesn't mean everyone can," he mumbled.

Lopez looked at him as a smile blossomed on her face.

"You really think that?"

Ethan sought to backtrack.

"Sometimes," he admitted. "But don't get used to compliments, I'm not paid to blow sunshine at you."

Lopez smiled again, but the moment passed. Ethan watched her for a moment, and then gestured to the car.

"We need to talk."

Lopez climbed into the car with Ethan and drove out of the hospital grounds, heading toward I-85. Ethan let the cool breeze funnel in through the open window onto his face as they drove.

"What's the news on Saffron?" Lopez asked.

"She's recovering well," Ethan replied, "but she'll be going to trial in Santa Fe as soon as she's well enough to stand."

"Doug's working on it," Lopez said. "They've been going insane trying to figure out who the eighth person is, but I don't think it's getting them anywhere. I'm surprised they haven't tried blackmailing us yet."

"With what?" Ethan smiled smugly, settling back into his seat. "They can hardly fire us as we don't work for them directly. Either they work it out for themselves or they uphold their end of the deal."

"Do you know who it is?" Lopez asked.

"I'm working on it."

"Are you tryin' to tease me?"

"Are you enjoying it?"

"You're an asshole, Warner."

"I'm not the one who sold out," Ethan replied.

Lopez kept her eyes on the road ahead for a long moment before she replied. "I took an educated gamble," she said finally. "It made sense at the time, okay? Everybody came out the other side of it."

"By the skin of our goddamned teeth, Nicola!" Ethan shot back, dragging a hand across his face. "Jesus, we're going up against some pretty unsavory people here and I don't know if I can trust you from one day to the next."

"You can trust me," Lopez said instantly, locking her dark eyes onto his.

"How can I when I don't know what you're doing behind my back?" Ethan replied. "That little scheme of yours could have cost us all our lives, and for what? A quarter million bucks? Is that what I'm worth to you?"

"That's not how it went down!" Lopez protested. "I didn't do anything behind your back. Jesus, we're not married, you know."

"Yes, we are," Ethan insisted, "ever since we went into business together."

"Hey, I'm not just the dutiful wife in this arrangement, cowboy," she shot back, the wind rippling her hair in a dark halo. "I'm not just a passenger here, your damsel in distress, okay? I can make my own calls, and I'll goddamned follow my own leads with or without you."

Ethan chuckled bitterly and shook his head.

"You didn't do this for your equality," he muttered. "You did it to make hard cash, and you risked my life along with everybody else's to pull it off. Where will it stop, Nicola? Breaking into jeeps is one thing: this is something else entirely. What will you do next?"

"There is no *next*," Lopez said, more reasonably. "I did what I set out to do, and now it's over."

"Is it?" Ethan asked. "I thought it was over when Doug Jarvis pulled us out two days ago. You said it was."

Lopez stared ahead as she drove, and sighed heavily.

"It's over," she repeated, and then looked at him. "I won't let you down, okay? Let's just drop it for now."

Ethan turned away from her to look out across the blistering deserts passing by.

75

Santa Fe Police Department
Camino Entrada

Claire Montgomery sat in the police department's interview room with Lieutenant Enrico Zamora facing her across a desk. Two Styrofoam cups sat between them and the unblinking black eye of a closed-circuit camera watched them from one corner of the room.

"He made you do that?" Zamora asked. "Every day?"

"Most days," Claire replied. "Jeb Oppenheimer liked his assistants to provide him with all and any services. He called them our 'targets.' Trouble for him, and fortunate for us, was that most days the old bastard couldn't get it up."

Zamora stifled a grin, nodding as he looked down at his notes.

"Okay, so you're sure that Tyler Willis was being held within the SkinGen building when he died."

"Definitely." Claire nodded. "I heard him on the intercom shouting something, but I had no idea what was happening down there in that theater. Christ, it just doesn't cross your mind that someone could be so evil. I mean, maybe Nazis or something, but not your boss."

Zamora nodded.

"Okay, well that's all we really needed to know, Montgomery. We were pretty sure that Tyler Willis died at the hands of Jeb Oppenheimer, and the only witness we have other than Lillian Cruz was one of Oppenheimer's bodyguards, who died out in New Mexico last week and . . ." Zamora paused as he noticed Claire's confused expression. "What?"

"You said Tyler Willis died at the hands of Jeb Oppenheimer?"

"Yes."

"That's not possible," Claire said. "He was alive when Oppenheimer left the building. Willis died hours later."

Zamora was about to reply when an urgent knock sounded at the door of the interview room, and a sergeant stepped quickly in with a sheet of paper in his hand.

"Results of the autopsy on Tyler Willis," he said to Zamora. "You need to see this."

Zamora took the piece of paper and scanned it, his eyes widening with each line. He stood up out of his chair.

"You're sure? Absolutely sure?"

The sergeant nodded.

"And that's not all," he said. "There's been another abduction."

76

Santa Fe
4:38 p.m.

Ethan sat on the edge of his bed in the hotel room and stared at the face of his cell phone, the Illinois number waiting there to be dialed taunting him in silence. The absurdity of his own reluctance, nervousness even, to just call the damned number wasn't lost on him. But through Ethan's moral compass he owed Saffron a promise, and despite himself he jabbed the dial button and listened to the phone ringing in his ear. To his surprise, it was picked up on the second ring.

"Ethan."

There was no detectable tone in the voice to suggest surprise, concern, or awareness that his father hadn't received a call from him for two years. Ethan found himself suddenly unable to think of anything to say.

"Hi, Dad. How's things?"

Lame. The towering rock that was Henry Warner would probably laugh in his face and cut the line off with a shake of his craggy head and . . .

"Not so bad, son. Haven't heard from you for a while, thought you'd forgotten what goddamned phones were for."

"I've been busy," Ethan replied, strangely feeling a little bit more relaxed. "I've started a new business."

"Doing what?"

"I work for the government, investigations and such like. It's going good."

"Where are you right now?"

"Santa Fe."

"Good for you. To what do I owe the honor of this unexpected communiqué?"

Ethan took a breath.

"I just thought I should call. We haven't spoken in a long time and I didn't want to leave it any longer."

"You didn't have to leave it at all, Ethan. Your mother's found this all very difficult, you know."

"So have I," Ethan replied.

A long silence enveloped the line as Ethan struggled to find something useful to say. Nothing came. He considered hanging up, but his hand refused to obey. Henry Warner's voice suddenly sounded down the line, startling him.

"I take it you've gotten over what happened in Palestine, Ethan, Joanna's disappearance and your . . . difficulties afterward."

"Diplomatic as always, Dad," Ethan responded.

"No sense in treading on eggshells," his father replied. "Ever since you resigned your commission your life hasn't followed a steady track."

"There's no such thing as a steady goddamned track, Dad. Life doesn't work like that."

"Your mother and I have done okay by living how I—"

"Mom just does as she's told because it's easier than having to negotiate with *you*!" Ethan shot back, unable to contain himself any further. "People are not machines, Dad! Families are not the Marine Corps, and home is not a barracks! When are you going to learn that?"

A long silence filled the line. Ethan realized that he was now holding his breath.

"The Corps did you proud, son, you should remember that. There was a time when you were afraid to climb up that tiny tree in our garden, let alone go to war for your country."

"People change as they grow up," Ethan said through gritted teeth. "You should try it someday. Is Mom there?"

"She's out visiting," came the response, then a long pause. "Has there been any word, about Joanna?"

Images of his long-vanished fiancée flickered through Ethan's mind. It had been so long now since she had disappeared from Gaza that he had begun to associate her memory with the four years he had spent searching for her, instead of the good times they had shared. Now he tried not to think about her at all.

"Nothing," Ethan said in a whisper. "Four years now. Whatever happened to Joanna, I don't think she'll be coming back."

"Then perhaps you can finally move on."

"Sure," Ethan muttered. "Sounds easy if you say it quickly enough."

He heard a heavy sigh in response, as though his father was already tiring of the conversation.

"I didn't mean it like that, Ethan, and you know it. Why don't you come by when you're back in the state, come and visit? Natalie's here too, on sabbatical."

Ethan was surprised to become aware of the broad smile breaking across his face. He hadn't seen his sister in two years. Last he'd heard, she was studying politics at college. He took a breath.

"Sure. I'll hopefully have some time off when we get back to Illinois."

"We?"

"My partner and I," Ethan replied. "That is, my *business* partner and I, at least for the moment."

"Well, you're welcome to bring her too. Just make sure you bring yourself, okay? It's been too long, son, you know that, don't you?"

Ethan avoided answering the question directly.

"Sounds great, Dad. I'll be there within a few days."

"Good. We'll all be glad to see you, Ethan, after all that's happened and despite our . . . differences. If there's one thing that's more important than anything else, it's family."

Ethan smiled again, and was about to reply when a sudden realization slammed into his field of awareness like a bullet through glass. Ethan's smile vanished as he stared blankly into nowhere and his jaw fell open as he heard his father's words echoing around in his mind. *If there's one thing that's more important than anything else, it's family.*

"Ethan?"

"Dad, I've got to go, something just turned up."

Ethan leaped up from the bed and dashed from the room, sprinting down the corridor outside and down a stairwell three at a time. He burst into the hotel foyer, where Lopez was casually leafing through a magazine.

"They've got away!" Ethan shouted.

Lopez dropped the magazine, jumped to her feet, and stared at him.

"Who's got away?"

"I can't believe I've been so stupid," Ethan said, holding his head in his hands.

"What the hell's going on?"

"Get the car," Ethan said. "We've got to go, right now."

Medical Investigator Facility
Albuquerque

Ethan opened the car door and stepped out even before Lopez had fully braked to a stop, the outside of the facility swarming with police cars. Ethan hurried to the police cordon and asked for Lieutenant Zamora, who emerged from the facility a minute or two later and waved Ethan and Lopez forward.

"I can't believe it," Zamora said.

Ethan looked at him.

"There's been another abduction, right?"

Lieutenant Zamora nodded, running a hand through his hair.

"We put a guard on the house just in case and extra security here, but it's like they just vanished into thin air. I can't understand how it happened."

Ethan closed his eyes. "I can."

Lopez grabbed Ethan's shoulder and squeezed it hard.

"Will you tell me what the hell is going on?!"

77 |

Hotel De Bilderberg
Oosterbeek
Holland
8:12 a.m. (European Time)

Gregory Hampton III sat in the plush surroundings of his

penthouse suite and looked out the window at the sumptuous grounds of the Hoge Veluwe National Park. A fine early-morning mist had enveloped the park, streetlights glowing like candle flames among the trees, and at this early hour there were no pedestrians. He stood and walked to the door of the suite, opening it at the appointed hour and moving into the corridor outside.

Hampton was not a man who was used to being recognized, but he saw clear recognition in the smartly dressed person striding confidently toward him, precisely on time.

Gregory Hampton had been born in Hampshire, England, in 1936, just before the world collapsed into the chaos and destruction of the Second World War. Sent away from his family to live in the West Country away from the blitzkrieg blasting England's major cities, he returned home an orphan and determined that he would never be subject to the whims of others again. Within fifteen years he was a millionaire magnate presiding over a booming property portfolio in London, profiting from a rebuilding frenzy in the aftermath of the war's destruction. Twenty years later, he was a billionaire. Another decade after that, he stopped even trying to count his fortune. He owned islands in the Pacific, a significant proportion of Dubai and Manhattan, and several cruise liners, but he prided himself on the fact that nobody would have known him if they passed him on the street.

The person who recognized him stopped at the entrance to his suite, wearing an immaculate dark gray suit and highly polished shoes, with hand extended.

"Thank you for agreeing to see me."

Hampton shook the proffered hand.

"I must ask: How did you know where to find me?"

His visitor smiled.

"I've watched you for many years, Mr. Hampton, long before you chose to recede into anonymity. I suspected that if it was not you that I needed to see, then you would at least know where to send me."

Hampton's eyes narrowed.

"And you are approaching me now, after all that has happened to you. Why?"

Again, the calm smile.

"Can we speak inside?"

Hampton nodded and gestured into the suite, following his guest in and closing the door. Inside the suite, four more men waited patiently, each wearing suits that cost as much as some cars and cautious expressions as they surveyed their visitor.

"I hope," one of them said to Hampton, "that this is worthwhile. We're taking an awful risk here."

"As am I," the visitor said, "after what happened to Donald Wolfe."

Gregory Hampton gestured to their guest.

"Gentlemen, please do me the courtesy of listening to what our new associate has to say. I feel certain that you will appreciate it."

"Who are you?" one of the men asked the guest.

The guest sat down, crossing one long leg over the other.

"My name is Lillian Cruz," she said, "and I was born in Montrose, Colorado, in the year 1824."

A silence descended upon the men in the suite as they looked at her.

"Go on," Gregory Hampton prompted her. "My associates here are familiar with the basic potential of human longevity."

Lillian Cruz regarded the men for a long moment before speaking.

"I am the last survivor of eight soldiers of the Union army who took sanctuary in a place called Misery Hole in New Mexico in 1862, just after the Battle of Glorieta Pass."

"What on earth were you doing in an army?" one of the younger men asked.

"I was one of many women who served alongside their countrymen in the Civil War," Lillian said hotly. "In my case, I met my husband within the ranks, an officer named Ellison Thorne. He died recently after being pursued for years by a man named Jeb Oppenheimer so that I might still live today."

Lillian took a moment before continuing.

"I have been alive for a hundred eighty-seven years, and in that time I have seen this country, and this planet, gradually destroyed by the parasite that we call humanity. We are the only species that consumes without replenishing, that takes more than our fair share without giving anything back. For almost one hundred fifty years, five of the seven men who had sheltered with me in Misery Hole lived in a small area of the New Mexico desert with barely anybody knowing they were even there. They took only what they needed, and in doing so were a part of their environment, not a predator upon it."

Lillian reached into the pocket of her jacket and produced a small vial filled with a clear liquid that caught the light streaming in through the suite's broad windows. She held it up to the men demonstratively.

"I have tested this serum," she said, "because we were dying of old age. The bacteria that infected us when we were in the caves began dying because it was unable to sustain itself indefinitely on the iron we consumed in our food, perhaps because of our physical size compared to other, smaller mammals such as bats, which I believe may have originally harbored it. I myself have not yet suffered any symptoms, possibly because females tend to live longer than males in many mammalian species. This serum has corrected the deficiency in the bacteria *Bacillus permians*, allowing them to sustain cellular senescence indefinitely within the bodies of large mammals such as ourselves."

"How?" one of the younger men demanded. "How can you be sure that the bacteria will work this time?"

"The degradation of our bodies was being caused by a hemoglobin deficiency," Lillian said, "itself caused by the bacteria's demands for iron, effectively making us permanently anemic. Hemoglobin deficiency decreases blood's oxygen-carrying capacity, causing loss of blood, nutritional deficiencies, bone-marrow problems, and kidney failure—the red blood cells in iron deficiency anemia become hypochromic and microcytic. Hemolysis, the accelerated breakdown of red blood cells, follows and jaundice is caused by the hemoglobin metabolite bilirubin, which can cause renal failure. The increased levels of bilirubin improperly degraded the hemoglobin and clogged small blood vessels, especially in the kidneys, causing kidney damage and muscular breakdown. We were essentially falling apart. The same quorum sensing that caused the infection to extend telomere life in us eventually also caused us to become anemic and then began to break down our bodies at the cellular level. The reason for all this was that our immune systems were finally overpowering the bacteria as they gradually died off over the years due to lack of iron. The bacterial population reduced sufficiently that quorum sensing ceased, and we aged rapidly. The bacteria within us simply had not evolved enough to exist in symbiotic harmony with human beings."

"I take it," one of the men asked, "that you have overcome this unfortunate flaw?"

She shook the vial in her hand. "This serum is the result of my work since: it is still not perfect, but it's already more effective than the naturally occurring bacteria. It no longer causes anemia and, as you know, lasts for at least a hundred fifty years. By that time, we'll have worked out how to make it last a lot longer, producing a bacterium that exists in perfect harmony with our own bodies."

Gregory Hampton looked at her for a long moment.

"Your colleagues died to protect that," he said. "Why would you now betray their memory?"

Lillian sighed.

"Because they had not really lived in the modern world," she said. "They hadn't seen what it has become. And because you are all already wealthy beyond avarice and I suspect your interest in this is not financial. I take it that you wish for there to be a world for humans to live in in a hundred years' time, a thousand?"

The men regarded her with respect for the first time. Hampton spoke softly for them.

"The Bilderberg Group has a greater cause than merely acting as an annual reservoir for political discourse," he said solemnly. "Unless there are major technological breakthroughs in the fields of energy generation, farming, and pharmaceuticals in the next ten years, it is highly likely that civilization as we know it will collapse, the burden of our population too great for even human ingenuity to support it. It has already begun, as fossil fuels are now running out and freshwater is becoming increasingly scarce. We exist as a think tank dedicated to preserving human endeavor after the coming apocalypse, and rebuilding it in the future."

Lillian nodded in understanding. "Then this serum could help extend your leadership, should the need arise."

"How much do you want for it?" one of them asked.

Lillian shook her head.

"I don't want money," she said. "What I want is to be protected so that I don't have to spend my life hiding from people like Jeb Oppenheimer. I want this serum to be used only by those who *earn* it, and I want the reduction of the population of this planet to be achieved humanely over time. If there's one thing that we absolutely agree on, it's that if either our population or our rate of consumption is not reduced then humanity is ultimately doomed. Any species that becomes too numerous eventually suffers a collapse, and I fear that ours is well and truly overdue. In that, at least, Jeb Oppenheimer was right."

Hampton looked at his accomplices, and they all nodded together.

"Security is the one thing that we absolutely can guarantee," he said to her finally. "You will be safe among us, I can assure you."

Lillian stood and handed the vial to Hampton. She fixed her gaze on his.

"This serum can be cultured and grown. As a bacteria, it will by its very nature divide and propagate without limitation. You have in your hands the fountain of youth," she said, "the elixir. Use it wisely."

78

Santa Fe
New Mexico

"Lillian Cruz is the eighth soldier?" Lopez said in disbelief.

Ethan nodded as he leaned against a squad car.

"It was staring us in the face all along and I never realized it," he said. "I even read about women serving as soldiers during the Civil War in the records office, but didn't make the connection."

"When did you figure it out?" Zamora asked.

Ethan shrugged.

"Too late. We were all pretty sure that Ellison and his men needed someone on the inside to handle things, but we assumed it would be somebody they had befriended. It crossed my mind only when I thought about Hiram Conley. He had accrued injuries throughout his life, some of them serious enough to have required hospitalization, maybe even surgery. I'd wondered how his great age could have been covered up if he'd died in some other way, in an operating theater surrounded by surgeons and nurses or similar. There had to be a plan in place to cover that eventuality."

Lopez caught on quickly.

"Like a coroner," she said. "Damn, somebody even mentioned that Lillian Cruz had worked there for longer than they could remember."

"Lillian could get them in and out of a laboratory without attracting attention," Ethan said. "She could administer medicines to them, possibly even perform surgery. Hard to make friends like that over decades as the faces would keep changing, and the secret would be too hard to keep. But if Lillian took on a medical role, an official position, then they'd be in good stead to survive just about anything. All she'd have to do is move occasionally, or change her name and such like, to avoid exposing herself."

"So she was the one who took the photograph?" Zamora asked.

"She must have fought alongside the men during the Civil War," Ethan said, "maybe been married to one of them. Lillian could have taken her maiden name when she took up residence in Santa Fe, to hide their connection. Lillian came back to work very soon after the gunfight at the caves, didn't she?"

"Yeah, within a day or two," Zamora said. "That's why we put a guard on her, just in case. How'd you know about that? You were in the hospital."

"Because she would have needed to use her laboratory to check that the bacterial samples she'd gotten from Lechuguilla Cave were alive," Ethan said. "She went after Jeb Oppenheimer when we were trapped underground, and when I caught up with her she'd tried to drown him in the water. She was drenched herself, and only Saffron Oppenheimer's arrival just before me had stopped her from killing him. But she didn't care whether Jeb Oppenheimer lived or died, she just needed a sample of the fluids in that pool."

Lopez frowned.

"But I thought that wasn't enough, that the infection doesn't last forever in the body?"

"No," Ethan agreed, "but by having a sample that she could test, and samples of the remains of both Hiram Conley and Lee Carson, she could experiment. Lillian must have known that it was something to do with iron. When Doug Jarvis's team went into the SkinGen labs they found Conley's and Carson's remains, but in both cases tissue had been removed from the area surrounding the bullets that killed them."

Enrico Zamora thought for a moment.

"The iron kept the tissues alive?" he asked.

"It stopped the tissue samples from decaying," Ethan said. "So by comparing the bacteria in the living tissue and the raw bacteria from the cave, Lillian may have been able to understand how to develop a method to genetically modify the bacteria to survive indefinitely within a human body. The perfect elixir."

Ethan had finally realized what had been happening all along, perhaps even before he and Lopez had arrived in New Mexico the previous week.

"Lillian hasn't been abducted—there's just no need for her to hang around here anymore. She's taken everything she's got now that Ellison and his men are dead, and has probably sold it off to the highest bidder."

Lopez shook her head thoughtfully.

"No way," she said. "Lillian may have taken off with whatever she's gotten from that cave, but I don't believe she would have done that, not after Ellison gave his life so we could escape from the caves."

"I wouldn't be so sure of that," Enrico Zamora said. "We got the autopsy reports back this morning for Tyler Willis. It turns out that he died not from a beating or at the hands of Jeb Oppenheimer. He died from blood loss caused by the severing of an artery in his left thigh, a tiny cut that was almost missed by the coroner."

"Oppenheimer had him killed though," Lopez said. "He was holding Willis hostage."

"Yes," Zamora agreed, "but Willis died when Oppenheimer was nowhere near him. Claire Montgomery, his assistant, confirmed that for us, and despite her earlier indiscretions she has no reason to lie now that Oppenheimer's dead. The only person in the room with Willis with the skill and knowledge to make an incision like that was Lillian Cruz, and Willis would never have known as she'd just hit him with shots of morphine after Oppenheimer had finished torturing him."

"Let me guess," Lopez said. "The morphine shot went into his leg."

"He bled out right there and then" Zamora nodded. "Never knew a thing about it."

"As don't you," Ethan said. "You stick to the nondisclosure policy from this point on, okay? I don't want the DIA hounding you out of a job down here or something."

"I ain't got nothing to say," Zamora replied. "Been a pleasure, in a weird kind of way."

Ethan shook Zamora's hand, and Lopez gave the cop a hug before he turned and walked back into the facility.

Ethan turned for their car, knowing deep inside that wherever Lillian had gone it was highly unlikely they would ever see her again.

"Well, whatever happens, I hope she knows what she's doing. It's likely the last time that she will ever hold the fountain of youth in her hands again."

Ethan climbed into the car, Lopez joining him in the passenger seat. They sat for a long moment in silence before he looked across at her.

"You ready to go home?"

Lopez stared into the middle distance for a long moment before replying.

"I'm not sure where home is right now."

Ethan glanced south toward the Pecos and the endless deserts beyond that led to the border with Mexico and her distant hometown, Guanajuato.

"Can I ask you something?" Ethan said finally.

"Shoot."

"Do you trust me?"

Lopez looked at him but did not respond immediately, and Ethan realized that she was thinking seriously about it.

"Yeah, I do," she said finally.

"Then why not confide in me?"

Lopez held his gaze for a long beat before turning away again. One hand played idly with her long black hair.

"Because I don't want to go through what happened to my last partner again," she said finally. "Ever."

"Lucas Tyrell," Ethan said, recalling Lopez's detective partner from Washington, D.C., who had died the previous year.

"He was a good man," Lopez said, "who didn't deserve to die the way he did."

Ethan leaned back in his seat. Fact was, he also had avoided close relationships since losing Joanna. The fear they both carried, veiled just beneath a thin veneer of normality, was that anybody could lose anyone at any time and never see them again. Be it by a bullet, or accident, or just plain stupidity, people died all the time. Only those left behind grieved for their loss, and all too often were unable to let go.

"I take it you got all the money that you swindled out of Jeb Oppenheimer," Ethan said. "And that now that he's dead, there'll be no comeback."

"No sense in crying over spilled milk," she said with a brief smile.

"Doug and the DIA won't have missed something like that," Ethan warned. "They'll know what you've pulled."

"Let them," Lopez said. "Every last dime's been taxed. There's nothing they can do about it. You're welcome to half, seeing as we're partners."

Ethan shook his head. He couldn't resist a wry sense of admiration for Lopez's sheer audacity, but it was that same recklessness that had almost gotten them both killed.

"I can't do this, Nicola," he said, "unless I know for sure that you've got my back."

Lopez nodded.

"I know."

"Then make your call. Are you in, or are you out?" he asked.

"Of what?"

"Of this, of Warner and Lopez Incorporated, bail bonds and investigations. Where's home, Nicola? It's your call, you can do anything you want. Do you want to head north back to Chicago with me, or south for Mexico?"

"What would you do if I went back over the border?" she asked.

"I'd carry on," he replied without hesitation. "Probably get more work done without you in the way, and Warner Incorporated rolls off the tongue nicely anyway."

Lopez smiled faintly.

"I just need time to figure everything out," she said. "It's been only a year since I left the police and we started this little venture. You lost Joanna four years ago, and you're still not quite who you were before, are you?"

Ethan couldn't bring himself to meet her gaze, but he shook his head briefly.

"Then let's just take this one step at a time, okay?" she suggested.

Ethan sighed again, knowing he could hardly pressure Lopez when he was barely over Joanna himself. Lopez and Tyrell hadn't, as far as he knew, been an item, but they'd been partners and sometimes that bond could be just as strong. Ethan had experienced just such camaraderie in the Marine Corps: men who had faced death together tended to face life together as well, a brotherhood forged in shared hardship that felt as though it could last for an eternity.

"Okay," he agreed finally. "I just don't want to be wearing dentures before I know I can rely on you."

Lopez smiled brightly. "Maybe we should have saved some of that elixir for your wrinkled old ass."

"Like hell," Ethan shot back as he started the car's engine. "Every crease has been earned."

He sat for a moment, thinking about Saffron Oppenheimer, and then looked across at Lopez again.

Lopez stared out to the south, toward the shimmering mountains and the hard blue skies, and then back at Ethan.

"I want to go home," she said. "North."

ACKNOWLEDGMENTS

Once again I owe an enormous debt of gratitude to my brilliant literary agent Luigi Bonomi, who upon our first-ever meeting suggested that I should pen a story based upon the search for immortality. He felt that nobody had ever attempted to concoct a potentially viable method by which eternal youth could be achieved. If he's still representing authors in the year 2176, we'll know he had an ulterior motive. My heartfelt thanks go to Stacey Creamer, Lauren Spiegel, and all at Simon & Schuster/Touchstone, who never fail to impress me and whom I count myself lucky to work with. I'd also like to thank so many of my friends who have been incredibly supportive and enthusiastic since the launch of my debut novel, *Covenant*—you all know who you are. Finally, as ever I owe my parents, Terry and Carolyn, everything, for without them I would not have such a love of the written word nor such an example in life to live up to.